BIOPROCESSING FOR VALUE-ADDED PRODUCTS FROM RENEWABLE RESOURCES

New Technologies and Applications

BIOPROCESSING FOR VALUE-ADDED PRODUCTS FROM RENEWABLE RESOURCES

New Technologies and Applications

edited by

SHANG-TIAN YANG

Department of Chemical Engineering
Ohio State University
Columbus, Ohio, U.S.A.

ELSEVIER

Amsterdam – Boston – Heidelberg – London – New York – Oxford – Paris
San Diego – San Francisco – Singapore – Sydney – Tokyo

Elsevier
Radarweg 29, PO Box 211, 1000 AE Amsterdam, The Netherlands
The Boulevard, Langford Lane, Kidlington, Oxford OX5 1GB, UK

First edition 2007

Library of Congress Cataloging-in-Publication Data
A catalog record for this book is available from the Library of Congress

British Library Cataloguing in Publication Data
A catalogue record for this book is available from the British Library

ISBN-13: 978-0-444-52114-9
ISBN-10: 0-444-52114-3

For information on all Elsevier publications
visit our website at books.elsevier.com

Transferred to digital print 2008
Printed and bound by CPI Antony Rowe, Eastbourne

Contents

Contributors

Sanem Argin-Soysal
Food Bioprocess Engineering Laboratory, Department of Nutrition and Food Science, University of Maryland, 3102 Marie Mount Hall, College Park, Maryland 20742, USA

Feng Chen
Department of Botany, The University of Hong Kong, Pokfulam Road, Hong Kong, China
South China Sea Institute of Oceanology, Chinese Academy of Sciences, Guangzhou, China

Rachel R. Chen
School of Chemical & Biomolecular Engineering, Georgia Insitute of Technology, Atlanta, GA 30332, USA

Shulin Chen
Biomass Processing and Bioproduct Laboratory, Department of Biological Systems Engineering, Washington State University, Pullman, WA 99164, USA

Wilfred Chen
Department of Chemical and Environmental Engineering, University of California, Riverside, CA 92521, USA

Catherine M-H. Cho
Department of Chemical and Environmental Engineering, University of California, Riverside, CA 92521, USA

Wolf-Dieter Deckwer
Group of TU-BCE, GBF-German Research Center for Biotechnology, Mascheroder Weg 1, 38124 Braunschweig, Germany

Hesham A. El-Enshasy
Bioprocess Development Department, Mubarak City for Scientific Research and Technology Applications, New Burg Al Arab, 21934 Alexandria, Egypt

King Wai Fan
Department of Botany, The University of Hong Kong, Pokfulam Road, Hong Kong, China

Xuan Guo
School of Chemical & Biomolecular Engineering, Georgia Insitute of Technology, Atlanta, GA 30332, USA

Lucita De Guzman
Food Technology Program, School of Technology, University of the Philippines in the Visayas, Miag-ao, Iloilo, Philippines

Chia-Chi Ho
Department of Chemical and Materials Engineering, University of Cincinnati, 497 Rhodes Hall, Cincinnati, OH 45221, USA

Chia-Hua Hsu
Food Bioprocess Engineering Laboratory, Department of Nutrition and Food Science, University of Maryland, 3102 Marie Mount Hall, College Park, Maryland 20742, USA

Hanjing Huang
Department of Chemical and Biomolecular Engineering, The Ohio State University, 140 W. 19th Avenue, Columbus, OH 43210, USA

Wei-Cho Huang
Bioprocessing Innovative Company, Inc., 4734 Bridle Path Ct., Dublin, OH 43017, USA

Hongfei Jia
Department of Chemical Engineering, The University of Akron, 200 E. Buchtel Commons, Akron, OH 44325, USA

Lu-Kwang Ju
Department of Chemical Engineering, The University of Akron, Akron, OH 44325, USA

Kung-Ta Lee
Department of Biochemical Science and Technology, National Taiwan University, Taipei 106, Taiwan

Wei Liao
Biomass Processing and Bioproduct Laboratory, Department of Biological Systems Engineering, Washington State University, Pullman, WA 99164, USA

Chuanbin Liu
Biomass Processing and Bioproduct Laboratory, Department of Biological Systems Engineering, Washington State University, Pullman, WA 99164, USA

Xiaoguang Liu
Department of Chemical and Biomolecular Engineering, The Ohio State University, 140 West 19th Avenue, Columbus, OH 43210, USA

Y. Martin Lo
Food Bioprocess Engineering Laboratory, Department of Nutrition and Food Science, University of Maryland, 3102 Marie Mount Hall, College Park, Maryland 20742, USA

Hongwu Ma
Group of Systems Biology, GBF-German Research Center for Biotechnology, Mascheroder Weg 1, 38124 Braunschweig, Germany

Ching-An Peng
Department of Chemical Engineering and Department of Materials Science, University of Southern California, Los Angeles, CA 90089, USA

Wei Qin
Department of Chemical Engineering, Tsinghua University, Beijing 100084, China

Peter J. Reilly
Department of Chemical and Biological Engineering, Iowa State University, Ames, Iowa 50011, USA

Sheryl B. Rubin-Pitel
Department of Chemical and Biomolecular Engineering, University of Illinois, Urbana, IL 61801, USA

Ellen C. San Nicolas
College of Engineering, Central Colleges of the Philippines, 52 Aurora Blvd., Quezon City, Philippines

Hyun-Dong Shin
School of Chemical & Biomolecular Engineering, Georgia Insitute of Technology, Atlanta, GA 30332, USA

Chin-Han Shu
Department of Chemical and Materials Engineering, National Central University, #300, Jungda Road, Jungli 32054, Taiwan, R.O.C.

Wei Wen Su
Department of Molecular Biosciences and Bioengineering, University of Hawaii, Honolulu, HI 96822, USA

Jibin Sun
Group of Systems Biology, GBF-German Research Center for Biotechnology, Mascheroder Weg 1, 38124 Braunschweig, Germany

I-Ching Tang
Bioprocessing Innovative Company, Inc., 4734 Bridle Path Ct., Dublin, OH 43017, USA

Bernie Y. Tao
Department of Agricultural and Biological Engineering, Purdue University, 745 Agricultural Mall Drive, West Lafayette, IN 47907, USA

Abdullatif Tay
Department of Chemical and Biomolecular Engineering, The Ohio State University, 140 W. 19th Avenue, Columbus, OH 43210, USA

Liping Wang
Department of Chemical and Biomolecular Engineering, The Ohio State University, 140 West 19th Avenue, Columbus, OH 43210, USA

Ping Wang
Department of Chemical Engineering, The University of Akron, Akron, 44325 USA

Si-Jing Wang
State Key Laboratory of Bioreactor Engineering, East China University of Science and Technology, 130 Meilong Road, Shanghai 200237, China

Wei Wang
Group of TU-BCE, GBF-German Research Center for Biotechnology, Mascheroder Weg 1, 38124 Braunschweig, Germany

Attaya Wasanasathian
Department of Chemical Engineering, University of Southern California, Los Angeles, CA 90089, USA

Zhiyou Wen
Biomass Processing and Bioproduct Laboratory, Department of Biological Systems Engineering, Washington State University, Pullman, WA 99164, USA
Department of Biological Systems Engineering, Virginia Tech, Blacksburg, VA 24061, USA

Zhinan Xu
Institute of Bioengineering, Department of Chemical and Biochemical Engineering, School of Material Science and Chemical Engineering, Zhejiang University, Hangzhou 310027, China

Shang-Tian Yang
Department of Chemical and Biomolecular Engineering, The Ohio State University, 140 West 19th Avenue, Columbus, OH 43210, USA

Jian Yu
Hawaii Natural Energy Institute, School of Ocean and Earth Science & Technology, University of Hawaii, 1680 East-West Rd. Honolulu, HI 96822, USA

Yali Zhang
Department of Chemical and Biomolecular Engineering, The Ohio State University, 140 West 19th Avenue, Columbus, OH 43210, USA

An-Ping Zeng
Group of Systems Biology, GBF-German Research Center for Biotechnology, Mascheroder Weg 1, 38124 Braunschweig, Germany

Huimin Zhao
Department of Chemical and Biomolecular Engineering, Department of Chemistry, and Center for Biophysics and Computational Biology, University of Illinois, Urbana, IL 61801, USA

Jian-Jiang Zhong
College of Life Science & Biotechnology, Shanghai Jiao Tong University, 800 Dong Chuan Road, Shanghai 200240, China

Ying Zhu
PDL Biopharma, 34801 Campus Drive, Fremont, CA 94555, USA

Preface

The rapid depletion of petroleum reserves and fossil fuels presents a challenging problem to the world, especially to countries whose industries are heavily reliant on petroleum-based feedstocks. A new industrial trend is to move from petroleum-based to bio-based products and manufacturing processes that can conserve the earth's natural resources and save the planet from industrial pollution. Meanwhile, many industrialized countries, such as the United States, also face the dual problems of surplus agricultural commodities with low economic returns for farmers and large amounts of food processing wastes with high disposal costs because of increasingly tight environmental regulations. The development of a bio-based product industry will offer an economical and environmentally friendly solution to the aforementioned problems.

The oil crisis in the 1970's and the birth of a new biotechnology industry in the early 1980's have given society the hope of becoming fully sustainable by using renewable resources. Since then, many new bio-based products and bioprocessing technologies have been developed. Until recently, however, the biotechnology industry has focused its effort mainly on recombinant protein drugs and transgenic plants and animals. Much less attention has been given to high-volume medium-value industrial products such as oxychemicals, drug intermediates, polymers, industrial solvents, fuels, and food ingredients and feed supplements, although many of these have traditionally been or are currently produced by fermentation from carbohydrate-based feedstocks. The slow progress in commercial development of these bio-based products in the past 20 to 30 years can be attributed to low investment return due to relatively inefficient bioprocesses and steep price competition from the petroleum-based products that dominate the market, and the difficulty of changing the corporate cultures of the traditionally segregated chemical and agricultural industries. This situation has been drastically changed in the last few years. Traditional agricultural companies, including Cargill and ADM, have expanded and transformed from being primarily commodity food/feed suppliers to major manufacturers of value-added products, including specialty chemicals and fuel ethanol. During the same period, several large chemical companies, including DuPont, Dow Chemical, and Monsanto, have also made major R&D investments in biotechnology-based manufacturing processes. This shift toward a bio-based economy has further accelerated as the price of crude oil has been doubled in two years and recently reached US $70 per barrel, making many bio-based products economically competitive and appealing to corporate and private investors.

This book provides a comprehensive review of the fundamentals of biotechnology and bioprocess engineering as well as industrial examples of new bio-based products and advancements in technology development that are important to the general field of sustainable bioprocessing for value-added industrial products from renewable resources. Critical enabling technologies, from genomics to metabolic and bioprocess engineering, are discussed, with some examples. Both fundamentals and novel developments in biotechnology and bioprocess engineering, and their applications to existing and new bio-based industrial products are described in sufficient detail to allow both experts and non-experts to comprehend recent progress in this field. This book first discusses the modern technologies in the fields of biotechnology and bioprocess engineering that are the cornerstones for building a new bio-based products industry. The second part of the book reviews different organisms, ranging from bacteria to algae, that are suitable for bioprocessing because of their unique characteristics, process requirements, products, and applications. The third part of the book comprises a variety of unconventional and novel bioprocesses currently in development. Finally, the book provides examples of the economical use of different renewable resources as feedstocks to produce industrial products.

We started this book more than two years ago. The invited contributing authors are leading experts in their respective research field from the USA and other countries. Without their contribution and editorial assistance from Kevin Yang, this book would not have been finished in time for publication.

Dublin, Ohio
September, 2006

Shang-Tian Yang (Editor)

Chapter 1. Bioprocessing – from Biotechnology to Biorefinery

Shang-Tian Yang

Department of Chemical and Biomolecular Engineering, The Ohio State University, 140 West 19th Avenue, Columbus, OH 43210, USA

1. INTRODUCTION

Recent advances in biotechnology and public concern about environmental pollution and the sustainability of natural resources have rapidly transformed the nation's many manufacturing industries, from chemical to pharmaceutical, to become more environmentally benign and bio-based. For example, almost all major pharmaceutical companies now dedicate more than 50% of their new drug development to biotech R&D, a trend away from traditional chemical synthesis. Likewise, large chemical companies, such as DuPont and Dow Chemicals, are aggressively developing new bio-based products to replace petrochemical ones. Meanwhile, rising energy demands and oil prices have prompted large petroleum companies such as Shell to explore biofuels as alternative energy sources. Together with the agricultural industry expanding its product portfolios beyond traditional food and feed is the birth of an emerging biorefinery industry that promises reduced dependence on fossil energy and a truly sustainable economy [1–3].

This chapter will introduce some important applications of biotechnology and recent developments in bioprocessing technologies for biomass utilization with a focus on the industrial bioconversion of renewable resources to fuels and chemicals. The concept and principles of integrated biorefineries to attain the sustainable production of food, energy, and industrial products are also presented.

2. INDUSTRIAL BIOTECHNOLOGY – HISTORY AND APPLICATIONS

Biotechnology has been described as the last great technological innovation of the twentieth century and has touched upon almost every aspect of human life, from healthcare to agriculture to the production of industrial products (Figure 1). Biotechnology, broadly defined, includes any technique that uses living organisms or parts of organisms to make or modify products, improves plants or animals, or develops microorganisms for specific uses. Based on this definition, mankind has a long history of using biotechnology; in 6000 B.C. our ancestors already knew how to make fermented foods and alcoholic beverages, although the process was not elucidated until 1857, when Pasteur proved fermentation was caused by microorganisms. In the 1910s, the fermentation industry was born and soon became the main force in the production of ethanol and solvents (mainly acetone and butanol from ABE fermentation by *Clostridium acetobutyricum*). The fermentation industry continued to develop, with citric acid being the first organic chemical and penicillin the first drug produced by fermentation in 1923 and 1944, respectively. However, with the discovery of oil and the rapid development of the petroleum industry in the 1950s, many of bulk chemicals and solvents, including ethanol, butanol, acetic acid, and lactic acid that previously had been predominately produced from sugars by fermentation were replaced with petroleum-based products produced by chemical synthesis [3]. Not until the first oil crisis in the 1970s did people start to realize that fossil fuels are exhaustible and that the oil-based economic development is not sustainable. Although there have been extensive efforts to develop renewable energy technologies, bio-based industrial products, with a few exceptions, have not been very successful because of relatively cheap oil. However, with recent crude oil prices exceeding $70 per barrel, bio-based products become increasingly attractive.

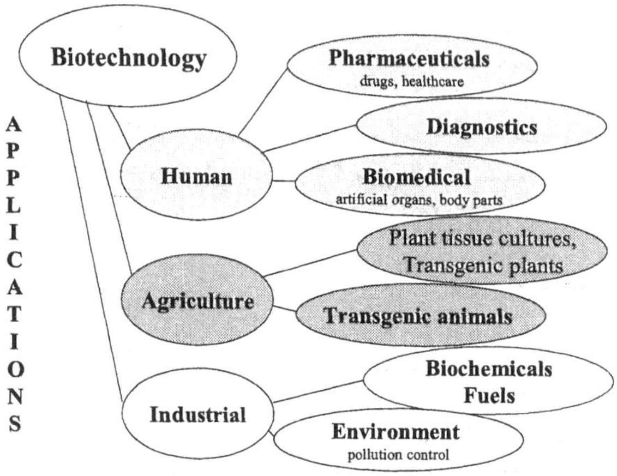

Fug. 1. Applications of biotechnology in various market sectors.

Table 1
Historical milestones in the development of biotechnology

Year	Historical Events
6000 B.C.	Alcoholic beverages, bread, and cheese made by fermentation
1857	Pasteur proves fermentation is caused by microorganisms
1910	Fermentation industry developed (fuel & solvent production)
1923	Citric acid produced by industrial fermentation
1944	Penicillin mass-produced for Normandy landings in WWII
1953	DNA structure elucidated
1973	Recombinant DNA makes genetic engineering possible
1982	First commercial recombinant protein product (human insulin)

Until now and with only a few exceptions, most of fermentation products are drugs, foods, or animal feeds. In terms of quantity, ethanol is the leading industrial product from fermentation. Table 2 lists some of the current industrial fermentation products and their estimated global annual production.

Table 2
Some major industrial fermentation products

	Production* (metric tons)	Microorganism	Separation method	Applications
Citric acid	1,200,000	*A. niger*	Extraction	Food
Ethanol	26,000,000	*S. cerevisiae*	Distillation	Fuel
Glutamate (MSG)	1,000,000	*C. glutamicum*	Crystallization	Flavoring
Lactic acid	400,000	*Lactobacillus sp.*	Extraction	Food, Plastics
Lysine	800,000	*C. glutamicum*	Crystallization	Feed
Penicillin	60,000	*P. chrysogenum*	Extraction	Drug
Xanthan gum	100,000	*X. campestris*	Precipitation	Food, Oil drilling

*2006 data from personal communication with industry sources.

2.1. Pharmaceutical industry

Following the elucidation of the structure of DNA by Watson and Crick in 1953, advances in molecular biology and the development of recombinant DNA technology (with the first demonstration of the transfer of heterologous genes via plasmids into *Escherichia coli* cells by Boyer in 1973) gave birth to the modern biotechnology industry. Genentech, the first biotechnology company, developed and launched, with the help of their licensing partner Eli Lilly, the first recombinant human protein (human insulin) for therapeutic applications in 1982. In the following two decades, the biotechnology industry continued to grow in the therapeutics sector to over $32 billion in annual sales in 2003 [4]. Today, the global market for biopharmaceuticals already exceeds $40 billion, which is about 10% of the total pharmaceutical market of over $400 billion. A large portion (more than 50%) of new drug development now comes from biotech R&D, a trend away from traditional chemical synthesis. In addition to the recombinant therapeutic proteins produced by fermentation, many small-molecule drugs and drug intermediates, especially chiral compounds, are produced by biocatalysis or biotransformation using enzymes or whole cells as the catalysts.

2.2. Agriculture and food

Biotechnology also has had a major impact on the U.S. agricultural and food industries. Transgenic plants and crops have contributed to increased farm productivities and are made into food and animal feed. For example, biotech varieties of corn increased to 52% of U.S. corn acres planted in 2005, and corn production yield increased from 129.3 bushels/acre in 2002 to 142.2 bushels/acre in 2003 and 160.4 bushels/acre in 2004 [5]. Major transgenic crops in the U.S. also include soybeans (81%), cotton (73%), and canola (70%) [6]. Transgenic crops were already valued at over $20 billion in 2002 and are expected to rapidly increase in value as transgenic plants are or can be used to produce pharmaceuticals, chemicals, and fuels [7]. In the U.S. dairy industry, milk and cheese production have also increased to 177 billion lbs and 9.1 billion lbs, respectively, in 2005 [8], largely due to the improved milk production per cow resulting from the use of recombinant bovine somatotropin (BSA). Increased agriculture productivity has not only increased total production but also kept prices low, providing an opportunity and need for their increased utilization in non-food areas. Large agricultural companies, including ADM and Cargill, have aggressively expanded their business into the value-added product market beyond the traditional food and animal feed markets.

2.3. Chemical industry

The chemical industry is huge, consisting of four major subsectors: commodity chemicals, specialty chemicals, consumer care products, and pharmaceuticals, with over $2 trillion on sales worldwide [3]. Biotechnology can offer both economic and environmental benefits to the chemical industry and thus has great potential to achieve the sustainable production of existing and new products from renewable feedstocks. With few exceptions, the chemical industry has been largely built on hydrocarbon feedstocks, with nearly $24 billion worth of them being used annually. As the production of non-renewable fossil energy has reached its limit and oil and natural gas prices have skyrocketed in the recent years, interest in biobased industrial products and bioenergy has grown. The potential for biotechnology applications in the chemical and energy industries is huge, although so far the impact of biotechnology on these market sectors is very small in terms of both sales and market share. This is expected to change in the next few years, as several major chemical and agricultural companies have developed novel technologies to economically produce biobased chemicals and industrial products to replace petrochemicals and products derived from fossil fuels.

The biotechnological production of chemicals is greener than chemical methods because biocatalysts (enzymes or cells) are highly selective, resulting in higher product yield with less or no byproducts, which are usually difficult to separate. Chemical synthesis often requires toxic solvents and generates large amounts of wastes, causing disposal and pollution problems. One example is the production of cephalexin, a semisynthetic antibiotic derived from cephalosporin C. Enzymatic and direct fermentation methods developed by DSM can reduce the process steps from 10 to 4 and wastes to less than one third of those from the chemical process [9]. An enzymatic process for acrylamide developed by Mitsubishi Rayon uses ~20% as much energy as the conventional process. It also requires milder conditions and achieves greater conversion and a higher final product concentration. DuPont and Genencor have co-developed a recombinant *E. coli* fermentation for the production of 1,3-propanediol (PDO) from corn [10]. The biobased PDO, which is a key ingredient in Sorona polymer, consumes 30% to 40% less energy. A joint venture with Tate & Lyle is building a $100 million, 100,000 tons/yr Bio-PDO plant expected to be in operation in 2006 [11]. NatureWorks, renamed from a previous Cargill-Dow joint venture and now solely owned by Cargill, produces polylactic acid (PLA) from lactic acid derived from corn at its 300-million-pound (140,000-metric-ton) capacity manufacturing plant and the world's largest lactic acid plant (400-million-pound or 182,000-metric-ton capacity) in Blair, Nebraska. The use of biodegradable PLA is expected to grow rapidly in the packaging material and textile fiber markets. Dow Chemical is developing vegetable-based polyol products, and ADM has announced a plan to build a polyol facility that will use carbohydrate and glycerol-based feedstocks. In addition, DSM is commercializing products derived from succinic acid produced from corn. Metabolix is developing a new generation of high-performance plastics based on polyhydroxyalkanoid (PHA) produced from renewable resources. These biopolymers can replace petroleum-based polymers and fibers, including polyesters, polyacrylics, polyamides, and polyurethanes, which have a worldwide production of 150 million tons per year [3].

2.4. Fuel and energy

Biomass has potential energy value both as a fuel for heat and power generation and as a feedstock for the production of chemicals and materials. Fuel and energy production from biomass thus represents another major market sector for biotechnology. Biofuels, including ethanol, biodiesel, butanol, methane, and hydrogen can be produced from renewable resources, replacing some fossil fuels. In the U.S., about 4.5 billion gallons of bioethanol are currently produced from corn and used along side with gasoline, which currently has an annual consumption rate of 140 billion gallons [12]. Brazil has increased its sugar cane-based ethanol output to an annual production of 15 billion liters (4 billion gallons) that can satisfy over 33% of the country's gasoline needs [13]. Domestic biofuels could eventually reduce U.S. dependence on foreign oil. In his 2006 state of the Union speech, President George W. Bush called for a 75% reduction in oil imports from the Middle East by 2025. The Energy Policy Act of 2005 mandates that 7.5 billion gallons of fuel ethanol be produced annually to replace 5% of gasoline by 2012. In fact, more than 2.2 billion gallons of annual capacity will be added in the U.S. by 2007 because demand for bioethanol is increasing rapidly as petroleum refiners phase out methyl-*tert*-butyl ether (MTBE) as an oxygenate in gasoline. The legislation also requires bioethanol use to increase to 30 billion gallons to replace 20% of gasoline by 2025. The production of biodiesel, mainly from soybean oil, also has increased rapidly from 500,000 gallons in 1999 to 75 million gallons in 2005. Biodiesel can replace petroleum-based diesel, and new U.S. legislation requires its use to increase to 250 million gallons in 2008 and 2 billion gallons in 2015.

Current bioethanol production consumes more than 12% of the corn produced in the U.S. It is clear that corn and soybeans alone will not be able to produce enough renewable fuels to displace a significant fraction of imported petroleum. Lignocellulosic materials are the most abundant renewable resources on earth, and new technologies are being developed to use them more economically as feedstocks for fuel and chemical production in the future [14]. In fact, Iogen in Canada has already built a cellulose-ethanol demonstration plant in 2004 that produces 800 liters of ethanol per day from wheat straw. Also, a Swedish ethanol plant started in summer 2005 uses sawdust as feedstock [13]. The commercial viability of cellulose-ethanol plants is promising as oil prices continue to rise and enzyme costs for cellulose hydrolysis continue to decrease as the result of new developments in this field. Biotechnology also can improve the growth yield and composition of energy crops, increasing their oil content for biodiesel production or carbohydrate content for bioethanol production, or decreasing lignin or changing cellulose crystallinity structure to facilitate faster hydrolysis. The biomass resources available for use represent 6–10 quadrillion Btu (quads) of energy.

Butanol is an important industrial solvent but also a potential liquid fuel that can be used directly to replace gasoline in current automobile engines [15]. The production of butanol totals about 350 million gal per year worldwide and 220 million gallons or 0.8 million tons in the U.S., all from petroleum. However, during World War I and until the 1950s, the "Weizmann" fermentation process was the main method for acetone and butanol production; this ceased in the U.S. and Europe during the early 1960s because of competition from petroleum-based solvents and the high price of sugar substrates [16]. The fermentation process is limited by the relatively low productivity and yield from sugars in the ABE fermentation process. The last commercial ABE fermentation plant in South Africa closed in 1981. However, recent advances in fermentation and separation technologies and rising energy prices will soon make biobutanol economically attractive again [17, 18]. Industrial ABE fermentation is now being considered in Austria [19, 20]. The extraction of butanol from the fermentation broth with biodiesel may generate a product with 18% ABE in biodiesel that can be used as a fuel without further cleaning [21].

Current technology for hydrogen production from biomass is far from economical, although hydrogen is a clean fuel and can be used in fuel cells. Biohydrogen can be produced by several routes: biophotolysis of water by algae and cyanobacteria, photodecomposition of organic compounds by photosynthetic bacteria, fermentative hydrogen production from organic compounds by anaerobic bacteria, and hybrid systems with both fermentative and photosynthetic bacteria [22]. However, none of these has the needed productivity and yield to be economically competitive at this time. Up to 12 moles of hydrogen can be obtained from glucose via ethanol fermentation followed with steam reforming of ethanol [23]:

$$\text{ethanol fermentation: } C_6H_{12}O_6 \rightarrow 2\ C_2H_5OH + 2\ CO_2$$

$$\text{steam reforming: } C_2H_5OH + 3\ H_2O \rightarrow 6\ H_2 + 2\ CO_2$$

The steam reforming process is highly endothermic. Thus, oxidative steam reforming with a reduced hydrogen yield (~5 mol) but minimal heat input is preferred. However, this fermentation-reforming process will be economical only when ethanol can be more economically produced from lignocellulosic biomass [24].

3. BIOPROCESSING – CURRENT STATUS AND DEVELOPMENT

A bioprocess usually consists of feedstock pretreatment, fermentation or biocatalysis, and downstream processing or separation for product recovery and purification (Figure 2). The

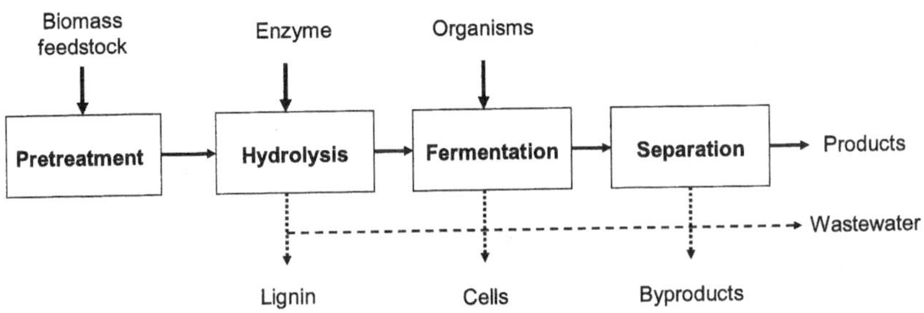

Fig. 2. A general bioprocess flowsheet.

actual bioprocess and required unit operation steps are largely dependent on the substrate and organisms used and the nature and applications of the final product. This section will briefly discuss organism choice, fermentation bioreactor design, and separation methods. More detailed discussions of various biomass feedstocks along with pretreatment methods and the hydrolysis of polysaccharides are given in the next section.

3.1. Organisms

As can be seen in Table 2, current industrial fermentation processes use all types of microorganisms: bacteria, yeasts, and filamentous fungi. The choice of microorganism for a fermentation process is usually based on the product, available substrate, and growth requirements, which affect the fermentation design and downstream processing. The final decision is dictated by economics. The general characteristics, products, and applications of unicellular bacteria and yeasts are discussed in Chapter 8, and those of filamentous fungi are given in Chapter 9. In addition to microorganisms, plant cell and hairy-root cultures also have gained increasing importance in industrial biotechnology [25]. They are discussed in Chapter 10. Furthermore, marine biotechnology offers immense potential for finding new biologically active compounds that have not been thoroughly explored. Chapter 11 gives some examples of high-value products of marine microalgae. Although still in their embryonic stage, marine sponge cultures offer a new type of bioprocess that one day could produce a wide spectrum of compounds for use as drugs [26, 27]. It is noted that animal cells play a very important role in the biopharmaceutical and biomedical industries, but are yet to find applications in industrial biotechnology with mostly small molecules as the products.

Strain development is an important part of industrial fermentation. In recent years, conventional random mutagenesis and screening methods have largely been replaced by more rational approaches that use modern genetic engineering tools, including directed evolution and metabolic engineering [28, 29]. Systems biotechnology or functional genomics using data from genomics, transcriptomics, proteomics, and metabolomics has also emerged as important tools for strain development and bioprocess analysis and optimization [30, 31]. Today, genome breeding, genome engineering, and genome shuffling allow effective evolutionary whole-cell engineering of industrial strains [32]. Also, high-throughput screening techniques can speed up the discovery of new biocatalysts, organisms, and biologically active compounds. Chapter 2 provides the systems biotechnology view of bioprocess development. The principles and applications of genetic and metabolic engineering are described in Chapters 3 and 4, respectively.

Biocatalysts, including both enzymes and whole cells, are important in the production of specialty chemicals, especially chiral compounds that are difficult to make by chemical synthesis [33]. Chapters 12 and 13 provide some important examples of biocatalysis in the industrial production of chemicals.

3.2. Fermentation and bioreactor engineering

The majority of today's fermentation processes can be classified as submerged or solid state fermentations, with the former dominating in the Western fermentation industry. Over the last few decades, extensive research has been focused on bioreactor engineering to improve mixing and heat and mass transfer inside large-scale reactor vessels. However, new bioreactor development efforts focus on microbioreactors that allow high-throughput screening for strain and media optimization. For submerged fermentation, Chapter 6 provides a comprehensive review of bioreactor engineering. Chapters 18 and 19 give some examples of solid state fermentation and algal photobioreactors, respectively. One key issue in industrial fermentation is improving productivity. Cell immobilization provides a viable solution to the productivity issue and has been extensively studied in the past three decades. Large-scale immobilized cell fermentation processes have been limited to industrial wastewater treatment, but will find important applications in fermentation that requires improved productivity and product titer in order to be economically competitive. A brief review on recent development in immobilized cell fermentation for the production of chemicals and fuels is given in Chapter 14. Oxygen transfer is often the rate limiting step in high cell density and viscous fermentations. A water-in-oil cultivation technology for improving viscous xanthan gum fermentation is described in Chapter 15.

3.3. Downstream processing

Except for recombinant protein therapeutics, most of industrial fermentation products are separated and purified by one of several separation techniques based on differences in their size, density, volatility, solubility, and partition coefficient in two different phases [34]. Downstream processing usually consists of 1) cell (particle) separation by filtration, centrifugation, or sedimentation, 2) primary separation or enrichment by extraction, adsorption, precipitation, or evaporation, 3) secondary separation or purification by crystallization, liquid chromatography, or distillation, and 4) polishing and product packaging. Drying is used for powder products. Membrane processing, including microfiltration for particle separation and liquid sterilization, ultrafiltration for concentrating macromolecules such as proteins and polysaccharides, reverse osmosis and diafiltration to remove inorganic salts from the liquid product stream, and electrodialysis to separate ions based on their electrical charge difference, is becoming increasingly important in bioprocessing.

Table 3
Some fermentation products and downstream-processing steps used in their recovery and purification

Product	Concentration (g/L)	Major steps used in downstream processing
Ethanol	70–120	Stripping, distillation
Organic acids	50–100	Precipitation / Solvent extraction, crystallization
Antibiotics (Penicillin)	10–30	Filtration, solvent extraction, crystallization, drying
Amino Acids	1–100	Filtration, precipitation, crystallization, drying
Xanthan gum	25–50	Alcohol precipitation, centrifugation, drying
Enzymes	2–25	Precipitation, adsorption, chromatography
Vitamin B_{12}	0.02–0.06	Flocculation, filtration, adsorption, crystallization

Adapted from [34].

Table 3 shows some fermentation products and their downstream processing steps. Chapter 7 describes some bioseparations using membranes with more detailed discussion on membrane fouling, which has been a major issue in biotechnological applications of membranes. Integrated fermentation and separation processes have additional advantages such as alleviating product inhibition and increasing reactor productivity [35]. Chapter 16 provides an example of extractive fermentation for carboxylic acid production.

4. BIOMASS FEEDSTOCKS

Biomass currently provides over 3% of the total energy consumed in the United States and is the largest domestic renewable energy source [36]. Biomass includes any organic matter that is available on a renewable or recurring basis. Because it is renewable and abundant, biomass has the potential to offer diverse supplies of reliable, affordable, and environmentally sound energy and chemicals to replace fossil fuels and petrochemicals. The U.S. Department of Energy and the Department of Agriculture envision that biomass will provide 5% of power (heat and electricity), 20% of liquid transportation fuels (ethanol and biodiesels), and 25% of industrial products (chemicals and materials) by 2030, representing thirty percent of the current US petroleum consumption, which would require over 1 billion dry tons of biomass feedstock annually [36]. Table 4 shows the potential production of various kinds of biomass, including dedicated energy crops and trees, agricultural crop residues, logging and wood processing residues, animal manures, and other waste materials.

Table 4
Annual biomass potential from agricultural and forest resources in the United States*

Agricultural resources	(10^6 dry tons)	Forest resources	(10^6 dry tons)
Crop residues	446	Logging residues	64
Grass and woody crops	377	Excess biomass thinning	60
Municipal solid waste	111	Fuel wood	51
Grains to fuels	87	Mill processing residues	145
Animal manures	44	Urban wood residues	48
Food processing residues	44		
Total	998	Total	368

*Including both currently available and potential growth in agricultural and forest lands [36].

4.1. Starch and sugar crops

Currently, starch and sugar from agricultural crops are the main fermentation feedstocks used in industry. This practice will not change until the cost of using lignocellulosic biomass as feedstock has been substantially lowered. Compared to cellulose, starch is much easier to hydrolyze to glucose by either chemical or enzymatic methods. The enzymatic method involves starch liquefying and saccharification enzymes (see Table 5) and is used in the corn refinery industry, which processes more than 22% of the 11.8 billion bushels (~300 million metric tons) of corn produced annually in the U.S. into high-fructose-corn-syrup, dextrose, starch, and fuel alcohol. Dextrose or glucose derived from starch is the main substrate for industrial fermentation in the U.S. About 4 billion gallons of fuel ethanol were produced from 1.4 billion bushels of corn in 2005 in the U.S. On the other hand, sucrose from sugar canes and sugar beets is the main fermentation substrate in other regions. The global production and compositions of some major starch and sugar crops are listed in Table 6. Grains and processed products from these crops are mainly consumed as foods and animal feeds. Only a small fraction, the wasted crops, may be collected and used to produce fuels, chemicals and other industrial products [37]. However, the global quantity of crop residues available as potential fermentation feedstock is huge, estimated at 204 million tons for corn stover, 731 million tons for rice straw, 354 million tons for wheat straw, and 180 million tons for sugar cane bagasse [37].

Table 5
Enzymes (amylases) for breaking down starch

Enzyme	**Reaction**
α-Amylase	Randomly cuts α-1,4-glycosidic bonds in starch molecules
β-Amylase	Cleaves maltose disaccharide from the non-reducing end
Glucoamylase	Cleaves glucose from the non-reducing end
Pullulanase	Cuts α-1,6-glycosidic bonds at the branching point in amylopectin

Table 6
Major starch and sugar crops – global annual production and compositions (dry basis, wt%)

Crop	Production* (10^6 metric tons)	Starch	Sugar	Protein	Oil	Fiber
Corn	695	72	-	10	5	13
Wheat	628	80	-	14	-	5
Rice	619	89	-	8	1	2
Soybeans	210	16	16^+	40	21	5
Sugar cane	1,290	-	55	-	-	45
Sugar beet	243	-	68	6	-	7
Sweet sorghum	59	-	50	-	-	50

*2005 data from FAOSTAT Database available at http://faostat.fao.org/
$^+$including sucrose and soluble oligosaccharides

4.2. Lignocellulosic biomass

Lignocelluloses are the most abundant biomass found in almost all plant-derived materials, from wood and grass to agricultural residues and municipal solid wastes. The major components of lignocelluloses are cellulose, hemicellulose, and lignin; however, their compositions vary greatly, depending on the type of plant, cultivation conditions, and the age of the plant. Table 7 shows the compositions of some lignocellulosic materials.

Table 7
Organic components of some lignocellulosic biomass (dry basis, wt%)

Feedstock	Cellulose	Hemicellulose	Lignin	Other
Bagasse	40	24	25	11
Corn stover	40	25	17	18
Corn cob	39	35	15	11
Corn fiber	15	35	8	42*
Rice straw	35	25	12	28
Wheat straw	38.2	21.2	23.4	17.2
Wheat chaff	38	36	16	11
Switch grass	45	31	12	12
Hard wood (hybrid poplar)	44.7	18.6	26.4	10.3
Soft wood (pine)	44.6	21.9	27.7	5.8
Waste paper	76	13	11	0

*including 23.7% starch. Sources of data: [24, 39, 40]

Cellulose, which usually makes up the major organic components (up to 50%) of a plant, is a linear polymer of D-glucose linked by β-glucosidic bonds. The linear cellulose polymers, called elemental fibrils, are linked together by hydrogen bonds and van der Waals forces to form microfibrils, which group together to constitute cellulose fiber and are usually covered by hemicellulose and lignin. Plant cellulose appears in either crystalline or amorphous form. The former is a highly ordered form that is difficult to degrade biologically. Hemicelluloses are a group of complex heteropolysaccharides made up of various sugars (D-xylose, D-glucose, D-mannose, D-galactose, and L-arabinose) and sugar acids (D-glucuronic and D-galacturonic acids), depending on the plant species. Unlike cellulose, hemicelluloses have branches with short lateral chains of different sugars, do not form aggregates, and are easily hydrolysable. Lignin, present in the cellular cell wall, is an amorphous heteropolymer consisting of phenylpropane units (coniferyl alcohol, sinayl alcohol, and coumaryl alcohol) joined together by different types of linkages [38]. These polymers are biodegradable but difficult to use directly as substrates in industrial fermentation.

The hydrolysis of lignocelluloses to fermentable sugars remains the greatest challenge in the development of economical plant biomass feedstock for the biorefinery industry [39]. The enzymatic hydrolysis of cellulose, which is expensive but nevertheless preferable to other methods, requires several cellulases – endoglucanases (endo-1,4-β-glucanases, EG), cellobiohydrolases (exo-1,4-β-glucanases, CBH), and cellobiase (β-glucosidase) (see Table 8). EG hydrolyze internal bonds, while CBH work from the existing ends of cellulose, releasing cellobiose molecules, which are further broken down to two molecules of glucose by β-glucosidase. To facilitate the enzymatic hydrolysis of cellulose, it is usually necessary to pretreat the lignocellulosic materials to partially remove or degrade hemicellulose and lignin and to break up or loosen crystalline cellulose and increase its surface areas accessible for enzyme adhesion.

Table 9 lists and compares commonly used pretreatment methods [40–42]. All the pretreatment methods increase the accessibility of crystalline cellulose for enzymatic hydrolysis. They also hydrolyze hemicellulose and lignin to different extents, depending on the treatment conditions. Dilute acid pretreatment with H_2SO_4 is the most often used method in industry, but it usually generates some toxic byproducts that need to be removed before yeast fermentation.

Table 8
Enzymes for breaking down cellulose and lignin

Enzyme	**Reaction**
Cellulases	
endo-1,4-β-glucanases (EG)	Hydrolyze internal β-1,4-glycosidic bonds in the amorphous region
cellobiohydrolases (CBH)	Release cellobiose disaccharides from the reducing end (CBH I) and the non-reducing end (CBH II) of the cellulose chain
β-glucosidase	Cleaves the cellobiose disaccharide to glucose
Lignin-degrading enzymes	
Peroxidases, Laccases	Depolymerize lignin by oxidizing phenolic compounds, but also non-phenolic compounds in the presence of mediators

Table 9
Some pretreatment methods for lignocellulosic biomass

Pretreatment methods	Conditions and performance
Steam explosion	Uses steam at 210–290°C, 20–50 bar for 2 min., followed with sudden pressure release; low xylose yield of 45–65%
Liquid hot water (LHW)	Uses compressed hot water 200–230°C for up to 15 min; high xylose yield (88–98%); requires recycling of water
Dilute acid pretreatment	Uses 0.5–1.5% H_2SO_4 or HCl at 160–220°C; good xylose yield: 75–90%; requires neutralization before cellulose hydrolysis; generates some toxic byproducts (acetic acid, furfural, phenolic compounds, etc.); current industrial method
Ammonia fiber explosion (AFEX)	Uses liquid ammonia (5–15%) and steam explosion (160–180 °C); enhances hydrolysis of (hemi)cellulose from grass, but not as effective for soft and hard woods that contain more lignin
Alkali pretreatment	Uses lime or NaOH at lower temperatures and pressures for a longer time (hours); removes all lignin but only some hemicellulose

4.3. Industrial waste

There is abundant biomass present as processing wastes requiring proper disposal to avoid pollution. One example is the corn refinery (wet milling) industry, which processes more than

20% of the 11.8 billion bushels (~300 million metric tons) of corn annually produced in the U.S. and generates more than 17 million metric tons of corn byproducts (corn fiber, etc.) that are currently of limited use and pose significant environmental problems. In the U.S. dairy industry, about 50% of the milk produced is used to produce cheese, generating ~80 billion lbs of cheese whey in 2005. Cheese whey contains about 7% total solids, of which ~70% is lactose and 13% is protein. Currently, less than 50% of the total amount of cheese whey produced in the U.S. is used to produce dry whey powder (469,000 metric tons), whey proteins, and lactose (300,000 metric tons), leaving more than 50% of the lactose in whey unused, which requires costly disposal because of its high biological oxygen demand (BOD). These abundant and inexpensive renewable resources can be readily used to produce chemicals and fuels by fermentation. The bioconversion of whey lactose to exopolysaccharides is discussed in Chapter 22. Chapter 23 describes the production of microbial polyhydroxyalkanoid (PHA) from renewable resources, including whey.

In addition, there are also large amounts of municipal solid wastes, waste sulfite liquor from the paper and pulp industry, and animal wastes available as inexpensive feedstocks. Chapter 25 discusses the use of animal manure to produce value-added products.

4.4. Lipids

Oils and fats are important raw materials for the production of oleochemicals, including free fatty acids, methyl esters, fatty alcohols and amines, and glycerol as a byproduct. Vegetable oils account for about 80% of the global oil and fat production, which was 105 million tons in 2000 [43]. Soybeans are the most important oil crop, followed by palm, rapeseed, and sunflower oils. About 15–17 million tons of vegetable oils are used by industry for the production of surfactants, lubricants, coatings, cosmetics, and other products. Fatty acid methyl esters have an important new use as liquid fuels. The production of biodiesel in the EU has steadily increased to 1.4 million tons in 2003, and the trend is continuing. As a result of increased biodiesel production, more glycerol, a byproduct of the production of fatty acids and esters from triglycerides, is expected to be available at a lower price in the future. Chapter 24 provides detailed discussion of vegetable oils and their industrial applications.

4.5. Proteins and nucleic acids

Proteins and nucleotides are nitrogenous compounds that are present in biomass at a relatively small percentage weight. Proteins are important nutrients in human and animal diets. Many fermentation processes require organic nitrogen sources for cell growth. Inexpensive nitrogen sources can be obtained from organic wastes containing proteins or amino acids. However, there is much less protein than carbohydrates available for non-food applications. Nucleic acids have never been considered as potential biomass feedstock. However, deoxyribonucleic acid (DNA) has unique applications as biophotonic materials [44, 45]. Although not considered as potential feedstock for chemical and fuel production, protein and DNA can be valuable products that may improve the economics of a biorefinery.

5. BIOREFINERIES

A biorefinery is a manufacturing facility that uses biomass as feedstock to produce fuels, power, and chemicals. It is analogous to today's petroleum refineries, which use petroleum-based feedstocks, mainly oil and natural gas, to produce multiple fuels, commodity chemicals, industrial products, and commercial goods. Figure 3 illustrates the biorefinery concept. There are two different routes or platforms for biomass conversion [46]. The syngas or thermochemical platform involves the gasification of biomass at 650–900°C by reacting with air, oxygen and steam to gaseous products (CO, CO_2, H_2, CH_4). In addition, liquefaction or pyrolysis of biomass at 450–500°C in the absence of any reactive compounds or oxidants can produce pyrolysis oil. All components of biomass, including lignin, which is resistant to biological conversion, can be converted to chemical building blocks. However, the thermochemical platform cannot compete economically with fossil fuels, especially coal. On the other hand, the sugar platform requires thermochemical pretreatment and enzymatic hydrolysis of lignocellulosic biomass, which at present is still too expensive because of high enzyme costs. Other economic barriers include relatively low sugar yields, low solid concentration (<20%), and impurities in the sugar solution. Advances in biotechnology one day will lead to new technologies that can produce better cellulases at lower costs [47], allowing the sugar platform to become economically attractive.

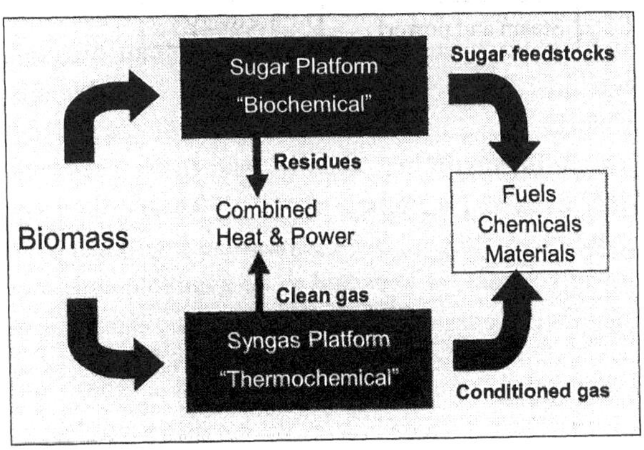

Fig. 3. Biorefinery concept. Adapted from [46].

Hydrocarbon feedstocks represent the largest share of raw materials purchased in the U.S. chemical industry, nearly $24 billion in 2001. The fossil-based feedstock energy used in the production of chemicals was ~3.2 quads, equivalent to over 700 million barrels of oil [48]. In addition, over 3.6 quads were used at petroleum refineries as feedstocks for non-fuel products, of which over 55% were sold as intermediates for plastics and other chemical products. By comparison, agricultural commodities purchased by the chemical industry for conversion into products totaled about $4.5 billion in 2001. Biorefineries will have a significant impact on both the chemical and petroleum industries if biomass can be economically used to supplant fossil-based feedstocks.

5.1. Integrated biorefineries

The DOE has envisioned that by 2030, at least one third of the present transportation fuels in the U.S. will be produced from biomass [48]. In addition, the DOE's biorefinery vision includes producing high-value chemicals and materials from biomass. High-value products may represent only a small fraction of total products, but may account for most of the profits. Integrating the production of higher value bioproducts into biorefinery fuel and power output improves the overall profitability and productivity of all products. Figure 4 illustrates the concept of an integrated biorefinery that utilizes all components of the biomass feedstock to produce energy, fuels, and chemicals [46].

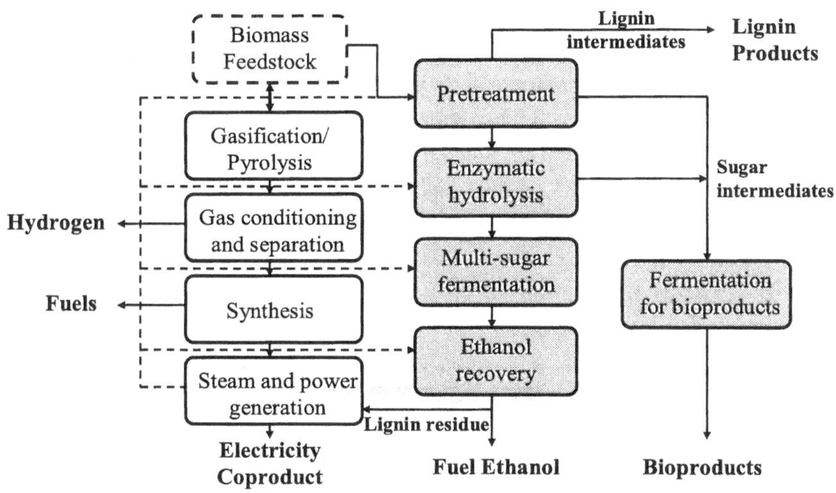

Fig. 4. An integrated biorefinery. Adapted from [46].

The DOE has also identified top chemical building blocks (see Figure 5) from biomass that have great potential to be used in current chemical synthesis processes to produce a cascade of intermediate and final industrial and consumer products [49]. These chemical building blocks are C3–C6 monomers exhibiting multiple functionalities suitable for further conversion as derivatives or molecular families. They are not already supercommodity chemicals, but can be produced from either starch or lignocellulosic biomass. These chemical building blocks include 1,4-dicarboxylic acids (succinic, fumaric, and malic acids), 2,5-furan dicaboxylic acid, 3-hydroxy propionic acid, aspartic acid, glucaric acid, glutamic acid, itaconic acid, levulinic acid, 3-hydroxybutyrolactone, glycerol, sorbitol, xylitol, and arabinitol. These chemical building blocks can be chemically converted to acrylic acid, adipic acid, acrylamide, 1,4-butanediol, and tetrahydrofuran, which are widely used in the chemical synthesis of polymers [50]. However, in order to use these chemical building blocks, their production costs from biomass need to be reduced to less than US$0.25/lb in order to compete with petroleum-based polymers. To meet this target, the cost of sugar needs to be reduced from $0.15/lb to $0.10/lb or less.

Among the chemical building blocks identified by the DOE, glutamic acid (monosodium glutamate) and itaconic acid are commercially produced by glucose fermentation [51, 52], sorbitol is produced by hydrogenizing glucose, and glycerol is produced as a byproduct of the transesterification of vegetable oils with alcohol. Succinic acid and fumaric acid also can be produced from glucose by fermentation [53, 54]. There is no known microorganism capable of producing 3-hydroxypropionic acid, which is an isomer of lactic acid (2- hydroxypropionic acid) but has very distinct and unique physiochemical properties. Through metabolic engineering, Cargill and Codexis are making good progress toward developing mutants that can produce this chemical at a significant titer. Xylitol and arabinitol can be produced by hydrogenizing or fermenting pentoses (xylose and arabinose, respectively) obtained from the hydrolysis of hemicellulose [55]. Dehydrating C6 sugars leads to the production of 2,5-furandicarboxylic acid and levulinic acid, among others. Glucaric acid and 3-hydroxybutyrolactone can be produced from the oxidation of starch. However, these chemical methods need to be further developed and optimized in their selectivity in order to be economical.

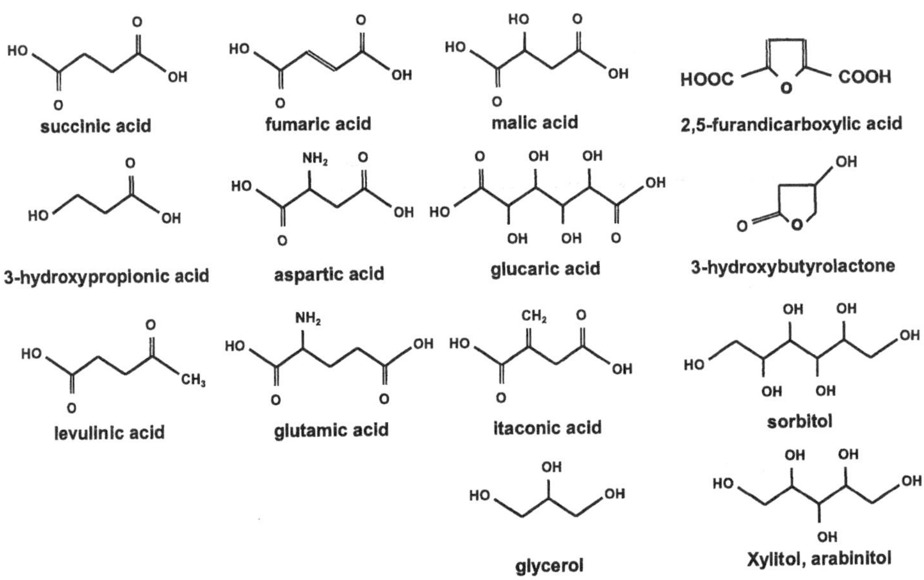

Fig. 5. Potential chemical building blocks from biomass identified by the DOE.

Hydrocarbons, isoprenoids (natural rubber), and novel non-biodegradable polymers, such as polythioesters, also can be produced from renewable resources by fermentation [56, 57]. However, direct microbial synthesis of these chemicals and polymers is far from practical application and requires much more development work.

5.2. Corn refineries (wet milling)

In the United States, corn and soybeans are the largest biomass resources for industrial products. About 1.5 billion bushels of corn and 3.4 billion pounds of soybeans are used annually for this purpose. In addition to corn oil, starch, and feed products, various bioproducts, including ethanol, lactic, citric and itaconic acids, amino acids, and xanthan gum are currently produced by microbial fermentation in large corn wet-milling plants (see Figure 6). The expanded corn refinery plant also may include chemical conversion of glucose to sorbitol via hydrogenation, production of industrial enzymes for the conversion of starch to maltodextrins and high fructose corn syrup (HFCS), and an on-site cogeneration system providing electricity and steam for various processes. The chemical products are used in foods, detergents, and plastics. The ethanol is used as a solvent or for transportation fuels. Lactic acid can be converted to polylactic acid and used as bioplastics for packaging and textile fibers. Lactic acid and ethanol can react to form ethyl lactate ester, which can be used as an industrial "green" solvent, replacing the petroleum-based solvents currently used in the semi-conductor industry.

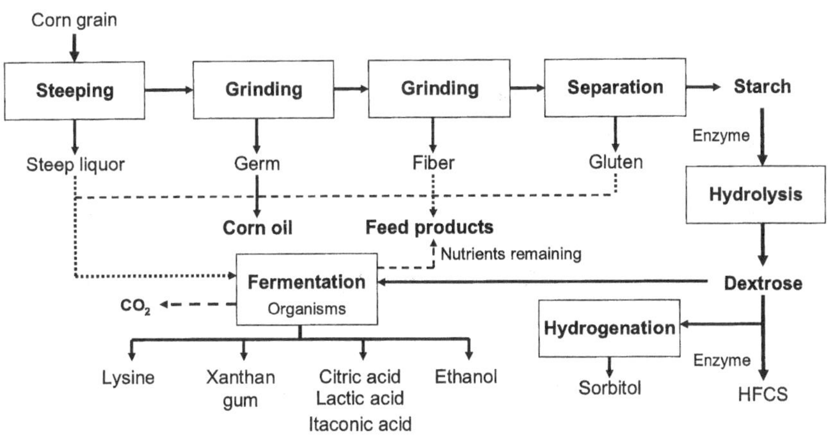

Fig. 6. Corn wet milling to various bioproducts.

In addition, 1,3-propanediol and succinic acid are chemical building blocks that can be produced from corn dextrose. An emerging corn biorefinery may also produce high-value biopolymers, such as PHA and poly-γ-glutamate [58, 59]. In future corn biorefineries, corn stover and other crop residues will also be used as feedstock.

5.3. Whey processing

The U.S. dairy industry processes more than 50% of the milk produced to cheese, generating more than 80 billion pounds of whey a year. Large dairy plants produce whey powder for human and animal food uses. Whey proteins and lactose are also marketable products of whey processing (Figure 7). However, not all whey or whey permeate can be economically used for foods, and they are available as low-cost feedstock for fermentation to produce fuels and chemicals. For example, the lactose in whey (permeate) can be converted to lactic acid, propionic acid, and acetic acid via anaerobic fermentation [60–62]. The mother liquor produced in why processing is high in lactose and salt, and difficult to use as animal feed. One possible economic use of this waste stream is to produce calcium magnesium acetate as a road deicer [63]. The ultimate goal of a whey biorefinery is to achieve total utilization of all whey components and zero emission.

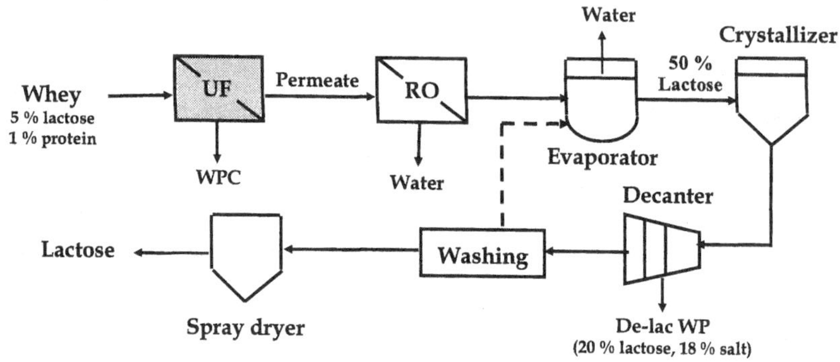

Fig. 7. A general flowsheet for whey processing to produce whey protein concentrate (WPC) and lactose from whey, which is a byproduct from cheese manufacturing. UF: ultrafiltration; RO: reverse osmosis; De-lac WP: de-lactosed whey permeate or mother liquor from crystallization.

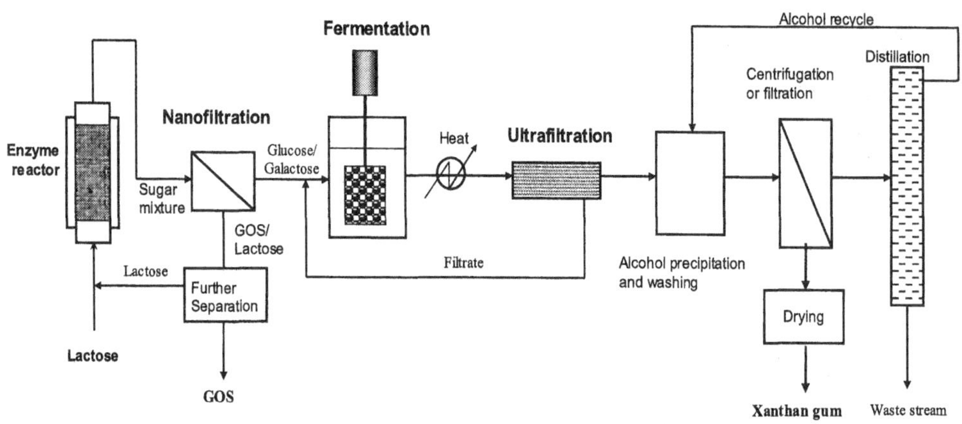

Fig. 8. An integrated process for GOS and xanthan gum production from whey lactose.

In an emerging whey processing plant, an immobilized enzyme reactor can be used to economically produce galacto-oligosaccharides (GOS) from lactose [64]. GOS contain 2 to 5 galactose units and one glucose. GOS are prebiotic, with wide applications in human and animal foods. As non-digestible dietary fibers, GOS are not susceptible to decomposition by human digestive enzymes, so they pass to the intestines, where they stimulate the growth of *Bifidobacteria* known to provide many health benefits, including increasing calcium absorption and the reduction of toxic metabolites and serum cholesterol. Current worldwide production of GOS is estimated at 25,000 tons per year, mainly in Japan and Europe.

However, in the GOS production process using whey lactose as the substrate, large amounts of glucose and galactose are also produced, which must be removed from GOS because these monosaccharides do not have and can reduce the prebiotic effect. It is thus desirable to convert these unwanted sugars to a high-value product, such as xanthan gum, to reduce waste while improving the process economics. Figure 8 shows an integrated bioprocess for the production of xanthan gum from glucose and galactose present in the waste sugar stream from the GOS production process. In the fermentation, a novel, rotating fibrous bed bioreactor can be used for cell immobilization to produce cell-free xanthan broth [65]. The centrifugal force generated from rotating the fibrous matrix separates the xanthan polymer from the immobilized cells, thus producing a cell-free xanthan broth that is then concentrated via ultrafiltration before further purification with alcohol precipitation. The permeate from ultrafiltration can be recycled and reused in the fermentation, reducing the raw material costs and the amount of spent medium. The integrated bioprocess for GOS and xanthan gum production from whey lactose not only can greatly increase the product value but also can reduce the amount of waste generated by the plant.

5.4. Lignocellulose biorefineries

Future biorefineries will use lignocellulosic materials from either crop residues or energy crops, such as switch grass, as feedstock. It has been estimated that there will be more than one billion tons of plant biomass available annually in the U.S. These lignocelluloses need to be converted to fermentable sugars in order to be used to produce fuels and chemicals. The cost of the hydrolytic enzymes is currently too high and must be reduced significantly for lignocellulose biorefineries to be economical. The production of ethanol from xylose and

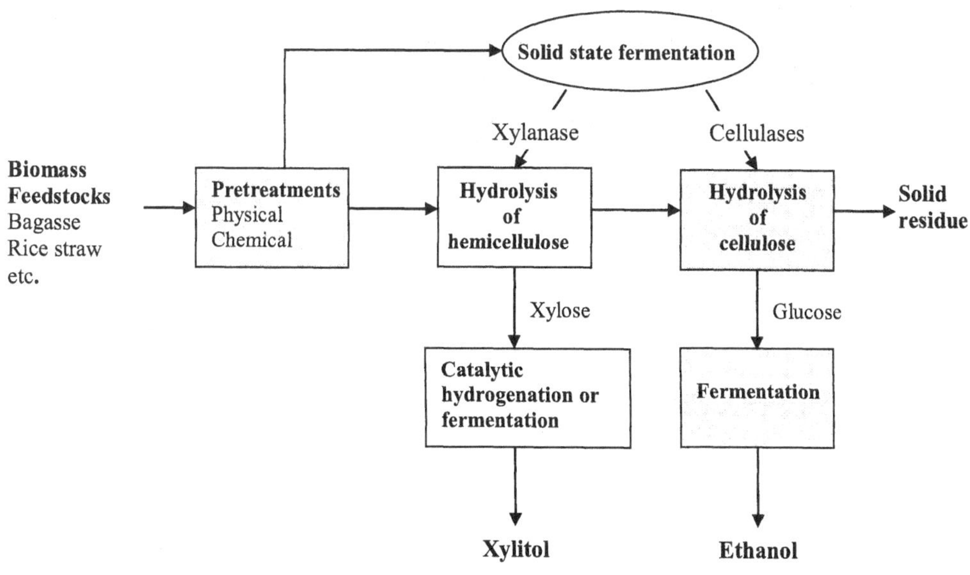

Fig. 9. An integrated biorefinery with solid state fermentation to produce xylanase and cellulase for the hydrolysis of hemicellulose and cellulose to xylose and glucose, which are then used to produce xylitol and ethanol, respectively.

other pentoses remains a challenge, although several mutants have been created for this purpose [66, 67]. It has been suggested that consolidating the bioprocessing of lignocellulosic biomass by integrating cellulase production, cellulose hydrolysis, and the fermentation of both glucose and xylose into one step, would significantly reduce the production cost of ethanol [14]. However, the anaerobic bacteria capable of fermenting cellulose do not produce ethanol or organic acids at sufficiently high rates or concentrations. On the other hand, it is difficult to metabolically engineer an industrial fermentation strain to use cellulose directly.

One way to reduce the cost of enzymes is to produce and use them on site. It is also important to produce the highest-value product possible from all components of the feedstock. As an example, Figure 9 shows an integrated biorefinery producing both xylanases and cellulases for the hydrolysis of hemicellulose and cellulose, respectively. The xylose obtained from the hydrolysis of xylan can be converted to xylitol, a high-value sweetener currently used in chewing gum, whereas the glucose obtained from the hydrolysis of cellulose can be fermented to ethanol and other chemical products. The lignin and other solid residues can be burned to generate heat or steam for processing use.

6. SUMMARY AND OUTLOOK

Bioprocessing and bioproducts have gained commercial interest because of the perceived "green" advantages of using biomass rather than fossil energy for the production of chemicals and industrial products [68]. Other key benefits in the move towards bioproducts include the sustainability of renewable biomass, replacing depleted fossil energy, and reducing greenhouse gas emissions from the present petroleum-based chemical and energy industries. Currently, only 10% of the one hundred million metric tons of fine, specialty, intermediate, and commodity chemicals produced annually in the U.S. are biobased. Today, hydrocarbons still dominate the U.S. economy. Currently, plant biomass provides only about 7% (by weight) of organic chemical products and 3% of transportation fuels and power. However, with rising oil prices and advances in industrial biotechnology, the potential of biomass to replace petroleum-based chemicals and fuels is huge. The potential for new bioproducts with improved performance to move into new and non-conventional markets is also substantial. The Department of Energy has predicted that the biotech-derived chemicals will exceed U.S. $100 billion in 2010 and 400 billion, representing about 50% of the market for organic chemicals, in 2030. However, only with competitive prices can bioproducts make significant inroads into markets historically dominated by petrochemical products.

There are still many challenges facing the emerging biorefinery industry. In the short term, starch and sugar present in crops and processing wastes will continue to be the main substrates for fermentation. In the long run, lignocellulosic biomass offers the greatest potential as feedstock, but the cost of fermentable sugars derived from lignocelluloses must be significantly reduced to less than $0.10/lb. In order to convert carbohydrates to chemical building blocks, continued efforts to discover and engineer new and more efficient biological routes and metabolic pathways are also needed. Biorefineries must also take an integrated approach to make use of all biomass components to produce energy, fuels, chemicals, and high-value products in order to be environmentally sound and cost-competitive with traditional products. With new bioprocessing technologies under development, the biofuels and biobased chemicals and polymers are promising and will lead to a more sustainable society [69].

ACKNOWLEDGEMENTS

The author's bioprocessing research program is supported by the National Science Foundation, the Department of Energy, the US Department of Agriculture, Environmental Protection Agency, the Consortium for Plant Biotechnology Research, Inc., and Dairy Management, Inc.

REFERENCES

[1] H. Ohara, Biorefinery. Appl Microbiol Biotechnol 62 (2003) 474–477.
[2] B. Kamm and M. Kamm, Principles of biorefineries. Appl Microbiol Biotechnol 64 (2004) 137–145.
[3] Th. Willke and K.-D. Vorlop, Industrial bioconversion of renewable resources as an alternative to conventional chemistry. Appl Microbiol Biotechnol 66 (2004) 131–142.
[4] A.K. Pavlou and J.M. Reichert, Recombinant protein therapeutics – success rates, market trends and values to 2010. Nat Biotechnol 22 (2004) 1513–1519.
[5] Corn Refiners Association, Corn, part of a healthy diet, Corn Refiners Association 2004 Annual Report.
[6] C.F Runge and B. Ryan, The economic status and performance of plant biotechnology in 2003: adoption, research and development in the United States. A report prepared for Council for Biotechnology Information (CBI), Washington, DC, December, 2003.
[7] M. Gavrilescu and Y. Chisti, Biotechnology – a sustainable alternative for chemical industry. Biotechnol Adv 23 (2005) 471–499.
[8] USDA National Agricultural Statistics Service, Milk Production, Disposition and Income, 2005 Summary, April 2006.
[9] E. Vandamme and C.G. Bienfait, Industrial biotechnology and sustainable chemistry. Brussels: Royal Belgian Academy Council of Applied Science; 2004. p. 32.
[10] J.V. Kurian, A new polymer platform for the future – Sorona® from corn derived 1,3-propanediol. J Polymers and the Environment, 13 (2005) 159–167.
[11] Anonymous, Feedstock change is hot and poppin, Chemical Processing, January 2006, 22–26.
[12] G. Hess, Push for biofuels seen in farm bill, Chem Eng News, May 22, 2006, 29–31.
[13] Y. Lin and S. Tanaka, Ethanol fermentation from biomass resources: current state and prospects. Appl Microbiol Biotechnol 69 (2006) 627–642.
[14] L.R. Lynd, W.H. van Zyl, J.E. McBride and M. Laser, Consolidated bioprocessing of cellulosic biomass: an update. Curr Opin Biotechnol 16 (2005) 577–583.
[15] Environmental Energy, Inc., Butanol is an oxygenate which can replace gasoline. 2006. http://butanol.com/
[16] D.T. Jones and D.R. Woods, Acetone-butanol fermentation revisited. Microbiol. REV. 50 (1986) 484-524.
[17] T.C. Ezeji, N. Qureshi, H.P. Blaschek, Acetone butanol ethanol (ABE) production from concentrated substrate: reduction in substrate inhibition by fed-batch technique and product inhibition by gas stripping. J Appl Microbiol 63 (2004) 653–658.

[18] N. Qureshi and H.P. Blaschek, ABE production from corn: a recent economic evaluation. J Ind Microbiol Biotechnol 27 (2001) 292–297.
[19] J.R. Gapes, The economics of acetone-butanol fermentation: theoretical and market considerations. J Mol Microbiol Biotechnol 2 (2000)27–32.
[20] J.R. Gapes, The history of the acetone-butanol project in Austria. J Mol Microbiol Biotechnol 2 (2000) 5–8.
[21] E. Crabbe, C. Nolasco-Hipolito, G. Kobayashi, K. Sonomoto, A. Ishizaki, Biodiesel production from crude palm oil and evaluation of butanol extraction and fuel properties. Process Biochem 37 (2001) 65–71.
[22] D. Das and T.N. Veziroglu, Hydrogen production by biological processes: a survey of literature. Int J Hydrogen Energy 26 (2001) 13–28.
[23] H. Agus, F. Sun, M. Naveen, A. Sushil, Current status of hydrogen production techniques by steam reforming of ethanol: A review. Energy & Fuels 19 (2005) 2098–2106.
[24] C.N. Hamelinck, G. van Hooijdonk, A.P.C. Faaij, Ethanol from lignocellulosic biomass: techno-economic performance in short-, middle- and long-term, Biomass and Bioenergy 28 (2005) 384–410.
[25] C.F. Runge and B. Ryan, The economic status and performance of plant biotechnology in 2003: adoption, research and development in the United States. A report prepared for Council for Biotechnology Information (CBI), Washington, DC, December, 2003.
[26] E.H. Belarbi, A.C. Gomez, Y. Chisti, F.G. Camacho, E.M. Grima, Producing drugs from marine sponges. Biotechnol Adv 21 (2003) 585–693.
[27] N.L. Thakur and W.E.G. Müller, Biotechnological potential of marine sponges. Curr Sci 86 (2004) 1506– 1512.
[28] M. Chartrain, P.M. Salmon, D.K. Robinson, and B.C. Buckland, Metabolic engineering and directed evolution for the production of pharmaceuticals. Curr. Opin. Biotechnol. 11 (2000) 209–214.
[29] H. Shimizu, Metabolic engineering – Integrating methodologies of molecular breeding and bioprocess systems engineering. J. Biosci. Bioeng. 94 (2002) 563–573.
[30] T. Hermann, Using functional genomics to improve productivity in the manufacture of industrial biochemicals. Curr. Opin. Biotechnol. 15 (2004) 444–448.
[31] S.Y. Lee, D.-Y. Lee, and T.Y. Kim, Systems biotechnology for strain improvement. TRENDS in Biotechnol. 23 (2005) 349–358.
[32] R. Petri and C. Schmidt-Dannert, Dealing with complexity: evolutionary engineering and genome shuffling. Curr. Opin. Biotechnol. 15 (2004) 298–304.
[33] A.S. Bommarius, Biocatalysis: fundamentals and applications. New York: Wiley, 2004.
[34] S. Basu and S.T. Yang, Bioseparations, in S. Lee (ed.), Encyclopedia of Chemical Processing, Marcel Dekker (2005), pp. 221–236.
[35] D. Stark and U. von Stockar, In situ product removal (ISPR) in whole cell biotechnology during the last twenty years. Adv. Biochem Eng/Biotechnol. 80 (2003) 150-175.
[36] Oak Ridge National Laboratory, Biomass as Feedstock for a Bioenergy and Bioproducts Industry: The Technical Feasibility for a Billion-Ton Annual Supply, U.S. Department of Energy, DOE/GO-201995-2135, 2005.
[37] S. Kim, B.E. Dale, Global potential bioethanol production from wasted crops and crop residues, Biomass and Bioenergy 26 (2004) 361–375.
[38] J. Perez, J. Munoz-Dorado, T. de la Rubia, J. Martinez, Biodegradation and biological treatments of cellulose, hemicellulose and lignin: an overview, Int Microbiol 5 (2002) 53–63.
[39] R.C. Brown, Biorenewable Resources: Engineering New Products from Agriculture, Iowa State Press, Ames, Iowa, 2003, Ch. 3.
[40] Badal C. Saha, Hemicellulose bioconversion, Ind Microbiol Biotechnol 30 (2003) 279–291.
[41] N. Mosier, C. Wyman, B. Dale, R. Elander, Y.Y. Lee, M. Holtzapple, M. Ladisch, Features of promising technologies for pretreatment of lignocellulosic biomass, Bioresource Technology 96 (2005) 673–686.

[42] Y. Sun and J. Cheng, Hydrolysis of lignocellulosic materials for ethanol production: a review, Bioresource Technology 83 (2002) 1–11.

[43] J.O. Metzger and U. Bornscheuer, Lipids as renewable resources: current state of chemical and biotechnological conversion and diversification. Appl Microbiol Biotechnol 71 (2006) 13–22.

[44] J.G. Grote, D.E. Diggs, R.L. Nelson, J.S. Zetts, F. K. Hopkins, et al., DNA photonics [Deoxyribonucleic acid]. Mol. Cryst. Liq. Cryst., 426 (2005) 3–17.

[45] L. Wang, J. Yoshida and N. Ogata, Self-assembled supramolecular films derived from marine deoxyribonucleic acid (DNA)-cationic surfactant complexes: Large-scale preparation and optical and thermal properties. Chem. Mater. 13 (2001) 1273-1281.

[46] Office of the Biomass Program, Multi Year Program Plan. U.S. Department of Energy, Office of Energy Efficiency and Renewable Energy, Washington, D.C. August 31, 2005.

[47] M.E. Himmel, M.F. Ruth, and C.E. Wyman, Cellulase for commodity products from cellulosic biomass. Curr. Opin. Biotechnol. 10 (1999) 358–364.

[48] M. Paster, J.L. Pellegrino, and T.M. Carole, Industrial Bioproducts: Today and Tomorrow. Report prepared by Energetics, Inc., Columbia, Maryland for the U.S. Department of Energy, Office of Energy Efficiency and Renewable Energy, Office of the Biomass Program, Washington, D.C. July 2003.

[49] T. Werpy and G. Petersen, Top Value Added Chemicals From Biomass, Volume I: Results of Screening for Potential Candidates from Sugars and Synthesis Gas. U.S. Department of Energy, Office of Energy Efficiency and Renewable Energy, Office of the Biomass Program, Washington, D.C. August 2004.

[50] S.Y. Lee, S.H. Hong, S.H. Lee, and S.J. Park, Fermentative production of chemicals that can be used for polymer synthesis. Macromol. Biosci. 4 (2004) 157–164.

[51] E. Kimura, Metabolic engineering of glutamate production, Adv. Biochem. Eng./Biotechnol. 79 (2003) 37–57.

[52] T. Willke and K.D. Vorlop, Biotechnological production of itaconic acid. Appl. Microbiol. Biotechnol. 56 (2001) 289–295.

[53] H. Songa and S.Y. Lee, Production of succinic acid by bacterial fermentation. Enz. Microb. Technol. 39 (2006) 352–361.

[54] N. Cao, J. Du, C.S. Gong, and G.T. Tsao, Simultaneous production and recovery of fumaric acid from immobilized *Rhizopus oryzae* with a rotary biofilm contactor and adsorption column. Appl. Environ. Microbiol. 62 (1996) 2926–2931.

[55] E. Winkelhausen and S. Kuzmanova, Microbial conversion of D-xylose to xylitol. J. Ferment. Bioeng. 86 (1998) 1–14.

[56] N. Ladygina, E.G. Dedyukhina, and M.B. Vainshtein, A review on microbial synthesis of hydrocarbons. Process Biochem. 41 (2006) 1001–1014.

[57] A. Steinbuchel, Non-biodegradable biopolymers from renewable resources: perspectives and impacts. Curr. Opin. Biotechnol. 16 (2005) 607–613.

[58] C.S.K. Reddy, R. Ghai, Rashmi, V.C. Kalia. Polyhydroxyalkanoates: an overview. Bioresource Technol. 87 (2003) 137–146.

[59] M. Ashiuchi and H. Misono, Biochemistry and molecular genetics of poly-γ-glutamate synthesis. Appl. Microbiol. Biotechnol. 59 (2002) 9–14.

[60] E.M. Silva and S.T. Yang, Continuous production of lactic acid from acid whey by *Lactobacillus helveticus* in a fibrous-bed bioreactor, J. Biotechnol., 41 (1995) 59–70.

[61] S.T. Yang, Y. Huang, and G. Hong, A novel recycle batch immobilized cell bioreactor for propionate production from whey lactose, Biotechnol. Bioeng., 45 (1995) 379–386.

[62] Y. Huang and S.T. Yang, Acetate production from whey lactose using co-immobilized cells of homolactic and homoacetic bacteria in a fibrous-bed bioreactor, Biotechnol. Bioeng., 60 (1998) 498–507.

[63] S.T. Yang, Y.L. Huang, Z. Jin, Yan Huang, H. Zhu, and W. Qin, Calcium magnesium acetate at lower-production cost: production of CMA deicer from cheese whey, Final report to Federal Highway Administration and New York State Energy and Development Authority; Publication No. FHWA-RD-98-174; April 1999.

[64] N. Albayrak and S.T. Yang, Production of galacto-oligosaccharides from lactose by *Aspergillus oryzae* β-galactosidase immobilized on cotton cloth, Biotechnol. Bioeng., 77 (2002) 8–19.

[65] S.T. Yang, Y.M. Lo, and D.B. Min, Xanthan gum fermentation by *Xanthomonas campestris* immobilized in a novel centrifugal fibrous-bed bioreactor, Biotechnol. Progress, 12 (1996) 630–637.

[66] C. Bro, B. Regenberg, J. Forster, and J. Nielsen, In silico aided metabolic engineering of *Saccharomyces cerevisiae* for improved bioethanol production. Metabolic Eng. 8 (2006) 102–111.

[67] A. Aristidou and M. Penttila, Metabolic engineering applications to renewable resource utilization. Curr. Opin. Biotechnol. 11 (2000) 187–198.

[68] EuropaBio, White biotechnology: Gateway to a more sustainable future. EuropaBio, Brussels, Belgium. (www.mckinsey.com/clientservice/chemicals/pdf/BioVision_Booklet_final.pdf)

[69] A.J. Ragauskas, C.K. Williams, B.H. Davison, G. Britovsek, J. Cairney, C.A. Eckert, W.J. Frederick Jr., J.P. Hallett, D.J. Leak, C.L. Liotta, J.R. Mielenz, R. Murphy, R. Templer, T. Tschaplinski, The path forward for biofuels and biomaterials. *Science* 311 (2006) 484–489.

Bioprocessing for Value-Added Products from Renewable Resources
Shang-Tian Yang (Editor)

Chapter 2. Application of Genomic and Proteomic Data for Bioprocess Analysis and Optimization

An-Ping Zeng[a], Jibin Sun[a], Wei Wang[b], Hongwu Ma[a] and Wolf-Dieter Deckwer[b]

[a]Group of Systems Biology, [b]Group of TU-BCE, GBF- German Research Center for Biotechnology, Mascheroder Weg 1, 38124 Braunschweig, Germany

1. INTRODUCTION

Since the first complete genome sequencing of the bacterium *Haemophilus influenzae* in 1995 [1] the genomes of more than 170 organisms have been completely sequenced and more than 190 sequencing projects were active by July 2004. The sequenced organisms include many biotechnologically important production strains such as *Escherichia coli, Bacillus subtilis, Saccharomyces cerevisiae, Corynebacterium glutamicum, Aspergillus niger* and lactic acid bacteria (http://www.ncbi.nlm.nih.gov/Genomes/index.html). Rapid advances in genome sequencing and methods for functional genomic studies (e.g. transcriptomic, proteomic and metabolomic analyses) make it possible to comprehensively characterize different production strains/mutants and even to reconstruct and analyze the genome-wide genetic and metabolic networks involved in cell growth and metabolism [29]. This opens up new horizons for the development and optimization of bioprocesses both at the molecular (i.e. via metabolic engineering of strains) [10, 11] and process engineering level (i.e. via improved medium design and feeding strategy). In this chapter we focus on the applications of genomic and proteomic analyses for bioprocess characterization and optimization. Transcriptomic and metabolomic analyses are also very promising tools for bioprocess development. The techniques for the generation and evaluation of transcriptomic and metabolomic data are less well established than the tools for genomic and proteomic analyses and their direct applications in bioprocess engineering are still very limited; however, their potential for the breakthrough to a target-oriented systems biotechnology is expected to be of immense value.

In the following, we will first address the use of genomic data for the construction and analysis of potential metabolic pathways and metabolic and regulatory networks. We will then briefly mention the general application of proteomics to bioprocess analysis. This is followed by demonstrating the practical use of genomic and proteomic approaches for the analysis and optimization of a real fermentation process in more detail.

2. APPLICATIONS OF GENOMIC AND PROTEOMIC ANALYSES

2.1. Reconstruction of metabolic pathways and networks from genome data

The starting point for the application of genomic data to the analysis of a bioprocess is to identify those genes that are potentially involved in information processing, metabolism, and other cellular processes related to cell growth and biosynthesis. From a metabolic engineering point of view, the potential metabolic pathways and network are of primary interest. Several methods and metabolic databases, such as EcoCyc, WIT and KEGG [2, 3, 5], are available to reconstruct the organism specific metabolic pathways and network from genome information. The KEGG map is an easy-to-use research platform for finding the bioreactions and pathways of organisms from genome data. The general scheme for constructing the potential metabolic pathways and network from the genome sequence is shown in Fig.1. The method is based on annotating a genome; this means identifying the open reading frame (ORF) or protein coding sequence (CDS) and its function assignment. For the reconstruction of the metabolic pathways and network, the identification of the CDSs with the appropriate EC numbers (enzyme) is compulsory. For an annotated organism the gene-enzyme relation can be extracted from databases, such as KEGG. By using a suitable reaction database, for instance KEGG LIGAND, one can obtain further information on relations among the enzymes and the reactions they catalyze. Thus, one achieves a list of reactions or pathways from the list of enzymes. Of course, there is no simple one-to-one relationship between an enzyme and a reaction. One enzyme may catalyze several different reactions and the same reaction may be catalyzed by different enzymes. For example, the enzyme fatty-acid synthase (2.3.1.85) catalyzes 31 reactions in the fatty acid synthesis pathway, while the reaction R04535:

$$\text{3-Hydroxydecanoyl-[acyl-carrier protein]} = \text{trans-Dec-2-enoyl-[acp]} + H_2O \quad (1)$$

is catalyzed by 5 different enzymes (2.3.1.85, 2.3.1.86, 4.2.1.58, 4.2.1.60, 4.2.1.61). Unfortunately, in most enzyme databases only the main reaction is listed for each enzyme. A complete metabolic network can not be reconstructed from such enzyme databases.

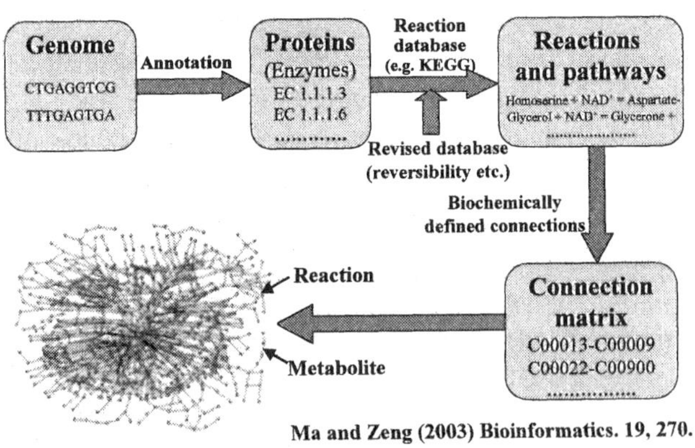

Fig.1. A general scheme to construct a metabolic pathway and network from genome data.

For computer processing the compound names in bioreactions are converted into compound indices and the reactions are converted into index representation. For example, the following reaction:

$$\text{L-Aspartate} + \text{ATP} = \text{4-Phospho-L-aspartate} + \text{ADP} \qquad (2)$$

is converted into:

$$\text{C00049} + \text{C00002} = \text{C03082} + \text{C00008} \qquad (3)$$

The metabolic network of a specific organism can be represented by a directed graph [7]. In this representation the directed connections (corresponding to irreversible reactions in a metabolic network) are called arcs, and undirected connections (reversible reactions) are called edges. For example, for the reversible reaction:

$$2\ \text{C00022} = \text{C00900} + \text{C00011} \qquad (4)$$

the corresponding edges are C00022-C00900 and C00022-C000011, where C00022 is the compound index of the substrate, C00900 and C00011 are the indices of the products. In such a way, we can write a list of arcs and edges that represents the metabolic network structure for every organism from its reaction constituents. The graph shown in Fig.1 is the so-called metabolic graph of the metabolic network. It can be also represented as a reaction graph in which reactions are defined as nodes and metabolites as connections between the nodes. A directed link from reaction node A to reaction node B is added if the product of reaction A is the substrate of reaction B [12]. Both the metabolic and reaction graphs can be used for further structural analysis of metabolic networks.

Information about the reversibility of bioreactions is also essential for the expression of a metabolic network as a directed graph. This kind of information is, however, often not included in databases of bioreactions. Although reaction direction is shown in the well-known KEGG metabolic maps, inconsistencies in different maps exist and some irreversible reactions exhibit opposite directions. Ma and Zeng [7] manually checked the reversibility information and corrected some of the mistakes in the KEGG database. The determination of reversibility for a reaction is sometimes a matter of debate. Ma and Zeng [7] suggested eleven kinds of reactions that can be considered as irreversible in metabolic networks. With the extensively revised KEGG LIGAND gene-enzyme-reaction database, Ma and Zeng [7] reconstructed the metabolic networks from genome information for more than 100 fully sequenced organisms. The results are saved in a binary matrix R for comparative studies, where $R_{ij} = 1$ if reaction i exists in the metabolic network of organism j.

As outlined before, the method described above is based on an annotated genome. The genome annotation information for any individual organism is continually updated. For most of the sequenced organisms, only about half of the open reading frames (ORFs) have been assigned to functions and errors may exist for certain annotated genes. However, most of the unknown genes or uncertain annotations are related to regulatory and signal transduction pathways. Most genes coding for metabolic enzymes can be accurately identified. This makes it possible to construct fairly complete metabolic networks for the fully annotated organisms. The KEGG databases are continually updated by incorporating the most up-to-date gene annotation information from various databases [13]. We can update the enzyme-gene and reaction-enzyme relations from the newest KEGG database files to construct strain specific metabolic networks with the newest information. The updating process can be carried out automatically by program. For newly sequenced and annotated organisms, the corresponding metabolic networks can also be constructed through these updating programs.

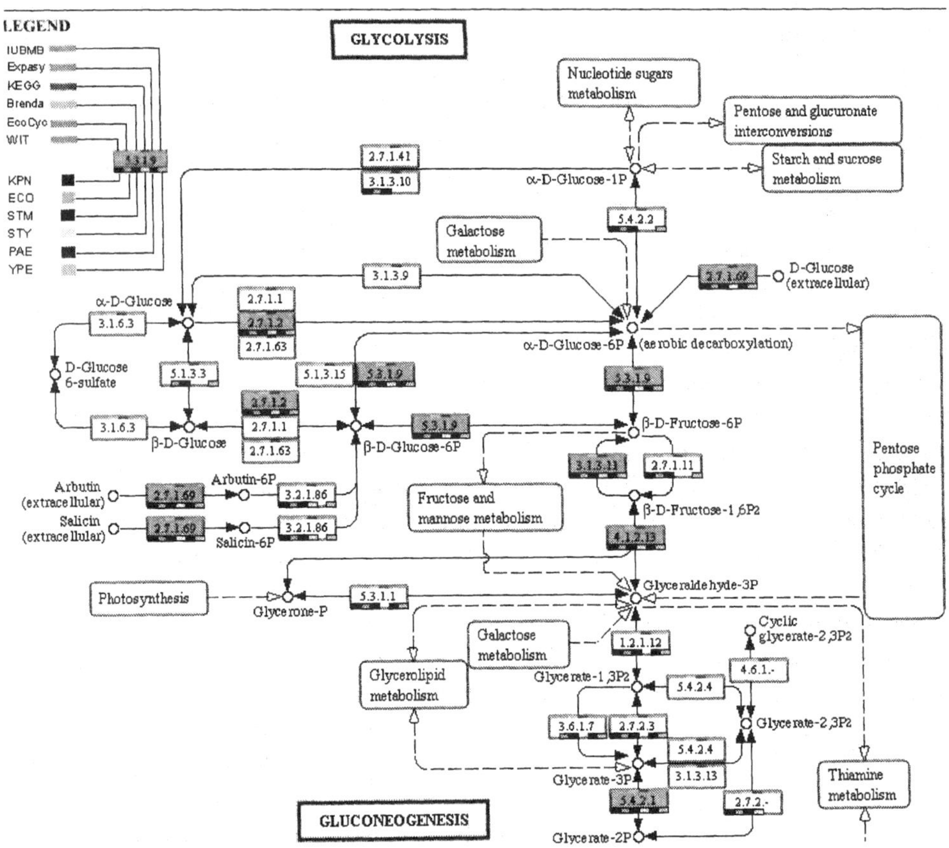

Fig. 2. Glycolysis pathways as an example for demonstration of metabolic comparisons among different organisms. KPN: *K. pneumoniae* MGH78578, ECO: *Escherichia coli* K-12 MG1655, STM: *Salmonella typhimurium* LT2, STY: *Salmonella typhi*, PAE: *Pseudomonas aeruginosa* PA01, YPE: *Yersinia pestis* strain CO92. Gray background with enzyme EC number means that this enzyme exists in all the compared organisms. Rectangles under the EC number box represent links to the strain-specific annotation of the corresponding enzyme. Bars above the EC number link the enzyme to the corresponding entry in different public databases. (For colour figure, refer to BMC Bioinformatics 2004, 5:112)

The annotation-based method described above has some drawbacks [14]. To have a reliable annotation, a genome sequence data set with more than 10 times coverage of the genome is normally needed. Despite many advances, annotation is still time and resource consuming. Available methods for annotation often encounter difficulties when dealing with unfinished, error-containing genome sequences. On the other hand, an early exploitation of the genome data is desirable for bioprocess development. In an effort to accelerate the use of raw genome sequence for functional studies, Sun and Zeng [14] developed a fast method for identifying coding sequences for proteins (particularly for metabolic enzymes) directly from unannotated low-coverage genomic data, which are then used to construct the metabolic network. Their method, based on a reversed approach to the conventional query procedure, uses extensive DNA and protein databases for predicting all the CDSs in a local database of the unannotated genome sequence of an organism. Functions are simultaneously assigned to the predicted CDSs. By using two genome sequence data sets of *Klebsiella pneumoniae* with different genome coverage, it was demonstrated that a 3.9-fold sequence coverage of the bacterial genome is sufficient for the *in silico* reconstruction of the entire metabolic network. Compared to other gene finding methods, such as CRITICA, the new method is more suitable for exploiting sequences with low genome coverage. Based on the new method, a program called IdentiCS (Identification of Coding Sequences From Unfinished Genome Sequences) has been created that combines the identification of the CDSs with the reconstruction, comparison and visualization of metabolic networks (Fig. 2) [14].

Following the reconstruction of the metabolic networks, the next challenge is the analysis of their structure and function. For bioprocess analysis and optimization, a network-level understanding of metabolic potential, maximum yield, optimal pathway(s), and the regulation and control of metabolic fluxes for the synthesis of the target product is desirable. Pathway analysis methods, such as metabolic flux balance [15, 16], elementary flux mode analysis, and extreme pathway analysis, have been shown to be useful tools for investigating the metabolic capacity and pathway structure of networks [17–19]. Some efforts have been made to use flux balance analysis for predicting the flux distribution in a genome-based metabolic network under certain physiological conditions [15, 16, 20]. Most of the methods are, however, hampered by the combinatorial explosion problem when applied to such large-scale networks as those reconstructed from genomic data. Hence, a decomposition of the network is necessary before performing functional analysis using the methods of pathway analysis mentioned above. Metabolic networks are organized in a modular and hierarchical structure [12, 21–23]. Methods for the rational decomposition of a metabolic network into relatively independent functional subsets are also essential in order to better understand the modularity and organization principles of large-scale, genome-wide networks.

Several methods for decomposing metabolic networks have been proposed [21, 22, 24]. Ravasz et al. [21] used a topological overlap matrix for network decomposition and classified the metabolites into different subsets. However, from a biological viewpoint, a subset of metabolites cannot sufficiently define a unique functional pathway or module. Reactions catalyzed by enzymes are the functional and regulatory units in a metabolic network. Therefore, a module should be a subset of reactions, not merely a subset of metabolites. Some metabolites can participate in reactions belonging to different modules. Schuster et al. [24] proposed a decomposition method that removed the highly connected metabolites. Using this method, they analyzed the metabolic network of the parasitic bacterium *Mycoplasma pneumoniae*. Compared to other organisms, *M. pneumoniae* has a relatively small metabolic network that includes less than 200 reactions [25]. Using the metabolic network of *E. coli* as an example, Gagneur et al. [23] extended the method of Schuster et al. [17] by serially grouping reactions in order by the degree of the metabolites they connect. Thus, a hierarchical organizational structure was also found. A potential drawback of these degree-based methods is that the global organization structure may be invisible from local features, such as connection degrees. Considering this drawback, Holme et al. [22] suggested a method for revealing the sub-network hierarchies of the network by successively removing reactions of high betweenness centrality. With this method, it is generally difficult to obtain subnetworks of similar size. Instead, a large subnetwork and many isolated nodes are often found.

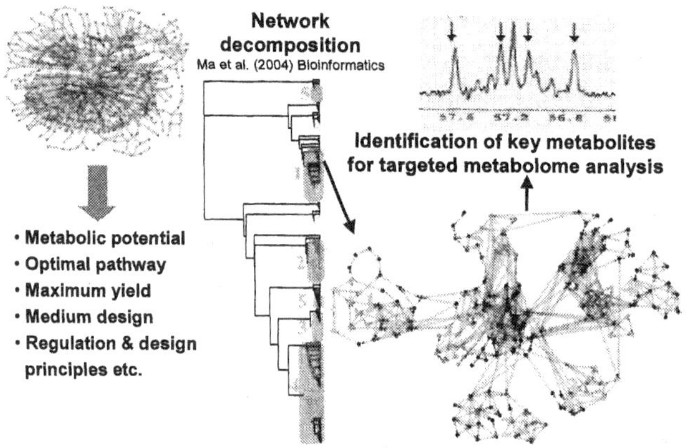

Fig. 3. From network to modules for functional study.

Ma and Zeng [7, 26] studied the global topological and connectivity structures of genome-wide metabolic networks from various organisms. Among others, they found that a bow-tie organizational structure consisting of three major subsets of metabolites and reactions, i.e. a fully connected sub-network, a substrate subset, and a product subset, exist in all organisms. This uncovered structure feature makes the structural and functional analysis of large-scale metabolic networks more feasible. Based on the bow-tie structure, Ma et al. [12] proposed a new decomposition method that allows for better identification of functional modules and the inherent organization of metabolic networks (Fig. 3). This method first calculates the path length between the nodes in giant strong component of the bow-tie structure and then uses it as a dissimilarity measure for the reaction nodes [27]. Based on this dissimilarity measure, hierarchical classification methods, such as neighbor joining, are used to calculate the hierarchical classification tree from which reaction modules are obtained. These modules are then extended to include reactions belonging to the IN and OUT subsets of the bow-tie structure. Investigations of the biological function of the reactions in each module showed that these structurally identified modules are really functionally related subsets. These new methods are being used for identifying key modules and metabolites for metabolomic studies of bioprocesses.

2.2. Genotype-phenotype mapping for metabolic engineering

Inverse metabolic engineering (IME) was proposed as a promising metabolic engineering strategy several years ago [28, 29]. The basic idea of IME is to determine the genotype for a desired phenotype and to endow the phenotype on another organism. Therefore, elucidating the relationships between genotypes and phenotypes is essential for IME in particular and for other areas of biotechnology and biomedicine in general. Various approaches have been proposed for genotype-phenotype mapping. Of these, methods based on the stoichiometric matrix of a metabolic network, such as extreme pathway analysis (EP), elementary mode analysis (EM), and flux balance analysis (FBA), have been shown to be very useful [30–35]. The only prerequisite for these methods is the metabolic stoichiometric information, which can be obtained at the genome level from the metabolic network reconstructed from genome data as described above. Using EM or EP, one can calculate all the possible metabolic phenotypes for a determined genotype and then study how the cell changes its phenotype under different environmental conditions, such as varying carbon sources [18]. FBA calculates the optimized state (may be multiple states [36]) for a desired phenotype (such as maximal growth or energy production) by using constrained optimization methods. It has been successfully used in phenotype prediction from the reconstructed metabolic networks of *E. coli* and *S. cerevisiae* [37–39] and in identifying the high-flux backbone of the *E. coli* metabolic network [40].

2.3. From genome to regulatory network

Though the methods described so far can successfully predict the metabolic phenotype, they can not provide explanations of how the cell controls fluxes in order to access a desired phenotype. Therefore, it is extremely important to integrate the regulatory information. Regulatory information provides new constraints on the metabolic network and thus can reduce the number of possible phenotypes [41, 42]. The regulatory information can be obtained from databases or directly from literature [43–45]. Compared with the information on metabolic networks, the available information on genome scale transcriptional regulatory networks is rather limited. Even for the best studied model microorganism, *E. coli,* the genome scale network reconstructed from RegulonDB contains only about one thousand genes, a mere one quarter of encoded genes in its genome [45]. Genome-wide expression data from microarray experiments offer a new possibility for reconstructing the regulatory network (so-called reverse engineering of biological networks) [46]. However, the reliability of the data from such high throughput experiments must be carefully investigated before being applied. Recently, Ma et al. [9] extended the *E. coli* transcriptional regulatory network by integrating information from different data sources: RegulonDB [45], EcoCyc [5], the paper of Shen-Orr et al. [47], and a literature survey, resulting in a network with 1278 genes and 2724 regulatory interactions. Among the 1278 genes in the network, ca. 500 are metabolic genes which code for 300 metabolic enzymes. This means that these enzymes are expressed only under specific conditions. Thus, we can construct condition specific metabolic networks which are much smaller than the network directly derived from the genome information. From these condition specific metabolic networks we can make more reasonable phenotype predictions.

Another more important use of regulatory information is the choosing of regulatory genes as targets for gene modification. The classical method of metabolic engineering, alleviating the network bottleneck by enzyme overexpression, has only limited practical success [48]. One possible reason is that the target gene is controlled by other genes, thus more gene copies does not mean more enzyme proteins expressed. The long history of strain improvement practice has shown that modifying the regulatory relationships (reducing or removing the inhibiting or activating effect) is more efficient in screening improved strains for the overproduction of objective products.

In order to select a regulatory gene as the target for strain improvement, we must know the regulatory relationships in the cell. Unfortunately, our knowledge of regulatory networks is very limited when compared with that of metabolic networks. However, recently developed high throughput technologies, such as genome sequencing, microarray, proteomics, and metabolomics, provide powerful tools for studying the gene regulatory relationships in the cell. A general strategy for using these high throughput techniques in process development can be: find all the gene differences between the wild type strain and the strain with the desired phenotype by DNA sequencing; use microarray techniques to measure the transcriptional profiling to discover how these gene differences affect the expression of other genes and then combine this with literature data and other experimental technologies to elucidate the relations between these genes; estimate the flux changes in the metabolic network (e.g. using NMR or metabolomics) and investigate how the gene difference affects the fluxes; establish mathematical models to describe the genotype and phenotype relations and, through analysis, find an efficient way to engineer the genotype to further improve the strain (model driven process development).

2.4. Applications of proteomic analysis

With the successful genome sequencing of more and more organisms, the last few years have seen extraordinary improvements in proteome analysis or proteomics. Protein separation by one- or two-dimensional gel electrophoresis or using gel-free techniques, such as liquid chromatography, coupled with protein identification by mass spectrometric techniques, such as matrix-assisted laser desorption ionization time-of-flight mass spectrometry (MALDI-TOF MS) and electrospray ionization quadrupole-time-of-flight mass spectrometry (ESI-QqTOF MS) are widely used for large-scale proteome analysis. The emerging and rapid development of proteomics has had a strong impact on bioprocess development [49]. One important application of proteomics is the quantitative and comparative analysis of cellular protein expression patterns representing different metabolic states defined by precisely controlled experimental conditions. Despite some drawbacks, two-dimensional gel electrophoresis (2-DE) is still the most powerful technique for the separation and visualization of complex protein mixtures and is especially useful for the study of cellular physiological behavior by directly comparing differences in protein expression patterns. Knowledge generated in this way, i.e. the characterization of metabolic and regulatory networks and the identification of key enzymes involved in the formation of desired products, is valuable for directed strain optimization by metabolic engineering and for the rational improvement of cultivation parameters in order to increase the efficiency and productivity of bioprocesses.

According to this strategy, proteome analysis has been employed – sometimes in combination with other techniques such as transcriptome analysis or enzyme activity assay – to obtain comprehensive views of the physiological and metabolic responses of microorganisms to changed environmental conditions or to heterologous biosynthesis. For example, *E. coli* has been extensively used as a model system for the development of bacterial proteome analysis. Recombinant strains of *E. coli* have been characterized by proteomic analysis for the overproduction of various value-added bioproducts as reviewed by Lee and Lee [50].

Peng and Shimizu [51] showed the usefulness of studying the global metabolic regulations in the central metabolic pathways of *E. coli* K12 in response to different carbon sources by 2-DE and enzyme activity measurement. Lee and coworkers analyzed the physiological behavior of *E. coli* during high cell density cultivation (HCDC) through the combined use of transcriptome and proteome profilings [52, 53]. They characterized metabolic changes in central metabolic pathways and elucidated phenomena observed during HCDC, such as the reduced specific productivity of recombinant proteins due to the decreased level of amino acid biosynthetic enzymes, or the alteration of outer membrane permeability at the increased cell density. A combination of transcriptome analysis and proteome analysis was also used for monitoring process-related stress responses in lab-scale and industry-scale fermentations, such as heat or osmotic shock, glucose excess or oxygen limitation in poorly mixed zones [54].

Comparative 2-DE proteome analysis has often been applied to study the metabolic burdens and cellular responses to the production of recombinant proteins in microorganisms, such as *E. coli* and *B. subtilis*, in high cell density fed-batch fermentations (see reviews [50, 53]). In general, overproduction of heterologous proteins triggers the heat shock-like stress response mechanisms of the host cells, leading to enhanced expression of protein chaperones, such as DnaK and GroEL and Tig, but reduced expression of some ribosomal proteins [55–60].

Two research groups have also used 2-DE to characterize the protein expression changes of recombinant *E. coli* before and after induction for producing polyhydroxybutyrate (PHB). They found that accumulation of PHBs in *E. coli* acted as a stress on the cells, resulting in significantly up-regulated expressions of heat shock proteins, such as GroEL, GroES and DnaK. They correlated the increased synthesis of some enzymes to increased demands for acetyl-CoA and NADPH caused by PHB synthesis [61, 62].

Information obtained from proteome analysis can be used to engineer production strains and thus to improve the productivity of a bioprocess. Han and Lee [53] carried out proteome profiling of *E. coli* in response to the overproduction of a serine-rich human leptin and found that the expression levels of some enzymes for amino acid biosynthesis decreased. Those involved in the biosynthesis of serine-family amino acids were most likely to show significant decreases. Based on this result, they successfully designed a strategy for enhancing the cell growth rate by two-fold and the specific leptin productivity by four-fold through the manipulation and coexpression of the cysteine synthase A encoding gene.

3. BIOPROCESS ANALYSIS AND OPTIMIZATION GUIDED BY GENOMIC AND PROTEOMIC ANALYSES: THE EXAMPLE OF MICROBIAL PRODUCTION OF 1,3-PROPANEDIOL

The microbial production of 1,3-propanediol (PDO) is an important new industrial bioprocess. PDO is the monomer for a very promising new polyester, polytrimethylene terephthalate, with wide applications [63]. In nature, PDO is produced from glycerol fermentation by such organisms as *K. pneumoniae* and *Clostridium butyricum* (Fig. 4). The substrate glycerol can be obtained as a byproduct in the processing of plant oil, such as rapeseed oil, for biodiesel production. Efforts and impressive progress have also been made to produce PDO from cheaper substrates, such as glucose, by using recombinant microorganisms. Most of the fundamental knowledge about this fermentation process has been gathered through investigations of the model organism *K. pneumoniae*. In the following we will demonstrate with examples from our recent studies how genomic and proteomic analyses have helped to identify new genes and pathways and to better understand the unusual dynamic behavior of glycerol fermentation by this bacterium.

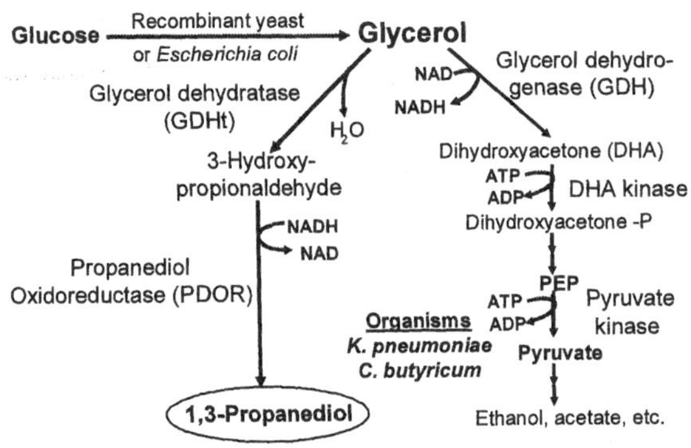

Fig. 4. Application example: microbial production of 1,3-propanediol.

3.1. Reconstruction and analysis of the *dha* regulon and related genes from genomic data

The enzymes for PDO production, namely glycerol dehydratase (GDHt) and 1,3-propanediol oxidoreductase (PDOR), are encoded for by genes found in the so-called dihydroxyacetone (*dha*) regulon (Figs. 4 and 5). Portions of the *dha* regulon in *K. pneumoniae* were first sequenced by Skraly et al. [64]. From the sequenced portions of the *dha* regulon, several authors have tried to use a genetic approach (metabolic engineering) to optimize the production of PDO from glycerol [65–68] and glucose [69–71]. Impressive progress has been made on producing PDO from glycerol or glucose from transformed *E. coli* that originally does not contain the *dha* regulon. However, the metabolic engineering of organisms that contain a native *dha* regulon often did not result in significant improvement of PDO formation. For example, Menzel [68] overexpressed GDHt and PDOR in *K. pneumoniae*. However, the recombinant strain did not produce more 1,3-propanediol than the wild type strain did under real fermentation conditions. Zeng [72] postulated a strong feedback inhibition and global regulations of gene expression of the *dha* regulon. The exact genetic basis and mechanisms for these kinds of feedback inhibition and global regulation are not yet clear.

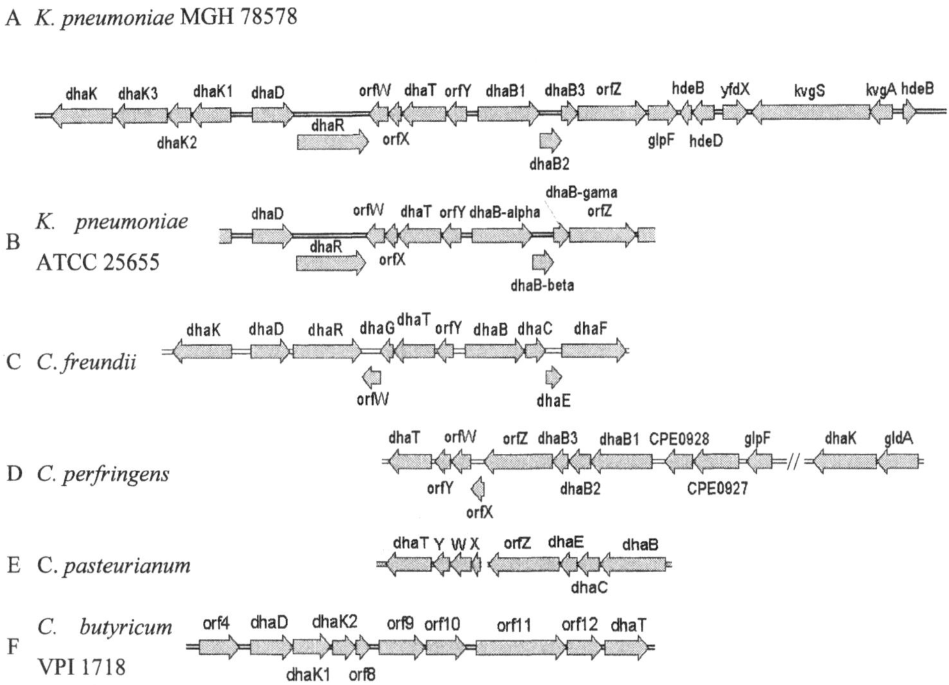

Fig. 5. Comparative genomic analysis of *dha* regulon and related genes in different organisms. A: *Klebsiella pneumoniae* MGH 78578; B: *K. pneumoniae* ATCC 25655; C: *C. freundii*; D: *C.* perfringens; E: C. *pasteurianum*; and F: *C. butyricum* VPI 1718. (From [73]. Reprinted with permission)

In 2002 the entire genome of *K. pneumoniae* strain MGH 78578 was sequenced and published (http://genome.wustl.edu). In order to gain more insights into the genetic organization and regulation of genes from the *dha* regulon, Sun et al. [73] reconstructed a larger and more complete *dha* regulon for *K. pneumoniae* from the genome data. Database searches of genomic data were also made in order to analyze and compare the *dha* regulon and related genes in different organisms with respect to gene organization, sequence similarity, and possible functions. This genomic analysis discovered another organism, *C. perfringens*, that contains a complete *dha* regulon bearing all the known enzymes and was later shown to be able to produce PDO from glycerol as well (own results not published). The components and their organization in the *dha* regulon of these two organisms differ considerably from each other and also from the previously partially sequenced *dha* regulons in *Citrobacter freundii*, *C. pasteurianum*, and *C. butyricum* (Fig. 5). Unlike all the other organisms, genes for the oxidative and reductive pathways of anaerobic glycerol metabolism in *C. perfringens* are located in two separate organization units on the chromosome. Overall, the genomic analysis resulted in the identification of several new genes and regulatory factors related to PDO production in different organisms. In particular, a novel dha kinase (DHAK II) that is phosphoenolpyruvate-dependent was identified in *K. pneumoniae* in addition to the known ATP-dependent *dha* kinase (DHAK I).

Dha kinase plays an important role in the oxidative pathway of glycerol metabolism which is coupled to the reductive pathway of glycerol for PDO production (Fig. 4). The discovered DHAK II has a very interesting property; namely, it simultaneously catalyzes the conversion of phosphoenolpyruvate to pyruvate (Fig. 6). This newly discovered metabolic pathway explains some hitherto not well understood experimental observations. For example, in our previous work with the glycerol metabolism of *K. pneumoniae,* it was observed that pyruvate kinase (PK) for the conversion of PEP to pyruvate had a quite low activity compared to the enzymes of oxidative glycerol dissimilation (e.g. GDH) and pyruvate metabolism (e.g. pyruvate: formate-lyase) [74, 75]. This led to the conclusion that pyruvate kinase (PK) is a limiting step in the oxidative pathway of glycerol utilization. Surprisingly, the measured *in vitro* activity of PK is in some cases even lower than the *in vivo* activity of this enzyme calculated from the flux of PEP to pyruvate, which appears to be unrealistic [74].

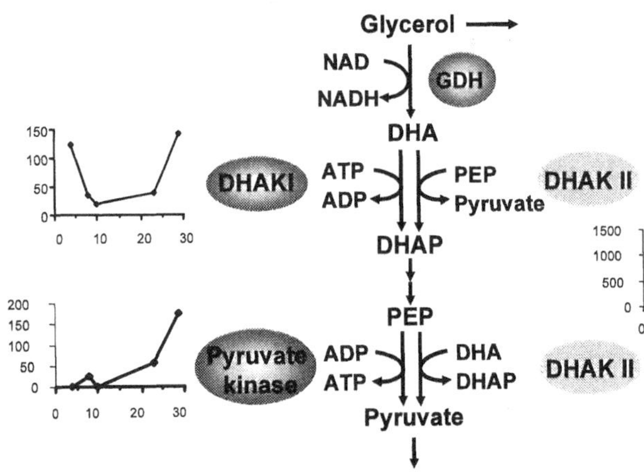

Fig. 6. Improved understanding of the oxidative pathway of glycerol metabolism. The time-course curves next to the enzymes show the spot intensity (normalized spot volume) change of the corresponding enzymes or one of their subunits detected by proteomic analysis.

Sun et al. [73] measured the activities of the enzymes DHAK I, DHAK II, PK, GDH, PDOR and GDHt in glycerol fermentation. It was shown that DHAK II had a higher activity than DHAK I and PK. The aforementioned seemingly unrealistic results can now be understood if we consider the contribution of the PEP-dependent DHAK II to the conversion of PEP to pyruvate. The above example demonstrates that genomic information can be very useful for understanding cellular metabolism. In general, information about the genetic makeup and regulation of genetic units in different organisms can guide the selection of the optimal organism and the desired genetic modification of metabolic pathways for developing efficient bioprocesses.

3.2. Proteomic analysis of glycerol fermentation

Whereas the genomic data gives important information about the metabolic potential of an organism, proteomic analysis of protein expression patterns under given experimental conditions can provide much information about the functionality and regulation of the cellular metabolism. Wang et al. [76] conducted a proteomic analysis of glycerol fermentation by *K. pneumoniae*. One of their goals was to understand the culture dynamics at the proteome level. The fed-batch fermentation of glycerol to PDO by *K. pneumoniae* displayed an unusual dynamic behavior that can be clearly divided into four distinct phases according to the cell growth and CO_2 evolution rate (Fig. 7).

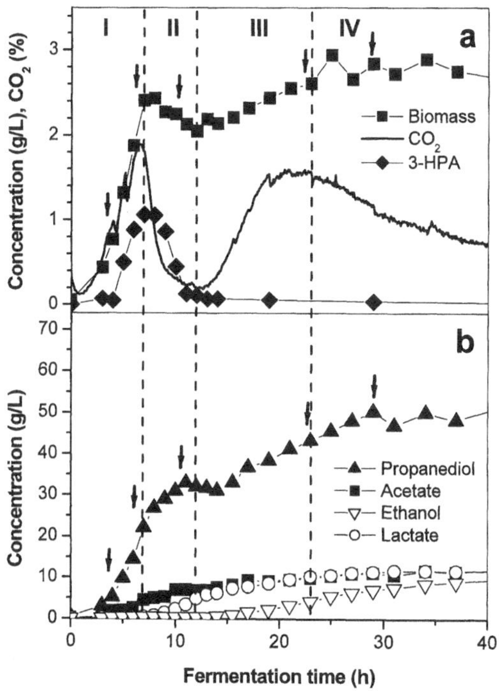

Fig.7. Time course of cell growth and product formation in fed-batch fermentation of glycerol by *Klebsiella pneumoniae*: arrows indicate sampling time for 2D-proteomic analysis. (From [76]. Reprinted with permission)

In Phase I, cells fast grew to a relatively high concentration almost without a lag phase and the CO_2 production reached a maximum. After that, cells ceased to grow and both the biomass concentration and CO_2 production reached a minimum (Phase II). In Phase III, cells grew again and the CO_2 production reached a second maximum. In Phase IV, the CO_2 production continuously declined and the biomass concentration remained relatively constant. These varied phases of cell growth and CO_2 production were also reflected in the product formation (Fig. 7b). The accumulation of some toxic metabolites, such as acetate and 3-hydroxypropionaldehyde (3-HPA), was considered as a possible explanation for the observed changes in cell growth and metabolism during this fermentation process. In fact, a significant accumulation of 3-HPA was observed in the first phase (Fig. 7a), which was then re-consumed in the second phase. It is known that 3-HPA is very toxic to the growth of many microorganisms. It may be possible that the accumulation of 3-HPA caused the metabolic changing from Phase I to Phase II. The re-consumption of 3-HPA allowed the cells to grow again (Phase III). The change from Phase III to Phase IV may be due to the accumulation of other major fermentation products (PDO, lactate, and ethanol). In order to understand the reasons for the metabolic changes and its regulation at the enzymatic level, Wang et al. [76] measured the *in vitro* activities of the three enzymes, GDHt, GDH and PDOR, encoded in the *dha* regulon. Surprisingly they showed apparently different patterns of expression despite belonging to the same regulon and thus, theoretically being controlled by the same regulators.

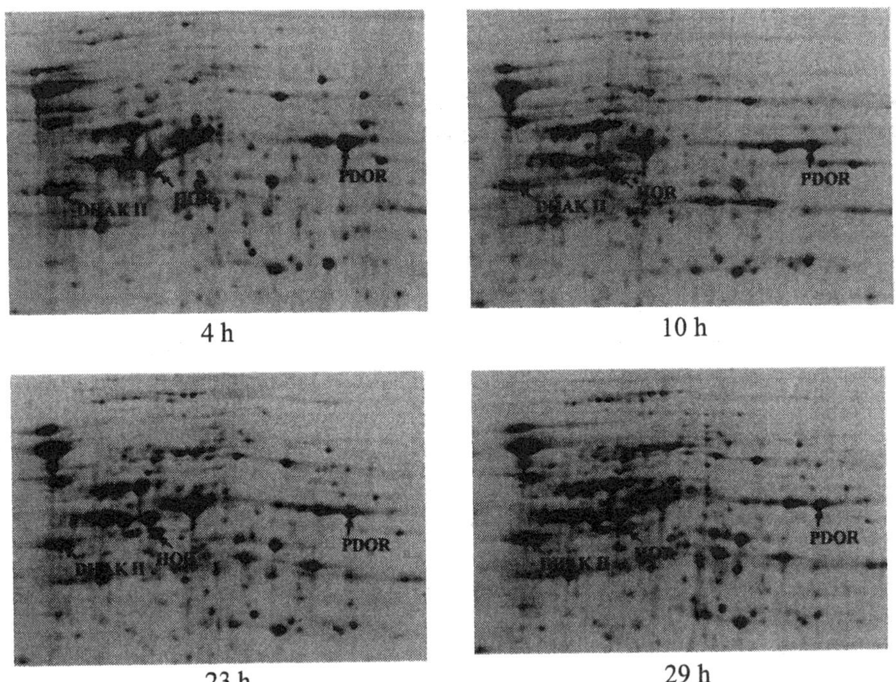

Fig. 8. Changes of protein expression measured by 2-DE as function of fermentation time. (From [76]. Reprinted with permission)

In order to understand the culture dynamics and patterns of enzyme formation at a more systemic level, the expression patterns of intracellular proteins of *K. pneumoniae* from different phases of the fed-batch fermentation were analyzed using 2-DE and other measurements. Using Immobiline DryStrip (pH 3 to 10 NL) for the first-dimensional separation, SDS-PAGE for the second-dimension separation and Coomassie staining, more than 200 spots can usually be detected. As shown in Fig. 8 by windows of the 2DE map, significant changes of the expression patterns can be ascertained at different time points of the fermentation.

One of the often encountered difficulties in proteomic analysis with 2-DE is the accurate identification of the spots on the gels. Peptide mass fingerprinting (PMF) by MALDI-TOF MS is the widely used technique for protein identification. This approach relies, however, on the presence of the proteins studied in publicly-accessible protein databases or the availability of annotated genome sequences for the organism. Quite often, the highly expressed proteins or proteins of interest (i.e. with significantly changed expression level) are identified as "hypothetical proteins" by conventional database searches. Wang et al. [77] investigated the reliability of using raw genome sequences for identifying proteins by PMF without the need for additional information, such as amino acid sequences. The method for proteomic analysis of glycerol fermentation by *K. pneumoniae* is demonstrated. For 197 spots excised from 2-DE gels and submitted for mass spectrometric analysis 164 spots were clearly identified as 122 individual proteins. 95% of the 164 spots can be successfully identified merely by using peptide mass fingerprints and a strain-specific protein database (ProtKpn) constructed from the raw genome sequences of *K. pneumoniae.* Cross-species protein searching in public databases mainly resulted in the identification of 57% of the 66 highly expressed protein spots in comparison to the 97% identified using the ProtKpn database. Ten *dha* regulon-related proteins that are essential for the initial enzymatic steps of anaerobic glycerol metabolism were successfully identified using the ProtKpn database, whereas none of them could be identified by cross-species searching. With this improved method of protein identification, the dynamic expression pattern changes of more than 120 proteins in glycerol fermentation were followed. The expression profiles for enzymes of the major fermentation pathways and some of the proteins of biosynthesis are depicted in Fig 9. These include all three known enzymes of the *dha* regulon mentioned above (PDOR, GDH, and two of the three subunits of GDHt) and the new enzyme DHAK II with its two subunits. It is worth mentioning that DHAK I, which had been considered responsible for the conversion of DHA to DHAP during glycerol fermentation, had a much lower expression level than DHAK II throughout the fermentation.

Proteomic analysis also led to the identification of an interesting spot that turned out to be a hypothetical oxidoreductase (HOR) and is potentially important for PDO production [76]. The expression level of this protein is higher than the expression of PDOR (Fig. 9). As shown in Fig. 10, the protein sequence alignment of the amino acid sequence of this protein is 89% identical to the hypothetical oxidoreductase of *E. coli (YqhD)*, 38% identical to the NADH-dependent butanol dehydrogenase A of several bacteria, and 25% identical to the PDOR of *C. freundii* and *K. pneumoniae.*

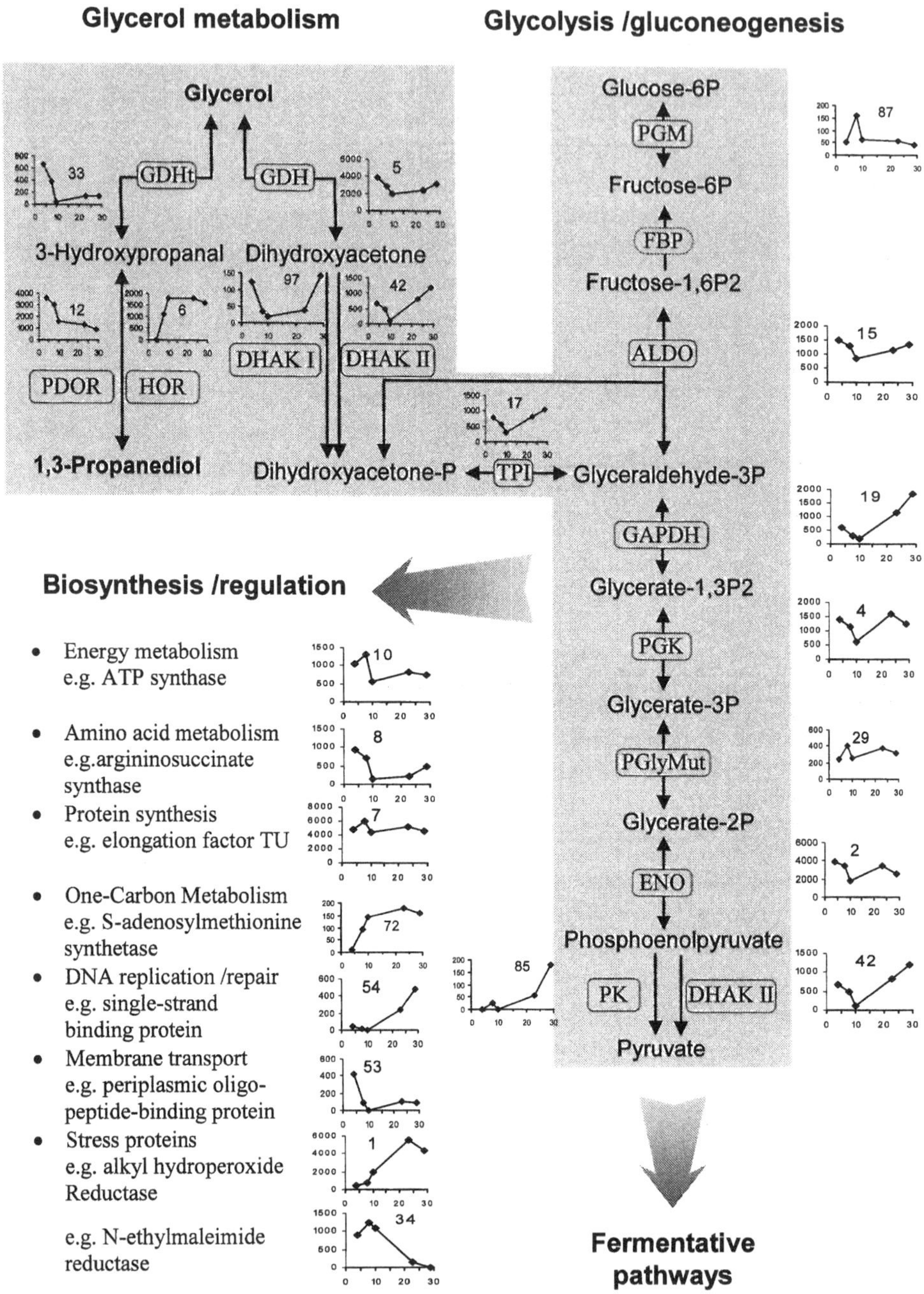

Fig. 9. Schematic presentation of identified proteins (enzymes) and their functions related to the metabolism of glycerol and biosynthesis. Also shown are the expression change patterns of these proteins during the fed-batch fermentation. X axis: fermentation time in hours; Y axis: the spot intensity (proportional to the quantity) of the corresponding protein or one of its subunits detected by proteomic analysis. (Adapted from [76]. Reprinted with permission)

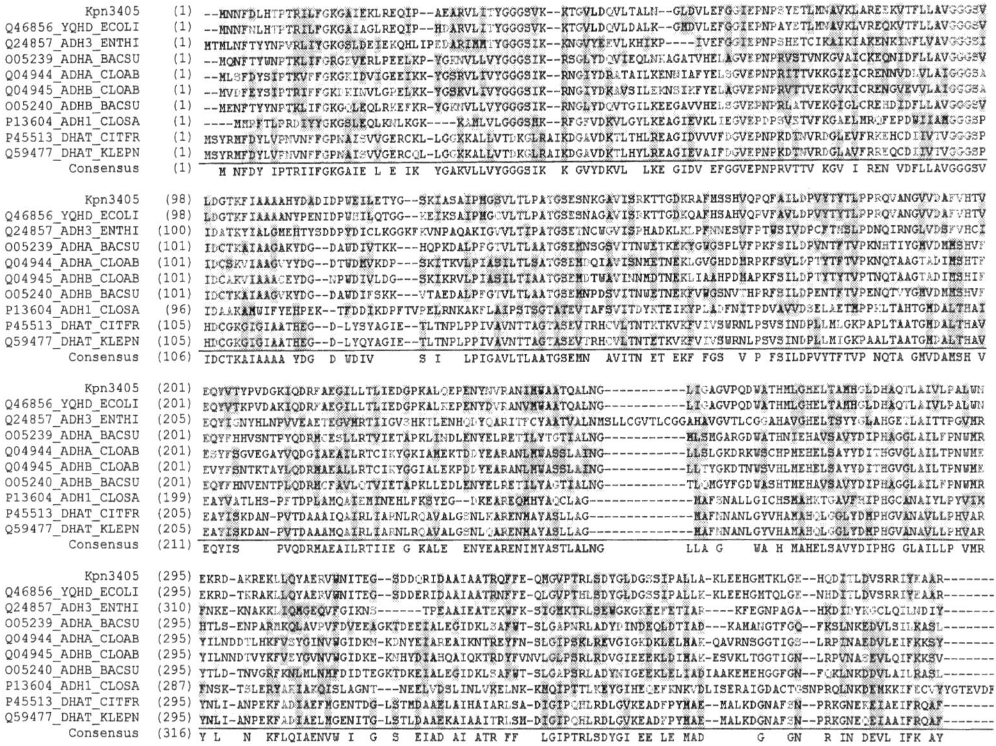

Fig. 10. Protein sequence alignments of a new putative 1,3-propanediol oxidoreductase Kpn3405 of *K. pneumoniae* with similar proteins in SWISS-PROT: Q46856_YQHD_ECOLI, hypothetical oxidoreductase YqhD of *E. coli*; Q24857_ADH3_ENTHI, alcohol dehydrogenase 3 of *Entamoeba histolytica*; O05239_ADHA_BACSU, probable NADH-dependent butanol dehydrogenase 1 of *Bacillus subtilis*; Q04944_ADHA_CLOAB, NADH-dependent butanol dehydrogenase A of *Clostridium acetobutylicum*; Q04945_ADHB_CLOAB, NADH-dependent butanol dehydrogenase B of *C. acetobutylicum*; O05240_ADHB_BACSU, probable NADH-dependent butanol dehydrogenase 2 of *Bacillus subtilis*; P13604_ADH1_CLOSA, NADPH-dependent butanol dehydrogenase of *Clostridium saccharobutylicum*; P45513_DHAT_CITFR, 1,3-propanediol oxidoreductase of *Citrobacter freundii*; Q59477_DHAT_KLEPN, 1,3-propanediol oxidoreductase of *K. pneumoniae*. (Adapted from [76]. Reprinted with permission.)

The HOR of *E. coli* was confirmed to be capable of replacing the function of PDOR in *E. coli* [78]. *E. coli*, which does not contain any native genes in the *dha* regulon, can produce a high amount of PDO after transformation with the *dha* regulon genes of *K. pneumoniae* except for the PDOR gene (*dhaT*). Interestingly, such transformants can even produce much higher amounts of PDO than strains transformed with the entire *dha* regulon. When the native gene *yqhD* was disrupted, the high producer completely lost its ability to produce PDO, proving that YqhD is directly responsible for reducing 3-HPA to PDO in the recombinant *E. coli* strain. Thus, the HOR of *K. pneumoniae*, which is strongly similar to YqhD, is most probably involved in PDO production in *K. pneumoniae* as well. Additional evidence for the existence of enzyme(s) other than PDOR for the formation of PDO in *K. pneumoniae* comes from a comparison of changes in PDOR determined by both 2D-gel analysis and its directly measured apparent activity by enzyme assay [76].

The genomic and proteomic analysis of genes and proteins involved in the glycerol metabolism revealed the existence and functionality of three parallel reactions at the key points of glycerol conversion and PDO production catalyzed by the pairs PDOR/HOR, DHAK I/DHAK II, and PK/DHAK II (Fig. 9). The observed unusual dynamic behavior of this fermentation process, as shown in Fig. 7, is obviously caused by the varied expression levels of these proteins at different times in the fermentation. The discovery of these parallel reactions may also help to understand the oscillation and multiplicity phenomena observed for this fermentation process [79, 80].

In addition to the key enzymes directly involved in glycerol metabolism and PDO formation the expression patterns of several proteins involved in amino acid metabolism and cofactor formations were also investigated by proteomic analysis. This hinted to a limited synthesis of a key amino acid in glycerol fermentation. By amending the limitation of this amino acid in a fed-batch process, both cell growth and PDO formation could be significantly improved (Fig. 11).

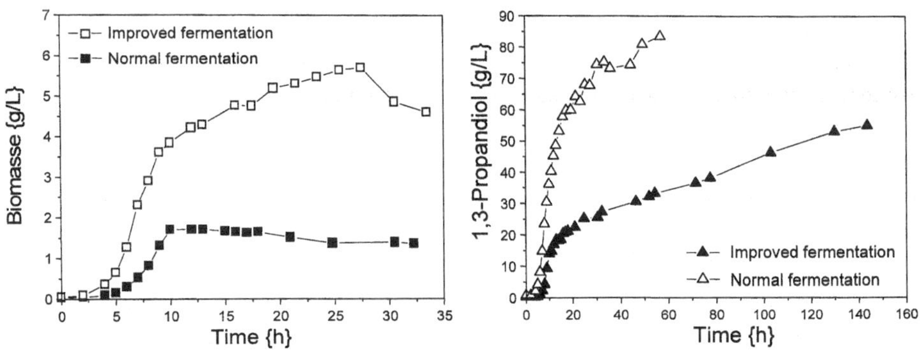

Fig. 11. Improved growth and product formation by amending limitation of an amino acid.

In general, genomic and proteomic analyses can be very useful for better defining targets for the metabolic engineering of bioprocesses. As demonstrated for glycerol fermentation by *K. pneumoniae*, the discovery and function confirmation of the PEP-dependent dihydroxyacetone kinase DHAK II make it clear that an overexpression of the previously assumed limiting enzyme pyruvate kinase will not lead to a significant improvement in the process since it is not really a limiting step. On the contrary, HOR appears to be a more promising target for metabolic engineering. One problem in glycerol fermentation is the accumulation of the intermediate 3-hydroxypropionaldehyde (3-HPA), which is a precursor of PDO and very toxic to cell growth. As shown by Ahrens et al. [74] PDOR from *K. pneumoniae* has a relatively high reaction activity for the reversed direction in the conversion of 3-HPA to PDO, resulting in a possibly high accumulation of 3-HPA when production of

PDO is increased. However, it is known that the activity of butanol dehydrogenase (BuDH) from *C. acetobutyricum*, which is well characterized [81, 82] and has a quite high protein sequence identity to HOR, is 50-fold lower in the reversed direction than in the forward direction [83]. As mentioned above, HOR from *E. coli* showed similar properties to BuDH and its overexpression is one key factor leading to the very high final PDO concentration (>130 g/l) in the conversion of glucose to PDO by recombinant *E. coli* strains.

4. CONCLUDING REMARKS AND OUTLOOK

The aim of this contribution is firstly to point out some of the presently available genome-based technologies and bioinformatics tools, and secondly to demonstrate how these data and tools can be utilized for bioprocess analysis and optimization. We showed that genomic data, even unfinished, raw genome sequences with relatively low genome coverage, can be converted into useful information about protein coding sequences and their potential assigned functions. This enables us to construct a strain specific metabolic network, which by itself is already an important result for functional genomic studies, such as proteomic analysis. By using genomic data and results from other functional genomic studies, such as gene expression profiling with microarrays, the gene regulatory network of a given microorganism can be inferred by using bioinformatics tools. However, what the process engineers are really looking for are the functional networks of cellular metabolism and gene regulation, which are dependent on the specific process conditions ("environome"). In this regard, one should not only think of selected nutrients and operational modes (e.g. batch, fed-batch, or continuous), but also of typical engineering quantities, such as mixing and aeration, which are sometimes influential, for instance in the cultivation of fungi showing either mycelium or pellet formation. Such phenotypic differences correspond to the functional levels of the gene regulatory and metabolic networks.

The functional networks, particularly the metabolic network, cannot yet be constructed *in silico* but instead require information from the various "omics" technologies. Proteomics (e.g. 2-DE plus mass spectrometry and its evaluation by bioinformatics tools) has advanced to become a well-established technology and has numerous applications in the laboratory and in industry; techniques for studying transcriptomics and metabolomics are also available but are still under fast development. Of particular importance for bioprocess engineers, however, is the functional metabolic network, or even better, the metabolic "fluxome," which reflects the dynamics of the cell and is directly linked to cell physiology under given environmental conditions. From a scientific point of view, the greatest challenge and opportunity in the use of genomic and functional genomic data for bioprocess analysis and development is the deciphering and purposeful manipulation of the interactions and dynamics within cellular networks at different molecular levels (e.g. metabolic, signaling and gene regulatory networks) which in the end determine the genotype and phenotype of the production strains but are very complex systems. New concepts and methods are desperately needed in order to achieve these goals. These needs gave rise to the birth of a "systems biotechnology" which could potentially revolutionize future bioprocess development.

REFERENCES

[1] Fleischmann RD, Adams MD, White O, Clayton RA, Kirkness EF, Kerlavage AR *et al.*: Whole-genome random sequencing and assembly of Haemophilus influenzae Rd. *Science* 1995, 269: 496-512.

[2] Ogata H, Goto S, Sato K, Fujibuchi W, Bono H, Kanehisa M: KEGG: Kyoto Encyclopedia of Genes and Genomes. *Nucleic Acids Res* 1999, 27: 29-34.

[3] Overbeek R, Larsen N, Pusch GD, D'Souza M, Selkov E Jr, Kyrpides N *et al.*: WIT: integrated system for high-throughput genome sequence analysis and metabolic reconstruction. *Nucleic Acids Res* 2000, 28: 123-125.

[4] Schilling CH, Covert MW, Famili I, Church GM, Edwards JS, Palsson BO: Genome-scale metabolic model of Helicobacter pylori 26695. *J Bacteriol* 2002, 184: 4582-4593.

[5] Karp PD, Riley M, Saier M, Paulsen IT, Collado-Vides J, Paley SM *et al.*: The EcoCyc Database. *Nucleic Acids Res* 2002, 30: 56-58.

[6] Forster J, Famili I, Fu P, Palsson BO, Nielsen J: Genome-scale reconstruction of the Saccharomyces cerevisiae metabolic network. *Genome Res* 2003, 13: 244-253.

[7] Ma HW, Zeng AP: Reconstruction of metabolic networks from genome data and analysis of their global structure for various organisms. *Bioinformatics* 2003, 19: 270-277.

[8] Herrgard MJ, Covert MW, Palsson BO: Reconstruction of microbial transcriptional regulatory networks. *Curr Opin Biotechnol* 2004, 15: 70-77.

[9] Ma HW, Kumar B, Ditges U, Gunzer F, Buer J, Zeng AP: An extended transcriptional regulatory network of *Escherichia coli* and its structural analysis. *Nucleic Acids Res* 2004, 32: 6643-6649.

[10] Bro C, Nielsen J: Impact of 'ome' analyses on inverse metabolic engineering. *Metab Eng* 2004, 6: 204-211.

[11] Gill RT: Enabling inverse metabolic engineering through genomics. *Curr Opin Biotechnol* 2003, 14: 484-490.

[12] Ma HW, Zhao XM, Yuan YJ, Zeng AP: Decomposition of metabolic network based on the global connectivity structure of reaction graph. *Bioinformatics* 2004, 20: 1870-1876.

[13] Kanehisa M: The KEGG database. *Novartis Found Symp* 2002, 247: 91-101.

[14] Sun J, Zeng AP: IdentiCS--identification of coding sequence and in silico reconstruction of the metabolic network directly from unannotated low-coverage bacterial genome sequence. *BMC Bioinformatics* 2004, 5: 112.

[15] Edwards JS, Covert M, Palsson B: Metabolic modelling of microbes: the flux-balance approach. *Environ Microbiol* 2002, 4: 133-140.

[16] Schilling CH, Edwards JS, Palsson BO: Toward metabolic phenomics: analysis of genomic data using flux balances. *Biotechnol Prog* 1999, 15: 288-295.

[17] Schuster S, Fell DA, Dandekar T: A general definition of metabolic pathways useful for systematic organization and analysis of complex metabolic networks. *Nat Biotechnol* 2000, 18: 326-332.

[18] Stelling J, Klamt S, Bettenbrock K, Schuster S, Gilles ED: Metabolic network structure determines key aspects of functionality and regulation. *Nature* 2002, 420: 190-193.

[19] Palsson BO, Price ND, Papin JA: Development of network-based pathway definitions: the need to analyze real metabolic networks. *Trends Biotechnol* 2003, 21: 195-198.

[20] Edwards JS, Covert M, Palsson B: Metabolic modelling of microbes: the flux-balance approach. *Environ Microbiol* 2002, 4: 133-140.

[21] Ravasz E, Somera AL, Mongru DA, Oltvai ZN, Barabasi AL: Hierarchical Organization of Modularity in Metabolic Networks. *Science* 2002, 297: 1551-1555.

[22] Holme P, Huss M, Jeong H: Subnetwork hierarchies of biochemical pathways. *Bioinformatics* 2003, 19: 532-538.

[23] Gagneur J, Jackson DB, Casari G: Hierarchical analysis of dependency in metabolic networks. *Bioinformatics* 2003, 19: 1027-1034.

[24] Schuster S, Pfeiffer T, Moldenhauer F, Koch I, Dandekar T: Exploring the pathway structure of metabolism: decomposition into subnetworks and application to Mycoplasma pneumoniae. *Bioinformatics* 2002, 18: 351-361.

[25] Ma HW, Zeng AP: Phylogenetic comparasion of metabolic capacities of organisms at genome level. *Molecular Phylogenetics and Evolution* 2003, 28.

[26] Ma HW, Zeng AP: The connectivity structure, giant strong component and centrality of metabolic networks. *Bioinformatics* 2003, 19: 1423-1430.

[27] Csete M, Doyle J: Bow ties, metabolism and disease. *Trends Biotechnol* 2004, 22: 446-450.

[28] Bailey JE, Sburlati A, Hatzimanikatis V, Lee K, Renner WA, Tsai PS: Inverse metabolic engineering: a strategy for directed genetic engineering of useful phenotypes. *Biotechnol Bioeng* 2002, 79: 568-579.

[29] Bulter T, Bernstein JR, Liao JC: A perspective of metabolic engineering strategies: moving up the systems hierarchy. *Biotechnol Bioeng* 2003, 84: 815-821.

[30] Papin JA, Stelling J, Price ND, Klamt S, Schuster S, Palsson BO: Comparison of network-based pathway analysis methods. *Trends Biotechnol* 2004, 22: 400-405.

[31] Klamt S, Stelling J: Two approaches for metabolic pathway analysis? *Trends Biotechnol* 2003, 21: 64-69.

[32] Schuster S, Dandekar T, Fell DA: Detection of elementary flux modes in biochemical networks: a promising tool for pathway analysis and metabolic engineering. *Trends Biotechnol* 1999, 17: 53-60.

[33] Price ND, Reed JL, Papin JA, Famili I, Palsson BO: Analysis of metabolic capabilities using singular value decomposition of extreme pathway matrices. *Biophys J* 2003, 84: 794-804.

[34] Wiback SJ, Palsson BO: Extreme pathway analysis of human red blood cell metabolism. *Biophys J* 2002, 83: 808-818.

[35] Ramakrishna R, Edwards JS, McCulloch A, Palsson BO: Flux-balance analysis of mitochondrial energy metabolism: consequences of systemic stoichiometric constraints. *Am J Physiol Regul Integr Comp Physiol* 2001, 280: R695-R704.

[36] Reed JL, Palsson BO: Genome-scale in silico models of E. coli have multiple equivalent phenotypic states: assessment of correlated reaction subsets that comprise network states. *Genome Res* 2004, 14: 1797-1805.

[37] Famili I, Forster J, Nielsen J, Palsson BO: Saccharomyces cerevisiae phenotypes can be predicted by using constraint-based analysis of a genome-scale reconstructed metabolic network. *Proc Natl Acad Sci U S A* 2003, 100: 13134-13139.

[38] Covert MW, Knight EM, Reed JL, Herrgard MJ, Palsson BO: Integrating high-throughput and computational data elucidates bacterial networks. *Nature* 2004, 429: 92-96.

[39] Edwards JS, Ibarra RU, Palsson BO: In silico predictions of Escherichia coli metabolic capabilities are consistent with experimental data. *Nat Biotechnol* 2001, 19: 125-130.

[40] Almaas E, Kovacs B, Vicsek T, Oltvai ZN, Barabasi AL: Global organization of metabolic fluxes in the bacterium Escherichia coli. *Nature* 2004, 427: 839-843.

[41] Covert MW, Palsson BO: Constraints-based models: regulation of gene expression reduces the steady-state solution space. *J Theor Biol* 2003, 221: 309-325.

[42] Covert MW, Palsson BO: Transcriptional regulation in constraints-based metabolic models of Escherichia coli. *J Biol Chem* 2002, 277: 28058-28064.

[43] Matys V, Fricke E, Geffers R, Gossling E, Haubrock M, Hehl R *et al.*: TRANSFAC: transcriptional regulation, from patterns to profiles. *Nucleic Acids Res* 2003, 31: 374-378.

[44] Munch R, Hiller K, Barg H, Heldt D, Linz S, Wingender E *et al.*: PRODORIC: prokaryotic database of gene regulation. *Nucleic Acids Res* 2003, 31: 266-269.

[45] Salgado H, Gama-Castro S, Martinez-Antonio A, Diaz-Peredo E, Sanchez-Solano F, Peralta-Gil M *et al.*: RegulonDB (version 4.0): transcriptional regulation, operon organization and growth conditions in Escherichia coli K-12. *Nucleic Acids Res* 2004, 32 Database issue: D303-D306.

[46] D'haeseleer P, Liang S, Somogyi R: Genetic network inference: from co-expression clustering to reverse engineering. *Bioinformatics* 2000, 16: 707-726.
[47] Shen-Orr SS, Milo R, Mangan S, Alon U: Network motifs in the transcriptional regulation network of Escherichia coli. *Nat Genet* 2002, 31: 64-68.
[48] Bailey JE: Lessons from metabolic engineering for functional genomics and drug discovery. *Nat Biotechnol* 1999, 17: 616-618.
[49] Mullner S: The impact of proteomics on products and processes. *Adv Biochem Eng Biotechnol* 2003, 83: 1-25.
[50] Lee PS, Lee KH: Escherichia coli--a model system that benefits from and contributes to the evolution of proteomics. *Biotechnol Bioeng* 2003, 84: 801-814.
[51] Peng L, Shimizu K: Global metabolic regulation analysis for Escherichia coli K12 based on protein expression by 2-dimensional electrophoresis and enzyme activity measurement. *Appl Microbiol Biotechnol* 2003, 61: 163-178.
[52] Yoon SH, Han MJ, Lee SY, Jeong KJ, Yoo JS: Combined transcriptome and proteome analysis of Escherichia coli during high cell density culture. *Biotechnol Bioeng* 2003, 81: 753-767.
[53] Han MJ, Lee SY: Proteome profiling and its use in metabolic and cellular engineering. *Proteomics* 2003, 3: 2317-2324.
[54] Schweder T, Kruger E, Xu B, Jurgen B, Blomsten G, Enfors SO *et al.*: Monitoring of genes that respond to process-related stress in large-scale bioprocesses. *Biotechnol Bioeng* 1999, 65: 151-159.
[55] Durrschmid K, Marzban G, Durrschmid E, Striedner G, Clementschitsch F, Cserjan-Puschmann M *et al.*: Monitoring of protein profiles for the optimization of recombinant fermentation processes using public domain databases. *Electrophoresis* 2003, 24: 303-310.
[56] Hoffmann F, Weber J, Rinas U: Metabolic adaptation of Escherichia coli during temperature-induced recombinant protein production: 1. Readjustment of metabolic enzyme synthesis. *Biotechnol Bioeng* 2002, 80: 313-319.
[57] Champion KM, Nishihara JC, Joly JC, Arnott D: Similarity of the Escherichia coli proteome upon completion of different biopharmaceutical fermentation processes. *Proteomics* 2001, 1: 1133-1148.
[58] Jurgen B, Hanschke R, Sarvas M, Hecker M, Schweder T: Proteome and transcriptome based analysis of Bacillus subtilis cells overproducing an insoluble heterologous protein. *Appl Microbiol Biotechnol* 2001, 55: 326-332.
[59] Jurgen B, Lin HY, Riemschneider S, Scharf C, Neubauer P, Schmid R *et al.*: Monitoring of genes that respond to overproduction of an insoluble recombinant protein in Escherichia coli glucose-limited fed-batch fermentations. *Biotechnol Bioeng* 2000, 70: 217-224.
[60] Rinas U: Synthesis rates of cellular proteins involved in translation and protein folding are strongly altered in response to overproduction of basic fibroblast growth factor by recombinant Escherichia coli. *Biotechnol Prog* 1996, 12: 196-200.
[61] Han MJ, Yoon SS, Lee SY: Proteome analysis of metabolically engineered Escherichia coli producing Poly(3-hydroxybutyrate). *J Bacteriol* 2001, 183: 301-308.
[62] Kabir MM, Shimizu K: Proteome analysis of a temperature-inducible recombinant *Escherichia coli* for poly-hydroxybutyrate production. *J Biosci Bioeng* 2001, 92: 277-284.
[63] Zeng AP, Biebl H: Bulk chemicals from biotechnology: the case of 1,3-propanediol production and the new trends. *Adv Biochem Eng Biotechnol* 2002, 74: 239-259.
[64] Skraly FA, Lytle BL, Cameron DC: Construction and characterization of a 1,3-propanediol operon. *Appl Environ Microbiol* 1998, 64: 98-105.
[65] Tong IT, Liao HH, Cameron DC: 1,3-Propanediol production by Escherichia coli expressing genes from the Klebsiella pneumoniae dha regulon. *Appl Environ Microbiol* 1991, 57: 3541-3546.

[66] Tong IT, Cameron DC: Enhancement of 1,3-propanediol production by cofermentation in Escherichia coli expressing Klebsiella pneumoniae dha regulon genes. *Appl Biochem Biotechnol* 1992, 34-35: 149-159.

[67] Sprenger GA, Hammer BA, Johnson EA, Lin EC: Anaerobic growth of Escherichia coli on glycerol by importing genes of the dha regulon from Klebsiella pneumoniae. *J Gen Microbiol* 1989, 135: 1255-1262.

[68] Menzel K: *Analyse der Stofffluesse und Metabolic Engineering der Glycerinvergaerung zu 1,3-Propandiol durch Klebsiella pneumoniae.* Ph.D thesis. Technical University of Braunschweig, Germany; 1999.

[69] Cameron DC, Altaras NE, Hoffman ML, Shaw AJ: Metabolic engineering of propanediol pathways. *Biotechnol Prog* 1998, 14: 116-125.

[70] Dunn-Coleman AU, Diaz TM, Chase MW, Trimbur D. Increased production of 1,3-propanediol by fermentation of inexpensive carbon sources. Genencor. [WO 9821341]. 1998.

[71] Mark HSLLPJaWG. Process for the biological production of 1,3-propanediol with high titer. Genencor. [World Patent 01/12833 A2, 2001.]. 2001.

[72] Zeng AP: *Quantitative Zellphysiologie, Metabolic Engineering und Modellierung der Glycerinfermentation zu 1,3-Propandiol.* Habilitationschrift, Technical University of Braunschweig, Germany; 2000.

[73] Sun J, van den HJ, Soucaille P, Qu Y, Zeng AP: Comparative genomic analysis of dha regulon and related genes for anaerobic glycerol metabolism in bacteria. *Biotechnol Prog* 2003, 19: 263-272.

[74] Ahrens K, Menzel K, Zeng AP, Deckwer WD: Kinetic, dynamic, and pathway studies of glycerol metabolism by Klebsiella pneumoniae in anaerobic continuous culture: III. Enzymes and fluxes of glycerol dissimilation and 1,3-propanediol formation. *Biotech Bioeng* 1998, 59: 544-552.

[75] Menzel K, Ahrens K, Zeng AP, Deckwer WD: Kinetic, dynamic, and pathway studies of glycerol metabolism by Klebsiella pneumoniae in anaerobic continuous culture: IV. Enzymes and fluxes of pyruvate metabolism. *Biotechnology and Bioengineering* 1998, 60: 617-626.

[76] Wang W, Sun J, Hartlep M, Deckwer WD, Zeng AP: Combined use of proteomic analysis and enzyme activity assays for metabolic pathway analysis of glycerol fermentation by Klebsiella pneumoniae. *Biotechnol Bioeng* 2003, 83: 525-536.

[77] Wang W, Sun J, Nimtz M, Deckwer WD, Zeng AP: Protein identification from two-dimensional gel electrophoresis analysis of Klebsiella pneumoniae by combined use of mass spectrometry data and raw genome sequences. *Proteome Sci* 2003, 1: 6.

[78] Emptage M, Haynie S, Laffend L, Pucci J, Whited G.. Process for the biological production of 1,3-propanediol with high titer. E.I.du Pont de Nemours and Co.and Genencor International, Inc. [U.S. patent WO 01/12833 A2.]. 2001.

[79] Menzel K, Zeng AP, Biebl H, Deckwer WD: Kinetic, dynamic, and pathway studies of glycerol metabolism by Klebsiella pneumoniae in anaerobic continuous culture .1. The phenomena and characterization of oscillation and hysteresis. *Biotechnology and Bioengineering* 1996, 52: 549-560.

[80] Zeng AP, Menzel K, Deckwer WD: Kinetic, dynamic, and pathway studies of glycerol metabolism by Klebsiella pneumoniae in anaerobic continuous culture: II. Analysis of metabolic rates and pathways under oscillation and steady-state conditions. *Biotechnology and Bioengineering* 1996, 52: 561-571.

[81] Youngleson JS, Jones WA, Jones DT, Woods DR: Molecular analysis and nucleotide sequence of the adh1 gene encoding an NADPH-dependent butanol dehydrogenase in the Gram-positive anaerobe Clostridium acetobutylicum. *Gene* 1989, 78: 355-364.

[82] Walter KA, Bennett GN, Papoutsakis ET: Molecular characterization of two Clostridium acetobutylicum ATCC 824 butanol dehydrogenase isozyme genes. *J Bacteriol* 1992, 174: 7149-7158.

[83] Welch RW, Rudolph FB, Papoutsakis ET: Purification and characterization of the NADH-dependent butanol dehydrogenase from Clostridium acetobutylicum (ATCC 824). *Arch Biochem Biophys* 1989, 273: 309-318.

Bioprocessing for Value-Added Products from Renewable Resources
Shang-Tian Yang (Editor)

Chapter 3. Directed Evolution Tools in Bioproduct and Bioprocess Development

Sheryl B. Rubin-Pitel[a], Catherine M-H. Cho[b], Wilfred Chen[b]*, and Huimin Zhao[a,c]*

[a]Department of Chemical and Biomolecular Engineering, University of Illinois, Urbana, IL 61801

[c]Department of Chemistry, University of Illinois, Urbana, IL 61801

[c]Center for Biophysics and Computational Biology, University of Illinois, Urbana, IL 61801

[b]Department of Chemical and Environmental Engineering, University of California, Riverside, CA 92521

1. INTRODUCTION

Darwinian evolution is often regarded as a negative process, with natural selection merely purging the less competitive among an uneven population. George Gaylord Simpson, in his 1944 treatise *Tempo and Mode in Evolution*, countered that selection does not "simply kill off or permit to live fixed types of organisms delivered to it...Selection also determines which among the millions of possible types of organisms will actually arise, and it is therefore a truly creative factor in evolution" [1, 2]. Man has harnessed the creative power of selection for thousands of years through the process of classical breeding, thereby molding a plethora of livestock, crops, and companion animals to fulfill collective needs or desires. Only in the past decades have researchers exploited the positive nature of selection at the scale of biological macromolecules or single cells rather than an entire organism. Evolutionary methods have been applied to achieve improved or novel characteristics in nucleic acids, proteins, viruses, and bacterial strains. The general strategy of mimicking natural evolution in the laboratory is termed "directed (molecular) evolution" or "*in vitro* evolution" [3]. Since it was first described in the 1970s, directed evolution has grown in popularity and found a wide range of applications across industry, academia, and medicine.

One of the earliest examples of "directed evolution" was *in vitro* evolution of nucleic acids carried out by Mills *et al* [4]. However, it was not until several decades later that the concept of directed evolution was applied for the *in vitro* engineering of proteins on the molecular level [5–7]. More recently, directed evolution techniques have been applied to the engineering of more complex subjects such as metabolic pathways, viruses, and bacterial genomes [8–14].

The method of directed evolution involves an iterative strategy. The procedure begins by determining a target biomolecule, metabolic pathway, or organism, and a desired phenotypic goal. A diverse library of mutants is generated *in vivo* or *in vitro* through methods that mirror the strategies of traditional evolution: introduction of random mutations in the genetic material and/or "sexual" gene recombination. A high-throughput screening or selection method is used to identify improved progeny among the library, which are subsequently used as parents in a second round of the cycle. The process is repeated until the phenotypic goal is achieved, or when no further improvement of the phenotype is observed despite repeated iterations.

Microorganisms and the enzymes they hold have been exploited by man for thousands of years, for example, in the production of food products through fermentation. Recent decades have seen an expanding role for enzymes and microbes in the development of bioproducts and bioprocesses in industry, organic synthesis, and medical therapies. While existing enzymes may hold great potential, their use is often hindered by the low stability, lack of specificity, and low catalytic efficiency encountered when exporting these highly evolved biological entities into non-natural environments and applications [3]. Directed evolution provides the means to enhance the performance of enzymes under requisite process conditions and customize the reactions they catalyze. Directed evolution tools have been used to improve synthesis yields of desired products, limit or expand substrate specificity, alter cofactor specificity, and improve stability over a wider range of temperature and pH.

The methods, applications, and achievements of directed evolution have been described in many recent review articles and books [3, 15–18]. This review will focus only on the strategies for diversity generation that are applicable to the development of bioproducts and bioprocesses via directed evolution. The application of directed evolution to functional nucleic acids is of limited relevance compared to the engineering of protein catalysts and improved strains, and so will not be addressed herein; interested readers are referred to several recent review articles [19–21]. Additionally, high-throughput screening and selection methods for sorting through diverse mutant libraries will not be discussed in this chapter.

2. DIRECTED EVOLUTION TOOLS FOR DIVERSITY GENERATION

By natural evolution, the Earth began with an ancient unicellular ancestor and filled its skies, land, and oceans with a vast array of organisms. Damage to genetic material by irradiation or oxidation, failures of DNA replication, recombination, or repair, and invasion by parasitic DNA elements led to substitutions, deletions, insertions, duplications, inversions, and translocation of DNA segments from one chromosome to another [22]. These events—predominantly accidents or mistakes—led fortuitously to the existence of human life and the amazing diversity we experience. It must be noted, however, that evolution is a creative but sluggish process. The *in vivo* mechanisms of evolution mentioned above are highly inefficient, producing negligible changes in gene structure or function after thousands or even millions of years. For organisms possessing more advanced DNA replication and repair machinery, it has been suggested that a typical protein (of 400 amino acids) would suffer a random amino acid change in the germline approximately once every 200,000 years [22]. Thus, while nature has created a bountiful variety of life, it should not be surprising what can be accomplished when one has four billion years to tinker.

To recreate evolution in the laboratory, the mechanisms of natural evolution must be accelerated such that meaningful diversity can be created and selected in a much shorter timeframe, mere days to weeks being favored. This defines the two-fold strategy of directed evolution: rapid generation of a functionally diverse collection of mutants, and rapid identification of the best performers among them [3]. The two natural evolutionary processes which have been adapted for *in vitro* evolution are gene recombination and random mutagenesis. Gene recombination refers to the exchange of blocks of genetic material among two or more DNA strands, and is often considered the "sexual" component of evolution. Recombination can be divided into four main types: (i) homologous recombination, where recombination occurs between two genes with high sequence identity, (ii) non-homologous recombination, where recombination occurs between two DNA sequences with little or no sequence identity, (iii) reciprocal recombination, in which a symmetrical exchange of genetic material occurs between two DNA strands, and finally (iv) site-specific recombination, in which specialized nucleotide sequences exhibiting some degree of target site specificity are moved between nonhomologous sites within a genome [22, 23]. Stemmer introduced DNA shuffling [5, 6], the first *in vitro* homologous recombination method, in 1994. Since that time, numerous other homologous recombination methods have been developed, as well as methods for recombination of genes lacking sequence identity.

Random mutagenesis refers to changes in the genome resulting from improper DNA replication or inadequate repair of DNA damage following events such as irradiation, exposure to oxidative or alkylating agents, and natural deamination of cytosine. Random mutation can be divided into five categories: (i) transitions, which involve substitution of a purine nucleotide by another purine, or a pyrimidine by a second pyrimidine, (ii) transversions, which involve substitution of a purine nucleotide by a pyrimidine, or vice-versa, (iii) deletions, in which one or more nucleotides are eliminated from a gene, (iv) insertions, in which one or more extra nucleotides are incorporated into a gene, and (v) inversions, which involve the 180° rotation of a double-stranded DNA segment of two base pairs or longer [3, 24]. *In vitro* random mutagenesis methods have been developed to generate substitutions, deletions, and insertions. One of the simplest and most popular directed evolution tools, error-prone polymerase chain reaction (PCR) takes advantage of the fallibility of DNA polymerase to generate random base pair substitutions. Similarly, mutator strains of *E. coli* exploit defective DNA repair machinery and also create random point mutations.

Random mutagenesis and gene recombination methods are compared in Fig. 1. Random mutagenesis methods use a single gene as a starting point, and introduce mutations along the entire gene or in predefined sites or regions. Nucleotides may be substituted randomly, generating point mutations, inserted into the sequence, or deleted. As many point mutations will be deleterious, a low mutation rate is necessary to preserve protein function. In contrast, gene recombination typically begins with a collection of parent molecules and exploits the

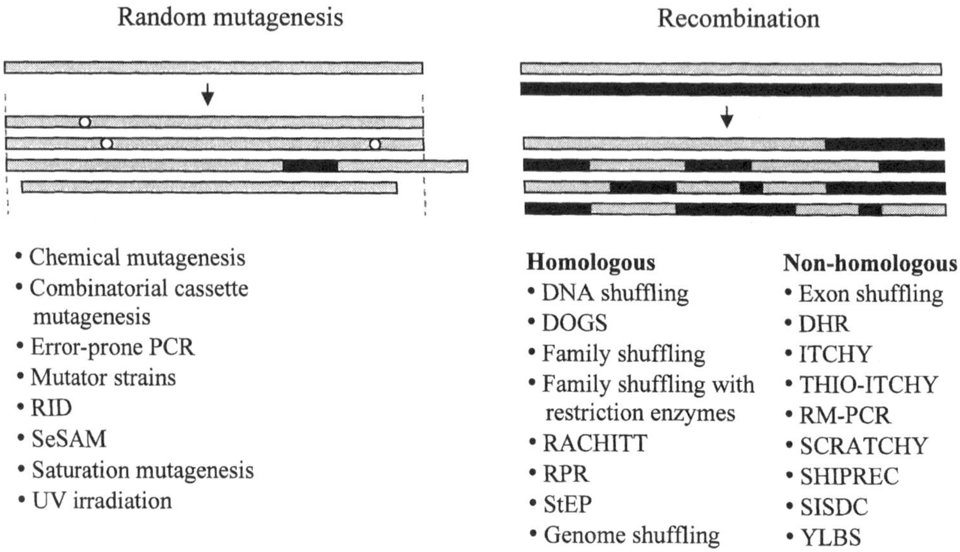

Fig. 1. Comparison of (a) random mutagenesis and (b) recombination strategies.

existing variation among them to create novel sequences. The pool of parent genes could be an assortment of mutant progeny resulting from random mutagenesis of a single parent (DNA shuffling), or a set of closely related genes from different strains or species (family shuffling). Typically the parent sequences are fragmented and the resulting short strands are pieced together into complete genes. The chimeric progeny are created with contributions from at least two parents. Unlike random mutagenesis, in which mutation events are restricted, maximal recombining of the genes, or crossover, may be desired. Some recently developed gene shuffling strategies merge gene recombination and random mutagenesis by using PCR to generate full-length progeny sequences from the gene fragments and also to amplify them [3]. In this way, misincorporations by DNA polymerase can provide additional diversity in the recombination library. As shown in Fig. 1, numerous experimental protocols have been formulated for each diversification strategy. These protocols will be described in further detail below.

2.1. *In vitro* mutagenesis methods

Random mutagenesis strategies are relatively simple and popular methods for generating molecular diversity. Early mutagenesis protocols involved creation of point mutations in a parent gene by damaging the DNA strand, for example by treatment with chemical mutagens including hydroxylamine [25], nitrous acid [25], methoxylamine [26], and sodium bisulfite [27], or by ultraviolet irradiation [25]. These methods tend to be inefficient, because they are typically discontinuous and can cause substantial cell damage if performed *in vivo* [28]. Point mutations can also be induced by error-prone PCR [29–32] or mutator strains of *E. coli* [28, 33–36]. The aforementioned mutagenesis methods will generate point mutations across the entire length of the parent gene. Other schemes have been developed that allow mutations to be focused in specific sites or regions of the parent DNA sequence. Some of the most common random mutagenesis methods are listed in Table 1.

Table 1
Random mutagenesis methods

Method	Advantages	Disadvantages	References
Chemical mutagenesis	Simplicity	Accumulates deleterious mutations Low mutation level Low efficiency Limited amino acid substitutions Cannot control mutation rate	[26, 27]
Mutator strains	Simplicity	Low mutation level Accumulates deleterious mutations Progeny must be transferred to DNA repair-competent strain for screening Limited amino acid substitutions Cannot control mutation rate	[28, 33–36]
Error-prone PCR	Simplicity	Accumulates deleterious mutations Limited amino acid substitutions Polymerase bias	[29, 30]
Saturation mutagenesis	Simplicity Mutate specific site(s) in a gene Access all 20 amino acids	Limited diversity generation Gene sequence required	[44]
Sequence saturation mutagenesis (SeSAM)	Overcomes polymerase bias Target a specific nucleotide in a sequence	Small fragments not mutagenized Four PCR reactions needed to remove bias Limited amino acid substitutions	[41]
Random insertion / deletion (RID)	Flexible Insert or remove an amino acid randomly Access all 20 amino acids	Point mutations may occur Time-consuming and technically challenging	[45]

2.1.1. Mutagenic strains

Propagating a gene of interest in a mutational strain represents the simplest method of random mutagenesis. Mutator strains of *E. coli* are deficient in one or more DNA repair genes, leading to single base substitutions at a rate of approximately 1 mutation per 1000 base pairs and mutation cycle [36]. This mutation rate is fairly low, and mutations may occur outside of the gene of interest, across the plasmid vector and bacterial genome. To generate a mutant library, the gene of interest is cloned into a plasmid or phagemid and propagated in mutator *E. coli* cells through a limited number of replications [33, 34]. The plasmid or phagemid library is then rescued from the mutator strains and stably expressed in a DNA-repair competent strain for amplification and selection of the mutant progeny; if necessary, the procedure of mutation, amplification, and selection is repeated until the desired phenotype is achieved [34]. The process is relatively easy, and commercial mutator strains such as XL1-Red (Stratagene, La Jolla, California) are available. Mutator *E. coli* strains find only modest use today, despite comparable methods being more time-consuming, difficult to implement, and expensive [33]. Rather, error-prone PCR is by far the most popular random mutagenesis method.

2.1.2. Error-prone polymerase chain reaction

Error-prone PCR relies on the misincorporation of nucleotides by DNA polymerase to generate point mutations in a gene sequence. The accuracy of DNA polymerase can be adjusted *in vitro* by addition of manganese ion into the PCR reaction mixture [37]. Additionally, PCR mutagenesis protocols have been designed which incorporate nucleotide analogs or use "mutagenic polymerases" [38–40]. Any one of these strategies, or a combination, can be incorporated into a PCR reaction to achieve a specific mutation rate. The relative simplicity and versatility of error-prone PCR have propelled it to become the most widely used mutagenesis strategy, but it suffers from several drawbacks. First, due to the redundancy of the genetic code, error-prone PCR methods are limited in their ability to create diversity at the protein level. From a single amino acid, an average of less than six other amino acids can be obtained, rather than all 19 possible substitutions [41]. Second, DNA polymerases used in PCR reactions have mutational biases that limit diversity. *Taq* polymerase and Mutazyme (Stratagene, La Jolla, California) will preferentially induce mutations at AT base pairs over GC base pairs [41]. Further, the majority of mutations are transitions, and amino acid substitutions, when present, tend to preserve the characteristics of the original residue [3, 41]. Third, in order to maintain adequate numbers of functionally active progeny, the mutation rate is kept low, generally only 1-3 mutations per 1000 base pairs [38]; these few mutations are unlikely to occur next to each other [41]. Finally, nucleotide analogs are not incorporated by DNA polymerases efficiently, and their incorporation tends to occur at certain favored sites [41]. Thus, nucleotide analog methods may result in low mutation frequencies, limited diversity, and low product yield [41–43].

2.1.3. Saturation mutagenesis

The limitations of error-prone PCR mutagenesis may be overcome by site-directed mutagenesis and saturation mutagenesis methods. Site-directed mutagenesis uses an oligonucleotide primer to introduce a single-base pair substitution at a specified position in a gene [46]. Saturation mutagenesis involves the substitution of all possible amino acids randomly at a predetermined residue or continuous series of residues in the protein of interest [3]. Several strategies of saturation mutagenesis have been developed, including combinatorial cassette mutagenesis [47, 48], recursive ensemble mutagenesis [49, 50], scanning saturation mutagenesis [51–53], and codon cassette mutagenesis [54, 55]. More recently, Wong *et al.* [41] described the method of sequence saturation mutagenesis (SeSaM), which is able to randomize a DNA sequence at every nucleotide position through use of a universal base.

2.1.4. Mutagenesis by random insertion or deletion

Random mutagenesis can also be accomplished by insertion or deletion of nucleotides from a target gene sequence. Random insertion or deletion leads to a net change in the length of the gene of interest, opening a new realm of diversity that cannot be reached by point mutation alone. In the past random insertion has been accomplished by exploiting naturally occurring transposable elements or by random elongation mutagenesis, in which peptide "tails" are fused to a gene [56–58]. Transposable elements have several advantages for random mutagenesis: transposons can be designed to carry selectable markers such as antibiotic resistance or phage immunity; the occurrence of transposon insertion can be controlled; mutagenesis is highly efficient; and the occurrence of secondary mutations is low [25]. However, transposons cannot be used to create random deletions. Random elongation mutagenesis can also create a functionally diverse library of mutants, but is limited to fusing additional peptides to the C-terminus of a protein, and also cannot facilitate random deletions. A more recent method developed by Murakami *et al.* [45] can introduce both insertions and deletions at any position in a gene sequence. Random insertion/deletion (RID) mutagenesis allows the deletion of up to 16 bases from random sites on the target gene and subsequent insertion of a random or predetermined sequence of any number of bases at the same position [45]. This method can be used to replace three randomly selected base pairs by a specific codon, a mixture of codons, a restriction site, or by four-base codons for non-natural amino acids [45]. Though a more versatile method, RID mutagenesis is also technically challenging, time consuming, requires a large amount of template DNA, and is difficult to iterate [3].

Because most mutations will be neutral or deleterious, a low mutation rate is maintained in random mutagenesis methods. As a result random mutagenesis uncovers diversity in a very small region of sequence space, and is unlikely to foster detection of synergistic effects of multiple beneficial mutations in a single gene [3]. Furthermore, the small evolutionary steps taken by random mutagenesis may not be sufficient to allow the wholesale changes required, for example, to evolve a novel activity in a target gene. Neutral or deleterious point mutations may also accumulate in a library of progeny. Such nonessential mutations may make the resulting protein immunogenic [6]. Finally, random mutagenesis methods are restricted by the use of a single parent as a starting point. Although it can be clearly defined as to which of a collection of existing enzymes has the most favorable characteristics, it is impossible to predict which enzyme has the greatest potential for improvement through directed evolution. Use of only a single parent represents a fundamental flaw of random mutagenesis methods, and limits the evolutionary potential of progeny [59]. This shortcoming of random mutagenesis is overcome by recombination methods.

2.2. *In vitro* homologous recombination methods

Homologous recombination methods mimic the "sexual" recombination of genetic material that rearranges maternal and paternal chromosomes in germ cell DNA. Such recombination increases the genetic variation among a population and is vital to the continued evolution of organisms in response to an ever-changing environment [22]. Unlike mutagenesis methods, which create novel diversity at the molecular level, recombination methods simply rearrange existing gene sequences to exploit the diversity that naturally exists among a population. While the results of random point mutations are unpredictable and often deleterious, recombination provides the advantage that all diversity present in a mutant sequence was drawn from folded and fully functional proteins. Recombination also makes it possible to remove neutral or deleterious mutations, which accumulate during random mutagenesis, by backcrossing progeny with excess parental or wild-type DNA [5]. Table 2 compares the advantages and disadvantages of various homologous recombination methods.

2.2.1. DNA shuffling and family shuffling

Stemmer introduced DNA shuffling, the first homologous recombination method, in 1994 [5, 6]. DNA shuffling involves the digestion of a gene by DNaseI into random fragments, and the reassembly of those fragments into a full-length gene by primerless PCR: the fragments prime on each other based on sequence homology, and recombination occurs when fragments from one copy of a gene anneal to fragments from another copy, causing a template switch, or crossover event. This method was used to fragment and recreate a single gene, to recombine a group of point mutants, and to recombine several related genes. The reassembly process introduces point mutations at a rate similar to error-prone PCR, due to misincorporations by the DNA polymerase. These mutations add to the diversity of the mutant library, and any unnecessary mutations can later be eliminated by backcrossing to parent or wild-type sequences. If necessary, use of a high fidelity DNA polymerase allows the rate of random point mutations to be reduced drastically [60]. Several years after the introduction of DNA shuffling, the method was applied to the recombination of a family of related genes from various species. This new method, termed family shuffling, applied DNA shuffling to a group of naturally occurring homologous genes rather than laboratory-created mutants. Crameri *et al* found that family shuffling significantly accelerated the rate of functional enzyme improvement in a single recombination-selection cycle [61]. Although they are powerful methods, DNA shuffling and family shuffling are not without limitations. Shuffling methods require the presence of zones of relatively high sequence homology surrounding regions of diversity [6]. Additionally, significant biases are found in where crossover events occur and in which parents are involved: crossover tends to occur in regions of higher homology, and among parents which share greater sequence identity [62]. Bias is also introduced by nonrandom gene fragmentation by the DNaseI enzyme [63]. All of these factors limit the diversity created in a shuffled library. In extreme cases, lack of homology among parents can lead to the majority of reconstructed "shuffled" sequences entirely representing a single parent [64].

Table 2
Homologous recombination methods

Method	Advantages	Disadvantages	References
DNA shuffling	Robust, flexible Back-crossing to parent removes non-essential mutations	Biased to crossovers in high homology regions Low crossover rate High percentage of parent	[5, 6]
Family shuffling	Exploit natural diversity Accelerates functional enzyme improvement	Biased to crossover in high homology regions Need high sequence homology in the gene family High percentage of parent	[61]
Family shuffling using restriction endonucleases	Lower representation of parent in a library	Point mutations Low crossover rate	[65]
DOGS	Reduced parental genes in a shuffled library Lower homology required Can bias representation of parent in library	Point mutations Frameshifts may occur Relatively low crossover rate	[64]
RACHITT	No parent genes in a shuffled library Higher rate of recombination Recombine genes of low sequence homology	Complex Requires synthesis and fragmentation of single-stranded complement DNA	[66]
RPR	Compatible with ssDNA DNase I-independent Removes sequence bias Independent of template length Less parent DNA needed	Need gene sequence Biased point mutations also occur	[67]
StEP	Simplicity	Need high homology Low crossover rate Need tight control of PCR	[68, 69]
Synthetic shuffling	Greater flexibility Increased diversity	Chemical synthesis of many degenerate oligonucleotides	[70]
Genome shuffling	Improve complex, poorly understood phenotypes Adapt to multiple phenotypic goals New strains not GMOs	Possibility of novel antibiotic resistance or pathogenicity Genome flexibility restricted by metabolic network rigidity	[10, 11, 13]

Numerous homologous gene recombination methods have been designed to address the limitations of family shuffling. Kikuchi *et al* described a method for gene shuffling using endonuclease digestion at restriction sites, rather than DNaseI digestion; however, sequence homology surrounding the digested restriction sites is still required for overlap extension to occur [64, 65]. Degenerate oligonucleotide gene shuffling (DOGS) utilizes a PCR reaction with degenerate-end, complementary primer pairs to shuffle genes with limited sequence similarity and G + C content [64]. Additionally, by modifying primer extension conditions the progeny can be biased towards one or more of the parent genes [64].

2.2.2. Oligonucleotide- and oligonucleotide primer-based methods

Several other alternatives to DNA shuffling have been established, including random-priming *in vitro* recombination (RPR) [67], the staggered extension process (StEP) [69], and synthetic shuffling [70] . Recombination by RPR utilizes elongation from random sequence primers to generate a collection of small DNA fragments complementary to different areas of the template sequence(s) [67]. The method of RPR is shown in Fig. 2. Similar to DNA shuffling, fragments prime each other based on sequence homology and are reconstructed into a full length sequence by DNA polymerase-catalyzed elongation [67]. StEP also utilizes primer elongation to generate small DNA fragments for recombination. In StEP recombination, flanking primers are annealed to a denatured template and allowed to extend for a very short time period; cycles of denaturation and short annealing/elongation are repeated [68, 69]. Crossover occurs when partially extended primers anneal randomly to different templates based on homology, and extend further [68, 69]. The cycle of denaturation/annealing/elongation is continued until full-length genes are created, and if necessary, a traditional PCR amplification can be used to increase the yield of chimeric progeny [68, 69]. In synthetic shuffling, the fragments to be shuffled are degenerate oligonucleotides that are chemically synthesized and encode all the variations in a family of homologous genes [70]. Compared to fragmentation-based DNA shuffling formats, synthetic shuffling is more flexible in the construction of permutated protein libraries and also introduces more diversity into these libraries. For example, this method does not require physical starting genes and can incorporate optimal codon usage or known beneficial mutations.

2.2.3. Random Chimeragenesis on Transient Templates (RACHITT)

In contrast to the above methods, RACHITT does not utilize thermocycling, strand switching, or staggered extension of primers [66]. Instead, a uracil-containing parent gene is made single-stranded to serve as a scaffold for the ordering of top-strand fragments of additional, homologous parent gene(s), and recombination occurs when fragments from different parent genes hybridize to the scaffold. *Pfu* DNA polymerase 3'-5' exonuclease activity removes the unhybridized 5' or 3' overhanging "flaps" created by fragment annealing, and also fills gaps between the annealed fragments using the transient scaffold as a template. The template strand is then eliminated by treatment with uracil-DNA-glycosylase before applying the template-chimera hybrid to PCR, resulting in amplification of double stranded, homoduplex chimerical gene sequences. The process of RACHITT recombination

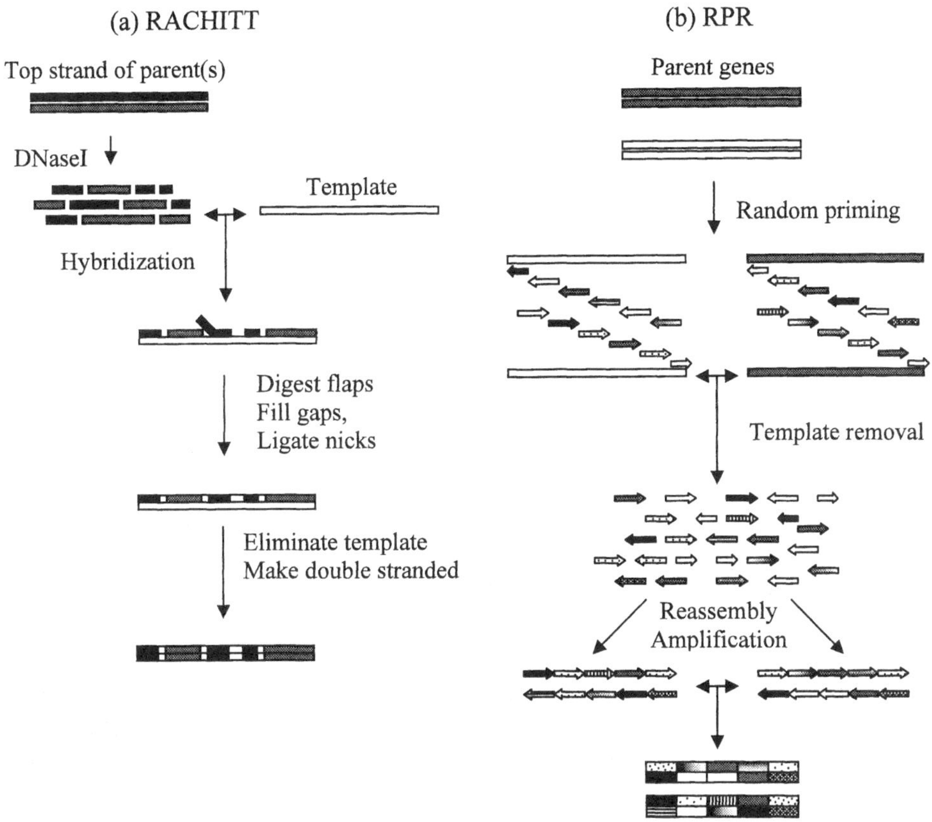

Fig. 2. Random homologous DNA recombination by (a) RACHITT and (b) RPR.

is illustrated in Fig. 2. RACHITT provides a significantly higher rate of crossover compared to other family shuffling methods, with an average of 14 crossovers per gene versus one to four crossovers for most other methods. RACHITT also generates 100% chimerical progeny with no duplications of recombination pattern in chimerical genes. Although the benefits of this method are obvious, its use may be limited by its complexity and the requirement to create single stranded gene fragments as well as single stranded, uracil-DNA template.

2.2.4. Genome shuffling

The technique of genome shuffling emerged recently as an alternative method for the optimization of industrial production strains [10–13]. Strain optimization is typically achieved by classical strain improvement techniques, which involve rounds of recombination and/or mutagenesis followed by screening for a desired phenotype, selective breeding, and rational schemes of metabolic engineering. Like other recombination methods, genome shuffling exploits the diversity that already exists among a population of organisms and allows back-crossing of progeny to parents to eliminate non-essential or deleterious gene changes that may accumulate during rounds of random mutagenesis. In genome shuffling, homologous recombination of genomes is achieved by protoplast fusion. The process of protoplast fusion in bacteria was reviewed by Gokhale *et al.* [71]. Protoplast fusion first involves the isolation of protoplasts from cells by digestion of the cell wall in the presence of osmotic stabilizers. Isolation of protoplasts from gram negative organisms is generally more difficult than gram positive due to their complex cell wall. Fusion is achieved by mixing of the parental protoplasts and addition of a fusogen, such as polyethylene glycol (PEG). PEG stimulates aggregation of protoplasts, and fusion events occur after the PEG is diluted or washed away. The PEG-treated protoplasts are subsequently plated onto appropriate media and the fused protoplasts are identified by selection. Protoplast fusion has also been described for the production of improved yeast strains [72].

The technique of genome shuffling by protoplast fusion offers several advantages. Protoplast fusion is a well-established technique that is applicable to an array of organisms including bacteria and both lower and higher eukaryotes. Protoplast fusion also provides simultaneous changes at different positions throughout the entire genome, without the requirement of genome sequence data [11]. This technique is therefore particularly applicable to the engineering of complex or poorly understood phenotypes, engineering of multiple phenotypic goals simultaneously, and engineering of organisms with limited availability of molecular biological tools and sequence information. Additionally, strains engineered by protoplast fusion, a form of natural homologous recombination, are not considered to be "genetically modified" [13], and therefore avoid the additional regulations and public distaste reserved for genetically modified organisms (GMOs). Genome shuffling by protoplast fusion has already shown promise in the improvement of industrial production strains. Zhang *et al.* showed the utility of genome shuffling to *Streptomyces* species [13], which are commonly employed in the commercial production of antibiotics. Genome shuffling of existing *Streptomyces fradiae* industrial strains was used to create a new strain with higher production of the polyketide antibiotic tylosin. By only two rounds of genome shuffling, strain improvement was equivalent to the results achieved after 20 rounds of classical strain improvement (CSI; sequential random mutagenesis and screening). Patnaik *et al* applied genome shuffling to the improvement of acid tolerance of *Lactobacillus* species, which are exploited in the commercial production of lactic acid [10]. The improved strain produced by genome shuffling showed faster growth and higher lactic acid production at a lower pH value, with tolerance to acidic pH approximately 5-fold higher than the wild type.

2.3. *In vitro* non-homologous recombination methods

The requirement of high sequence identity among parent genes limits the application of homologous recombination methods. In many situations it may be desirable to shuffle genes with low or even no evident sequence identity. The increasing availability of protein structures has also indicated that many enzymes with little or no sequence homology can have high protein structural homology; it may also be useful to shuffle such proteins, but would be inefficient with homologous recombination methods [73]. The intron-exon organization of eukaryotic genomes also facilitates non-homologous gene recombination [74]. A single exon or a collection of exons often encodes a distinct protein domain, and it is advantageous to swap domains and create combinatorial libraries of proteins. By recombining genes within non-homologous introns, exchange of protein domains is permitted while still ensuring the integrity of the coding DNA sequence, the exons. Such "exon shuffling" reflects a mechanism of natural evolution which swapped exons among unrelated genes, and led to existent proteins from distant families sharing conserved functional domains.

Table 3
Non-homologous recombination methods

Method	Advantages	Disadvantages	References
Exon shuffling	Preserves exon function	Requires known intron-exon organization of target gene Limited diversity	[73]
ITCHY	Eliminate recombination bias Structural knowledge not needed	Limited to two parents Significant fraction of progeny out-of-frame Complex, labor-intensive	[74]
THIO-ITCHY	Same advantages as ITCHY Combines recombination and random mutagenesis Simplified ITCHY method	Same disadvantages as ITCHY Incorporated dNTP analogs may complicate further experimentation	[75]
SCRATCHY	Eliminate recombination bias Structural knowledge not needed	Limited to two parents Significant fraction of progeny out-of-frame Complex, labor-intensive	[74]
DHR	High recombination rate Eliminate recombination bias	Synthesize numerous complementary oligonucleotides Gene sequence needed	[76]
RM-PCR	Unbiased incorporation of variable size DNA fragments	Frame shifts may occur Mutants may be longer or shorter than expected	[77]
SHIPREC	Crossovers occur at structurally related sites	Limited to two parents Single crossover per gene	[78]
SISDC	Recombines fragments without bias Ligates fragments in a desired order	Gene sequence needed Must engineer endonuclease sites into parent genes Must synthesize numerous oligonucleotide primers	[79]
YLBS	Recombines variable size DNA fragments Shuffles large fragments such as exons or domains	Non-stoichiometric incorporation of DNA fragments Frame shifts may occur Low product recovery	[80]

2.3.1. Exon shuffling

The method of *in vitro* exon shuffling has been described by Kolkman and Stemmer [73]. The general scheme of this recombination method is shown in Figure 3. Exon shuffling requires the creation of DNA fragments containing exons or combinations of exons that encode a protein domain. The exon fragments are amplified with a mixture of synthetic chimeric oligonucleotides, causing the fragments to be spliced together randomly. These spliced fragments are then assembled by primerless PCR, where individual fragments prime against each other to recreate a full-length gene. Recombination occurs when a chimeric oligonucleotide connects an exon from one parent gene to a second exon from a different parent gene. The diversity in an exon shuffling library is controlled by the number of modules which are recombined, and the number of homologs that are included for each module; in some cases, the availability of homologous domains may limit the creation of a shuffled library. The diversity of an exon shuffling library can also be controlled experimentally through the design of the chimeric oligonucleotides, facilitating certain connections between domains but not others, or by modifying the molar ratio of domain-encoding fragments to control the stoichiometry of the individual domains in the progeny. As with other recombination methods, additional diversity can be created in the library by introducing random point mutations, insertions, or deletions. Rearranging the order of domain-encoding exons also creates novel diversity.

2.3.2. Incremental truncation methods

Several non-homologous recombination methods have been designed to facilitate the shuffling of genes with insufficient sequence identity for homologous shuffling techniques. Ostermeier *et al* introduced the technique of incremental truncation for the creation of hybrid enzymes (ITCHY) [74], in which random fusion of domains from two parent enzymes is used to generate novel chimeras. Because it is difficult to predict at what locations two protein domains should be fused for maximal performance or novel activity, ITCHY libraries contain every combination. This is achieved through controlled digestion of DNA by exonuclease III to generate a collection of all possible truncated fragments of the parent genes, followed by blunt-end ligation of the fragments to form hybrid proteins. Tight control of exonuclease activity is required in addition to frequent removal of digested fragments and quenching of the reaction, in order to collect a variety of fragment lengths. Thus, ITCHY becomes a time-consuming and laborious method. ITCHY is also limited by other factors, including that only two parents can be used, gene length is not conserved by random fusion of fragments, recombination predominantly occurs at sites which are not structurally related, and only a fraction of crossover events connect fragments from two parent genes at sites where the sequences align [78]. A modified incremental truncation method, termed THIO-ITCHY, introduces a simpler procedure for creating fragment libraries from the parent genes [75]. THIO-ITCHY entails the random, low-frequency incorporation (spiking) of α-phosphothioate nucleotide analogs into the parent genes. The α- phosphothioate nucleotides protect the DNA from exonuclease activity, and so ensure the desired variation in truncation length without timed removal and quenching of digestion aliquots. If a DNA polymerase is used to incorporate nucleotide analogs, then random mutagenesis can also be integrated into the library via error-prone PCR conditions. Additional diversity can also be created by shuffling of two ITCHY libraries. This method, termed SCRATCHY, was described by Ostermeier *et al* [74].

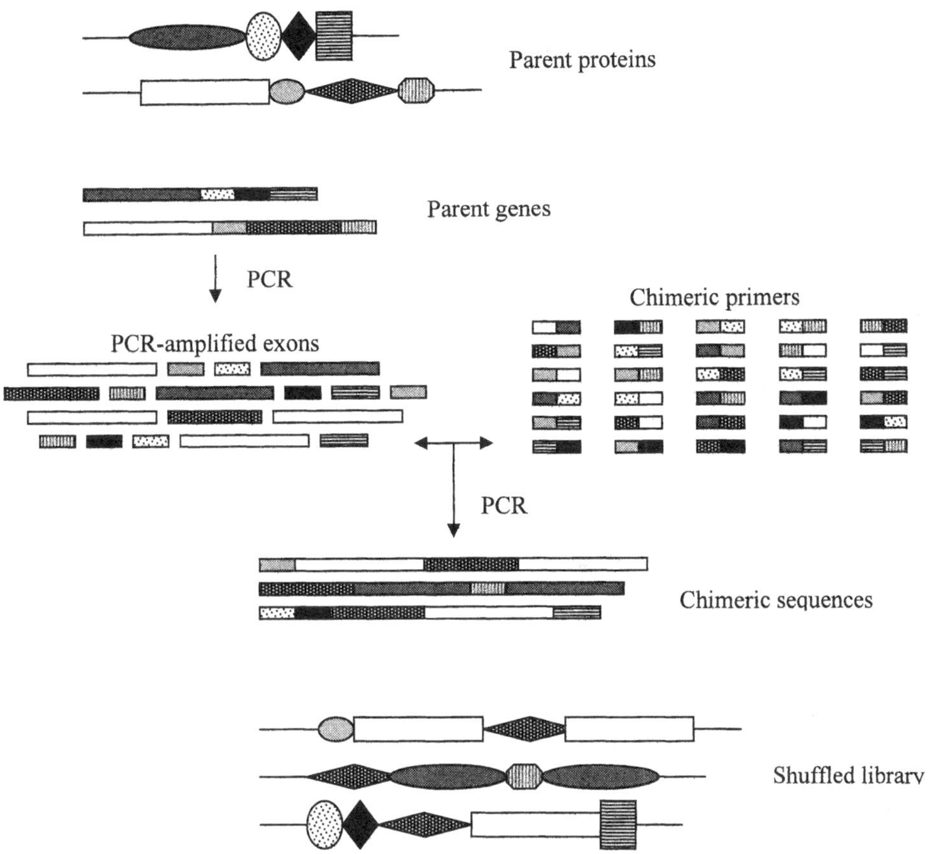

Fig. 3. Method of non-homologous recombination by exon shuffling.

2.3.3. Sequence Homology-Independent Protein Recombination (SHIPREC)

Another method conceptually similar to ITCHY is sequence homology-independent protein recombination (SHIPREC), which was used by Sieber *et al.* to create a library of interspecies hybrids from a membrane-bound human cytochrome P450 and a soluble bacterial P450 from *Bacillus megaterium* [78]. SHIPREC also involves the fusion of two parent genes and creation of a library of random length fragments. Two parent genes are joined in the first step, with a linker between them containing a unique restriction site. The fusion product is then digested with DNase I to form a library of random fragments, and fragments of length corresponding to the size of either parent gene are isolated and treated with S1 nuclease to produce blunt ends. The fragments are then circularized by blunt-end ligation and relinearized by digestion at the restriction site within the linker sequence; by this method, the gene at the 5' end of the dimer will now be at the 3' end and provide the C-terminus of the hybrid protein. SHIPREC is superior in its ability to create fusion hybrids where sequence alignment is maintained, but is limited to only one crossover event and also permits only two parent genes. Other methods for recombination of genes with limited sequence identity include degenerate homoduplex recombination (DHR) [76], random multirecombinant PCR (RM-PCR) [77], sequence independent site-directed chimeragenesis (SISDC) [79], and Y-ligation based shuffling (YLBS) [80]. A comparison of the advantages and disadvantages of these methods is provided in Table 3.

3. APPLICATIONS OF DIRECTED EVOLUTION TOOLS

3.1. Applications in enzyme engineering

Enzyme biocatalysis is increasingly viewed as a competitive and cost-effective alternative for the manufacturing of fine chemicals, pharmaceuticals, and agrochemical intermediates. Enzymes have major appeal for catalysis because of their high turnover number and refined level of selectivity, particularly in the synthesis of single-enantiomer compounds. Until recently, most of the successful industrial applications of enzymes have been limited to hydrolytic enzymes such as lipases, esterases, acylases, and hydantoinases. This situation is changing with the emergence of enzymes that perform a wide range of transformations, including asymmetric reduction, oxidation, and carbon–carbon bond formation [81–84].

Historically, microbial culture has been the most important route for enzyme discovery, even though only a small fraction of all microbes can be sampled by this method [85]. This classical strategy has rapidly been replaced by high-throughput methods based on genomic sequence discovery [86]. However, even these strategies are limited by the natural ability of enzymes to perform only a well-defined set of transformations. Directed evolution has been used with great success in recent years for the diversification of gene sequences and optimization of enzyme phenotypes [15, 87]. By surveying the available gene sequence space, specific traits are created through screening of libraries consisting of 10^4–10^{10} individuals. In all cases, optimal assay development is critical to the success in optimizing the fitness landscape of these enzymes.

3.1.1. Improving catalytic activity/stability

One of the most popular applications of directed evolution is to improve enzyme activity or stability under well-defined process conditions. By screening for initial activity and residual activity at an elevated temperature, both the thermostability and activity of mesophilic subtilisin E [88] and *p*-nitrobenzyl esterase [89] were significantly increased. Similarly, a directed evolution approach was successfully used to enhance the specific activity of a thermophilic 3-isopropylmalate dehydrogenase at lower temperatures [90], demonstrating the flexibility of this method in tailoring desirable enzymatic traits. In addition to thermal properties, enzymes with enhanced activity have also been created. In one example, directed evolution was used to improve the hydrolysis rate of organophosphorus hydrolase for several poorly degraded pesticides (25 to 700 fold) [91, 92], suggesting that this approach may be useful in generating other variants that could rapidly decontaminate structurally similar chemical warfare agents. Directed evolution approaches have also been used to enhance catalytic activities in non-natural environments such as organic solvents, for organic-phase syntheses. Moore and Arnold [93] created several *p*-nitrobenzyl esterase variants that were up to 60-fold more active in 30% dimethylformamide. Another recent work using error-prone PCR was described to achieve a five-fold improvement in the amylase activity at pH 10, an alkaline pH required for the paper industry and as a detergent additive [94].

3.1.2. Expanding specificity

Another application of directed evolution is to fine-tune the specificity of enzymes. Many successful examples have been demonstrated that are useful for the production of important industrial products. The *E. coli* D-2-keto-3-deoxy-6-phosphogluconate (KDPG) aldolase, which catalyzes the highly specific reversible aldol reaction on D-configurated KDPG substrates, was subjected to DNA shuffling and screening, and one variant was isolated capable of accepting both D- and L-glyceraldehyde as substrates in a non-phosphorylated form [95]. More recently, the P450 BM-3 monoxygenase, normally specific for medium-chain fatty acids, has been evolved to accept small hydrocarbon substrates and convert them at very high rates [96].

Perhaps the most dramatic success in this area is the use of directed evolution to create novel specificity and activity. Sun *et al.* [97] used combinatorial mutagenesis to change the substrate specificity of galactose oxidase to use glucose as a substrate. One variant (with only three point mutations) exhibited activity against D-glucose and oxidized other primary and secondary alcohols. Family shuffling of two homologous biphenyl dioxygenases created several variants with enhanced substrate specificity towards *ortho*-substituted polychlorinated biphenyls [98] and other aromatic compounds such as benzene [99], suggesting the feasibility to expand the biodegradability of other highly recalcitrant pollutants.

In addition to substrate specificity, product specificity can also be altered by directed evolution. Wild-type toluene 4-monooxygenase (T4MO) of *Pseudomonas stutzeri* OX1 oxidizes toluene to *p*-cresol (96%) and oxidizes benzene sequentially to phenol, catechol, and 1,2,3-trihydroxybenzene. To synthesize novel dihydroxy and trihydroxy derivatives of benzene and toluene, DNA shuffling of the alpha-hydroxylase fragment of T4MO (TouA) and saturation mutagenesis of the TouA active site residues were used to generate random mutants [100]. Several variants were isolated to form 4-methylresorcinol, 3-methylcatechol, and methylhydroquinone from *o*-cresol, whereas wild-type T4MO formed only 3-methylcatechol. These variants also formed catechol, resorcinol, and hydroquinone from phenol, whereas wild-type T4MO formed only catechol. These reactions show the potential synthesis of important intermediates for pharmaceuticals.

3.1.3. Changing stereo- and enantio-selectivity

Often the production of enantiomerically pure compounds is of extreme importance, particularly in the pharmaceutical industry. In this respect, directed evolution has been useful in creating enzymes with desirable enantioselectivity. May *et al.* were the first to demonstrate the feasibility to invert the enantioselectivity of D-hydantoinase to generate an enzyme that has enhanced selectivity towards L-5-(2-methylthioethyl)hydantoin [101]. Similarly, inversion of enantioselectivity of a lipase was achieved towards (R)-selectivity with E =30 (comparing to E = 1.1 for the wild type enzyme) [100]. Perhaps the best industrial success was demonstrated with the synthesis of *cis*-(1S, 2R)-indandial, a key precursor of an inhibitor of HIV protease, by toluene dioxygenase [102]. In three rounds of screening, several variants with up to three-fold decrease in production of the undesirable 1-indenol (only 20% from 60%) were obtained. In addition to enantioselectivity, the steroselectivity can be easily altered by directed evolution. Williams *et al.* [103] demonstrated that stereospecificity of tagatose-1,6-bisphosphate aldolase can be altered by 100-fold via three rounds of DNA shuffling and screening. The resulting mutant catalyzes the formation of carbon-carbon bonds with unnatural diastereoselectivity, where the >99:<1 preference for the formation of tagatose 1,6-bisphosphate was switched to a 4:1 preference for the diastereoisomer, fructose 1,6-bisphosphate.

3.2. Applications in pathway engineering

Metabolic pathway engineering is a rapidly growing area with great potential to impact industrial biocatalysis [104]. As enzymes are the central components in metabolic pathways, the strategy for the generation of sequence diversity and the screening/selection methods can be readily applied for pathway engineering. Directed evolution can be used to optimize an existing pathway, but the ability of this evolutionary approach to create new pathways that are capable of synthesizing novel compounds may be the most promising aspect for the future.

Carotenoids are important antioxidants and food additives that have been attracting commercial attention in recent years. Unfortunately, the synthesis of useful quantities from conventional chemical routes or from natural microorganisms is often costly and limited. The colorful nature of carotenoids makes them easy to detect via high-throughput screening. As a result, gene clusters for carotenoid synthesis have been introduced into *E. coli* and by performing directed evolution on two phytoene desaturases and two lycopene cyclases, several novel carotenoids were produced [105]. More recently, the C30 carotene synthase CrtM from *Staphylococcus aureus* was subjected to one round of mutagenesis and screening, and variants capable of synthesizing C-40 carotenoids were identified [106]. This plasticity of CrtM with respect to its substrate and product range highlights the potential in creating further new carotenoid backbone structures. As a result, previously unknown C-45 and C-50 carotenoid backbones were obtained from the appropriate isoprenyldiphosphate precursors [107]. Similar strategies have been applied successfully to evolve pathways for porphyrin synthesis [108].

Polyketides belong to a second class of important bioactive compounds and efforts have been directed towards the generation of novel structures for uses as antibiotics or anti-cancer agents. The modular nature of the polyketide synthases (PKS) renders polyketide synthesis inherently amenable to directed evolution strategy, particularly in the engineering of novel polyketide structures. Typically a given PKS can generate only one product. However, Shen *et al.* [109] reported that a minimal PKS from *Streptomyces coelicolor* is capable of generating more than 30 different structures, suggesting the flexibility in engineering a large number of useful structures by a single PKS. By systematically deleting domains of the erythromycin PKS or exchanging domains with other PKS modules, several variants were obtained that are capable of generating more than 50 different polyketides [110]. These examples imply the feasibility of creating entirely novel products via directed evolution of metabolic pathways.

Directed evolution can also be used as a powerful tool in optimizing an entire metabolic pathway. Functional evolution of an arsenic resistance operon has been accomplished by three rounds of shuffling and selection, resulting in cells that grew in 0.5 M arsenate, a 40-fold increase in resistance [111]. Ten mutations were located in *arsB*, encoding the arsenite membrane pump, resulting in a 4-fold to 6-fold increase in arsenite resistance. While *arsC*, the arsenate reductase gene, contained no mutations, its expression level was increased, and the rate of arsenate reduction was increased 12-fold.

Directed evolution has also been shown to enable the construction of artificial networks of transcriptional control elements in living cells [112]. By applying directed evolution to genes comprising a simple genetic circuit, a nonfunctional circuit containing improperly matched components can evolve rapidly into a functional one. Such an approach is likely to result in a library of genetic devices with a range of behaviors that can be used to construct more complex genetic circuits.

4. ALTERNATIVES TO DIRECTED EVOLUTION

4.1. Rational approaches to enzyme evolution

In addition to combinatorial approaches to enzyme evolution, many different methods for rational protein design have been devised. The strengths of directed evolution and rational design are highly complementary and may be combined to provide significant advantages over the use of a single approach.

4.1.1. Rational computational design

Computational methods [113] represent a widely used approach for rational protein design. These methods rely on the use of a force field to identify amino acid sequences that are optimal for stabilizing a protein backbone. The major drawback of these methods is that the number of possible sequences will often exceed what could be exhaustively searched by existing computing power. However, recent developments in powerful search algorithms have generated new excitement in this area [114]. Specifically, experimental data are incorporated to iteratively improve the empirical force field calculations [115]. In one example, a novel active site for activated ester hydrolysis was computationally designed into the scaffold of thioredoxin [116]. Ranking of different active site designs on the basis of substrate binding resulted in an enzyme with the ability to catalyze the predicted reaction. Even though the activity is quite modest, this example demonstrates the utility of the computational approach in designing proteins with the desired catalytic functions.

Computational methods have also been used to guide experimental design for directed evolution. There have been several studies to optimize the mutation or recombination rate with respect to the number of mutants that can be screened [117]. Other methods focused on algorithms that target the diversity of regions that will preserve structures during the evolutionary process. Inverse folding algorithms were used to predict the protein sequences that are amenable to mutagenesis without perturbing the overall protein structure [118].

4.1.2. Site-directed mutagenesis

Rational design by site-directed mutagenesis has enjoyed some success in the past, mostly in term of engineering enzyme specificity. Our ability to redesign enzyme mechanisms or completely new reactions, however, remains a difficult task. With the ever increasing knowledge of protein structure and function, site-directed mutagenesis could become a powerful complementary approach to directed evolution.

Active site substitution by site-directed mutagenesis based on structural information has been the conventional approach in protein engineering. This strategy has been used successfully in reshaping substrate or co-factor specificity and reactivity [119]. Another powerful tool in enzyme design is based on the use of structural homology to graft the desired properties from one enzyme into another via site-directed mutagenesis. Very often, new catalytic residues are introduced to alter enzyme mechanism and function. The introduction of a Ser-His-Asp triad into a peptidyl-prolyl isomerase resulted in a remarkably efficient proline-specific endopeptidase [120]. In addition, homology-driven design provides information for more accurate and detailed physical models for future rational enzyme design.

The ability to carry out novel, unique chemistries has been achieved by incorporating new catalytic groups using related proteins with similar folding structures. For example, four substitutions were enough to confer an oleate-hydroxylase activity on an oleate-desaturase [121].

4.2. Semi-rational approaches to enzyme evolution

4.2.1. SISDC: Sequence-independent site directed chimeragenesis

Conventional methods for gene shuffling are useful only if the parental genes share high levels of sequence identity (usually 70%). This means that including a relatively diverse pool of parental sequences, potentially allowing the exploration of more vast sequence space, is often difficult to accomplish. As mentioned above, several methods have been reported for creating chimeric protein libraries independent of homology, such as ITCHY, SCRATCHY, and SHIPREC. These methods, however, generate large numbers of non-functional sequences, and therefore diverse libraries of functional proteins have not been demonstrated convincingly. Recently, the Arnold group presented a simple and general method called sequence-independent site-directed chimeragenesis (SISDC) that allows for recombination of distantly related proteins at multiple discrete sites with little sequence bias and in which all targeted fragments were recombined in the desired order [79]. If desired, various modifications, such as insertion, deletion, and rearrangement, can be incorporated easily. A complementary computational algorithm called SCHEMA was also developed to estimate the disruption by the inheritance of amino acid from different parents upon recombination [122].

4.2.2. GSSM: Gene site saturation mutagenesis

Gene site saturation mutagenesis (GSSM) technology is a unique method for rapid laboratory evolution of proteins whereby each amino acid of a protein is replaced with each of the other 19 naturally occurring amino acids [123]. This is accomplished at the genetic level through the use of degenerate primer sets, comprising either 32 or 64 codon variants, for each amino acid residue. Subsequent use of standard methods for DNA replication generates a library of genes possessing all codon variations required for saturation mutagenesis of the original gene. A unique application of this method was demonstrated to evolve a nitrilase as a process-scale enantioselective biocatalyst [124]. Comprehensive mutagenesis and screening using LC-MS resulted in a nitrilase variant with high enantiomeric excess (*ee*) at high substrate concentrations. The essential mutation required two base changes in a single codon, which is difficult to achieve through other random mutagenesis methods.

5. CONCLUSION

Directed evolution tools have been increasingly used to engineer new or improved enzymes, metabolic pathways, and whole genomes for various bioprocessing applications. In the past decade, numerous molecular biology techniques have been developed to create genetic diversity through random mutagenesis and/or homologous or non-homologous recombination in the target genes, pathways and genomes. Coupled with the development of powerful high-throughput screening or selection methods, these evolutionary techniques have been successfully used to solve challenging problems in protein engineering and metabolic engineering. For the foreseeable future, directed evolution will not only remain a powerful tool for bioproduct and bioprocess development, but also a powerful research tool for solving fundamental biological problems such as the protein structure-function relationship and protein folding. In addition, directed evolution is highly complementary to rational design which capability is rapidly growing due to recent advances of structural genomics and computational biology. It seems that the combination of directed evolution and rational design represents the most powerful tool for protein engineering and metabolic engineering, and will likely become a fertile ground for innovations in the coming years.

ACKNOWLEDGEMENTS

We thank the Office of Naval Research (N000140210725 to H.Z.), National Science Foundation (BES-0348107 to H.Z. and Graduate Research Fellowship to S.R.-P.), and National Institute of Health (CMBTG Fellowship to S. R.-P.) for supporting our work on development and applications of new directed evolution tools for protein engineering and metabolic engineering.

REFERENCES

[1] W.M. Fitch and F.J. Ayala, Proc Natl Acad Sci U S A, 91(1994) 6717.
[2] G.G. Simpson, Tempo and Mode in Evolution, Columbia University Press, New York, N.Y., 1944.

[3] H. Zhao and W. Zha, (A. Svendsen), Evolutionary Methods for Protein Engineering, Marcel Dekker, Inc., New York, 2004, 353.
[4] D.R. Mills, R.L. Peterson, and S. Spiegelman, Proc Natl Acad Sci U S A, 58(1967) 217.
[5] W.P. Stemmer, Nature, 370(1994) 389.
[6] W.P. Stemmer, Proc Natl Acad Sci U S A, 91(1994) 10747.
[7] K. Chen and F.H. Arnold, Proc Natl Acad Sci U S A, 90(1993) 5618.
[8] N.W. Soong, et al., Nat Genet, 25(2000) 436.
[9] M. Kramer, et al., Metab Eng, 5(2003) 277.
[10] R. Patnaik, et al., Nat Biotechnol, 20(2002) 707.
[11] R. Petri and C. Schmidt-Dannert, Curr Opin Biotechnol, 15(2004) 298.
[12] G. Stephanopoulos, Nat Biotechnol, 20(2002) 666.
[13] Y.X. Zhang, et al., Nature, 415(2002) 644.
[14] M. Kleerebezem, et al., J Biotechnol, 98(2002) 199.
[15] F. Valetti and G. Gilardi, Nat Prod Rep, 21(2004) 490.
[16] H. Tao and V.W. Cornish, Curr Opin Chem Biol, 6(2002) 858.
[17] D.E. Robertson and B.A. Steer, Curr Opin Chem Biol, 8(2004) 141.
[18] K.A. Powell, et al., Angew Chem Int Ed, 40(2001) 3948.
[19] D.S. Wilson and J.W. Szostak, Annu Rev Biochem, 68(1999) 611.
[20] A. Jaschke and B. Seelig, Curr Opin Chem Biol, 4(2000) 257.
[21] G.F. Joyce, Annu Rev Biochem, 73(2004) 791.
[22] B. Alberts, et al., Molecular Biology of the Cell, Garland Science, New York, 2002.
[23] F.E. Wurgler, Mutat Res, 284(1992) 3.
[24] W.-H. Li and D. Graur, Fundamentals of Molecular Evolution, Sinauer Associates, Inc, Sunderland, MA, 1991.
[25] D. Botstein and D. Shortle, Science, 229(1985) 1193.
[26] J.T. Kadonaga and J.R. Knowles, Nucleic Acids Res, 13(1985) 1733.
[27] D. Shortle and D. Botstein, Methods Enzymol, 100(1983) 457.
[28] O. Selifonova, F. Valle, and V. Schellenberger, Appl Environ Microb, 67(2001) 3645.
[29] D.W. Leung, E. Chen, and D.V. Goeddel, Technique, 1(1989) 11.
[30] R.C. Cadwell and G.F. Joyce, PCR Methods Appl, 3(1994) S136.
[31] M. Fromant, S. Blanquet, and P. Plateau, Anal Biochem, 224(1995) 347.
[32] R.A. Beckman, A.S. Mildvan, and L.A. Loeb, Biochemistry, 24(1985) 5810.
[33] X. Lu, et al., Journal of Virological Methods, 94(2001) 37.
[34] G. Coia, P.J. Hudson, and R.A. Irving, J Immunol Methods, 251(2001) 187.
[35] E. Henke and U.T. Bornscheuer, Biol Chem, 380(1999) 1029.
[36] U.T. Bornscheuer, J. Altenbuchner, and H.H. Meyer, Biotechnol Bioeng, 58(1998) 554.
[37] H. Zhao, et al., (J.E. Davies), Methods for optimizing industrial enzymes by directed evolution, ASM Press, Washington DC, 1999, 597.
[38] J. Cline and H.H. Hogrefe, Strategies, 13(2000) 157.
[39] O.P. Kuipers, Methods Mol Biol, 57(1996) 351.
[40] M. Zaccolo, et al., J Mol Biol, 255(1996) 589.
[41] T.S. Wong, et al., Nucleic Acids Res, 32(2004) e26.
[42] J.H. Spee, W.M. de Vos, and O.P. Kuipers, Nucleic Acids Res, 21(1993) 777.
[43] M.A. Innis, et al., Proc Natl Acad Sci U S A, 85(1988) 9436.
[44] R.M. Myers, L.S. Lerman, and T. Maniatis, Science, 229(1985) 242.
[45] H. Murakami, T. Hohsaka, and M. Sisido, Nat Biotechnol, 20(2002) 76.
[46] R.B. Wallace, et al., Nucleic Acids Res, 9(1981) 3647.
[47] J.A. Wells, M. Vasser, and D.B. Powers, Gene, 34(1985) 315.
[48] J.F. Reidhaar-Olson and R.T. Sauer, Science, 241(1988) 53.
[49] S. Delagrave, E.R. Goldman, and D.C. Youvan, Protein Eng, 6(1993) 327.
[50] A.P. Arkin and D.C. Youvan, Proc Natl Acad Sci U S A, 89(1992) 7811.
[51] J.A. Maynard, et al., Methods Mol Biol, 182(2002) 149.
[52] G. Chen, et al., Protein Eng, 12(1999) 349.
[53] E.A. Burks, et al., Proc Natl Acad Sci U S A, 94(1997) 412.
[54] D.M. Kegler-Ebo, G.W. Polack, and D. DiMaio, Methods Mol Biol, 57(1996) 297.

[55] D.M. Kegler-Ebo, C.M. Docktor, and D. DiMaio, Nucleic Acids Res, 22(1994) 1593.
[56] B. Hallet, D.J. Sherratt, and F. Hayes, Nucleic Acids Res, 25(1997) 1866.
[57] Y. Kang, et al., J Bacteriol, 186(2004) 4921.
[58] T. Matsuura, et al., Nat Biotechnol, 17(1999) 58.
[59] J.E. Ness, et al., Nat Biotechnol, 17(1999) 893.
[60] H. Zhao and F.H. Arnold, Nucleic Acids Res, 25(1997) 1307.
[61] A. Crameri, et al., Nature, 391(1998) 288.
[62] J.M. Joern, P. Meinhold, and F.H. Arnold, J Mol Biol, 316(2002) 643.
[63] J. Sambrook, E.F. Fritsch, and T. Maniatis, Molecular Cloning: A Laboratory Manual, Cold Spring Harbor Laboratory Press, Cold Spring Harbor, NY, 1989.
[64] M.D. Gibbs, K.M. Nevalainen, and P.L. Bergquist, Gene, 271(2001) 13.
[65] M. Kikuchi, K. Ohnishi, and S. Harayama, Gene, 236(1999) 159.
[66] W.M. Coco, et al., Nat Biotechnol, 19(2001) 354.
[67] Z. Shao, et al., Nucleic Acids Res, 26(1998) 681.
[68] H. Zhao, Methods Enzymol, 388(2004) 42.
[69] H. Zhao, et al., Nat Biotechnol, 16(1998) 258.
[70] J.E. Ness, et al., Nat Biotechnol, 20(2002) 1251.
[71] D.V. Gokhale, U.S. Puntambekar, and D.N. Deobagkar, Biotech Advs, 11(1993) 199.
[72] S. Scheinbach, Biotech Advs, 1(1983) 289.
[73] J.A. Kolkman and W.P. Stemmer, Nat Biotechnol, 19(2001) 423.
[74] M. Ostermeier, A.E. Nixon, and S.J. Benkovic, Bioorg Med Chem, 7(1999) 2139.
[75] S. Lutz, M. Ostermeier, and S.J. Benkovic, Nucleic Acids Res, 29(2001) E16.
[76] W.M. Coco, et al., Nat Biotechnol, 20(2002) 1246.
[77] T. Tsuji, M. Onimaru, and H. Yanagawa, Nucleic Acids Res, 29(2001) E97.
[78] V. Sieber, C.A. Martinez, and F.H. Arnold, Nat Biotechnol, 19(2001) 456.
[79] K. Hiraga and F.H. Arnold, J Mol Biol, 330(2003) 287.
[80] K. Kitamura, et al., Protein Eng, 15(2002) 843.
[81] H.E. Schoemaker, D. Mink, and M.G. Wubbolts, Science, 299(2003) 1694.
[82] A. Schmid, et al., Nature, 409(2001) 258.
[83] O. Kirk, T.V. Borchert, and C.C. Fuglsang, Curr Opin Biotechnol, 13(2002) 345.
[84] S. Panke and M.G. Wubbolts, Curr Opin Biotechnol, 13(2002) 111.
[85] N.J. Turner, Trends Biotechnol, 21(2003) 474.
[86] M. Lafferty and M.J. Dycaico, Methods Enzymol, 388(2004) 119.
[87] H. Zhao, K. Chockalingam, and Z. Chen, Curr Opin Biotechnol, 13(2002) 104.
[88] H. Zhao and F.H. Arnold, Protein Eng, 12(1999) 47.
[89] A. Gershenson, et al., Biochemistry, 39(2000) 4658.
[90] T. Suzuki, et al., Protein Eng, 14(2001) 85.
[91] C.M.H. Cho, A. Mulchandani, and W. Chen, Appl Environ Microb, 68(2002) 2026.
[92] C.M.H. Cho, A. Mulchandani, and W. Chen, Appl Environ Microb, 70(2004) 4681.
[93] J.C. Moore and F.H. Arnold, Nat Biotechnol, 14(1996) 458.
[94] C. Bessler, et al., Protein Sci, 12(2003) 2141.
[95] S. Fong, et al., Chem Biol, 7(2000) 873.
[96] A. Glieder, E.T. Farinas, and F.H. Arnold, Nat Biotechnol, 20(2002) 1135.
[97] L. Sun, et al., Chembiochem, 3(2002) 781.
[98] F. Bruhlmann and W. Chen, Biotechnol Bioeng, 63(1999) 544.
[99] H. Suenaga, et al., J Bacteriol, 183(2001) 5441.
[100] G. Vardar and T.K. Wood, Appl Environ Microbiol, 70(2004) 3253.
[101] O. May, P.T. Nguyen, and F.H. Arnold, Nat Biotechnol, 18(2000) 317.
[102] N. Zhang, et al., Metab Eng, 2(2000) 339.
[103] G.J. Williams, et al., Proc Natl Acad Sci U S A, 100(2003) 3143.
[104] L. Rohlin, M.K. Oh, and J.C. Liao, Curr Opin Microbiol, 4(2001) 330.
[105] C. Schmidt-Dannert, D. Umeno, and F.H. Arnold, Nat Biotechnol, 18(2000) 750.
[106] D. Umeno, A.V. Tobias, and F.H. Arnold, J Bacteriol, 184(2002) 6690.
[107] D. Umeno and F.H. Arnold, J Bacteriol, 186(2004) 1531.
[108] S.J. Kwon, et al., Chembiochem, 5(2004) 1069.

[109] Y. Shen, et al., Proc Natl Acad Sci U S A, 96(1999) 3622.
[110] R. McDaniel, et al., Proc Natl Acad Sci U S A, 96(1999) 1846.
[111] A. Crameri, et al., Nat Biotechnol, 15(1997) 436.
[112] Y. Yokobayashi, R. Weiss, and F.H. Arnold, Proc Natl Acad Sci U S A, 99(2002) 16587.
[113] A.G. Street and S.L. Mayo, Structure Fold Des, 7(1999) R105.
[114] J.R. Desjarlais and N.D. Clarke, Curr Opin Struct Biol, 8(1998) 471.
[115] D.B. Gordon, S.A. Marshall, and S.L. Mayo, Curr Opin Struct Biol, 9(1999) 509.
[116] D.N. Bolon and S.L. Mayo, Proc Natl Acad Sci U S A, 98(2001) 14274.
[117] C.A. Voigt, S. Kauffman, and Z.G. Wang, Adv Protein Chem, 55(2000) 79.
[118] C.A. Voigt, et al., Proc Natl Acad Sci U S A, 98(2001) 3778.
[119] F. Cedrone, A. Menez, and E. Quemeneur, Curr Opin Struct Biol, 10(2000) 405.
[120] E. Quemeneur, et al., Nature, 391(1998) 301.
[121] P. Broun, et al., Science, 282(1998) 1315.
[122] M.M. Meyer, et al., Protein Sci, 12(2003) 1686.
[123] J.M. Short, Patent US6171820 (2001).
[124] G. DeSantis, et al., J AM Chem Soc, 125(2003) 11476.

Bioprocessing for Value-Added Products from Renewable Resources
Shang-Tian Yang (Editor)

Chapter 4. Metabolic Engineering – Applications, Methods, and Challenges

Shang-Tian Yang, Xiaoguang Liu, and Yali Zhang

Department of Chemical and Biomolecular Engineering, The Ohio State University, 140 West 19th Avenue, Columbus, OH 43210, USA

1. INTRODUCTION

A living cell is a complex chemical reactor in which more than 1000 independent but highly coupled enzyme-catalyzed reactions and selective membrane transport occur. Classical strain improvement (CSI) uses random mutagenesis to accumulate genomic alterations and then screens for phenotypes with desirable characteristics. This technique has been successfully used for many decades in industrial fermentation to improve the production of penicillins, amino acids, citric acid, and many other industrial products. However, CSI is time consuming and only works in some cases. Modern recombinant DNA technology provides a powerful tool for the directed improvement of cellular properties through the modification of specific biochemical reactions or the introduction of new ones that can not only provide new capabilities but also greatly speed up the strain development process [1]. Metabolic engineering, as first defined by Baily in 1981 [2], is "the improvement of cellular activities by manipulating enzymatic, regulatory, and transport functions of the cell with the use of recombinant DNA technology." Since then, metabolic engineering has emerged as a highly multidisciplinary scientific field, with many excellent review articles [3–5] and books published [6–8]. Today, metabolic engineering is not only widely applied in industrial fermentation for strain improvement and metabolite overproduction, but has also found many important applications in functional genomics, biological research (e.g., signal transduction), and medical research (e.g., drug discovery and gene therapy) [9, 10]. In this chapter, we will focus on industrial biotechnology.

Metabolic engineering focuses on the metabolic pathways leading to metabolites or biomolecules that are desired fermentation products. Glucose or another carbohydrate is usually the starting material or substrate. In a typical cell, glucose is first catabolized to pyruvate usually by one of the three glycolysis pathways – Embdon-Meyerhof-Parnas (EMP), pentose phosphate (PP) or hexose monophosphate (HMP), or Entner-Doudoroff (ED). Under aerobic conditions, pyruvate is converted to acetyl-CoA, which then enters the tricarboxylic acid (TCA) cycle. The electrons present in the cofactors (mainly NADH and NADPH) generated in glycolysis and the TCA cycle are then transferred down the electron transport chain to oxygen, yielding ATP or energy and regenerating the cofactors (NAD^+ and $NADP^+$). The intermediate metabolites, also referred to as primary metabolites, produced in these pathways are used as precursors for the biosynthesis of amino acids, ribonucleotides, deoxyribonucleotides, hexosamines, fatty acids, vitamins, and secondary metabolites, such as antibiotics. Finally, these small biomolecules are used to synthesize macromolecules or biopolymers, including proteins, lipids, nucleic acids, polysaccharides, polyalkanoates, etc. These macromolecules are the building blocks of cell structural components and organelles. However, under some conditions when cell growth is not "optimized," some metabolites and biosynthetic products may be overproduced. Metabolic engineering is about making cells overproduce a desirable metabolite or biosynthetic product at a high rate and yield from glucose and other affordable carbon sources.

Therefore, metabolic engineering can be said to be pathway engineering. However, there are many ways to engineer metabolic pathways: directly by manipulating the genes encoding the enzymes catalyzing the reactions in the pathways, or indirectly by altering the regulatory pathways affecting gene expression and enzyme activity. Then, one has to decide where in the pathways to target the reactions in order to get the desirable results. This will require either a good knowledge of the metabolic networks or a good strategy to select the gene target for re-engineering. Much of the field of metabolic engineering also focuses on developing a good metabolic engineering strategy and methods for selecting targets and analyzing and predicting the consequences of genetic modifications and changes in the metabolic pathways.

In this chapter, we will first discuss several industrial fermentation examples, followed by the methodologies used in metabolic engineering and then the challenges and approaches complementary to metabolic engineering for industrial strain and bioprocess development.

2. APPLICATIONS AND SOME EXAMPLES

Generally speaking, the main purpose of metabolic engineering is to optimize a biotechnologically important process carried out by organisms by using genetic manipulation to change the distribution of intracellular chemical reactions (flux). The application of metabolic engineering to industrial biotechnology has mainly focused on improving cell metabolism to increase productivity – higher product yield, production rate, and cell growth efficiency (energy efficiency), and to eliminate or reduce undesirable byproducts. It has also been applied to eliminate or reduce feedback inhibition. By recruiting heterologous activities, a partial pathway in an organism can be completed for the production of the desirable product. Also, hybrid or new metabolic pathways can be developed to produce new products, extend substrate range, or enhance processing characteristics. All of these can be found in various examples discussed in this section. It is not possible and unnecessary to review every metabolic engineering study that can be found in the literature. Here we have chosen the following products because of their industrial importance from various metabolite groups to illustrate how various metabolic engineering strategies and approaches can be used to improve existing or create new production strains and complexities and problems that may arise in the process.

2.1. Amino acids

Amino acids are the basic building blocks of proteins and have many important applications in food and biomedical industries. The essential amino acids, such as lysine, methionine, threonine, and phenylalanine are important nutritional supplement in animal feed, MSG (monosodium glutamate) is used as a flavoring agent, and tryptophan is an important pharmaceutical ingredient. Since the discovery of glutamate-overproducing *Corynebacterium glutamicum* in 1957, traditional strain development and, more recently, metabolic engineering have been widely studied and applied to enhance the production of various L-amino acids from sugars by fermentation. Today, except for methionine and a few others, many important amino acids are commercially produced from sugars by microbial fermentation in large quantities with high product titers and yields (Table 1) [11–13].

Table 1
Some important amino acids produced by fermentation [11, 13]

	Annual Production (metric tons)	Fermentation Organism	Concentration (g/L)	Product yield (g/g)
Glutamate (MSG)	1,200,000	*C. glutamicum*	150	0.5
L-Lysine	600,000	*C. glutamicum*	150	0.5
L-Threonine	40,000	*E. coli*	100	0.4 – 0.5
L-Phenylalanine	13,000	*E. coli*	50	0.2 – 0.25
L-Tryptophan	1,200	*E. coli*	58	0.2 – 0.25

Amino acids are produced in the biosynthetic pathways of most bacteria. However, their production in the cell is affected by the central metabolism and is highly regulated. Figure 1 shows the pathways involved in the biosynthesis of some amino acids. Historically, intensive strain development through sequential mutations and screening processes has been used to obtain high producers of some amino acids. Glutamic acid is overproduced by some industrial strains of *C. glutamicum* when its fatty acid biosynthesis pathway is hindered by either *DtsR* mutation, biotin limitation, or inhibition by penicillin or surfactant in the growth medium [14]. It was later discovered that the overproduction of glutamate by *C. glutamicum* was not related to the membrane structure or permeability but resulted from the carbon flux shift from fatty acid biosynthesis towards the TCA cycle. Also, the activity level of ODHC (2-oxoglutarate dehydrogenase complex) affected the flux distribution at the branch point of 2-oxoglutarate and glutamate synthesis, primarily by reductive amination in the presence of ammonia and NADPH. Increasing glutamate dehydrogenase (GDH) has no effect on glutamate production, suggesting that its secretion is limited by the export system.

To overproduce amino acids in the aspartate group, a change in the metabolic flux resulting from mutations affecting the repression or feedback inhibition of the key biosynthetic enzymes is usually required [11]. For example, lysine synthesis in *C. glutamicum* is controlled by aspartate kinase (AsK), which is allosterically inhibited by lysine and threonine; this feedback control is eliminated by mutations in the β-subunit of the kinase or relieved due to a low threonine concentration resulting from low homoserine dehydrogenase (HDH) activity.

Threonine synthesis in corynebacteria is regulated by the activity of HDH, which is strongly inhibited by threonine. In addition, the genes coding for homoserine dehydrogenase (*hom*) and homoserine kinase (*thrB*) are on the same operon, which is repressed by methionine [12]. Because corynebacteria do not convert homoserine to threonine effectively or secrete threonine well, efforts to breed a corynebacterial threonine producer have not been as successful as those with *E. coli*, which has threonine-insensitive AsK and HDH [15]. The diversity of the anaplerotic enzymes present in *C. glutamicum* exert a complicated control over carbon flux that affects the biosynthesis of lysine [16]. The biosynthesis of methionine in microorganisms requires energy (ATP) and is highly regulated. Consequently, no methionine overproducing microorganism is available for industrial fermentation, and methionine is currently produced either by chemical synthesis or by hydrolyzing proteins [17].

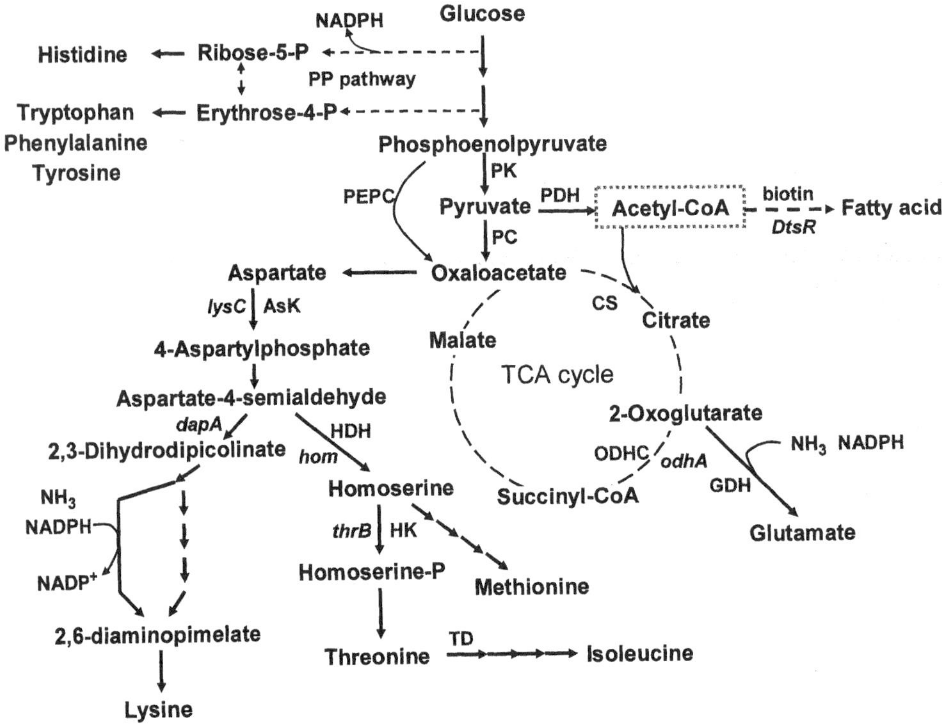

Fig. 1. Metabolic and biosynthetic pathways for amino acids production from glucose.
AsK: aspartate kinase; CS: citrate synthase; GDH: glutamate dehydrogenase; HDH: homoserine dehydrogenase; HK: homoserine kinase; ODHC: 2-oxoglutarate dehydrogenase complex; PC: pyruvate carboxylase; PDH: pyruvate dehydrogenase; PEPC: phosphoenolpyruvate carboxylase; PK: pyruvate kinase; TD: threonine dehydratase.

Many different metabolic engineering strategies have been studied and applied to enhance amino acid production. They include the metabolic engineering of biosynthetic and central metabolic pathways and transport engineering.

2.1.1. Engineering biosynthetic pathways

The first target for metabolic engineering is the terminal pathway leading to the synthesis of a desired amino acid. Successful strategies include:

1. Amplification of the gene coding for the rate-limiting enzyme in the biosynthetic pathway to eliminate the bottleneck and thus increase the amino acid production. For example, overexpression of the dihydrodipicolinate synthase gene *dapA* increases lysine accumulation [18]
2. Amplification of the gene coding for the branch-point enzyme to redirect metabolic flow from the common intermediate to another amino acid. For example, increasing *hom* for HDH shifted the metabolic flux from lysine to threonine production [19].
3. Introducing heterologous enzymes with different control architectures allowing them to bypass the regulatory step (e.g., feedback inhibition or repression) in the biosynthetic pathway. For example, expressing *E. coli* threonine dehydratase insensitive to feedback inhibition increases isoleucine production in *C. glutamicum* [20]
4. Introducing heterologous enzymes with different catalytic mechanisms that are functionally or energetically advantageous. An example is the construction of an alanine-producing strain of *C. glutamicum* by replacing its alanine dehydrogenase with the one from *Arthrobacter oxydans*; the mutant produced more alanine under limited oxygen conditions due to a better balanced redox potential [11].
5. Amplification of the first enzyme in the pathway diverging from the central metabolism to increase carbon flow down the biosynthetic pathway. This strategy has been applied to a tryptophan-producing *C. glutamicum* strain [21], but it usually also requires to sequentially remove bottlenecks discerned by the accumulation of intermediates.

2.1.2. Engineering central metabolism

Central metabolism supplies precursors and energy that may limit amino acid biosynthesis in overproducing strains whose bottlenecks in the terminal pathways have been removed. Engineering the central metabolism is complicated because it is subjected to extensive and global regulation, which is still not fully understood. Anaplerotic carboxylation reactions are responsible for the synthesis of oxaloacetate, a key intermediate in amino acid biosynthesis. Deletion of phosphoenolpyruvate carboxykinase has been shown to lead to increased threonine productivity by more than 40% in *E. coli* model strains [22]. Simultaneous overexpression of pyruvate carboxylase (PC) and aspartate kinase increased lysine productivity by 250% in *C. glutamicum* [23]. Decreasing the level of the *dtsR* gene product triggered the overproduction of glutamate because of the decreased activity of ODHC [14]. The pentose phosphate pathway is responsible for supplying the NADPH, ribose-5-phosphate, and erythrose 4-phosphate required for histidine and aromatic amino acids. Amplifying the transketolase gene increases the supply of erythrose 4-phosphate to the non-oxidative PP pathway, leading to the creation of a tryptophan-hyperproducing strain of *C. glutamicum* [24]. On the other hand, better histidine production was achieved by a deficiency of transketolase, as this increased the supply of ribose 5-phosphate to the oxidative PP pathway [25].

2.1.3. Transport engineering

Mutant strains with altered transport systems can maintain a low intracellular level of the product amino acid and thus are not subject to feedback control; this has been shown to increase product yield in tryptophane and threonine fermentations with *C. glutamicum* and *E. coli*, respectively [26, 27]. Mutants with an active efflux system and impaired uptake of the amino acid can overproduce the amino acid without the complete deregulation of the relevant biosynthetic pathways [11].

2.1.4. Whole-cell engineering

Central metabolism plays a critical role in cellular physiology and is subject to extensive global regulation. Although the pathways and major fluxes in central metabolism have been extensively studied in some microorganisms, flux regulation and its role in global physiology is still not fully understood for most microorganisms of industrial interest. Even with advanced engineering modeling and metabolic network analysis, it is still difficult to predict the outcome of redirecting central metabolism. The junction between glycolysis and the TCA cycle is particularly crucial for the global regulation of amino acids biosynthesis. Moreover, pathway flux control is usually shared and not localized at any particular enzyme. Therefore, inverse metabolic engineering and systems biotechnology approaches have become popular in recent years [28]. An effective new metabolic engineering approach is to use data from functional genomics to carry out whole-cell engineering and genomic breeding, which improves strains by removing the unwanted mutations acquired during numerous mutation steps [29]. The minimally mutated strains obtained in "genomic breeding" by only introducing beneficial point mutations into the wild-type *C. glutamicum* genome resulted in remarkable lysine titers and productivity [30].

2.2. Antibiotics

β-lactam antibiotics such as penicillins are among the oldest and largest industrial products produced by fermentation [31]. Through classical strain improvement programs and fermentation process optimization, industrial production of penicillin has been improved over 1000-fold. However, metabolic engineering or directed genetic modifications through recombinant DNA technology can provide direct and more efficient approaches to strain improvement, and has recently been extensively studied and used to further improve the production of β-lactam antibiotics, including penicillins, cephalosporins, and cephamycins (see recent reviews by Gumarson et al. [32] and Thykaer and Nielsen [33]).

The biosynthetic pathways of penicillins, cephalosporins and cephamycins are well characterized (see Figure 2). The key genes involved in the biosynthesis of these antibiotics are often clustered on the same chromosome. Therefore, increasing antibiotic production can be achieved by overexpressing single genes or entire gene clusters in the biosynthetic pathways. For example, penicillin production in *P. chrysogenum* was increased significantly when additional copies of the *pcbC* and *penDE* genes were introduced [34]. The transformant with the amplified whole penicillin gene cluster showed a 176% increase in specific penicillin productivity, equivalent to that achieved from classical strain improvement over a period of five years from 1972 to 1977 [32]. Similarly, cephalosporin C production in *A. chrysogenum* was increased 2 to 3-fold by overexpressing the *cefG* gene alone [35, 36]. Improved penicillin titers in industrial fermentation have been found to be highly correlated to the cluster copy number in the production strains [37].

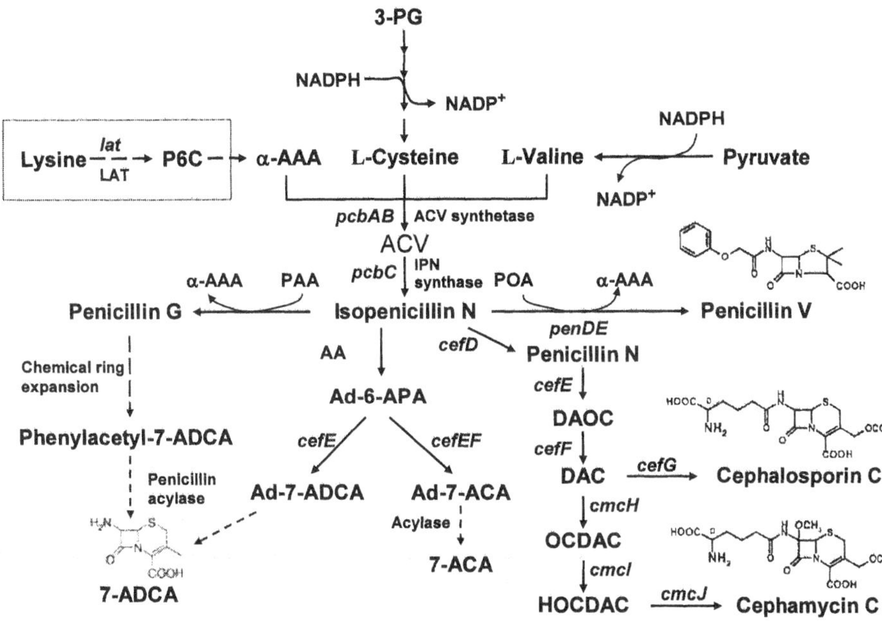

Fig. 2. Biosyntheses of the β-lactam penicillins, cephalosporin C and cephamycin C. Arrows with dashed lines indicate the *in vitro* enzymatic or chemical reactions carried out after fermentation. AA: adipic acid; α-AAA: L-α-aminoadipate; 7-ACA: 7-aminocephalosporanic acid; ACV: α-L-aminoadipyl- L-cysteinyl-D-valine; Ad-: adipoyl-; 7-ADCA: 7-aminodeacetoxycephalosoranic acid; DAC: deacetylcephalosporin C; DAOC: deacetoxycephalosporin C; HOCDAC: 7-α-hydroxy-O-carbamoyldeacetylcephalosporin C. IPN: isopenicillin N; OCDAC: O-carbamoyldeacetyl-cephalosporin C; PAA: penaylacetic acid; 3-PG: 3-phosphoglycerate; POA: phenoxyacetic acid.

However, further increasing the gene dosage and transcription levels of the entire penicillin biosynthesis gene cluster in industrial strains may not be effective because there appears to be an upper limit on the number of copies of the gene cluster that can be inserted while still increasing productivity [33]. This is because the precursor supply from the primary metabolism also limits the biosynthesis of β-lactam antibiotics. It has been suggested that the supply/regeneration of NADPH from the pentose phosphate pathway for L-cysteine biosynthesis could limit the formation of penicillin from glucose [38, 39]. Chary et al. increased cephamycin C production by overexpressing the enzyme lysine-6-aminotransferase (LAT), which catalyzes the deamination of lysine to L-α-aminoadipate (α-AAA) [40]. Increasing the available activated side chain precursor phenylacetic acid (PAA) by overexpressing PAA activating CoA ligase (PCL) in *P. chrysogenum* also improved penicillin G production [41].

Metabolic engineering has also been used to extend the pathway in well-characterized β-lactam production organisms [33, 42]. Traditional chemical methods for the production of 7-aminocephalosporanic acid (7-ACA) and 7-aminodeacetoxycephalosoranic acid (7-ADCA) involve environmentally harmful organic solvents [43]. When fed adipic acid, *P. chrysogenum* strain expressing *S. clarvuligerus cefE* or *A. chrysogenum cefEF* produced Ad-7-ADCA or Ad-7-ACA, respectively [44]. These cephalosporins with adipoyl side chains can be enzymatically converted to 7-ADCA and 7-ACA, respectively (see Figure 1). This new process replaced the old chemical process and is now used for the production of 7-ADCA by *P. chrysogenum* [45].

In addition to β-lactams, metabolic engineering has also been applied to other types of antibiotics, including polyketides and glycopeptides [32]. Polyketides are a class of bioactive secondary metabolites that are assembled from the coenzyme A esters of short-chain fatty acids by polyketide synthases (PKS) [46]. After assembly by the PKS, the polyketide backbone can be modified by other enzymes to form an extremely diverse class of antibiotics [47]. Metabolic engineering has been used to explore the modular nature of PKS, generating hybrid enzymes and producing novel compounds by combining various precursors [48]; this is so called "combinatorial engineering." Improving the supply of precursor molecules is critical to the production of polyketides [32]. By decreasing flux through the pentose phosphate pathway, biosynthesis of the polyketides actinorhodin and undecylprodigiosin in *Streptomyces lividans* can be improved because more glucose is catabolized through glycolysis, which provides the precursor of the antibiotic [49]. Butler *et al.* created mutants with deleted *zwf1* and *zwf2* genes encoding two isozymes of glucose-6-phosphate dehydrogenase and showed that the mutants produced twice as much antibiotics as the control strains did [50].

2.3. Indigo

Fermentative production of indigo with a recombinant *E. coli* strain carrying the *Pseudomonas putida* genes encoding the enzyme naphthalene dioxygenase (NDO) was first reported in 1983 by Ensley et al. [51]. The native *E. coli* tryptophanase converted tryptophan in the culture medium to indole, which was then converted to indoxyl by NDO, followed by spontaneous oxidation reactions forming indigo (Figure 3). Later, a mutant with an inactivated *trpB* gene encoding the β-subunit of tryptophan synthase, which catalyzes the terminal step in the biosynthesis of tryptophan from indole and serine, was cloned with the NDO genes to produce indigo directly from glucose [52]. However, indigo production was found to be limited at the first step of aromatic biosynthesis, the formation of 3-deoxy-D-arabino-heptulosonate 7-phosphate (DAHP) from PEP and E4P, because DAHP synthase (encoded by *aro*G^{fbr}) was inactivated by indigo [53]. To overcome this limitation, Berry *et al.* increased the dosage of the *aro*G^{fbr} gene and the availability of the enzyme substrates, which were found to protect the enzyme from inactivation by indigo, by amplifying the *tkt*A gene (encoding transketolase) and inactivating *pykA* and *pykF* genes (encoding the two isozymes of pyruvate kinase) [53]. The strain with all these mutations produced 18 g/L of indigo in 72 hours, a 60% increase over the control strain.

One problem in large scale denim dyeing processes using biologically produced indigo is the undesirable red cast caused by indirubin, which is a structural isomer of indigo formed from indoxyl and isatin, a byproduct of the spontaneous oxidation reaction that converts indoxyl to indigo. A mutant expressing the enzyme isatin hydrolase [54], which hydrolyzes isatin to isatic acid, reduced the indirubin content of the final indigo product by 50%. The biological indigo thus produced was found to be equivalent to chemically produced indigo in its dyeing performance [53]. This work illustrated that metabolic engineering can be applied to solve problems encountered in the production process and final application of the product.

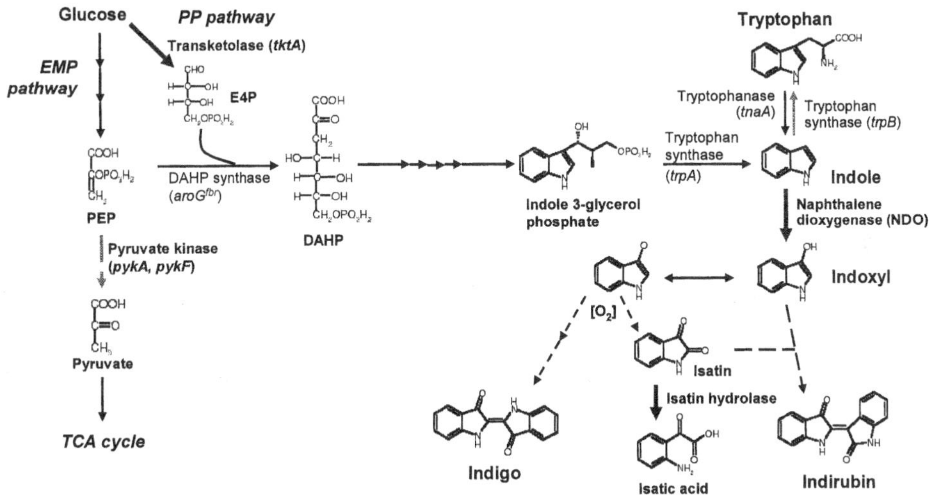

Fig. 3. Metabolic and biosynthetic pathways for indigo production from glucose. Bold arrows indicate the reactions with heterologous genes and dashed arrows indicate spontaneous chemical reactions.

2.4. 1,3-Propanediol

There has been a high interest in producing 1,3-propanediol as an industrial feedstock chemical from biomass by fermentation [55–57]. Recently, DuPont has developed a new polymer fiber based on the reaction between 1,3-propanediol and terephthalic acid [58]. The new polymer fiber, Sorona®, has many desirable properties and can replace current synthetic fibers for use in large-volume markets including apparel and carpet [58]. Naturally, a large number of anaerobic bacteria, including *Citrobacter, Clostridium, Enterobacter, Klebsiella* and *Lactobacillus* species, can produce 1,3-propanediol from glycerol, whereas yeasts such as *Saccharomyces cerevisiae* can produce glycerol from glucose [57]. Directed towards a more economical single-step fermentation process utilizing the lower cost feedstock glucose, DuPont and Genencor International have undertaken an ambitious metabolic engineering effort to design and develop a recombinant *E. coli* strain for the direct conversion of glucose to 1,3-propanediol [59–62]. The original *E. coli* K12 strain cannot produce 1,3-propanediol and has only a weak capacity to produce glycerol from glucose. The metabolic engineering work thus involved the cloning of glycerol 3-phosphate dehydrogenase (DAR1) and glycerol 3-phosphate phosphatase (GPP2) genes from *Saccharomyces cerevisiae* to divert carbon from dihydroxyacetone phosphate (DHAP) to glycerol [63, 64] and glycerol dehydratase (*dhaB1*, *dhaB2*, *dhaB3*) and its reactivating factors (*dhaBX*, *orfX*), obtained as a gene cluster from *K. pneumoniae*, to convert glycerol to 3-hydroxy-propionaldehyde (3-HPA) [59, 60], which was then converted to 1,3-propanediol using a previously uncharacterized oxidoreductase endogenous to *E. coli* (*yqhD*). Instead of using the heterologous 1,3-propanediol oxidoreductase (*dhaT*) from the bacteria, the *E. coli* strain using YqhD produced a high 1,3-propanediol concentration of ~130 g/L that was higher than that from the same strain using DhaT. Such high concentrations have never been obtained in glycerol-fed fermentations using natural 1,3-propanediol-producing bacteria [62]. The higher titer was attributed to the difference in cofactor balance: YqhD utilized NADPH whereas DhaT used NADH.

The host strain was further modified to optimize its energetic (Pi) and redox demands. Glucose transport by the PEP-dependent phosphotransferase system (PTS) in *E. coli* was eliminated and replaced with the more energetically efficient ATP-dependent phosphorylation. Concomitant with the modulation of triosphosphate isomerase (*tpi*), glyceraldehyde 3-phosphate dehydrogenase (*gap*) was downregulated to provide an improved flux control point. In addition, glycerol kinase (*glpK*) and glycerol dehydrogenase (*gldA*) were deleted to prevent glycerol from re-entering central carbon metabolism [65]. Figure 4 illustrates the metabolic pathways engineered to produce 1,3-propanediol from glucose. In fed-batch aerobic fermentations, metabolically engineered *E. coli* produced 1,3-propanediol

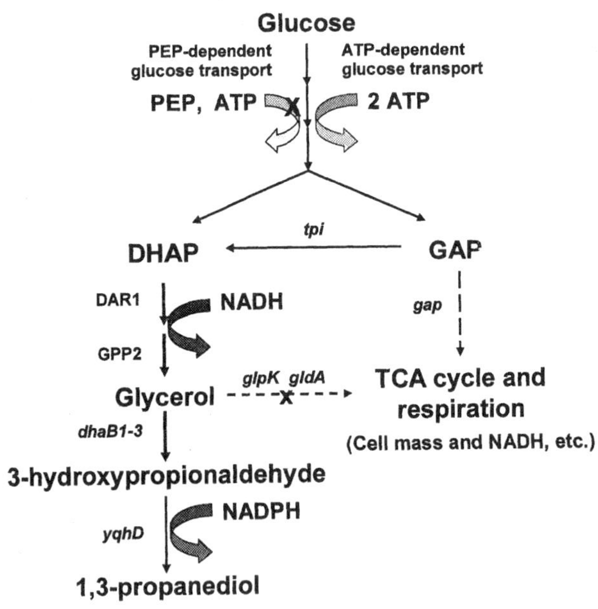

Fig. 4. Metabolic and biosynthetic pathways for 1,3-propanediol production from glucose. Bold arrows indicate reactions with heterologous genes. "X" indicates gene knock-out.

from glucose at a productivity of 3.5 g/L·h, a titer of 135 g/L, and a weight yield of 51%. By contrast, traditional anaerobic fermentations of glycerol typically give maximum values of 3.0 g/L·h, 78 g/L and 55%, respectively [57]. This is a successful example of using metabolic engineering to establish an industrial strain for commercial production of feedstock chemicals from renewable resources. DuPont and Tate & Lyle (formerly A.E. Staley) have completed their pilot fermentation process study and begun building their first plant, with a capacity of 100,000 tons/yr. The demand for polymers from 1,3-propanediol is estimated to be 1–2 billion pounds per year in ten years.

2.5. Lactic acid

Lactic acid is an important specialty chemical widely used in industry. In the last ten years, fermentative production of optically pure L(+)-lactic acid has gained large interest because of the increased industrial interest in biodegradable polylactic acid [66]. Lactic acid can be produced from sugars by anaerobic fermentation with homolactic acid bacteria (e.g., *Lactobacillus* and *Lactococci*) [67–69] and aerobic fermentation with the filamentous fungus *Rhizopus oryzae* [70]. However, these organisms have their respective disadvantages for use in large scale industrial fermentation. *R. oryzae* can produce high purity L(+)-lactic acid from sugars but it has a complex filamentous morphology that is usually difficult to control in stirred-tank fermentors. On the other hand, homofermentative lactic acid bacteria (LAB) usually produce a mixture of L(+)- and D(−)-lactic acid. Although mutants of *Lactobacillus* spp. with knocked-out D(−)-lactate dehydrogenase gene *ldhD* can produce pure L(+)-lactic acid at high yields [71], these bacteria require complex media and a relatively high pH value for growth [72], which cause difficulties in product recovery and purification.

In order to reduce the fermentation cost for the production of optically pure L(+)-lactic acid, much effort has been focused on the development of metabolically engineered yeasts in the last ten years [73–79]. Yeasts, which naturally produce ethanol but not much lactic acid, were given a heterologous lactate dehydrogenase (LDH) gene to shift the glycolytic flux toward the production of lactic acid [73, 74]. The metabolically engineered yeasts expressing LDH are better for lactic acid production in terms of their better acid tolerance and simpler nutrient requirements than lactic acid bacteria. A high concentration of L(+)-lactic acid with greater than 99.9% optical purity was produced by genetically engineered yeast (*Saccharomyces cerevisiae*) containing the bovine *ldh* gene [76]. However, the lactic acid yield from sugars using metabolically engineered *S. cerevisiae* was low, only about 60%, because of the co-production of ethanol [77]. Attempts to eliminate ethanol production by knocking out pyruvate decarboxylase (PDC) in *S. cerevisiae* were not successful because this yeast has two active structural *pdc* genes and PDC activity is critical to its growth ability [78].

Alternatively, Crabtree-negative yeast *Kluyveromyces lactis* possessing only one PDC gene was engineered to produce only lactic acid by knocking out the *pdc* gene and introducing the bovine *ldh* gene under the control of the inducible promoter of *KlPDC1* gene [78]. However, the lactic acid yield from glucose by this mutant was still low, only about 0.58 g/g, because this yeast actively transports acetyl-coenzyme A (CoA) from mitochondria to the cytosol and pyruvate is largely channeled into the tricarboxylic acid cycle by the pyruvate dehydrogenase

(PDH) complex under aerobic conditions. Further engineering the yeast to delete the PDH E1α subunit gene greatly improved the lactic acid production to a yield as high as 0.85 g/g in a fed-batch fermentation [79]. Figure 5 shows the metabolic pathway and genes involved in engineering the yeast for lactic acid production. This example illustrates the multiple engineering of metabolic pathways in a production strain that is suitable for industrial applications.

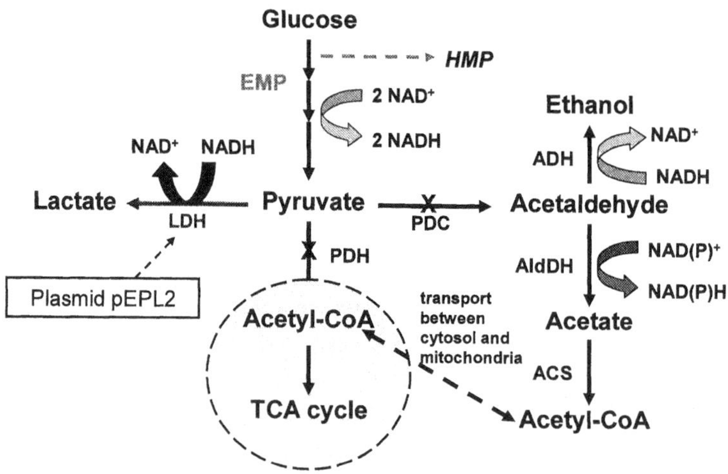

Fig. 5. Metabolic engineering of pyruvate metabolism and pyruvate bypass in *K. lactis* for the production of L(+)-lactic acid from glucose. Enzymes: ACS, acetyl-CoA synthetase; ADH, alcohol dehydrogenase; AldDH, aldehyde dehydrogenase; LDH, lactate dehydrogenase; PDC, pyruvate decarboxylase; PDH, pyruvate dehydrogenase. The area encircled with the dotted line represents the mitochondrial compartment. The yeast was transformed with a heterologous *LDH* gene to add the metabolic step for lactic acid production from pyruvate. Meanwhile, PDC and PDH were deleted to eliminate the flux toward ethanol production and TCA cycle, resulting in a homolactic acid proudcing yeast with a high lactate yield from glucose.

In addition, the central fermentation metabolism of *E. coli* has also been engineered to produce optically pure lactic acid from glucose and xylose. *E. coli* normally produces acetate, formate, D(–)-lactate, succinate, and ethanol from glucose under anaerobic conditions (see Figure 6). A *pta* mutant of *E. coli* deficient in phosphotransacetylase (PTA) was made to produce only D(–)-lactate by inactivating the phosphoenolpyruvate carboxylase (PPC) gene [80]. The *pta ppc* mutant, however, required dicarboxylic acids or complex nutrients for growth. By combining mutations in four genes: pyruvate formatelyase (*pflB*), acetate kinase (*ackA*), alcohol dehydrogenase (*adhE*), and fumarate reductase (*frdBC*), Zhou et al. developed a mutant strain that produced high yields of D(–)-lactate from sugars in a mineral salt medium without additional nutrients [81].

E. coli capable of producing pure L(+)-lactic acid as its major fermentation product was also developed by introducing plasmids containing heterologous L(+)-lactate dehydrogenase (*ldhL*) genes into *pta ldhA* or *pfl ldhA* double mutant lacking D(–)-lactate dehydrogenase [80,

82]. As much as 70 g/L of L(+)-lactic acid with a yield of 0.8 g/g of glucose was produced using complex media. Similarly, by replacing part of the chromosomal *ldhA* coding region with *ldhL*, Zhou et al. developed a stable mutant with five chromosomal deletions (*focA-pflB*, *frdBC*, *adhE*, *ackA*, *ldhA*) that produced L(+)-lactic acid in a mineral salt medium with high yields of ~95% from glucose and 93% from xylose [83]. The product's optical purity was 99.5%, and only small amounts of acetate and succinate were produced as byproducts in the fermentation. However, the volumetric productivity was low, less than 0.6 g/L·h.

Additional studies of metabolic engineering approaches for lactic acid production in various microorganisms can be found in a recent review article [84].

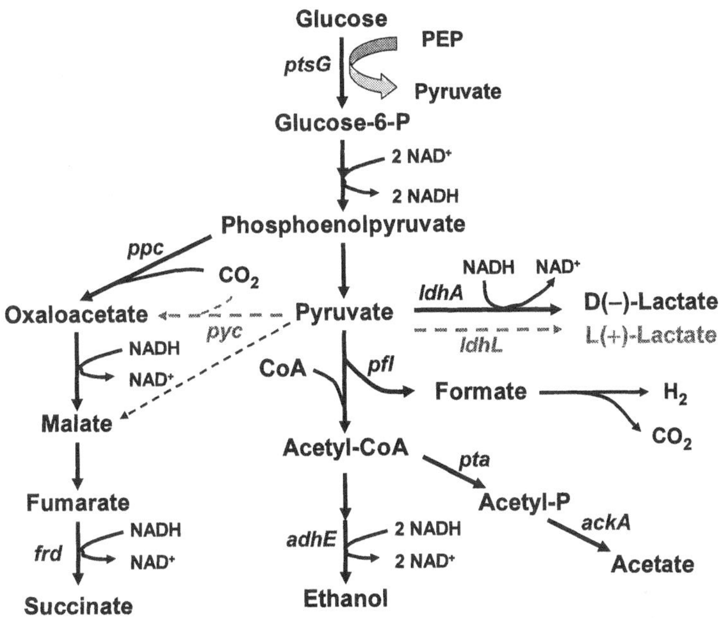

Fig. 6. Anaerobic fermentation pathways in *E. coli* with genes that have been mutated (knocked out or inserted) in order to shift the metabolic flux to overproduce a desirable fermentation product from glucose. Indigenous genes: *ackA*, acetate kinase; *adhE*, alcohol dehydrogenase; *frd*, fumarate reductase; *ldhA*, lactate dehydrogenase; *pfl*, pyruvate formatelyase; *ppc*, phosphoenolpyruvate carboxylase; *pta*, phosphotransacetylase; *ptsG*, glucose phosphotransferase. Heterologous genes: *ldhL*, lactate dehydrogenase; *pyc*, pyruvate carboxylase.

2.6. Succinic acid

Succinic acid is a four-carbon dicarboxylic acid with wide industrial applications [85]. It can be used as a precursor for many commercially important chemicals, including 1,4-butanediol. Like lactic acid and 1,3-propanediol, succinic acid can be used as a monomer for polymer synthesis [86]. Fermentative production of succinic acid from renewable resources has gained large interest in the past 10 years [87]. Obligate anaerobes such as *Anaerobiospirillum succiniciproducens* and *Actinobacillus succinogenes* produce high concentrations (up to 110 g/L) of succinic acid from various carbon sources, with an apparent yield of 1.2 mol/mol or 0.78 g/g glucose consumed [88]. These strict anaerobes, however, are difficult to use in large-scale industrial fermentations because of their high sensitivity to oxygen and relatively slow growth rates.

Many metabolically engineered mutants of *E. coli* have been developed as more favorable organisms for fermentative production of succinic acid from glucose [89–96]. Under anaerobic conditions, *E. coli* produce a mixture of organic acids, including succinic acid (see Fig. 6). In order to increase succinic acid production, many metabolic engineering approaches and mutants have been studied, including: 1) overexpressing native phosphoenolpyruvate (PEP) carboxylase to direct more pyruvate towards succinic acid [89], 2) overexpressing fumarate reductase to further improve the conversion of fumarate to succinate, which appeared to be a bottleneck in the pathway [90], 3) inactivating the pyruvate formate lyase (*pfl*) and lactate dehydrogenase (*ldhA*) genes to shut down competing pathways while overexpressing the malic enzyme [91], 4) introducing heterologous genes for pyruvate carboxylase (*pyc*) to increase the availability of oxaloacetate [92], and 5) inactivating the *ptsG* gene and the glucose phosphotransferase system (PTSG) to increase the availability of phosphoenolpyruvate [93]. An *E. coli* strain with mutations in *pfl*, *ldhA*, *pyc* and *ptsG* was

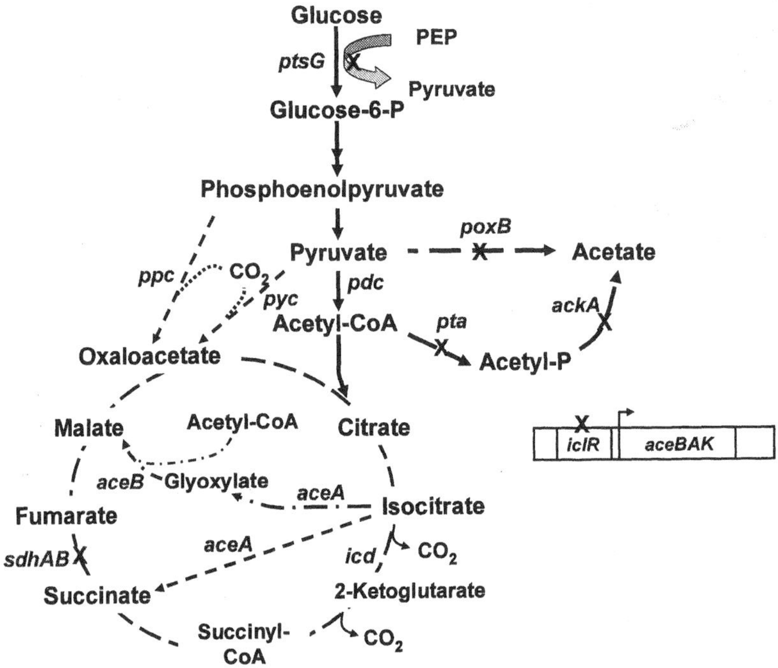

Fig. 7. Metabolic engineering of aerobic fermentation pathways in *E. coli* with genes that have been mutated in order to shift the metabolic flux to overproduce succinic acid from glucose [100]. "x" indicates genes knock out. *ackA*, acetate kinase; *icd*, isocitrate dehydrogenase; *iclR*, *aceBAK* operon repressor; *pdc*, pyruvate dehydrogenase; *poxB*, pyruvate oxidase; *ppc*, phosphoenolpyruvate carboxylase; *pta*, phosphotransacetylase; *ptsG*, glucose phosphotransferase; *pyc*, pyruvate carboxylase; *sdhAB*, succinate dehydrogenase.

able to produce succinic acid from glucose with a high final concentration (99.2 g/L), productivity (1.3 g/L·h), and yield (1.1 g/g) in a dual-phase (aerobic for cell growth followed by anaerobic for production) fed-batch fermentation using a complex medium containing yeast extract and tryptone [94, 95]. The higher than 1.0 g/g succinate yield was probably due to CO_2 incorporation and additional carbon sources present in the complex medium.

Fermentative production of succinic acid by recombinant *E. coli* is limited by the cofactor NADH and poor cell growth and slow production under anaerobic conditions [97, 98]. An aerobic succinate production system has been designed that allows *E. coli* to produce and accumulate succinate under aerobic conditions [99]. Multiple mutations in the tricarboxylic acid cycle (*sdhAB*, *icd*, *iclR*) and acetate pathways (*poxB*, *ackA-pta*) redirect the carbon flux toward succinic acid through two pathways: 1) the glyoxylate cycle when *icd* is also knocked out, and 2) the oxidative branch of the TCA cycle when *icd* is not knocked out (see Figure 7). The *sdhAB* knock-out allows the accumulation of succinate in the mutants, which normally does not happen in *E. coli* under aerobic conditions. With *ptsG* inactivation and overexpression of a malate feedback inhibition resistant PEP carboxylase, the mutants are able to produce the maximum theoretical succinate yield of 1.0 mol/mol (0.66 g/g) glucose consumed. This is the first aerobic succinate production system based on the creation of a new aerobic central metabolic network in *E. coli* [100]. However, the succinate productivity was low, less than 0.27 g/L·h, and there was also substantial accumulation of pyruvate and TCA cycle C6 intermediates in the fermentation. More metabolic engineering improvements need to be made to this system in order to make aerobic succinate production more efficient and economical.

2.7. Polyhydroxyalkanoates

Polyhydroxyalkanoates (PHAs) are natural polyesters produced as carbon and energy reserve materials by many microorganisms under unfavorable growth conditions in the presence of excess carbon sources [101–106]. These bioplastics have gained large commercial interest because they are biodegradable, can be made from renewable resources, and have a variety of chemical and structural properties [107, 108], which can be easily modulated by varying the substrates and feeding regimes used in the biosynthetic process [109]. Although poly(3-hydroxybutyrate) (P3HB) is the most frequently occuring microbial PHA, there are over 150 known PHAs [104], which are usually classified into two categories according to their constituent monomer chain length: short chain (C4 and C5) and medium chain (≥ C6) length polyhydroxyalkanoates (PHA_{SCL}, PHA_{MCL}) [110, 111]. In addition to homopolymers, block copolymers such as poly(3-hydroxybutyrate-*co*-3-polyhydroxyvalerate) and unnatural polythioesters (PTEs) also can be synthesized by bacterial PHA synthases, which have a broad substrate range [112, 113].

PHA can be biosynthesized from a variety of carbon sources by a wide range of naturally occurring bacteria, including *Ralstonia eutropha*, *Pseudomonas pudita*, *P. oleovorans*, *Aeromonas punctata*, *Allochromatium vinosum*, and some methylotrophs [101]. Figure 8 summarizes common biosynthetic pathways of PHAs from sugars and fatty acids. For the biosynthesis of short chain length PHAs, such as P(3HB) in *R. eutropha*, three enzymatic reactions are required: 1) the condensation of two acetyl coenzyme A (acetyl-CoA) molecules into acetoacetyl-CoA by 3-ketoacyl-CoA thiolase (PhaA), 2) the reduction of acetoacetyl-CoA to (R)-3-hydroxybutyryl-CoA by an NADPH-dependent acetoacetyl-CoA reductase (PhaB), and 3) polymerization of (R)-3-hydroxybutyryl-CoA monomers to P(3HB) by PHA polymerase or synthase (PhaC). For medium chain length PHAs, biosynthesis (such as in *P. oleovorans*) is closely associated with fatty acid metabolism (both de novo synthesis and β-oxidation) and is more complicated [114]. In this case, the monomer (R)-3-hydroxyacyl-CoA can be supplied from fatty acids via three different intermediates in the β-oxidation cycle: trans-2-enoyl-CoA by the action of hydratases (PhaJ), (S)-3-hydroxyacyl-CoA by the action of epimerases, and 3-ketoacyl-CoA by the action of reductases (PhaB). (R)-3-hydroxyacyl-CoA also can be produced from the intermediate (R)-3-hydroxyacyl ACP in the fatty acid de novo synthesis cycle by the action of transacylases (PhaG). Depending on the monomer(s) and degree of polymerization, PHAs with various molecular weights, structures and properties can be produced, including a variety of block copolymers. The size of PHA granules produced in *Ralstonia eutropha* was found to be inversely related to the concentration of a PHA-binding protein, PhaP, which is always granule bound but not necessary for PHA biosynthesis [101, 115].

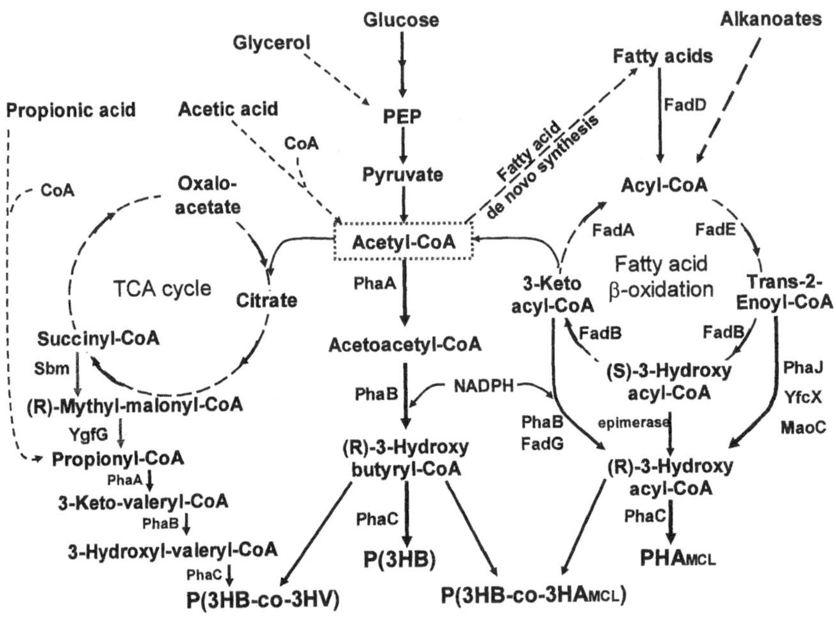

Fig. 8. Biosynthesis pathways for polyalkanoates production from various substrates. The major enzymes involved in PHA biosynethesis are: PhaA, 3-ketoacyl-CoA thiolase; PhaB, 3-ketoacetyl-CoA reductase, PhaC, PHA polymerase or synthase; PhaJ, (R)-specific enoyl-CoA hydratase; FadB, enoyl-CoA hydratase; FadD, acyl-CoA synthetase; FadE, acyl-CoA dehydrogenase; FadG, reductase; PhaG, (R)-3-hydroxyacyl ACP:CoA transacylase (not shown). Also shown are some heterologous enzymes added to the pathways to improve PHA production: Sbm, sleeping beauty mutase; YgfG, (R)-methyl-malonyl-CoA carboxylase; MaoC, YfcX, enoyl-CoA hydratase homologous emzyme.

The biosynthetic genes *phb*A, *phb*B, and *phb*C in *R. eutropha* are clustered and organized in one *pha*CAB operon, which makes it simple to clone the PHA synthesis genes. Similarly, the genes for PhaP, PhaC, and PhaJ in *A. punctata* are clustered on the *phaPCJ* operon. Metabolic engineering has been widely applied to improve PHA production by natural producers and heterologous hosts, including *E. coli* and some eukaryotes (*Saccharomyces cerevisiae* and plants) [116, 117]. Natural PHA-producing bacteria usually have a long generation time and relatively low PHA production yields. *E. coli* is a better host for PHA production because it grows fast, does not degrade PHA, is easy to lyse for the recovery of PHA granules, and can utilize various carbon sources, including glucose, sucrose, lactose and xylose, and cheap substrates such as molasses, whey, and hemicellulose hydrolysate [105]. Oilseed crops are considered good targets for seed-specific PHA production because both PHA and fatty acids are derived from acetyl-CoA. Metabolic engineering of plants for the diversion of acetyl-CoA towards PHA accumulation can be more directly achieved in the seeds of crops having a naturally high flux of carbon through acetyl-CoA. Therefore, PHA can potentially be produced in plants at a cost competitive with that of the petroleum based plastics [117].

Metabolic engineering strategies for improving PHA production in microorganisms have been extensively studied and include:

1. *Increasing the gene dosage for PHA synthase* – This strategy can potentially increase PHA production, but has had mixed success, being dependent on the limiting factors for PHA biosynthesis in the organism [101].

2. *Introducing heterologous genes for precursor production from inexpensive carbon sources* – Economical production of PHA is limited by the carbon sources available as precursors for PHA biosynthesis. Most of the desirable monomer precursors for PHAs cannot be produced by natural organisms and thus must be supplied to the fermentation. These precursors (e.g., propionic acid) not only are more expensive than sugars but also can be toxic to the producing organisms. Instead of feeding propionic acid, propionyl-CoA can be formed from succinyl-CoA, an intermediate in the TCA cycle, after introducing two genes encoding the enzymes sleeping beauty mutase (Sbm) and (R)-methyl-malonyl-CoA carboxylase (YgfG) (see Figure 8). P(3HB-co-3HV) was produced from glycerol when the host's gene encoding 2-methylcitrate synthase was also deleted to channel propionyl-CoA to 3HV in the mutant co-expressing phaBCA [118].

3. *Increasing acetyl-CoA and NADPH available for PHA biosynthesis* – The fluxes through the PP and ED pathways in *E. coli* have been increased by cloning the *zwf* gene to overexpress glucose-6-phosphate dehydrogenase, resulting in increased production of NADPH required in PHA synthesis [119].

4. *Inhibiting fatty acid β-oxidation pathway to increase medium chain length monomers for PHA synthesis* – Short chain length PHAs can be efficiently synthesized in *R. eutropha* and recombinant *E. coli* with a high productivity (up to 5 g/L·h) and yield (up to 90% cell dry weight) [105, 114]. However, it is usually more problematic to synthesize medium chain length PHAs. It is desirable to modulate the composition of PHA with increased medium chain length monomers in the polymers or block copolymers. By adding the β-oxidation

inhibitor acrylate, wild type *R. eutropha* accumulated P(3HP-co-3HB-co-3HHX-co-3HO) during growth on octanoate [120]. Amplifications of *fadD* and *fadE* in a *fadA* and/or *fadB* mutant of recombinant *E. coli* harboring a PHA synthase increased 3-hydroxyalkanoate monomers for PHA synthesis [114, 121]. Amplification of (R)-specific enoyl-CoA hydratase (PhaJ) and 3-ketoacyl-CoA reductase (PhaB, FadG) can also increase alkanoate monomers supplied from fatty acids, thus producing PHA with an altered monomer composition [114].

5. *Constructing non-natural biosynthetic pathways* – By introducing *C. acetobutyricum*'s butyrate kinase (*buk*) and phosphotransbutyrylase (*ptb*) genes and *Thiocapsa pfennigii*'s PHA synthase (PhaEC) gene into *E. coli*, non-natural PHAs such as P(4HB) and PTEs can be produced from appropriate substrates (4HB, 3-mercaptoalkanoates) [104, 112]. In this novel biosynthesis process, the coenzyme A released by PHA synthase is recycled (see Figure 9).

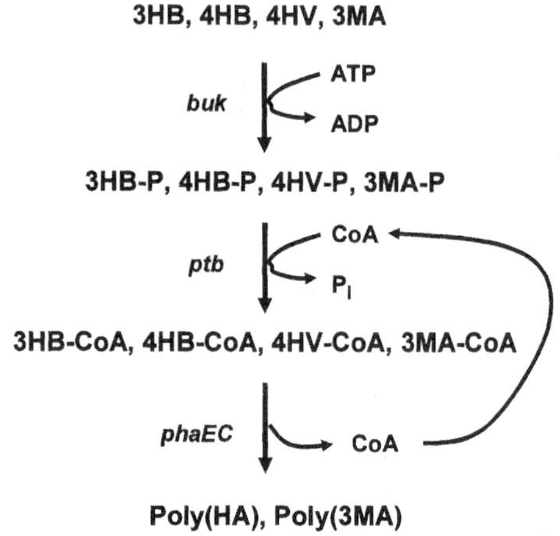

Fig. 9. A non-natural pathway with butyrate kinase (*buk*), phosphotransbutyrylase (*ptb*) and PHA synthase (*phaEC*) for biosynthesis of various polyhydroxy-alkanoates (PHAs) and polymercaptoalkanoates (PMAs). Abbreviations: 3MV, 3-mercaptoalkanoates; 4HV, 4-hydroxyvalerate; 3HB, 3-hydroxybutyrate.

Fermentative production of PHAs from sugars and fatty acids provides an attractive method for producing biopolymers with tunable properties that can potentially replace many current commercial plastics. It has been shown that PHA concentrations of up to 72 g/L and productivity of 1.91 g/L·h can be achieved in a high-cell-density fermentation of *P. putida* [122]. Metabolic engineering coupled with protein engineering and process engineering has the potential to further improve PHA production from low-cost carbon sources, including waste sugars and fatty acids. Commercial production of these biopolymers (e.g., Biopol™ by Metabolix; Nodax™ by Protor and Gamble) is expected to increase as the market is developed in the near future.

2.8. Ethanol

Ethanol fermentation is one of the oldest and most important fermentation processes used in the biotechnology industry. In the U.S. alone, about 4.5 billion gallons of ethanol are produced annually from corn and used as a transportation fuel. The annual production of bioethanol in the U.S. is expected to grow to more than 7.5 billion gallons in the next few years and reach 30 billon gallons by 2025. Many microorganisms, including bacteria and yeasts, can produce ethanol as the major fermentation product from carbohydrates [123]. Current industrial ethanol fermentation is mainly carried out with the yeast *Saccharomyces cerevisiae* because of its hardiness (low pH and high ethanol tolerance), although the bacterium *Zymomonas mobilis* has a higher specific ethanol productivity and yield from glucose and sucrose.

Metabolic engineering of *S. cerevisiae*, *Z. mobilis*, and *E. coli* for improved ethanol fermentation has been extensively studied [124–128]. Most efforts have been focused on the creation of efficient xylose-fermenting mutant strains because neither *S. cerevisiae* nor *Z. mobilis* can use xylose, which is the second most abundant sugar (next to glucose) present in plant biomass (hemicellulose). Since yeasts can grow on and ferment xylulose, the heterologous expression of bacterial xylose isomerase (XI) appeared to be a reasonable approach to engineer *S. cerevisiae* for xylose assimilation. However, all earlier efforts using this approach failed even though the cloned gene *XylA* from *Thermus thermophilus* and *Piromyces* sp E2 produced active xylose isomerase in *S. cerevisiae*. The failure was partially because xylose isomerase is strongly inhibited by xylitol and the isomerization equilibrium favors xylose formation. More recently, a mutant capable of growing anaerobically on xylose to produce a high yield of ethanol (0.42 g/g xylose) was obtained via evolutionary engineering of the genetically engineered strain expressing the *Xyl*A gene of the anaerobic fungus *Piromyces* sp E2 [129]. This work demonstrated that the metabolic pathway can be better engineered through the combination of genetic modification and directed evolution without the need to search for genetic targets for modification.

Introducing *XYL1* and *XYL2* genes, encoding for xylose reductase (XR) and xylitol dehydrogenase (XDH), respectively, from xylose-fermenting yeasts, such as *Pichia stipitis*, has been the major strategy used to metabolically engineer *S. cerevisiae* for ethanol production from xylose [128]. However, this strategy does not work well either, largely because the recombinant yeast, which converts xylose to xylulose by the combined action of NADPH-dependent xylose reductase and NAD-linked xylitol dehydrogenase, cannot sustain its anaerobic growth due to a redox imbalance of NAD^+/NADH, which also results in the excretion of xylitol. Yeasts cannot interconvert NADPH and NADH because they lack a transhydrogenase. Overexpressing transhydrogenase in yeasts does not improve the situation. Although the excess NADH can be effectively removed via aeration, with oxygen being an external electron receptor, over aeration may shift cell metabolism from fermentation to respiration and inhibit ethanol production. To solve the redox imbalance problem thus requires protein engineering to exactly match the coenzyme specificities of the two oxidoreductases in the XR/XDH pathway. Modulating the redox metabolism by either increasing NADPH-producing glucose 6-phosphate dehydrogenase activity or altering ammonia assimilation to become NADH, rather than NADPH, dependent has been shown to increase ethanol production from xylose [130].

Increasing xylitol dehydrogenase activity relative to xylose reductase in *S. cerevisiae* resulted in less xylitol and more ethanol production. However, overexpressing *XYL2* led to xylulose secretion, indicating that xylulokinase (XK) limits xylose metabolism in *S. cerevisiae*. When xylulokinase (encoded by *XYL3* or *XKS1*) was overexpressed along with *XYL1* and *XYL2* in yeast, the mutant *Saccharomyces* sp. 1400 (pLNH33) fermented both glucose and xylose to ethanol (50 g/L) with a high ethanol yield (~0.46 g/g or 90% of theoretical yield) and productivity (~1.4 g/L·h) [131, 132], which is the highest from glucose/xylose mixtures reported for a recombinant yeast to date. Since 2004, this *Saccharomyces* mutant strain has been used by Iogen in their pilot demonstration plant that produces 800 liters of ethanol per day from wheat straw. However, economical scale up of this process may require further strain improvement. Recombinant yeast cells cannot use xylose as efficiently as glucose and still produce a significant amount of xylitol. In the fermentation, glucose utilization usually precedes xylose utilization when both are present in the medium [133]. The poor xylose utilization in the presence of glucose is attributed to the lack of a specific xylose transportor in yeast. The yeast's pentose phosphate pathway may also limit the utilization of xylose. Overexpressing transketolase (*TKL1*) and transaldolase (*TAL1*) in the *XYL1-XYL2* containing *S. cerevisiae* strains considerably improved cell growth on xylose, but the effect on ethanol production was not as significant.

Figure 10 illustrates the metabolic pathways for glucose and xylose utilization in recombinant *S. cerevisiae*. Glucose is converted to pyruvate via the Embden-Meyerhof-Parnas pathway, whereas xylose, after conversion to xylulose, is metabolized through the pentose phosphate cycle.

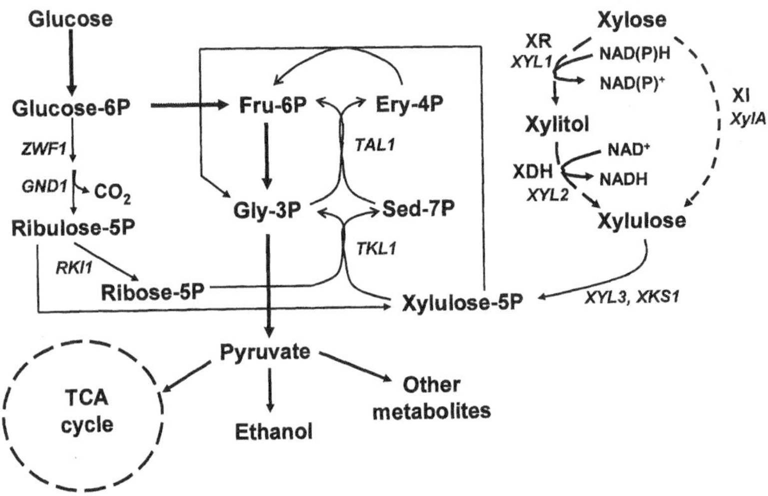

Fig. 10. Metabolic pathways in recombinant *S. cerevisiae* engineered for xylose fermentation. The Embden-Meyerhof-Parnas pathway is indicated by thick lines and the pentose phosphate pathway is indicated by thin lines. Important genes in the pathways are GND1: phosphogluconate dehydrogenase, RKI1; ribose-5-phosphate isomerase, *TAL1*: transaldolase, *TKL1*: transketolase, *XKS1*: xylulokinase, *XYL1*: xylose reductase (XR), *XYL2*: xylitol dehydrogenase (XDH), *XYL3*: xylulokinase, *XylA*: xylose isomerase (XI), *ZWF1*: G-6-P 1-dehydrogenase. XI, XR, and XDH are heterologous enzymes.

In order to maintain redox balance, yeast cells produce glycerol as a byproduct in ethanol fermentation, which reduces the ethanol yield to significantly lower than the theoretical value of 0.51 g/g glucose. Replacing NADPH-dependent glutamate dehydrogenase (encoded by

GDH1) with NADH-dependent glutamate dehydrogenase (encoded by *GDH2*) resulted in a 30% reduction in glycerol production. Overexpressing both the glutamate and glutamine synthase genes (*GLT1* and *GLN1*) in the *GDH1*-deleted mutant reduced glycerol production by 38% and increased ethanol yield by 10% [134]. Increasing the ethanol yield to close to its theoretical maximum value will have a significant impact on the process economics as the substrate (glucose) accounts for ~50% of the final product cost.

Like native *S. cerevisiae*, *Z. mobilis* can produce a large amount of ethanol from glucose and sucrose, but not pentoses. By introducing two operons encoding xylose assimilation and pentose phosphate pathway enzymes into *Z. mobilis*, the metabolically engineered strain can ferment both xylose and glucose to ethanol [135]. However, the recombinant *Z. mobilis* can produce ethanol only at a lower concentration level and is sensitive to inhibitors present in hemicellulose hydrolysates, both of which limit its potential for industrial application.

An alternative to developing a pentose-fermenting ethanologenic strain is to introduce the genes for ethanol production into host cells that can utilize pentoses. A metabolically engineered *E. coli* strain KO11 with *Z. mobilis* pyruvate decarboxylase (*pdc*) and alcohol dehydrogenase (*adh*B) genes integrated into its chromosome and a mutation in fumarate reductase that eliminates succinate production can produce ~40 g/L ethanol in 48 hours from hemicellulose hydrolysate supplemented with corn steep liquor as a nutrient source, with a typical ethanol yield of ~0.46 g/g sugar or 90% of the theoretical yield [136]. Further mutation of this strain with a defective phosphoenolpyruvate-dependent phosphotransferase system (PTS) produced mutants that were unable to use glucose, but some of them produced almost 60 g/L ethanol from 120 g/L xylose in 48 hours [137]. Ethanol fermentation with recombinant *E. coli* KO11 has been successfully demonstrated at the 10,000-L scale. However, its commercial application is limited because of the neutral pH required for bacterial fermentation, its lower ethanol tolerance, and issues related to the disposal of spent media containing enteric bacterial cell biomass, which cannot be used as animal feed as yeast can.

A similar approach has also been used to engineer *Klebsiella oxytoca*, which can ferment cellobiose and cellotriose, to produce ethanol from cellulose. This can potentially reduce the cost of cellulase enzymes required in the hydrolysis of lignocellulosic materials for ethanol fermentation.

2.9. Other products and applications

Metabolic engineering also has been applied to many other fermentation products, including citric acid [138], pyruvate [139], propionic acid [140], butyric acid [141], butanol [142], 1,2-propanediol [143], vitamin B12 [144], riboflavin [145], flavonoids [146], astaxanthin [147], carotenoids [47, 148], lycopene [149], indene [150], steroids [151], exopolysaccharides [152], and recombinant proteins. In industrial production of heterologous protein by recombinant *E. coli*, it is critical to grow the cells to a high cell density by controlling or minimizing the production of acetate, which inhibits cell growth. Acetate is produced when cell growth is limited by oxygen and pyruvate or acetyl-CoA accumulates due to an oversupply of glucose. Several metabolic engineering strategies for acetate reduction have been successfully used to improve recombinant protein production in *E. coli*, including eliminating the pathway leading to acetate formation by knocking out *pta* and/or *ack* genes (see Figure 6), reducing the glucose uptake rate by introducing a mutation in *ptsG* (encoding the PEP:glucose phosphotransferase system), and redirecting carbon flow from acetate to acetoin by overexpressing the *Bacillus subtilis* acetolactate synthase (ALS) gene [153].

Expressing a *Vitreoscilla* hemoglobin gene (*vgb*) in *E. coli* also improved its growth because of the improved oxygen transfer mediated by the hemoglobin [154]. Another classical example of improving cell growth is by cloning glutamate dehydrogenase in *Methylophilus methylotrophus* to alter its nitrogen assimilation pathway, leading to improved energy efficiency and a higher yield of single cell protein from methanol [155]. By cloning the human-derived glucose transporter gene (*glut1*) into *Phaeodactylum tricornutum*, the microalga was engineered from an obligate photoautotroph to grow solely on glucose in the absence of light [156]. Metabolic engineering also can be applied to control the morphology of filamentous fungi [157] and in bioremediation [158].

3. METABOLIC ENGINEERING STRATEGIES AND LIMITATIONS

General speaking, metabolic engineering uses genetic manipulations to affect the distribution of intracellular chemical reactions (flux) among the metabolic pathways to optimize a biotechnologically important process. The strategies used in matabolic engineering depend on the strain/process development goals (see Table 2), and their degree of success varies with host organisms and the products of interest, as illustrated in various examples discussed in the previous section.

Table 2
Some common metabolic engineering goals and strategies

Strategies for increasing productivity and/or yield
• Amplify the gene (cluster) coding for the rate-limiting enzyme to eliminate the bottleneck
• Amplify the gene coding for the branch-point enzyme to redirect metabolic flow
• Delete key genes in the branched pathways to eliminate non-productive reactions
• Modulate enzyme expression levels to avoid the accumulation of intermediates
• Introduce genes/enzymes with different control architectures to avoid feedback inhibition or repression
• Increase the supply of precursor
• Change cofactor dependency of key enzymes to balance redox and/or improve energy and growth efficiency
• Engineer the transport systems to improve product efflux and/or substrate uptake
Strategies for eliminating/reducing byproducts
• Delete key genes in the branched pathways leading to undesirable byproducts
• Amplify the gene coding for the branch-point enzyme to increase metabolic flow towards the desired product
Strategies for extending substrate range
• Introduce heterologous genes encoding for the enzymes needed to assimilate the substrate
• Introduce the genes encoding for the enzymes needed to convert the metabolites to the desired product in non-native producers with the ability to use the substrates

Other common metabolic engineering goals include improving cellular properties and extending product spectrum or synthesizing novel products. Although there are many successful examples of microorganisms metabolically engineered for industrial biotechnology applications, numerous attempts and multiple and iterative engineering efforts are usually required before obtaining an improved strain suitable for industrial fermentation [5]. Quite often, the initial targets fail to generate the much-anticipated results. This is largely because the organisms have evolved to their current physiological state (phenotype) with their genetic makeup (genotype) adapted to their native environment. Any alteration to their genetic makeup may cause unexpected genes to turn on or off, sendng a cascade of effects down entire regulatory, metabolic, and physiological networks to compensate for or offset the "suboptimal" genetic change or mutation [159]. In other words, the central metabolism is tightly regulated at various levels – genomic, transcriptomic, proteomic, and metabolomic [160], as illustrated in Figure 11. It is thus quite difficult to predict the consequence of a genetic modification on the microorganism's phenotype or physiological behavior.

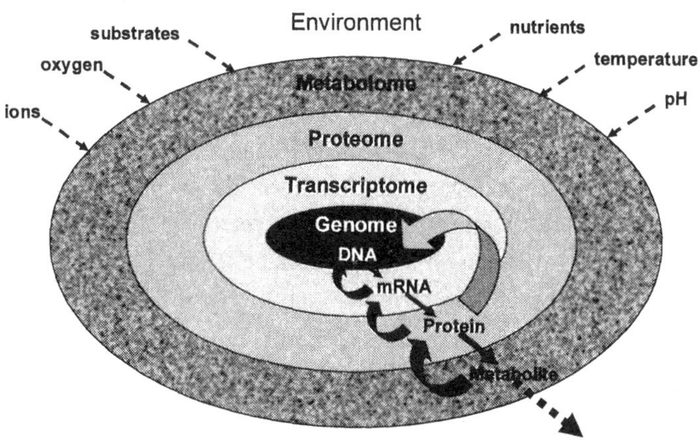

Fig. 11. Cellular processes from genome to metabolome involving complicated interactions among genes (DNA), mRNA, proteins, and metabolites within a cell and between the cell and environment.

A successfully metabolically engineered strain usually requires simultaneous multiple mutations on multiple genes [23]. Overexpressing a single gene (cluster) coding for the "rate-limiting" enzyme may not produce any effect because there are usually multiple flux control points in the pathway and the rate limiting reaction step may be shifted from one to another after a mutation [161].

In addition, there may be other (unknown) pathways or isoenzymes present in the cells that do not function under normal situations. Inactivating one pathway or enzyme may activate other pathways or enzymes that can restore the lost functions. This is especially true for the production of primary metabolites in the central metabolism, which is tightly regulated and controlled.

For the same reason, it may be difficult to extend the substrate range of an industrial strain, such as in the case of making yeast *Saccharomyces* sp. produce ethanol from pentoses [127]. Even after extensive trials and multiple mutations, no recombinant yeast strain can produce ethanol from pentoses as efficiently as from glucose. On the other hand, it is usually easier to introduce a new pathway into non-native producing cells to produce the desired product because the heterologous genes and enzymes are not regulated in the host cells. However, the non-native hosts usually do not have as high a tolerance to the end product, which at high concentrations is usually toxic. Increasing tolerance to the toxic product is not a straightforward metabolic engineering task, and to date there is no known successful industrial example. Classical strain development through random mutagenesis and/or adaptation and evolutionary engineering coupled with high throughput screening appears to be more effective in this regard [162]. The combination of metabolic engineering with other strain development methods will be discussed at the end of this chapter. In order to do metabolic engineering more efficiently, many engineering analytical methods and experimental characterization tools have been developed, which are discussed next in this chapter.

4. METABOLIC ENGINEERING METHODOLOGIES AND TOOLS

As illustrated in Figure 12, the first step in metabolic engineering is to choose the gene and enzyme to modify [162]. Once the target gene has been identified, genetic modifications will be done to create new mutant strains, which will then be characterized using various experimental tools. Finally, the mutant's metabolic behaviors as the consequence of the genetic changes will be analyzed. If the original goal was not met, then a new target will be

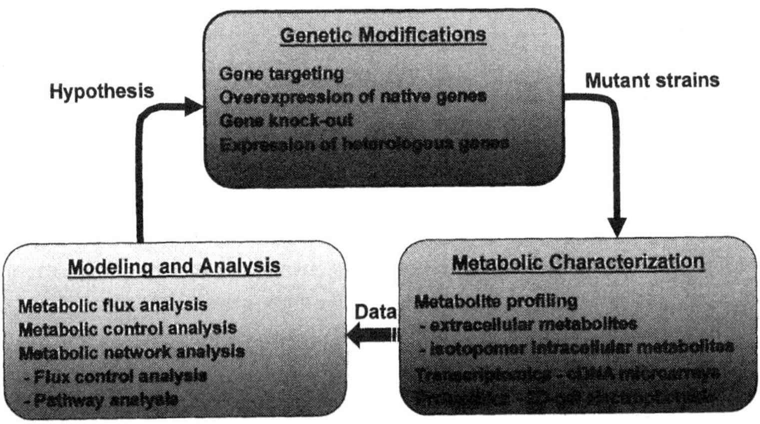

Fig. 12. Metabolic engineering ususally starts from genetic modifications, followed with characterization and analysis of the mutants and consequences of the changes in metabolic pathways. Inverse metabolic engineering starts by first characterizing existing mutants to find and understant beneficial mutations before selecting the gene targets for further genetic modifications. With mass genomic and proteomic data available to several microorganisms, one can do *in silico* pathway engineering to help selecting gene targets before doing metabolic engineering.

selected for a new round of genetic modifications. However, it is difficult to select target genes for metabolic engineering because it is hard to predict the consequences of changes to metabolic pathways. Therefore, engineering modeling and computational methods have often been used to analyze and predict the consequences of and guide the selection of target genes for metabolic engineering [163, 164]. Various genetic engineering techniques, including random mutagenesis and recombination methods, are available to genetically modify cells. For this topic, the readers are referred to Chapter 3 in this book.

Over the past 15 years, metabolic engineering has developed into a scientific field that applies engineering principles and methodologies (mainly mathematical modeling) to guide the choice of useful genetic alteration and predict the consequences. Many metabolic engineering modeling and computational tools based on classical chemical engineering principles, including mass balance, energy balance, and reaction kinetics, have been developed. These metabolic engineering tools are widely used to evaluate the system properties of metabolic networks, with emphases on metabolic flux and its control. In addition, many high-throughput experimental methods have also been developed and used to generate massive amounts of data on genomes, transcriptomes, and proteomes that can be used in metabolic engineering as well as fuctional genomics research. These engineering and experimental tools are discussed in this section.

4.1. Metabolic flux analysis (MFA)

MFA, an engineering method based on stoichiometric equations representing reactions in metabolic pathways, is widely used to determine the metabolic flux distribution that reflects or represents cell physiology [4]. In a fermentation process, cell productivity and product yield are predicated by metabolic flux distribution, which is mainly affected by the fermentation conditions but can also be changed by genetic modifications. In fact, the main purpose of metabolic engineering is to alter metabolic flux distribution so that cellular properties and/or metabolite overproduction in fermentation can be optimized. Based on the principle of mass balance, stoichiometric equations have long been used to estimate the maximum or theoretical yield of a metabolite from a particular substrate under certain conditions (constraints) [165]. MFA uses a similar approach to estimate the intracellular flux distribution and to find possible reaction bottlenecks in the metabolic pathways [166].

In MFA, the reaction network is represented by a set of stoichiometric equations in a matrix such that $\boldsymbol{S} \times \boldsymbol{v} = 0$, where the stoichiometric coefficients are collected in the matrix $\boldsymbol{S}$ and $\boldsymbol{v}$ is the vector containing the fluxes or the rates at which input metabolites are processed to form output metabolites in the metabolic pathways. The intracellular fluxes ($\boldsymbol{v}_i$) for intermediate metabolites and cofactors in the pathway network are usually difficult to measure directly, but can be estimated from the fluxes of extracellular metabolites ($\boldsymbol{v}_o$) on the basis of mass conservation at a steady state, at which there is no net accumulation of intracellular metabolites. That is $\boldsymbol{v}_i = \boldsymbol{S}_o \times \boldsymbol{v}_o$ where $\boldsymbol{S}_o$ contains only the stoichiometric coefficients associated with the extracellular metabolites. If the number of unknown (intracellular) fluxes is greater than the number of measured (extracellular) fluxes, the unknown intracellular fluxes cannot be uniquely determined but are restricted to a constrained solution space. One can search the solution space to find the optimal solution for a given objective function, such as maximum cell growth or product yield. The determined flux distribution indicates the cell capability associated with the objective function within the given constraints.

MFA can be used to choose an appropriate metabolic engineering strategy and as a tool for the physiological characterization of a metabolically engineered strain. MFA is most useful in cases where a relatively large fraction of the carbon is directed towards the product, such as primary metabolites in the central metabolism (glycolysis and TCA cycle), or product precursors for secondary metabolites or biosynthetic products. By comparing the flux distributions in different mutants, it is also possible to obtain information about regulation at the branch point in a metabolic pathway. MFA has been successfully used in metabolic engineering studies of many industrially important fermentation products, including amino acids and antibiotics [166–170].

MFA using mass balance and measurements of extracellular metabolites to analyze intracellular flux distribution has limitations in elucidating complex metabolic pathways involving split pathways that converge at another point of the network, metabolic cycles, and unknown network structure and cofactor utilization [4]. To overcome these limitations, ^{13}C-labelled carbon sources, such as [1-^{13}C] glucose, have been used in tracer experiments to measure the fluxes of intracellular metabolites by gas chromatography-mass spectrometry (GC-MS) and nuclear magnetic resonance (NMR) [171–174].

4.2. Metabolic control analysis (MCA)

Identifying which enzyme(s) exert flux control in a given metabolic pathway is critical to the success of metabolic engineering but also important because flux control tends to be distributed among multiple enzymes, instead of controlled by just one "rate limiting" step. MCA allows one to assess the relative importance of different enzymes in the control of metabolic fluxes by measuring the "control coefficients" of enzymes or calculating them from known kinetic properties or transient metabolite concentrations. Flux control coefficients (FCC) are defined as the relative effects of small (infinitesimal) changes in enzyme activities on flux changes at a steady state and can be represented by the following equation [4]:

$$C_i^k = \frac{\partial v_k / v_k}{\partial e_i / e_i}$$

Where C_i^k is the control coefficient and e_i and v_k are the enzyme activity of reaction i and the flux of reaction k, respectively, in the pathway. In principle, the enzyme with the highest control coefficient exerts the greatest control over flux through the pathway and should be amplified in order to increase the overall flux or rate through the pathway. However, in many cases where this was done, the results were not satisfactory.

Unlike MFA, which only requires the stoichiometric coefficients and simple modeling and experimental methods, MCA or dynamic simulation requires additional information on the kinetic parameters (rate constants, elasticity, etc.), which are usually unavailable and difficult to obtain in larger networks. So far, the effectiveness of MCA has been confirmed in only a few studies where FCCs were experimentally determined by a large perturbation on enzyme activities [175, 176]. However, MCA generally cannot be used to predict large (finite) changes because the control would be shifted away from the enzyme that was increased in concentration. Also, flux control may reside outside the pathway proper [8]. In order to increase flux, simultaneous modulations of the activities of all the enzymes in the pathway, including transport of the desired product through the cell membrane, may be required. For example, an earlier attempt to increase flux in the glycolytic pathway of *S. cerevisiae* by individually over-expressing every gene in the pathway was unsuccessful [177].

4.3. Genome-scale metabolic network models and pathway analysis

The complexity of metabolic networks, which consist of metabolic reactions, substrate and product transport, gene regulation, signal transduction, energy supply, and redox/cofactor balancing, make it very difficult to predict the effects of genetic modifications on the resulting phenotype. Although MFA and MCA have been used with some success in this respect, more often they have failed in finding the right targets or generating the predicted results. In recent years, it has become clear that metabolism needs to be analyzed from a global and holistic perspective. The development of genome-scale metabolic models thus has become the new trend in metabolic engineering [159, 162, 164, 178], and several have been developed and used to study the metabolic networks in *E. coli* [179] and *S. cerevisiae* [180]. These genome-scale metabolic models are basically stoichiometric models similar to those used in MFA but with much more detailed metabolic networks involving hundreds of reactions, metabolites and cofactors [181]. For example, the metabolic network model for *S. cerevisiae* has 1212 reactions and 601 metabolites [182].

A metabolic network model can be rapidly developed by reconstructing information on annotated genes and metabolic pathways, much of which is available in online databases (see Table 3). These models are used mainly to quantify flux distribution and predict phenotypic behavior under various conditions. As discussed in MFA, the stoichiometric equations, which are set up in matrix form with mass balance under a steady state assumption, are presented as an underdetermined linear model that can be analyzed by two different approaches.

The first approach uses linear programming to find a unique solution for the stoichiometric model under the given conditions. In MFA, the external fluxes of the extracelluar metabolites are experimentally measured and used to determine the unknown (intracellular) fluxes, whereas metabolic network analysis (MNA) uses isotope-labelled substrates to quantify some intracellular fluxes. In flux balance analysis (FBA), which is also referred to as constraints-based or *in silico* modeling, flux quantification is entirely based on metabolite balancing and solving the linear model with a given objective function (e.g., max. cell growth) and additional constraints (e.g., positive ATP generation, product formation, and substrate consumption rates) [183]. An *in silico* model reconstructed from existing knowledge of genome sequences, biochemistry, and cell physiology can be analyzed to generate new hypotheses and to characterize emergent properties that arise from the integrative functions of the network. For example, *in silico* modelling can demonstrate that wild-type *E. coli*'s metabolic network is optimized to maximize growth, which is consistent with experimental results [184]. Gene deletion analysis using FBA correlates well with experimental data, and its success rate of predicting the essential genes for growth is high, ~86% for *E. coli* and ~88% for *S. cerevisiae* [164]. Therefore, a combination of *in silico* and experimental biology can be used to obtain a quantitative genotype-phenotype relationship for metabolism in bacterial cells [181]. The network structure defined by the stoichiometric model also can be used to deduce some regulatory information [185].

In the second approach, all possible steady-state solutions or pathways are enumerated and analyzed via so called extreme pathways (EP) or elementary flux modes (EFM) [186, 187]. Both methods use convex analysis to characterize all the possible steady-state flux distributions of a biochemical network. In EFM, each mode consists of the minimum number of reactions that it needs to exist as a functional unit (i.e., genetic independence and non-decomposability) or the minimal sets of enzymes that can each generate valid steady states. Similarly, EP specifies that each pathway consists of the minimum number of reactions needed to exist as a functional unit. Extreme pathways are the systemically independent subset of elementary modes, and thus, the interpretation of EPs can lead to questionable results because of missing genetically independent pathways. In general, pathway analysis can be used to help optimize conversion yields and predict the effects of the insertion or deletion of enzymes. However, the pathway analysis approach will have a very limited use in genome-scale models because of the combinatorial complexity of large metabolic networks [188]. By introducing additional regulatory constraints through the use of Boolean operators, a large number of pathways can be eliminated to reduce the complexity of the problem [189, 190].

In general, metabolic pathway analysis can predict optimal cell growth in wild-type *E. coli* and *S. cerevisiae* [179, 182]. However, the metablic network of genetically engineered *E. coli* strains did not operate in an optimum growth regime [191], and their phenotypic behavior suggested that they underwent minimal redistribution of fluxes with respect to the wild type [164]. In other words, *in silico* modeling and pathway analysis generally work well in predicting cell growth but not necessary for metabolite overproduction. By using energy balance analysis to eliminate thermodynamically infeasible results from FBA, the flux distribution in gene-deletion *E. coli* mutants can be better predicted [192]. However, applying metabolic network modeling and analysis to metabolically engineering industrial strains is limited by the little knowledge we have about cellular regulation. Furthermore, these stoichiometric reaction pathway models cannot tell anything about the reaction rates or provide kinetic information in a dynamic environment. Nevertheless, metabolic network analysis coupled with "omics" can be a very useful tool for functional genomics, which has become an important part of metabolic engineering as well as of systems biology [193].

4.4. Kinetic modelling and Silicon Cells

Kinetic models have long been used in studying fermentation kinetics but rarely used in metabolic flux analysis because the kinetic parameters of all enzymes in a pathway are rarely available [194]. Furthermore, it is very difficult to measure the kinetic parameters under conditions closely resembling intracellular physiological states or *in vivo* conditions. Most of the metabolic databases, such as KEGG and MetaCyc (see Table 3), rarely contain the kinetic parameters required for kinetic modelling. However, a simplified kinetic model useful in studying cellular processes could be developed by reverse engineering [195]. The Silicon Cells (www.siliconcell.net) concept presents a greater challenge to systems biology but may one day become a powerful tool for metabolic engineering using whole cell simulation at the molecular level (www.e-cell.org).

Table 3
Some useful online resources and databases available for metabolic engineering research

Database	Name	Web site
Genomics	DDBJ	www.ddbj.nig.ac.jp
	EMBL	www.ebi.ac.uk/embl.html
	Entrez Genomes	www.ncbi.nlm.nih.gov/entrez/query.fcgi?dbZGenome
	TIGR	www.tigr.org/tdb/mdb/mdbcomplete.html
Microarray	ArrayExpress	www.ebi.ac.uk/arrayexpress
	GEO	www.ncbi.nlm.nih.gov/geo/
Proteomics	COG	www.ncbi.nlm.nih.gov/COG
	GELBANK	gelbank.anl.gov
Metabolic pathways	BioCarta	www.biocarta.com/genes/allPathways.asp
	BioSilico	biosilico.kaist.ac.kr
	BRENDA	www.brenda.uni-koeln.de
	KEGG	www.genome.jp/kegg/
	MetaCyc	metacyc.org

4.5. Omics and high-throughput tools

The genome sequences of many microorganisms, including *E. coli*, *S. cerevisiae*, *C. glutamicum*, *Bacillus subtilis*, *Lactococcus lactis*, and *C. acetobutylicum*, have been completed and many more are in progress. These genome sequences are available for functional genomics and metabolic engineering research. Besides genomic sequence information, transcriptomic, proteomic and metabolomic data can be generated at an ever-increasing rate from high-throughput technologies, such as DNA sequencers, microarrays (gene chips), two-dimensional gel electrophoresis combined with tandem mass spectrometry, and isotopic label distributions probing metabolic phenotypes (see Table 4). Genome libraries can be analyzed to identify the genetic basis of relevant phenotypes [161, 196]. DNA microarrays can be used to efficiently analyze gene disruption/molecularly barcoded mutant libraries to identify genes essential to a particular phenotype [197, 198]. Transcriptional profiling or the analysis of genome-wide gene expression (transcriptome) can differentiate between genes with altered expression levels, either as the result or cause of the phenotype of interest [199, 200]. Proteomics enables quantitative profiling of cellular proteins using two-dimensional gel electrophoresis or chromatography coupled with mass spectrometry [201, 202]. High-throughput quantification of metabolites (metabolome) by sophisticated NMR, gas chromatography-mass spectrometry (GC-MS), liquid chromatography-mass spectrometry (LC-MS), and MALDI-TOF MS enables the comparative analysis of metabolite profiles under genetic and environmental perturbations [203, 204]. Intracellular fluxes can then be obtained by metabolite balancing using computational methods, such as MFA and FBA, and isotopomer experiments using ^{13}C-labelled substrates [171–173]. These fluxomic data allow us to have a better understanding of physiological states and phenotypic behaviors and the relationship between genes and their functions, also referred to as phenomics [205].

In the post-genomic era, the vast genome sequence information can be used to manipulate the metabolism of the organism, resulting in more efficient production strains [206, 207]. Through comparative analysis of wild-type and recombinant strains, genomics can be used to identify gene targets [208], and transcriptomics have been used to optimize fermentation conditions [209] and understand regulatory mechanisms [199, 210]. Today, functional genomics, which combines transcriptomic, proteomic, and metabolomic data, can provide insights on cellular metabolism that are difficult to obtain with traditional approaches. Based on results from functional genomic studies, new metabolic pathways that are expressed under different conditions or stress can be identified, and new strategies for rationally engineering metabolic pathways and cellular properties can be developed [160–162].

Table 4
Omic data and high-throughput experimental tools useful to metabolic engineering

Omics	Data	Tools	References
Genomics	the DNA sequence of the genome	DNA sequencer	
Transcriptomics	the abundance of all mRNA's of a genome	cDNA microarray	[196]
Proteomics	the presence or absence of all proteins of the genome	2D electrophoresis Mass spectrometry Protein microarray	[201, 202]
Metabolomics	intracellular concentrations of metabolites	GC-MS, LC-MS NMR	[203, 204]
Fluxomics	the steady-state rates at which extracellular metabolites are produced	Flux & isotopomer balance	[171–173]

5. CHALLENGES AND NEW APPROACHES FOR METABOLIC ENGINEERING

Metabolic engineering has the general goal of altering cell metabolism to improve the productivity and yield of metabolites or to produce novel biochemical compounds by changing genes encoding key pathway enzyme(s) or regulatory proteins using recombinant DNA technology. Classical rational metabolic engineering often also uses metabolic models to guide the selection of gene targets. Recent rapid development in high-throughput technologies have also generated various kinds of -omic information that allow for a more efficient "inverse metabolic engineering" approach to strain and bioprocess development [211]. However, many issues and challenges remain in this field.

5.1. Biological compelxity

The biological network is very complex not only at the ecological but also at the cellular level (see Figure 11). Cells are robust, with a high degree of redundancy in their biochemical networks that are insensitive to many mutations, especially those affecting critical core activities [212]. Changes in the relative level of a particular gene, transcript (mRNA), or protein do not imply a corresponding change in the level of its transcript, protein activity, or reaction rate in the cell because of the complicated interactions among the metabolic and regulatory networks, many of which are still unknown and thus cannot be considered in rational metabolic engineering approaches. It is thus not a surprise that knocking out main genes or introducing heterologous enzymes and pathways often have almost no effect on many flux ratios. On the other hand, because of the global, genome-wide regulation in the cell, any perturbation to a single gene or enzyme usually results in multiple transcriptomic and proteomic responses that are not predictable with our current knowledge.

So far most of the metabolic engineering studies have focused on the manipulation of metabolic network stoichiometry, which has been successful in cases ranging from improving cellular properties to overproducing primary (e.g., ethanol and lactic acid) and secondary (e.g., antibiotics) metabolites and even complex biomolecules (e.g., carotenoids) and biopolymers (e.g., PHA). As discussed before, these successful cases usually were achieved through increasing or redirecting the metabolic flux towards desirable products by relieving metabolic bottlenecks, increasing precursor supply, balancing cofactors, deleting product inhibition, or some combination thereof. A useful strain usually requires "compensatory" mutations to ameliorate adverse phenotypic changes caused by the initial genetic modifications. Nevertheless, there are many more unsuccessful metabolic engineering attempts that failed largely because regulatory pathways were rarely considered due to the lack of knowledge in this field.

However, there are a few successful cases involving the manipulation of regulatory systems [213]. For example, disrupting *E. coli*'s carbon storage regulatory protein (*Csr*A) increased phenylalanine biosynthesis by up regulating gluconeogenic and glycogen synthesis enzymes while downregulating glycolytic enzymes, which increased the level of phosphoenolpyruvate, the precursor for the aromatic amino acid [214]. Another new approach to improve strain performance is to construct non-native regulatory circuits to modulate or expand regulatory networks, which was demonstrated in the production of lycopene by *E. coli* [215].

5.2. Functional genomics

Functional genomics is currently the most effective approach for increasing knowledge at the molecular level of metabolic and adaptive processes in cells, and has become a new scientific field providing powerful tools for unraveling and understanding biological complexity. However, it has major limitations. For example, about 40% of the open reading frames (ORFs) of the known genomic sequences of many microorganisms have no known function. A complete genomic sequence is required in order to use DNA microarrays for transcriptomic studies, although shotgun DNA microarrays have been constructed and used to study the metabolism and gene functions of several microorganisms [216].

Also, there are large errors and uncertainties associated with transcriptomic and proteomic data obtained from microarrays and 2D gel electrophoretic maps. Currently, all omic data are comparative in nature and their interpretation can be difficult, especially when the difference is not large. Transcriptomes and proteomes are not only dependent on genomes, but also change with environmental or experimental conditions, which can vary greatly between *in vivo* and *in vitro* setups and even between two different process scales (e.g., flasks vs. fermentors). For unicellular organisms such as bacteria and yeasts, there are several thousands genes, a similar number of different types of mRNAs, and probably 10 times more different types of proteins in each cell. Today, a DNA microarray can cover several thousand genes or the entire genome of a bacterial or yeast cell, but a 2D gel map can only give several thousand protein spots, of which many are still unknown. Data mining is one of the greatest challenges in bioinformatics.

For metabolomics, one would expect a similarly large, if not larger, number of small-molecule metabolites in a cell. However, the number of metabolites known or estimated in genome-scale *in silico* metabolic models is much smaller than the numbers of genes and reactions [196]. Apparently, there are many more metabolites still unknown to us or difficult to detect due to a lack of analytical methods and/or chemical standards. True whole-cell metabolome profiling is far from realization. Flux analysis is also limited by our limited knowledge of cell metabolism and physiology. Knocking out the key genes in a metabolic pathway often failed to eliminate the production of the targeted metabolite [140, 141, 217]. Even for the most studied and best understood microorganisms, *E. coli* and *S. cerevisiae*, there are probably unknown metabolic pathways waiting to be uncovered.

Finally, there is a missing link between functional genomics and industrial biotechnology. For industrial applications, metabolically engineered mutants must be tested for their fermentation performance under industrial fermentation conditions. The development of high-throughput microbioreactors that can generate kinetic data comparable to those from industrial-scale bioreactors is thus needed [218].

5.3. Modulation of gene expression and enzyme activity

Knock-out and knock-in mutants are so much different from the initial (wild type) organism. In order to survive, these mutants may activate an entirely new set of genes or undergo secondary mutations to minimize or compensate for any physiological changes. Therefore, modulating production organisms around their physiological states by subtly under- or over-expressing genes of potential interest around their normal expression levels may be more effective in inducing desirable phenotypic changes [219]. Modulating or fine-tuning gene expression is also important to the understanding of gene regulation and its effects on the metabolism. Gene expression can be modulated by using inducible promoters or promoter cassettes and synthetic promoter libraries (SPL). Modulated gene expression has been applied to metabolic engineering. For example, xylose utilization in recombinant *S. cerevisiae* was improved by several fold when a YRP promoter library was used to fine-tune the ZWF1 gene encoding glucose-6-phosphate dehydrogenase, reducing its activity by 1% to 13%, instead of a complete disruption, which would reduce xylose consumption by more than 80%.

Modulating enzyme activities may have similar benefits. This can be done by titrating enzyme activities with inhibitors, varying the dose of the gene encoding the enzyme, and point mutations in the gene encoding the protein. Directed enzyme evolution can be used to generate enzymes with better processing capabilities and is thus a promising technique for metabolic engineering as well [220]. In fact, enzyme engineering is an effective strategy to eliminate feedback inhibition as already discussed in amino acid biosynthesis. Antisense technology using antisense RNA and small interfering RNA, which inhibit the expression of a target gene, also can be used to downregulate enzymes controlling undesirable pathways [221].

5.4. Genome breeding and evolutionary engineering

Rational metabolic engineering, including inverse metabolic engineering, requires a good knowledge of the functional genomics in order to select the right targets for genetic modifications. In the genomic era, it is possible to do genome breeding and genome engineering to improve industrial strains for the production of biochemicals [22]. Genome breeding can improve strains by removing unwanted mutations acquired during numerous mutation steps. Genome engineering or removing unnecessary genes or operons that are genomic burdens on the cell is another promising approach for developing new production strains [222]. A large number of genes are redundant or have no function and their removal does not seem to adversely affect normal physiology. Removing the dispensable genomic regions thus may pave the way towards optimized cell factories since the "minimized" strain may require less energy or produce fewer byproducts.

Biological complexity and our limited knowledge of functional genomics greatly impede practical applications of rational metabolic engineering. Evolutionary engineering, based on random mutagenesis and selection, is thus an appealing approach for strain development [223]. *In vivo* recombination or genome shuffling has been successfully demonstrated in several studies [224–227]. Compared to classical strain selection, genome shuffling can decrease the number of rounds of mutation and screening by an order of magnitude. Likewise, *in vitro* evolution or molecular breeding that combines biosynthetic genes from different organisms can produce a wider spectrum of complex biomolecules such as carotenoids [228]. Combinatorial engineering and a hybrid of evolutionary and rational metabolic engineering have also been used to improve production strains [229]. It is also possible to create a large random knock-out and/or knock-in mutant library and then select for better producing strains using high-throughput screening. The advantage of evolutionary engineering, as opposed to rational metabolic engineering, is that no prior knowledge is required. However, a high-throughput screening method for selecting and characterizing useful but unknown mutants is critical to its successful application. The development of novel biosensors for this application is thus an important area for future research and development.

5.5. Process engineering

Before metabolic engineering, genetic engineering, and recombinant DNA technologies were developed, process engineering and classical strain development programs using random mutagenesis and screening were the main approaches to improving industrial fermentation. As discussed before, cellular metabolism and phenotypes are strongly influenced by the growth environment and thus can be manipulated by changing the medium composition or process conditions. Therefore, interactions between cells and the environment should not be overlooked in metabolic engineering. As we have learned from antibiotic resistant bacterial strains and extremophiles, microorganisms have tremendous capabilities to adapt to almost any adverse environment through various mechanisms, which may shed a light on how to develop an effective metabolic engineering strategy for strain engineering [230]. A common problem in producing carboxylic acids using anaerobic acidogens is strong end-product inhibition and poor acid tolerance. While there has been no known metabolic engineering to improve cell's acid tolerance, in-process adaptation has been successfully used to produce acid tolerants that can produce acids at two to three-fold higher final concentrations in the fermentation media [231]. Extractive fermentation also has been successfully used to alleviate end product inhibition, leading not only to a higher final product concentration but also to a higher product yield due to altered flux distribution [232]. Thus, any effect on intracellular flux distribution should be looked at from the process engineering perspective as well.

Metabolic engineering using recombinant DNA technology also has practical difficulties in generating mutants capable of carrying out complex reactions that naturally require several microbial species growing in an interactive community. One example is the anaerobic digestion process that produces biogas (methane and carbon dioxide) from organic matter involving at least three metabolically diverse microorganisms – hydrolytic, acidogenic, and methanogenic bacteria. Learning from nature, one can design a better biomethanation process by coordinating the metabolic pathways from three different microbial species [233]. Similarly, by culturing two species together, a better or more complete fermentation process can be obtained to convert an otherwise unfermentable substrate to the desirable product [234].

Metabolic engineering thus should not be limited to designing or engineering the metabolic pathways in just one cell or species using recombinant DNA technology. Any means, whether biological, chemical, or physical, on the molecular (gene) or process level, should be considered in future metabolic engineering as long as it can achieve the goal of improving the industrial production of biochemicals and biofuels.

6. SUMMARY

Classical metabolic engineering changes cell metabolism by changing its pathway enzyme(s) or regulatory protein(s) using recombinant DNA technology, thus improving the productivity and yield of an industrial fermentation product or producing novel biochemical compounds. However, classical metabolic engineering rarely considers gene regulation, and it is thus difficult to guide the selection of target genes for modification. Inverse metabolic engineering takes advantage of the knowledge learned from mutants or existing strains with improved performance. Functional genomics using high-throughput approaches provides more efficient metabolic engineering design and analysis, but still has its limitations. Future metabolic engineering should consider every biochemical network in the cell and account for cellular responses through its own regulation hierarchy to the genetic changes and environmental stimuli. Until we have a complete knowledge of the cell and its complex metabolic and regulatory networks, evolutionary engineering and classical process engineering approaches will continue to contribute to both strain and bioprocess development.

REFERENCES

[1] G. Stephanopoulos and J. Vallino, Network rigidity and metabolic engineering in metabolite overproduction. Science, 252 (1991) 1675–1681.
[2] J. Bailey, Towards a science of metabolic engineering. Science, 252 (1991) 1668–1674.
[3] J. Nielsen, Metabolic engineering. Appl Microbiol Biotechnol, 55 (2001) 263–283.
[4] H. Shimizu, Metabolic engineering – integrating methodologies of molecular breeding and bioprocess systems engineering, J Biosci Bioeng. 94 (2002) 563–573.
[5] R.M. Raab, K. Tyo, and G. Stephanopoulos, Metabolic engineering, Adv Biochem Eng/Biotechnol. 100 (2005) 1–17.
[6] G. Stephanopoulos, A. Aristidou, and J. Nielsen, Metabolic Engineering Principles and Methodologies. London, UK: Academic Press; 1998.
[7] S.Y. Lee and E.T. Papoutsakis (eds.), Metabolic Engineering, Marcel Dekker, Inc., New York, 1999.
[8] B.N. Kholodenko and H.V. Westerhoff (eds.), Metabolic Engineering in the Post Geonmic Era, Horizone Bioscience, Wymondham, UK, 2004.
[9] G. Chotani, T. Dodge, A. Hsu, M. Kumar, R. LaDuca, D. Trimber, W. Weyler, and K. Sanford, The commercial production of chemicals using pathway engineering. Biochim Biophys Acta, 1543 (2000) 434–455.
[10] M. Chartrain, P.M. Salmon, D.K. Robinson, and B.C. Buckland, Metabolic engineering and directed evolution for the production of pharmaceuticals. Curr Opin Biotechnol. 11 (2000) 209–214.
[11] M. Ikeda. Amino acid production processes, Adv Biochem Eng/Biotechnol. 79 (2003) 1–35.
[12] D. Kumar and J. Gomes, Methionine production by fermentation, Biotechnol Adv. 23 (2005) 41–61.
[13] W. Pfefferle, B. Möckel, B. Bathe, A. Marx, Biotechnological manufacture of lysine, Adv Biochem Eng/Biotechnol. 79 (2003) 59–112.
[14] E. Kimura, Metabolic engineering of glutamate production, Adv Biochem Eng/Biotechnol. 79 (2003) 37–57.
[15] V.G. Debabov, The threonine story. Adv Biochem Eng/Biotechnol. 79 (2003) 113–136.
[16] A.A. de Graaf, L. Eggeling, H. Sahm, Metabolic engineering for L-lysine production by *Corynebacterium glutamicum*, Adv Biochem Eng/Biotechnol. 73 (2001) 9–29.
[17] D. Kumar and J. Gomes, Methionine production by fermentation. Biotechnol Adv. 23 (2005) 41–61.
[18] J. Cremer, L. Eggeling, and H. Sahm, Control of the lysine biosynthesis sequence in *Corynebacterium glutamicum* as analyzed by overexpression of the individual corresponding genes. Appl Environ Microbiol. 57 (1991) 1746–1752.
[19] R. Katsumata, T. Mizukami, Y. Kikuchi, K. Kino, Threonine production by the lysine producing strain of *Corynebacterium glutamicum*with amplified threonine biosynthetic operon. In: M. Alacevic, D. Hranueli, Z. Toman (eds) Genetics of Industrial Microorganisms. Yugoslavia (1986) p. 217.
[20] S. Guillouet, A.A. Rodal, G. An, P.A. Lessard, A.J. Sinskey, Expression of the *Escherichia coli* catabolic threonine dehydratase in *Corynebacterium glutamicum* and its effect on isoleucine production. Appl Environ Microbiol. 65 (1999) 3100–3107

[21] M. Ikeda, K. Nakanishi, K. Kino, R. Katsumata, Fermentative production of tryptophan by a stable recombinant strain of *Corynebacterium glutamicum* with a modified serine-biosynthetic pathway. Biosci Biotechnol Biochem. 58 (1994) 674

[22] T. Hermann, Using functional genomics to improve productivity in the manufacture of industrial biochemicals. Curr Opin Biotechnol. 15 (2004) 444–448.

[23] M.A.G. Koffas, G.Y. Jung, and G. Stephanopoulos, Engineering metabolism and product formation in *Corynebacterium glutamicum* by coordinated gene overexpression, Metabol Eng. 5 (2003) 32–41.

[24] M. Ikeda M and R. Katsumata, Hyperproduction of tryptophan by *Corynebacterium glutamicum* with the modified pentose phosphate pathway, Appl Environ Microbiol 65 (1999) 2497–2502.

[25] M. Ikeda, K. Okamoto, R. Katsumata, Cloning of the transketolase genen and the effect of its dosage on aromatic amino acid production in *Corynebacterium glutamicum*, Appl Microbiol Biotechnol 51 (1999) 201–206.

[26] M. Ikeda M and R. Katsumata, Tryptophan production by transport mutants of *Corynebacterium glutamicum*. Biosci Biotechnol Biochem 59 (1995) 1600–1602.

[27] K. Okamoto, K. Kino, M. Ikeda, Hyperproduction of L-threonine by an *Escherichia coli* mutant with impaired L-threonine uptake. Biosci Biotechnol Biochem 61 (1997) 1877–1882.

[28] M. Koffas and G. Stephanopoulos, Strain improvement by metabolic engineering: lysine production as a case study for systems biology, Curr Opin Biotechnol. 16 (2005) 361–366.

[29] R. Petri and C. Schmidt-Dannert, Dealing with complexity: evolutionary engineering and genome shuffling. Curr Opin Biotehnol. 15 (2004) 298–304.

[30] J. Ohnishi, S. Mitsuhashi, M. Hayshi, S. Ando, H. Yokoi, K. Ochiai, and M. Ikeda, A novel methodology employing *Corynebacterium glutamicum* genome information to generate a new L-lysine-producing mutant. Appl Microbiol Biotechnol. 58 (2002) 217–223.

[31] A.L. Demain and R. Elander, The β-lactam antibiotics: past, present, and future. Antonie van Leeuwenhoek 75 (1999) 5–19.

[32] N. Gunnarssonl, A. Eliasson, J. Nielsen, Control of fluxes towards antibiotics and the role of primary metabolism in production of antibiotics, Adv Biochem Eng/Biotechnol. 88 (2004) 137–178.

[33] J. Thykaer and J. Nielsen, Metabolic engineering of β-lactam production, Metabolic Eng. 5 (2003) 56–69.

[34] H.Aa. Theilgaard, M. van den Berg, C. Mulder, R.A.L. Bovenberg, J. Nielsen, Quantitative analysis of *Penicillium chrysogenum* Wis54-1255 transformants overexpressing the penicillin biosynthetic genes. Biotechnol Bioeng. 72 (2001) 379–388.

[35] L. Mathison, C. Soliday, T. Stepan, T. Aldrich, J. Rambosek, Cloning, characterization, and use in strain improvement of the *Cephalosporium acremonium* gene *cefG* encoding acetyl transferase. Curr Genet. 23 (1993) 33–41.

[36] S. Gutiérrez, J. Velasco, A.T. Marcos, F.J. Fernández, F. Fierro, J.L. Barredo, B. Diez, J.F. Martin, Expression of the *cefg* gene is limiting for cephalosporin biosynthesis in *Acremonium chrysogenum*. Appl Microbiol Biotechnol. 48 (1997) 606–614.

[37] R.W. Newbert, B. Barton, P. Greaves, J. Harper, G. Turner. Analysis of a commercially improved *Penicillium chrysogenum* strain series: involvement of recombinogenic regions in amplification and deletion of the penicillin gene cluster. J Ind Microbiol Biotechnol. 19 (1997) 18–27.

[38] W.M. van Gulik, W.T. de Laat, J.L. Vinke, J.J. Heijnen, Application of metabolic flux analysis for the identification of metabolic bottlenecks in the biosynthesis of penicillin-G. Biotechnol Bioeng. 68 (2000) 602–618.

[39] B. Christensen, J. Thykaer, J. Nielsen, Metabolic characterization of high- and low-yielding strains of *Penicillium chrysogenum*. Appl Microbiol Biotechnol. 54 (2000) 212–217.

[40] V.K. Chary, J.L. de la Fuente, A.L. Letiäso, P. Liras, J.F. Martin, Overexpression of the *lat* gene in *Nocardia lactamdurans* from strong heterologous promoters results in very high levels

of lysine-6-aminotransferase and up to two-fold increase in cephamycin C production. Appl Microbiol Biotechnol. 53 (2000) 282–288.

[41] B. Minambres, H. Martinez-Blanco, E.R. Olivera, B. Garcia, B. Diez, J.L. Barredo, M.A. Moreno, C. Schleissner, F. Salto, J.M. Luengo. Molecular cloning and expression in different microbes of the DNA encoding *Pseudomonas putida* U phenylacetyl-CoA-ligase. J. Biol. Chem. 271 (1996) 33531–33538.

[42] J. Velasco, J.L. Adrio, M.A. Moreno, B. Diez, G. Soler, J.L. Barredo, Environmentally safe production of 7-aminodeacetoxycephalosporanic acid (7-ADCA) using recombinant strains of *Acremonium chrysogenum*. Nature Biotechnol. 18 (2000) 857–861.

[43] A. Bruggink, P.D. Roy. Industrial synthesis of semisynthetic antibiotics. In: Bruggink, A. (Ed.), Synthesis of β-lactam antibiotics, chemistry biocatalysis and process integration. Kluwer Acadenic Publishers, Dordrecht, The Netherlands, 2001, pp. 13–56.

[44] L. Crawford, A.M. Stepan, P.C. McAda, J.A. Rambosek, M.J. Confer, V.A. Vinci, C.D. Reeves. Production of cephalosporin intermediates by feeding adipic acid to recombinant *Penicillium chrysogenum* strains expressing ring expansion activity. Bio/Technol. 13 (1995) 58–62.

[45] E.J.A.X. van de Sandt and E. de Vroom. Innovations in cephalosporin and penicillin production: Painting the antibiotics industry green. Chim. Oggi–Chem. Today 18 (2000) 72–75.

[46] B.A. Pfeifer, S.J. Admiraal, H. Gramajo, D.E. Cane, and C. Khosia, Biosynthesis of complex polyketides in a metabolically engineered strain of *E. coli*. Science 291 (2001) 1790–1792.

[47] B.N. Mijts and C. Schmidt-Dannert, Engineering of secondary metabolite pathways. Curr Opin Biotechnol. 14 (2003) 597–602.

[48] Y.J. Yoon, B.J. Beck, B.S. Kim, H.Y. Kang, K.A. Reynolds, and D.H. Sherman, Generation of multiple bioactive macrolides by hybrid modular polyketide synthases in Streptomyces venezuelae. Chem Biol. 9 (2002) 203–214.

[49] C. Avignone Rossa, J. White, A. Kuiper, P.W. Postma, M. Bibb, and M.J. Teixeira de Mattos, Carbon flux distribution in antibiotic-producing chemostat cultures of *Streptomyces lividans*. Metab Eng. 4 (2002) 138–150.

[50] M.J. Butler, P. Bruheim, S. Jovetic, F. Marinelli, P.W. Postma, and M.J. Bibb, Engineering of primary carbon metabolism for improved antibiotic production in *Streptomyces lividans*. Appl Environ Microbiol. 68 (2002) 4731–4739.

[51] B.D. Ensley, B.J. Ratzkin, T.D. Osslund, M.J. Simon, L.P. Wackett and D.T. Gibson. Expression of naphthalene oxidation genes in *Escherichia coli* results in the biosynthesis of indigo. Science 222 (1983) 167–169.

[52] D. Murdock, B. D. Ensley, C. Serdar and M. Thalen. Construction of metabolic operons catalyzing the de novo biosynthesis of indigo in *Escherichia coli*. Bio/Technology 11 (1993) 381–386.

[53] A. Berry, T.C. Dodge, M. Pepsin and W. Weyler. Application of metabolic engineering to improve both the production and use of biotech indigo. J Ind Microbiol Biotechnol. 28 (2002) 127–133.

[54] W. Weyler, T.C. Dodge, J.J. Lauff and D.J. Wendt. Microbial production of indigo. US Patent 5,866,396, 1999.

[55] W.-D. Deckwer. Microbial conversion of glycerol to 1,3-propanediol. FEMS Microbiol Rev, 16 (1995) 143–149.

[56] D. C. Cameron, N.E. Altaras, M.L. Hoffman, A.J. Shaw. Metabolic engineering of propanediol pathways. Biotechnol Prog, 14 (1998) 116–125.

[57] A.-P. Zeng and H. Biebl: Bulk chemicals from biotechnology: the case of 1,3-propanediol production and the new trends. Adv Biochem Eng Biotechnol, 74 (2002) 239–259.

[58] J.V. Kurian. A new polymer platform for the future – Sorona® from Corn derived 1,3-Propanediol. J. Polymers and the Environment 13 (2005) 159–167.

[59] L.A. Laffend, V. Nagarajan, C.E. Nakamura, Bioconversion of a fermentable carbon source to 1,3-propanediol by a single microorganism. United States Patent 1997, No 5,686,276.
[60] R.V. Nair, M.S. Payne, D.E. Trimbur, F. Valle, Method for the production of glycerol by recombinant organisms. PCT Int Application 1999, Publication No WO 99/28480.
[61] M. Diaz-Torres, N.S. Dunn-Coleman, M.W. Chase, D. Trimbur, Method for the recombinant production of 1,3-propanediol. United States Patent 2000, No 6,136,976.
[62] M. Emptage, S.L. Haynie, .L.A. Laffend, J.P. Pucci, G. Whited, Process for the biological production of 1,3-propanediol with high titer. United States Patent 2003, No 6,514,733.
[63] K. Larsson, R. Ansell, P. Eriksson, L. Adler, A gene encoding sn-glycerol 3-phosphate dehydrogenase (NAD^+) complements an osmosensitive mutant of *Saccharomyces cerevisiae*. Mol Microbiol 10 (1993) 1101–1111.
[64] J. Norbeck, A-K. Pahlman, N. Akhtar, A. Blomberg, L. Adler, Purification and characterization of two isoenzymes of DL-glycerol-3-phosphatase from *Saccharomyces cerevisiae*. J Biol Chem. 271 (1996) 13875–13881.
[65] C.E. Nakamara, G..M. Whited, Metabolic engineering for the microbial production of 1,3-propanediol. Curr Opin Biotechnol. 14 (2003) 454–459.
[66] J. Lunt, Large-scale production, properties and commercial applications of polylactic acid polymers. Polymer Degrad Stabil. 59 (1997) 145–152.
[67] W.M. de Vos and J. Hugenholtz. Engineering metabolic highways in Lactococci and other lactic acid bacteria. Trends Biotechnol. 22 (2004) 72–79.
[68] M. Kleerebezem and J. Hugenholtz. Metabolic pathway engineering in lactic acid bacteria. Curr Opin Biotechnol. 14 (2003) 232–237.
[69] J. Hugenholtz and M. Kleerebezem. Metabolic engineering of lactic acid bacteria: overview of the approaches and results of pathway rerouting involved in food fermentations. Curr Opin Biotechnol. 10 (1999) 492–497.
[70] A. Tay and S.T. Yang. Production of L(+)-lactic acid from glucose and starch by immobilized cells of *Rhizopus oryzae* in a rotating fibrous bed bioreactor. Biotechnol Bioeng., 80 (2002) 1–12.
[71] K. Kylä-Nikkilä, M. Hujanen, M. Leisola, and A. Palva. Metabolic engineering of *Lactobacillus helveticus* CNRZ32 for production of pure L-(+)-lactic acid. Appl Environ Microbiol. 66 (2000) 3835–3841.
[72] K. Hofvendahl and B. Hahn-Hägerdal. Factors affecting the fermentative lactate acid production from renewable resources. Enz Microb Technol. 26 (2000) 87–107.
[73] D. Porro, L. Barmbilla, B. M. Ranzi, E. Martegani, and L. Alberghina. Development of metabolically engineered *Saccharomyces cerevisiae* cells for the production of lactic acid. Biotechnol Prog. 11 (1995) 294–298.
[74] E. Adachi, M. Torigoe, S. Sugiyama, J. Nikawa, and K. Shimizu. Modification of metabolic pathways of *Saccharomyces cerevisiae* by the expression of lactate dehydrogenase and deletion of *pyruvate decarboxylase* genes for the lactic acid fermentation at low pH value. J Ferment Bioeng. 86 (1998) 284–289.
[75] C. D. Skory. Lactic acid production by *Saccharomyces cerevisiae* expressing a *Rhizopus oryzae* lactate dehydrogenase gene. J Ind Microbiol Biotechnol. 30 (2003) 22–27.
[76] S. Saitoh, N. Ishida, T. Onishi, K. Tokuhiro, E. Nagamori, K. Kitamoto, and H. Takahashi. Genetically engineered wine yeast produces a high concentration of L-lactic acid of extremely high optical purity. Appl Environ Microbiol. 71 (2005) 2789–2792.
[77] N. Ishida, S. Saitoh, K. Tokuhiro, E. Nagamori, T. Matsuyama, K. Kitamoto, and H. Takahashi. Efficient production of L-lactic acid by metabolically engineered *Saccharomyces cerevisiae* with a genome-integrated L-lactate dehydrogenase gene. Appl Environ Microbiol. 71 (2005) 1964–1970.
[78] D. Porro, M. M. Bianchi, L. Brambilla, R. Menghini, D. Bolzani, V. Carrera, J. Lievense, C. L. Liu, B. M. Ranzi, L. Frontali, and L. Alberghina. Replacement of a metabolic pathway for large

scale production of lactic acid from engineered yeasts. Appl Environ Microbiol. 65 (1999) 4211–4215.

[79] M. M. Bianchi, L. Brambilla, F. Protani, C. L. Liu, J. Lievense, and D. Porro. Efficient homolactic fermentation by *Kluyveromyces lactis* strains defective in pyruvate utilization and transformed with the heterologous *LDH* gene. Appl Environ Microbiol. 67 (2001) 5621–5625.

[80] D.-E. Chang, H.-C. Jung, J.-S. Rhee, and J. Pan. Homofermentative production of D- or L-lactate in metabolically engineered *Escherichia coli* RR1. Appl Environ Microbiol. 65 (1999) 1384–1389.

[81] S. Zhou, T. B. Causey, A. Hasona, K. T. Shanmugam, and L. O. Ingram. Production of optically pure D-lactic acid in mineral salts medium by metabolically engineered *Escherichia coli* W3110. Appl Environ Microbiol. 69 (2003) 399–407.

[82] B. S. Dien, N. N. Nichols, and R. J. Bothast. Recombinant *Escherichia coli* engineered for production of L-lactic acid from hexose and pentose sugars. J Ind Microbiol Biotechnol. 27 (2001) 259–264.

[83] S. Zhou, K. T. Shanmugam, and L. O. Ingram. Functional replacement of the *Escherichia coli* D-(−)-lactate dehydrogenase gene (*ldhA*) with the L-(+)-lactate dehydrogenase gene (*ldhL*) from *Pedioccus acidilactici*. Appl Environ Microbiol. 69 (2003) 2237–2244.

[84] S.K. Singh, S.U. Ahmed, and A. Pandey, Metabolic engineering approaches for lactic acid production. Process Biochem. 41 (2006) 991–1000.

[85] J.G. Zeikus, M.K. Jain, P. Elankovan. Biotechnology of succinic acid production and markets for derived industrial products. Appl Microbiol Biotechnol. 51 (1999) 545–552.

[86] S.Y. Lee, S. H. Hong, S.H. Lee, S.J. Park. Fermentation production of chemicals that can be used for polymer synthesis. Macromol Biosci. 4 (2004) 157–164.

[87] H. Song and S.Y. Lee, Production of succinic acid by bacterial fermentation. Enz Microb Technol. 39 (2006) 352–361.

[88] M.J. Van der Werf, M.V. Guettler, M.K. Jain, J. G. Zeikus. Environmental and physiological factors affecting the succinate product ratio during carbohydrate fermentation by *Actinobacillus* sp. 130Z. Arch Microbiol. 167 (1997) 332–342.

[89] C.S. Millard, Y. Chao, J.C. Liao, M.I. Donnelly. Enhanced production of succinic acid by overexpression of phosphoenolpyruvate carboxylase in *Escherichia coli.* Appl Environ Microbiol. 62 (1996) 1808–1810.

[90] I. Goldberg, K. Lonberg-Holm, E.A. Bagley, B. Stieglitz. Improved conversion of fumarate to succinate by *Escherichia coli* strains amplified for fumarate reductase. Appl Environ Microbiol. 45 (1983) 1838–1847.

[91] L. Stols, M.I. Donnelly. Production of succinic acid through overexpression of NAD^+-dependent malic enzyme in an *Escherichia coli* mutant. Appl Environ Microbiol. 63 (1997) 2695–2701.

[92] R.R. Gokarn, M.A. Eiteman, E. Altman. Expression of pyruvate carboxylase enhances succinate production in *Escherichia coli* without affecting glucose uptake. Biotechnol Lett. 20 (1998) 795–798.

[93] R. Chatterjee, C.S, Millard, K. Champion, D.P. Clark, M.I. Donnelly. Mutation of the *ptsG* gene results in increased production of succinate in fermentation of glucose by *Escherichia coli.* Appl Environ Microbiol. 67 (2001) 148–154.

[94] G.N. Vemuri, M.A. Eiteman, and E. Altman. Effects of growth mode and pyruvate carboxylase on succinic acid production by metabolically engineered strains of *Escherichia coli.* Appl Environ Microbiol. 68 (2002) 1715–1727.

[95] G.N. Vemuri, M.A. Eiteman, and E. Altman. Succinate production in dual-phase *Escherichia coli* fermentations depends on the time of transition from aerobic to anaerobic conditions. J Ind Microbiol Biotechnol. 28 (2002) 325–332.

[96] S.H. Hong and S.Y. Lee. Metabolic flux distribution in a metabolically engineered *Escherichia coli* strain producing succinic acid. J Microbiol Biotechnol. 10 (2000) 496–501.

[97] H. Lin, G.N. Bennett, K.-Y. San. Effect of carbon sources differing in oxidation state and transport route on succinate production in metabolically engineered *Escherichia coli*. J Ind Microbiol Biotechnol. 32 (2005) 87–93.
[98] K.-Y. San, G.N. Bennett, S.J. Berríos-Rivera, R.V. Vadali, Y.-T. Yang, E. Horton, F.B. Rudolph, B.Sariyar, K. Blackwood. Metabolic engineering through cofactor manipulation and its effects on metabolic flux redistribution in *Escherichia coli.* Metabolic Eng. 4 (2002) 182–192.
[99] H. Lin, G.N. Bennett, K.-Y. San. Genetic reconstruction of the aerobic central metabolism in *Escherichia coli* for the absolute aerobic production of succinate. Biotechnol Bioeng. 89 (2005) 148–156.
[100] H. Lina, G.N. Bennettb, K.-Y. San. Metabolic engineering of aerobic succinate production systems in *Escherichia coli* to improve process productivity and achieve the maximum theoretical succinate yield. Metabolic Eng. 7 (2005) 116–127.
[101] L.L. Madison, G.W. Huisman. Metabolic engineering of poly(3-hydroxyalkanoates): from DNA to plastic. Microbiol Mol Biol Rev. 63 (1999) 21–53.
[102] A. Steinbüchel, Perspectives for biotechnological production and utilization of biopolymers: metabolic engineering of polyhydroxyalkanoate biosynthesis pathways as a successful example, Macromol Biosci. 1 (2001) 1–24.
[103] I.S. Aldor, J.D. Keasling. Process design for microbial plastic factories: metabolic engineering of polyhydroxyalkanoates. Curr Opin Biotechnol. 14 (2003) 475–483.
[104] A. Steinbuchel, T. Lutke-Eversloh. Metabolic engineering and pathway construction for biotechnological production of relevant polyhydroxyalkanoates in microorganisms. Biochem Eng J. 16 (2003) 81–96.
[105] C.S.K. Reddy, R. Ghai, Rashmi, V.C. Kalia. Polyhydroxyalkanoates: an overview. Bioresource Technol. 87 (2003) 137–146.
[106] S.Y. Lee, J. Choi. Production of microbial polyester by fermentation of recombinant microorganisms. In: A. Steinbuchel, W. Babel (Eds.) Biopolyesters. Adv Biochem Eng Biotechnol, Berlin: Springer-Verlag; 71 (2001) 183–207.
[107] J. Asrar, K.J. Gruys, Biodegradable polymer (Biopol®), in: Y. Doi, A. Steinbüchel (Eds.), Biopolymers, vol. 3a,Wiley/VCH,Weinheim, 2002, pp. 53–90.
[108] A.Steinbüchel, H.E. Valentin, Diversity of bacterial polyhydroxyalkanoic acids, FEMS Microbiol Lett. 128 (1995) 219–228.
[109] S.Y. Lee, S.J. Park, Fermentative production of SCL-PHAs, in: Y. Doi, A. Steinbüchel (Eds.), Biopolymers, vol. 3a, Wiley/VCH, Weinheim, 2002, pp. 263–290.
[110] R.A. Weusthuis, B. Kessler, M.P.M. Dielissen, B. Witholt, G. Eggink, Fermentative production of medium-chain-length poly(3-hydroxyalkanoate), in: Y. Doi, A. Steinbüchel (Eds.), Biopolymers, vol. 3a, Wiley/VCH, Weinheim, 2002, pp. 291–316.
[111] A.S. Kelley, N.V. Mantzaris, P. Daoutidis, F. Srienc. Controlled synthesis of polyhydroxyalkanoic (PHA) nanostructures in *R. eutropha*. Nano Lett. 1 (2001) 481–485.
[112] T. Lütke-Eversloh, A. Steinbüchel. Microbial Polythioesters. Macromol Biosci. 4 (2004) 165–174.
[113] T. Lutke-Eversloh, A. Fischer, U. Remminghorst, J. Kawada, R.H. Marchessault, A. Bogershausen, M. Kalwei, H. Eckert, R. Reichelt, S.-J. Liu, et al. Biosynthesis of novel thermoplastic polythioesters by engineered *Escherichia coli*. Nat Mater, 1 (2002) 236–240.
[114] S.J. Park, J. Choi, S.Y. Lee. Engineering of *Escherichia coli* fatty acid metabolism for the production of polyhydroxyalkanoates. Enzyme Microb Technol. 36 (2005) 579–588.
[115] M. Pötter, M.H. Madkour, F. Mayer, A. Steinbüchel, Regulation of phasin expression and polyhydroxyalkanoate (PHA) granule formation in *Ralstonia eutropha* H16, Microbiol. UK 148 (2002) 2405–2412.
[116] R. Carlson, D. Fell, F. Srienc, Metabolic pathway analysis of a recombinant yeast for rational strain development, Biotechnol Bioeng. 79 (2002) 121–134.

[117] Snell KD, Peoples OP: Polyhydroxyalkanoate polymers and their production in transgenic plants. Metab Eng. 4 (2002) 29–40.

[118] I.S, Aldor, S. W. Kim, K.L.J. Prather, J.D. Keasling. Metabolic engineering of a novel propionate-independent pathway for the production of poly(3-hydroxybutyrate-co-3-hydroxyvalerate) in recombinant *Salmonella enterica* serovar Typhimurium. Appl Environ Microbiol. 68 (2002) 3848–3854.

[119] J.C. Choi, H.D. Shin, Y.H. Lee. Modulation of 3-hydroxyvalerate molar fraction in poly(3-hydroxybutyrate-3-hydroxyvalerate) using *Ralstonia eutropha* transformant co-amplifying phbC and NADPH generation-related *zwf* genes. Enzyme Microb Technol. 32 (2003) 178–185.

[120] P.R. Green, J. Kemper, L. Schechtman, L. Guo, M. Satkowski, S. Fiedler, A. Steinbuchel, B.H.A. Rehm. Formation of short chain length/medium chain length polyhydroxyalkanoate copolymers by fatty acid β-oxidation inhibited *Ralstonia eutropha*. Biomacromolecules 3 (2002) 208–213.

[121] S.J. Park, J.P. Park, S.Y. Lee, Y. Doi. Enrichment of specific monomer in medium-chain-length poly(3-hydroxyalkanoates) by amplification of fadD and fadE genes in recombinant *Escherichia coli*. Enzyme Microb Technol. 33 (2003) 62–70.

[122] S.Y. Lee, H.H. Wong, J. Choi, S.H. Lee, S.C. Lee, C.S. Han. Production of medium-chain-length polyhydroxyalkanoates by high-cell.density cultivation of *Pseudomonas putida* under phosphorus limitation. Biotechnol Bioeng. 68 (2000) 466–470.

[123] Y. Lin and S. Tanaka, Ethanol fermentation from biomass resources: current state and prospects. Appl Microbiol Biotechnol 69 (2006) 627–642.

[124] A. Aristidou and M. Penttila, Metabolic engineering applications to renewable resource utilization. Curr Opin Biotechnol. 11 (2000) 187–198.

[125] L.O. Ingram, P.E. Gomez, X. Lai, M. Moniruzzaman, B.E. Wood, L.P. Yomano, S.W. York, Metabolic engineering of bacteria for ethanol production. Biotechnol Bioeng. 58 (1998) 204–214.

[126] S. Ostergaard, L. Olsson, and J. Nielsen, Metabolic Engineering of *Saccharomyces cerevisiae*. Microbiol Mol Biol Rev. 64 (2000) 34–50.

[127] B. Hahn-Hägerdal, C.F. Wahlbom, M. Gárdonyi, W.H. van Zyl, R.R. Cordero Otero, L.J. Jönsson, Metabolic engineering of *Saccharomyces cerevisiae* for xylose utilization. Adv Biochem Eng/Biotechnol. 73 (2001) 53–84.

[128] T.W. Jeffries and Y.-S. Jin, Metabolic engineering for improved fermentation of pentoses by yeasts. Appl Microbiol Biotechnol. 63 (2004) 495–509.

[129] M. Kuyper, A.A. Winkler, J.P. van Dijken, J.T. Pronk, Minimal metabolic engineering of *Saccharomyces cerevisiae* for efficient anaerobic xylose fermentation: a proof of principle. FEMS Yeast Res. 4 (2004) 655–664.

[130] T. Grotkjaer, P. Christakopoulos, J. Nielsen, L. Olsson, Comparative metabolic network analysis of two xylose fermenting recombinant *Saccharomyces cerevisiae* strains. Metabol Eng. 7 (2005) 437–444.

[131] N.W. Ho, Z. Chen, and A.P. Brainard, Genetically engineered *Saccharomyces* yeast capable of effective cofermentation of glucose and xylose. Appl Environ Microbiol. 64 (1998) 1852–1859.

[132] M.S. Krishnan, N.W. Ho, and G.T. Tsao, Fermentation kinetics of ethanol production from glucose and xylose by recombinant *Saccharomyces* 1400 (pLNH33). Appl Biochem Biotechnol. 78 (1999) 373–388.

[133] J. Zaldivar, C. Roca, and C. Le Foll, Ethanolic fermentation of acid pre-treated starch industry effluents by recombinant *Saccharomyces cerevisiae* strains. Bioresource Technol. 96 (2005) 1670–1676.

[134] C. Bro, B. Regenberg, J. Forster, and J. Nielsen, In silico metabolic engineering of *Sacharomyces cerevisiae* for improved bioethanol production. Metabol Eng. 8 (2006) 102–111.

[135] M. Zeng, C. Eddy, K. Deanda, M. Finkelstein, and S. Picataggio, Metabolic engineering of a pentose metabolism pathway in ethanologenic *Zymomonas mobilis*. Science 267 (1995) 240–243.

[136] K. Ohta, D.S. Beall, J.P. Mejia, K.T. Shanmugam, and L.O. Ingram, Genetic improvement of *Escherichia coli* for ethanol production: chromosomal integration of *Zymomonas mobilis* genes encoding pyruvate decarboxylase and alcohol dehydrogenase II. Appl Environ Microbiol. 57 (1991) 893–900.

[137] S.E. Lindsay, R.J. Bothast, and L.O. Ingram, Improved strains of recombinant *Escherichia coli* for ethanol production from sugar substrates. Appl Microbiol Biotechnol. 43 (1995) 70–75.

[138] H. Pedersen, B. Christensen, C. Hjort, and J. Nielsen, Construction and characterization of an oxalic acid nonproducing strain of *Aspergillus niger*. Metabol Eng. 2 (2000) 34–41.

[139] Q. Wang, P. He, D. Lu, A. Shen, and N. Jiang, Metabolic engineering of *Torulopsis glabrata* for improved pyruvate production. Enz Microb Technol. 36 (2005) 832–839.

[140] S. Suwannakham, Y. Huang and S.T. Yang, Construction and characterization of *ack* knock-out mutants of *Propionibacterium acidipropionici* for enhanced propionic acid fermentation, Biotechnol Bioeng. 94 (2006) 383–395.

[141] Y. Zhu, X. Liu and S.T. Yang, Construction and characterization of *pta* gene deleted mutant of *Clostridium tyrobutyricum* for butyric acid fermentation, Biotechnol Bioeng. 90 (2005) 154–166.

[142] E.T. Papoutsakis and G.N. Bennett, Molecular regulation and matabolic engineering of solvent production by *Clostridium acetobutyricum*. In S.Y. Lee and E.T. Papoutsakis, eds., Metabolic Engineering. Marcel Dekker, New York, 1999, pp.253-280.

[143] N.E. Altaras and D.C. Cameron, Enhanced production of (R)-1,2-propanediol by metabolically engineered *Escherichia coli*. Biotechnol Prog. 16 (2000) 940–946.

[144] Y. Piao, M. Yamashita, N. Kawaraichi, R. Asegawa, H. Ono, and Y. Murooka, Production of vitamin B12 in genetically engineered *Propionibacterium freudenreichii*. J Biosci Bioeng. 98 (2004) 167–173.

[145] W. Sybesma, C. Burgess, M. Starrenburg, D. van Sinderen, and J. Hugenboltz, Multivitamin production in *Lactococcus lactis* using metabolic engineering. Metab Eng. 6 (2004) 109–115.

[146] H. Jiang, K.V. Wood, and J.A. Morgan, Metabolic engineering of the phenylpropanoid pathway in *Saccharomyces cerevisiae*. Appl Environ Microbiol. 71 (2005) 2962–2969.

[147] H. Visser, A.J.J. van Ooyen, and J.C. Verdoes, Metabolic engineering of the astaxanthin-biosynthetic pathway of *Xanthophyllomyces dendrorhous*. FEMS Yeast Res. 4 (2003) 221–231.

[148] J. Maury, M.A. Asadollahi, K. Moller, A. Clark, and J. Nielsen, Microbial isoprenoid production: an example of green chemistry through metabolic engineering. Adv Biochem Eng/Biotechnol. 100 (2005) 19–51.

[149] S.-W. Kim and J.D. Keasling, Metabolic engineering of the nonmevalonate isopentenyl diphosphate synthesis pathway in *Escherichia coli* enhances lycopene production. Biotechnol Bioeng 72 (2001) 408–415.

[150] D.E. Stafford, K.S. Yanagimachi, and G. Stephanopoulos, Metabolic engineering of indene bioconversion in *Rhodococcus* sp. Adv Biochem Eng/Biotechnol. 73 (2001) 85–102.

[151] F.M. Szczebara, C. Chandelier, C. Villeret, A. Masurel, S. Bourot, C. Duport, S. Blanchard, A. Groisillier, E. Testet, P. Costaglioli et al., Total biosynthesis of hydrocortisone from a simple carbon source in yeast. Nat Biotechnol. 21 (2003) 143–149.

[152] I. Sa-Correia, A.M. Fialho, P. Videira, L.M. Moreira, A.R. Marques, and H. Albano, Extracellular polysaccharides review, Gellan gum biosynthesis in *Sphingomonas paucimobilis* ATCC 31461: genes, enzymes and exopolysaccharide production engineering. J Ind Microbiol Biotechnol. 29 (2002) 170–176.

[153] Y.-T. Yang, G.N. Bennett, and K.-Y. San, Genetic and metabolic engineering. Electronic J. Biotechnol. 1 (1998) 134–141.

[154] C. Khosla and J.E. Bailey, Heterologous expression of a bacterial haemoglobin improves the growth propoerties of recombinant *Escherichia coli.* Nature 331 (1988) 633–635.

[155] J.D. Windass, M.J. Worsey, E.M. Pioli, D. Pioli, P.T. Barth, K.T. Atherton, E.C. Dart, D. Byrom, K. Powell, and P.J. Senior, Improved conversion of methanol to single-cell protein by *Methylophilus methylotrophus*. Nature 287 (1980) 396–401.

[156] L.A. Zaslavakaia, J.C. Lippmeier, C. Shih, D. Ehrhardt, A.R. Grossman, K.E. Apt, Trophic conversion of an obligate photoautotrophic organism through metabolic engineering. *Science* 292 (2001) 2073–2075.

[157] M. McIntyre, C. Müller, J. Dynesen, and J. Nielsen, Metabolic engineering of the morphology of *Aspergillus*. Adv Biochem Eng/Biotechnol. 73 (2001) 103–128.

[158] H.-S. Kim and J.-Y. Lee, Molecular bioremediation: metabolic engineering for biodegradation of recalcitrant pollutants. In S.Y. Lee and E.T. Papoutsakis, eds., Metabolic Engineering. Marcel Dekker, New York, 1999, pp.391-404.

[159] K.R. Patil, I. Rocha, J. Förster, and J. Nielsen, Evolutionary programming as a platform for *in silico* metabolic engineering. BMC Bioinformatics 6 (2005) 308.

[160] K. Sanford, P. Soucaille, G. Whitted and G. Chotani, Genomics to fluxomics and physiomics – pathway engineering. Curr Opin Microbiol. 5 (2002) 318–322.

[161] R.T. Gill, Enabling inverse metabolic engineering through genomics, Curr Opin Biotechnol. 14 (2003) 484–490.

[162] M.J. van der Werf, Towards replacing closed with open target selection strategies. Trends Biotechnol. 23 (2005) 11–16.

[163] Torres and Voit, Pathway Analysis and Optimization in Metabolic Engineering, Cambridge University press, London, 2002.

[164] K.R. Patil, M. Akesson, and J. Nielsen, Use of genome-scale microbial models for metabolic engineering. Curr Opin Biotechnol. 15 (2004) 64–69.

[165] E.T. Papoutsakis and C.L. Meyer, Equations for propionic acid bacteria and production of assorted oxychemicals from various sugars. Biotechnol Bioeng. 27 (1985) 67–80.

[166] J.J. Vallino and G. Stephanopoulos, Metabolic flux distributions in *Corynebacterium glutamicum* during growth and lysine overproduction. Biotechnol Bioeng. 41 (1993) 633–646.

[167] S. Takac, G. Cahk, F. Mavituna and G. Dervakos, Metabolic flux distribution for the optimized production of L-glutamate. Enz Microb Technol. 23 (1998) 286–300.

[168] H.S. Jorgensen, J. Nielsen and J. Villadsen, Metabolic flux distribution in *Penicillium chrysogenum* during fed-batch cultivations. Biotechnol Bioeng. 46 (1995) 117–131.

[169] H. Shimizu, N. Shimizu and S. Shioya, Role of glucose and acetate as carbon sources in L-histidine production revealed by metabolic flux distribution analysis. Biotechnol Bioprocess Eng. 7 (2002) 171–177.

[170] T. Shiraia, A. Nakatoa, N. Izutania, K. Nagahisab, S. Shioyaa, E. Kimurac, Y. Kawarabayasid, A. Yamagishif, T. Gojoborig, H. Shimizu, Comparative study of flux redistribution of metabolic pathway in glutamate production by two coryneform bacteria. Metabol Eng 7 (2005) 59–69.

[171] W. Wiechert, ^{13}C flux analysis. Metab Eng. 3 (2001) 195–206.

[172] M. Dauner and U. Sauer, GC-MS analysis of amino acids rapidly provides rich information for isotopomer balancing. Biotechnol Prog. 16 (2000) 642–649.

[173] C. Rollin, V. Morgant, A. Guyonvarch and J.L. Guerquin-Kern, ^{13}C NMR studies of *Corynebacterium melassecola* metabolic pathways. Eur J Biotechnol. 227 (1995) 488–493.

[174] C. Yang, Q. Hua and K. Shimizu, Quantitative analysis of intracellular metabolic fluxes using GC-MS and two-dimensional NMR spectroscopy. J Biosci Bioeng. 93 (2002) 78–87.

[175] T. Simpson, H. Shimizu and G. Stephanopoulos, Experimental determination of group flux control coefficients in metabolic network. Biotechnol Bioeng. 58 (1998) 149–153.

[176] H. Shimizu, H. Tanaka, A. Nakato, K. Nagahisa and S. Shioya, Metabolic control analysis in glutamate synthetic pathway, In M.R. Marten, T.H. Park and T. Nagamune (ed.) Biological

systems analysis. ACS Symposium series, vol. 830. Oxford University Press, Washington, D.C. (2002) pp. 39–52.
[177] I. Schaaff, J. Heinisch and F.K. Zimmermann, Overproduction of glycolytic enzymes in yeast. Yeast 5 (1989) 285–290.
[178] E.J. Smid, D. Molenaar, J. Hugenholtz, W.M. de Vos and B. Teusink, Functional ingredient production: application of global metabolic models. Curr Opin Biotechnol. 16 (2005) 190–197.
[179] J.L. Reed, T.D. Vo, C.H. Schilling and B.O. Palsson, An expanded genome-scale model of *Escherichia coli* K-12 (iJR904 GSM/GPR). Genome Biol. 4 (2003) R54.
[180] I. Famili, J. Forster, J. Nielsen and B.O. Palsson, *Saccharomyces cerevisiae* phenotype can be predicted by using constraint-based analysis of a genome-scale reconstructed metabolic network. Proc Natl Acad Sci USA 100 (2003)13134–13139.
[181] J.S. Edwards, R.U. Ibarra and B.O. Palsson, *In silico* predictions of *Escherichia coli* metabolic capabilities are consistent with experimental data. Nat Biotechnol. 19 (2001) 125–130.
[182] J. Forster, I. Famili, P. Fu, B.O. Palsson and J. Nielsen, Genome-scale reconstruction of the *Saccharomyces cerevisiae* metabolic network. Genome Res. 13 (2003) 244–253.
[183] K.J. Kauffman, P. Prakash and J.S. Edwards, Advances in flux balance analysis. Curr Opin Biotechnol. 14 (2003) 491–496.
[184] R.U. Ibarra, J.S. Edwards and B.O. Palsson, *Escherichia coli* K-12 undergoes adaptive evolution to achieve *in silico* predicted optimal growth. Nature 420 (2002) 186–189.
[185] J. Stelling, S. Klamt, K. Bettenbrock, S. Schuster and E.D. Gilles, Metabolic network structure determines key aspects of functionality and regulation. Nature 420 (2002) 190–193.
[186] S. Klamt and J. Stelling, Two approaches for metabolic pathway analysis? Trends Biotechnol. 21 (2003) 64–69.
[187] J.A. Papin, J. Stelling, N.D. Price, S. Klamt, S. Schuster and B.O. Palsson, Comparison of network-based pathway analysis methods. Trends Biotechnol. 22 (2004) 400–405.
[188] S. Klamt and J. Stelling, Combinatorial complexity of pathway analysis in metabolic networks. Mol Biol Rep. 29 (2002) 233–236.
[189] M.W. Covert and B.O. Palsson, Constraints-based models: regulation of gene expression reduces the steady-state solution space. J Theor Biol. 221 (2003) 309–325.
[190] M.W. Covert and B.O. Palsson, Transcriptional regulation in constraints-based metabolic models of *Escherichia coli*. J Biol Chem. 277 (2002) 28058–28064.
[191] D. Segre, D. Vitkup, G.M. Church, Analysis of optimality in natural and perturbed metabolic networks. Proc Natl Acad Sci USA 99 (2002) 15112–15117.
[192] D.A. Beard, S.D. Liang and H. Qian, Energy balance for analysis of complex metabolic networks. Biophys J. 83 (2002) 79–86.
[193] B.O. Palsson, Systems biology, properties of reconstructed networks. Cambridge University Press, New York, NY. 2006.
[194] H.V. Westerhoff and B.O. Palssom, The evolution of molecular biology into systems biology. Nat Biotechnol. 22 (2004) 1249–1252.
[195] D. Visser and J.J. Heijnen, Dynamic simulation and metabolic redesign of a branched pathway using linlog kinetics. Metab Eng. 5 (2003) 164–176.
[196] R.T. Gill, S. Wildt, T. Yang, S. Ziesman and G. Stephanopoulos, Genome wide screening for trait conferring genes using DNA microarrays. Proc Natl Acad Sci USA 99 (2002) 7033–7038.
[197] G. Glaever, A.M. Chu, L. Ni, C. Connelly, L. Riles, S. Veronneau, S. Dow, A. Lucau-Danila, K. Anderson, B. Andre, et al., Functional profiling of the *Saccharomyces cerevisiae* genome. Nature 418 (2002) 387–391.
[198] V. Badarinarayana, P. Estep, J. Shendure, J. Edwards, S. Tavazoie, F. Lam and G. Church, Selection analyses of insertional mutants using subgenic resolution arrays. Nat Biotechnol. 19 (2001) 1060–1064.

[199] S.B. Tummala, S.G. Junne, C.J. Paredes and E.T. Papoutsakis, Transcriptional analysis of product concentration driven changes in cellular programs of recombinant *Clostridium acetobutylicum* strains. Biotechnol Bioeng. 84 (2003) 842–854.

[200] M. Karos, C. Vilarino, C. Bollschweiler and J.L. Revuelta, A genome-wide transcription analysis of a fungal riboflavin overproducer. J Biotechnol. 113 (2004) 69–76.

[201] S.D. Patterson and R.H. Aebersold, Proteomics: the first decade and beyond. Nat Genet. 33 (2003) 311–323.

[202] W.P. Blackstock and M.P. Weir, Proteomics: quantitative and physical mapping of cellular proteins. Trends Biotechnol. 17 (1999) 121–127.

[203] E. Fischer and U. Sauer, Metabolic flux profiling of *Escherichia coli* mutants in central carbon metabolism by GC-MS. Eur J Biochem. 270 (2003) 880–891.

[204] C. Wittmann and E. Heinzle, Application of MALDI-TOF MS to lysine producing *Corynebacterium glutamicum*. Eur J Biochem. 268 (2001) 2441–2455.

[205] U. Sauer, High-throughput phenomics: experimental methods for mapping fluxomes. Curr Opin Biotechnol. 15 (2004) 58–63.

[206] S.Y. Lee, D.-Y. Lee, and T.Y. Kim, Systems biotechnology for strain improvement, Trends Biotechnol. 23 (2005) 349–358.

[207] C. Bro and J. Nielsen, Impact of 'ome' analyses on inverse metabolic engineering. Metab Eng. 6 (2004) 204–211.

[208] C. Rückert, A. Pühler and J. Kalinowski, Genome-wide analysis of the L-methionine biosynthetic pathway in *Corynebacterium glutamicum* by targeted gene deletion and homologous complementation. J Biotechnol. 104 (2003) 213–228.

[209] J. Ohnishi, M. Hayashi, S. Mitsuhashi and M. Ikeda, Efficient 40 °C fermentation of L-lysine by a new *Corynebacterium glutamicum* mutant developed by genome breeding. Appl Microbiol Biotechnol. 62 (2003) 69–75.

[210] J.H. Lee, D.-E. Lee, B.-U. Lee and H.-S. Kim, Global analyses of transcriptomes and proteomes of a parent strain and an L-threonine-overproducing mutant strain. J Bacteriol. 185 (2003) 5442–5451.

[211] J.E. Bailey, A. Sburlati, V. Hatzimanikatis, K. Lee, W.A. Renner, and P.S. Tsai, Inverse metabolic engineering: a strategy for directed genetic engineering of useful phenotypes. Biotechnol Bioeng. 79 (2002) 568–579.

[212] .E. Bailey, Lessons frm metabolic engineering for functional genomics and drug discovery. Nat Biotechnol. 17 (1999) 616–618.

[213] T. Butler, J.R. Bernstein and J.C. Liao, A perspective of metabolic engineering strategies: moving up the systems hierarchy. Biotechnol Bioeng. 84 (2003) 815–821.

[214] M. Tatarko and T. Romeo, Disruption of a global regulatory gene to enhance central carbon flux into phenylalanine biosynthesis in *Escherichia coli*. Curr Microbiol. 43 (2001) 26–32.

[215] W.R. Farmer and J.C. Liao, Improving lycopene production in *Escherichia coli* by engineering metabolic control. Nat Biotechnol. 18 (2000) 533–537.

[216] A.Zaigler, S.C. Schuster and J. Soppa, Construction and usage of a onefold-coverage shotgun DNA microarray to characterize the metabolism of the archaeon *Haloferax volcanii*. Mol Microbiol. 48 (2003) 1089–1105.

[217] J. Aarnikunnas, N. von Weymarn, K. Rönnholm, M. Leisola and A. Palva, Metabolic engineerng of *Lactobacillus* fermentum for production of mannitol and pure L-lactic acid or pyruvate. Biotechnol Bioeng. 82 (2003) 653–663.

[218] G.T. John, I. Klimant, C. Wittmann and E. Heinzle, Integrated optical sensing of dissolved oxygen in microtiter plates: a novel tool for microbial cultivation. Biotechnol Bioeng. 81 (2003) 829–836.

[219] B.J. Koebmann, J. Tornoe, B. Johansson and P.R. Jensen, Experimental modulation of gene expression. In B.N. Kholodenko and H.V. Westerhoff (eds.), Metabolic Engineering in the Post Geonmic Era, Horizone Bioscience, Wymondham, UK, 2004. pp. 155–179.

[220] E.T. Farinas, T. Bulter and F.H. Arnold, Directed enzyme evolution. Curr Opin Biotechnol. 12 (2001) 545–551.

[221] L.K. Lee and C.M. Roth, Antisense technology in molecular and cellular engineering. Curr Opin Biotechnol. 14 (2003) 505–511.

[222] H. Westers, R. Dorenbos, J.M. van Dijl, J. Kabel, T. Flanagan, K.M. Devine, F. Judge, S.J. Seror, A.C. Beekman, et. al. Genome engineering reveals large dispensable regions in *Bacillus subtilis*. Mol Biol Evol. 20 (2003) 2076–2090.

[223] U. Sauer, Evolutionary engineering of industrially important microbial phenotypes. Adv Biochem Eng/Biotechnol. 73 (2001) 1305–169.

[224] G. Stephanopoulos, Metabolic engineering by genome shuffling. Nat Biotechnol. 20 (2002) 666–668.

[225] Y. Zhang, K. Perry, V. Vinci, K. Powell, W. Stemmer, S. delCardayre, Genome shuffling leads to rapid phenotypic improvement in bacteria. Nature 415 (2002) 644–646.

[226] M. Ikeda and S. Nakagawa, The *Corynebacterium glutamicum* genome: features and impacts on biotechnological processes. Appl Microbiol Biotechnol. 62 (2003) 99–109.

[227] R. Patnaik, S. Louie, V. Gavrilovic, K. Perry, W. Stemmer, C. Ryan and S. delCardayre, Genome shuffling of *Lactobacillus* for improved acid tolerance. Nat Biotechnol 20 (2002) 707–712.

[228] C. Schmidt-Dannert, D. Umeno and F. Arnold, Molecular breeding of carotenoid biosynthetic pathways. Nat Biotechnol 18 (2000) 750–753.

[229] L. Rohlin, M.-K. Oh and J.C. Liao, Microbial pathway engineering for industrial processes: evolution, combinatorial biosynthesis and rational design. Curr Opin Microbiol. 4 (2001) 330–335.

[230] J. Bonomo and R.T. Gill, Antibiotic resistance as a model for strain engineering. Comp Chem Eng. 29 (2005) 509–517.

[231] Y. Zhu and S.T. Yang, Adaptation of *Clostridium tyrobutyricum* for enhanced tolerance to butyric acid in a fibrous-bed bioreactor. Biotechnol Prog. 19 (2003) 365–372.

[232] Z. Jin and S.T. Yang, Extractive fermentation for enhanced propionic acid production from lactose by *Propionibacterium acidipropionici.* Biotechnol Prog. 14 (1998) 457–465.

[233] S.T. Yang, I.C. Tang, and M.R. Okos, Defined bacterial culture developments for methane generation from lactose. Biotechnol Bioeng. 32 (1988) 8–37.

[234] Y. Huang and S.T. Yang, Acetate production from whey lactose using co-immobilized cells of homolactic and homoacetic bacteria in a fibrous-bed bioreactor. Biotechnol Bioeng. 60 (1998) 498–507.

Bioprocessing for Value-Added Products from Renewable Resources
Shang-Tian Yang (Editor)

Chapter 5. Amylase and Cellulase Structure and Function

Peter J. Reilly

Department of Chemical and Biological Engineering, Iowa State University, Ames, Iowa 50011, U.S.A.

1. INTRODUCTION

Amylases and cellulases hydrolyze α- and β-(1,4) glycosidic bonds, respectively, in oligo- and polysaccharides containing primarily D-glucosyl residues. In this chapter a somewhat wider range of enzymes that includes those hydrolyzing α- and β-(1,4) bonds in glucose-containing disaccharides, as well as those in two phosphorylase families, will be considered. These hydrolases are among the most important found in nature, as α-(1,4) bonds are found in starch and glycogen, the main food storage components in most plants and animals, and β-(1,4) bonds occur in cellulose, the chief structural material in plants. Enzymes mainly active on other bonds or on oligo- and polysaccharides that contain sugars instead of or in addition to glucose will not be discussed here.

1.1. Amylases

In industry, α-amylase (1,4-α-D-glucan glucanohydrolase, EC 3.2.1.1) and glucoamylase (1,4-α-D-glucan glucohydrolase, EC 3.2.1.3) are extensively used to hydrolyze starch solubilized by steam treatment first to maltooligosaccharides and then to glucose, and they are in tonnage terms along with proteases the most used of all industrially-employed enzymes. Their use increased greatly in the 1970s as high-fructose syrup was developed as a substitute sweetener for sucrose.

In the first of two enzymatic steps that produce high-glucose syrup as an intermediate to high-fructose syrup and other products, bacterial α-amylase is employed at pH 6–7 and about 105°C to hydrolyze starch purified by wet milling to roughly DE (dextrose equivalent) 10, DE being the percent of reducing end-groups in the aqueous sugar mixture relative to those in pure glucose of the same concentration. This indicates that the average maltooligosaccharide produced has a DP (degree of polymerization) of 10, although the mixture has a very wide DP range. DE 30 or more can be attained if the reaction proceeds to completion, but DE 10 is chosen to minimize the probability of pseudo-crystallization at lower DEs and the extent of alkaline-catalyzed isomerization of the reducing-end glucosyl residue at higher DEs.

In the second step, at pH 4.3–4.5 and roughly 60°C, fungal glucoamylase and a small amount of pullulanase (pullulan 6-glucanohydrolase, EC 3.2.1.41), the latter used to rapidly hydrolyze the α-1,6 bonds from the original starch feed, convert the maltooligosaccharide mixture to approximately 96% glucose (dry basis). The remainder is composed of byproducts from α-amylase hydrolysis plus mainly isomaltose [α-glucopyranosyl-(1,6)-glucose] and isomaltotriose from the glucoamylase-catalyzed condensation of glucose.

The glucose made by these two steps is further converted to a near-equimolar mixture of glucose and fructose by immobilized glucose isomerase (actually a xylose isomerase). Fructose is chromatographically enriched and then backblended with glucose to yield an approximately 55% (dry weight) fructose–42% glucose aqueous solution, the remaining 3% being composed of the substances mentioned above. This is indistinguishable in sweetness, taste, and mouth feel from a sucrose solution of equal concentration. Other food uses of the products of α-amylase- and glucoamylase-catalyzed starch hydrolysis are various malto-oligosaccharides and crystalline and aqueous glucose preparations. Furthermore, fungal α-amylase, which hydrolyzes starch and long maltooligosaccharides to primarily maltose and maltotriose, and β-amylase (1,4-α-D-glucan maltohydrolase, EC 3.2.1.2), which mainly produces maltose from the same starting materials, find substantial use.

Enzymatic starch conversion to low-molecular-weight sugars is a mature industry, at least in the United States, where it has already captured over half of the natural sweetener market from sucrose. Sucrose retains some uses because it can be crystallized, whereas high-fructose syrup cannot. Furthermore, sucrose production is protected in many countries by either import quotas or tariffs. Interestingly, this has led, especially in the United States, to increased starch hydrolysis, as products from the latter can nearly always be marketed at lower cost than domestic or imported sucrose.

Starch has been fermented by yeast since prehistoric times to make beverages. In the last 30 years, public subsidies or mandates in some countries have spurred the development of ethanol for use as a transportation fuel. Unlike the high-glucose and high-fructose mixtures described above, which are produced from highly pure starch made by wet milling, starch for ethanol is usually made in lower purity but lower cost by dry milling. Solubilized starch is converted by α-amylase and glucoamylase to glucose, maltose, and maltotriose before fermentation. Sucrose is the starting material in those countries, such as Brazil, where it can be produced cheaply. Ethanol for fuel is separated from water and minor products to near-purity by several distillation steps.

1.2. Cellulases

Enzymatic hydrolysis of β-1,4-linked cellulose is a completely different story than hydrolysis of α-1,4-linked starch, as the former is a structural material, designed to withstand degradation, while the latter is designed for food storage. Enzymatic access to cellulose is difficult, as in plant tissues it is nearly always encased in hemicelluloses and lignin, which themselves require many enzymes to be degraded. Furthermore, cellulose is often crystalline, unlike starch. Finally, β-1,4 glycosidic bonds are intrinsically more resistant to hydrolysis by both acid and enzymes than are α-1,4 glycosidic bonds. These factors have prevented the large-scale commercial use of cellulases [endo-1,4-β-glucanase (1,4-β-D-glucan glucanohydrolase, EC 3.2.1.1, EG); cellobiohydrolase (1,4-β-D-glucan cellobiohydrolase, EC 3.2.1.91, CBH); and β-glucosidase (EC 3.2.1.21)] to produce cellooligosaccharides, cellobiose, and glucose. A solution to this problem would unlock the very large supply of cellulosic biomass available for conversion to sugars, far greater than the available starch supply, and intensive research has been devoted to this aim for many years.

2. AMYLASE AND CELLULASE CLASSIFICATION AND MECHANISMS

2.1. Protein structure

Proteins are classified into four structures of successively larger scope:

- Primary – their amino acid sequences;
- Secondary – local α-helices, β-strands, loops, and other small structures;
- Tertiary – the overall protein folding pattern;
- Quaternary – in some cases, the association of two or more tertiary structures, either the same or different, into a larger structure.

Primary structures determine the larger structures. After proteins are produced, they fold into secondary, tertiary, and sometimes quaternary structures by themselves or with the help of protein chaperonins. The proper fold allows specific amino acid residues to be positioned in the active site so that some bind the reactant (the substrate in biochemical terms), generally by hydrogen binding and hydrophobic interactions, and so that others are catalytically active.

2.2. Glycoside hydrolase mechanisms

The catalytic domains of nearly all glycoside hydrolases have one amino acid residue that acts as a catalytic nucleophile, being mainly negatively charged at the pH of maximal enzyme activity, and a second residue that acts as a catalytic proton donor, being mainly undissociated at that pH. These may be two aspartate residues, two glutamate residues, or one of each. In the case of one aspartate and one glutamate catalytic residue, the former is more likely to be the catalytic nucleophile and the latter the catalytic proton donor. With phosphorylases, a phosphate is the catalytic nucleophile and either aspartate or glutamate is the catalytic proton donor.

Glycoside hydrolases employ mainly two catalytic mechanisms. The first, called the inverting mechanism because the newly-produced reducing-end glycosyl residue has the opposite configuration than before its glycosidic bond was cleaved, features a single-displacement reaction (Fig. 1) [1, 2]. Here the two torsional angles surrounding the glycosidic-bond oxygen atom (O4') are sometimes rotated during substrate binding, the proton donor donates a proton to O4', the ring of the residue to the nonreducing side of the bond (the glycon) is flattened and its C1–O5 bond gains partial double-bond character, the hydrolytic water molecule is activated by the catalytic nucleophile and donates a hydroxyl group to C1, the C1–O4' bond breaks, and both the leaving group O4' and the catalytic nucleophile gain a proton. The distance between proton donor and nucleophile is characteristically 9–10 Å.

In the retaining mechanism, the new reducing-end glycosyl residue retains its original configuration through a double-displacement reaction (Fig. 2) [1, 2]. Here the nucleophile forms a covalent bond with C1. A proton is donated to the leaving group, and water donates a hydroxyl group to C1, breaking its covalent bond with the nucleophile. The distance between proton donor and nucleophile is 5.5–7 Å.

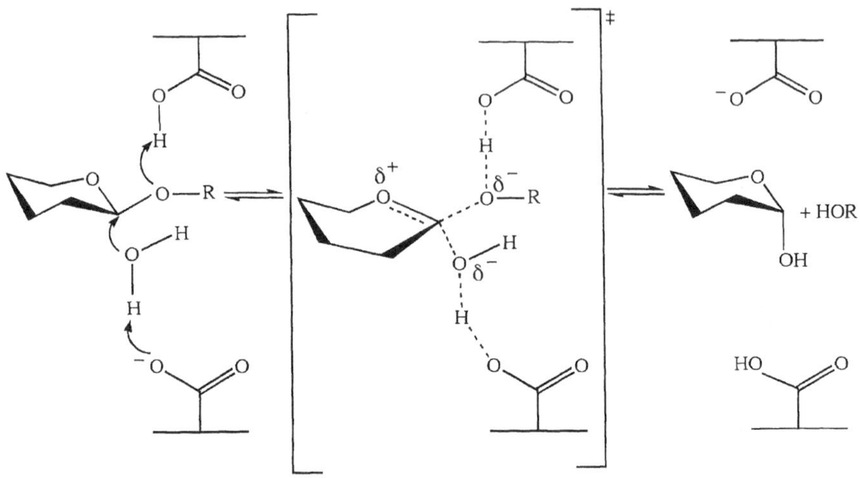

Fig. 1. Mechanism of glycoside hydrolase-catalyzed inverting reaction.

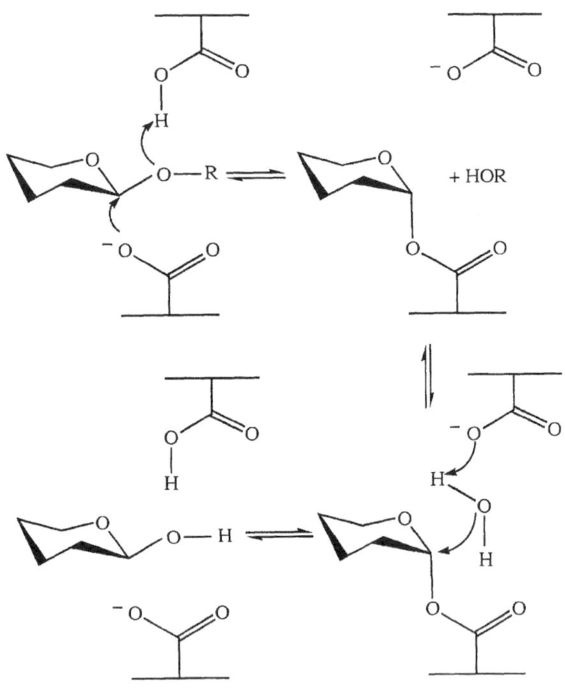

Fig. 2. Mechanism of glycoside hydrolase-catalyzed retaining reaction.

2.3. Glycoside hydrolase classification

The CAZy (Carbohydrate-Active enZymes) database [3] classifies glycoside hydrolase/transglycosidase, glycosyltransferase, polysaccharide lyase, carbohydrate esterase, and carbohydrate-binding module structures. Amylases, cellulases, and related enzymes are part of the glycoside hydrolase/transglycosidase classification, which encompasses over one hundred families whose primary structures bear either no or very little similarity to each other. Individual families are defined by the amino acid sequences of their protein members, which are sufficiently similar in any family that it is clear that all its members are descended from a common ancestor. Different proteins within the family may be produced by organisms from different genera or species.

Table 1, derived from the CAZy database, lists the ten families whose members hydrolyze α-1,4 glycosidic bonds. A further fourteen families contain enzymes that hydrolyze β-1,4 bonds. The mechanisms, catalytic nucleophiles and proton donors, tertiary structure folds, and clans of some families are much better known than those of others.

A very common fold in cellulases and amylases is a $(\beta,\alpha)_y$ barrel (Figs. 3–7), where y β-strands form an inner shell and an additional y α-helices form an outer shell. A somewhat common fold is an $(\alpha,\alpha)_x$ barrel (Figs. 8–10), where x α-helices form an inner shell about the active site, and where a further x α-helices form an outer shell. Although other β-strands and α-helices may be present in these folds, they are generally sufficiently peripheral that they are not included in the fold designation. Two cellulase families have β-jelly roll structures (Figs. 11, 12). All members of the same family have the same fold designation. In many cases different families having the same fold designation can be grouped into a clan, all of whose members have the same distant ancestor [3].

Glycoside hydrolases with $(\alpha,\alpha)_6$ barrels all have inverting mechanisms; however, some cleave α-1,4 while others cleave β-1,4 bonds. Hydrolases with $(\beta,\alpha)_y$ barrels can have either inverting or retaining mechanisms and catalyze either α- or β-1,4 bond hydrolysis. Families having the same folds do not necessarily belong to the same clan.

It is immediately clear from Table 1 that glycoside hydrolases with the same names and EC numbers can be found in different families and can have different tertiary structures, as well as different mechanisms and catalytic nucleophiles and proton donors. For instance, EGs are found in eleven different families and CBHs are found in four, immediately implying that convergent evolution has occurred to produce different structures that act on the same substrates to yield the same products. This observation further implies that the EC classification system, which is based on types of reactions catalyzed and types of bonds broken, is insufficiently detailed to recognize that these enzymes are intrinsically different. This has come about because the system was developed when protein tertiary structures were rare, and when there were not enough structures of enzymes of any one EC number to realize their differences.

A further conclusion from Table 1 is that two or more different enzyme types can be found in the same family. EGs and CBHs are found together in four families. In glycoside hydrolase family 6 (GH6), one of the two having the largest number of these two enzymes, the cellulose chain proceeds, non-reducing end first, through a short tunnel in CBHs [14]. A cellobiose molecule with inverted configuration is formed when the chain passes over the catalytic nucleophile and proton donor, both aspartate residues. GH6 EGs do not have the tunnel, presumably by divergent evolution from CBHs, and therefore cellulose chains can approach the active site nonprocessively, allowing a wide range of cellooligosaccharides to be formed [15]. In GH7 CBHs, cellulose chains proceed, reducing ends first, through a long tunnel, and are cleaved to form cellobiose molecules with retained configuration by the catalytic

Table 1
Families of enzymes hydrolyzing α- and β-1,4 glycosidic bonds

Family	Principal enzymes	Mechanism	Nucleophile	Proton donor	Fold	Clan
1	β-Glucosidase, many others	R	Glu	Glu	$(\beta,\alpha)_8$	GH-A
3	β-Glucosidase, several others	R	Asp	Glu	$(\beta,\alpha)_8$	—
4	α-Glucosidase, several others	*	—	—	$(\beta,\alpha)_8$	—
5	EG, many others	R	Glu	Glu	$(\beta,\alpha)_8$	GH-A
6	CBH, EG	I	Asp	Asp	Distorted β,α barrel	—
7	CBH, EG	R	Glu	Glu	β-Jelly roll	GH-B
8	EG, chitosanase, xylanase	I	Asp	Glu	$(\alpha,\alpha)_6$	GH-M
9	EG, CBH (few)	I	Asp	Glu	$(\alpha,\alpha)_6$	—
12	EG	R	Glu	Glu	β-Jelly roll	GH-C
13	α-Amylase, CGTase, pullulanase, many others	R	Asp	Glu	$(\beta,\alpha)_8$	GH-H
14	β-Amylase	I	Glu	Glu	$(\beta,\alpha)_8$	—
15	Glucoamylase, glucodextranase	I	Glu	Glu	$(\alpha,\alpha)_6$	GH-L
31	α-Glucosidase, some others	R	Asp	—	—	—
44	EG	I	—	—	—	—
45	EG	I	Asp	Asp	β-Barrel	—
48	CBH, EG	I	—	Glu	$(\alpha,\alpha)_6$	GH-M
57	α-Amylase, amylopullulanase	R	Glu	—	$(\beta,\alpha)_7$	—
61	EG	—	—	—	—	—
63	α-Glucosidase	—	—	—	—	GH-G
65	Maltose phosphorylase, others	I	P	Glu	$(\alpha,\alpha)_6$	GH-L
74	EG, some others	I	Asp	Asp	7-fold β-propeller	—
77	4-α-Glucanotransferase	R	Asp	Glu	$(\beta,\alpha)_8$	GH-H
94	Cellobiose phosphorylase, others	I	P	Asp	$(\alpha,\alpha)_6$	—
97	α-Glucosidase	—	—	—	—	—

Abbreviations: Asp: aspartate; CBH: cellobiohydrolase; CGTase: cyclomaltodextrin glucanotransferase; EG: endo-1,4-β-glucanase; GH: glycoside hydrolase; Glu: glutamate; I: inverting; P: phosphate ion; R: retaining. * Involves NAD^+.

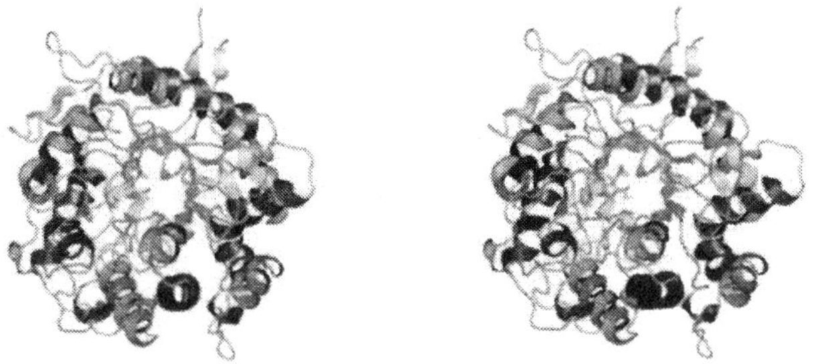

Fig. 3. Three-dimensional view of glycoside hydrolase family 1 (GH1) β-glucosidase from *Paenibacillus polymyxa* (Protein Data Base accession number 1BGG) [4].

Fig. 4. Three-dimensional view of GH3 exo-1,3-1,4-glucanase from *Hordeum vulgare* (1IEX) [5].

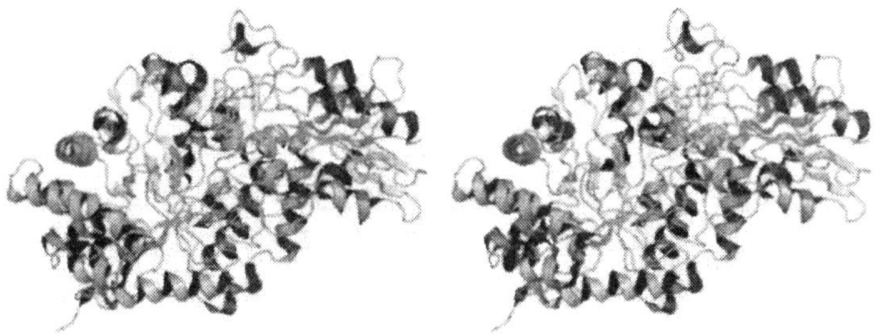

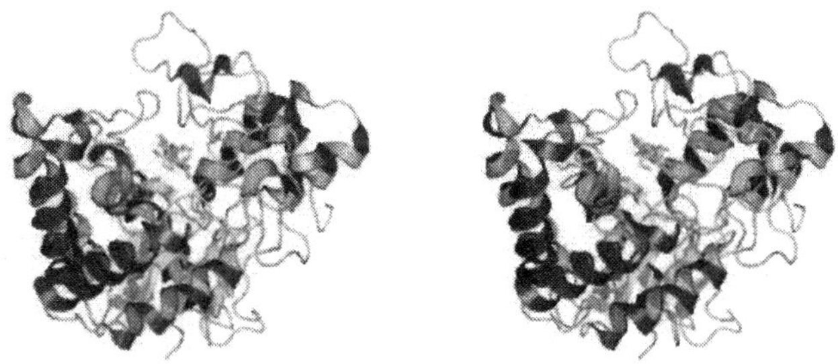

Fig. 5. Three-dimensional view of GH6 CBH from *Humicola insolens* (1OCN) [6]. The view is from the side to show the active-site tunnel.

Fig. 6. Three-dimensional view of GH13 pancreatic α-amylase from *Homo sapiens* (1B2Y) [7].

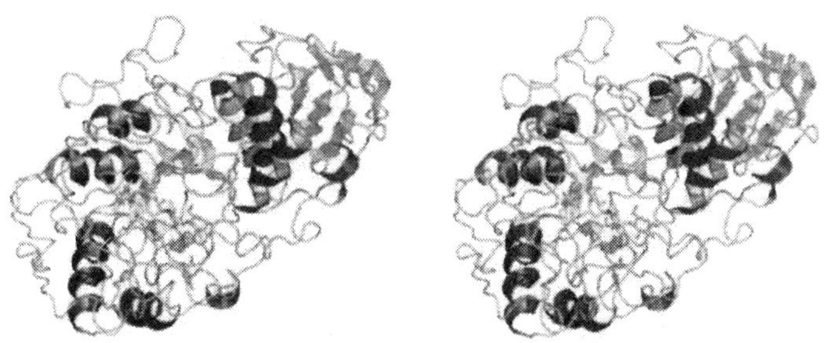

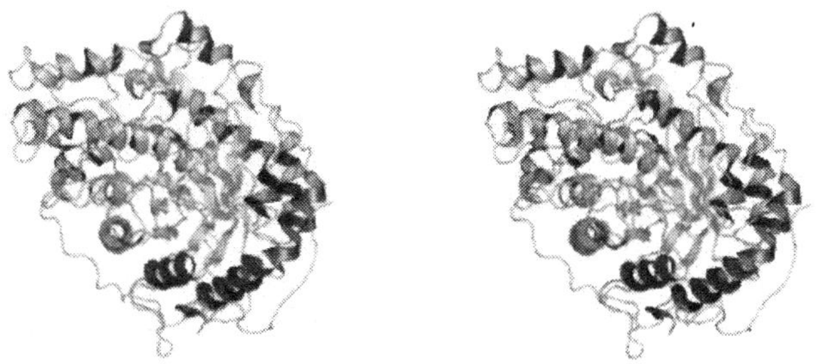

Fig. 7. Three-dimensional view of GH14 β-amylase from *Hordeum vulgare* (1B1Y) [8].

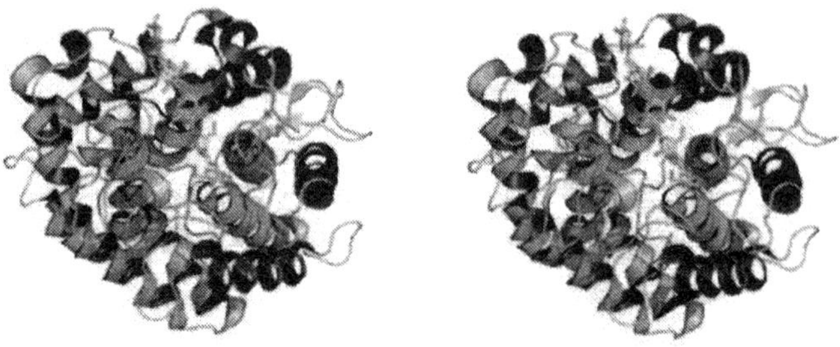

Fig. 8. Three-dimensional view of GH8 EG from *Clostridium thermocellum* (1KWF) [9].

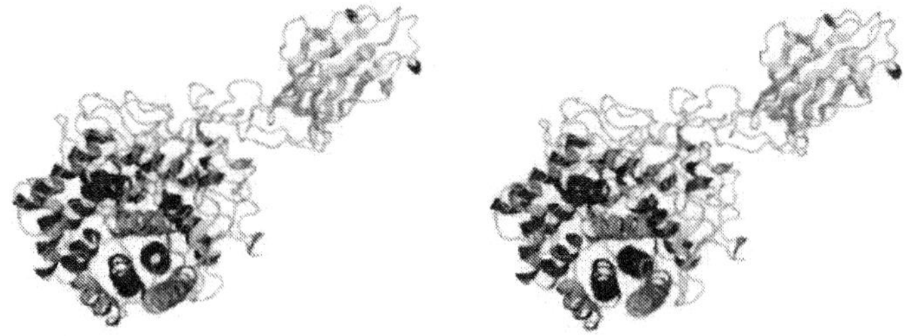

Fig. 9. Three-dimensional view of GH9 endo/exo-cellulase from *Thermobifida fusca* with a CBM3 to the upper right (4TF4) [10].

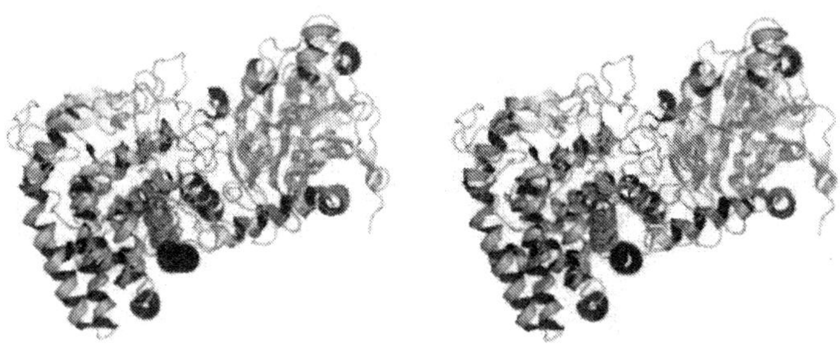

Fig. 10. Three-dimensional view of GH15 glucoamylase from *Thermoanaerobacterium thermosaccharolyticum* with a domain of unknown function to the right (1LF9) [11].

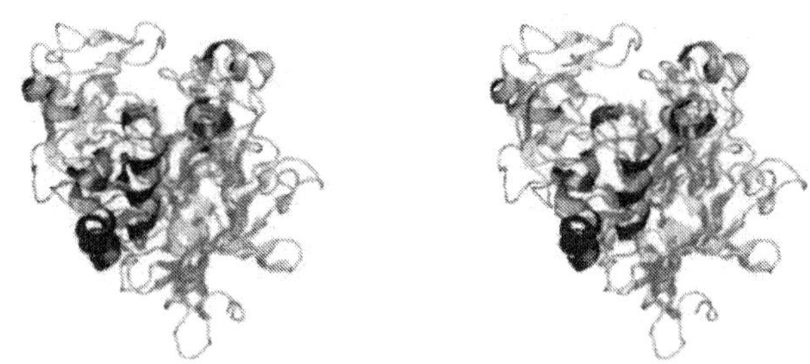

Fig. 11. Three-dimensional view of GH7 CBH from *Hypocrea jecorina* (1Q2E) [12]. The view is from the side to show the active-site tunnel.

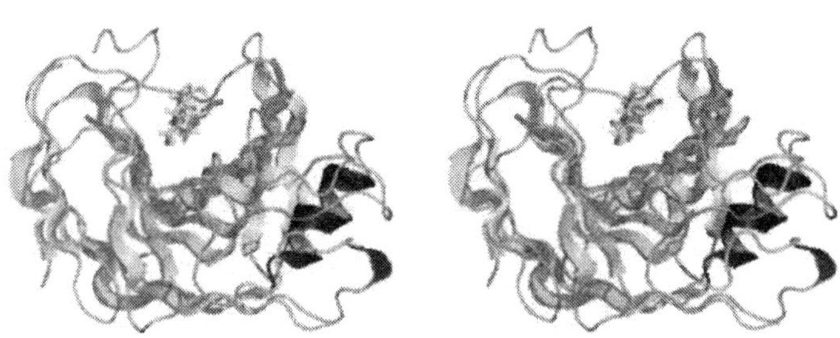

Fig. 12. Three-dimensional view of GH12 EG from *Streptomyces lividans* (2NLR) [13].

nucleophile and proton donor, both glutamate residues [16, 17]. As in GH6, GH7 EGs have no tunnels and therefore produce a range of cellooligosaccharides [18].

In some families, very small changes in the active-site structure can yield substantial differences in substrate specificity. Such occurs with GH1 hydrolases, all of which attack β-glycosidic bonds and nearly all of which are more active on small molecules than on larger ones. However, enzymes more or less specific to many different glycons are part of the family, and the member enzymes are not designated by the configuration of the glycosidic bond to the leaving group or in most cases by leaving group identity. To illustrate, GH1 is composed of sixteen different enzymes, fifteen of which are sufficiently well established to have EC numbers [3]. Glycons for which individual family members are mainly specific include the β-forms of fucose, galactose, glucose, glucuronate, mannose, 6-phosphogalactose, and 6-phosphoglucose, although cross-specificity for other glycons is common. Enzymes such as phlorizin hydrolase, strictosidine β-glucosidase, prunasin β-glucosidase, oligoxyloglucan β-glycosidase, and raucaffricine β-glucosidase indicate their substantial specificity to individual leaving groups, but in general this is not the case with GH1 members.

Yet a further conclusion from Table 1 is that members of different glycoside hydrolase families can be members of the same clan, having the same fold and implying distant common ancestry, and yet act very differently. For instance, it is now clear that GH15 glucoamylases are distantly related to the GH65 kojibiose, maltose, and trehalose phosphorylases and trehalases and the GH94 cellobiose, cellodextrin, and chitobiose phosphorylases and cyclic β-1,2-glucan synthases [11, 19, 20], each having $(\alpha,\alpha)_6$ folds and the first two at least being members of Clan GH-L, and with all three having inverting mechanisms. Yet GH15 enzymes have glutamate nucleophiles and proton donors, while GH65 enzymes have phosphate nucleophiles and glutamate proton donors and GH94 enzymes have phosphate nucleophiles and aspartate proton donors.

2.4. Carbohydrate-binding modules

Many glycoside hydrolases, and specifically amylases and cellulases, have not only a catalytic domain by which they are identified, but also other domains, some of which are second catalytic domains. However, it is more likely that these additional domains bind solid polysaccharides. CAZy has classified these carbohydrate-binding domains (CBMs) in much the same fashion as the glycoside hydrolase classification (Table 2). At present, almost forty CBM families differentiated by primary structure exist. Of these, four CBM families bind starch and a further eleven bind cellulose. Cellulose-binding CBMs often can bind other β-linked nonglucan polysaccharides such as xylan and chitin. Furthermore, a substantial number of CBMs do not bind glucans at all.

In general, CBMs are not associated with specific catalytic domains, such as GAs or EGs. Furthermore, catalytic domains in single glycoside hydrolase families may be associated with several different CBMs. For instance, GH6 members are found in eight subfamilies [21]. One composed mainly of actinobacterial EGs has a few forms with CBM2s and CBM4s, a second of chytridiomycotal fungal EGs and CBHs has CBM1s and CBM10s, a third of actino- and proteobacterial CBHs has mainly CBM2s, a fourth of chytridiomycotal fungi is associated only with CBM10s, two others, one of ascomycotal fungal CBHs and the other of basidiomycotal fungal CBHs, have CBM1s, and two others have virtually no CBMs [20]. Even within single glycoside hydrolase subfamilies, CBMs may be found either to the N-terminal or C-terminal sides of the catalytic domains.

Despite the many different families of CBMs, their different specificities, and the different catalytic domains to which they are attached, all whose folds are known are composed of β-sheets or β-sandwiches (Table 2).

Table 2
Carbohydrate-binding modules that bind starch and cellulose

Family	# of Res.	Origin	Ligands	Fold	Parent catalytic domains
1	~40	F, Pl (1), oomycete (1), virus (1)	C, chitin (1 case)	β-Sheet	CBH, EG, EX, others
2	~100	B (many), An, Pr, Ar	C, xylan, chitin	β-Sheet	EG, chitinase, CBH, EX, others
3	~150	B	C, chitin (1)	β-Sandwich	EG, CBH, others
4	~150	B	Xylan, non-cell. glucans, C	β-Sandwich	EG, CBH, EX, others
5	~60	B, Ar (1)	Chitin, C	β-Sheet	Chitinase, EG
6	~120	B, F (1), An (1)	Xylan, C	β-Sheet	EX, agarase, EG, others
8	—	Pr	C	—	EG
10*	~50	B	Mannan, xylan, C	β-Sheet	EM, EG, EX
11	~190	B	C, non-cell. glucan	—	EG
17⁺	~200	B	C	β-Sandwich	EG
20	~110	B, F, An, Pl, Pr, Al	Starch	β-Sheet	α-Amylase, GA, β-amylase, CGTase, others
21	~100	F, An	Starch	—	GA, α-amylase, others
25/26	~85	B	Starch	—	α-Amylase, β-amylase, G_5- and G_6-forming amylases
30	~100	B	C	—	EG
34	~120	B, Ar	Starch	β-Sheet	Amylases, CGTases, and pullulanases

*Also known as dockerin D2.

⁺ Combined with Family 28.

Abbreviations: Al: algae; An: animals; Ar: archaea; B: bacteria; C: cellulose; CBH: cellobiohydrolase; CGTase: cyclomaltodextrin glucanotransferase; EG: endo-1,4-β-glucanase; EM: endo-β-1,4-mannanase; EX: endo-1,4-β-xylanase; F: fungi; GA: glucoamylase; Pl: plants; Pr: protozoa.

3. CONCLUSIONS

Polysaccharide hydrolysis is an important part of Nature's effort to recycle components produced and used by living organisms. As has been seen here, even hydrolysis of starch and cellulose, the two most important but not by any means the only polysaccharides, is a complicated procedure. Two main mechanisms and many different enzymes have evolved to carry out the task, a clear case of divergent evolution. These enzymes often have two domains, with different catalytic domains in the same family often having different CBMs, demonstrating the ubiquity of gene shuffling and transfer. Different enzymes within single families can be active on different substrates, apparently caused by quite minor changes in their active site structures. Different families of catalytic domains and CBMs have very different tertiary structures but often the same oligo- and polysaccharide specificities and product patterns, a case of convergent evolution.

It is hoped that this description of the structures and functions of amylases and cellulases allows a more informed use of these enzymes in processes to use and convert renewable resources.

ACKNOWLEDGMENT

The author thanks Anthony Hill (Iowa State University) for his help in producing the three-dimensional enzyme views.

REFERENCES

[1] M.L. Sinnott, Chem. Rev., 90 (1991) 1170.
[2] J.D. McCarter and S.G. Withers, Curr. Opin. Struct. Biol., 4 (1994) 885.
[3] P.M. Coutinho and B. Henrissat. Carbohydrate-Active enZymes server at URL: http://afmb.cnrs-mrs.fr/CAZY/, 1999–2006.
[4] J. Sanz-Aparicio, J.A. Hermoso, M. Martinez-Ripoll, J.L. Lequerica, and J. Polaina, J. Mol. Biol., 275 (1998) 491.
[5] M. Hrmova, J.N. Varghese, R. Degori, B.J. Smith, H. Driguez, and G.B. Fincher, Structure, 9 (2001) 1005.
[6] A. Varrot, J. Macdonald, R.V. Stick, G. Pell, H.J. Gilbert, G.J. Davies, direct submission (2003).
[7] M. Qian, R. Haser, G. Buisson, E. Duee, and F. Payan, Biochemistry, 33 (1994) 6284.
[8] B. Mikami, H. J. Yoon, and N. Yoshigi, J. Mol. Biol., 285 (1999) 1235.
[9] D.M. Guerin, M.B. Lascombe, M. Costabel, H. Souchon, V. Lamzin, P. Beguin, and P.M. Alzari, J. Mol. Biol., 316 (2002) 1061.
[10] J. Sakon, D. Irwin, D.B. Wilson, and P.A. Karplus, Nat. Struct. Biol., 4 (1997) 810.
[11] A.E. Aleshin, P.-H. Feng, R.B. Honzatko, and P.J. Reilly, J. Mol. Biol., 327 (2003) 61.
[12] I. Von Ossowski, J. Stahlberg, A. Koivula, K. Piens, D. Becker, H. Boer, R. Harle, M. Harris, C. Divne, S. Mahdi, Y. Zhao, H. Driguez, M. Claeyssens, M. L. Sinnott, and T. T. Teeri, direct submission (2003).

[13] G. Sulzenbacher, L.F. Mackenzie, K.S. Wilson, S.G. Withers, C. Dupont, and G.J. Davies, Biochemistry, 38 (1999) 4826.

[14] V. Harjunpää, A. Teleman, L. Ruohonen, T.T. Teeri, O. Teleman, and T. Drakenberg, Eur. J. Biochem., 240 (1996) 584.

[15] M. Spezio, D.B. Wilson, and P.A. Karplus, Biochemistry, 32 (1993) 9906.

[16] M. Vrsanska and P. Biely, Carbohydr. Res., 227 (1992) 19.

[17] B.K. Barr, Y.L. Hsieh, B. Ganem, and D.B. Wilson, Biochemistry, 35 (1996) 586.

[18] G. Sulzenbacher, H. Driguez, B. Henrissat, and M. Schülein. Biochemistry, 35 (1996) 15280.

[19] M.-P. Egloff, J. Uppenberg, L. Haalck, and H. van Tilbeurgh, Structure, 9 (2001) 689.

[20] H. Hidaka, Y. Honda, M. Kitaoka, S. Nirasawa, K. Hayashi, T. Wakagi, H. Shoun, and S. Fushinobu, Structure, 12 (2004) 937.

[21] B. Mertz, R.S. Kuczenski, R.T. Larsen, A.D. Hill, and P.J. Reilly, Biopolymers, 79 (2005) 197.

Bioprocessing for Value-Added Products from Renewable Resources
Shang-Tian Yang (Editor)

Chapter 6. Bioreactor Engineering

Si-Jing Wang[a] and Jian-Jiang Zhong[a,b]

[a]State Key Laboratory of Bioreactor Engineering, East China University of Science and Technology, 130 Meilong Road, Shanghai 200237, China

[b]College of Life Science & Biotechnology, Shanghai Jiao Tong University, 800 Dong Chuan Road, Shanghai 200240, China

1. INTRODUCTION

By definition, a bioreactor is a vessel in which a biological reaction or change takes place. The biological systems involved include enzymes, microorganisms, animal cells, plant cells, and tissues. The bioreactor is a place where an optimum external environment is provided to meet the needs of the biological reaction system so that a high yield of the bioprocess is achieved. Obviously, there are complicated interactions between the biological system and the physical and chemical aspects of this process. To design an appropriate bioreactor for a particular bioprocess, intensive studies on the biological system, such as cell growth and metabolism, genetic manipulation, and protein or other product expression are needed to understand the cells' requirement on their physical and chemical environment. A variety of bioreactor types and configurations have thus been exploited and developed along with the advances in the understanding of biological systems. In addition, it is necessary to control the bioreactor's operating parameters in order to favor the desired functions of the living cells or enzymes. Dissolved oxygen concentration, pH, temperature, mixing, and supplementation of nutrients all need to be controlled and optimized. Because two distinct bodies of knowledge, namely, molecular biology and process engineering, are involved and the bioreactor is the core of the bioprocess, a systematic science-based approach to studying bioreactors is needed and the term "bioreactor engineering" becomes more appropriate than the term "bioreactor" or the much earlier used "fermentor." Fig. 1 shows a simplified schematic representation of the process and scope of bioreactor engineering.

As shown in Fig. 1, the bioreactor actually is the core of a number of biological processes. To consider a bioreactor system, the final objective of this biological process must be identified, which is often determined by the market demand for a certain product or beneficial biotransformation process. Because of the rapid advances in recombinant DNA technology and genome sequencing, the same product or biological process may be achieved by different biological systems: microorganisms, plant cells, animal cells, or enzymes. Their genetic expressions, metabolic manipulation, and bioreaction pathways all need to be understood. The

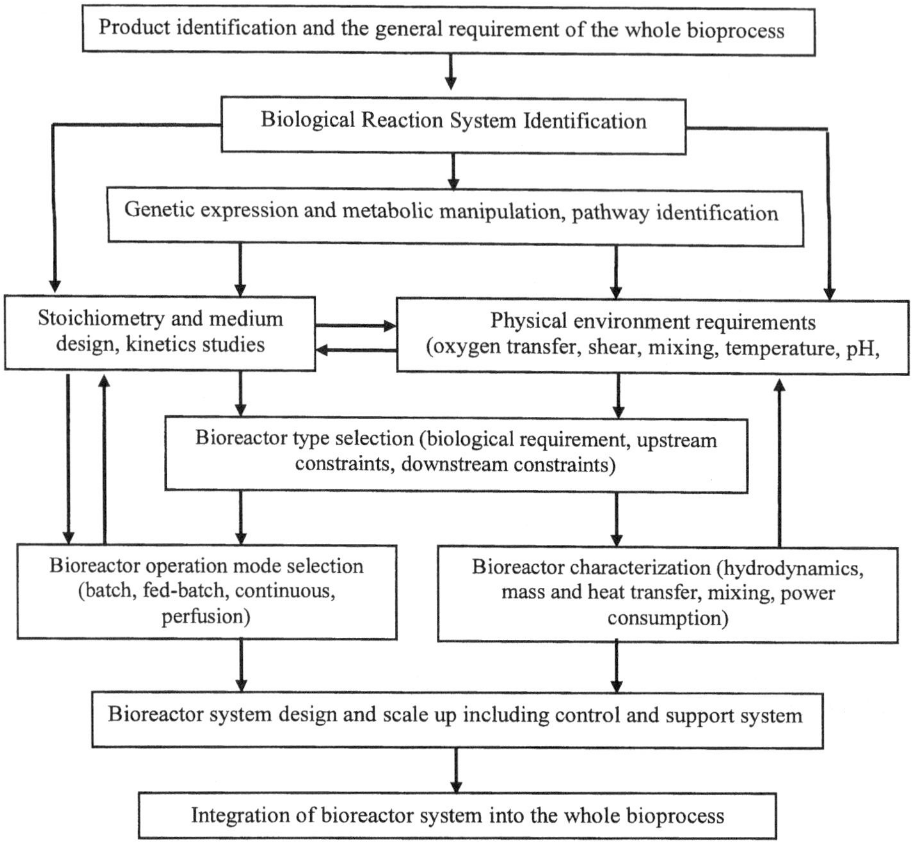

Fig. 1. Schematic representation of the process and scope of bioreactor engineering.

next step is to identify the medium requirements for the efficient performance of the chosen biological system. The media design and optimization can be based on a basic knowledge of stoichiometry and experimental data, including monitoring the composition changes of the media, intermediates, products, and nutrients. Stoichiometric calculations provide quantitative relationships between yields of biomass and product synthesis, maintenance requirement and energy production. Complementing stoichiometric data for the design of a bioreactor, a kinetics study will reveal the biological reaction rates, including cell growth, substrate consumption, product synthesis and by-product formation rates. Many enzymatic reactions are involved, and inhibitions caused by products, byproducts, or even substrate at high concentrations are often observed. On the other hand, the physical environment directly affects the biological performance. Shear stress, mass transfer, mixing, pH and temperature are all interrelated and can influence biological reactions. Also, equally important factors to be considered are downstream process requirements. With this understanding of the biological system and its requirements on its physical and chemical environment, a proper bioreactor type can be selected. Among the bioreactor types available for a certain bioprocess, it is important to have a balanced consideration of many factors, including oxygen transfer, mixing, shear, operational stability and reliability, scale-up, and cost. The chosen bioreactor should be further characterized and the operational mode should be optimized. The bioreactor's characteristics and operational mode also greatly affect the biological performances. An efficient bioreactor system relies greatly on its control and support systems. No matter how important the bioreactor system is, it must be closely and efficiently integrated into the whole production system. Therefore, other process requirements and constraints should also be considered.

This chapter provides a brief description of a variety of bioreactor systems, with an overview of the latest advances in their design, control, and applications. While we try to categorize them by their distinct attributes, it is obvious that there are some overlapping characteristics. Since bioreactor engineering covers such an extensive area, any attempt to cover all of it in a short chapter would be simply impossible.

2. VARIOUS TYPES OF BIOREACTORS

In general, most biological reaction systems can be classified into two main groups: suspension systems and immobilization systems. Stirred tank, air-lift and bubble column bioreactors are mainly for suspension cultures; membrane, packed bed, and fluidized bed bioreactors are mainly for cultivating attached cells or immobilized enzymatic reactions. Obviously there are some bioreactors that can be applied in both of these two categories. For example, with the appropriate carriers, the immobilized cells or enzymes on carriers can be suspended in stirred tank bioreactors or air-lift/bubble column bioreactors.

The design and selection of each type of bioreactors is unique but some basic fundamental principles are followed. Nutrients must be effectively provided to the cells, and waste products must be removed. Cell growth and product formation kinetics should be assessed so that an optimal environmental condition can be defined and an operational mode can be determined. Transport phenomena, including mixing, shear force, and oxygen transfer, should be studied in order to define the criteria for bioreactor design and scale-up. Operating parameters, such as temperature, pH, dissolved oxygen concentration and substrate concentrations should be easy to control and monitor. In addition, the bioreactor should be as simple and inexpensive as possible and it should easily operate free of contamination with microorganisms. In the biopharmaceutical industry, bioreactor design and selection should also consider cGMP compliance. Most often it is difficult or impossible to meet all the requirements, and so some compromise must be made. For example, it is very important to give a balanced consideration between mixing and mass transfer requirements and the shear sensitivity of cells in the design of large-scale bioreactor systems [1].

Some basic types of bioreactors widely used in industrial fermentation are briefly discussed in the following sections. Bioreactors mainly used for solid state fermentation and photobioreactors for algal cultures are included in Chapters 18 and 19 of this book, respectively, and they are not repeated in this chapter. Bioreactors such as roller bottles and wave bioreactors that are used in cell cultures also will not be discussed in this chapter.

2.1. Stirred-tank bioreactor

One of the most conventional bioreactors is the stirred-tank bioreactor. Fig. 2 shows a schematic diagram of a typical stirred-tank bioreactor. The core component of the stirred tank bioreactor is the agitator or impeller, which performs a wide range of functions: heat and mass transfer, aeration, and mixing for homogenization. Two types of impellers are widely used in the conventional fermentation industry: axial and radial flow impellers. Over time, vast and valuable research endeavors have illustrated the transport phenomena of these impellers: oxygen transfer, heat transfer, power consumption, and fluid dynamics. These greatly facilitate the design, installation and optimization of these impellers in conventional fermentation, and so the standard stirred-tank bioreactor is used almost universally in the fermentation industry. Besides the impeller type, there are a number of geometric specifications important for the performance of the stirred tank reactor; these include the impeller off-bottom clearance, the impeller size, the baffles and their width, the sparger type and position, the ratio of liquid height to tank diameter and so on. For large-scale vessels, multiple impellers are often been installed in order to provide sufficient mixing and mass transfer. A number of researchers have studied the hydrodynamics of multiple impeller systems [2–4].

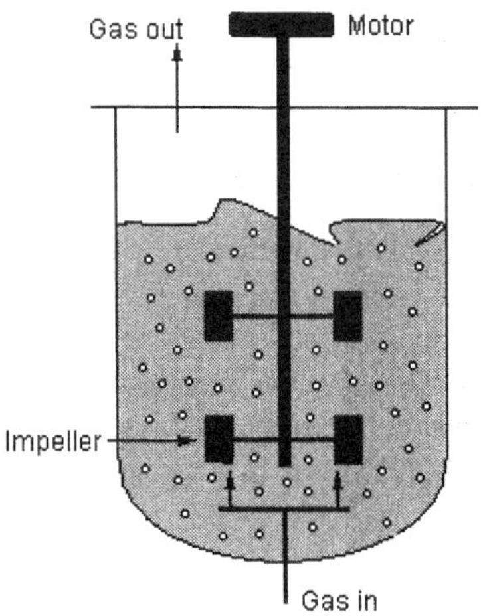

Fig. 2. A typical stirred-tank bioreactor.

For shear-sensitive biological systems, such as animal and plant cell cultures, conventional impellers that produce high shear stress cannot be directly applied. Because of their fragile cellular structure, animal cells are very sensitive to shear and bubble damage in the bioreactor environment. The two principal mechanisms that can lead to physical cell damage are hydrodynamic shear force induced by agitation, and air bubble damage caused by unprotected gas sparging. Many investigators have worked hard to eliminate or reduce cell damage by these two mechanisms. In general, these endeavors can be summarized by the following three methods: developing a new oxygenation device in order to reduce shear caused by bubbles [5–8], exploiting different protective agents [9–12], and modifying existing impellers and designing new types of agitators [8, 13].

Along with the installation of a proper oxygenation device, such as bubble-free aeration, gas basket, and cage-aeration, and the addition of appropriate protective agents (Pluronic 68 is the most widely used), many modifications of the marine impeller have been proposed in order to provide more efficient mixing at lower impeller tip speeds [13]. A number of high-flow, low-power-number impellers such as Intermig, Lightnin, Prochem Maxflow T and Scaba 6SRGT have been developed to provide improved performance [14]. In addition to the modification of existing conventional impellers, some new types of agitation impellers are developed for shear-sensitive cell culture processes, including cell-lift [15] and centrifugal impellers [16, 17].

Plant cells are also sensitive to shear stress, but not as much as animal cells are. As such, many of the stirred tank bioreactor systems used for plant cell suspension cultures are modifications of those used for microbial systems. With greater power input to the broth by mechanical agitation, stirred tanks are generally more suitable for viscous liquids containing suspended particles such as plant cells. As such, they have been successfully applied to industrial-scale plant cell cultures [18–20]. The impellers used for plant cell cultures range from the standard Ruston turbine, curved-blade disk turbine, and hydrofoil impellers to semiconventional agitators, such as helical ribbon and centrifugal impellers [21]. Wang and Zhong designed a novel centrifugal impeller bioreactor for shear-sensitive biological systems [16, 17], which has been demonstrated to be very successful in high cell density plant cell cultures [22–25]. As shown in Fig. 3, this impeller is designed based on the principle of a centrifugal pump. The rotation of the centrifugal impeller creates an area of negative pressure in the center of impeller, drawing liquid and cells from the reactor bottom through the draft tube and centrifuging them toward the bulk liquid. Compared to the flat turbine impeller, this impeller has reduced shear force, much better mixing performance [16], and improved oxygen transfer capacity [17].

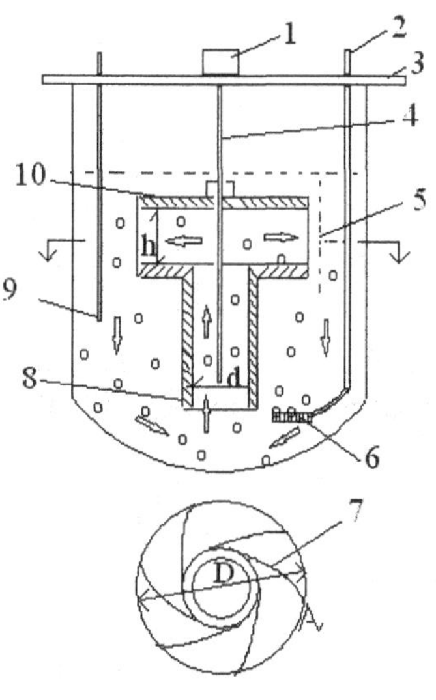

Fig. 3. Schematic diagram of a centrifugal impeller bioreactor. 1-Stirrer, 2-gas in, 3-head plate, 4-shaft, 5-measuring points for liquid velocity, 6-sparger, 7-blade, 8-draft tube, 9-DO probe, and 10-rotating pan (Reprinted from Ref. [16] with permission of John Wiley & Sons, Inc.).

In general, the stirred tank bioreactor has several advantages for the cultivation of shear-sensitive cells: existing industrial capacity, proven performance, ease of maintaining homogeneous conditions, and ease of scale-up and control. Thus, currently in the biopharmaceutical industry, the stirred-tanks are the most widely used bioreactors for GMP production of monoclonal antibodies (MAb) therapeutics and other biologicals using animal cell cultures. Several biopharmaceutical manufacturers have implemented stirred tank bioreactors at the 10,000 to 20,000 liter scale for large-scale animal cell cultures [26].

2.2. Pneumatically agitated bioreactors

There are two main types of pneumatically agitated bioreactors: air-lift and bubble-column bioreactors. As shown in Fig. 4, the main difference between them is that the air-lift bioreactors contain a draft tube (internal loop) or an external loop. The draft tube or the external loop gives the air-lift bioreactor a number of advantages: preventing bubble coalescence by directing them in one direction; distributing shear stresses more evenly throughout the reactor, thus providing a more favorable environment for cell growth; enhancing the cyclical movement of fluid, thus increasing mass and heat transfer rates.

In a typical air-lift bioreactor with an internal loop (as shown in Fig. 4a), air is fed through a sparger ring into the bottom of a central draft tube, which directs the circulation of both air bubbles and liquid. Air bubbles flow up inside the central draft tube; some of them coalesce and exit at the top of the column while other bubbles follow the degassed liquid and circulate down from the area outside the draft tube. Some air-lift bioreactors use an external-loop. The air-lift reactor with an external riser sparges bubbles into the section which is outside the draft tube. Turbulence is generally greater in the riser rather than the downcomer section of an air-lift reactor. Since the heating/cooling jacket is located on the walls of the air-lift reactor, a reactor with an external riser will have the advantage of having greater turbulence near the jacket and thus better heat transfer efficiency. It is also believed that reactors with external risers foam less than those with internal risers.

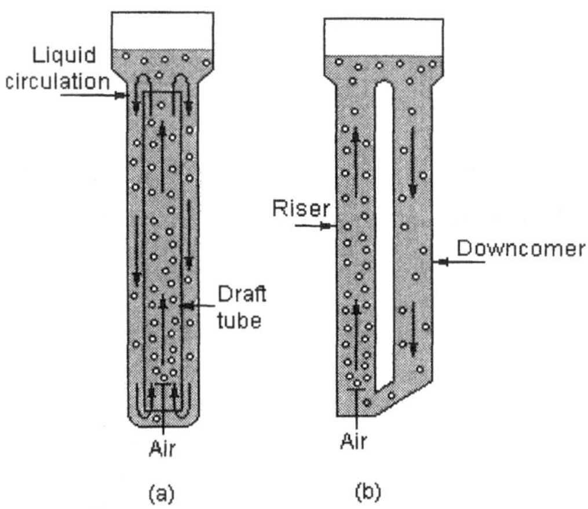

Fig. 4. Air-lift and bubble-column bioreactors. (a) Air-lift with internal loop; (b) Air-lift with external loop; (c) Bubble-column.

Air-lift bioreactors with various configurations have been constructed for use in a variety of fermentation processes, cell cultures, and biological wastewater treatment. The air-lift bioreactor is the second type that is well documented and characterized, but is less so than the stirred tank bioreactor. Much experimental and modeling work has been done to illustrate the transport phenomena, such as liquid circulation, mixing, and oxygen transfer. A variety of designs for the air-lift bioreactors have also been proposed and tested. Several column designs with vertical circulation have been tested by Viestures et al. [27, 28]. Several researchers have investigated air-lift bioreactors with multiple draft tubes for their solid circulation, hydrodynamics, mixing, and oxygen transfer characteristics [29–32]. Recently, Wei et al. [33, 34] investigated the hydrodynamics and mass transfer of an internal-loop air-lift reactor with a convergence-divergence draft tube. Other designs include a simple split-column [35, 36], a propeller-driven loop [37] and varieties in sparging devices [37, 38].

Air-lift bioreactors have been widely used in filamentous fermentation, biological wastewater treatment, production of single cell proteins, and plant and animal cell cultures. Because they provide a low-shear environment and good mass transfer, air-lift bioreactors are often been preferred in filamentous fermentations [39, 40]. Compared to stirred tank reactors, one of their major advantages is that the cost associated with agitation and aeration can be substantially reduced [41, 42]. Other advantages of using air-lift bioreactors include: ease of scale-up, low shear characteristics, no moving parts, high O_2 transfer efficiency, and predictable flow patterns.

Compared to the air-lift bioreactor, standard bubble columns have some considerable disadvantages: backmixing in the continuous liquid phase and the decrease of interfacial area due to bubble coalescence in the viscous liquids. To overcome these disadvantages, some modifications to the standard design have been proposed. For example, multistage bubble column reactors sectionalized by perforated plates have been used. It has been commonly acknowledged that sectionalizing bubble column reactors can significantly improve their mass transfer characteristics and, at the same time, substantially reduce the degree of backmixing in contacted phases [43, 44].

There are many recent research reports [45–47] on the fluid dynamic characteristics of bubble column reactors, but their application to biological processes, such as microbial fermentation and cell cultures, is quite limited. Hu et al. [48] scaled up a *Panax notoginseng* cell culture process from shake flasks to a 1.0-L bubble column reactor and concentric-tube air-lift reactor. Both the maximum cell density and productivity of ginseng saponin in the batch culture were found to be higher than in shake flasks but lower than in air-lift reactors. Barbosa et al. [49] studied the effect of hydrodynamic stress on two different microalgae strains, *Dunaliella tertiolecta* and *D. salina*, cultivated in bench-scale bubble columns. In this type of bubble column reactor, it was found that bubble rising and bubble bursting were not causing cell death. Instead, bubble formation at the gas sparger was found to be mainly responsible for cell death.

In general, the reports on the development of new types of pneumatically agitated bioreactors, including air-lift and bubble column reactors, are few while research on the application of the existing pneumatically-agitated bioreactors (with some modifications in configuration or operating conditions) is abundant. This is because the only way to evaluate a bioreactor is to apply it to a particular biological reaction system.

2.3. Membrane bioreactors

A membrane bioreactor, broadly defined, is a flow reactor within which membranes are used to separate cells or enzymes from the feed or product streams. The most important feature of membrane bioreactors are that cells or enzymes are retained within the reactors, so the reactors can be continuously perfused without concerns about washing out the cells or enzymes. Sometimes, membranes are used for *in situ* separation of cells or enzymes, thus integrating the production and separation into a single step.

Membranes are made from a variety of materials, including cellulose, acetate and nitrate, polyvinylidene difluoride, polysulfone, polypropylene, polytetrafluoroethylene (PTFE), and polyacrylonitrile. Also, other types of membranes are used, including ceramic [50], silicone rubber [51], and ion exchange membranes [52]. Microfiltration and ultrafiltration membranes are most commonly used. Microfiltration membranes have pore sizes between 0.1 and 0.5 μm and can be used to confine cells within a reactor while imposing little restriction on the passage of soluble nutrients and products. Ultrafiltration membranes typically have pore sizes between 20 and 1000 Å and are used to retain or exclude macromolecules. Before membrane separation can be used, methods of economically and efficiently packaging large areas of membrane are required. These packages are called membrane modules. In general, membranes are packed into one of the following modules: plate-and-sheet modules, tubular modules, spiral-wound modules, and hollow-fiber modules. The most commonly used geometry for membrane bioreactor is the hollow fiber.

Membrane reactors have been used for an enormous variety of applications. In general, these applications can be categorized into four major areas: biocatalysis, fermentation, cell cultures, and wastewater and waste gas treatments.

In biocatalysis, an enzymatic membrane reactor (EMR) couples a membrane separation process with an enzymatic reaction. Enzyme molecules are either freely circulated on the retentate side or are immobilized onto the membrane surface or inside its porous structure. Compared to other immobilization techniques, membrane entrapment of the enzyme is perhaps the gentlest approach, as no chemical agents or harsh conditions are employed. The decay of enzyme activity can therefore be attenuated in membrane bioreactors. For example, Kuo et al. [53] studied the hydrolysis of chitosan in a continuous enzymatic membrane reactor and found that the enzyme activity was not greatly altered when used in the membrane reactor and that the membrane reactor gave a higher productivity than that of the conventional batch reactor. A widespread and classical application of the enzymatic membrane reactor is in the hydrolysis of macromolecules for food and pharmaceutical applications [54, 55]. An integral combination with a membrane for enzyme retention and as a downstream process was applied in the enzymatic production of glycosides [56, 57] and for enzymatic cellulose hydrolysis [58]. Recently, enzymatic membrane reactors were also used in wastewater treatment. Akay et al. [59] and Erhan et al. [60] employed a cross-flow enzyme-immobilized membrane reactor for the removal of phenol and catechol from water. A crude enzyme extract from a species of *Pseudomonas syringae* was chemically immobilized onto a flat-sheet polyamide membrane with a nominal pore size of 0.2 μm. Their results showed that the reaction rate was diffusion controlled. The immobilized enzyme showed better stability than free enzymes in solution and retained 70% of its initial activity for about 100 hours.

A very interesting and important fact in applying membrane bioreactors to microbial fermentation is that the retention of cells within a membrane reactor allows for a high cell density to be maintained in the reactor. Kamoshita et al. [61] exploited a stirred ceramic membrane reactor for the rapid fermentation of lactic acid by *Lactococcus lactis*, which reached a high cell density of 140 g L^{-1}. A 4.3-fold increase in productivity was obtained in a bioreactor coupled to a microfiltration module for the production of an enzyme, superoxide dismutase, by *Streptococcus lactis* [62]. More significant improvements (over conventional suspension) in cell density and volumetric productivity have also been reported [63]. Chung et al. [64] reported bacterial concentrations of more than 500 g cell dry wt L^{-1} in a hollow-fiber reactor. This concentration was even higher than the solid content of normal microorganisms, indicating significant dewatering of cells by membrane entrapment.

The main attractions of using membrane bioreactors in animal cell cultures are also high cell density and high volumetric productivity. In a simple stirred-tank bioreactor, a typical animal cell density is on the order of 10^6 cells/mL [65]. In a hollow-fiber bioreactor, on the other hand, a high cell density of $>10^8$ cells/mL can be achieved [66–68]. Another advantage of using a membrane bioreactor in animal cell cultures is that the shear stress problem often encountered in stirred-tank reactors can be eliminated in a membrane reactor because the cells are sequestered in a relatively quiescent region wherein they are protected from mechanical damage and are not in direct contact with air bubbles. It should be noted, however, that applying membrane bioreactors to large-scale cell cultures can cause many operating problems resulting in poor cell viability, poor process stability, product heterogeneity, and diffusion gradients. Therefore their use in animal cell culture process is mainly limited to small-to-medium scales.

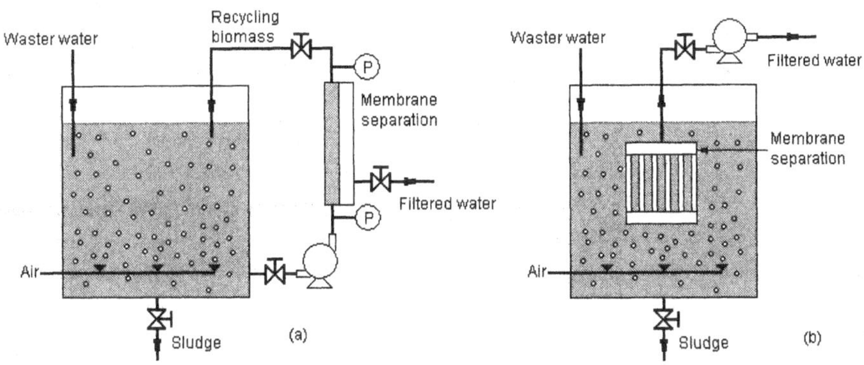

Fig. 5. Typical membrane bioreactors for biological wastewater treatment. (a) Bioreactor with external membrane separation; (b) Submerged membrane bioreactor.

Most successful applications of the membrane bioreactor can be found in biological wastewater treatment. Fig. 5 shows two typical membrane bioreactors for wastewater treatment: one with an external membrane separation module (Fig. 5a) and one with a submerged membrane module (Fig. 5b). The complete retention of sludge by a membrane allows operation at much higher biomass concentrations. As a direct consequence of the high biomass concentration obtained in membrane reactors with complete sludge retention, the microorganisms utilize a growing portion of the carbon content of the feed for maintenance purposes and much less for cell growth. When the ratio of feed to the microorganism concentration (F/M) becomes low enough, no or almost no excess sludge is produced [69–71]. Rosenberger et al. [72] studied the aerobic treatment of municipal wastewater in a membrane bioreactor for 535 days. When the feed to microorganism (F/M) ratio was decreased to as low as 0.07 kg COD (kg MLSS)$^{-1}$ d^{-1}(MLSS: mixed-liquor suspended solids), no net sludge was produced. It was found that the treatment performance was very stable and on a high level over a long time. Since membrane fouling can be reduced to an acceptable level by appropriate backwashing and air scotching, membrane bioreactors have been widely used in large-scale wastewater treatment plants [73]. In addition to wastewater treatment, the membrane bioreactor also has great potential in waste-gas treatment [74]. The use of membrane bioreactors in both wastewater and waste-gas treatment has been well documented in the literature, and a number of literature review papers or books have been published [74–78].

2.4. Fixed bed bioreactors

Fixed (packed) bed reactors are one of the most frequently employed types of bioreactor for immobilization systems. This type of bioreactor has the advantages of simplicity of operation and high reaction rates. Enzymes or cells are immobilized in appropriate carriers, which are packed in the fixed reactors, resulting in high solid-liquid specific interfacial contact areas, and the velocity of liquid creeping over the static solid particles substantially alleviates the film resistance to mass transfer. The major disadvantages of the fixed bed bioreactor are their relatively poor mass and heat transfer coefficients due to low liquid velocities. For aerobic biological systems, efficient gas-liquid contact and carbon dioxide removal are very critical. A fixed bed reactor often accumulates stagnant gas pockets, causing gas flooding and producing poor liquid distribution. It has therefore not been widely used in aerobic microbial fermentation processes. To avoid gas accumulation, Shiotani and Yamane [79] proposed a shallow horizontal packed bed reactor for ethanol fermentation. In the horizontal reactor, there is a free space above the packed bed, so carbon dioxide gas can be easily released upward into the free space by its buoyancy. Compared to a vertical packed bed reactor, the ethanol productivity was enhanced by 1.5 times.

Most applications of the packed bed reactors are found in the treatment of wastewater and waste gases. Biotrickling filters, which are based on the principle of the fixed bed bioreactor, have been widely applied for the biodegradation of many pollutants over the past two decades. A pollutant-degrading biofilm is established on the surface of the packed bed carriers, and wastewater or contaminated air is passed through the packed bed for treatment. A variety of packing materials, such as lava rock [80], plastic packings [81], activated carbon [82] and sand [83], have been used in biotrickling filters. The most important requirements for the packing materials include high porosity, large specific surface area, and high chemical stability and structural strength.

Fixed bed bioreactors also have some applications in animal cell cultures, as animal cells have a much lower oxygen transfer requirement than microorganisms. Since animal cells are immobilized within a carrier matrix, they are less sensitive to shear force than when suspended. In addition, the microenvironment created for the immobilized cells within the carrier matrix can be more favorable than that in the bulk liquid. With the appropriate selection of carriers for cell immobilization, high cell density and high productivity can be obtained. Yang et al. [84] compared different culture systems for Monoclonal Antibody production. A packed bed bioreactor shows advantages in achieving high cell density and MAb concentration. Using a non-woven polyester fibrous matrix to immobilize cells in a fibrous-bed bioreactor, a high viable cell density of $3x10^8$ cells/mL packed bed with a high volumetric MAb productivity of 1 g L^{-1} day^{-1} under continuous feed conditions was obtained.

In addition, it was found that the fibrous matrix could selectively retain healthy, non-apoptotic cells for long-term cultures [84].

2.5. Fluidized bed bioreactors

The fluidized bed reactor is another widely employed bioreactor for immobilization systems. A fluidized bed reactor can provide a degree of mixing intermediate between the two extremes of the packed bed reactor and stirred tank reactor. On the one hand, the upward movement of fluid carries immobilized cells upwards; the particles rise until the force of gravity causes the particles to fall. Fluidization is achieved by the combined upward and downward movement of particles. Compared to the fixed bed reactor, the major advantages of the fluidized bed bioreactor are listed below:

1. The system is homogeneous and it is therefore easier to monitor and control the operating parameters such as temperature, pH, and dissolved oxygen concentration.
2. Good mixing is achieved so that gradients do not occur across the reactor.
3. Higher mass transfer and heat transfer rates are expected between the bulk fluids and the particles as a result of the free movement of the particles and the high specific surface area of small particles.
4. Easy particle sampling and replacement of the active fractions, even during operation.
5. Scale-up can be achieved without increasing concentration gradients. As the scale increases, the advantages become more apparent.

However, there are also operational difficulties. The major problem is that it is not easy to predict the back-mixing and fluidization patterns. Since fluidized beds have a narrow range of optimum operating conditions at relatively high bed expansion and low stability levels, it is quite difficult to maintain nonfluctuating operation. These problems have hampered the efforts to scale up this type of bioreactors.

The application of the fluidized bed reactor with immobilized cells has been primarily achieved in wastewater treatment [85]. In fluidized-bed bioreactors for wastewater treatment, cells are either immobilized in carriers or self-granulated, resulting in biomass retention in the reactor and improved reactor volumetric conversion capacity. Use of the fluidized bed of biomass can be traced back to 1940 in the UK while the development of particle-supported biofilm reactors began in the early 1970s [85]. The fluidized particles (carriers) provide a large specific surface area for cell growth and allow biomass concentrations in the 10–40 kg m^{-3} range to develop [86]. Fluidized bed bioreactors have been used in almost all areas of wastewater treatment processes, including both aerobic and anaerobic treatments of industrial effluents [87, 88] and domestic waste water [85, 89, 90].

Fluidized reactors have also been employed for microcarrier cultures [91, 92]. Recently Durrschmid et al. [91] compared two cell culture systems for the production of recombinant protein from CHO cells. Using a Cytopilot Mini fluidized bed bioreactor (FBR, Vogelbusch-Amersham Biosciences, Austria), CHO cells were cultivated as adherent cells attached on Cytoline macroporous microcarriers. In comparison, the same cell line was cultivated in suspension using a stirred tank bioreactor equipped with an ultrasonic resonator based cell separation device. It was found that both systems were equally well-suited for stable, long-term, high cell density perfusion cell cultures and can provide industrial scalability and high yields.

3. EFFECTS OF PROCESS PARAMETERS ON BIOLOGICAL PERFORMANCES

The analysis of bioreactors is central to the successful design and operation of biotechnical processes. The main objective of bioreactor selection, design, and control is to provide the optimal environment for a biological reaction system. The bioreactor should provide optimum conditions (e.g., temperature, pH, oxygen transfer, mixing, and substrate concentration), in addition to its basic function of containment. For example, the ability to control the substrate concentration is an important function of the bioreactor. The substrate concentration can be subjected to spatial variation – advertently or inadvertently – and may also change with time in batch or fed-batch operation. Cellular metabolism depends on local concentrations in the reactor, as well as on the physiological status of the cell [93]. In order to understand bioreactor operation, cellular metabolism must be considered together with the flow profile and the mass transfer characteristics of the bioreactor because they closely interact with each other.

3.1. Temperature

Temperature is one of the most critical parameters to be closely controlled in a bioreactor. Microorganisms are often classified according to their growth temperature as either thermophiles (growth temperature: >50°C), mesophiles (growth temperature: from 20°C to 50°C), or psychrophiles (growth temperature: <20°C) [94]. Regardless of the microorganism type, microorganisms always have a quite narrow optimal temperature range for growth. If grown at a temperature below the optimum, growth occurs slowly resulting in a reduced rate of cellular production and product synthesis. On the other hand, if the growth temperature is too high, not only will death occur, but protein expression or metabolite synthesis will also be seriously affected, lowering product yield or affecting product quality.

The effect of temperature on chemical or enzymatic reactions is typically modeled using the Arrhenius equation. This has also been used to describe the effect of temperature on the specific cell growth or cell death of the microbial system [94]. Fig. 6 shows a typical growth rate curve as a function of the temperature. The cell growth rate increases when the temperature is increased toward the optimum. When the temperature exceeds the optimum, the growth rate decreases and thermal death occurs. The net cell growth rate is proposed as follows:

$$\frac{dX}{dt} = (\mu - k_d)X \tag{1}$$

where X and t are cell concentration and time, μ and k_d are cell growth and cell death rate, respectively.

Both μ and k_d can be expressed as functions of temperature following the Arrhenius equation, as follows:

$$\mu = A \cdot e^{(-E_a/RT)} \tag{2}$$

$$k_d = A' \cdot e^{(-E_d/RT)} \tag{3}$$

where E_a and E_d are activation energies for cell growth and thermal death, respectively. Equation (2) represents the increase in specific growth rate with temperature, and the value of E_a is typically in the range of 10 to 20 kcal/mol. Equation (3) represents the thermal death rate, which is substantially increased with temperature as E_d is much higher than E_a. E_d is typically in the range of 60 to 80 kcal/mol [94].

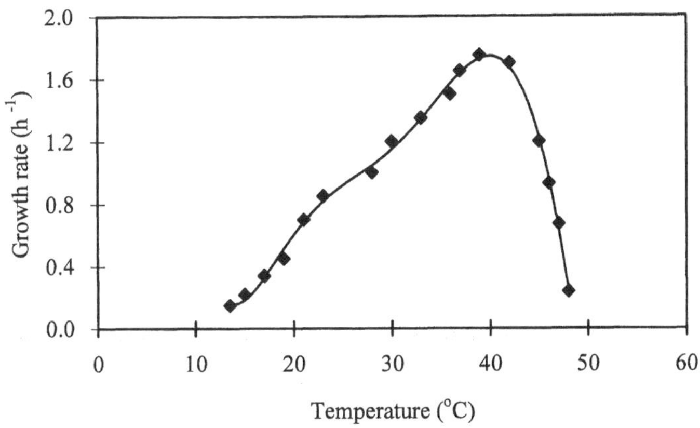

Fig. 6. Effect of temperature on cell growth rate of *E. coli* (adapted from Figure 6.7 of Ref. [94]).

There are a number of reports on the effect of temperature on microbial growth rates and product formation rates for different fermentation processes [94, 95]. In a conventional microbial fermentation process, once the optimal temperature is determined, it will normally be maintained throughout the whole fermentation process. This, however, may not always be the case for mammalian cell culture processes.

In mammalian cell culture process, a large portion of the protein product is synthesized during the post growth phase. Since the cell viability drops quickly after the cell density approaches maximum, the cultivation of cells at reduced temperatures has been proposed to improve batch culture performances. It has been consistently reported that a decrease in cultivation temperature leads to prolonged culture viability [96, 97]. However, a culture temperature below 37°C normally inhibits cell growth [98]. A concept of two stages is therefore proposed: a growth phase and a production phase. During the first stage, the temperature favoring cell proliferation (e.g. 37°C) is used to obtain a high cell density. In the second stage, temperature is reduced in order to decelerate the drop in cell viability. This strategy, however, is not straightforward, as temperature is also a very critical parameter for protein synthesis. The effect of reduced temperature on heterologous protein production of mammalian cells varied among different studies. For example, a shift in the cultivation temperature from 37°C to 28°C greatly increased the specific productivity of a CHO cell line for the production of Fab antibody fragments [99]. Also, up to 1.7-fold increase in the specific human secreted alkaline phosphatase (SEAP) productivity by a CHO cell line was recorded when the culture temperature was reduced from 37°C to 30°C [100]. In contrast, the specific monoclonal antibody productivity of murine hybridoma CC9C10 decreased by 21% at 33°C compared with controls at 37°C [101]. In another study [98] on the effect of reduced temperature using a recombinant CHO cell line to produce a C-terminal α-amidating enzyme in the temperature range of 37°C to 30°C, the maximum productivity was achieved at 32°C, but a further reduction in temperature to 30°C resulted in an obvious decline in productivity. Clearly, an optimal temperature exists for each individual cell culture process.

3.2. Effects of pH

Different biological systems have different optimal pH ranges. Most microorganisms grow best between pH 5 and 7. During fermentation, pH can change. As the cells grow, metabolites are released into the medium; substrate consumption also causes pH change. For example, ammonia is a common nitrogen source. When ammonia is utilized during fermentation, pH decreases. Therefore, the pH of the medium must be monitored and adjusted by base or acid addition in order to constantly maintain an optimum pH.

A number of researchers have investigated the effect of pH on the growth kinetics of microorganisms, enzymatic activities, and product synthesis [102–105]. For example, Elmahdi et al. [106] investigated the influence of various pH control strategies on growth and erythromycin synthesis by *Saccharopolyspora erythraea* CA340. A two-fold increase in erythromycin biosynthesis was achieved by pH control. In addition, a pH monitoring and control strategy was applied to microscale fermentation and a similar enhancement was obtained. In recent years, there has been increasing interest in researching fermentation optimization and process development carried out in a microtitre (microwell) plate format [107]. Compared to the conventional shake flask approach, process optimization and development can be greatly facilitated because only small volumes are required and a large number of microplates can be run in parallel.

In animal cell culture processes, culture pH is often controlled by the addition of an alkaline reagent, such as $NaHCO_3$ or NaOH, to neutralize the acidic effects of lactate and CO_2 production during cell growth [108]. Another scheme for pH control in animal cell culture process is CO_2 addition. CO_2 is added to a sodium bicarbonate-containing medium in order to control the pH via the following reaction:

$$CO_{2\,(aq)} + H_2O \rightleftharpoons H^+ + HCO_3^- \tag{4}$$

In general, using CO_2 to control pH is simple and efficient. However, it may cause the following problems: in high cell density cell cultures, a high rate of CO_2 production will limit controllability by CO_2 addition, CO_2 sparging can decrease the oxygen supply and upset DO control, and in the case of high lactate concentrations or during periods of rapid lactate production, the limited buffering capacity of the bicarbonate system may become inadequate [109].

Animal cells are more sensitive to changes in pH than microorganisms. The media pH is one of the most important process parameters in mammalian cell culture; its effects on cell growth, metabolism, recombinant protein synthesis, and protein quality have been extensively studied. A small change in culture pH can greatly influence the cell growth, metabolism, and production synthesis. In hybridoma cultures, pH 7.2 was found to be optimal for cell growth. When pH was decreased from 7.2 to 6.9, a two-fold decrease in specific growth rate, specific glucose consumption rate, and specific lactate production rate was observed [110]. On the other hand, Wayte et al. [108] reported that the final antibody concentration of two different murine hybridoma cell lines was increased 1.5-fold when the bioreactor pH was slightly reduced from 7.2 to 7.1. Xie et al. [111] studied PER.C6® cell growth, metabolism, and adenovirus production in stirred bioreactors under different pH conditions. It was found that cell metabolism in both infected and uninfected cultures was very sensitive to culture pH, causing dramatic changes in glucose/glutamine consumption and lactate/ammonium production under different pH conditions. A more than 2-fold increase in adenovirus productivity was observed by reducing pH from 7.6 to 7.3. Sauer et al. [112] investigated the effect of pH on the fed-batch process for six Sp2/0-derived cell lines (A, B, F, G, H, I). It was found that the bioreactor pH set point significantly affected cell growth, cell metabolism and culture productivity. The culture pH can also affect product quality. Effects of pH on protein glycosylation in different cell lines, including CHO [113–115] and HL60 [116], have also been reported.

3.3. Mixing

In bioreactors, adequate mixing is essential in order to ensure the adequate supply of nutrients and to prevent the accumulation of toxic metabolites. For a bioreactor designed for a suspension system, mixing time is a critical parameter to be studied and evaluated. The fluid hydrodynamics, fluid rheology, impeller type, power input, and vessel size can all influence the mixing conditions. Generally, the following equation can be used to describe the effects of different parameters on the mixing time in a stirred tank bioreactor [117]:

$$t_m = f(\frac{d}{D}, N, \eta, \rho, \frac{V_a}{V}, \frac{P}{P_a}, \varepsilon_T) \tag{5}$$

where t_m is the mixing time (s), d is the stirrer diameter (m), D is the bioreactor diameter (m), N is the impeller rotational speed (rpm), V is the medium volume (m^3), V_a is the volumetric air flow rate ($m^3\ s^{-1}$), P is the power consumption for mixing non-aerated broth (W), P_a is the power consumption for mixing aerated broth (W), η is the apparent viscosity (cP), ρ is the density (kg m^{-3}), and ε_T is the energy dissipated (W m^{-3}).

Numerous equations have been proposed in the literature for calculating mixing time. However, caution should always be exercised when applying these model equations directly to a specific bioreactor system because many variables are involved. Oniscu et al. [117] studied and modeled the mixing time for non-aerated suspensions of bacteria (*Propionibacterium shermanii*), yeasts (*Saccharomyces cerevisiae*), and fungi (*Penicillium chrysogenum*, free mycelia and mycelial aggregates) at different concentrations in a laboratory bioreactor with a double-turbine impeller. It was found that the presence of biomass significantly reduced the mixing efficiency, even at low broth viscosity levels. The magnitude of this effect depends on the type of biomass and its concentration and morphology. The mixing time increases in the following order: fungal free mycelia > fungal pellets > yeasts > bacteria. At the same concentration and under the same operational conditions, the mixing time for fungal cell suspensions was significantly higher due to their high viscosity and non-Newtonian behavior.

In fermentation or cell culture processes, mixing has often been evaluated in terms of biological performance, such as cell growth rate and productivity. The control of temperature, pH, and substrate concentration are all dependent on good mixing in the bioreactor. Although it is easy to maintain a homogeneous condition in a small-scale reactor, mixing often becomes one of the constraints during scale-up. In large-scale bioreactors, poor mixing often leads to undesirable concentration gradients and a decrease in mass transfer efficiency. In shear-sensitive biological systems, such as animal and plant cell cultures and filamentous fungal fermentation, mixing cannot be enhanced simply by increasing agitation intensity because excessive agitation can cause mechanical damage to living cells. There are numerous reports on the effect of mixing on biological performance in the literature; the following are some of the latest.

Toma et al. [118] investigated the effect of mixing on glucose fermentation by *Zymomonas mobilis* in a stirred tank bioreactor. At higher stirrer speeds, the biomass yield and ethanol productivity were enhanced while the byproduct synthesis was reduced. In plant cell cultures, Zhong et al. [24] studied the effect of mixing time on taxoid production in a centrifugal impeller bioreactor. In the agitation intensity range where no damage was observed on the cultured cells, two different mixing times (5 s and 10 s) were applied by adjusting the impeller agitation speed. A higher cell density and taxoid productivity were obtained under the shorter mixing time. Poor mixing limited oxygen transfer and led to the formation of larger cell aggregates.

As mentioned before, animal cells are very sensitive to pH changes. A fast adjustment of bioreactor pH relies on the mixing condition of the bioreactor. Langheinrich and Nienow [119] studied macromixing conditions on pH control in a large-scale free suspension cell culture bioreactor. When Na_2CO_3 was added at or near the liquid surface to control pH, the added Na_2CO_3 could not be quickly mixed well with the bulk liquid, and very poor homogenization was observed [120]. When the addition position was changed from above the liquid surface to the impeller region, there was no Na_2CO_3 accumulation and the pH value was raised in a continuous and smooth manner, minimizing the danger of contact between cells and the alkali.

3.4. Oxygen transfer

Oxygen transfer is always a concern in aerobic biological systems. Most nutrients required for cellular growth and metabolism are highly soluble in water; sufficient and timely supply of these nutrients can be achieved in a well-mixed bioreactor. However, oxygen transfer often becomes a limiting step to the optimal performance of biological systems and also for scale-up because oxygen is only sparingly soluble in aqueous solutions. When the supply of oxygen is limited, both cell growth and product formation can be severely affected. For example, it was reported that ceasing aeration in the medium during penicillin fermentation for just a few minutes seriously impacted the ability of the cells to produce the antibiotic [121]. In a well-mixed suspension system, the oxygen mass balance is written as:

$$\frac{dC_o}{dt} = k_L a(C_o^* - C_o) - Q_o X \tag{6}$$

where k_La, C_o^*,C_o and Q_o are the volumetric mass transfer coefficient, saturated oxygen concentration, the oxygen concentration in the liquid, and the specific oxygen uptake rate, respectively.

At steady state, the above equation can be solved to obtain the oxygen concentration in the liquid:

$$C_o = C_o^* - \frac{Q_o X}{k_L a} \tag{7}$$

Since C_o^* is constant at a fixed air pressure, C_o is determined by three factors: the specific oxygen uptake rate Q_o, which is determined by the biological system, cell concentration X, and the volumetric mass transfer coefficient, k_La. For a given biological system (bacteria, yeast, animal or plant cells), a serious shortage of oxygen can be expected at a high cell density. Aggravating this problem, high cell density often causes the oxygen transfer coefficient to deteriorate. Since k_La is so important in supplying oxygen to the medium, a very critical aspect of bioreactor design is to achieve a sufficiently high oxygen transfer coefficient, k_La, which is affected by many factors, including the geometrical and operational characteristics of the reactor vessel, agitation speed, aeration rate, fluid hydrodynamics, media composition, cell type, morphology and concentration, and biocatalyst properties. It was estimated that oxygen transfer management accounts for about 15–20% of all operating costs for aerobic fermentation [122].

There are numerous reports studying the effects of oxygen concentration or oxygen transfer on microbial fermentation. For example, a number of researchers reported the effects of oxygen limitation on cell growth, metabolism and product formation in L-lysine fermentation. Ensari and Lim [123] investigated the effects of bioreactor operating variables, including aeration, agitation, dissolved oxygen, and dilution rate, on L-lysine fermentation by *Corynebacterium lactofermentum* ATCC 21799 in a continuous culture. It was found that L-lysine production was strongly influenced by the dissolved oxygen level; the specific growth rate, substrate consumption, product formation, and oxygen uptake rate all depended on the dissolved oxygen concentration in the reactor. In order to maximize lysine production, they suggested that the fermentation should be carried out at 50% dissolved oxygen or above. In another study, Hadj Sassi et al. [124] found that both the substrate consumption rate and L-lysine yield were decreased by oxygen limitation. Compared with cultures grown under 15–20% dissolved oxygen, a 20% increase in L-lysine production was obtained when the dissolved oxygen level was increased to 30–35%. On the other hand, Hua et al. [125] applied metabolic flux analysis to microaerobic lysine fermentation using *C. glutamicum* ATCC 21253. Their results showed that the activities of TCA cycle enzymes decreased with the decrease in oxygen supply. As a result, a 30% increase in lysine yield due to increased phosphoenol pyruvate (PEP) carboxylation was achieved for the microaerobic culture (5% DO) as compared with aerobic fermentation (20–80% DO).

In filamentous fungal fermentations, it has often been observed that the high apparent viscosities and the non-Newtonian behavior of the broths require a strong agitation intensity in order to provide adequate mixing and oxygen transfer. On the other hand, the stirrer speed can strongly influence mycelial morphology, cell viability and productivity [126–128]. Amanullah et al. [126, 129] investigated the effect of agitation intensity on growth, mycelial morphology and amyloglucosidase (a recombinant protein) production in cultures of *Aspergillus oryzae* in chemostat and fed-batch cultures. It was found that the mycelial morphology was significantly affected by agitation intensity. However, protein production was not found to be affected by changes in agitation intensity in constant-mass chemostat cultures where the dissolved oxygen level was maintained at 75% of air saturation. In fed-batch cultures using the same genetically modified industrial strain, they found that the biomass concentration and protein secretion increased with increasing agitation speed when the dissolved oxygen level was controlled at 50% of air saturation. However, when the dissolved oxygen fell below 40% due to the enhanced viscosity of the broth, the protein production stopped. These studies indicate that the agitation intensity must be manipulated so that it meets process requirements in terms of dissolved oxygen levels and bulk mixing. With proper control of the process parameters, such as dissolved oxygen and agitation intensity, recombinant protein productivity can be sustained.

Although the oxygen consumption of plant and animal cells is lower than that of microorganisms, limitation in oxygen transfer is also often a constraining factor for cell cultures at high cell density. Maintaining a suitable oxygen concentration in the culture broth is equally important. The optimal dissolved oxygen concentration may be different for cell growth and product formation in animal cells [130, 131]. Chotigeat et al. [132] studied the role of environmental conditions, including the dissolved oxygen concentration and the level of sodium butyrate, on the expression levels, glycoform pattern, and the levels of sialytransferase for human follicle stimulating hormone (hFSH) production by recombinant CHO cells. In steady-state perfusion cultures in a stirred tank bioreactor at a range of different dissolved oxygen concentrations, it was found that both the specific productivity of hFSH and specific activity of sialyl transferase were increased from 0.7 to 2.6 ng (10^6cells)$^{-1}$ h^{-1} and from 1.0 to 4.9 mol (mg protein)$^{-1}$ h^{-1}, respectively, when the dissolved oxygen was increased from 10% to 90% of the air saturation level. The number of viable cells was found to be relatively constant, ranging from 4.5–5.7 x10^6 cells/ml over the dissolved oxygen levels studied. In another report, Donaldson [133] reported that elevating O_2 to 80% saturation resulted in a significant decrease in SEAP production by BTI Tn5B1-4 cells. In plant cell cultures, the deleterious effects of an over-supply of oxygen were demonstrated by Smart and Fowler [134–136]. In a suspension culture of *Catharanthus roseus* in air-lift bioreactors, a maximum biomass density of 14.3 g dry wt. per liter was achieved at a k_La value of 14.5 h^{-1}. When k_La was increased to 39 h^{-1}, only 8.9 g (dry weight) per liter was obtained.

4. INDUSTRIAL APPLICATIONS OF BIOREACTORS

Bioreactors play an important role in many industries, including fermentation, food, pharmaceuticals, and wastewater treatment. For example, a membrane bioreactor was recently applied to the treatment of foul condensates from Kraft pulp mills at high temperatures, and it showed technical feasibility and good potential for industrial application [137]. Also, industrial wastewater bioreactors are rich sources of novel microorganisms for biotechnology. Because microorganisms exist in nature as members of complex, mixed communities, the microbial communities in industrial wastewater bioreactors can be used as model systems to study the evolution of new metabolic pathways in natural ecosystems [138]. In the following, recent studies of industrial bioreactors are briefly discussed.

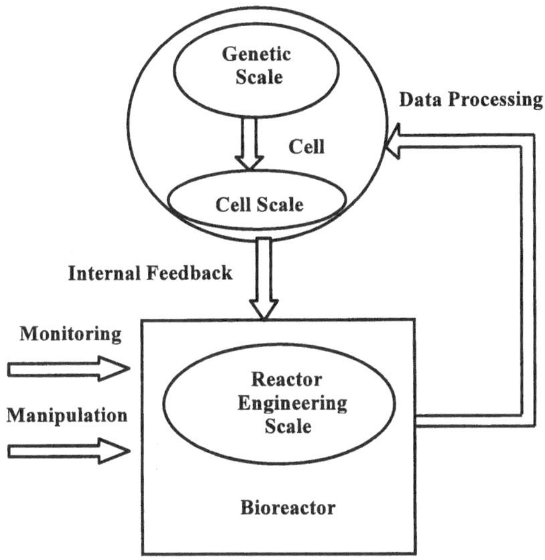

Fig. 7. Network of relationships between the different scales in a bioreactor.

4.1. Multi-scale study of industrial bioreactors and bioprocesses

Qualitative and quantitative descriptions of a production process through the analysis of various parameters by automatic or manual methods are necessary for process control and optimization. A multi-scale approach to study industrial fermentation processes was recently proposed (Fig. 7). The objects of process monitoring can be the environmental status or the varied values of operational variables. Through analysis, the cellular or engineering problems of a bioreactor on different scales can be identified. Inter-scale observation and operation is crucial in bioprocess optimization. Based on parameter correlations and the scale-up technique for the regulation of multiple parameters in bioprocesses, an optimization methodology for the study of multi-scale problems in fermentation processes was proposed through investigations on typical industrial fermentation processes for penicillin, erythromycin, chlortetracycline, inosine, and guanosine [139].

4.2. Measurement of parameters in industrial-scale bioreactors

Measurement and analysis of bioprocess parameters are very important for understanding industrial-scale bioreactor behaviors. A lack of models and sensors for describing and monitoring large-scale solid substrate cultivation (SSC) bioreactors has hampered the industrial development and application of this type of process. An indirect dynamic measurement model for water content in a 200-kg-capacity fixed-bed SSC bioreactor under periodic agitation was presented [140]. For the growth of the filamentous fungus *Gibberella fujikuroi* on wheat bran, the model uses CO_2 production rate and inlet air conditions to estimate average bed water content and average bed temperature. The model adequately reproduces the evolution of the average bed water content and can therefore be used as an on-line estimator in pilot-scale SSC bioreactors. It may prove useful in establishing advanced model-based operational and control strategies [140]. In industrial high-density animal cell cultures, dielectric spectroscopy was applied and used to on-line monitor the concentration of CHO cells immobilized on macroporous microcarriers in a stirred tank bioreactor and in a packed-bed of disk carriers [141]. The cell concentration predicted from the spectroscopic data was in excellent agreement with off-line cell counting data for both processes. Turker [142] attempted the measurement of metabolic heat in an industrial-scale bioreactor using continuous and dynamic heat balance calorimetry. The contributions of individual heat sources influencing the temperature of the broth were evaluated and the magnitude of metabolic heat was calculated from the general energy balance. Good correlations were obtained between the oxygen uptake rate and metabolic heat. Heat balance in an industrial bioreactor can be simplified by accurately identifying individual heat sources, as opposed to laboratory bioreactors, where the contribution of each source can have a significant impact. This reduces the number of measurements for accurate heat balance and makes heat balance feasible on a large scale [142]. Wahl et al. reported serial C-13-based flux analysis of an L-phenylalanine-producing *E. coli* strain under industry-like conditions in a 300-liter bioreactor [143]. Based on the NMR labeling analysis data, three subsequent flux patterns were successfully derived by monitoring the L-Phe formation. Linear programming was performed to identify optimal flux patterns for L-Phe formation. Additionally, flux sensitivity analysis was used to identify the most promising metabolic engineering target [143].

4.3. Modeling and simulation

Various models and tools have been proposed for modeling and simulating large-scale bioreactors. A networks-of-zones analysis of mixing and mass transfer was conducted in three different industrial fermenters: 3 and 31 m^3 triple-impeller stirred tank reactors and a 236 m^3 bubble column reactor [144]. A structured unsegregated cybernetic model able to simulate the growth of baker's yeast in any possible condition in multistage industrial production was developed. The kinetic and mass transfer model developed allows us to find and maintain the optimal conditions of biomass growth in industrial fed-batch bioreactors [145].

A specially designed model reactor based on a 42-L laboratory fermenter was equipped with six stirrers (Rushton turbines) and five cylindrical disks for the parallel exponential fed-batch cultivation. In this model reactor, the mixing time, Θ_{90}, turned out to be 13 times longer (Θ_{90} = 130 seconds) compared with the 42-L standard laboratory fermentor fitted with two Rushton turbines and four wall-fixed longitudinal baffles (Θ_{90} = 10 seconds). The suitability of the model reactor for scale-down studies of mixing-time-dependent processes was proven in a scaled-down industrial L-lysine fed-batch fermentation process. The model reactor represents a valuable tool to simulate the conditions of poor mixing and inhomogeneous substrate distribution in industrial scale bioreactors [146].

In industrial fed-batch bioreactors, imperfect mixing coupled with the biological consumption of nutrients causes temporal and spatial concentration gradients leading to the formation of zones very rich in substrate close to the feed port and low or even depleted regions further from it. The direct consequence is that cells experience a changing environment during the cultivation process and, thus, respond differently than in laboratory cultivation, where a good degree of homogeneity can be assumed throughout the reactor. A drastic decline in the performance of the bioprocess is often observed in large-scale reactors due to this nonhomogeneity. Modeling of the performance of industrial bioreactors with a dynamic microenvironmental approach is illustrated in Fig. 8 [147].

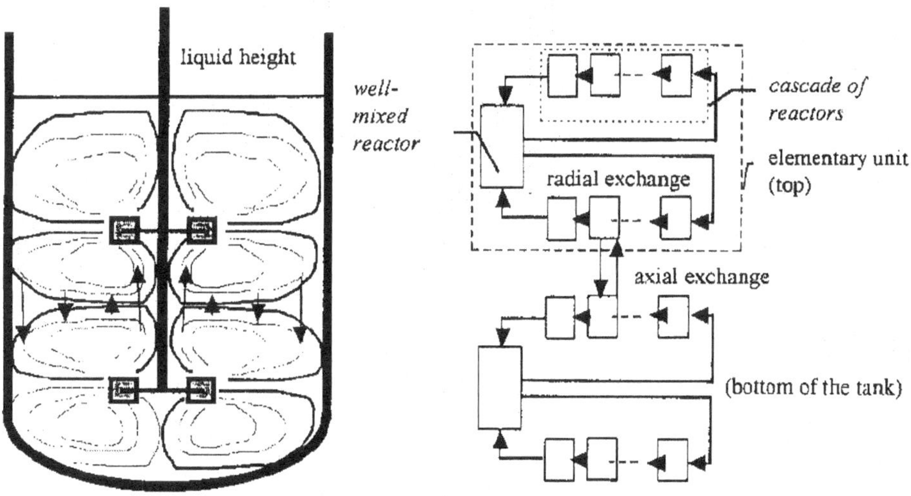

Fig. 8. Compartment mixing model: schematic of the concept (Reprinted from Ref. [147] with permission of Wiley-VCH).

Studies related to the scale-up of high-cell-density *E. coli* fed-batch fermentations using multi-parameter flow cytometry have been carried out. A changing microenvironment with respect to substrate (glucose) concentration and the dissolved oxygen tension (DOT) has a profound effect on cell physiology and hence on viable biomass yield in fermentations at both production (20 m^3) and bench (5×10^{-3} m^3) scales. The relatively poorly mixed conditions in the large-scale fermenter led to a low biomass yield, but, surprisingly, a high cell viability throughout the fermentation was achieved. Similar results were obtained in the small-scale fermentation that most closely mimicked the large-scale heterogeneity (i.e., a region of high glucose concentration and low DOT analogous to a feed zone). At the larger scale, and to differing degrees in scaled-down simulations, cells periodically encounter regions of higher glucose concentrations [148]. Further studies related to the scale-up of high cell density *E. coli* fed-batch fermentations were carried out in order to address the additional effect of a changing microenvironment when aqueous ammonia was used to control pH. It was demonstrated that in a 20-m^3 industrial fed-batch fermentation, the biomass yield of *E. coli* W3110 was lower and final cell viability was higher than those found in the equivalent well-mixed 5 L laboratory scale case. However, by using a combination of the well-mixed 5 L stirred tank reactor (STR) with a suitable plug flow reactor (PFR) to mimic the changing microenvironment at the large scale, very similar results to those in the 20-m^3 reactor were obtained [149].

5. TRENDS IN BIOREACTOR ENGINEERING

Bioreactor engineering science is experiencing rapid progress. In recent years, microbioreactors have received great interest. With the tremendous progress in functional genomics, metabolic engineering and systems biology, there is a great potential for a single cell working as a super bioreactor. It is also very exciting to see more and more achievements using plants and animals as integrated bioreactor systems.

5.1. Microbioreactor

Low-cost microbioreactors have been designed for use in high-throughput bioprocessing. An optical sensing system was used for continuous measurements of pH, dissolved oxygen, and optical density in a microbioreactor with 2-mL working volume [150]. When used for *Escherichia coli* fermentation, the microbioreactor showed similar pH, dissolved oxygen, and optical density profiles as those in a standard 1-L bioreactor. This work provided a basis for developing a multiple-bioreactor system for high-throughput bioprocess optimization. Recently, Keasling and his colleagues [151] demonstrated a scalable array for the parametric control of high-throughput cell cultivations. The technology makes use of commercial printed circuit board technology, integrated circuit sensors, and an electrochemical gas generation system. Growth data are presented for *E. coli* cultured in the array of eight 250-μL microbioreactors with varying microaerobic conditions using electrochemically generated oxygen.

Zanzotto et al. fabricated a microbioreactor, with microliters volume, out of poly(dimethylsiloxane) (PDMS) and glass (Fig. 9) [152]. Aeration was done through a gas-permeable PDMS membrane. Sensors were integrated for on-line measurement of optical density (OD), dissolved oxygen (DO), and pH, all of which were measured based on optical methods. Bacterial fermentations carried out in the microbioreactor under well-defined conditions were found to be comparable to the fermentation in a 500-mL bench-scale bioreactor. The behavior of the bacteria in the microbioreactor was similar to that in the larger bioreactor. Furthermore, it was demonstrated that the sensitivity and reproducibility of the microbioreactor system were such that statistically significant differences in the time evolution of the OD, DO, and pH could be used to distinguish between different physiological states.

To improve primary adult rat hepatocyte cultures, two types of PDMS microbioreactors containing a membrane used as a scaffold for cell attachment were built: one with a commercially-available polyester membrane, the other with a PDMS membrane (5×5 μm hole

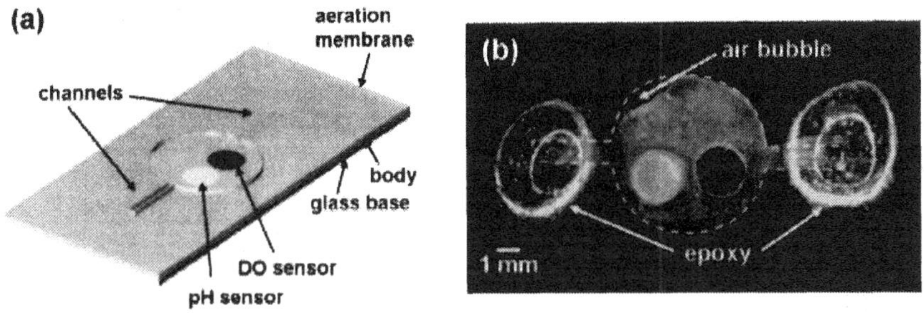

Fig. 9. Microbioreactor built of three layers of PDMS on top of a layer of glass. (a) Solid model drawn to scale; (b) photograph of microbioreactor at the end of a run (Reprinted from Ref. [152] with permission of John Wiley & Sons, Inc.)

size) made in the laboratory. These new membrane-based PDMS microbioreactors, which closely mimic the *in vivo* liver architecture, revealed themselves to be very promising tools for future applications in drug screening and liver tissue engineering [153].

In an effort to develop microbioreactor device for animal cell culture processing, Hung et al. [154, 155] recently designed a 10 × 10 microfluidic array for continuous perfusion culture. The 10 × 10 array was fabricated on a 2 × 2 cm device, consisting of a circular microfluidic chamber, a set of narrow perfusion channels surrounding the main chamber, and four ports for fluidic access. Human carcinoma (HeLa) cells were cultured inside the device, and successful operation of the continuous perfusion culture was verified over 16 days. The device functioned well for repeated cell growth/passage cycles, reagent introduction, and real-time optical analysis [155].

5.2. Cell as a super bioreactor

Many different kinds of commercially important products are derived from the cell factory, and metabolic engineering serves as an integrated approach to design new cell factories by providing rational design procedures and valuable mathematical and experimental tools [156]. For example, lactic acid bacteria were metabolically engineered to produce important compounds, including diacetyl, alanine, and exopolysaccharides [157]. As a consequence of large sequencing programs, the complete genomic sequence has become available for an increasing number of organisms. This has resulted in substantial research efforts in assigning functions to all identified open reading frames – referred to as functional genomics. In both metabolic engineering and functional genomics, there is a trend towards the application of a macroscopic view to cell function, leading to an expanded role for the classical approach in microbial physiology. With the increased understanding of molecular mechanisms, it will be possible to describe the interaction between all the components in a cellular system (the cell) at the quantitative level. This is the goal of systems biology, and would significantly facilitate studies on microbial physiology and metabolic engineering [158].

It is very interesting to engineer the plant cell factory for secondary metabolite production, because plants synthesize an extensive array of secondary metabolites that can be used as drugs, dyes, flavors, and fragrances. These plant metabolites often have highly complex structures. Currently, most pharmaceutically important secondary metabolites are isolated from wild or cultivated plants because their chemical synthesis is not economically feasible. To increase secondary metabolite production, different strategies may be adopted, such as overcoming rate limiting steps, reducing flux through competitive pathways, reducing catabolism, and overexpressing regulatory genes [159]. Our limited knowledge of secondary metabolite pathways and the genes involved is one of the main bottlenecks. However, advances in plant genomics and metabolite profiling offer unprecedented possibilities for exploring the extraordinary complexity of plant biochemical capacity. State-of-the art genomics tools can be used to enhance the production of known target metabolites or to synthesize entire novel compounds by so-called combinatorial biochemistry in cultivated plant cells [160]. Plant cell cultures combine the merits of whole-plant systems with those of microbial and animal cell cultures and already have an established track record for the production of valuable therapeutic secondary metabolites. Although no recombinant proteins have yet been produced commercially using plant cell cultures, there have been many proof-of-principle studies and several companies are investigating the commercial feasibility of such production systems [161].

The heterogeneity of plant secondary metabolites is an extremely interesting and important issue because these structurally similar natural products have different biological activities. For example, Rg_1 stimulates the central nervous system, whereas Rb_1 tranquilizes it and Rc inhibits it. It is very advantageous to intentionally manipulate the heterogeneity of secondary metabolites in cell cultures by altering or stimulating their genome and/or the subsequent processes, resulting in the enzymatic biosynthesis of secondary metabolites and allowing the production of secondary metabolites with a high degree of chemical diversity from the existing plant cell culture library. The main strategy for manipulating the production of individual ginsenosides is to intentionally change external environmental factors in cell cultures.

Our group has used chemically synthesized 2-hydroxyethyl jasmonate (HEJA) to induce ginsenoside biosynthesis and to manipulate the product heterogeneity in suspension cultures of *P. notoginseng* [162]. Interestingly, it was found that HEJA could stimulate ginsenosides biosynthesis and change their heterogeneity more efficiently than methyl jasmonate (MJA) and that the activity of Rb_1 biosynthetic enzyme, *i.e.* UDPG-ginsenoside Rd glucosyltransferase (UGRdGT), was also higher in the former case. Our results suggest that MJA and HEJA may induce ginsenoside biosynthesis via induction of endogenous JA biosynthesis and key enzymes in the ginsenoside biosynthetic pathway such as UGRdGT.

This valuable information is useful for the hyper-production of plant-specific heterogeneous products. It is expected that the dream of manipulating plant cells in order to directly produce high-value-added secondary metabolites will come true with the advancement of functional genomics and plant metabolic engineering.

5.3. Plant and animal as powerful protein-producing bioreactors

The limited capacity of current bioreactors has led the biopharmaceutical industry to investigate alternative protein expression systems. The use of whole plants for the synthesis of recombinant proteins has recently received a great deal of attention from industry as a natural bioreactor for the production of industrial and chemical products because of advantages in economy, scalability, and safety over traditional microbial and mammalian production systems. Useful expression systems based on promoters which optimize transgene expression in plant cells hold the key to maximizing the potential of this concept of molecular-farming or industrial plants. The use of plants, which are natural bioreactors, for heterologous protein production has received increasing attention [163].

The high-level expression and efficient recovery of recombinant proteins are two main critical factors that determine the use of transgenic plants as natural bioreactors to produce foreign proteins for industrial applications. The potential of a new strategy involving chloroplast transformation, GUS-fusions and affinity-tag based chromatography to overexpress and purify a human therapeutic protein, interferon gamma (IFN-γ), in tobacco plants was demonstrated by Leelavathi and Reddy [164]. The IFN-γ accumulation reached up to 6% of total soluble protein when expressed as a GUS-fusion protein in tobacco chloroplasts. Addition of His-tag simplified the downstream process and the recombinant protein yields were high (~360 μg/g fresh leaf tissue). Using plants as 'natural bioreactors', the new strategy has a tremendous potential for the large-scale production of proteins from heterologous sources, independent of their physio-chemical and biological properties.

Transgenic animals are ready to become industrial bioreactors for the preparation of pharmaceuticals in milk and probably in the future, in egg white. The milk of transgenic cattle may provide an attractive vehicle for the large-scale production of biopharmaceuticals. The production of recombinant human lactoferrin, an iron-binding glycoprotein involved in innate host defense, at gram per liter concentrations in bovine milk was reported [165]. The results illustrate the potential of transgenic cattle in the large-scale production of biopharmaceuticals. Park et al. reported the expression of a recombinant version of human α-fetoprotein (a 68 kDa glycoprotein, rhAFP) in the milk of transgenic goats. After purification and characterization, the results demonstrate that an active form of rhAFP can be produced on an industrial scale by expression in transgenic goat milk [166]. The generation of a transgenic rabbit producing recombinant human erythropoietin (rhEPO) in the lactating mammary gland was also reported [167]. Transgenic individuals are viable, fertile and transmit the rhEPO gene to the offspring. The level of rhEPO secretion in the founder female, measured in the period of lactation, varied in the range of 60–178 and 60–162 mIU/ml in the milk and blood plasma, respectively. The biological activity of the milk rhEPO was confirmed by a standard [H-3]-thymidine incorporation test. The model of a rhEPO-transgenic rabbit, valuable for studies of rhEPO glycosylation and function, can be useful for the development of transgenic approaches designed for the preparation of recombinant proteins by alternative biopharmaceutical production.

ACKNOWLEDGMENTS

J.J.Z. thanks the National Science Fund for Distinguished Young Scholars (NSFC project no. 20225619) and the Cheung Kong Scholars Program of the Ministry of Education of China.

REFERENCES

[1] J. Varley and J. Birch, Cytotechnology, 29 (1999) 177.
[2] D.G. Cronin and A.W. Nienow, Proceedings of the Third NEL Bioreactor Research Symposium Paper No. 1-C (1989) 17.
[3] A.W. Nienow, Chem. Eng. Prog., 86(1990) 61.
[4] M. Whitton, 2nd International Conference on Bioreactor Fluid Dynamics, King R (Ed), Elsevier Applied Science, Oxford, 1988, p135.
[5] J. Lehmann., J. Vorlop and H. Buntemayer, In: Spier R.E. and Griffiths J.B. (eds) Animal Cell Biotechnology Vol. 3. Academic Press, London, 1988, p 221.
[6] A. Handa-Corrigan, In: Spier R.E. and Griffiths J.B. (eds) Animal Cell Biotechnology Vol. 4. Academic Press, London, 1990, p123.
[7] W.S. Hu, J. Meier and D.I.C. Wang, Biotech. Bioeng., 28 (1986) 122.
[8] S.-J.Wang , J.-J.Zhong, Y. Chen and J. Yu., J. Ferment. Bioeng., 80 (1995) 71.
[9] N. Kioukia, A.W. Nienow, A.N. Emery and M. Al-Rubeai, Trans. I. Chem. E.,70 C (1992) 143.
[10] L.A. Van der Pol, I. Beeksma and J. Tramper, Enzyme Microb.Technol., 17 (1995) 401.
[11] L. Van der Pol, D. Bonarius, G. Vandewouw and J. Tramper, Biotechnol. Prog., 9(1993) 504.
[12] E.T. Papoutsakis, Trends Biotechnol., 9 (1991), 316.
[13] A.W. Nienow, C. Langheinrich, N.C. Stevenson, A.N. Emery, T.M. Clayton and N.K.H. Slater, Cytotechnology, 22 (1996) 87.
[14] M.Reuss, Stirred Tank Bioreactors. In: Bioreactor System Design, Asenjo JA & Merchuk JC (Eds). Marcel Dekker, New York, 1995, p207.
[15] Y. Shi, D.Y. Dewey and H. Park, Biotechnol. Bioeng., 69 (1992) 260.
[16] S.J. Wang and J.J Zhong, Biotechnol. Bioeng. 51 (1996) 511.
[17] S.J. Wang and J.J Zhong, Biotechnol. Bioeng. 51 (1996) 520
[18] M. E. Curtin, Bio/Technology, 1 (1983) 649.
[19] T. Hashimoto, S. Azechi, S. Sugita and K. Suzuki, Large scale production of tobacco cells by continuous cultivation. In: Plant Tissue Culture 1982; Fujiwara, A., Ed.; Japanese Association for Plant Tissue Culture, Tokyo, 1982, p 403.
[20] E..Rittershaus, J.Ulrich and K. Westphal, Int. Assoc. Plant Tissue Cult. Newsl., 61 (1990) 2.
[21] P.M. Doran, Biotechnol. Prog., 15 (1999) 319.
[22] Z.Y. Zhang and J.J. Zhong, Biotechnol. Prog., 20 (2004) 1076.
[23] J.J. Zhong, J. Biosci. Bioeng., 94 (2002) 591.
[24] J.J. Zhong, Z.W. Pan, Z.Y. Wang, J.Y. Wu, F. Chen, M. Takagi and T. Yoshida, J. Biosci. Bioeng., 94 (2002) 244.

[25] J.J. Zhong, F. Chen and W.W. Hu, Process Biochem., 35 (2000) 491.
[26] D. M. Marks, Cytotechnology 42 (2003) 21.
[27] U.E. Viesturs, I.A. Sturmanis, V.V. Krikis and V.D. Prokopenko, Biotechnol. Bioeng., 22 (1980) 799.
[28] U.E. Viesturs, I.A. Sturmanis, V.V. Krikis and V.D. Prokopenko, Biotechnol. Bioeng., 23 (1981) 1171.
[29] A. Margaritis and J.D. Sheppard, Biotechnol. Bioeng. 23 (1981) 2117.
[30] W.A.M. Bakker, H.J.L. van Can, J. Tramper and C.D. Der Gooijer, Biotechnol. Bioeng., 42 (1993) 994.
[31] F.A. Figen and B.Z. Uysal, J. Chem. Technol. Biotechnol., 72 (1998) 143.
[32] H. Kawasaki, H. Hirano and H. Tanaka, J. Chem. Eng. Jpn., 27 (1994) 669.
[33] C. Wei, B. Xie and H. Xiao, Chem. Eng.Technol., 23 (2000) 38.
[34] C. Wei, B. Xie, H. Xiao and D. Wang, Chem. Eng. Technol., 23 (2000) 597.
[35] V. Lazarova, J. Meyniel, L. Duval and J. Manem, Chem. Eng. Sci., 52 (1997) 3919.
[36] Y. Chisti, F. Wenge and M. Moo-Young, Chem. Eng. J., 57 (1995) B7.
[37] H. Blenke, Adv. Biochem. Eng., 13 (1979) 121.
[38] R.D. Fraser and G.A. Hill, Can. J. Chem. Eng. 71 (1993) 419.
[39] B. Konig and K. Schügerl, Biotechnol. Bioeng., 24 (1982) 259.
[40] D.J. Pollard, A.P. Ison, P.A. Shamlou and M.D. Lilly, Biotechnol. Bioeng., 58 (1998) 453.
[41] N.Y. Chen, E.F. Kondis and S. Srinivasan, Biotechnol. Bioeng., 24 (1987) 414.
[42] C. Sola and F. Godia, Bioreactor System Design, Asenjo JA and Merchuk JC (Eds) Marcel Dekker, New York, 1995, pp. 511–552.
[43] Y.T. Shah, B.G. Kelkar, S.P.Godbole and W.-D.Deckwer, A.I.Ch.E. J. 28 (1982) 353.
[44] H. Tsuge, S. Otatsume, K. Kobayashi, K. Terasaka, M. Hayasaki and H. Kobayashi, J. Chem. Eng. Jpn., 37 (2004) 941.
[45] S.D. Vlaev and M. Fialova, Can. J. Chem. Eng. 81 (2003) 535.
[46] F.C. Rubio, A.S. Miron, M.C.C. Garcia, F.G.. Camacho, E.M. Grima and Y. Chisti, Chem. Eng. Sci. 59 (2004) 4369.
[47] A.S. Miron, M.C.C Garcia, F.G. Camacho, E.M. Grima EM and Y. Chisti, Chem. Eng. Res. Des. 82 (2004) 1367.
[48] W.W. Hu, H. Yao and J.J. Zhong, Biotechnol. Prog., 17 (2001) 838.
[49] M.J. Barbosa, Hadiyanto and R.H. Wijffels, Biotechnol. Bioeng., 85 (2004) 78.
[50] M.M. Resende, S.M. Ratusznei, C.A.T. Suazo, R.C. Giordano, Proc. Biochem. 37 (2002) 1297.
[51] J.Y. Xin, S.B. Li, Y. Xu, J.R. Chui, C.G. Xia, J. Chem. Technol. Biotech., 76 (2001) 579.
[52] S. Velizarov, J.G. Crespo, M.A. Reis, Biotechnol. Prog., 18 (2002) 296.
[53] C.H. Kuo, C.C. Chen and B.H. Chiang, J. Food Sci., 69 (2004) E332.
[54] D.M.F. Prazeres and J.M.S. Cabral, Enzyme Microb. Technol., 16 (1994) 738.
[55] B.M. de Roode, M.C.R. Franssen, A. vam der Padt and R.M. Boom, Biotech. Prog., 19 (2003) 1391.
[56] O. Gaouar, C. Aymard, N. Zakhia and G.M. Rios, J. Chem. Technol. Biotechnol. 69 (1997) 367.
[57] R. Lopez-Ulibarri and G.M. Hall, Enzyme Microb. Technol., 21 (1997), 398.
[58] Q. Gan, S.J. Allen and G. Taylor, Biochem. Eng. J., 12 (2002) 223.
[59] G. Akay, E. Erhan, B. Keskinler and O.F. Algur, J. Membr. Sci., 206 (2002) 61.

[60] E. Erhan, B. Keskinler, G. Akay and O.F. Algur, J. Membr. Sci., 206 (2002) 361.
[61] Y. Kamoshita, R. Ohashi and T. Suzuki, J. Ferment. Bioeng., 85 (1998) 422.
[62] M. Taniguchi, I. Nakagawa, K. Hoshino, T. Itoh, K. Ohno and M. Fujii, Agric. Biol. Chem., 53 (1989) 2447.
[63] D.S. Inloes, W.J. Smith, D.P. Taylor, S.N. Cohen, A.S. Michaels and C.R. Robertson, Biotechnol. Bioeng., 25 (1983) 2653.
[64] B.H. Chung, H.N.Chang and I.H. Kim, Enzyme Microb. Technol. 9 (1987) 345.
[65] M. Scholz and W.-S. Hu, Cytotechnoplogy, 4 (1990) 127.
[66] L.R. Jackson, L.J. Trudel, J.G. Fox and N.S. Lipman, J. Immunol. Methods 189 (1986) 217.
[67] T. Stoll, C. Perregaux, U. Von Stockar and I.W. Marison, Cytotechnology 17 (1995) 53.
[68] R. Valdes, N. Ibarra, M. Gonzalez, T. Alvarez, J. Garcia, R. Llambias, C.A. Perez, O. Quintero and R. Fischer, Cytotechnology 35 (2001) 145.
[69] E.W. Low and H.A. Chase, Water Res., 33(1999) 847.
[70] S. Chaize and A. Huyard, Water Sci. Technol., 23 (1991) 1591.
[71] E.B. Muller, A.H. Stouthamer, H.W. van der Verseveld and D.H. Eikelboom, Water Res., 29 (1995) 1179.
[72] S. Rosenbergera, U. Kr.ugera, R. Witzigb, W. Manzb, U. Szewzykb and M. Kraume, Water Res., 36 (2002) 413.
[73] A. Tazi-Pain, J.C. Schrotter, G. Bordb, M. Payreaudeaub and H. Buisson, Desalination, 146 (2002) 439.
[74] S.J. Ergas. Membrane Bioreactors. In: Bioreactors for Waste Gas Treatment (Kennes and Veiga (eds.), Kluwer Academic Publishers, Dordrecht. 2001, p163.
[75] T. Stephenson, S. Judd, B. Jefferson and K. Brindle, Membrane Bioreactors for Wastewater Treatment. 2000, IWA Publishing, London.
[76] M.W. Reij, J.T.F Keurentjes and S. Hartmans, J. Biotechnol., 59 (1998) 155.
[77] L.M. Freitas dos Santos, P. Pavasant, L.F. Strachan, E.N. Pistikopoulos, A.G. Livingston, Pure Appl. Chem. 69 (1997) 2459.
[78] E.Casey, B. Glennon and G. Hamer, Resources, Conservation and Recycling, 27 (1999) 203.
[79] T. Shiotani and T. Yamane, Eur. J. Appl. Microbiol. Biotechnol., 13 (1981) 96.
[80] N.Y. Fortin and M.A. Deshusses, Environ. Sci. Technol., 33 (1999) 2980.
[81] S.L. Kazenski and K.A. Kinney, Biofiltration of paint spray booth emissions: packing and media considerations and VOC interactions. IN: Proceedings of the 93rd Annual Meeting & Exhibition of the Air & Waste Manage. Assoc., 2000, Pittsburgh, PA.
[82] R. Yan, Y.L. Ng, X.G. Chen, A.L. Geng, W.D. Gould, H.Q. Duan, D.T. Liang and L.C.C. Koe, Wat. Sci. Technol., 50 (2004) 299.
[83] J.H. So, D.A. Webster, B.C. Stark and K.R. Pagilla, Biodegradation, 15 (2004) 161.
[84] S-T. Yang, J. Luo and C. Chen, Adv. Biochem. Eng. Biotechnol., 87 (2004) 61.
[85] C. Nicolella, M.C.M. van Loosdrecht and J.J. Heijnen, J. Biotechnol. , 80 (2000) 1.
[86] P.F. Cooper and P.M. Sutton, P.M., Chem. Eng. 90 (1983), 392.
[87] S. Sen and G.N. Demirer, Wat. Res., 37 (2003) 1868.
[88] G. Gonzalez, G. Herrera, M.T. Garcia, M. Pena, Bioresource Technol., 80 (2001) 137.
[89] A.E.F. Botrous, M.F. Dahab and P. Mihaltz, Wat. Sci. Technol., 49 (2004) 65.

[90] M. An and K.V. Lo, J. Environ. Sci. Health Part A-Toxic/Hazardous Substances & Environ. Eng., 36 (2001) 819.
[91] M.P. Durrschmid, K. Landauer, G. Simic, H. Klug, T. Keijzer, F. Trampler, A. Oudshoorn, M. Groschl, D. Muller, O. Doblhoff-Dier, Biotechnol. Prog., 19 (2003) 1045.
[92] Y.J. Kwon and C.A. Peng, Biotechniques, 33 (2002) 212.
[93] G. Liden, Bioproc. Biosys. Eng., 24 (2002) 273.
[94] M.E. Shuler and F. Kargi, Bioprocess Engineering: Basic Concepts, Prentice Hall, Upper Saddle River, NJ, 2002, p 169-173.
[95] J.A. Asenjo, J.L. Spencer and V. Phuvan, Ann. N.Y. Acad. Sci., 469 (1986) 404.
[96] P. Ducommun, P.-A. Rueux, A. Kadouri, U. von Stockar and I.W. Marison, Biotechnol. Bioeng., 77 (2002) 838.
[97] A. Moore, J. Mercer, G. Dutina, J. Donahue, K. Bauer, J. Mather, T. Etcheverry and T. Ryll, Cytotechnology, 23 (1997) 47.
[98] K. Furukawa and K. Ohsuye, Cytotechnology, 26 (1998) 153.
[99] S.M. Schatz, R.J. Kerschbaumer, G. Gerstenbauer, M. Kral, F. Dorner and F. Scheiflinger, Biotechnol. Bioeng., 84 (2003) 433.
[100] H. Kaufmann, X. Mazur, M. Fussenegger and J.E. Bailey, Biotechnol. Bioeng., 63 (1999) 573.
[101] N. Barnabé and M. Butler, Biotechnol. Bioeng. 44 (1994) 1235.
[102] I.C. Tang, M.R. Okos and S.T. Yang, Biotechnol. Bioeng., 34 (1989) 1063.
[103] D. Weuster-Botz, J. Altenbach-Rehm and M. Arnold, Biochem. Eng. J. 7 (2001) 163.
[104] A. Amanullah, C.M. McFarlane, A.N. Emery and A.W. Nienow, Biotechnol. Bioeng., 73 (2001) 390.
[105] M. Elibol, Process Biochem., 37 (2002) 1381.
[106] I. Elmahdi, F. Baganz, K. Dixon, T. Harrop, D. Sugden, G.J. Lye, Biochem. Eng. J., 16 (2003) 299.
[107] G.J. Lye, P. Ayazi-Shamlou, F. Baganz, P. Dalby and J.M. Woodley, Trends Biotechnol., 21 (2003) 29.
[108] J. Wayte, R. Boraston, H. Bland, J. Varley and M. Brown, Gen. Eng. Biotechnol., 17 (1997) 125.
[109] M.W.Glacken, Bioreactor Control and Optimization. In: Animal Cell Bioreactors (Edited by C.S. Ho and D.I.C. Wang). Butterworth-Heinemann, MA, USA. 1991. P 379.
[110] S.S. Ozturk and B.O. Palsson, Biotechnol. Prog. 7 (1991) 481.
[111] L. Xie, W. Pilbrough, C. Metallo, T. Zhong, L. Pikus, J. Leung, J.G. Aunins, W. Zhou, Biotechnol. Bioeng., 80 (2002) 569.
[112] P.W. Sauer, J.E. Burky, M.C. Wesson, H.D. Sternard and L. Qu, Biotechnol. Bioeng., 67 (2000) 585.
[113] M.C. Borys, D.H. Linzer and E.T. Papoutsakis, Bio/Technology, 11(1993) 720.
[114] M.C. Borys, D.H. Linzer and E.T. Papoutsakis, Biotechnol. Bioeng., 43 (1994) 505.
[115] J.A. Zanghi, A.E. Schmelzer, T.P. Mendoza, R.H. Knop and W.M. Miller, Biotechnol. Bioeng., 65 (1999) 182.
[116] C.L. McDowell and E.T. Papoutsakis, Biotechnol. Prog., 14 (1998) 567.
[117] C. Oniscu, A.I. Galaction, D. Cascaval and F. Ungureanu, Biochem. Eng. J., 12 (2002) 61.

[118] M.M. Toma, U. Kalnenieks, A. Berzins, A. Vigants, M. Rikmanis and U. Viesturs, Process Biochem., 38 (2003) 1347.
[119] C. Langheinrich and A.W. Nienow, Biotechnol. Bioeng., 66 (1999) 171.
[120] A.W. Nienow, S.M. Drain, A.P. Boyes and K.J. Carpenter, Ind. Eng. Chem. Res., 36 (1997) 2984.
[121] J. Nielsen and J. Villadsen, Bioreaction Engineering Principles. New York: Plenum Press, 1994.
[122] B.H. Junker, M. Stanik, C. Barna, P. Salmon and B.C. Buckland, Bioprocess Eng., 19 (1998) 403.
[123] S. Ensari and H.C. Lim, Process Biochem., 38 (2003) 1531.
[124] A. Hadj Sassi, A.M. Deschamps and J.M. Lebeault. Process Biochem. 31 (1996) 493.
[125] Q. Hua, P-C. Fu, C. Yang and K.Shimizu, Biochem. Eng. J., 2 (1998) 89.
[126] A. Amanullah, L.H. Christensen, K. Hansen, A.W. Nienow and C.R. Thomas, Biotechnol. Bioeng., 77 (2002) 815.
[127] P. Ayazi Shamlou, H.Y. Makagiansam, A.P. Ison and M.D. Lilly, Chem. Eng. Sci., 49 (1994) 2621.
[128] J.Nielsen, O.L. Johansen, M. Jacobsen, P. Krabben and J.Villadsen, Biotechnol. Prog., 11 (1995) 93.
[129] A. Amanullah, R. Blair, A.W. Nienow and C.R. Thomas, Biotechnol. Bioeng., 62 (1999) 434.
[130] M. M. William, R. W. Charles, W. B. Harvey, J. Cell. Physiol., 132 (1987) 524.
[131] S. Reuveney, D.Velez, J. D. Macmillan and L. Miller, J. Immunol. Methods, 86 (1986) 53.
[132] W. Chotigeat, Y. Watanapokasin, S. Mahler and P.P. Gray, Cytotechnology, 15 (1994) 217.
[133] M. Donaldson, H. A. Wood, P. C. Kulakosky and M.L. Shuler, Biotechnol. Bioeng. 63 (1999) 255.
[134] N.J. Smart and M.W. Fowler, Biotechnol. Lett., 3 (1981) 171.
[135] N.J. Smart and M.W. Fowler, Appl. Biochem. Biotechnol., 9 (1984) 209.
[136] N.J. Smart and M.W. Fowler, J. Exp. Bot., 35 (1984) 531.
[137] J.C.T. Dias, R.P. Rezende, C.M. Silva and V.R. Linardi, Process Biochem., 40 (2005) 1125.
[138] M.G. Bramucci and V. Nagarajan, Trends Biotechnol., 18 (2000) 501.
[139] S.L. Zhang, J. Chu and Y.P. Zhuang, Adv. Biochem. Eng. Biotechnol., 87 (2004) 97
[140] M.P.Y. Lillo, R. Perez-Correa, E. Agosin and E. Latrille, Biotechnol. Bioeng., 76 (2001) 44.
[141] P. Ducommun, A. Kadouri, U. von Stockar and I.W. Marison, Biotechnol. Bioeng., 77 (2002) 316.
[142] M. Turker, Chem. Eng. Comm., 190 (2003) 573.
[143] A. Wahl, M. El Massaoudi, D. Schipper, W. Wiechert and R. Takors, Biotechnol. Prog., 20 (2004) 706.
[144] J. Zahradnik, R. Mann, M. Fialova, D. Vlaev, S.D. Vlaev, V. Lossev and P. Seichter, Chem. Eng. Sci., 56 (2001) 485.
[145] M. Di Serio, R. Tesser and E. Santacesaria, Chem. Eng. J., 82 (2001) 347.
[146] B.M. Schilling, W. Pfefferle, B. Bachmann, W. Leuchtenberger and W.D. Deckwer, Biotechnol. Bioeng., 64 (1999) 599.
[147] F. Guillard and C. Tragardh, Chem. Eng. Technol., 22 (1999) 187.

[148] C.J. Hewitt, G. Nebe-Von Caron, B. Axelsson, C.M. McFarlane and A.W. Nienow, Biotechnol. Bioeng., 70 (2000) 381.
[149] H. Onyeaka, A.W. Nienow and C.J. Hewitt, Biotechnol. Bioeng., 84 (2003) 474.
[150] Y. Kostov, P. Harms, L. Randers-Eichhorn and G. Rao, Biotechnol. Bioeng., 72 (2001) 346.
[151] M.M. Maharbiz, W.J. Holtz, R.T. Howe and J.D. Keasling, Biotechnol. Bioeng., 85 (2004) 376.
[152] A. Zanzotto, N. Szita, P. Boccazzi, P. Lessard, A.J. Sinskey and K.F. Jensen, Biotechnol. Bioeng., 87 (2004) 243.
[153] S. Ostrovidov, J.L. Jiang, Y. Sakai and T. Fujii, Biomed Microdevices 6 (2004) 279.
[154] P.J. Hung, P.J. Lee, P. Sabounchi, N. Aghdam, R. Lin and L.P. Lee, Lab Chip, 5 (2005) 44.
[155] P.J. Hung, P.J. Lee, P. Sabounchi, R. Lin and L.P. Lee, Biotechnol. Bioeng. , 89 (2005) 1.
[156] K.R. Patil, M. Akesson and J. Nielsen, Curr. Opinion Biotechnol., 15 (2004) 64.
[157] M. Kleerebezem, P. Hols and J. Hugenholtz, Enzyme Microb. Technol., 26 (2000) 840.
[158] J. Nielsen and L. Olsson, FEMS Yeast Res., 2 (2002) 175.
[159] R. Verpoorte, R. van der Heijden and J. Memelink, Transgenic Res., 9 (2000) 323.
[160] K.M. Oksman-Caldentey and D. Inze, Trends Plant Sci., 9 (2004) 433.
[161] S. Hellwig, J. Drossard, R.M. Twyman and R. Fischer, Nat. Biotechnol., 22 (2004) 1415.
[162] W. Wang, Z.J. Zhao, Y.F. Xu, X.H. Qian and J.J. Zhong, Appl. Microbiol. Biotechnol., 70 (2006) 298.
[163] K. Yoshida and A. Shinmyo, J. Biosci. Bioeng., 90 (2000) 353.
[164] S. Leelavathi and V.S. Reddy, Mol Breeding 11 (2003) 49.
[165] P.H.C. van Berkel, M.M. Welling, M. Geerts, H.A. van Veen, B. Ravensbergen, M. Salaheddine, E.K.J. Pauwels, F. Pieper, J.H. Nuijens and P.H. Nibbering, Nat. Biotechnol., 20 (2002) 484.
[166] M.H. Park, E. Birck-Wilson, G. Allard, N. Masiello, M. Day, K.P. Murphy, V. Paragas, S. Silver and M.D. Moody, Protein Exp Purif 38 (2004) 177.
[167] T. Mikus, M. Poplstein, J. Sedlakova, V. Landa, G. Jenikova, P. Trefil, J. Lidicky and P. Maly, Transgenic Res., 13 (2004) 487.

Bioprocessing for Value-Added Products from Renewable Resources
Shang-Tian Yang (Editor)

Chapter 7. Membranes for Bioseparations

Chia-Chi Ho

Department of Chemical and Materials Engineering, University of Cincinnati, 497 Rhodes Hall, Cincinnati, OH 45221-0012 USA

1. INTRODUCTION

Membranes have always played an integral part in downstream processing in the biotechnology and pharmaceutical industries. The advantages of using membranes for bioseparations include: they can operate at room temperature and without a phase change; they are easy to scale up; and they operate with low energy consumption. These properties are particularly important for the purification of enzymes and proteins, which are easily deactivated or denatured under extreme conditions. In this chapter, membrane processes used in bioseparations are briefly reviewed in Section 2 while membrane materials and modules commonly used in biotechnology industries for concentration or separation of biochemical compounds are summarized in Section 3. Membrane fouling, the most critical issue in applying membrane processes to biological methods, is emphasized in Section 4, with discussions on the factors affecting fouling phenomena and mathematical modeling of the fouling process. This is followed by Sections 5 and 6 which discuss some key applications and outlook on membrane separations in biotechnology.

2. MEMBRANES IN BIOSEPARATIONS

Membrane processes are conveniently categorized based on the pore size of the membrane, as shown in Table 1. Pervaporation is a unique membrane process where the mechanism of separation is not size-based, but rather based solely on the solubility and diffusivity of the permeate through the membrane. In pervaporation processes, the feed side is usually a liquid, while the permeate side is a vapor. Pervaporation is typically used for the removal of volatile organic compounds from water or the removal of water from organic solvents, e.g., recovery of ethanol from fermentation processes [1]. Unlike other membrane processes, a phase transition occurs and the heat of vaporization of the permeate must be continuously supplied to maintain a fix temperature.

Reverse osmosis membranes retain solutes with molar masses below 1 kDa, such as salts and amino acids, and are used to separate small salts or organic molecules (e.g., sugars [2] and organic acids [3]) from water. Reverse osmosis was first designed for the desalination of sea water and is now used for reclaiming or concentrating salts, amino acids, and sugar solutions. In reverse osmosis processes, high pressure, in excess of the osmotic pressure of the feed, is applied to force the solvent to diffuse through the membrane into the dilute compartment. Ideal reverse osmosis membranes allow only the solvent to pass through via the solution-diffusion mechanism in which transport is set by the solubility and diffusivity of the penetrant in the membrane.

Nanofiltration membranes retain polypeptides and other solutes with molar mass ranging between 1 kDa and 3 kDa. Nanofiltration processes were originally referred to as low pressure reverse osmosis because of the lower pressure (10 to 30 bar) required. Nanofiltration is used to retain sugars [4], polypeptides, antibiotics [5], and bivalent and higher valent ions [6]. Two mechanisms are involved in separation during nanofiltration, size-based molecular sieving and Donnan exclusions, which arise from the repulsion of charged species from membrane of similar charge.

Ultrafiltration membranes have a pore size ranging from 10 nm to 0.1 μm and are typically used for concentration of proteins, DNA or polysaccharides, buffer exchange, clarifying antibiotics, and virus clearance. Microfiltration membranes have a pore size between 0.1 μm and 10 μm and are commonly used to remove bacteria, cells, and other large particles from fermentation broths, beverages, and water. These membrane processes are used throughout the downstream purification of important biological products, with ultrafiltration and microfiltration being applied most frequently in bioseparations.

Excellent overviews of membrane processes for bioseparations are available in books by Ho and Sirkar [7], McGregor [8], and Zeman and Zydney [9].

Table 1
Membrane filtration processes

Filtration Type	Pore Size/Nominal MW cut-off	Permeating Species	Retained Species
Microfiltration	0.1 to 10 μm	viruses, colloids, proteins, nucleic acids, sugars, salts	yeast, bacteria intact cells, cell debris
Ultrafiltration	10 to 100 nm 1 to 1,000 kDa	Proteins, oligosaccharides, nucleic acids, surfactants, sugars, salt	viruses, proteins DNA, polysaccharides
Nanofiltration	1 to 10 nm 1 to 3 kDa	salt, water sugar, organic acids	polypeptides antibiotics
Reverse Osmosis	< 1 nm < 1 kDa	water	salts, amino acids sugars
Pervaporation	not based on size	water organic solvent	water or organic solvent

3. MICROFILTRATION AND ULTRAFILTRATION PROCESSES

3.1. Membrane Materials

Microfiltration and ultrafiltration membranes are made from a variety of materials, including polymers (e.g., polyethersulfone, polyethylene, polytetrafluoroethylene, polyvinylidene fluoride, nylon, polyester, polycarbonate, cellulose acetate, and regenerated cellulose), ceramics (aluminum and zirconium oxide), glasses (borosilicate glass fiber), and metals (silver and stainless steel). Regenerated cellulose, polyethersulfone, and polyvinylidene fluoride membranes are most commonly used for bioseparations due to their low protein binding characteristics. The underlying pore morphology of these materials varies significantly, depending on the technique used to prepare the membrane. Most polymeric microfiltration membranes consist of an isotropic network of polymer fibers resulting in a highly interconnected pore structure. These membranes tend to have a fairly broad pore size distribution throughout the membrane. Metallic membranes generally consist of an array of sintered metal particles or spheroids, giving an isotropic structure with more uniform pores in the interstices between the metal particles. Many ultrafiltration membranes have an asymmetric structure consisting of an ultrathin skin (approximately 0.5 μm thick) which determines the sieving characteristics of the membrane, a porous substructure, and a porous matrix which provides the membrane with its structural integrity. Although less commonly used for large-scale separations, there are a variety of membranes with nearly uniform straight through pores. For example, track-etched polycarbonate membranes have uniform cylindrical pores which are generated by etching away the damaged regions within a polymer film that has been exposed to the radioactive decay fragments from a trans-uranium element. The pore size rating of microfiltration membranes typically refers to the size of species that has a retention coefficient (R), defined as follows, of 0.9.

$$R = 1 - \frac{C_{\text{filtrate}}}{C_{\text{feed}}} \tag{1}$$

Most ultrafiltration membranes are rated by nominal molecular weight cut-offs indicating the molar mass of the species with a specified retention coefficient. The exact retention coefficient used to define the pore size rating of membranes varies throughout the membrane industry.

3.2. Membrane Modules

In addition to the membrane itself, the physical unit that houses the membrane can also have significant effects on the performance of the membrane device. Two general classes of filtration modules are used commercially. In dead-end filtration, the feed flow is directed perpendicular to the membrane surface, while in cross flow filtration, the feed flows tangentially across the membrane with a fraction of the feed driven through the membrane by the applied transmembrane pressure drop. The main advantage of dead end filtration is its simplicity. It is typically used to filter relatively small volumes or solutions with very low levels of impurities. The disadvantage of dead-end filtration is that all solutes and particles are carried directly to the membrane surface, leading to rapid particle deposition and fouling. These adverse effects are minimized in cross flow filtration since the tangential flow sweeps the particles along the membrane surface. A variety of configurations are commercially available, including plate and frame, spiral wound, hollow fiber, tubular, and vortex flow devices (Fig. 1).

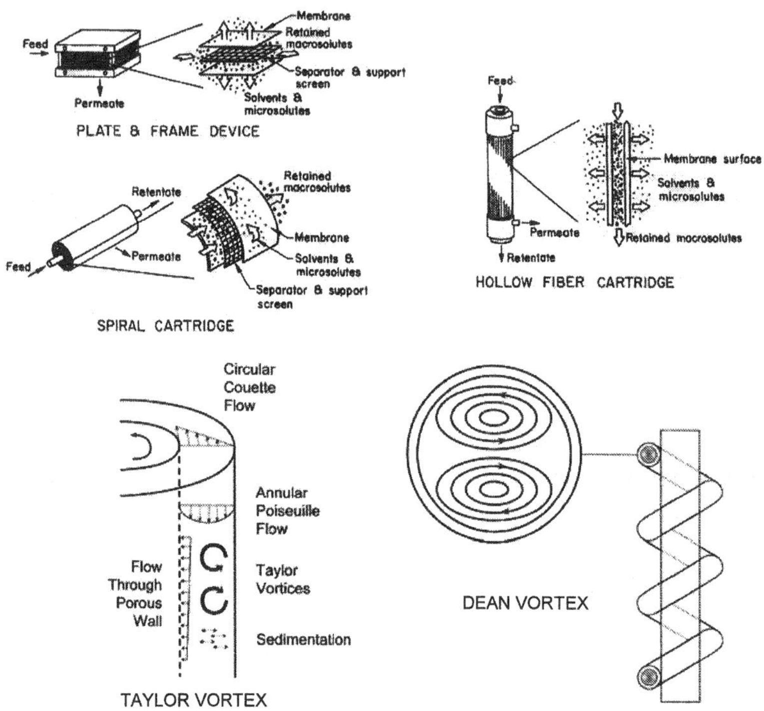

Fig. 1. Examples of different membrane modules. (Plate & frame, spiral, and hollow-fiber cartridge schematics adapted from R. S. Tutunjian in M. Moo-Young, ed., Comprehensive Biotechnology, Vol. 2., Elsevier Science, London, 1985.)

Plate and frame modules consist of several flat sheet membranes stacked together; the details of the design can be found in the book by RF Madsen [10]. Spiral wound modules also consist of flat sheet membranes separated by a porous mesh sheet in a sandwich configuration. The three sandwiched layers are sealed along three edges to form a pocket for permeate flow. The open side of the pocket is connected to a central tube which collects the permeate, and the membrane stack is spirally wound around the center tube and fitted into a tubular cartridge. In hollow fiber modules, bundles of fibers with diameters ranging from 200 to 2500 μm are potted at the ends in an epoxy or polyurethane resin and cut open to expose the open bores (lumens) of the fibers. The feed typically flows on the outside of the fibers while the permeate passes through the lumens and exits from the permeate port. Tubular devices are similar to hollow fiber modules but with the fibers replaced with large diameter tubes (typically 0.3 to 2.5 cm). Vortex devices take advantage of secondary flows present in rotating annular devices (Taylor vortices) and spiral wound tubes (Dean vortices) to improve bulk mass transfer in membrane modules.

The development of membrane processes for bioseparations is similar to that of membrane systems for non-biological applications. However, one critical factor that needs to be considered in using membranes for bioseparations is the significantly increased fouling of membranes by biological macromolecules (e.g., proteins, cells, and bacteria).

4. MEMBRANE FOULING

4.1. Membrane Fouling

Membranes are used extensively for protein purification. This includes the sterile filtration of therapeutic proteins prior to final formulation, the clarification of protein solutions from harvested cell culture media, and plasma collection from whole blood for therapeutic and commercial uses. One of the critical factors governing the performance of microfiltration or ultrafiltration processes is the irreversible alteration of the membrane caused by specific interactions between the membrane and various proteins in the process stream. Protein fouling usually manifests itself as a decay in filtrate flux and/or an alteration in membrane selectivity. Fouling often continues throughout the filtration process and eventually requires the membrane be cleaned or replaced.

It is convenient to examine irreversible fouling in terms of two distinct phenomena: protein adsorption, which describes the interaction between proteins and the membrane polymer that occurs in the absence of any convective flow through the membrane, and protein deposition, which refers to any additional protein that becomes associated with the membrane during filtration. Most studies of protein adsorption have found approximately monolayer adsorption throughout the internal area of the microfiltration membrane [11]. For example, monolayer adsorption has been reported for BSA adsorption on 0.16 μm polyethersulfone [12] and 0.22 μm aluminum oxide [13] membranes and hemoglobin adsorption on 0.2 μm alumina membranes [14]. This type of monolayer adsorption generally has little effect on the filtrate flux during microfiltration processes since the pore size of these membranes (0.1–1 μm) is orders of magnitude larger than the protein size (1–10 nm) [15]. For ultrafiltration processes, protein adsorption and bulk mass transfer limitations can lead to more significant flux decline. Protein deposition has been shown to cause significant fouling, with the flux declining by more than a factor of ten during microfiltration of cell-free protein solutions. Marshall *et al.* [16] have provided an excellent review of the literature on protein adsorption and deposition and their effects on both ultrafiltration and microfiltration.

Opong and Zydney [12] examined the effect of BSA adsorption and deposition on the hydraulic resistance of 30,000 and 100,000 nominal molecular weight cut-off (MWCO) ultrafiltration and 0.16 μm pore size microfiltration polyethersulfone (PES) membranes. The hydraulic resistance was evaluated from data for the saline flux in the absence of any protein. Protein adsorption caused a large increase in resistance for the partially retentive 100,000 MWCO membrane, but had minimal effect on the resistance of either the fully retentive 30,000 MWCO or the very large pore 0.16 μm membranes. The hydraulic resistance of the three membranes after BSA filtration was quite similar, with BSA deposition causing an increase of almost two orders of magnitude in the resistance of the microfiltration membrane. Kelly and Zydney [17] attributed the increase in hydraulic resistance of the microfiltration membranes to the formation of a thick protein deposit on the membrane surface. This was confirmed by SEM images which showed a deposit consisting of large protein aggregates surrounded by a more amorphous protein matrix. Lee and Merson [18] also found that BSA and β-lactoglobulin formed sheet-like deposits on both microfiltration and ultrafiltration membranes, while immunoglobulin G formed deposits that were composed of large protein granules. These protein deposits are generally about 1 μm thick, although deposits as thick as 30 μm have been reported [19].

Most of the research on the mechanism of protein fouling during microfiltration has focused on the effects of denatured or aggregated proteins. These protein aggregates can be present in the bulk protein solution or they may be formed during the filtration process due to pumping or to high shear rates near the membrane surface or in the membrane pores. Chandavarkar [20] demonstrated that the flux decline during BSA filtration through 0.2 μm polycarbonate membranes was due primarily to the deposition of particles (presumably BSA aggregates) on the membrane surface. Quasi-elastic light scattering was used to show that repeatedly pumping the BSA solutions caused a steady increase in the concentration of these large aggregates. Chandavarkar hypothesized that protein aggregation was a consequence of shear denaturation of the protein molecules during pumping, with these denatured proteins forming aggregates through strong intermolecular interactions.

Bowen and Gan [13] measured the flux decline during stirred cell filtration of BSA through 0.2 μm aluminum oxide membranes. They hypothesized that the high shear rate at the entrance to the membrane pores caused shear induced protein denaturation, with the subsequent deposition of these denatured proteins on the pore walls leading to the large flux decline observed experimentally. Stirring, elevated temperatures (33°C), and high cross flow velocity have also been shown to enhance protein aggregation, and in turn fouling [21, 22].

Kelly *et al.* [21] examined the effects of BSA aggregates on the fouling of 0.16 μm PES membranes. The results demonstrated that different commercial BSA preparations can have dramatically different fouling characteristics. For example, heat shock precipitated BSA caused a much more rapid flux decline than cold alcohol precipitated BSA due to the greater amount of BSA aggregates formed during the heat shock precipitation. These protein aggregates can be removed by prefiltering the BSA solution through 100,000 MWCO membranes; the prefiltered BSA solutions show little to no flux decline.

Kelly and Zydney [17] studied the mechanism of BSA aggregation and showed that the aggregates arose primarily through the formation of intermolecular thiol-disulfide bonds. BSA has a single free sulfhydryl group and 17 internal disulfide bonds, two of which are shown schematically in Figure 2. The free sulfhydryl can become ionized at high pHs, and it can then serve as a nucleophile and attack an existing disulfide linkage in another BSA molecule. This forms a dimer which has two free sulfhydryl groups available for further reactions to form large protein aggregates. These disulfide linkages have been identified in the aggregation of a wide range of proteins [23]. Kelly and Zydney [17] showed that capping this free sulfhydryl group with either a cysteinyl group (forming an –S–S– bond) or a

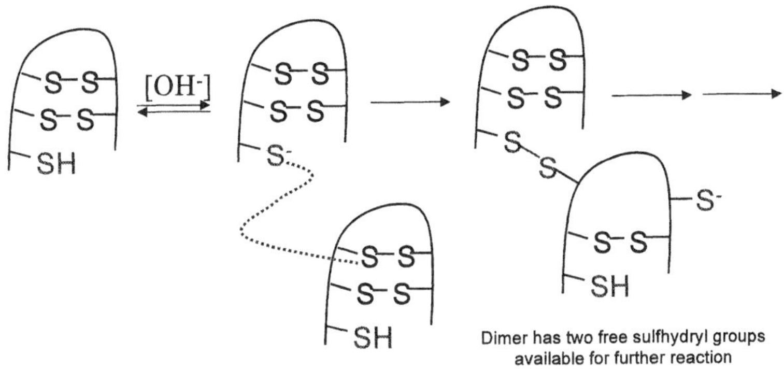

Fig. 2. Schematic representation of the thiol-disulfide interchange reaction. Only two of the 17 disulfide linkages in BSA are shown explicitly.

carboxymethyl group (forming an $-S-CH_2-COOH$ group) completely eliminates the flux decline seen during BSA microfiltration by blocking these thiol-disulfide interchange reactions. They also showed that the extent of BSA aggregation, and thus BSA fouling, could be substantially reduced by the addition of a metal chelator like EDTA or citrate, both of which reduce the catalytic activity of any divalent metal cations present in solution.

4.2. Factors Affecting the Rate of Fouling

4.2.1. Solution Environment

The solution environment can affect the rate and extent of protein aggregation as well as the nature of the interactions (hydrophobic and electrostatic) between the protein and the membrane surface. Bansal *et al.* [14] found that the flux decline during microfiltration of hemoglobin solutions was greatest at the protein isoelectric point, i.e. under conditions where the protein has no net electrical charge. Similar results were obtained by Palecek and Zydney [24] during microfiltration of BSA. This behavior was attributed to the reduction in electrostatic repulsion between protein molecules, which increased the extent of protein deposition and reduced the permeability of the protein cake.

Palecek and Zydney [24] developed a simple model to describe the quasi-steady state flux during microfiltration based on the physical situation that at steady state the drag force on the protein associated with the filtrate flux is exactly balanced by the intermolecular repulsive force between the proteins in the bulk solution and those already deposited on the membrane surface. The steady state flux was given as:

$$J_{ss} = J_{pI} + \frac{\sigma^2}{3\mu\kappa\varepsilon_0\varepsilon'} e^{-\kappa d_s} \tag{2}$$

where J_{pI} is the flux at the isoelectric point, σ is the protein surface charge density, κ^{-1} is the Debye length, d_s is the characteristic separation distance between the bulk protein and the deposit when the proteins first attach to the deposit, and ε' and ε_0 are the dielectric constant of the medium and permittivity of the vacuum, respectively. The model was in good agreement with data for a variety of proteins, with J_{pI} found to be nearly identical for BSA, lysozyme, immunoglobulin G, ribonuclease, and hemoglobin.

Kelly and Zydney [17] studied the effect of pH on the rate of flux decline during the filtration of BSA solutions. A much higher fouling rate was seen at pH 8.4 than at pH 6.9, which they attributed to the greater degree of BSA aggregation associated with the increased ionization of the free thiol groups at the higher pH.

McDonogh *et al.* [25] used radioactively labeled serum albumin to investigate effects of pH on protein adsorption and concentration polarization in cross-flow ultrafiltration. The amount of retained protein, either held dynamically in the polarization layer or adsorbed to the membrane, was greatest at the protein isoelectric point, which they attributed to the higher rate of protein aggregation under these conditions.

Pincet *et al.* [26] examined the effects of pH on the molecular interactions between cellulose acetate films (CA) and two proteins (ribonuclease A and human serum albumin) using a surface force apparatus. Both proteins retained their native conformation on interacting with the CA film at high pH, but significant disturbances in the tertiary structure were seen at pH at or below the isoelectric point.

The effects of the solution's ionic strength on protein fouling are somewhat more complicated. Palecek and Zydney [27] observed a significant increase in the quasi-steady flux at low ionic strengths due to the increase in intermolecular repulsion, consistent with Equation 2. In contrast, Kelly and Zydney [17] observed greater fouling at low ionic strengths, which they attributed to changes in the conformation of the BSA molecules due to the intramolecular interactions. Turker and Hubble [28] observed a 66% reduction in BSA adsorption to 10,000 MWCO hollow fiber polyethersulfone membranes as the ionic strength was increased from 0.01 to 0.06 M, which they attributed to the increase in BSA solubility at high ionic strengths.

Although most studies of the solution environment have focused on the effect of pH and ionic strength, other small molecules and large species can also have a significant effect on fouling. Kelly and Zydney [17] showed that BSA fouling during microfiltration can be substantially reduced by adding metal chelators, like EDTA or citrate, while the rate of fouling was increased in the presence of divalent metal cations, like copper. Guell *et al.* [29] and Kuberkar and Davis [30] showed that adding yeast to the protein solution enhances protein transmission and reduces the flux decline. The yeast cells apparently act as a secondary membrane that traps protein aggregates, thereby reducing the rate of fouling.

4.2.2. Device Hydrodynamics

The fluid mechanics within the membrane device can also affect the rate of protein fouling. Numerous studies have explored techniques to reduce fouling by exploiting hydrodynamic interactions. For example, Jeffree *et al.* [31] and van der Waal *et al.* [32] used rough channels to induce fluid mixing at the membrane solution interface. Thomas [33] placed protuberances directly on the membrane surface at defined distances to induce periodic unsteady flows in the concentration boundary layer. Vera *et al.* [34] and Lee *et al.* [35] injected gas into the cross-flow stream to reduce fouling.

Rodgers and Sparks [36] examined the effects of rapid transmembrane pressure pulsing on the flux. Back pressures of several psi at frequencies of several Hertz increased the solute (protein) flux by as much as two orders of magnitude in cross-flow ultrafiltration. They subsequently showed that the reduction in fouling was due to the disturbance of the concentration boundary layer caused by translation of the body force through the membrane.

Illias and Govind [37] developed a mathematical model to evaluate the performance of a tubular membrane module under oscillatory flow conditions. The analysis accounted for the effects of osmotic pressure, axial pressure variation, and convective and diffusive mass transport. The model was in good agreement with experimental data obtained by Kennedy *et al.* [38] for the filtrate flux during reverse osmosis of 10 wt% sucrose solutions. Illias and Govind [37] concluded that that the increase in flux associated with the oscillatory flow more than compensates for the increased power consumption.

Lopez-Leiva [39] and Lieberherr [40] suggested the use of Taylor vortices to reduce solute build up at the membrane surface. Taylor vortices are generated by centrifugal instabilities in an annular system when the inner cylinder rotates. Kroner and coworkers [41, 42] demonstrated that the filtrate flux and solute permeation rate were both increased in the presence of Taylor vortices. The disadvantages of this rotating device include higher energy consumption, inadequate sealing, and non-linear scale up. Winzeler and Belfort [43] and Brewster *et al.* [44] proposed using Dean vortices resulting from the centrifugal instabilities generated by flow in a curved channel to improve membrane performance. Brewster *et al.* [44] analyzed the design of a spiral channel configuration for both narrow and wide gaps, accounting for the stabilizing effect of the wall flux and the variation in channel curvature within the spiral.

4.2.3. Membrane Properties

Studies have shown that membrane surface chemistry can affect protein adsorption characteristics. Protein adsorption is somewhat reduced on more hydrophilic membranes, with less than monolayer adsorption for BSA [13] and β-lactoglobulin [11] on hydrophilic polyvinylidene fluoride membranes. Most attempts to reduce fouling by modifying the surface chemistry have focused on increasing the membrane hydrophilicity and/or increasing intermolecular repulsive interactions between the solutes and the membrane surface. Specific techniques include chemical modification [45], adsorption of hydrophilic polymers [46], ultraviolet irradiation [47], and low temperature plasma activation [48]. Although these studies do show a reduction in protein adsorption, the improvement in flux during microfiltration has generally been quite small. For example, Mueller and Davis [49] found very similar flux decline profiles during constant pressure filtration of bovine serum albumin (BSA) solutions through 0.5 μm pore size polyethylene membranes both before and after treatment with polyvinyl alcohol, which rendered the membrane surface more hydrophilic. This behavior is consistent with previous studies showing that fouling occurs primarily by the physical deposition of large protein aggregates on the membrane surface. The reduction in protein adsorption on hydrophilic membranes has relatively little effect on the filtrate flux.

The membrane pore size, porosity, and pore morphology can also have a significant effect on protein fouling. Bowen and Gan [13] obtained data for the flux decline during constant pressure filtration of bovine serum albumin solutions through 0.2 μm pore size polycarbonate, anodized aluminum oxide (Anopore), and polyvinylidene fluoride (PVDF) membranes. The flux decline was much more pronounced for the polycarbonate membrane, which the authors attributed to the lower porosity (9% compared to 50% for the Anopore and 75% for the PVDF).

Mueller and Davis [49] examined the fouling characteristics of 0.2 μm pore size polycarbonate, polysulfone, cellulose acetate, and polyvinylidene fluoride membranes during constant pressure microfiltration of BSA. The extent of fouling was smallest for the cellulose acetate membrane, which they attributed to the much higher porosity of this microfiltration membrane. However, the flux decline for the polysulfone membrane was more rapid than that for the polycarbonate membrane, even though the polysulfone membrane had a slightly greater surface porosity (12% vs. 9%) and a much greater bulk porosity (65–80% compared to only 9% for the track-etch polycarbonate membrane). Mueller and Davis hypothesized that this difference might be related to the greater thickness and lower hydrophilicity of the polysulfone membrane, but no quantitative analysis of these effects was provided.

Ho and Zydney [50] demonstrated that the membrane pore interconnectivity can have a large effect on the rate of flux decline. Data obtained on isotropic membranes with interconnected pore structures showed much smaller rates of flux decline than seen with membranes having straight pores. This reduction in fouling is directly due to the interconnected pore structure, which allows fluid to flow under and around any pore blockage on the upper surface of the membrane. The initial rate of flux decline was also a function of the membrane thickness, with thicker membranes fouling less rapidly since the surface blockage only affects filtrate flow through a relatively thin penetration distance into the membrane pore structure.

Ho and Zydney [51] provided the first means for directly evaluating the extent of pore connectivity. Unlike traditional techniques for quantifying membrane pore structure, (e.g., bubble point, solute rejection, gas adsorption, gas diffusion, thermoporometry, and SEM or TEM image analysis), this technique evaluates the pore connectivity by measuring the fluid flow rate (or diffusive solute flux) in both directions within the membrane. This is done by partially covering both the upper and lower surfaces of the membrane to force the fluid to flow in the directions normal and parallel to the membrane surface, with the relative contributions of the two flow directions determined by the magnitude of the overlap and the ratio of the permeabilities (or diffusivities).

4.3. Flux Decline Models

Most studies of protein fouling have interpreted the observed differences in flux decline during protein filtration using the classical fouling models: standard pore blockage, intermediate pore blockage, pore constriction, and/or cake filtration [9, 52]. The mathematical analysis of the flux decline is developed by assuming that the membrane is composed of a uniform array of parallel cylindrical pores. The flux is evaluated using a Darcy's law expression:

$$Q = \frac{\Delta P - \sigma_0 \Delta \pi}{\mu(R_m + R_P)} A_0 \tag{3}$$

where μ is the solution viscosity, ΔP is the transmembrane pressure, R_m is the resistance of the clean membrane, R_P is the resistance of the protein deposit or cake that forms on the external surface of the membrane, and σ_0 and $\Delta\pi$ are the osmotic reflection coefficient and the osmotic pressure difference across the membrane, respectively. The osmotic reflection coefficient is a measure of the permselectivity of the membrane to the protein. It varies from one for a fully retentive membrane to zero for a nonretentive membrane. For a clean microfiltration membrane, the osmotic reflection coefficient is essentially equal to zero since the protein size is so much smaller than that of the membrane pores. Thus, the osmotic pressure term in Eq. 3 is negligible during the initial stages of the filtration process. This term can become important at longer filtration times, although it has generally been neglected in previous models for flux decline during microfiltration.

4.3.1. Complete Pore Blockage Model

In the complete pore blockage model, the volumetric flow rate declines as the available membrane area (A_{open}) decreases due to pore blockage:

$$Q_{open} = \frac{\Delta P}{\mu R_m} A_{open} \tag{4}$$

Cake formation is assumed to be negligible, i.e. $R_P = 0$ in Eq. 3. The rate of pore blockage is assumed to be proportional to the convective flow rate of protein aggregates to the membrane surface:

$$\frac{dA_{open}}{dt} = -\alpha_1 Q_{open} C_b \tag{5}$$

where C_b is the bulk protein concentration and α_1 is a pore blockage parameter which is equal to the membrane pore area blocked per unit mass of protein convected to the membrane surface. Substitution of Eq. 4 into Eq. 5 gives a first order ordinary differential equation for A_{open} that can be integrated to give:

$$A_{open} = A_0 \exp\left(-\frac{\alpha_1 \Delta P C_b}{\mu R_m} t\right) \tag{6}$$

The volumetric flow rate through the partially fouled membrane is given directly by Eq. 4 with A_{open} evaluated from Eq. 6. Q_{open} thus decreases exponentially with time at a rate which is proportional to the bulk protein concentration:

$$\frac{Q_{open}}{Q_0} = \exp(-\frac{\alpha_1 \Delta P C_b}{\mu R_m} t) \tag{7}$$

where Q_0 is the initial filtrate flow rate through the clean membrane. Eq. 7 can be integrated to evaluate the total collected filtrate volume $V = \int Q_{open} dt$. The filtrate flux ($J = Q_{open}/A_0$) is conveniently expressed as a linearly decreasing function of V as:

$$J = J_0 - \frac{\alpha_1 \Delta P C_b}{\mu R_m A_0} V \tag{8}$$

where $J_0 = Q_0/A_0$ is the initial filtrate flux through the clean (unfouled) membrane.

4.3.2. Intermediate Pore Blockage Model

The intermediate pore blockage model accounts for the possibility that particles land on top of other particles when they deposit on the membrane surface. In this case the rate of surface coverage is assumed to be proportional to the fractional area of the membrane remaining uncovered (A_{open}/A_0):

$$\frac{dA_{open}}{dt} = -\alpha_1 Q_{open} C_b \left(\frac{A_{open}}{A_0}\right) \tag{9}$$

Eq. 9 is combined with Eq. 4 to give the following expression for the normalized flux:

$$\frac{Q_{open}}{Q_0} = \left(1 + \frac{\alpha \Delta P C_b}{\mu R_m} t\right)^{-1} \tag{10}$$

4.3.3. Pore Constriction Model

In the pore constriction model, the particles or aggregates are assumed to deposit uniformly on the pore walls throughout the internal membrane volume. The rate of change in the pore volume is again assumed to be proportional to the rate of particle convection to the membrane:

$$\frac{d}{dt}(N_0 \pi r_p^2 \delta_m) = -\alpha_{pore} Q C_b \tag{11}$$

where the parameter α_{pore} equals the volume of foulant deposited in the pore interior per unit mass of protein filtered through the membrane. The flow rate Q is evaluated as a function of the pore radius (r_p) using the Hagen-Poiseuille equation for laminar flow in a cylinder:

$$Q = \frac{N_0 \pi r_p^4 \Delta P}{8 \mu \delta_m} \tag{12}$$

where N_0 is the total number of pores and δ_m is the membrane thickness. The normalized flux is evaluated by integrating Eqs. 11 and 12 to give:

$$\frac{Q}{Q_0} = \left(1 + \frac{\alpha_{pore} Q_0 C_b}{\pi r_0^2 \delta_m} t\right)^{-2} \tag{13}$$

where r_0 is the initial pore radius. Eq. 13 can be integrated over time to evaluate the total filtration volume (V), with the results conveniently put in the following linear form:

$$\frac{t}{V} = \frac{\alpha_{pore} C_b}{\pi r_0^2 \delta_m} t + \frac{1}{Q_0} \tag{14}$$

4.3.4. Cake Filtration Model

In the cake filtration model, it is assumed that proteins deposit on the upper surface of the membrane, forming a uniform cake layer. The hydraulic resistance provided by the particle cake is assumed to be proportional to the mass per unit area of the cake layer (m_p):

$$R_p = R' m_p \tag{15}$$

where R′ is the specific cake resistance. The rate of particle deposition is directly proportional to the rate of particle convection to the membrane:

$$\frac{dm_p}{dt} = f' J C_b \tag{16}$$

where f′ is the fraction of protein convected to the membrane that actually adds to the growing deposit. Substitution of Eqs. 3 and 15 into 16 gives the following equation upon integration:

$$\frac{Q}{Q_0} = \left(1 + \frac{2 f' R' Q_0 C_b}{A_0 R_m} t\right)^{-\frac{1}{2}} \tag{17}$$

Eq. 17 can be recast in linear form in terms of the filtrate volume as:

$$\frac{t}{V} = \frac{1}{Q_0} + \frac{\mu R' f' C_b}{2 \Delta P A_0^2} V \tag{18}$$

A more detailed discussion of the underlying assumptions and mathematical development of these models are provided by Hermia [52] and Zeman and Zydney [9].

4.3.5. Combined Pore Blockage and Cake Filtration Model

Several studies have instead demonstrated that membrane fouling typically occurs on the upper surface of the membrane, resulting in both pore blockage and the formation of a dense deposit or cake layer [9]. Ho and Zydney [53] recently developed a new fouling model that describes the flux decline due to simultaneous pore blockage and cake formation. The normalized flow rate at time t is expressed as a convolution integral over the time (t_p) at which each region of the membrane surface is first blocked by a protein aggregate or particle:

$$\frac{Q}{Q_0} = \exp(-\beta J_0 t) + \int_0^t \frac{\beta R_m}{R_m + R_p} \exp(-\beta J_0 t_p) dt_p \tag{19}$$

where β is the pore area blocked per unit volume of filtrate and R_m is the resistance of the clean (unfouled) membrane. The resistance of the cake layer or deposit (R_p) increases with time during filtration [53]:

$$R_m + R_p = \left(R_m + R_{p0}\right)\sqrt{1 + \frac{2R'R_m J_o}{\left(R_m + R_{p0}\right)^2}\left(t - t_p\right)} \tag{20}$$

where R_{po} is the initial resistance of the deposit (i.e., the resistance of the first particle or protein aggregate that deposits on the membrane surface) and R' is the specific resistance of the growing cake layer multiplied by the concentration of protein aggregates present in the feed solution (C_{agg}). Note that Eq. (20) accounts for the fact that the cake layer only grows over the time interval $t-t_p$ since for $t < t_p$ that region of the membrane remains free of any deposit. Ho and Zydney [53] have shown that the convolution integral in Eq. 19 can be approximated by assuming that the deposit resistance over the fouled region of the membrane is uniform at its maximum value:

$$\frac{Q}{Q_o} = \exp(-\beta J_o t) + \frac{R_m}{R_m + R_p}\left[1 - \exp(-\beta J_o t)\right] \tag{21}$$

with R_p evaluated using Eq. 20 with $t_p = 0$.

4.3.6. Other Fouling Models

Many other fouling models have also appeared in the literature. A number of these were developed as empirical modifications of the classical fouling models described in the last section. For example, Wu *et al.* [54] suggested that the rate of flux decline is proportional to the filtrate flux at short times and approaches a steady state at long times. They expressed this mathematically as:

$$\frac{dJ}{dt} = -k_p e^{-k_f t} J \tag{22}$$

where k_p and k_f are empirical constants. At short times, Eq. 22 reduces to the complete pore blockage model while at long times the flux approaches a constant value:

$$J = J_0 e^{-k_p/k_f} \tag{23}$$

Other investigators have modified the cake filtration model to account for the back transport of particles away from the membrane. This is particularly important in cross flow systems where hydrodynamic interactions are significant. For example, Green and Belfort [55] evaluated the rate of cake growth as:

$$\frac{dm_{cake}}{dt} = f'(Q - Q^*)C_b / A_0 \tag{24}$$

where Q^* is an effective flow rate describing the back transport of particles. Green and Belfort [55] assumed that Q^* was related to hydrodynamic lift forces while Cohen and

Probstein [56] suggested that the back transport was due to electrostatic repulsion between the particles in the bulk suspension and those deposited on the membrane surface. Suki *et al.* [57] proposed that particle removal from the membrane was due to tangential shear stresses near the membrane surface with Q* proportional to the mass of the deposit. Field *et al.* [58] defined Q* as the critical flux below which there is no fouling, with the value of Q* determined by both hydrodynamic and intermolecular interactions.

Koltuniewicz [59] evaluated the flux using a surface renewal model. They assumed that the instantaneous permeate flux J for each element of the membrane was given as:

$$J = (J_0 - J^*)e^{-qt} + J^* \tag{25}$$

where J* is the terminal or steady-state flux for a dead end experiment, q is a parameter describing the rate of particle deposition, and t is the time measured from the instant at which the surface was "renewed". Equation 25 reduces to the classical pore blockage model when $J^* = 0$. Equation 25 was applied to cross flow filtration by assuming that each element of the membrane surface was periodically renewed by the cross-flow. The final result is:

$$J = (J_0 - J^*)\left(\frac{s}{q+s}\right)\left(\frac{1-e^{-(q+s)t}}{1-e^{-st}}\right) + J^* \tag{26}$$

where the parameter s describes the surface renewal process. Model calculations were in good agreement with experimental data obtained during filtration of BSA and kaolin suspensions using the same values of J* and q in both dead end and cross-flow filtration.

An alternative approach that has been used to describe the flux in cross-flow filtration is based on the concentration polarization model. In this case, the quasi-steady flux is calculated by balancing the convective transport of particles towards the membrane with the diffusive transport back into the bulk suspension. A simple stagnant film model is typically used to evaluate the flux in terms of the bulk particle concentration (C_b) and the particle concentration at the membrane surface (C_w):

$$J = k_m \ln\left(\frac{C_w - C_f}{C_b - C_f}\right) \tag{27}$$

where k_m is the mass transfer coefficient, equal to the ratio of the free solution diffusivity (D_∞) to the boundary layer thickness, and C_f is the particle concentration in the filtrate solution. For laminar flow, the length averaged mass transfer coefficient can be estimated using the Leveque solution for a thin boundary layer [60] to give the following expression for the length averaged permeate flux:

$$\langle J \rangle = 0.81\left(\frac{\gamma_0 D_\infty^2}{L_z}\right)^{1/3} \ln\left(\frac{C_w - C_f}{C_b - C_f}\right) \tag{28}$$

where L_z is the channel length and γ_0 is the shear rate at the membrane surface.

Although Eqs. 27 and 28 provide a reasonably accurate description of the filtrate flux in protein ultrafiltration, the predicted flux in microfiltration systems using the Brownian diffusivity are typically one or two orders of magnitude less than that measured experimentally [55]. Zydney and Colton [61] hypothesized that the Brownian diffusivity in Equation 28 should be replaced with the shear induced hydrodynamic diffusivity which arises from particle-particle interactions in the shear flow of a concentrated suspension. This effect was first measured by Eckstein *et al.* [62] by tracking the motion of individually labeled particles. For particle volume fractions between 0.2 and 0.45, the shear induced diffusivity D_s can be expressed as:

$$D_s = 0.3\gamma_0 a^2 \tag{29}$$

where a is the particle radius. Substitution of Eq. 29 into Eq. 28 gives a flux which varies linearly with the shear rate and with the particle radius to the 4/3 power, a dependence which is in good agreement with filtrate flux data obtained during cross-flow filtration of a concentrated suspension of red blood cells.

Altena and Belfort [63] proposed that the back transport of particles away from the membrane was due primarily to inertial forces associated with hydrodynamic interactions between the particle and the boundary. The inertial lift velocity (v_L) for a dilute suspension of spherical particles under laminar flow is:

$$v_L = \frac{b\rho_0 a^3 \gamma_0^2}{16\mu} \tag{30}$$

where ρ_0 is the fluid density and b is a function of the dimensionless distance from the wall. Altena and Belfort [63] hypothesized that the flux continues to decline until the filtration velocity exactly equals the inertial lift velocity, giving a dependence on γ_0^2 and a^3.

Baralla et al. [64] used a completely different approach to model membrane fouling. The porous structure of an inorganic membrane was simulated using a 2-dimensional Voronoi tessellation which divides the space into an array of irregular convex polygons. The edges of the polygons were used to represent the pore space, with the geometry of each pore specified by the pore body diameter, the pore throat diameter, and the pore length. The pore body diameter was evaluated from the experimentally determined porosity. Pore throat diameters were randomly assigned to each pore segment based on a previously specified pore size distribution. The model was used to simulate the filtration of a particle suspension with a given particle size distribution. Particles larger than the pore bodies were assumed to deposit on the membrane surface, leading to cake formation, while particles small enough to penetrate the matrix followed a path that depended on the geometry and connectivity of the pore structure. Although this framework can provide useful insights into the fouling phenomena, the complex numerical calculations are very poorly suited to describing the filtrate flux as a function of time.

5. APPLICATIONS IN BIOTECHNOLOGY INDUSTRIES

5.1. High Performance Tangential Flow Filtration

Membrane processes are now used widely in biotechnology industries for the purification of blood components, recombinant proteins, natural protein products, industrial enzymes, antibiotics, and amino acids. Traditional processes have been limited to separating species that differ in size by at least ten fold. The recent development of High Performance Tangential Flow Filtration (HPTFF) has expanded the application of ultrafiltration to protein-protein separation where there are very small size differences. This is made possible by four new developments that have been implemented in cross-flow filtration processes. (1) To minimize the variation in transmembrane pressure and flux throughout the module, HPTFF uses co-current feed and filtrate flow to balance the feed-side pressure drop throughout the module. This allows the system to operate at a specific flux for optimal separation. (2) The solution environment is adjusted to maximize the *effective* size differences between the two species. Charged proteins have a larger effective volume due to the diffuse electrical double layer surrounding the protein. Adjusting the pH to the PI of one protein allows it to pass through the membrane while the charged proteins remain excluded due to their larger effective size. Differences in the diffusive double layer thickness are further enhanced by reducing the solution's ionic strength. (3) Membranes are chosen to have like charge as the excluded species to provide greater rejection for species with like charge. (4) A diafiltration mode is applied to wash out the impurity from the retentate through the continuous addition of buffer to the retentate during filtration.

With HPTFF, a very high degree of separation can be obtained. For example, purification factors for the separation of bovine serum albumin (MW68,000) from an antigen binding fragment (Fab) derived from a recombinant DNA antibody (MW45,000) can exceed 800 fold [65]. More recently, Ebersold and Zydney [66] demonstrated that membrane systems can be used to separate a negatively charged myoglobin variant (differing from native myoglobin by just one amino acid residue) from native myoglobin on the basis of electrostatic interactions with the membrane. Using a two stage diafiltration process, they achieved greater than nine-fold purification and 90% yield of native myoglobin. This is the first example of membrane systems being used to separate proteins that differ by only a single amino acid residue.

5.2. Use of V_{max} Analysis for System Scale Up

Normal flow filtration is carried out by passing the feed solution directly through the membrane, without any tangential flow, in what is often referred to as a dead-end mode. As the filtration proceeds, the resistance to flow increases due to fouling. Fouling causes a decay in flow rate for constant pressure operation and it increases the pressure for constant flow filtration. The filter capacity is defined as the volume of feed solution that can be processed before the flow rate falls below a specified value (for constant pressure operation) or before the pressure differential exceeds a specified limit (for constant flow rate operation)

The simplest approach to sizing normal flow filters is the flow decay method, in which the cumulative filtrate volume is measured through a small-area test filter until the flow rate drops to 10% (or 20%) of its initial value. An attractive alternative to the flux decay method is the

V_{max} analysis [67, 68]. In this case, flux decay data are only obtained over a short filtration time (typically 10 to 15 minutes), with the data extrapolated to longer filtration times using the linearized form of the pore constriction model [52] as shown in Equation 14, identifying V_{max} with the group of parameters:

$$V_{max} = \left(\frac{\pi r_o^2 \delta_m}{\alpha_{pore} C_b} \right) \tag{31}$$

V_{max} is the maximum volume of fluid that can be filtered before the membrane is completely plugged by foulant. V_{max} is evaluated directly from the flux decay data as the inverse of the slope on a plot of t/V as a function of t (Eq. 14). The system capacity is then calculated using the expression for the flow rate as a function of time for the pore constriction model [52] shown in Eq. 13 to give:

$$V_{capacity} = V_{max} \left[1 - \sqrt{Q_{min} / Q_o} \right] \tag{32}$$

where Q_{min} is the minimum specified flow rate (typically 10 or 20% of Q_o).

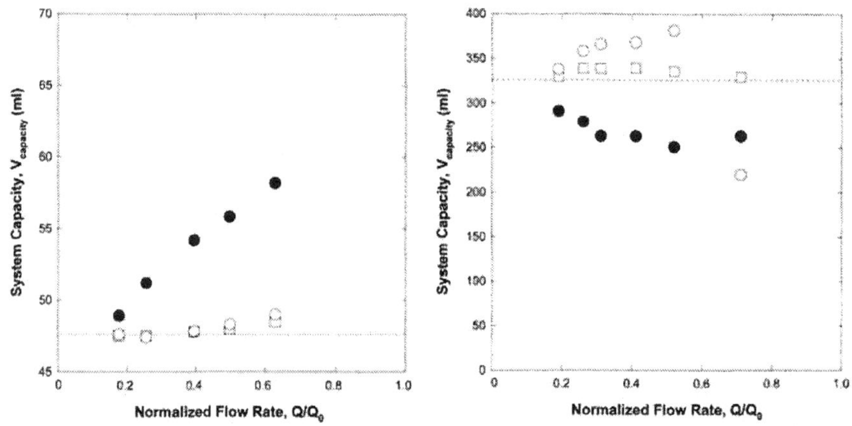

Fig. 3. System capacity for BSA (left panel) and lysozyme (right panel) as a function of the normalized flow rate range used to evaluate the model parameters. Filled circles represent results from the V_{max} method. Open squares (full model) and open circles (approximate solution) represent results for the pore blockage–cake filtration model. Dotted line is the actual system capacity [69].

Although the V_{max} method is used quite extensively for filter sizing, the assumption that membrane fouling is due to the uniform constriction of cylindrical pores is rarely met in practice. Fig. 3 shows the results given by the V_{max} analysis (filled circles), the full convolution integral (open squares, Eq. 19), and the approximate solution (open circles, Eq. 21) for BSA (left panel) and lysozyme (right panel) [69]. The calculations were done using data over a range of Q/Q_0 values with V_{max} or model parameters determined using flux decline data down to the value of Q/Q_0 (shown in the x axis of Fig. 3). For BSA, the system capacity predicted by the pore blockage – cake filtration model is in excellent agreement with the true capacity (shown as the dotted line) for both the full model and the approximate solution. This is particularly true for data obtained down to $Q/Q_o = 0.4$, but even results using data only for $Q/Q_o > 0.63$ gave a capacity ($V = 49.0$ mL for the approximate solution and 48.7 mL for the full model) that was within 5% of the actual value. Thus, the pore blockage – cake filtration model allows one to estimate the capacity for BSA using data obtained over only 3 min (corresponding to $V = 21$ ml), compared to the 11 min required for the direct experimental determination of the capacity using the flux decay method. In contrast, the capacity estimated using the V_{max} model is 23 % larger than the actual capacity for data obtained using $Q/Q_o > 0.63$. The error in the V_{max} calculations does decrease as one uses data over a wider range of flux decay, approaching the actual capacity as one conducts the filtration out to $Q/Q_o = 0.2$. However, the use of such long filtration runs completely eliminates the potential time and fluid savings that are the basis for using the V_{max} method.

For lysozyme, the full convolution integral (Eq. 19) provides a much better estimate of capacity than the approximate solution. This is due to the relatively small value of R_{po} for lysozyme, which causes the contributions from both pore blockage and cake filtration to be important at the same time. The full convolution integral provides an accurate estimate of the capacity using data for $Q/Q_o > 0.71$ corresponding to less than 9 min of filtration. Again, accurate estimates of the capacity (within 10% of the actual value) could only be obtained using data for Q/Q_o down to 0.2, corresponding to 90 min of filtration. This is 10 times longer than the time required using the full pore blockage – cake filtration model and 50 percent larger than the time required using the approximate analytical solution (Eq. 21).

5.3. OTHER APPLICATIONS

The development of affinity membranes was motivated by affinity chromatography wherein the desired product is isolated through specific binding to the stationary phase. Affinity membrane processes are highly effective in recovery of fusion proteins [70, 71]. In comparison to affinity resins used in chromatography, the use of affinity membranes eliminates the slow diffusion mass transfer and high pressure drop along the column.

Supported liquid membranes consisting of an organic solvent immobilized within the pores of hydrophobic microfiltration membranes have been widely used for the removal of ions, amino acids, or organic acids from aqueous or gas mixtures [72–74]. The organic solvent typically contains a carrier that selectively binds to a class of compounds from the feed stream and transports them across the membrane. Compared to classical liquid-liquid extraction, the use of supported liquid membranes significantly reduces the amount of extractant while maintaining high selectivity. Despite the potential of supported liquid membranes, the loss of the carrier or solvent during the process limits its long-term stability. Approaches to overcoming this limited stability include encapsulating the liquid by skinning both surfaces of the membrane [75, 76] or using a gel that can be grafted to the porous support [77, 78].

Unlike supported liquid membranes, membranes used in membrane distillations are typically not wetted by the bulk liquid phase. In typical applications to separate water from salt [79], sugar [80], and other non-volatile solutes, only water vapor from the heated feed passes through the hydrophobic membrane and condenses on the cold permeate side.

6. OUTLOOK

Membrane processes should continue to be an integral part of downstream bioprocessing applications. Microfiltration is used extensively for initial harvesting of therapeutic products from mammalian, yeast, and bacterial cell cultures. Ultrafiltration has been the method of choice for protein concentration and buffer exchange. With the development of HPTFF, membranes can now be applied to high resolution separations. New advances in membrane materials and process operations to minimize fouling and enabling protein purification using affinity membranes or the development of membrane chromatography will lead to the broader use of membranes in bioseparations.

REFERENCES

[1] L.M. Vane, J. Chem. Technol. Biotechnol., 80 (2005) 603.
[2] S.S. Madaeni, K. Tahmasebi and S.H. Kerendi, Eng. Life Sci., 4 (2004) 187.
[3] G. Laufenberg, S. Hausmanns and B. Kunz, J. Mem. Sci., 110 (1996) 59.
[4] N. Aydogan, T. Gurkan and L. Yilmaz, Sep. Sci. Technol., 39 (2004) 1059.
[5] L. Tessier, P. Bouchard and M. Rahni, J. Biotechnol., 116 (2005) 79.
[6] L.M. Ortega, R. Lebrun, I.M. Noel and R. Hausler, Sep. Pur. Technol., 44 (2005) 45.
[7] W.S.W. Ho and K.K. Sirkar, Membrane Handbook, Chapman & Hall, New York, 1992.
[8] W.C. McGregor, Membrane Separations in Biotechnology, Marcel Dekker, New York, 1986.
[9] L.J. Zeman and A.L. Zydney, Microfiltration and Ultrafiltration: Principles and Applications, Marcel Dekker, New York, 1996.
[10] R.F. Madsen, Hyperfiltration and Ultrafiltration in Plate-andFrame Systems, Elsevier, Amsterdam, 1977.
[11] K.M. Persson, G. Capannelli, A. Bottino and G. Tragardh, J. Mem. Sci., 76 (1993) 61.
[12] W.S. Opong and A.L. Zydney, J. Colloid Interface Sci., 142 (1991) 41.
[13] W.R. Bowen and Q. Gan, Biotechnol. Bioeng., 38 (1991) 688.
[14] A. Bansal, Y.H. Ma and W.M. Clark, Key Eng. Materials, 61 (1991) 505.
[15] G. Belfort, R.H. Davis and A.L. Zydney, J. Mem. Sci., 96 (1994) 1.
[16] A.D. Marshall, P.A. Munro and G. Tragardh, Desalination, 91 (1993) 65.
[17] S.T. Kelly and A.L. Zydney, Biotechnol. Bioeng., 44 (1994) 972.
[18] D.N. Lee and R.L. Merson, J. Dairy Sci., 58 (1974) 1423.
[19] F.A. Glover and B.E. Brooker, J. Dairy Sci., 41 (1974) 89.
[20] A.S. Chandavarkar, Dynamics of Fouling of Microporous Membranes by Proteins, Ph.D. Dissertation, Cambridge, MA, Massachusetts Institute of Technology, 1990.
[21] S.T. Kelly, W.S. Opong and A.L. Zydney, J. Mem. Sci., 80 (1993) 175.
[22] M. Meireles, P. Aimar and V. Sanchez, Biotechnol. Bioeng., 38 (1991) 528.
[23] W.R. Liu, R. Langer and A.M. Klibanov, Biotechnol. Bioeng., 37 (1991) 177.
[24] S.P. Palecek and A.L. Zydney, Biotechnol. Prog., 10 (1994) 207.
[25] R.M. McDonogh, H. Bauser, N. Stroh and H. Chmiel, Desalination, 79 (1990) 217.
[26] F. Pincet, E. Perez and G. Belfort, Langmuir, 11 (1995) 1229.
[27] S.P. Palecek and A.L. Zydney, J. Mem. Sci., 95 (1994) 71.
[28] M. Turker and J. Hubble, J. Mem. Sci., 34 (1987) 267.
[29] C. Guell, P. Czekaj and R.H. Davis, J. Mem. Sci., 155 (1999) 113.
[30] V.T. Kuberkar and R.H. Davis, Biotechnol. Prog., 15 (1999) 472.
[31] M.A. Jeffree, J. Peacock, I.J. Sobey and B.J. Bellhouse, Clinical and Experimental Dialysis and Apheresis, 5 (1981) 373.
[32] M.J. van der Waal and I.G. Racz, J. Mem. Sci., 40 (1989) 243.
[33] D.G. Thomas, Ind. Eng. Chem. Fundamentals, 12 (1973) 396.
[34] L. Vera, R. Villarroel, S. Delgado and S. Elmaleh, J. Mem. Sci., 165 (2000) 47.
[35] C.K. Lee, W.G. Chang and Y.H. Ju, Biotechnol. Bioeng., 41 (1993) 525.

[36] V.G.J. Rodgers and R.E. Sparks, J. Mem. Sci., 68 (1992) 149.
[37] S. Illias and R. Govind, Sep. Sci. Technol., 25 (1990) 1307.
[38] T.J. Kennedy, R.L. Merson and B.J. McCoy, Chem. Eng. Sci., 29 (1974) 1927.
[39] M. Lopez-Leiva, Ultrafiltration in a Rotary Annular Filter, Ph.D. Dissertation, Lund, Lund University, 1979.
[40] J. Lieberherr, Scerfiltration im Ringspalt, Ph.D. Dissertation, Zurich, ETH, 1978.
[41] K.H. Kroner and V. Nissinen, J. Mem. Sci., 36 (1985) 85.
[42] K.H. Kroner, V. Nissinen and H. Ziegler, Bio-Technol., 5 (1987) 921.
[43] H.B. Winzeler and G. Belfort, J. Mem. Sci., 80 (1993) 35.
[44] M.E. Brewster, K.Y. Chung and G. Belfort, J. Mem. Sci., 81 (1993) 127.
[45] M.J. Steuck and N. Reading, Porous Membrane Having Hydrophilic Surface and Process, US Patent No. 4 618 533 (1986).
[46] K.J. Kim, A.G. Fane and C.J.D. Fell, Desalination, 70 (1988) 229.
[47] M. Nystrom and P. Jarvinen, J. Mem. Sci., 60 (1991) 275.
[48] P.W. Kramer, Y.S. Yeh and H. Yasuda, J. Mem. Sci., 46 (1989) 1.
[49] J. Mueller and R.H. Davis, J. Mem. Sci., 116 (1996) 47.
[50] C.C. Ho and A.L. Zydney, J. Mem. Sci., 155 (1999) 261.
[51] C.C. Ho and A.L. Zydney, J. Mem. Sci., 170 (2000) 101.
[52] J. Hermia, Transactions of the Institution of Chemical Engineers, 60 (1982) 183.
[53] C.C. Ho and A.L. Zydney, J, Colloid and Interface Sci., 232 (2000) 389.
[54] D. Wu, Howell, J.A., Turner, N.M., Trans. Inst. Chem. Eng. (London), 69 (1991) 77.
[55] G. Green and G. Belfort, Desalination, 35 (1980) 129.
[56] R.D. Cohen and R.F. Probstein, J. Colloid and Interface Sci., 114 (1986) 194.
[57] A. Suki, A.G. Fane and C.J.D. Fell, J. Mem. Sci, 21 (1984) 269.
[58] R.W. Field, D. Wu and J.A. Howell, J. Mem. Sci, 100 (1995) 259.
[59] A.B. Koltuniewicz, J. Mem. Sci., 68 (1992) 107.
[60] M.C. Porter, Ind. Eng. Chem. Prod. Res. Dev., 11 (1972) 233.
[61] A.L. Zydney and C.K. Colton, Chem. Eng. Comm., 47 (1986) 1.
[62] E.C. Eckstein, P.G. Bailey and A.H. Shapiro, J. Fluid Mech., 79 (1977) 191.
[63] F.W. Altena and G. Belfort, Chem. Eng. Sci., 39 (1984) 343.
[64] G. Baralla, M. Mattea and V. Gekas, Sep. Pur. Technol., 22 (2001) 489.
[65] R. Van Reis and A.L. Zydney, Current Opin. Biotech., 12 (2001) 208.
[66] M.F. Ebersold and A.L. Zydney, Biotechnol. Prog., 20 (2004) 543.
[67] F. Badmington, E. Honig, M. Pyane and R. Wilkins, Pharm. Tech., 19 (1995) 64.
[68] E.S. Honig and P.D. Schwartz, Filtration and Separation, 1 (1997) 73.
[69] A.L. Zydney and C.C. Ho, Desalination, 146 (2002) 75.
[70] F. Cattoli and G.C. Sarti, Sep. Sci. Technol., 37 (2002) 1699.
[71] D.W. Wood, V. Derbyshire, W. Wu, M. Chartrain, M. Belfort and G. Belfort, Biotechnol. Prog., 16 (2000) 1055.
[72] R. Chiarizia, J. Mem. Sci, 55 (1991) 39.
[73] J. Gega, W. Walkowiak and B. Gajda, Sep. Pur. Technol., 22-3 (2001) 551.
[74] R. Molinari, L. Debartolo and E. Drioli, J. Mem. Sci, 73 (1992) 203.
[75] M.C. Wijers, M. Jin, M. Wessling and H. Strathmann, J. Mem. Sci, 147 (1998) 117.
[76] Y.C. Wang and F.M. Doyle, J. Mem. Sci, 159 (1999) 167.
[77] A.M. Mika, R.F. Childs, J.M. Dickson, B.E. McCarry and D.R. Gagnon, J. Mem. Sci, 108 (1995) 37.
[78] M. Ulbricht, Reactive & Functional Polymers, 31 (1996) 165.
[79] K. Schneider, W. Holz, R. Wollbeck and S. Ripperger, J. Mem. Sci., 39 (1988) 25.
[80] S. Nene, S. Kaur, K. Sumod, B. Joshi and K. Raghavarao, Desalination, 161 (2004) 305.

Bioprocessing for Value-Added Products from Renewable Resources
Shang-Tian Yang (Editor)

Chapter 8. Bacterial and Yeast Cultures – Process Characteristics, Products, and Applications

Wei-Cho Huang and I-Ching Tang

Bioprocessing Innovative Company, Inc., 4734 Bridle Path Ct., Dublin, Ohio 43017, USA

1. INTRODUCTION

Paintings on the walls of Egyptian tombs depict the brewing of beer, which uses microorganisms in the fermentation process. However, the existence of bacteria did not become known until the development of sufficiently powerful microscopes in the late 1600s. Today, microorganisms provide us a valuable tool with which to produce useful chemicals and fuels. They are widely applied in biotechnology because their high metabolic rates and specialized enzymes, which allow for the specific degradation of substrates and synthesis of products. In nature, organic substances are broken down into simpler compounds through digestion by bacteria, yeasts, filamentous fungi and other microorganisms. These chemical changes, called fermentation, are generally accompanied by the generation of heat and gases. Meanwhile, they obtain energy for metabolism by means of organic or inorganic electron donors and acceptors. Various microorganisms produce different metabolites via different metabolic pathways. Furthermore, the same species may release different products when grown under different environmental conditions as a result of different metabolic pathways. Recombinant DNA technology has been widely used to aid the development of economical bioprocesses in the food and pharmaceutical industries in the past two decades. Recent bioprocessing research emphasizes using genetic and metabolic engineering techniques to manipulate the metabolic pathway to produce new bio-based products. These powerful techniques can make it economically feasible to produce various bio-based value-added chemicals with less energy and waste.

All organisms have different capacities to acclimate to environmental stresses, such as acidity, alkalinity, cold, heat, or high pressure. Many bacteria and some yeasts can survive in extreme environments through changes in the enzymes or other proteins that they produce. These adjustments enable bacteria to grow in a variety of conditions. Gradual exposure to the stress may enable bacteria to synthesize new enzymes that allow them to live under severe conditions or that enhance their ability to cope with the stressing agent. For example, some bacteria can pump out acid when grown in a low pH environment. Osmophilic yeasts excrete glycerol in order to balance the osmotic pressure. These adaptations offer us useful tools for overproducing value-added products from renewable resources.

Compared to other organisms, bacteria and yeasts have many advantages as producing organisms for industrial products. For examples, bacteria and yeasts can grow in a variety of environments with wide ranges of temperature and pH conditions. They replicate rapidly with a short lag phase by taking various kinds of nutrients. Bacteria and yeasts are easy to genetically manipulate for strain improvement and usually have no limitations to transport oxygen or nutrients into cells. They can be easily separated from media and product solutions with low energy costs. High product yields and productivities can usually be attained. Compared to plant and animal cell cultures, bacteria and yeasts cell cultures have the benefits of faster growth, cheaper media, and higher gene expression levels. Thus, using bacteria and yeasts as the producing microorganisms for industrial fermentation products is usually more economical and efficient than using other organisms.

The environmental pollution caused by petroleum-derived wastes and the global shortage of fossil fuel are long-term issues that remain unsolved. Renewable, bio-based products have the potential to overcome these problems and reduce the greenhouse effect. Therefore, microbial production of biodegradable polymers and alternative energy provides a solution to improve the environment and enhance quality of life.

This chapter will review bacterial and yeast cultures, their fermentation products and process characteristics, and challenges in large-scale fermentation for production of industrial bio-based products from renewable resources.

2. BACTERIA

2.1. General characteristics

Bacteria are single-celled microorganisms visible only through a microscope that magnifies them to at least 500 times their actual size (0.5~3 μm). They have different shapes, such as spheres, cylinders, or spirals. They are the most diversed group of organisms, living almost everywhere on Earth, including the deepest parts of the ocean. They are in the air, in the soil, in food, and in other living organisms. Anywhere there is life, it includes bacterial life. Even our bodies are home to many different kinds of bacteria. Our lives are closely intertwined with them, and the health of our planet depends very much on their activities.

Bacteria are truly noteworthy in their adaptations to extreme environments and their ability to survive in parts of Earth that are inhospitable to other forms of life. Extremophiles surviving in extreme environments such as deep ocean, underground ice in Greenland, hot springs, and dry desert soil. Some extremophiles can survive at high pressure (>100 atm), high temperature (>110 °C), or extreme pH (above pH 10 or below pH 2).

Table 1 lists major characteristics of some representative bacteria. Bacteria are usually classified by their morphology and by means of a technique called Gram's stain (invented by Hans Christian Gram in 1884). A gentian violet dye is applied to stain cells. After the staining procedure, gram-positive bacteria with thick peptidoglycan walls appear purple and gram-negative bacteria with thin peptidoglycan walls and an outer membrane appear colorless or reddish. The gram stain method reveals intrinsic differences in the cell wall structure.

Table 1

Characteristics of common bacteria [1–3]

Species	*Bacillus subtilis*	*Clostridium* spp.	*Escherichia coli*	*Lactobacillus* spp.	*Pseudomonas* spp.	*Streptomyces* spp.
Gram-stain	+	+	–	+	–	+
Sporulation	Endospore	Endospore (some)	Nonspore	Nonspore	Nonspore	Conidiospore
Generation time	20~30 min	4.5~5 h	22 min	25 min ~1.4 h	15~96 min	21 h
Growth media, carbon source	Sucrose, glucose, starch	Cellulose, lactose, starch, xylan	Glucose, lactose, xylose, acetate	Lactose, glucose, fructose, xylose	Lactic acid, lipid, xylose, sucrose	Starch, glucose, chitin, cellobiose
Growth temperature	Mesophilic, thermophilic	Mesophilic, thermophilic	Mesophilic	Mesophilic, thermophilic	Mesophilic	Mesophilic, thermophilic
pH range	5.5~8.5 (7.5~8)[1]	4~8 (6.5~7)	4.4~7	2.5~6.5 (5.5~5.8)	5~6	4.5~8
O_2 demand	Aerobic, facultative anaerobic	Strictly anaerobic	Facultative anaerobic	Facultative anaerobic	Obligate aerobic	Aerobic
Genetic modification	Easy	Somewhat difficult	Very easy	Easy	Easy	Average
GRAS*	Yes	No	No	Yes	No	Some
Others		Some pathogenic		Probiotic		Filamentous

[1]Numbers in the parentheses are optimal pH range for growth.

*GRAS: generally recognized as safe

Bacteria are also classified according to their growth temperature, pH, and nutrient requirements. Psychrophiles, psychrotrophs, mesophiles, and thermophiles are referred to bacteria that grow optimally at 10–15, 15–30, 30–40, and 50–85 °C, respectively. Some extreme thermophiles, such as *Pyrobacterium brockii*, can grow at as high as 115 °C. Bacteria usually live in neutral or weakly acidic environments. However, acidophiles, such as *Sulfolobus acidocaldarius*, can survive at pH as low as 1.0 and alkaliphiles grow at pHs between 8 and 11.

The availability of water or water activity (a_w) is critical to the growth of all cells. Most bacteria require a water activity of higher than 0.9, but halophiles and osmophiles can grow at a_w as low as 0.75. Oxygen or air is another important factor affecting bacterial growth. Most bacteria belong to either obligate aerobes, facultative anaerobes, or obligate anaerobes. All bacteria that can live in the presence of O_2 contain superoxide dismutase, which converts superoxide (O_2^-) to hydrogen peroxide (H_2O_2). Most of them then use catalase to decompose H_2O_2. Certain aerotolerant bacteria, such as lactic acid bacteria, decompose H_2O_2 by means of peroxidase. Obligate anaerobes lack of these enzymes, and therefore cannot live with oxygen but thrive only in oxygen-free environments, such as our intestines or the ooze at the bottom of swamps.

Bacterial growth requires a variety of nutrients, including carbon, energy, and nitrogen sources, and vitamins, minerals, and trace elements. Most bacteria and some archaea use organic compounds for their energy and carbon sources. These chemoheterotrophs usually use glucose and other sugars as the substrate for their growth, but methylotrophs can use simple one-carbon compounds, such as methanol and methane. Many archaea, such as methanogens, and a few bacteria use inorganic compounds (e.g. H_2, NH_3, NO_2, H_2S) as the energy source and CO_2 as the carbon source. Cyanobacteria and some purple and green bacteria are photoautotrophs, using light and CO_2 as the energy and carbon source, respectively.

In general, bacteria have a relatively short generation time. For example, *Escherichia coli*, one of the most studied microorganisms, can divide every 22 minutes under favorable conditions. *E. coli* can easily be genetically modified to produce many kinds of useful products. *Bacillus subtilis*, which is often used as the Gram-positive equivalent of *E. coli*, has been widely used in chemical, food, and pharmaceutical industries because it is generally recognized as safe and its ease of genetic modification, metabolic diversity, and long-term storage (its endospore can survive extreme conditions). Members of genus *Clostridium* are Gram-positive, spore-forming anaerobes. They are ubiquitous in nature. Although some of them are pathogenic, many *Clostridium* spp. can produce alcohols, solvents, and organic acids from sugars and cellulose. Lactic acid bacteria (LAB), including some *Lactobacillus* species, are used industrially for the production of yogurt, sauerkraut, pickles, silage, and other fermented foods. They are Gram-positive facultative bacteria and can convert lactose and other simple sugars to lactic acid. Lactobacilli are unique in that they do not require iron for growth and have a high hydrogen peroxide tolerance. They have small genomes of ~2–3 Mb, many of which have been sequenced. They are models for rational metabolic engineering because of their simple metabolism and low G-C content. Pseudomonades are Gram-negative, aerobic rods capable of metabolizing a variety of diverse substrates and forming biofilms to thrive in harsh conditions. They are widely used in bioremediation for environmental clean up. *Streptomyces* are non-motile, filamentous, Gram-positive bacteria. They are unique in the way they form mycelia and spores, and their ability to produce numerous antibiotics and pigments. These diverse capabilities make bacteria ideal for many industrial applications.

2.2. Industrial applications

Bacteria play a major role in the environment and have many industrial applications (*see* Table 2). In nature, they recycle many chemical elements and compounds. Without such bacterial activities as the recycling of CO_2, life on Earth would be impossible. Moreover, we would drown in garbage and wastes if bacteria did not decompose dead plant and animal matters.

Table 2

Industrial applications of bacteria

Applications	Examples
Bio-energy	Hydrogen, electricity, methane, ethanol
Biocatalysis	Enzymes, organic solvent tolerant bacterial cells
Bioleaching	Heavy metals extraction from ores or crude oil (biomining)
Bioremediation	Pollution control, toxic waste clean-up, wastewater treatment
Chemicals	Organic acids, polysaccharides, bio-surfactants, butanol, 1,3-propanediol
Food and beverages	Dairy products: yogurt, cheese; beverages: cider, wine; vinegar
Health-care	Human therapeutic proteins, antibiotics

Bacteria are of major importance to the food industry. Some of the most common are lactic acid bacteria (LAB), which are used in many fermentation processes involving milk, meats, cereals, and vegetables because of their unique metabolic characteristics [1]. LAB contribute to the fermentation of many dairy products. Yogurt, widely considered a healthy food, is produced by bacterial fermentation of milk. The bacteria produce lactic acid, which turns the milk sour, retards the growth of disease-causing bacteria, and gives a desirable flavor to the resulting yogurt. Other foods fermented by bacteria include cheese, pickles, soy sauce, olives, and sausages and other cured meats. In most of these fermentations, bacteria that produce lactic acid play major roles. Even though alcohol-producing yeasts are the primary strains in the manufacture of beer and wine, lactic-acid bacteria are also involved in making wine or cider. Furthermore, bacteria that produce acetic acid convert alcoholic beverages to vinegar.

In the 1970s, scientists used information about the replication of viruses and bacteria and DNA synthesis to begin the genetic engineering of bacterial cells. Recombinant DNA technology was born as scientists combined human DNA with bacterial DNA. Bacteria, such as recombinant *E. coli*, became factories for churning out human proteins, such as the human insulin. Because they multiply so rapidly, bacteria produce many copies of proteins in a short time. The growth of genetic engineering has opened the way to the even greater use of bacteria in large-scale industrial manufacturing and environmentally friendly processes.

Since the 1980s, bacteria have gained importance in the production of many bulk chemicals, including ethanol made from fermented corn. In addition, bacteria play a role in the production of specialty chemicals, certain plastics, enzymes used in laundry detergents, and many antibiotics, such as streptomycin and tetracycline. Bacteria have been at the center of recent advances in biotechnology and create products for human benefit.

The problem of environmental pollution caused by plastic wastes is recognized throughout the world. Conventional plastics that are chemically synthesized from petroleum are not biodegradable, and are considered environmentally harmful [2]. In contrast, biodegradable plastics, such as polyhydroxyalkanoate (PHA) and poly-D-3-hydroxybutyrate (PHB), can be produced by decomposing the sludge from a wastewater treatment plant, which also alleviates pollution. Because of the increasing demand for energy and the decreasing availability of fossil resources, fermentation of waste biomass has been applied to make use of renewable resources to produce bio-energy and useful chemicals while reducing wastes [3]. Recovering energy from renewable materials can reduce the cost of sewage treatment and reduce dependence on fossil fuels [4]. For example, methanogenic bacteria (archaea) can break down sludge to produce methane gas, which can then be used as fuel to power the treatment facility. Many *Pseudomonas* species can grow on n-alkanes as their sole source of carbon. These procedures for processing harmful wastes are also called bioremediation.

Extremophiles have been exploited as a source of enzymes for biocatalytic processes under extreme conditions, such as at 110°C (thermophiles), pH 1.0 (acidophiles), high salts (halophiles), or under 800 MPa (barophiles) [5]. The use of whole cells of extremophiles is one of the direct applications in bioprocesses. For example, *Pseudoalteromonas haloplanctis*, functioned as a cold-adapted biocatalyst, can operate at low temperatures for food processing, which offers an alternative for energy-efficient industrial biocatalysis [6]. Extremophiles, such as *Dunaliella salina*, are used to extract heavy metals (copper, cobalt, uranium, etc.) from ores, which is called bioleaching (biomining) [7]. In commercial bioleaching process, strains with improved characteristics can be applied to bioremediation for degradation of hydrocarbons released from crude oil [8]. In addition, *Acidithiobacillus* and *Sulfolobus* spp. (acidophiles) are used to produce soluble ferric or copper salts from gold-bearing ores (e.g. pyrite, or FeAsS) in the biomining industry [6].

A large number of substrates used in the chemical industry are dissolved in organic solvents. However, most bacteria and their enzymes are inactivated in the presence of organic solvents. Organic solvent tolerant bacteria, such as *Bacillus*, *E. coli*, *Pseudomonas* and *Rhodococcus* strains circumvent these toxic effects by various adaptations, such as solvent efflux pumps, rapid membrane repair, lower cell membrane permeability, increased membrane rigidity, decreased cell surface hydrophobicity, etc., and provide a good tool for biocatalysis and bioremediation [9]. For example, the water-insoluble compounds suspended in the wastewater will take a longer time to complete the bioconversion. Solvent tolerant bacteria survive in organic solvents and accelerate the decomposition of organic solvents.

Table 3

Characteristics of common yeasts [1–3]

Species	*Candida* sp.	*Kluyveromyces* sp.	*Pichia* sp.	*Saccharomyces cerevisiae*	*Torulopsis glabrata*	*Yarrowia lipolytica*
Generation time	2 h	3.5~4.5 h	2~3 h	1.5~2 h	5~14 h	1.5~2 h
Growth media, carbon source	n-Alkane, methanol	Glucose, lactose, whey	Glucose, whey, methanol	Glucose, maltose, molasses, sucrose	n-Alkane, polyols, methanol, glucose	Sucrose, n-decane, alcohols, fatty acids
Growth temperature (°C)	0~48 (25~30)[1]	20~28	37~42	0~40 (28~35)	23~43 (37)	16~38 (26)
pH range	2.5~8 (5~7)	4.2~9	6~12	4.5~7.5 (6.0)	2~8	3.1~9 (4.2~5.8)
O_2 demand	Aerobic	Aerobic	Aerobic	Aerobic	Aerobic	Aerobic
Genetic modification	Easy	Average	Easy	Very easy	Average	Easy
GRAS	Most	Yes	Yes	Yes	Some	Yes
Others	SCP*		Recombinant proteins	Ethanol production	SCP*	Preservative-resistant

[1]Numbers in the parentheses are optimal temperature or pH range for growth.

*SCP: single-cell protein

3. YEAST

3.1. General characteristics

Yeasts are one-celled fungi, 5~10 μm in size. Yeast cells are usually spherical, cylindrical, or oval and are important for their ability to ferment the carbohydrates within various substances. They are widespread in nature, existing in soil and on plants. Yeasts have been used since prehistoric times in the making of breads and wines, but their cultivation and use in large quantities only started in the 19th century. Today, they are used industrially in a wide range of fermentation processes, as feeds and foodstuffs, as a source of vitamins, and to produce various antibiotics and steroid hormones. Characteristics of some commonly used yeasts are list in Table 3. Yeasts can grow in a wide range of pH. For instance, *Candida* spp., *Torulopsis glabrata*, and *Yarrowia lipolytica* survive in the pH range from 3 to 8. Generally speaking, yeasts prefer to live at temperatures between 25 and 35°C under aerobic conditions. Pure yeast cultures are grown in a medium of sugars, nitrogen sources, minerals, and water. In anaerobic environments, yeast transforms simple sugars, such as glucose and sucrose, into ethanol and carbon dioxide.

Most cultivated yeasts belong to the genus *Saccharomyces*; those known as brewer's yeasts are strains of *S. cerevisiae*, which have been widely used for ethanol production. *S. cerevisiae* is the eukaryotic model organism in molecular and cell biology, similar to *E. coli* as the model prokaryote. Yeasts usually divide every few hours, though they have longer generation times than bacteria. Most yeasts are generally recognized as safe (GRAS), easy to be genetically modified, and easy to separate in downstream processing because of their relatively large size. *S. cerevisiae* is the most studied of simple eukaryotes. It is the first eukaryote with its genome completely sequenced and its genetics and physiology thoroughly characterized. The completion of the entire genome sequence of *S. cerevisiae* in 1996 was a milestone in the fundamental understanding of its physiology and will undoubtedly accelerate developments in the genetic improvement of *S. cerevisiae* and other yeasts.

Candida spp. are methanol-utilizing yeasts. They produce lipases of commercial interest, can grow on paraffin oil, fatty acids, triglycerides, and n-alkanes, and thus are widely used in bioprocessing and bioremediation. *Kluyveromyces* spp. can grow either as single cells or in filaments, which provide larger surface area and thus increase the product yield for industrial applications. *Pichia pastoris* can be used in the production of enzymes and recombinant proteins because it can grow on methanol to a high cell density. However, the heat generated from its fermentation must be removed due to the highly exothermic process. *Torulopsis glabrata* can decompose n-alkanes, polyols, and methanol. Its cells as well as *Candida* spp. are used for SCP production. *Yarrowia lipolytica* is well known for its ability to decompose fatty acids, hydrocarbons, and alcohols via the glyoxylate pathway. It has been considered as a preservative-resistant yeast with strong production of extracellular lipases and proteases.

Yeasts have the advantages of rapid growth and ease of genetic manipulation. Other advantages of employing yeasts as hosts for fermentation are their abundance of metabolic activities and safety. These characteristics have brought yeasts many applications in chemical, food, and pharmaceutical industries.

3.2. Industrial applications

Yeasts have many applications in industrial food and beverage production (Table 4). Industrial yeasts are suppliers of enzymes, proteins, and chemicals. The commercial importance of yeasts extends to their application to the treatment of industrial wastes and effluents. For example, *Kluyveromyces marxianus* can remove heavy metals from waste stream; some *Candida* spp. can detoxify and remove pollutants from wastewater.

Table 4

Industrial applications of yeasts

Applications	Examples
Baking and brewing	Bread, beer, wine, spirits
Bio-based fuels	Bio-ethanol from sucrose, glucose, and xylose
Bioremediation	Heavy metal removal, wastewater treatment
Chemicals	Glycerol, bio-surfactants, enzymes, organic acids, amino acids
Health-care	Human therapeutic proteins, steroid hormones
Nutrition and animal feed	Biomass, polysaccharides, vitamins, single cell proteins

Alcoholic fermentation is the oldest known biological reaction. The German chemist Eduard Buchner (1897) discovered that a cell-free extract of yeast can induce alcoholic fermentation. Beer is an alcoholic beverage made from cereal grains, usually barley but also corn, rice, wheat, and oats, by yeast fermentation that consumes sugars in the grain and produces alcohol and CO_2. Two yeasts, *S. cerevisiae* and *S. bayanus*, are used to make wine by fermenting grapes. Yeast is responsible for the presence of both positive and negative odors in wine. For example, yeast may produce hydrogen sulfide when stressed. Adding nutrients to the fermentation tank can avoid this undesirable quality. The time of fermentation also determines wine character. Above all, subtle differences in ingredients determine the unique characteristics of each brewing process. Yeast fermentation is also used to make leavened breads. The main function of baker's yeast (*S. cerevisiae*) in bread dough is to produce CO_2 from sugars. The dough is placed in a warm and moist environment, enabling the yeast to multiply, and CO_2 produced during fermentation causes the dough to rise. Alcohol produced during fermentation contributes to the aroma of the bread. Secondary fermentation produces organic acids that also add to the flavor of the bread. In the making of wines, beers, spirits, and industrial alcohol, the fermented medium after separation and purification is the desired product, and the yeast itself is used in animal feeds. Yeast biomass is a rich source of proteins, nucleic acids, vitamins, and minerals.

Furthermore, yeasts contribute as hosts for expressing foreign genes not shared by prokaryotic cells in modern recombinant DNA technology. They are used to produce human proteins in spite of plasmid instability and the economic costs of providing growth medium. Gene expression is better in *S. cerevisiae* than in *E. coli* because *S. cerevisiae* is more capable of excreting and post-translationally modifying genetic products [10]. *Pichia pastoris* has also been widely used with commercially available expression systems. Yeast RNA polymerase recognizes many animal promoters, and yeast utilizes inexpensive carbon sources. Recombinant yeasts take less time, reach higher yields, and are more genetically stable and cheaper than the insect and mammalian cell systems. Besides, yeast cultures are non-pathogenic, stable, and easy to operate and scale-up. In addition, stable mutants exist that enhance productivity.

Yeasts are also widely used to produce fuels and chemicals from biomass, which is discussed in the next section along with bacteria.

4. FERMENTATION PRODUCTS FROM BACTERIA AND YEASTS

Fermentation is a bioprocess of chemical reactions catalyzed by the enzymes produced from bacteria, molds, yeasts and other microorganisms. For example, lactase, a ferment produced by bacteria, hydrolyzes lactose to glucose and galactose. Various fermentation products of milk, such as acidophilus milk and yogurt, are consumed as food. The most famous example is ethanol fermentation, in which the zymase (complex enzymes) obtained from yeast converts simple sugars, such as glucose and fructose, into ethanol and CO_2. Most fermentation occurs naturally, e.g. butyric acid is formed when butter becomes rancid, and acetic acid is the product when wine turns to vinegar. Today, bacteria and yeasts are widely applied in the fermentation industry to produce value-added products from renewable resources. Table 5 lists some important fermentation products, including alcohols, biofuels, bio-polymers, bio-surfactants, specialty chemicals, materials, polysaccharides, enzymes, and vitamins.

4.1. Alcohols

Over the past three decades, ethanol production in the United States has increased over ten-fold. Bio-ethanol is a clean and sustainable alternative to petroleum. It has lower toxicity and is easily biodegradable, soluble in water, harmless to the environment and does not generate greenhouse gases [11]. A yeast strain with a high specific growth rate and ethanol productivity at high ethanol concentrations is extremely demanded in the fermentation industry. Hack and Marchant [12] isolated a novel thermotolerant strain of *Kluyveromyces marxianus* and improved ethanol productivity with cell recycling. Immobilizing the yeast *P. stipitis* in Ca-alginate beads increased the cell density and made the process of producing ethanol from glucose and xylose feasible [13]. Among the common fermenting yeasts, *S. cerevisiae* is the most frequently used. Co-fermentation of glucose and xylose using recombinant microorganisms is one of the most promising methods for producing bio-ethanol from lignocellulosic biomass, an attractive feedstock for ethanol fuel production. However, the diauxic behavior is unfavorable to xylose consumption. Sedlak and Ho [14] demonstrated a stable recombinant *Saccharomyces* yeast, 424A (LNH-ST), that contains cloned xylose-metabolizing genes integrated into the yeast chromosome and can ferment glucose and xylose in hydrolysates from cellulosic biomass to ethanol. This recombinant strain has the potential of making ethanol production from cellulosic biomass more profitable. Bio-ethanol also can be produced directly from cellulosic materials by a thermophilic bacterium, *Clostridium thermocellum*, in anaerobic conditions [15], and the production rate is enhanced at elevated pressures due to the effects on membrane fluidity during continuous culturing [16].

Glycerol is used in almost all chemical industries due to its particular combination of physical and chemical properties [17]. The majority goes into the manufacturing of synthetic resins and ester gums, drugs, cosmetics, and toothpastes because glycerol is a good solvent of many compounds. One of the biochemical processes that produce glycerol is aerobic fermentation with osmophilic yeast. Glycerol is accumulated in yeast as a compatible solute during adaptation to high osmotic pressures or high sugar concentrations (*see* Fig. 1). *S. cerevisiae* uses glycerol as its sole compatible osmolyte. The process usually decreases the specific growth rate because of the limited oxygen transfer rates (OTRs) of industrial bioreactors. *Candida krusei* is another osmophilic yeast which can ferment glucose into glycerol. Huang *et al.* [18, 19] suggested that a higher glycerol yield and productivity can be obtained when the oxygen transfer rate is enhanced by pressure pulsation. However, Liu *et al.* [20] reported that oxygen limitation improved enzyme activity, shifting the metabolic flux to produce more glycerol.

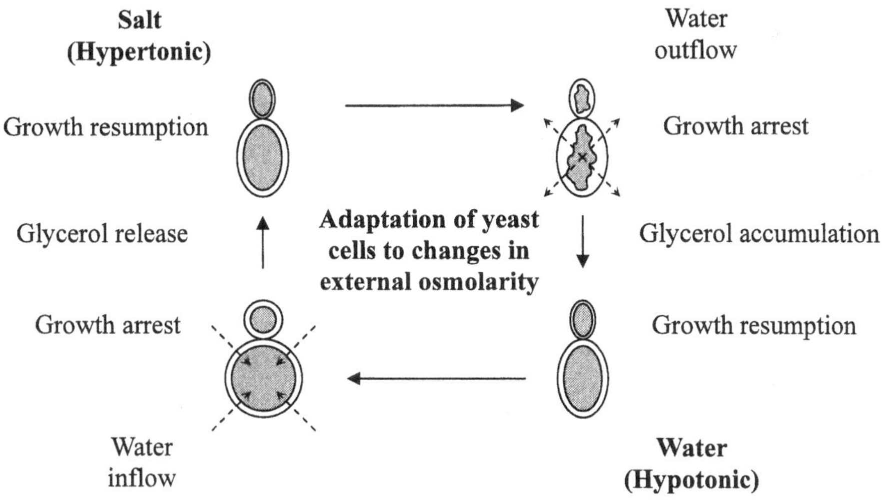

Fig. 1. The impact of osmotic stress on yeast cells and the basic response mechanisms. (adapted from Yeast stress responses / S. Hohmann and W. H. Mager [21]).

DuPont Corporate and Genencor have engineered biosynthetic pathways into an industrial strain of *E. coli* to directly convert glucose to 1,3-propanediol, a route not previously available in a single microorganism [22]. 2,3-Butanediol has potential use as an antifreeze agent, polymeric substance, and fuel additive. *Aerobacter aerogenes*, *Bacillus polymyxa*, and *Klebsiella pneumoniae* have been employed to produce 2,3-butanediol because of the low product yield of the alternative yeast fermentation. Converting cellulosic and hemicellulosic materials to alcohols has made this process economical [23].

Table 5
Current and potential industrial fermentation products from bacteria and yeasts

Fermentation product	Typical species	Carbon source	References
Alcohols, biofuels and bio-energies			
1,3-Propanediol	*E. coli, Clostridium butyricum, Klebsiella pneumoniae, Citrobacter freundii*	Glucose, glycerol	[22–24]
2,3-Butanediol	*Aerobacter aerogenes, Bacillus polymyxa, Klebsiella pneumoniae*	Xylose, glucose	[22–23]
n-Butanol	*Clostridium acetobutylicum*	Glucose	[25]
Ethanol	*Saccharomyces cerevisiae, Pichia stipitis, Clostridium thermocellum, E. coli, Zymomonas mobilis, Kluyveromyces matxianus*	Sucrose, glucose, xylose, lactose, cellulose	[11–13, 15, 26–29]
Glycerol	*Candida krusei, S. cerevisiae, Pichia farinosa*	Glucose, fructose	[18–20]
Hydrogen	*Enterobacter aerogenes, Clostridium butyricum, Caldicellulosiruptor saccharolyticus*	Propionic acid, xylose, glucose	[30–33]
Electricity	*Enterobateriaceae, Geobacter* sp.	Wastewater	[4]
Methane (biogas: methane and CO_2)	*Methanosarcina barkeri, Methanococcus mazei, Streptococcus lactis*	CO_2-H_2, acetate, methanol	[34–37]
Biopolymers and biosurfactants			
Biopolymer	*Rhizobium* sp.	Mannitol	[38]
Poly-glutamic acid (PGA)	*Bacillus subtilis, Bacillus licheniformis*	Glutamic acid, fructose	[2, 42]
Polyhydroxy-alkanoate (PHA)	*Alcaligenes eutrophus, E. coli* *Ralstonia eutropha, Azotobacter vinelandii*	Glucose, fatty acids, alkanes	[43–44]
Bio-surfactant	*Azotobacter vinelandii, Bacillus subtilis, Pseudomonas* sp.	Potato, cassava, glucose	[45–49]
Microbial polysaccharides			
Cellulose	*Acetobacter xylinum*	Molasses	[50]
Dextran	*Leuconostoc mesenteroides*	Molasses	[51]
Mannan	*Hansenula, Pichia, Pachysolen* sp.	Sucrose, glucose	[52]
Pullulan	*Azotobacter pullulans*	Starch syrup	[53]
Xanthan gum	*Xanthomonas campestris*	Glucose, sucrose	[54–55]
Carboxylic acids			
Acetic acid	*C. thermoaceticum, C. formicoaceticum, Gluconobacter oxydans, Acetobacter* sp.	Glucose, fructose, ethanol	[3–4, 56–57]
Butyric acid	*C. tyrobutyricum, C. butyricum*	Glucose, xylose	[58–59]
Citric acid	*Yarrowia lipolytica, Candida lipolytica*	Glycerol, glucose	[60–61]
Gluconic acid	*Gluconobacter suboxydans, Pseudomonas* sp.	Starch, glucose	[62–66]
Lactic acid	*Lactococcus lactis, Lactobacillus delbrueckii, L. helveticus, Kluyveromyces lactis, E. coli*	Glucose, cheese whey, food wastes	[67–70]
Propionic acid	*Propionibacterium acidipropionici*	Lactose, glucose	[71–72]
Pyruvic acid	*Torulopsis glabrata, E. coli, Candida* sp.	Glucose	[73–74]
Succinic acid	*Actinobacillus succinogenes, E. coli*	Glucose	[75]

Table 5 (Continued)
Current and potential industrial fermentation products from bacteria and yeasts

Fermentation product	Typical species	Substrates	References
Amino acids			
L-Alanine	*E. coli, Bacillus sphaericus*	Sucrose, glucose	[76–77]
L-Aspartic acid	*E. coli, Corneybacterium glutamicum*	Glucose	[78–79]
L-Glutamic acid	*C. glutamicum, Brevibacterium* sp.	Molasses, acetate	[80–81]
L-Lysine	*C. glutamicum, E. coli*	Glucose, acetate	[82–83]
L-Phenylalanine	*E. coli, B. subtilis*	Glucose	[78, 84]
L-Threonine	*E. coli, C. glutamicum*	Glucose, sucrose	[85–89]
L-Tryptophan	*E. coli, B. subtilis*	Glucose, indole	[90–91]
Antibiotics			
Aminoglycosides	*Streptomyces griseus, S. bikiniensis*	Glucose, starch	[83, 92]
Bacitracin	*Bacillus licheniformis*	Soybean meal	[93]
Bacteriocin	*B. licheniformis, B. cereus, Staphylococci* sp.	Cheese whey, thioglycollate	[94–95]
β-Lactam	*Streptomyces clavuligerus*	Glucose	[96–98]
Nisin	*Bacillus subtilis, Lactococcus lactis*	Cheese whey	[68, 96]
Pediocin	*Pediococcus acidilactic*	Glucose	[99]
Tetracyclines	*Streptomyces aureofaciens, S. rimosus*	Peanut shells	[100–101]
Enzymes			
Alkaline proteases	*B. licheniformis, B. amyloliquefaciens.*	Lactose, sucrose	[102–103]
α-Amylases	*B. subtilis, B. amyloliquefaciens*	Starch	[104–105]
Glucose isomerase	*Streptomyces rubiginosus, Thermus thermophilus.*	Starch, fructose	[106–110]
β-Lactamase	*B. subtilis, B. licheniformis*	Fructose, glucose	[111]
Lipases	*S. cerevisiae, Candida Antarctica*	Glucose, starch	[111–112]
Pullulanase	*Bacillus* sp., *Klebsiella pneumoniae*	Starch, flours	[114–118]
Vitamins			
β-carotene	*Candida utilis, Pichia farinosa*	Starch , glucose	[79–80]
Provitamin D_2	*Saccharomyces cerevisiae*	Glucose, molasses	[80, 83]
Vitamin B_{12} (Riboflavin)	*B. megaterium, Propionibacterium shermanii, Pseudomonas deniftrificans, B. subtilis, Candida flareri*	Glucose, methanol, n-decane, sugar beet molasses	[83, 119–123]
Nucleotides and others			
5'-IMP, 5'-GMP, 5'-XMP	*B. subtilis, Brevibacterium ammoniagenes, Corynebacterium ammoniagenes*	Glucose, fructose	[83, 124]
Lipid compounds	*Saccharomyces cerevisiae*	Fatty acids	[125–127]

Butanol is used as a solvent and in plasticizers, amino resins and butylamines. It can also be used as an alternative fuel and is more energy efficient than ethanol. Butanol was produced primarily by anaerobic fermentation with *Clostridium acetobutylicum* before World War II. However, the petroleum-based route to butanol replaced the lower yield biochemical route as oil prices declined. Nevertheless, U.S. legislation to produce strategic chemicals, fuels, and energy from domestic renewable resources and the need to lessen the dependence on diminishing petroleum supplies have resulted in the renaissance of the fermentation process as a possible source of solvents [25]. A few researchers are trying to increase butanol yield with improved bacterial strains, advanced reactor technology, and separation to reduce product inhibition. These may make it cost-competitive with the petrochemical route. Genetic and metabolic engineering of *C. acetobutylicum* have been studied as promising methods for overcoming the aforementioned problems in ABE fermentation [128]. Despite all these efforts, the best results ever obtained for ABE fermentations are still less than 20 g/L in butanol concentration from fermentation, 4.5 g/L·h in butanol productivity, and a butanol yield of less than 25 % (w/w) from glucose.

4.2. Fuels and energy

The realization that fossil fuel reserves are limited and that their use pollutes the environment has pushed research for alternative energy sources. One of the most well known alternatives to fossil fuels is ethanol, a renewable energy source produced from fermentation. In addition to ethanol and butanol, biomass also can be used to produce other fuels and energy, including hydrogen, methane, and electricity.

Hydrogen is a fuel that can reduce air pollution and the greenhouse effect. It is clean, efficient, and can be used in fuel cells to generate electricity [30]. Hydrogen can be generated via the fermentative conversion of organic substrates. To make hydrogen a more sustainable and economic source of energy, it should be produced from renewable resources, especially organic wastes. Also, hydrogen made from renewable energy resources provides clean and CO_2-free energy. Hydrogen can be biologically produced either photosynthetically by algae [129–130] and photosynthetic bacteria, such as *Rhodobacter sphaeroides* [131–132], or non-photosynthetically by fermentation with anaerobic bacteria, such as *Enterobacter aerogenes* and *Clostridium butyricum* [32]. *Rhodopseudomonas capsulate*'s fermentation rate is higher when propionic acid is used as a substrate than when other organic acids found in wastewater are used [31]. *Caldicellulosiruptor saccharolyticus*, an extreme thermophile, can use paper sludge as a cheap, renewable feedstock for hydrogen fermentation [33]. The hydrogen yields from fermentation range from 0.35 to 7.2 mol/mol glucose, depending on the fermentation conditions and bacteria used in the process.

Methane is usually produced as a mixture with carbon dioxide by bacterial degradation of organic matters and the biogas can be used as a fuel [34–35, 36–37]. Methane fermentation of lignocellulosic materials is an effective method for providing both a renewable energy source and a means to reduce the volume of municipal solid wastes, although the fermentation process has generally been observed to be slow and incomplete.

Microbial fuel cells (MFCs) have been suggested as an alternative to biohydrogen production in the waste treatment process. MFCs generate electricity using anaerobic bacteria that consume organic wastes and transfer electrons to an electrode. Many anaerobic bacteria, such as *Enterobateriaceae*, *Geobacter*, and *Shewanella* spp., have successfully generated electricity by fermentation [4].

4.3. Polymers and biodegradable plastics

Bio-based polymers include various synthetic polymers derived from renewable resources, biopolymers (nucleic acids, polyamides, polysaccharides, polyesters, and polyphenols), their derivatives, and their blends and composites. They are applied in the food, pharmaceutical chemical, and petroleum industries, and are used as emulsifying agents, stabilizing agents, flocculating agents, and lubricants [38]. Recently, lactic acid produced from fermentation has been used to synthesize biodegradable plastic (polylactic acid).

Biodegradable plastics have a high demand because they are thermoplastic, environmentally degradable, and help to reduce the disposal problem of non-degradable plastics. Several polyesters with properties comparable to conventional plastics, such as poly-butylene succinate (PBS), polyester carbonate (PEC), poly-D-3-hydroxybutyrate (PHB), poly-propiolactone (PPL), and poly-L-lactide (PLA) [133], are used as biodegradable plastics. Many of these biopolymers, such as PLA and polyglycolic acid, have been accepted for use in the medical industry as medical devices or cell culture matrices [134–135]. Poly-glutamic acid (PGA), produced by the genus of *Bacillus*, can be used as the basis in drug delivery applications for cancer therapy. PGA-conjugation can provide more stable and water-soluble drugs, which control drugs exposure to tumor cells [39–41].

Polyhydroxyalkanoate (PHA) is one of the largest groups of thermoplastic polyesters synthesized by numerous bacteria as an intracellular carbon and energy storage compound and accumulated as granules in the cytoplasm [42]. PHA is regarded as a potentially useful alternative to petroleum-derived thermoplastics because it is biodegradable and biocompatible. This makes PHA a good absorbable material for implantable medical applications. PHA has been industrially produced by pure cultures of *Alcaligenes latus*, *Azotobacter vinelandii*, *Pseudomonas oleovorans*, *Ralstonia eutropha*, recombinant *Alcaligenes eutrophus*, and recombinant *E. coli*. *E. coli* has proven to be a high-yield microbial species in the production of PHA. Metabolix Inc. applied metabolic engineering at the genomic level to produce several PHA copolymers in a single bacterial system. The composition of PHA is dependent on the metabolic capability of the microorganism and the substrate specificity of PHA synthase [44]. These properties make PHA a renewable and environmentally friendly alternative to synthetic plastics, adhesives, coatings, extruded products, fibers, and film.

4.4. Surfactants

Bio-surfactants produced by microorganisms, such as *Azotobacter vinelandii*, *B. subtilis*, *Pseudomonas* sp., and *Rhodococcus* sp., reduce the interfacial tension between two phases and can be used in textiles, environmental bioremediation, and fossil fuel recovery [45–46], as well as in cosmetic, pharmaceutical, and food industries. Bio-surfactants are membrane-associated metabolites, biodegradable, and can be produced from renewable resources by many bacteria. Surfactin, produced by *B. subtilis*, is also an effective antimicrobial and antiviral agent, which transport cations across membranes by inducing the formation of ionic pores in phospholipid bilayers [47–49].

4.5. Microbial polysaccharides

Microbial polysaccharides have been widely applied in chemical, food and pharmaceutical industries although their production costs are higher than those of traditional polysaccharides, such as cornstarch and cellulose-derived products that dominate the market. Microbial polysaccharides are water soluble biopolymers produced by many bacteria. Because of their rheological characteristics, microbial polysaccharides are used as binders, coagulants, emulsifiers, film formers, gelling agents, lubricants, stabilizers, and thickening and suspension agents [53]. Xanthan gum has the largest microbial polysaccharide market because of its rheological features over a wide range of temperatures and pH. It is used for salad dressings, syrups, starch-based products, beverages, abrasives, texturized coatings, and enhanced oil recovery [54–55]. Dextrans are employed in the manufacturing of molecular sieves. They are primarily produced by *Leuconostoc mesenteroides* and *Leuconostoc dextranicum* [51]. *Hansenula*, *Pichia*, and *Pachysolen* spp. produce extracellular phosphorylated mannans that are resistant to microbial attacks [52]. Pullulan produced by *Azotobacter pullulans* is used as a film-wrap food packaging material in Japan. Bacterial cellulose (BC) is produced by *Acetobacter xylinum* BPR2001 using molasses as a carbon source [50]. Because these microbial polysaccharides are resistant to digestive enzymes, they can be used as substitutes for starch in low-calorie foods [114].

4.6. Carboxylic acids

Carboxylic acids are widely used as additives in the food industry and as chemical feedstocks [60]. High yields of carboxylic acids can be obtained from anaerobic bacterial fermentations and in the TCA cycle of aerobic microorganisms. Acetic acid has long been produced as vinegar from ethanol in an aerobic fermentation by *Acetobacter* spp. However, a higher yield of acetic acid can be produced directly from sugars by anaerobic fermentation with homoacetogens [56]. Lactic acid bacteria (LAB) are regarded as safe organisms for the production and preservation of fermented foods. They convert certain sugars into organic acids. Lactic acid and acetic acid, the two main types of acids produced, also provide aroma [136]. Lactic acidity plays a direct role in the flavor of the bread. Lactic acid is widely used in food and chemical industries [70, 137].

Propionic acid is used in the manufacture of herbicides, chemical intermediates, artificial fruit flavors, pharmaceuticals, cellulose acetate propionate, and preservatives for food, animal feed, and grains [71]. Butyric acid has been widely used in beverage, foodstuff, and pharmaceutical industries. Its derivatives play an important role in plastic and textile industries. Butyric acid can be extracted from butter, but this method is too expensive. These acids are mainly produced by petrochemical routes, but there is increasing interest in producing them from biomass via. anaerobic fermentation [58]. Propionic acid is the main products from sugar fermentation by *Propionibacterium acidipropionici* and *P. shermani*, whereas butyric acid is the main product from *Clostridium tyrobutyricum and C. butyricum* [59, 72].

Citric acid is produced aerobically in industrial fermentation processes involving either *Aspergillus niger*, *Candida oleophila*, or *Yarrowia lipolytica* [60–61, 138]. Citric acid is widely used in food, beverages, and pharmaceuticals for its acidity, flavor, and salt-formation capability. Citrate salts have been used as sequestering agents and as an anti-coagulant blood preservative [61]. Their antioxidant properties can reduce metal-catalyzed oxidation in fats and oils. Gluconic acid and its salts have wide potential uses in food, pharmaceutical and medical industries [139]. Its strong chelating property also makes it a good cleaner for glass and metal washing in industry. In addition to fungal fermentation, gluconic acid can be derived from glucose by oxidative fermentation by some *Gluconobacter* [62–64] and *Pseudomonas* strains [65–66].

Pyruvate is one of the most important metabolites in the central metabolism of living cells. It is a precursor for amino acid synthesis and plays a pivotal role as an intermediate of glycolysis. Pyruvic acid and its salts are important chemicals used in agrochemical, cosmetic, pharmaceutical and food industries [74]. Because it costs less than the chemical method, a multi-vitamin auxotrophic yeast *Torulopsis glabrata* has been used in the commercial production of pyruvate [73]. Succinic acid and its salts have the potential to be commodity chemicals. It can be used as an intermediate to produce 1,4-butanediol and pyrrolidinones, which are important in the manufacture of plastics and solvents. Succinic acid can be produced from glucose through fermentation by obligate anaerobes, such as *Anaerobiospirillum succiniciproducens* and *Actinobacillus succinogenes*, and recombinant *E. coli* [75].

4.7. Amino acids

Amino acids have been widely used as food additives, medicine, cosmetics, and other materials in the chemical industry. They are used to enhance flavors in the food industry. *Corynebacterium*, *Brevibacterium* and their mutants are widely used to produce L-glutamic acid [80–81]. L-Lysine is one of the essential amino acids for daily nutrition. A fermentation based on glucose with a *gnd* mutant of *C. glutamicum* led to 15% increased production of L-lysine [82]. L-Threonine, as well as L-lysine, is supplemented to improve the nutritive value of animal feeds or used as a precursor of several flavoring agents [85]. *E. coli* has been genetically modified for commercial production of L-threonine, L-phenylalanine, and L-tryptophan [84–89]. L-Phenylalanine (L-Phe) and L-tryptophan are aromatic amino acids. They have many applications in food and pharmaceutical industries. L-Aspartic acid and L-phenylalanine can be used to synthesize aspartame, which is the most widely used low-calorie sweetener [78–79]. L-Tryptophan combined with L-histidine acts as an antioxidant to preserve powered milk. L-Alanine is a food additive generated through aerobic fermentation and the conversion of pyruvate by *E. coli* or *Bacillus* species [76–77]. These amino acids also can be used to synthesize, for example, polyalanine fibers, lysine isocyanate resins, and surfactants [83, 90–91].

Not all essential amino acids can be economically produced by fermentation. For example, methionine is commercially produced by either extraction/hydrolysis from food stuffs or by chemical synthesis, which employs toxic raw materials and generates racemic mixtures, complicating the purification process. Biosynthesis of these amino acids by *E. coli* or other microorganisms is thus an attractive alternative but requires further research and development.

4.8. Antibiotics

Antibiotics, products of secondary metabolism, inhibit the growth of other microorganisms even at low concentrations. Antibiotics have many applications, especially in the treatment of certain tumors, the control of plant diseases, and as animal growth promoters. Lactic acid bacteria (LAB) can produce bacteriocins, which are antimicrobial peptides used as biopreservatives in the food industry [95] and as one alternative to antibiotic therapy. *Bacillus* strains produce many kinds of peptide antibiotics, and have been safely used in the food industry [94]. One of the most industrially relevant bacteriocins is nisin, which has been applied in pharmaceutical, veterinary and microbiological industries. *Lactococcus lactis* subsp. *lactis* utilizes cheese whey as substrate to simultaneously produce nisin and lactic acid [68, 96]. The bioprocess not only produces value-added products, but also solves the pollution problem – disposal of cheese whey – and reduces the production cost. β-Lactam antibiotics inhibit bacterial cell wall (peptidoglycan) synthesis [97]. β-Lactam antibiotics and cephalosporins are two of the most effective antibiotics in control infectious diseases [98]. Bacitracin, which is a peptide antibiotic, is used as a topical antibiotic and as a growth promoter in animal feeds [93]. Aminoglycosides are used against Gram-negative bacteria to relieve severe infections. For instance, streptomycin is primarily used to treat tuberculosis [92]. Tetracyclines inhibit protein synthesis in both Gram-positive and Gram-negative bacteria [100]. Tecracyclines are the third most widely used antibiotics after β-Lactam antibiotics and cephalosporins [101]. *Pediococci* (pediocins) is used as a food biopreservative, providing effective measures to control pathogens [99].

4.9. Industrial enzymes

Amylases as well as amyloglucosidases are industrial enzymes used in starch-saccharification – converting starch to glucose. α-Amylases, endoenzymes, are produced by *Bacillus subtilis*, *B. amyloliquefaciens* and *B. licheniformis* [104–105]. Pullulanase, known as limit dextrinase, is able to specifically hydrolyze the β-1,6-glycosidic linkages of pullulan to generate maltotriose and break up the 1,4-linkages in amylopectin, which leads to the formation of maltose and glucose [114–115]. Therefore, this enzyme, mainly produced by *Bacillus* sp., plays a key role in the brewing process and starch hydrolysis [116–117]. Glucose isomerase catalyzes the reversible isomerization of D-glucose to D-fructose or D-xylose to D-xylulose [106]. It is used to produce high-fructose-corn-syrup, a low calorie sweetener. Whole cells of *Streptomyces* sp., immobilized in a bioreactor, have been continuously used for the production of glucose isomerase with heat treatment inactivating contamination of other enzymes in the cells [107–110]. The whole cells were employed as non-viable biocatalysts for biotransformation. In general, single-activity, immobilized cell systems with permeabilizing treatments with heat, surfactants or solvents are required [106].

Many bacteria, such as *B. licheniformis* and *B. amyloliquefaciens*, excrete alkaline proteases [102–103]. Proteases are used primarily in the detergent and dairy industries. They are also used for medicine. Lipases have been used for therapeutic purposes as digestive enzymes in the dairy industry. Lipases from *Candida cylindraceae* are used to hydrolyze oils in the soap industry [112]. In addition, lipases from *C. rugosa* can perform enantioselective esterification to synthesize esters from acids and alcohols in the pharmaceutical industry [140]. *Bacillus* species produce β-lactamase, an industrial enzyme that catalyzes the hydrolysis of the β-lactam ring in β-lactam antibiotics. β-Lactamase has been used in medicine for the specific assay of penicillins [111].

4.10. Vitamins

Vitamins (β-carotene, riboflavin, pyridoxine, etc.) are produced by yeasts or bacteria, such as *Candida utilis*, *Pichia farinosa*, *S. cerevisiae*, and *Mycobacterium smegmatis* [78–79]. Provitamin D_2 (ergosterol) can be produced by *Saccharomyces* strains at concentrations as high as 0.1-10% of the cell weight [79, 83]. Vitamin B_{12} is synthesized exclusively by microorganisms in nature; humans can obtain it only from food [83, 119–121]. Commercial production is currently carried out completely by fermentation. *Pseudomonas deniftrificans* is one of the most productive species for vitamin B_{12}. Riboflavin, as well as thiamine and nicotinic acid, are frequently added to flour to produce vitamin-enriched bread. Mutants of *Bacillus pumilus* can transform glucose to D-ribose for subsequent chemical conversion to riboflavin. High-yield microorganisms such as *C. acetobutylicum* and *Candida flareri* can produce riboflavin directly from glucose [122–123].

4.11. Nucleosides

Nucleosides and nucleotides, guanylic acid (5'-GMP), inosinic acid (5'-IMP) and xanthylic acid (5'-XMP), have widely used as flavor-enhancers for food. Nucleosides, nucleotides and related compounds are being tested for therapeutic purposes, such as cancer chemotherapy [124]. The Ajinomoto Company has greatly enhanced the amount of inosinic acid produced with an IMP production process that used a mutant of *Brevibacterium ammoniagenes* [83].

4.12. Other products

Sterols, phospholipids, fatty acids and ceramides are pharmaceutical lipid compounds of high commercial interest [125]. The lipid biosynthetic pathways of yeasts have been well characterized, and *S. cerevisiae* has been genetically modified for the synthesis of many lipid-derived compounds, including sterols, steroid hormones and polyunsaturated fatty acids [126, 127]. In addition, several yeasts are used to produce therapeutic proteins and single cell proteins (SCP), a highly nutritional food supplement. Many other applications to produce value-added bio-based products with bacteria and yeasts are not described here owing to the page limit.

5. FERMENTATION PROCESSES

As discussed in the previous section, bacteria and yeasts can be used to produce various useful products. However, the fermentation performance is affected by the culture used and many physical and chemical factors, and process conditions affecting small- or laboratory-scale processes are quite different on a larger scale. Scale-up studies are thus necessary for the industrialization of microbial processes. A basic understanding of potential alternative processes and knowledge of the manufacturing technology for a particular product play crucial roles in the process development and scale up. The culture, medium, and cultivation method used in a fermentation process influence not only the economics, but also the consistency of the final product. Fermentation byproducts may be reduced by only intermittently adding substrate or by using different carbon sources. Properly formulating the medium and choosing the cultivation system for a large-scale fermentation process are thus important. Design factors and challenges in fermentation process scale up include oxygen transfer, mixing, power consumption, heat transfer, and sterilization, all of which depend on the reactor design and affect cell density, productivity, and final product concentration and purity. Issues such as end-product inhibition, foaming, contamination, and process/product stability also need to be addressed.

Figure 2 illustrates the important technical areas for a fermentation process, including culture selection and genetic manipulation, raw materials and their sterilization, bioreactor design and operation, and down-stream processing for product recovery, purification, and formulation. The important issues in these areas are discussed in this section.

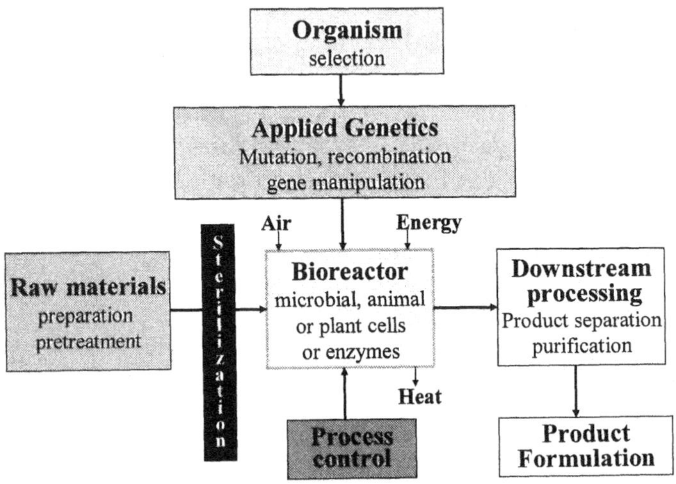

Fig. 2. Schematic overview of a fermentation process.

5.1. Culture characteristics affecting fermentation

The bacterial or yeast culture used in a fermentation process would determine the productivity, yield, and purity, which are also dependent on the operating conditions. The media composition (mainly carbon and nitrogen sources) and fermentation conditions (pH, temperature, mixing, aeration, etc.) are the common factors to be considered and need to be optimized based on the culture used in the fermentation. Furthermore, process design, including the reactor operation mode, down-stream processing, and even waste treatments, is also highly dependent on the culture. It is thus very important to select suitable cultures and processing conditions for economical industrial production. The selection of the culture for production of ethanol and lactic acid will be discussed as two examples.

As shown in Tables 6 and 7, each culture has its own pros and cons when used in the fermentation. The final selection thus will be based mainly on the economical factors, including raw material costs, productivity, yield, recovery costs, and waste disposal. For example, *S. cerevisiae* is widely used in industrial ethanol fermentation even though *Zymomonas mobilis* can produce ethanol from glucose at a higher productivity and yield [141, 142]. This is because the yeast cell is hardy and easy to separate from the fermentation broth. In order to use xylose for ethanol production, recombinant strains of *S. cerevisiae, Z. mobilis*, and *E. coli* have been developed. Although *E. coli* appears to be a better organism in simultaneously converting glucose and xylose to ethanol, its low ethanol tolerance, neutral pH for growth, and disposal after fermentation are unfavorable to its industrial application [28]. Yeast biomass generated in ethanol fermentation is used in animal feed and thus does not pose a disposal problem.

Table 6
Comparison of ethanol fermentations by bacterial and yeast cultures

	G(-) Bacteria	Yeast	G(+) Bacteria
Species	*E. coli*	*S. cerevisiae*	*Zymomonas mobilis*
Substrates	Glucose, xylose	Glucose, sucrose, xylose	Glucose, fructose, sucrose; xylose
Medium	Simple medium	Simple medium	Simple medium
Culture pH	pH 6~8	pH 5	pH 7
Product yield	0.46 g/g glucose 0.46 g/g xylose	0.47 g/g glucose 0.43 g/g xylose	0.49 g/g glucose nil from xylose
Product concentration	~50 g/L	120 g/L from glucose	120 g/L from glucose
Productivity	0.83 g/L·h	>1.4 g/L·h	>2 g/L·h with glucose

Table 7
Comparison of lactic acid fermentations by bacterial, yeast, and filamentous fungal cultures

	Bacteria	Yeast	Filamentous fungi
Species	*Lactobacillus* spp.	*Kluyveromyces lactis*	*Rhizopus oryzae*
Substrates	Glucose, lactose, sucrose; can't use starch	Glucose, lactose; can't use starch	Glucose, starch, xylose
Medium	Require complex growth nutrients	Relative simple industrial media	Simple medium with only trace minerals
Growth conditions	Anaerobic, pH >5	Anaerobic, pH 4.5	Aerobic, pH > 4
Products	Mixtures of L(+) and D(-)-lactic acids	Pure L(+)-lactic acid	Pure L(+)-lactic acid
Product yield	0.85 ~ 0.95 g/g glucose	~0.85 g/g glucose	~0.85 g/g glucose
Product concentration	Up to 150 g/L	60–109 g/L	Up to 120 g/L
Productivity	as high as 60 g/L·h	0.12–0.91 g/L·h	Up to 6 g/L·h

Food and pharmaceutical grades of lactic acid are mainly produced by LAB, such as *Lactobacillus* spp. because of the high productivity and yield. However, industrial LAB strains usually produce a mixture of L-(+) and D-(−)-lactic acids, which is not suitable for the synthesis of polylactic acid. LAB also require a complex medium for their growth. Tate and Lyle developed a recombinant yeast *Kluyveromyces lactis* that can produce pure L-(+)-lactic acid from glucose in a simple medium but has a lower productivity and yield. The filamentous fungus *Rhizopus oryzae* also can produce pure L-(+)-lactic acid from glucose, starch, and xylose in a simple medium, but its growth requires aeration and its filamentous morphology can be difficult for scale up [143].

5.1.1. Media formulation and feeding mode

Culture media contain all the necessary nutrients to maintain cell growth and to generate products. Nutrients include carbon, nitrogen, oxygen, hydrogen, sulfur, phosphorus, trace elements, vitamins, growth factors, and metabolic precursors. Anti-foaming agent and buffering chemicals also may be added in the medium. Media formulation is determined by the nature of the desired fermentation products and the culture used. Two types of growth media are usually used in fermentation: synthetic (with a well defined composition) and complex media. Defined media have specific amounts of pure chemical compounds and an identifiable chemical composition. Complex (enriched) media contain natural compounds whose precise chemical composition is not known. For example, a medium containing yeast extract, peptone, molasses, or corn steep liquor is a complex medium, which provides necessary nutrients and generates higher cell yields than defined media. However, complex media can vary from one batch to another, and thus affect the process reproducibility; defined media are more reproducible, giving the operator better control of the fermentation, but are more expensive than complex media. For economic reasons, complex media are widely used in industrial fermentations. Glucose and sucrose are often used as the carbon source, but they may be supplied in an industrial substrate such as molasses, which provide not only the carbon source but also nitrogenous substances, vitamins, and trace elements. The required nitrogen source is usually supplied in yeast extract or corn steep liquor, an inexpensive industrial byproduct from corn refining. These materials constitute an excellent source of nitrogen as well as other growth factors. For some bacteria, urea and inorganic nitrogenous compounds, such as ammonia and ammonium sulfate, also have been used.

The choice of the carbon source, which is usually the major raw material cost, is dependent on the cultures. For example, amylase-producing microbes can utilize starch and dextrins, instead of the more expensive glucose. Genetically modified yeasts can use xylose, which is a major sugar component in hemicellulose. *Kluyveromyces* and *Torulopsis* spp. can consume the lactose in cheese whey as a cheap carbon source. *Yarrowia lipolytica* is able to degrade lipids, proteins and n-paraffins as sole carbon source. *Candida* can assimilate n-alkanes and fatty acids as carbon sources. Methylotrophic bacteria, such as *Methylobacterium extorquens*, grow on methanol utilizing the serine cycle for carbon assimilation. Some yeasts, including *Candida*, *Hansenula*, *Pichia,* and *Torulopsis*, also can use methanol as the carbon source.

Batch, fed-batch, and continuous cultures are three common ways to grow microorganisms. In batch cultures, cells are initially inoculated into a fresh medium and no further nutrient added until the target product is produced. In fed-batch cultures, growth medium is added at various intervals, while effluent is removed discontinuously. High cell density is usually attained in fed-batch cultures, since nutrients are added as required to maintain higher cell growth and to prolong the growth phase of the fermentation process. In continuous cultures, fresh growth medium is continually added throughout the whole process, and cells and spent medium are removed simultaneously. By and large, growth and uniform product formation are maintained for a longer period in continuous cultures. Although batch cultures are the most used bioprocesses, a fed-batch system combined with the features of continuous culture and batch growth is also widely used in commercial plants.

5.1.2. Cell density and immobilized cell cultures

The reactor productivity is usually proportional to the viable cell density in the fermenter. Cell recycling has been used to increase the cell density and product concentration of continuous cultures. It can also reduce the formation of inhibitory end-products resulting from lower concentrations of substrates due to the higher dilution rate applied. Filtration, centrifugation, sedimentation, and immobilization are the representative methods used to retain a high cell density. In addition, high cell density fermentations can be achieved using fed-batch technology, as it overcomes substrate and product inhibition. Cell immobilization is the most efficient among these techniques and contributes to high cell growth rates and long-term stability in a bioreactor. For example, butanol production in ABE fermentation can be improved by cell recycling and immobilization to increase cell density and reactor productivity [144–145].

Fig. 3. SEM micrographs of *C. acetobutylicum* cells immobilized on fibers in a FBB. (Magnification: A: 500×; B: 2500×; C: 3500×; D: 5000×)

Immobilized cell cultures have many advantages over suspension cultures for large-scale fermentation. They provide high cell density, eliminate expensive cell recovery, alleviate cell washout at high dilution rates, and protect against shear damage. Thus, cell immobilization enhances productivity in continuous processes. The adsorption of cells onto inert support surfaces has been widely used for cell immobilization (readers are referred to Chapter 14). For example, the fibrous bed bioreactor (FBB) improved several organic acid fermentations with significantly enhanced productivity, yield, and product concentration [146–147]. The fibrous bed allows for good multiphase flows and provides renewable surfaces for cell immobilization [148], resulting in a high cell density (*see* Fig. 3). FBBs provide efficient, continuous operation with high cell density and economic downstream processing in large-scale fermentations [149].

5.1.3. Metabolism and metabolic engineering

Many industrial applications exploit the specific capability of microorganisms to make a variety of products. Increasing the productivity and yield of certain primary or secondary metabolites has become the objective of many biotechnologists [150]. To explore the full industrial potential of bacterial and yeast cells, it is necessary to understand their growth and metabolic pathways, which are linked to the successful commercial exploitation of fermentation products.

Bacterial and yeast cells produce various products as a consequence of different metabolisms. However, all cells generate energy to drive vital functions via catabolism and synthesize biological compounds through anabolic pathways. Almost all cells share similar metabolic pathways, which perform enzymatic reactions to transform substrates into end products (amino acids, lipids, or polysaccharides). In general, aerobic catabolism consists of the Embden-Meyerhof-Parnas (EMP) pathway: fermentation of glucose to pyruvate, the tricarboxylic acid (TCA) cycle, which converts pyruvate to CO_2 and NADH, and the electron transport chain, which forms ATP by electron transfer from NADH. The hexose-monophosphate (HMP) pathway, which can be shuttled into glycolysis, converts glucose-6-phosphate into carbon skeletons and reducing power for direct use in biosynthesis. Entner-Doudoroff (ED) pathway converts glucose to pyruvate and glyceraldehyde-3 phosphate by producing 6-phosphogluconate and then dehydrating it.

Recently, metabolic engineering techniques have been applied to enhance the production of industrial bio-based products. Modified genes can lead microorganisms to new metabolic pathways for novel products or achieve higher efficiencies in metabolite overproduction through alterations in metabolic flux distribution. Recent progress in metabolic engineering has provided many significant applications for engineered strains and will continue to expand into new areas of applications as more and more genes become available. These metabolically engineered cells either overexpress controlling genes (or key enzyme) or block undesired metabolic pathways to overproduce value-added products. Therefore, an increasing number of microorganisms will become more amenable to genetic manipulation and facilitate the design and control of metabolic pathways.

Metabolic modeling aids quantitative study of the metabolism and provides new insights into microorganisms. One of the most popular methods is metabolic flux analysis (MFA), which provides a measure of the change in overall cellular functions and metabolic processes [151]. MFA measures the inputs and outputs of a cell and uses knowledge of the metabolic pathways to calculate the fluxes through these pathways. Stoichiometric mass balance is a popular and readily applied method to determine metabolic flux distribution in the central metabolism under pseudo-steady-state assumptions. It requires neither enzymatic kinetics nor expensive equipment and statistical calculations for isotope tracers but provides significant metabolic information [43]. A metabolic network with carbon fluxes provides a clear picture of carbon distribution or *in vivo* fluctuations that affect yield and productivity. For instance, flux analysis of yeast described ATP requirements for biomass synthesis and intermediate metabolite transportation in carbon-limited chemostat cultures and identified the critical pathway in the metabolism of recombinant *S. cerevisiae* for the production of ethanol [11]. These metabolic engineering tools greatly minimize experimental efforts for process optimization and develop effective strains for reducing production costs. More details about metabolic engineering can be found in Chapter 4.

5.2. Fermentation process characteristics and challenges

The fermentation processes can be divided into two groups: aerobic and anaerobic fermentations. In general, aerobic fermentations are faster but require higher power inputs for aeration and agitation, and cooling capacity, which could pose difficulties in bioreactor scale up. On the other hand, anaerobic fermentations usually require low energy inputs but are much slower. The general characteristics and challenges in aerobic and anaerobic fermentations are discussed in this section. Also, problems associated with foaming, end-product inhibition, and downstream processing are discussed along with possible solutions.

5.2.1. Aerobic fermentation, oxygen transfer and mixing

Aerobic fermentation occurs in the presence of oxygen. It usually occurs at the beginning of the fermentation process. Aerobic fermentation is usually a shorter and more intense process than anaerobic fermentation. Oxygen limitation is a major problem in aerobic fermentations because oxygen has a low solubility in water. Dissolved oxygen (DO) concentration is generally kept as high as possible by increasing the oxygen transfer rate (OTR). To keep the culture at a high cell density, a high agitation and/or aeration rate is the most commonly used strategy to improve OTR [152–153]; they offer a larger driving force, gas-liquid interfacial area, and longer residence time of gas bubbles in the liquid for oxygen transfer into the liquid. There are also some studies on the effects of DO [154] and periodic changes in pressure on bioprocesses [155–157]. In recent years, the utilization of an increased air/O_2 pressure [158], such as pressure pulsation [18] and oscillating dissolved oxygen tension [159] have been applied to biological systems in order to enhance the OTR or DO level in fermenting liquid to produce more desired products (metabolites).

In large-scale bioreactors, the specific surface area (surface to volume ratio) is dramatically decreased. Namely, the relative surface aeration for gas exchange decreases, and thus can become a limiting factor of the process. An adequate oxygen supply is very important to the achievement of a high cell density, especially for the production of secondary metabolites in large-scale fermentations, because oxygen has a very low solubility in liquids. When the dissolved oxygen level is under the critical DO concentration, the growth rate is dependent on the DO concentration, which is thus a limiting factor in aerobic fermentations. Conventional stirred-tank fermenters provide high oxygen transfer rates with increased mechanical agitation, high gas flow rate, or the dispersion of size-reduced gas bubbles with air-sparging devices. The stirrer provides enough agitation to disperse gas bubbles, breaks them into smaller ones, and lengthens their residence time. The agitator also provides homogeneity and mixing. However, at low agitation speeds, it may not generate enough turbulent flow to disrupt and disperse bubbles. Higher stirrer speeds and higher gas flow rates improve the oxygen transfer capacity. However, the higher tip speed of a large-diameter impeller may destroy cells because of the large shear stress. Choosing the appropriate agitation speed and impeller shape will promote both oxygen transfer and mixing.

The ALSA (Air Lift with Side Arms) fermenter is an alternative to replace stirred tank fermenters because they reduce the cost of mechanical agitation. Figure 4 demonstrates the circulation pattern of an ALSA fermenter [160]. A side arm is attached to an air lift unit. There is an opening at the top such that the liquid can circulate between the side arm and the main body of the fermenter. The liquid at the top of the side arm can flow downward into the annular space and circulate throughout the system according to Bernoulli's Equation. ALSA fermenters also can reduce foaming, since no mechanical agitation is needed, and provide higher oxygen transfer with an increased frequency of gas-liquid contact by circulation.

Increasing pressure in the head space of the bioreactors can increase the driving force of oxygen transfer because of the increased gas solubility in the liquid phase according to Henry's Law. Recently, periodic changes in pressure, pressure pulsation [18], and oscillating dissolved oxygen tension [159] have been applied to biological systems in order to enhance the oxygen transfer rate or dissolved oxygen level in fermenting liquids. Pressure changes in bioreactors change the volume of bubbles. These techniques provide a means to improve the cell density and fermentation capacity not only by increasing the gas solubility but also by continuously creating new surfaces for oxygen transfer.

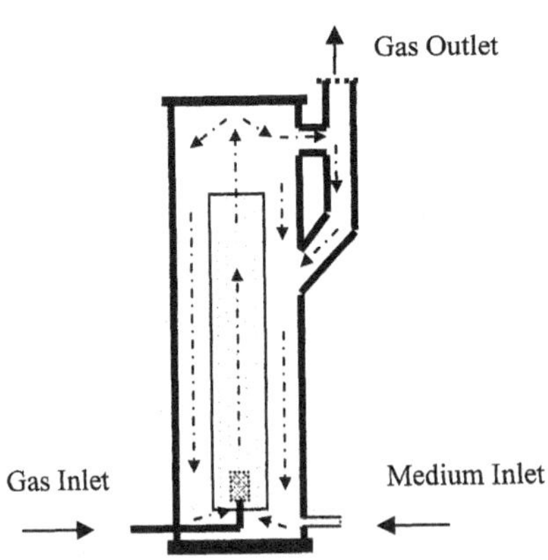

Fig. 4. Liquid circulation pattern in an air-lift-with-side-arms (ALSA) fermenter.

5.2.2. Anaerobic fermentation

Anaerobic fermentation occurs in the fermentation vessel once the oxygen is discharged and replaced with N_2, CO_2, or another by-product of the fermentation process. Anaerobic fermentation is usually a slower process. In the mid-1850s, the French chemist Louis Pasteur produced anaerobiosis by boiling the medium to drive out oxygen and then introducing inert gas for cultivation. He showed that a microorganism, probably *Clostridium butyricum*, was responsible for butyric acid fermentation. In the 1960s and 1970s, anaerobic chambers were invented that allowed the cultivation of numerous anaerobic cultures for certain strictly anaerobic organisms, including *C. botulinum*. During World War I, industrial anaerobic fermentation was further demonstrated by Perkins and Weizmann, who worked on acetone-butanol-ethanol (ABE) fermentation with *C. acetobutylicum*.

Anaerobes may grow under the unfavorable conditions used to minimize contamination during fermentations because they have unusual enzymes and catabolic pathways. Most anaerobic fermentations require little energy to keep cells in suspension. Because less biomass is produced in anaerobic fermentations, more carbon can be converted to the end product, and a higher product yield is attained. Anaerobes can utilize a wide range of substrates, including agricultural waste streams. This reduces the overall cost of the fermentation process. Anaerobic fermentation has been applied to many important industrial fermentations, such as ethanol production by yeasts, lactic acid preservation of foods, anaerobic digestion of organic matters in ruminant cultivation and waste treatment. The most important industrial fermentation is the anaerobic production of ethanol by *S. cerevisiae* and other yeasts.

However, mixed-culture processes in anaerobic fermentation are also difficult to study and model. The microbial communities are usually unstable, varying with environmental changes and the availability of nutrients. Obligate anaerobes need specialized media and apparatus. They are deactivated by exposure to oxygen. Hence, special skills and meticulous methods are required for the cultivation and manipulation of strictly anaerobic microorganisms. Compared to aerobic organisms, there is little known about methods for genetic manipulation and to express desired genes or biosynthetic pathways.

5.2.3. Foaming

Foaming is another major problem often found in large-scale fermentations. Foaming may cause the loss of fermenting broth and asepsis problems, increase the pressure drop, prevent adequate gas or liquid circulation in continuous bioreactors, and even terminate the whole process. If the foam provides a route for contaminating cells to enter the fermenter, this will ruin the product and may require that the culture be restarted from the beginning, especially if absolute sterility is required.

Complex media or high protein levels promote foam development. A mechanical foam breaker or antifoam agents (surfactants) can reduce foaming. A mechanical foam breaker can not work well in a heavy foaming situation. Antifoams are effective in reducing foam formation, but they lower the oxygen transfer rate, inhibit cell growth and lead to low cell density. A head space of 25% of the total fermenter volume provides room for gas to release from the broth, and reduces opportunities for contamination. In addition, a limited agitation speed and air flow rate can decrease the volume of foam. For example, the aforementioned non-agitated ALSA fermenter has a side arm acting as a foam breaker and a foam reducer.

5.2.4. Downstream processing

Product recovery and purification, such as centrifugation, chromatography, crystallization, dialysis, drying, electrophoresis, filtration, precipitation, etc., are essential finishing steps to any commercial fermentation process. Because of complex fermentation broths and high-purity requirements for some products, large-scale fermentation studies often involve sophisticated downstream processing, which may account for up to 90% of the total production cost [161]. For instance, the separation of organic compounds is carried out at the industrial level using distillation, which consumes large amounts of energy [162]. To overcome the problem of product inhibition, several integrated fermentation/product recovery technologies have been investigated for solvent removal and have improved fermentation performance in the last decade. These techniques include pervaporation [163–165], liquid-liquid extraction [166–169], perstraction [170–171], reverse osmosis [172], adsorption [145, 173], and gas-stripping [174–175]. Higher yields or productivities were obtained by means of product recovery techniques. Extractive fermentation is one example of simultaneous fermentation and product recovery.

5.2.5. Product inhibition and extractive fermentation

Yield, productivity, and final concentration of the product are the three most important parameters making industrial fermentation processes more economical to compete with petroleum-based chemical processes. However, product inhibition limits the desired product concentration in the bioreactor, and it may be necessary to separate the end products, which can result in feedback inhibition. For example, in conventional ABE fermentations, the butanol yield from glucose is low, typically ~15% (w/w), and rarely exceeds 25%. The production of butanol is also limited by severe product inhibition. Butanol at a concentration of 10 g/L can significantly inhibit cell growth and fermentation. Consequently, butanol titers in conventional ABE fermentations are usually lower than 15 g/L. Although several genetic engineering approaches are developing and the complete genome sequence of the type strain (*C. acetobutylicum* ATCC 824) is available (www.ncbi.nih.gov), no genetically engineered Clostridia can meet the requirements for industrial butanol production use. The low butanol yield and butanol concentration make butanol production from glucose by ABE fermentation uneconomical compared to chemical synthesis from petroleum feedstocks [176].

Extractive fermentation is one of the efficient product recovery processes extensively studied for alcohol fermentation. Despite extensive research, extractive fermentation has not been used in industrial fermentation because past effort using conventional extractants with severe solvent toxicity to cells and low separation efficiencies at the fermentation pH made the process inefficient. The conventional extractive fermentation process is also limited by inefficient phase separation and solvent regeneration. The advantages for extractive fermentation include increased reactor productivity, ease in reactor pH control without requiring base addition, and the possible use of high-concentration substrate for the process feed, reducing process wastes and production costs. The use of extractive fermentation also allows the engineer to produce and recover a desired fermentation product in one continuous step.

Recently, some researchers have successfully overcome these problems and demonstrated the feasibility and advantages of using extractive fermentation for the production of several organic acids [71, 177]. For example, an extractive fermentation process using an amine extractant and a hollow-fiber membrane extractor to selectively remove propionic acid from the fermentation broth has been developed to produce propionate from lactose. Compared to conventional batch fermentation, extractive fermentation had a much higher productivity (~1 g/(L·h)), higher propionate yield (up to 0.66 g/g), higher final product concentration (75 g/L), and higher product purity (~90%). The extractive fermentation process consists of a novel fibrous bed bioreactor (FBB) for immobilized cell fermentation that protects cells from direct contact with solvents (thus eliminating the solvent toxicity problem), and hollow-fiber membrane extractors containing an amine-based extractant for efficient product separation and continuous solvent regeneration. A three to five-fold increase in reactor productivity was usually attained in extractive fermentation. For instance, extractive fermentation with in-situ butanol removal from the fermentation broth has been shown to improve the productivity by two-fold as well as improve the butanol yield [178]. A few researchers have shown that the product inhibition has been overcome by extractive fermentation and much higher solvent concentrations than conventional ABE fermentation were obtained by an integrated system with gas stripping [179–180]. In addition, fermentation with pervaporation and gas stripping has been studied extensively for butanol recovery [176, 181].

Other issues important to large-scale fermentations, including heat removal, sterilization (asepsis), in-place cleaning, power consumption, and reactor design, are not discussed here. There are handbooks and encyclopedias of bioprocesses, biotechnology, and fermentation technology that provide further information [161, 182–186].

6. CONLUSION AND OUTLOOK

Bacteria and yeasts have been widely used to benefit our life. Several factors, such as cell characteristics, cell culture and fermentation processes, determining a successful and economical production are described in this chapter. Bacteria and yeast transform sugars from renewable resources into a variety of value-added chemicals, solvents, and fuels as alternatives to petroleum-based chemicals. Bacterial and yeast fermentations have provided us sustainable, cost-competitive, and biocompatible products from renewable resources.

In last two decades, scientists have engineered bacterial genes to improve the production of value-added substances, such as fine chemicals, biodegradable plastics, bio-fuels, and vitamins. The difficulties in converting biomass to desired products have been ameliorated by genetic manipulation. Metabolic engineering has been applied to improve and change the existing metabolic activities of several bacteria and yeasts for the production of industrial chemicals. These tools have enhanced utilization of biomass and reduced the cost of bioprocesses. Biotechnology will continue enabling us to exploit the potential of these microorganisms in agricultural, chemical, and pharmaceutical industries and benefit humankind.

REFERENCES

[1] E. Caplice and G.F. Fitzgerald, Food fermentations: role of microorganisms in food production and preservation. Int. J. Food Microbiol., 50 (1999) 131.

[2] M. Suresh Kumar, S.N. Mudliar, K.M.K. Reddy, and T. Chakrabarti, Production of biodegradable plastics from activated sludge generated from a food processing industrial wastewater treatment plant. Bioresource Technol., 95 (2004) 327.

[3] W.N. Chan and M.T. Holtzapple, Conversion of Municipal Solid Wastes to Carboxylic Acids by Thermophilic Fermentation. Appl. Biochem. Biotechnol., 111 (2003) 93.

[4] L.T. Angenent, K. Karim, M.H. Al-Dahhan, B.A. Wrenn, and R. Domíguez-Espinosa, Production of bioenergy and biochemicals from industrial and agricultural wastewater. Trends in Biotechnol., 22 (2004) 477.

[5] P. Hendry, Extremophiles : There's More to Life. Environ. Chem., 3 (2006) 75.

[6] A.A.H. Pakchung, P.J.L. Simpson, and R. Codd, Life on Earth. Extremophiles continue to move the goal posts. Environ. Chem., 3 (2006) 77.

[7] M. Podar and A.-L. Reysenbach, New opportunities revealed by biotechnological explorations of extremophiles. Curr. Opin. Biotechnol., 17 (2006) 250.

[8] I.M. Head, D.M. Jones, and W.F.M. Roeling, Marine microorganisms make a meal of oil. Nat. Rev. Microbiol., 4 (2006) 173.

[9] T.N. Sardessai and S. Bhosle, Industrial Potential of Organic Solvent Tolerant Bacteria. Biotechnol. Prog., 20 (2004) 655.

[10] J. Chu, S. Zhang, and Y. Zhuang, Fermentation Process Optimization of Recombinant *Saccharomyces cerevisiae* for the Production of Human Interferon-α2a. Appl. Biochem. Biotechnol., 111 (2003) 129.

[11] T. Çakir, K.Y. Arga, M.M. Altintas, and K.Ö. Ülgen, Flux analysis of recombinant *Saccharomyces cerevisiae* YPB-G utilizing starch for optimal ethanol production. Process Biochem., 39 (2004) 2097.

[12] C.J. Hack and R.Marchant, Characterization of a novel thermotolerant yeast, *Kluyveromyces marxianus* var *marxianus*: development of an ethanol fermentation process. J. Ind. Microbiol. Biotechnol., 20 (1998) 323.

[13] I. De Bari, D. Cuna, F. Nanna, and G. Braccio, Ethanol Production in Immobilized-Cell Bioreactors from Mixed Sugar Syrups and Enzymatic Hydrolysates of Steam-Exploded Biomass. Appl. Biochem. Biotechnol., 113-116 (2004) 539.

[14] M. Sedlak and N.W.Y. Ho, Production of Ethanol from Cellulosic Biomass Hydrolysates Using Genetically Engineered *Saccharomyces* Yeast Capable of Cofermenting Glucose and Xylose. Appl. Biochem. Biotechnol., 114 (2004) 403.

[15] R. Balusu, R.M.R. Paduru, G. Seenayya, and G. Reddy, Production of Ethanol From Cellulosic Biomass by *Clostridium thermocellum* SS19 in Submerged Fermentation: Screening of Nutrients Using Plackett-Burman Design. Appl. Biochem. Biotechnol., 117 (2004) 133.

[16] G.D. Bothun, B.L. Knutson, J.A. Berberich, H.J. Strobel, and S.E. Nokes, Metabolic selectivity and growth of *Clostridium thermocellum* in continuous culture under elevated hydrostatic pressure. Appl. Microbiol. Biotechnol., 65 (2004) 149.

[17] R.E. Kirk and D. Othmer, Glycerol, in M. Howe-Grant (eds.), Encyclopedia of Chemical Technology, 4th ed., John Wiley & Sons, New York, vol. 12, 1991, pp. 681-694.

[18] W.-C. Huang, C.S. Gong, and G.T. Tsao, Enhancement of Oxygen Transfer by Pressure Pulsation in Aqueous Glycerol Fermentation. Appl. Biochem. Biotechnol., 98-100 (2002) 909.

[19] W.-C. Huang, C.S. Gong, and G.T. Tsao, Effects of Pressure Pulsation on Oxygen Transfer Rate Measured by Sulfite Method. Appl. Biochem. Biotech., 106 (2003) 471.

[20] H.J. Liu, D.H. Liu, and J.J. Zhong, Oxygen limitation improves glycerol production by *Candida* krusei in a bioreactor. Process Biochem., 39 (2004) 1899.

[21] S. Hohmann and W. H. Mager (eds.), Yeast Stress Responses, Chapman & Hall, New York, 1997.
[22] C.E. Nakamura and G.M. Whited, Metabolic engineering for the microbial production of 1,3-propanediol. Curr. Opin. Biotechnol., 14 (2003) 454.
[23] N. Garg and Y.D. Hang, Microbial production of organic acids from carrot processing waste, J Food Sci. Technol., 32 (1995) 119.
[24] T.M. Carole, J. Pellegrino, and M.D. Paster, Opportunities in the Industrial Biobased Products Industry. Appl. Biochem. Biotechnol., 115 (2004) 871.
[25] W.-C. Huang, D.E. Ramey, and S.-T. Yang, Continuous Production of Butanol by *Clostridium acetobutylicum* Immobilized in a Fibrous Bed Bioreactor. Appl. Biochem. Biotech., 113-116 (2004) 887.
[26] Y. Kourkoutas, C. McErlean, M. Kanellaki, C.J. Hack, R. Marchant, I.M. Banat, and A.A. Koutinas, High-temperature wine making using the thermotolerant yeast strain *Kluyveromyces marxianus* IMB3. Appl. Biochem. Biotechnol., 112 (2004) 25.
[27] L.O. Ingram, H.C. Aldrich, A.C. Borges, T.B. Causey, A. Martinez, F. Morales, A. Saleh, S.A. Underwood, L.P. Yomano, S.W. York, J. Zaldivar, and S. Zhou, Enteric bacterial catalysts for fuel ethanol production. Biotechnol. Prog., 15 (1999) 855.
[28] Y. Lin and S. Tanaka, Ethanol fermentation from biomass resources: current state and prospects. Appl. Microbiol. Biotechnol., 69 (2006) 627.
[29] N.W. Ho, Z. Chen, and A.P. Brainard, Genetically engineered *Saccharomyces* yeast capable of effective cofermentation of glucose and xylose. Appl. Environ. Microbiol., 64 (1998) 1852.
[30] K.S. Lee, J.F. Wu, Y.S. Lo, Y.C. Lo, P.J. Lin, and J.S. Chang, Anaerobic hydrogen production with an efficient carrier-induced granular sludge bed bioreactor. Biotechnol. Bioeng., 87 (2004) 648.
[31] X.Y. Shi and H.Q. Yu, Hydrogen Production From Propionate by *Rhodopseudomonas capsulata*. Appl. Biochem. Biotechnol., 117 (2004) 143.
[32] N. Kataoka, A. Miya, and K. Kiriyama, Studies on hydrogen production by continuous culture system of hydrogen producing anaerobic bacteria. Water Sci. Tech., 36 (1997) 41.
[33] Z. Kadar, T. De Vrije, G.E. van Noorden, M.A.W. Budde, Z. Szengyel, K. Reczey, and P.A.M. Claassen, Yields from glucose, xylose, and paper sludge hydrolysate during hydrogen production by the extreme thermophile *Caldicellulosiruptor saccharolyticus*. Appl. Biochem. Biotechnol., 113-116 (2004) 497.
[34] V.C. Kalia and A.P. Joshi, Conversion of waste biomass (pea-shell) into hydrogen and methane through anaerobic digestion. Bioresource Technol., 53 (1995) 165.
[35] S.T. Yang, I.C. Tang, and M.R. Okos, Defined Bacterial Culture Developments for Methane Generation from Lactose, Biotechnol. Bioeng., 32(1988) 28.
[36] S.T. Yang and M.R. Okos, Kinetic Study and Mathematical Modeling of Methanogenesis of Acetate Using Pure Cultures of Methanogens, Biotechnol. Bioeng, 30 (1987) 661.
[37] S.T. Yang and M.N. Guo, Kinetics of Methanogenesis from Whey Permeate in Packed Bed Immobilized Cells Bioreactor, Biotechnol. Bioeng., 36 (1990) 427.
[38] F.P. Duta, F.P. De França, E.F.C. Sérvulo, L.M.D.A. Lopes, A.C.A. Da Costa, and A. Barros, Effect of Process Parameters on Production of a Biopolymer by *Rhizobium* sp. Appl. Biochem. Biotechnol., 114 (2004) 639.
[39] H.-F. Liang, T.-F. Yang, C.-T. Huang, M.-C. Chen, and H.-W. Sung, Preparation of nanoparticles composed of poly(gamma-glutamic acid)-poly(lactide) block copolymers and evaluation of their uptake by HepG2 cells. J. Control. Release, 105 (2005) 213.
[40] A. Richard and A. Margaritis, Rheology, oxygen transfer, and molecular weight characteristics of poly(glutamic acid) fermentation by *Bacillus subtilis*. Biotechnol. Bioeng., 82 (2003) 299.
[41] I.-L.Shih, Y.-T. Van, and M.-H. Shen, Biomedical applications of chemically and microbiologically synthesized poly(glutamic acid) and poly(lysine). Mini-Rev. Med. Chem., 4 (2004) 179.

[42] K.H. Law, P.L. Chan, W.S. Lau, Y.C. Cheng, Y.C. Leung, W.H. Lo, H. Lawford, and H.F. Yu, Construction of Recombinant *Escherichia coli* Strains for Production of Poly-(3-hydroxybutyrate-co-3-hydroxyvalerate). Appl. Biochem. Biotechnol., 114 (2004) 361.

[43] J. Yu and Y. Si, Metabolic Carbon Fluxes and Biosynthesis of Polyhydroxyalkanoates in Ralstonia eutropha on Short Chain Fatty Acids. Biotechnol. Prog., 20 (2004) 1015.

[44] I.J. Park and S.Y. Lee, Biosynthesis of Poly(3-hydroxybutyrate-co-3-hydroxyalkanoates) by Metabolically Engineered *Escherichia coli* Strains. Appl. Biochem. Biotechnol., 114 (2004) 335.

[45] M. Wainwright, An introduction to fungal biotechnology, J. Wiley & Sons, New York, 1992.

[46] K.D. Schaller, S.L. Fox, D.F. Bruhn, K.S. Noah, and G.A. Bala, Characterization of Surfactin from *Bacillus subtilis* for Application as an Agent for Enhanced Oil Recovery Appl. Biochem. Biotechnol., 115 (2004) 827.

[47] Y.-H. Wei, L.-F. Wang, and J.-S. Chang, Optimizing Iron Supplement Strategies for Enhanced Surfactin Production with *Bacillus subtilis*. Biotechnol. Prog., 20 (2004) 979.

[48] M. Nitschke and G.M. Pastore, Biosurfactant Production by *Bacillus subtilis* Using Cassava-Processing Effluent. Appl. Biochem. Biotechnol., 112 (2004) 163.

[49] D. Levišauskas, D.; V. Galvanauskas, G. Zunda, and S. Grigiškis, Model-based optimization of biosurfactant production in fed-batch culture *Azotobacter vinelandii*. Biotechnol. Lett., 23 (2004) 1141.

[50] S. Bae and M. Shoda, Bacterial Cellulose Production by Fed-Batch Fermentation in Molasses Medium. Biotechnol. Prog., 20 (2004) 1366.

[51] J. Behravan, B.S. Fazly Bazzaz, and Z. Salimi, Optimization of dextran production by Leuconostoc mesenteroides NRRL B-512 using cheap and local sources of carbohydrate and nitrogen. Biotechnol. Appl. Biochem., 38 (2003) 267.

[52] K. Pavlova, L. Koleva, M. Kratchanova, and I. Panchev, Production and Characterization of an Exopolysaccharide by Yeast. World J Microbiol. Biotechnol., 20 (2004) 435.

[53] A. Margaritis and G.W. Pace, Microbial Polysaccharides, in M. Moo-Young (eds.), Comprehensive Biotechnology: the Principles, Applications, and Regulations of Biotechnology in Industry, Agriculture, and Medicine, Pergamon, New York, vol. 3, 1985, pp. 1005–1044.

[54] M. Chaitali, M. Kapadi, G.K. Suraish Kumar, and R.D. Gudi, Productivity improvement in xanthan gum fermentation using multiple substrate optimization. Biotechnol. Prog., 19 (2003) 1190.

[55] S.T. Yang, Y.M. Lo, and D.B. Min, Xanthan Gum Fermentation by *Xanthomonas campestris* Immobilized in a Novel Centrifugal Fibrous-Bed Bioreactor, Biotechnol. Prog., 12 (1996) 630.

[56] S.T. Yang, I.C. Tang, and M.R. Okos, Kinetics of homoacetic fermentation of lactate by *Clostridium formicoaceticum*, Appl. Environ. Microbiol., 53 (1987) 823.

[57] I.C. Tang, S.T. Yang, and M.R. Okos, Acetic acid production from whey lactose by the coculture of *Streptococcus lactis* and *Clostridium formicoaceticum*, Appl. Microbiol. Biotechnol., 28 (1988)138.

[58] J. Zigova and E. Sturdik, Advances in biotechnological production of butyric acid. J. Ind. Microbiol. Biotechnol., 24 (2000) 153.

[59] Y.L. Huang, Z. Wu, L. Zhang, C.M. Cheung, and S.T. Yang, Production of carboxylic acids from corn meal by immobilized cell fermentation in a fibrous-bed bioreactor, Bioresource Technol., 82 (2002) 51.

[60] T.E. Arzumanov, N.V. Shishkanova, and T.V. Finogenova, Biosynthesis of citric acid by *Yarrowia lipolytica* repeat-batch culture on ethanol. Appl. Microbiol. Biotechnol., 53 (2000) 525.

[61] M. Mattey, The production of organic acids. Critical Rev. Biotechnol., 12 (1992) 87.

[62] G. Biagi, A. Piva, M. Moschini, E. Vezzali, F.X. Roth, Effect of gluconic acid on piglet growth performance, intestinal microflora, and intestinal wall morphology. J. Anim. Sci., 84 (2006) 370.

[63] K. Gao and D. Wei, Asymmetric oxidation by Gluconobacter oxydans. Appl. Microbiol. Biotechnol., 70 (2006) 135.

[64] T.Doneva, C. Vassilieff, and R. Donev, Catalytic and biocatalytic oxidation of glucose to gluconic acid in a modified three-phase reactor. Biotechnol. Lett., 21 (1999) 1107.

[65] N. Mori, K. Jitsunari, M. Yoshida, Y. Kitamoto, and Y. Ichikawa, Pyruvic acid production from gluconic acid by *Pseudomonas stutzeri*. Tottori Daigaku Nogakubu Kenkyu Hokoku, 40 (1987) 15.

[66] R. Kaur, J. Macleod, W. Foley, and M. Nayudu, Gluconic acid : An antifungal agent produced by *Pseudomonas* species in biological control of take-all. Phytochemistry, 67 (2006) 595.

[67] C. Liu, Y. Liu, W. Liao, Z. Wen, and S. Chen, Simultaneous production of nisin and lactic acid from cheese whey: Optimization of fermentation conditions through statistically based experimental designs. Appl. Biochem. Biotechnol., 113-116 (2004) 627.

[68] S.K. Singh, S.U. Ahmed, and A. Pandey, Metabolic engineering approaches for lactic acid production. Process Biochem. 41 (2006) 991.

[69] M. M. Bianchi, L. Brambilla, F. Protani, C. L. Liu, J. Lievense, and D. Porro. Efficient homolactic fermentation by *Kluyveromyces lactis* strains defective in pyruvate utilization and transformed with the heterologous LDH gene. Appl. Environ. Microbiol. 67 (2001) 5621.

[70] S. Zhou, T. B. Causey, A. Hasona, K. T. Shanmugam, and L. O. Ingram. Production of optically pure D-lactic acid in mineral salts medium by metabolically engineered *Escherichia coli* W3110. Appl. Environ. Microbiol. 69 (2003) 399.

[71] Z. Jin and S.T. Yang, Extractive fermentation for enhanced propionic acid production from lactose by Propionibacterium acidipropionici. Biotechnol. Prog., 14 (1998) 457.

[72] S. Suwannakham and S.-T. Yang. Enhanced propionic acid fermentation by Propionibacterium acidipropionici mutant obtained by adaptation in a fibrous-bed bioreactor, Biotechnol. Bioeng., 91 (2005) 325.

[73] Y. Li, J. Chen, and S.-Y. Lun, Biotechnological production of pyruvic acid. Appl. Microbiol. Biotechnol., 57 (2001) 451.

[74] B. Zelic, D. Vasic-Racki, C. Wandrey, and R. Takors, Modeling of the pyruvate production with *Escherichia coli* in a fed-batch bioreactor. Bioprocess Biosyst. Eng., 26 (2004) 249.

[75] H. Song and S.Y. Lee, Production of succinic acid by bacterial fermentation. Enz. Microb. Technol., 39 (2006) 352.

[76] M. Lee, G.M. Smith, M.A. Eiteman, and E. Altman, Aerobic production of alanine by *Escherichia coli* aceF ldhA mutants expressing the *Bacillus* sphaericus alaD gene. Appl. Microbiol. Biotechnol., 65 (2004) 56.

[77] G. Ravot, B. Ollivier,M.-L. Fardeau, B.K.C. Patel, K.T. Andrews, M. Magot, and J.-L. Garcia, L-Alanine production from glucose fermentation by hyperthermophilicmembers of the domeins Bacteria and Archaea: a remnant of an ancestral metabolism? Appl. Environ. Microbiol., 62 (1996) 2657.

[78] B. Atkinson and F. Mavituna, Biochemical engineering and biotechnology handbook, Nature Press, New York, 1983.

[79] G.M. Walker, Yeast physiology and biotechnology, J. Wiley & Sons, New York, 1998.

[80] R. Bona and A. Moser, Modeling of the L-glutamic acid production with *Corynebacterium glutamicum* under biotin limitation. Chem. Biochem. Eng. Quarterly, 12 (1998) 25.

[81] K.M. Nampoothiri and A. Pandey, Solid state fermentation for L-glutamic acid production using Brevibacterium sp. Biotechnol. Lett., 18 (1996) 199.

[82] J. Ohnishi, R. Katahira, S. Mitsuhashi, S. Kakita, and M. Ikeda, A novel gnd mutation leading to increased L-lysine production in *Corynebacterium glutamicum*. FEMS Microbiol. Lett., 242 (2005) 265.

[83] W. Crueger and A. Crueger, Biotechnology : a textbook of industrial microbiology, Science Tech., Madison, WI, 1984.

[84] J.L. Báez-Viveros, J. Osuna, and G. Hernández-Chávez, Metabolic engineering and protein directed evolution increase the yield of L-phenylalanine synthesized from glucose in *Escherichia coli*. Biotechnol. Bioeng., 87 (2004) 516.

[85] M.-H. Lee, H.-W. Lee, J.-H. Park, J.-O. Ahn; J.-K. Jung, and Y.-I. Hwang, Improved L-threonine production of *Escherichia coli* mutant by optimization of culture conditions. J. Biosc. Bioengr., 101 (2006) 127.

[86] E. Shimizu, T. Oosumi, H. Heima, T. Tanaka, J.Kurashige, H. Enei, K. Miwa, and S. Nakamori, Culture conditions for improvement of L-threonine production using a genetically self-cloned L-threonine hyperproducing strain of *Escherichia coli* K-12. Biosci. Biotech. Biochem., 59 (1995) 1095.

[87] K. Okamoto, K. Kino, and M. Ikeda, Hyperproduction of L-threonine by an *Escherichia coli* mutant with impaired L-threonine uptake. Biosci. Biotech. Biochem., 61 (1997) 1877.

[88] S. Furukawa, A. Ozaki, Y. Kotani, and T. Nakanishi, Breeding of L-threonine hyper-producer of *Escherichia coli* W. Appl. Microbiol. Biotechnol., 29 (1998) 253.

[89] Y. Ogawa-Miyata, H. Kojima, and K. Sano, Mutation analysis of the feedback inhibition site of aspartokinase III of *Escherichia coli* K-12 and its use in L-threonine production. Biosci. Biotechnol. Biochem., 65 (2001) 1149.

[90] K. Kawasaki, A. Yokota, and F. Tomita, L-Tryptophan production by a pyruvic acid-producing *Escherichia coli* strain carrying the *Enterobacter aerogenes* tryptophanase gene. J. Ferment. Bioeng., 82 (1996) 604.

[91] O. Kurahashi, M. Noda-Watanabe, Y. Toride, T. Takenouchi, K. Akashi, Y. Morinaga, and H. Enei, Production of L-tryptophan by azaserine-, 6-diazo-5-oxo-L-norleucine- and cinnamate-resistant mutants of *Bacillus subtilis* K. Agr. Biol. Chem. 51 (1987) 1791.

[92] P.G. Berwick, β-Lactam and aminoglycoside production from streptomycetes. J. Appl. Bacteriol. 64 (1988) 9.

[93] K. Farzana, S.N.H. Shah, F.B. Butt, and S.B. Awan, Biosynthesis of bacitracin in solid-state fermentation by *Bacillus licheniformis* using defatted oil seed cakes as substrate. J. Pharm. Sci. 18 (2005) 55.

[94] F. Cladera-Olivera, G.R. Caron, and A. Brandelli, Bacteriocin production by *Bacillus licheniformis* strain P40 in cheese whey using response surface methodology. Biochem. Eng. J., 21 (2004) 53.

[95] D. Bizani and A. Brandelli, Influence of media and temperature on bacteriocin production by *Bacillus cereus* 8A during batch cultivation. Appl. Microbiol. Biotechnol., 65 (2004) 158.

[96] X. Liu, Y.-K. Chung, S.-T. Yang, and A.E. Yousef, Continuous nisin production in laboratory media and whey permeate by immobilized *Lactococcus lactis*. Process Biochem., 40 (2005) 13.

[97] J. Thykaer and J. Nielsen, Metabolic engineering of β-lactam production. Metabol. Eng. 5 (2003) 56.

[98] J. Robin, S. Bonneau, D. Schipper, H. Noorman, and J. Nielsen, Influence of the adipate and dissolved oxygen concentrations on the β-lactam production during continuous cultivations of a *Penicillium chrysogenum* strain expressing the expandase gene from *Streptomyces clavuligerus*. Metabol. Eng. 5 (2003) 42.

[99] H.Y. Cho, A.E. Yousef, and S.T. Yang, Continuous production of pediocin by immobilized *Pediococcus acidilactici* PO2 in a packed-bed bioreactor, Appl. Microbiol. Biotechnol., 45 (1996) 589.

[100] A.E. Asagbra, A.I. Sanni, and O.B. Oyewole, Solid-state fermentation production of tetracycline by *Streptomyces* strains using some agricultural wastes as substrate. World J. Microbiol. Biotechnol. 21 (2005) 107.

[101] L.H. Hansen, B. Ferrari, A.H. Sorensen, D. Veal, and S.J. Sorensen, Detection of oxytetracycline production by *Streptomyces* rimosus in soil microcosms by combining whole-cell biosensors and flow cytometry. Appl. Environ. Microbiol. 67 (2001) 239.

[102] J.-H. Koo, I.-J. Choi, H.-S. Nam, H.-J. Lee, Z.-I. Shin, and T.-K. Oh, Medium optimization for production of thermostable alkaline proteases from *Bacillus licheniformis* NS70. Sanop Misaengmul Hakhoechi 25 (1997) 207.

[103] G. Calik, N. Pehlivan, N. Kalender, T.H. Oezdamar, and P. Calik, Utilization of pretreated molasses for serine alkaline protease production with recombinant *Bacillus* species. Chem. Eng. Comm. 190 (2003) 630.

[104] H. Huang, D. Ridgway, T. Gu, and M. Moo-Young, Enhanced amylase production by *Bacillus subtilis* using a dual exponential feeding strategy. Bioprocess Biosyst. Eng. 27 (2004) 63.

[105] U.I. Beshay and H.A. El-Enshasy, Effect of cultivation conditions on the kinetics of α-amylase production by *Bacillus amyloliquefaciens*. Deutsche Lebensmittel-Rundschau 98 (2002) 5.

[106] M. Mezghani, M.A. Borgi, R. Kammoun, H. Aouissaoui, and S. Bejar, Construction of new stable strain over-expressing the glucose isomerase of the *Streptomyces* sp. SK strain. Enz. Microb. Technol., 37 (2005) 735.

[107] R.K. Manhas and S. Bala, Thermostable glucose isomerase production by some bacteria and *Streptomyces*. Indian J. Microbiol., 44 (2004) 129.

[108] S. Mukhopadhyaya and S.K. Majumdar, Immobilization of whole cells of *Streptomyces kanamyceticus* containing glucose isomerase by a combination of heat treatment in the presence of minerals and entrapment in calcium alginate gels. Folia Microbiol., 41 (1996) 502.

[109] A. Converti and M.D. Borghi, Kinetics of glucose isomerization to fructose by immobilized glucose isomerase in the presence of substrate protection. Bioproc. Engr., 18 (1998) 27.

[110] S.H. Bhosale, M.B. Rao and V.V. Deshpande, Molecular and industrial aspects of glucose isomerase. Microbiol. Rev., 60 (1996) 280.

[111] E. Celik and P. Calik, Bioprocess Parameters and Oxygen Transfer Characteristics in β-Lactamase, Production by *Bacillus* Species. Biotechnol. Prog. 20 (2004) 491.

[112] G. Ciafardini, B.A. Zullo, and A. Iride, Lipase production by yeasts from extra virgin olive oil. Food Microbiol. 23 (2006) 60.

[113] J.-h. Liu, J.-w. Xu, F.-y. Wang, R.-l. Wu, F.-k. Yang, and J.-l. An, Studies on fermentation process for lipase production by *Candida antarctica*. Qingdao Keji Daxue Xuebao, Ziran Kexueban 26 (2005) 101.

[114] M. Doman-Pytka and J. Bardowski, Pullulan degrading enzymes of bacterial origin. Crit. Rev. Microbiol., 30(2) (2004) 107.

[115] H.S. Ling, L.T. Chuan, R. Mohammad, A. Ariff, and S. Shioya, Pullulanase type II from *Bacillus* spp H1.5: Isolation, production and characterization of the enzyme. Biotechnol. Sustain. Util. Biol. Res. Tropics, 17 (2004) 48.

[116] T. Kuroiwa, H. Shoda, S. Ichikawa, S. Sato, and S. Mukataka, Immobilization and stabilization of pullulanase from *Klebsiella pneumoniae* by a multipoint attachment method using activated agar gel supports. Process Biochem., 40(8) (2005) 2637.

[117] D. Malle, T. Itoh, W. Hashimoto, K. Murata, S. Utsumi, and B. Mikami, Overexpression, purification and preliminary X-ray analysis of pullulanase from *Bacillus subtilis* strain 168. Acta Crystallogr. F: Struct. Biol. Crystal. Commun., F62(4) (2006) 381.

[118] B. Ramesh, P.R.M. Reddy, G. Seenayya, and G. Reddy, Effect of various flours on the production of thermostable-amylase and pullulanase by *Clostridium thermosulfurogenes* SV2. Bioresource Technol., 76 (2001) 169.

[119] A. Quesada-Chanto, A.C. Schmid-Meyer, A.G. Schroeder, M.F. Carvalho-Jonas, I. Blanco, and R. Jonas, Effect of oxygen supply on biomass, organic acids and vitamin B12 production by *Propionibacterium shermanii*. World J. Microbiol. Biotechnol. 14 (1998) 843.

[120] Y. Tsuchiya, N. Nishio, and S.Nagai, Vitamin B12 production from methanol by continuous culture of *Pseudomonas* AM-1. J. Ferment. Technol. 58 (1980) 485.

[121] S.S. Marwaha, J.F. Kennedy, and R.P. Sethi, Vitamin B12 production from whey and simulation of optimal cultural conditions. Process Biochem. 18 (1983) 24.

[122] K.-P. Stahmann, J.L. Revuelta, and H. Seulberger, Three biotechnical processes using *Ashbya gossypii*, *Candida famata*, or *Bacillus subtilis* compete with chemical riboflavin production. Appl. Microbiol. Biotechnol. 53 (2000) 509.

[123] C. Olczyk, n-Alkanes as a substratum for riboflavin production. I. Investigations of the dynamics of the flavinogenesis in chosen yeasts of the genus *Candida*. Polish J. Pharmacol. Pharm. 30 (1978) 83.
[124] K. Tomita and Y. Kuratsu, Influence of carbon sources on the stimulatory effect of L-proline on 5'-inosinic acid production. J. Ferment. Bioeng. 74 (1992) 406.
[125] G. Daum, N.D. Lees, M. Bard, and R. Dickson, Biochemistry, cell biology and molecular biology of lipids of *Saccharomyces cerevisiae*. Yeast, 14 (1998) 1471.
[126] J.M. Dyer, D.C. Chapital, J.W. Kuan, R.T. Mullen, and A.B. Pepperman, Metabolic engineering of *Saccharomyces cerevisiae* for production of novel lipid compounds. Appl. Microbiol. Biotechnol., 59 (2002) 224.
[127] M. Veen and C. Lang, Production of lipid compounds in the yeast *Saccharomyces cerevisiae*. Appl. Microbiol. Biotechnol., 63 (2004) 635.
[128] M. Young, N.P. Minton, and W.L. Staudenbauer, Recent advantages in the genetics of the clostridia. FEMS Microbiol. Rev., 63 (1989) 301.
[129] T. Akano, Y. Miura, K. Fukatsu, H. Miyasaka, Y. Ikuta, H. Matsumoto, A. Hamasaki, N. Shioji, T. Mizoguchi, K. Yagi, and I. Maeda, Hydrogen production by photosynthetic microorganisms. Appl. Biochem. Biotechnol., 57-58 (1996) 677.
[130] V.B. Borodin, A.A. Tsygankov, K.K. Rao, and D.O. Hall, Hydrogen production by *Anabaena variabilis* PK84 under simulated outdoor conditions. Biotechnol. Bioeng., 69 (2000) 478.
[131] A.S. Fedorov, A.A. Tsygankov, K.K. Rao, and D.O. Hall, Hydrogen photoproduction by *Rhodobacter sphaeroides* immobilized on polyurethane foam. Biotechnol. Lett., 20 (1998) 1007.
[132] S.A. Markov, P.F. Weaver, and M. Seibert, Spiral tubular bioreactor for hydrogen production by photosynthetic microorganisms. Appl. Biochem. Biotechnol., 63-65 (1997) 577.
[133] Y. Tokiwa and B.P. Calabia, Review Degradation of microbial polyesters. Biotechnol. Lett., 26 (2004) 1181.
[134] D.D. Wright-Charlesworth, D.M. Miller, I. Miskioglu, and J.A. King, Nanoindentation of injection molded PLA and self-reinforced composite PLA after in vitro conditioning for three months. J. Biomed. Mater. Res. A, 74A(3) (2005) 388.
[135] X.-D. Li, H.-Q. Geng, J. Pan, H. Xie, and F. Chen, Amniotic fluid : a novel source for tissue engineering. Zhonghua yi xue za zhi, 85 (2005) 464.
[136] J. Hugenholtz and M. Kleerebezem, Metabolic engineering of lactic acid bacteria : overview of the approaches and results of pathway rerouting involved in food fermentations. Curr. Opin. Biotechnol., 10 (1999) 492.
[137] F. Leroy and L. De Vuyst, Lactic acid bacteria as functional starter cultures for the food fermentation industry. Trends in Food Sci. Technol., 15 (2004) 67.
[138] S. Anastassiadis, A. Aivasidis, and C. Wandrey, Citric acid production by *Candida* strains under intracellular nitrogen limitation. Appl. Microbiol. Biotechnol., 60 (2002) 81.
[139] O.V. Singh and R.P. Singh, Bioconversion of grape must into modulated gluconic acid production by *Aspergillus niger* ORS-4.410. J. Appl. Microbiol., 100 (2006) 1114.
[140] W.-C. Huang, S.-W. Tsai, and C.-S. Chang, Enantioselective Esterification of (S)-Naproxen from Racemic Naproxen by Alkylated Lipase in Organic Solvents. J. Chin. Inst. Chem. Engrs., 29 (1998) 153.
[141] Y. Lin and S. Tanaka, Ethanol fermentation from biomass resources: current state and prospects. Appl. Microbiol. Biotechnol., 69 (2006) 627.
[142] G.A. Sprenger, Carbohydrate metabolism in *Zymomonas mobilis*: a catabolic highway with some scenic routes. FEMS Microbiol. Lett., 145 (1996) 301.
[143] A. Tay and S.T. Yang, Production of L(+)-lactic acid from glucose and starch by immobilized cells of *Rhizopus oryzae* in a rotating fibrous bed bioreactor. Biotechnol. Bioeng., 80 (2002) 1.
[144] N. Qureshi and I.S. Maddox, Continuous production of acetone-butanol-ethanol using immobilized cells of *Clostridium acetobutylicum* and integration with product removal by liquid-liquid extraction. J. Ferment. Bioeng., 80 (1995) 185.

[145] X. Yang and G.T. Tsao, Enhanced acetone-butanol fermentation using repeated fed-batch operation coupled with cell recycle by membrane and simultaneous removal of inhibitory products by adsorption. Biotechnol. Bioeng., 47 (1995) 444.
[146] Y. Huang and S.-T. Yang, Acetate production from whey lactose using co-immobilized cells of homolactic and homoacetic bacteria in a fibrous-bed bioreactor. Biotechnol. Bioeng., 60 (1998) 498.
[147] Y. Zhu, Z. Wu, and S.T. Yang, Butyric acid production from acid hydrolysate of corn fiber by *Clostridium tyrobutyricum* in a fibrous-bed bioreactor. Process Biochem., 38 (2002) 657.
[148] E.M. Silva and S.-T. Yang, Kinetics and stability of a fibrous-bed bioreactor for continuous production of lactic acid from unsupplemented acid whey. J. Biotechnol., 41 (1995) 59.
[149] Y.L. Huang, K. Mann, J. Novak, and S.T. Yang, Acetic acid production from fructose by *Clostridium formiaceticum* immobilized in a fibrous-bed bioreactor. Biotechnol. Prog., 14 (1998) 800.
[150] H. Kacser and L. Acerenza, A universal method for achieving increases in metabolite production. Eur. J. Biochem., 216 (1993) 361.
[151] H. Zhang, K. Shimizu, and S. Yao, Metabolic flux analysis of *Saccharomyces cerevisiae* grown on glucose, glycerol or acetate by ^{13}C-labeling experiments. Biochem. Eng. J., 16 (2003) 211.
[152] R.K. Finn and J.E. Nowrey, A note on the stability of clostridia when held in continuous culture. Appl. Microbiol., 7 (1958) 29.
[153] A.W. Hixson and J.R.E.L. Gaden, Oxygen transfer in submerged fermentation. Ind. Eng. Chem., 42 (1950) 1792.
[154] D.G. Kilburn and F.C. Webb, The cultivation of animal cells at controlled dissolved oxygen partial pressure. Biotechnol. Bioeng., 10 (1968) 801.
[155] H. Kataoka, S. Sato, S. Mukataka, A. Namiki, K. Yoshimura, and J. Takahashi, Effects of periodic change in pressure and dissolved-oxygen concentration on the incubation characteristics of *Pseudomonas aeruginosa*. Biotechnol. Bioeng., 28 (1986) 663.
[156] M. Trager, G.N. Qazi, R. Buse, and U. Onke, Influence of constant and oscillating dissolved oxygen concentration on *Aspergillus niger*. J. Ferment. Bioeng., 74 (1992) 282.
[157] M. Rhiel and D.W. Murhammer, The effect of oscillating dissolved oxygen concentrations on the metabolism of a *Spodoptera frugiperda* IPLB-SF21-AE clonal isolate. Biotechnol. Bioeng., 47 (1995) 640.
[158] F. Campelo Ana and I. Belo, Fermentative capacity of baker's yeast exposed to hyperbaric stress. Biotechnol. Lett., 26 (2004) 1237.
[159] M.A. Trujillo-Roldan, C. Pena, O.T. Ramirez, and E. Galindo, Effect of oscillating dissolved oxygen tension on the production of alginate by *Azotobacter vinelandii*. Biotechnol. Prog., 17 (2001) 1042.
[160] J.M. Dominguez, N. Cao, C.S. Gong, and G.T. Tsao, Ethanol production from xylose with the yeast *Pichia stipitis* and simultaneous product recovery by gas stripping using a gas-lift loop fermentor with attached side-arm (GLSA). Biotechnol. Bioeng., 67 (2000) 336.
[161] N. Jens, Fermentation Monitoring, Design and Optimization. in M.C. Flickinge and S.W. Drew (eds.), The Encyclopedia of Bioprocess Technology: Fermentation, Biocatalysis, and Bioseparation, Wiley, New York, vol. 2, 1999, pp. 1149.
[162] D.K. Jitesh, V.G. Pangakar, and K. Niranjan, Pervaporative stripping of acetone, butanol and ethanol to improve ABE fermentation. Bioseparation, 9 (2000) 145.
[163] W.J. Groot and K.C.A.M. Luyben, Continuous production of butanol from a glucose/xylose mixture with an immobilized cell system coupled to pervaporation. Biotehnol. Lett., 9 (1987) 867.
[164] A. Friedl, N. Qureshi, and I.S. Maddox, Continuous acetone-butanol-ethanol (ABE) fermentation using immobilized cells of *Clostridium acetobutylicum* in a packed bed reactor and integration with product removal by pervaporation with product removal by pervaporation. Biotechnol. Boieng., 38 (1991) 518.

[165] N. Qureshi, M.M. Meagher, and J. Huang, Acetone butanol ethanol (ABE) recovery by Pervaporation using silicalite-silicone composite membrane from fed-batch reactor of *Clostridium acetobutylicum*. J. Membrane Sci., 187 (2001) 93.
[166] A.M. Dadgar and G.L. Foutch, Improving the acetone-butanol fermentation process with liquid-liquid extraction. Biotechnol. Prog., 4 (1988) 36.
[167] G. Eckert, and K. Schugerl, Continuous acetone-butanol production with direct product removal. Appl. Microbiol. Biotechnol., 27 (1987) 221.
[168] S.R. Roffler, H.W. Blanch, and C.R. Wilke, In-situ recovery of butanol during fermentation. Part 1: Batch extractive fermentation. Bioprocess Eng., 2 (1987) 1.
[169] W.J. Groot, H.S. Soejak, P.B. Donck, R.G.J.M. van der Lans, K.C.A.M. Luyben, and J.M.K. Timmer, Butanol recovery from fermentation by liquid-liquid extraction and membrane solvent extraction. Bioprocess Eng., 5 (1990) 203.
[170] W.J. Groot, J.M.K. Timmer, and K.C.A.M. Luyben, Membrane Solvent Extraction for In-situ Butanol Recovery in Fermentations, in O.M. Meijssel, R.R. van der Meer, and K.Ch.A.M. Luyben (eds.), Proceedings of the 4th European Congress on Biotechnology, Elsevier, Amsterdam, vol. 2, 1987, pp. 564–566.
[171] Y.J. Jeon and Y.Y. Lee, In situ product separation in butanol fermentation by membrane-assisted extraction. Enz. Microb. Technol., 11 (1989) 575.
[172] A. Garcia, E.L. Ianotti, and J.L. Fischer, Butanol fermentation liquor production and separation by reverse osmosis. Biotechnol. Bioeng., 28 (1986) 785.
[173] B.M. Ennis, N. Qureshi, and I.S. Maddox, In-line toxic product removal during solvent production by continuous fermentation using immobilized *Clostridium acetobutylicum*. Enz. Microb. Technol., 9 (1987) 672.
[174] T.C. Ezeji, N. Qureshi, and H.P. Blaschek, Acetone butanol ethanol (ABE) production from concentrated substrate: reduction in substrate inhibition by fed-batch technique and product inhibition by gas stripping. Appl. Microbiol. Biotechnol. 63 (2004) 653.
[175] M.C. Dale, M.R. Okos, and P.C. Wankat, An immobilized cell reactor with simultaneous product separation. I. Reactor design and analysis. Biotechnol. Bioeng., 27 (1985) 932.
[176] I.S. Maddox, The acetone-butanol-ethanol fermentation: recent progress in technology. Biotechnol. Genet. Eng. Rev., 7 (1989) 189.
[177] Z. Wu and S.T. Yang, Extractive fermentation for butyric acid production from glucose by *Clostridium tyrobutyricum*. Biotechnol. Bioeng., 82 (2003) 93.
[178] N. Qureshi and H.P. Blaschek, Production of acetone butanol ethanol (ABE) by a hyper-producing mutant strain of *Clostridium beijierinckii* BA101 and recovery by pervaporation. Biotechnol. Prog., 15 (1999) 594.
[179] T.C. Ezeji, N. Qureshi, and H.P. Blaschek, Production of acetone, butanol and ethanol by *Clostridium beijerinckii* BA101 and in situ recovery by gas stripping. World J. Microbiol. Biotechnol., 19 (2003) 595.
[180] T.C. Ezeji, P.M. Karcher, N. Qureshi, and H.P. Blaschek, Improving performance of a gas stripping-based recovery system to remove butanol from *Clostridium beijerinckii* fermentation. Bioproc. Biosys. Engr. 27 (2005) 207.
[181] N. Qureshi, X.-L. Li, S. Hughes, B.C. Saha, M.A. Cotta, Butanol Production from Corn Fiber Xylan Using *Clostridium acetobutylicum*. Biotechnol. Prog. 22 (2006) 673.
[182] M. Moo-Young, Comprehensive Biotechnology: The Principles, Applications, and Regulations of Biotechnology in Industry, Agriculture, and Medicine, Pergamon, New York, 1985.
[183] R.E. Spier, Encyclopedia of cell technology, John Wiley, New York, 2000.
[184] R.K. Robinson, C.A. Batt, and P.D. Patel (eds.), Encyclopedia of food microbiology, Academic Press, San Diego, 2000.
[185] J.G. Holt (ed.), Bergey's manual of systematic bacteriology, Williams & Wilkins, Baltimore, 1984.

[186] P. Singleton and D. Sainsbury, Dictionary of microbiology and molecular biology, Wiley, Chichester, New York, 2001.

Bioprocessing for Value-Added Products from Renewable Resources
Shang-Tian Yang (Editor)

Chapter 9. Filamentous Fungal Cultures – Process Characteristics, Products, and Applications

Hesham A. El-Enshasy

Bioprocess Development Department, Mubarak City for Scientific Research and Technology Applications, New Burg Al Arab, 21934 Alexandria, Egypt

1. INTRODUCTION

The success of filamentous fungi for the industrial production of biotechnological products is largely due to the metabolic versatility of this group of microorganisms. Filamentous fungi are known to produce many organic acids, polysaccharides, enzymes, plant growth regulators, alkaloids, pigments, mycotoxins, and antibiotics. This group of microorganisms also has a long history of antibiotic production and has thus saved countless lives since the discovery of penicillin in the mid 1940s. However, antibiotic production is not restricted to β-lactam from *Penicillium* and *Cephalosporium* species. Nowadays, different types of antibiotics are produced by fungi from other groups, such as cyclosporine by *Tolypocladium inflatum* and fusidic acid by *Acremonium fusidiodes.*

In general, fungal cells are characterized by an extraordinary ability to secrete large amounts of proteins, metabolites, and organic acids into their growth medium. However, the industrial importance of fungal cells is not limited to their wide range of products but also includes the development and commercialization of new products derived from genetic engineering. Therefore, during the last few years, numerous studies have been presented on the cultivation of *Aspergillus niger*, the most important fungal species for the production and secretion of different types of proteins. The employment of fungal cells as host organisms for the production and secretion of homologous and heterologous proteins increased the importance of this group of microorganisms. During the submerged cultivation of fungi, growth morphology can vary from discrete compact pellets of hyphae to homogeneous suspension of dispersed mycelia. These morphological differences are associated with significant differences in growth kinetics and physiology.

This chapter reviews the process characteristics of fungal growth in submerged cultures and its relationship to the kinetics of product formation with a focus on the unique growth morphology of filamentous fungi in submerged cultures. The factors affecting fungal morphology and their effects on fungal fermentation are discussed in detail. Immobilization of fungal cells and its benefits are also discussed. Finally, the future of and challenges in using filamentous fungi as biofactories of recombinant proteins are briefly discussed as well.

2. FUNGAL CELLS AS BIOFACTORIES

Filamentous fungi are typically saprophytic microorganisms which secrete a wide array of enzymes involved in the decomposition and recycling of complex biopolymers from both plant and animal tissues. The majority of these enzymes are hydrolytic and play an important role in fungal nutrition, releasing carbon and nitrogen locked in insoluble macromolecules obtained from the metabolic activities of other organisms. For more than a century, fungi have been known to produce and secrete different types of enzymes in large quantities, which has resulted in an increasing interest in studying and using filamentous fungi in industrial processes. Both hyper-production and hyper-secretion are desirable characteristics of organisms with eventual industrial applications. The production of fungal proteins, either homologous or heterologous, by filamentous fungi is usually very efficient, and production levels of grams per liter are within reach. Moreover, organisms such as *A. niger* and *A. oryzae* have a long history of usage in the fermentation industry and are generally regarded as safe (GRAS) in accordance with the Food and Drug administration (FDA). Therefore, the development of an expression system in these microorganisms is desirable.

This makes filamentous fungi attractive hosts for the production of secreted heterologous proteins [1–3]. Nowadays, the production of enzymes is an important and rapidly growing sector of the fermentation industry. Several of these enzymes have been developed for a variety of commercial uses, for example, in textile processing, leather manufacturing, paper and pulp processing, detergent production, and food processing. Some fungal strains used to produce enzymes in industrial processes are capable of secreting large amounts of their respective products. For example, *A. niger* usually produces glucoamylase at 0.5 g L^{-1}, but with mutation as well as medium development and optimized fermentation conditions, the yield increased 40-fold reaching about 20 g L^{-1} [4]. *Trichoderma reesei* produces cellulases at 30 g L^{-1}. The major component of complex cellulases is cellobiohydrolase I, which is heat stable. However, these cellulases produce cellobiose as an end product, which has feedback inhibition on the enzyme. The cellulase productivity of *T. reesei* was improved more than four-fold by strain improvement programs [5]. Production of homologous and heterologous proteins by filamentous fungi has been reviewed by many authors [3, 6–9]. The capacity of filamentous fungi for high-level protein secretion was one of the key features in considering them as potential hosts for producing high value recombinant therapeutic proteins [10]. One strategy to improve the production of heterologous protein in recombinant microorganisms is the development of secretion systems [11]. Exporting the produced protein from the host cell reduces the risk of protein degradation by intracellular proteases, allows glycosylation and protein folding to occur, simplifies down-stream purification, and reduces the effect of any feedback inhibition mechanisms present in the production pathway. Product secretion is also desired if the protein is toxic to the host cell [12]. Also, the production of chymosin using *A. niger* var *awamori* was extensively studied by Dunn-Coleman *et al.* [13]. Bovine chymosin production increased to 1 g L^{-1} after gene expression in *A. niger*. The range of commercially important native and recombinant fungal products is diverse. Table 1 list some industrially important products made by fermentations with filamentous fungi.

Table 1
Different types of industrially important filamentous fungal products
(Adapted from Gibbs *et al.* [14] with some modifications)

Product	Microorganism
Antibiotics	
Penicillins G and V	*Penicillium chrysogenum*
Cephalosporin C	*Cephalosporium acremonium*
Griseofulvin	*Penicillium patulum*
Penicillin N	*Emericellopsis sp.*
Pleuromutilin	*Pleurotus mutilus*
Cyclosporin A	*Tolepocladium inflatum*
Cyclosporin A and B	*Cylinrocarpum lucidum*
Enzymes	
Glucose oxidase, pectinase and phytase	*Aspergillus niger*
Xylanase and invertase	*Aspergillus awamori*
α-Amylase and glucoamylase	*Aspergillus oryzae*
Cellulase and hemicellulase	*Trichoderma reesei*
Mycotoxins	
Aflatoxins, citrinin and ochratoxin	*Aspergillus sp.*
Trichothecenes and zearalanone	*Fusarium sp.*
Citrinin, ochratoxin	*Penicillium sp.*
Other native fungal products	
Riboflavin	*Ashbya gossypii*
Citric and gluconic acid	*Aspergillus niger*
Kojic acid and biotin	*Aspergillus oryzae*
Itaconic acid	*Aspergillus terreus*
Pullulan	*Aureobasidium pullulans*
Biotin	*Fusarium culmorum*
Ergot alkaloid	*Claviceps purpurea*
Gibberellic acid	*Giberella fujikuroi*
Linoleic acid	*Martierella isabellina*
β-carotene	*Phycomyces blakesleanus*
Recombinant heterologous proteins	.
Human interleukin-6	*Aspergillus niger*
Tissue plasminogen activator (tPA)	*Aspergillus niger*
Human interleukin-6	*Aspergillus nidulans*

3. HYPHAL GROWTH AND PROTEIN SECRETION

The relationship between the cytology of hyphal growth and protein excretion in filamentous fungi has long been observed and reported. Protein excretion by filamentous fungi is mainly restricted to the tips of growing hyphae. The hyphal tips are free of all organelles, except for a large number of vesicles of varying size; and some of these appear to be in the process of fusing with the plasma membrane [15]. This observation led to the idea that the vesicles were involved in the transport of materials to the surface of the plasma membrane at the hyphal tip as well as in membrane growth. Using immuno-cytochemical methods, it was shown that glucoamylase secretion in *A. niger* occurs predominantly at the growing hyphal tip [16]. On the other hand, the secretion of enzymes involved in lignin degradation by *Phanerochaete chrysosporium*, a process associated with the non-growing phase in the physiological cycle of fungal culture, was also shown to be associated with the tips of newly formed hyphal branches. Generally, in other eukaryotic cells, it is presumed that the apical vesicles are the final step of the intracellular secondary pathway that begins at the endoplasmic reticulum (ER) and proceeds via a Golgi system. To date, a Golgi system has only been described in Oomycetes. However, other types of fungi are thought to have organelles equivalent to the Golgi structure [15]. The available evidence clearly points to protein secretion being a highly polarized process involving the movement of protein containing vesicles to the hyphal tip. The secretory vesicles appear to be associated with microtubules, and they probably move together via an ATP dependent process. A hypothetical secretory pathway in the hyphae of filamentous fungi is shown in Fig. 1. In eukaryotic cells, the desired proteins for secretion are synthesized on ribosomes of the endoplasmic reticulum. The secretion process is then initiated by sequencing the nascent extracellular portion into the lumen of the rough endoplasmic reticulum (RER). This process is determined by the information of the signal sequence attached to the protein molecule. In general, signal sequences of different organisms share a common feature: they comprise 13–30 amino acids with a basic N-terminal region and more polar C-terminal region, which is the cleavage site.

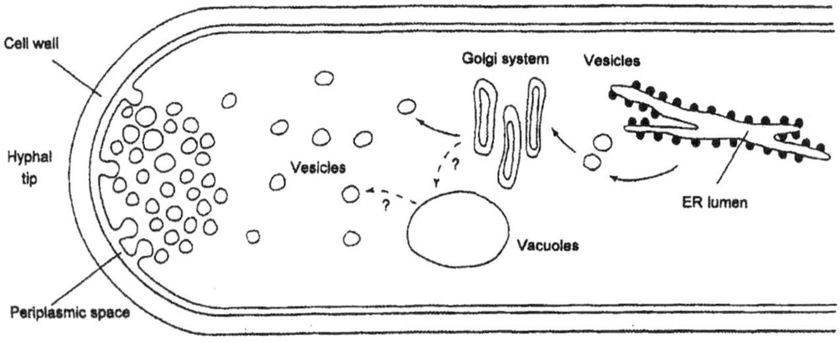

Fig. 1. A hypothetical secretory pathway in filamentous fungi. (Reprinted from [15] with permission)

The hypothesis implicating protein-conducting channels in the ER membrane was described by High [17]. Hence, the protein is transported through the organelles of the secretory pathway, and the signal peptide sequence is cleaved from the nascent protein by an endopeptidase contained within the ER lumen (Fig. 2).

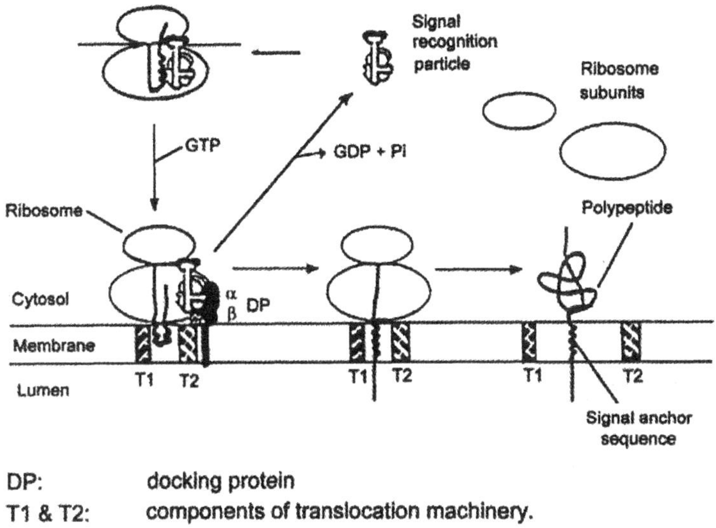

Fig. 2. Targeting of a secreted protein to the endoplasmic reticulum membrane. (Reprinted from [17] with permission)

Before proteins are secreted, they undergo several post-translational modifications. This starts in the ER and continues as the protein pass through the secretory vesicles. Three changes to a protein molecule may occur:

1- Proteolytic cleavage removes the signal sequence and other peptide sequences, if present.

2- A folding process involving the formation of disulphide bonds develops the tertiary and quaternary structure of the protein, where the disulphide bonds stabilize the molecule.

3- Glycosylation.

These maturation processes involve several different enzymes present in the ER.

3.1. Cell wall and protein excretion

The fungal cell wall fulfills several functions connected with the interaction between the cell and their environment. Some of these are:

1- Formation of a rigid, mechanical barrier on the surface of the protoplast, which also determines the cell shape.

2- Protection from osmotic stress on the protoplast.

3- Acting as a carrier of specific antigen characteristics of the particular cell and playing an important role in cell recognition in various cell interactions.

4- Acting as the site of various extracellular enzymes engaged in the exchange of nutrients and products of metabolism and the hydrolysis of cell wall components.
5- Acting as a reservoir of carbohydrates, which can be reutilized under limiting conditions or in certain stages of the life cycle.

The chemical composition of the cell wall is closely correlated with the taxonomic classification of fungi. In general, fungal cell walls share a common chemical structure composed of homo- and heteropolysaccharides, protein, protein-polysaccharide complexes, lipids, melanin, and polysaccharide chains of chitin. The unique mechanical, chemical and biological properties of the fungal cell wall are determined not only by their chemical composition but also by the spatial arrangement of the individual polymers. The layering of the cell wall components is one of the most characteristic ultra-structures of fungal cell walls. The general picture is that the skeletal, microfibrillar wall components, such as β-glucan, chitin, and/or cellulose, are embedded in an amorphous polysaccharide and protein-polysaccharide matrix.

The outer surface of the wall is usually smooth or slightly rough, whereas the skeletal polysaccharide microfibrils are more prominent on the inner surface of the wall [18]. For example, the hyphal walls of *N. crassa* consist of coaxial layers of individual wall components, and the chitin microfibrils in the innermost wall layer are covered by proteinaceous material and glycoprotein reticulum.

The outermost wall layer is smooth, composed of mixed α- and β-glucans [19]. Until now, there had been little information available about the nature of the linkages between the different components in fungal walls. The existence of covalent linkages between chitin and glucan has been described in *A. niger* [20].

In eukaryotic cells, the cell wall plays an important role as a bio-barrier for nutrient uptake and excretion. During excretion, the determination of the molecular threshold of cell walls suggests that the size limit is around 20,000 Dalton [15]. The location of the enzyme after its release from the surface of the cytoplasmic membrane is not clearly defined and the excretion processes are highly dependent on the porosity of the cell wall. In wild type *N. crassa*, invertase remains in the periplasmic space, whereas in the mutant form it is excreted outside the cell due to the increased porosity of the cell wall [21]. As the porous and nascent apical walls of fungi are transformed to the less porous lateral wall during growth, some exoenzymes are trapped and become bound within the cell wall. The hypothesis conflicting excretion and retention of exoenzyme by the wall is based on the structural and physiological differences between the apical and lateral walls of hyphal fungi as described by Chang and Trevithick [22]. Proceeding to the apical region within 2 μm of the hyphal tip are zones called (α, β and γ). Of particular interest are the β zone of maximum intussusceptions of new wall materials and the highly elastic and extendable γ zone. Both of these zones are mechanically weak. The rest of the lateral hyphal wall, the δ zone, is rigidified by secondary wall substances.

The transformation of the apical into the lateral region may be responsible for the fraction of exo-enzymes retained in the walls. During the process of rigidifying the pores in the wall (δ zone), which are initially large enough to release macromolecules from the intramural or periplasmic space, the pores of this zone become smaller due to the addition of secondary wall material. Consequently, this portion of exoenzyme becomes trapped during the transit and corresponds to the wall bound fraction. An alternative hypothesis is offered by the "bulk-flow" hypothesis, which assumes that proteins excreted by the very tip are pushed through the wall to the outside of the wall by the accretion of plastic wall polymers during apical wall growth [23]. Therefore, factors which increase the extent of hyphal branching may improve the yield of enzyme secretion in filamentous microorganisms [24]. The external/internal ratio of enzyme was found to be strain dependent, in accordance with differences in cell wall composition. For example, invertase is excreted by both *A. niger* and *A. nidulans*. In *A. niger*, the distribution of the enzyme is 70% cell bound and 30% excreted, but in *A. nidulans* the enzyme is distributed more equally. In all cases, 70% of the cell bound enzymes are external to the plasma membrane [15]. If the cell wall is impaired by using mutant strain such as mutant *N. crassa* [25] or removed, as in the protoplast of *A. nidulans*, the level of invertase excretion reaches around 90%. Moreover, the cell wall rigidity can be controlled by inhibiting chitin synthesis, the most important skeletal structure of the fungal cell wall, through the depletion of divalent cations in the cultivation medium. Among different divalent cations tested, Ca^{2+} and Co^{2+} increase the activity of chitin synthetase in *Phycomyces blakesleeanus* whereas Mg^{2+}, which is the most efficient divalent cation for stimulating enzyme catalysis, proved ineffective in the activation process [26].

4. FUNGAL GROWTH IN SUBMERGED CULTURE

The morphological growth forms of filamentous organisms in aerobic submerged cultivation may lead to suspension characteristics quite different from those of bacterial and yeast cultures. The macro-morphological features of the filamentous microorganisms, which have a significant effect on the rheological properties of the cultivation medium, are reflected directly in the production and excretion of different microbial metabolites. In submerged cultivation involving filamentous organisms, the morphology can vary from discrete compact pellets of hyphae to a homogeneous suspension of dispersed mycelia. These morphological differences are associated with significant differences in growth kinetics and physiology. The growth of dispersed mycelia is effectively equivalent to that of unicellular cells with a homogeneous distribution of biomass, substrate, and products and exponential growth at a constant specific rate in batch cultures where substrates are in excess.

The filamentous form of mycelial hyphae easily causes entanglement, and the cultivation broth thus becomes very viscous. The rheological behaviour is usually non-Newtonian, leading to relatively low viscosities in regions with high shear rates (near the impeller) and very high viscosities in region with low shear rates (near the wall). The high viscosity and pseudo-plasticity of the suspension cause many problems during cultivation, decreasing the mass transfer, heat transfer, and requiring more power input for mixing. In this case, only a small part of the bioreactor around the impeller is maintained at the optimal condition. Increasing the agitation rate improves the overall homogeneity, but this also raises the power consumption and the high shearing often damages the cells [27, 28]. On the other hand, the pellet form can be attractive for the cultivation of fungi. The most important advantages are the decrease in the viscosity of the cultivation broth and that the rheological properties become Newtonian. Newtonian fluid is characterized by good mass and heat transfer properties; moreover, the pellet formation facilitates the separation of fluid in down-stream processes [29]. The macroscopic features of cultivation media containing fungal pellets show homogeneity in their rheological properties. On the other hand, the microscopic scale of the pellets shows some heterogeneity due to zonation in accordance with the different hyphal densities inside the pellet. The zonation process has been discussed by many authors [30–33]. As long as there is a sufficient supply of oxygen to all cells within the pellet, it grows in density as well as in size. After some time, the oxygen concentration in the centre of the pellet drops to almost zero, restricting cell growth to an area near the pellet surface. The limiting of the oxygen supply and the removal of metabolic products can lead to an alteration of the cellular metabolism and enzyme excretion kinetics [34]. In studies of penicillin production using *P. chrysogenum* pellets, the transfer resistance inside the fungal pellet is high in large pellets. This causes an oxygen deficiency and autolysis of cells at the centre of the pellet. However, the thickness of the layer which contains the living cells remains constant, regardless of the pellet size. The resistance of the gas/liquid interface outside and inside the pellets are equal for pellets 400–500 μm in diameter [35, 36]. Pellets smaller than 400 μm in diameter consist of a metabolically active layer only, and all the cells are supplied with sufficient oxygen. For cell growth in the pellet form, there are two extreme cases. In pellets consisting of densely packed hyphae, growth is restricted by the diffusion of material from the liquid phase to the pellet centre, while unrestricted growth is limited to the hyphae in an outer peripheral shell. Thus, in batch culture, biomass (M) increases as a cubic function of time according to the following equation:

$$M^{1/3} = kt + M_0^{1/3} \qquad \textbf{(1)}$$

where M_0 represents the initial biomass and k is a rate constant. If a culture is assumed to consist of n spherical pellets of equal radius r and density ρ with an active outer mycelial shell of thickness w growing at a specific rate μ, then the rate constant k can be determined as follows:

$$k = \left(\frac{4}{3}\pi\rho n\right)^{1/3} \mu w \qquad (2)$$

When a pellet exceeds a certain size it is assumed that growth is limited to a peripheral zone of thickness w, through the limitations of the penetration rate of the growth-limiting nutrient. Although cubic-root growth kinetics has been observed in fungal cultures, experimental data do not always allow it to be distinguished from other models. In addition, the cubic-root law fails to consider the effects of mass transfer and substrate concentration on growth. Attempts have been made to consider oxygen uptake, consumption, and limitation within a pellet [37]. However, most do not take into account the variations in pellet size, density, and micro-morphology which occur in liquid culture. Such variation is of significance because it is the pellet size frequency distribution which defines the amount of mycelium in contact with the growth medium, the proportion of biomass which is growing, and the rate of substrate utilization. On the other hand, if the pellets consist of loose, open, more filamentous mycelia, agitation allows nutrients and oxygen to reach all the constituent hyphae and supports exponential growth in the entire biomass. The latter type of growth is also more easily controlled because, in the ideal state, all of the hyphae are growing exponentially and all are in contact with a well-stirred medium, so all can respond rapidly to manipulation of the medium. These advantages have to be traded off against increased viscosity caused by the filamentous growth. Growth morphology also has a significant role in the mixing characteristics and phase interactions inside the cultivation system. The influence of the growth morphology of *Trichoderma harzianum* on oil-air dispersion in fermentation broth was studied by Lucatero *et al.* [38]. Larger oil drop sizes were obtained with dispersed mycelia than with pellets as a result of the high apparent viscosity of the broth, which caused a drop in power drawn, reducing oil drop break-up. Unexpectedly, bubble sizes observed with dispersed mycelia were smaller than with pellets, a phenomenon which can be explained by the segregation occurring at high biomass concentrations with the dispersed mycelia. In this system, very complex oil drops were produced containing air bubbles and a high number of structures likely consisting of small water droplets. Bubble location was influenced by biomass morphology. The percentage (in volume) of oil-trapped bubbles increased from 32% to 80% as the dispersed mycelia concentration increased. A practically constant 32% of oil-trapped bubbles was observed with pelleted morphology at all biomass concentrations. This study evidenced the high complexity of phase interaction and the importance of mycelial morphology in such processes. Thus, the ability to control the morphology of a fungus in submerged culture is important, since morphology can affect the product yield.

4.1. Micro- and Macro-morphology in submerged cultures

The growth morphology of filamentous microorganisms was of interest to many scientists with many publications. Many reviews concerning the growth of filamentous microorganisms have been published [28, 39–42]. A detailed description of morphology is of interest because a correlation between morphology and productivity can then be found and used for process optimization. However, a quantitative experimental determination of morphology and therefore a systemic approach to solving this problem became possible only a few years ago with modern image analysis systems [43, 44]. Recently, a new method using an online laser probe for particle size determination was applied to study the fungal spore's aggregation in submerged cultures [45]. In general, fungal cell growth in submerged cultures undergoes three main phases:

Phase 1 - micromorphological growth: spore swelling, spore germination, hyphal cell extension, and branching.

Phase 2 - macromorphological growth: the formation of either the hyphal network or the fungal biopellet.

Phase 3 - fungal cell autolysis.

Different factors affecting each of these phases are described in more detail below.

4.1.1. Micromorphology

The micromorphological characterization of fungi includes the progress of spore swelling, germination, and germ tube elongation and branching. These are the main steps of micromorphological growth and sometimes determine whether further macro-morphological growth is in pellet or filamentous form. In general, a short germination phase is desired in biotechnological processes in order to decrease the cultivation time. Fermentation efficiency is also conditional on the synchronicity of the germination process and on the number of germination tubes formed by each spore [46].

4.1.1.1. Spore swelling and germination

Fungal spores are considered the beginning and the end of the development cycle of fungi. Like the seeds of plant species, fungal spores are characterized by a dormant state, in which their metabolic activities decrease to a large extent although they still retain some respiratory activity and use some functional links of the metabolic chain. Hence, a number of researchers denote fungal spores as dormant. Fungal spores aer surrounded by an outer rodlet layer of hydrophobic proteins and melanin. This works as a barrier against mass transfer from the medium to the inside of the spore, resulting in a reduction in swelling and germination [47, 48]. Moreover, this layer plays a significant role in aggregation between spores. More recently, the surface properties of *A. oryzae* spores were investigated by atomic force microscopy [49]. It was observed that the dormant spores were covered with a discontinuous layer of about 35 nm thickness. High resolution deflection images showed that this layer consisted of rodlets (10 ± 1 nm in diameter) and was assembled in parallel to form fascicles interlaced with different orientations. The germinating spore surface was much rougher and showed streaks oriented in the scanning direction. Recent studies by Fuchs *et al.* showed a direct relationship between the presence of two hydrophobic genes and microconidial chain formation in *Fusarium verticillioides* [50].

In general, fungal spores display two types of dormancy: exogenous (superficial) and endogenous (constitutive). Constitutive dormancy requires a specialized cell barrier (restricting the entry of nutrients) and the presence of auto-inhibitory compounds. These are spore-dependent factors and vary from species to species and between different types of spores within the same species. Endogenous dormancy in some spore species can be broken by the activation of the cAMP pathway and by trehalose breakdown [51]. It was also observed that this type of dormancy is regulated by some genes. In *A. nidulans,* the mitogen-activated protein kinase (MAP kinase) gene *mpkA* is involved in the germination of conidial spores and polarized growth [52].

When spores in exogenous dormancy are placed in a suitable environment, the germination process starts. During spore swelling and germination, the dormant spore shifts from low to high metabolic activity. This process starts automatically when the spore is placed in a suitable environment. Dormancy also can be broken by an activation process such as heat shock or by chemical treatment. Whatever the method of initiation, three basic structural changes during germination can be recognized by microscopic observation: spore swelling, germ tube emergence, and germ tube elongation [53, 54]. With the onset of germination, the spore begins to swell to several times its dormant diameter and a germ tube emerges. Usually, a spore is considered germinated if the length of this germ tube reaches one-half of the largest dimension of the spore. The swelling and germ tube emergence in the germination process constitute a major part of the lag phase in fermentation inoculated with spores. The processes of spore swelling and germination are regulated by different exogenous environmental factors such as temperature, humidity, carbon dioxide, and many other factors. Besides these exogenous factors, spore concentration and pre-treatment before inoculation also play a significant role in the germination time [55].

The early study by Anderson and Smith was focused on the influence of temperature on spore swelling and germination in *A. niger* during submerged cultivation [56]. The incubation temperature has a marked effect on the spherical growth rate of spores and their eventual size. Of the large range of temperatures tested (30–47°C), the highest spherical growth rate (swelling rate) was observed at 38°C. The size of the swollen spore reached about 20 μm in the temperature range of 30–44°C, although the time taken to reach this size varied depending on the spherical growth rate. Spores did not swell in high temperature cultures of 47°C even after an extended cultivation period. On the other hand, the optimal conditions for germination of *Rhizopus oligosporus* sporangiospores were 42°C and pH 4.0 [57]. In the same study, germination kinetics was also highly influenced by the type of carbohydrate used and by amino acid supplementation. Spore germination in *A. niger* is stimulated by the addition of a nitrogen-containing substrate and is a temperature-dependent process [46]. Temperature causes significant changes to the lipid bilayer and the neutral lipid composition of conidia, and thus influences spore germination. The temperature viability threshold does not exceed 45°C. The combined effect of chitosan and temperature on spore germination in *A. niger* was recently studied [58]. The optimal temperature for spore germination was 30°C and the addition of chitosan to the cultivation medium decreased germination. Chitosan inhibits spore swelling by a direct interaction with the spore cell wall, and the effect was directly proportional to its concentration.

Van Suijdam and Metz showed that the absence of carbon dioxide causes much slower spore germination [27]. In the absence of CO_2, the lag phase of *P. chrysogenum* spore was 99 ± 21 h, compared to a period of 44 ± 14 h when CO_2 was present under the same cultivation conditions.

The origin of the fungal spore, either from a surface or submerged culture, also has a significant effect on spore swelling and germination. *Trichoderma harzianum* aerial spores show higher hydrophobicity than those obtained by submerged culture. The latter are easily wettable and germinate at a higher rate. However, spores produced in aerial mycelia show higher resistance to stress than those produced under submerged culture conditions [59]. Spore germination in *Verticillium lacanii* is also influenced by the type of spore used. Aerial spores had a tendency to have rough, brittle surface characteristics; however, the submerged spores appeared smooth on the surface. During submerged cultivation in rich media, aerial spores did not show germ tubes until 8 h of incubation, while the submerged spores showed many germ tubes earlier. However, a spore germination percentage of over 90% was reached for both types of spores at 18 h of incubation [60].

In this respect, spore hydrophobicity might contribute to the maintenance of the dormant state in aerial spores. Spores produced in submerged cultures have facilitated interactions with the external medium, such as quick water and nutrient uptake; this results in a rapid response to external conditions, allowing a shorter germination time. In general, the germination time of most industrial fungi belonging to the genuses *Aspergillus* and *Penicillium* is about 6–8 h under favorable cultivation conditions [61–63]. An image analysis system was first used to study spore viability and swelling during the germination process by Paul *et al.* [53]. Since then, several papers have been published on using image analysis for measuring the germination characteristics of fungal spores during submerged cultivation [64].

An acridine orange (AO) staining method was successfully applied for the study of spore swelling under fluorescence microscopy. Metabolically active spores have a red fluorescence, which indicates the presence of higher amounts of single-stranded RNA (ssRNA) and minor quantities of DNA; whereas latent or dormant spores have a green color (which indicates the presence of a high amount of DNA and a minor quantity of ssRNA). This method was successfully used to follow the kinetics of spore swelling and germination in *A. awamori* and *A. niger* [63, 65]. Fig. 3 shows the different stages of spore swelling and germination in *A. niger* spores in submerged culture as observed under phase contrast and under fluorescence microscopy after AO staining [63]. The emergence of a germ tube for the fungal spore stained red only is in agreement with the idea that conidial germination in *Aspergillus* requires RAS signaling and protein synthesis [66]. They also observed that conidial germination in *A. nidulans* requires protein synthesis and that the initiation of translation is linked, through a signalling cascade that includes *rasA*, to a carbon-source-sensing apparatus. However, the transformation of the physiological status of spores from resting to growing is usually concomitant with a significant increase in cell metabolism with regard to enzyme production. Significant changes in fatty acid composition and the differential expression of desaturase genes are also observed during spore germination [67].

4.1.1.2. Germ tube formation, elongation and hyphal branching

Hyphal extension in fungi is an extreme example of polarized cell growth since it is localized in a small region at the hyphal apex and can attain high extension rates (up to 100 $\mu m\ min^{-1}$). The existence of ion gradients in such tip growing systems has been proposed as a mechanism for the establishment and maintenance of cell polarity [68, 69].

Calcium ions also play a crucial role in the polarized extension of fungal cells. It have been observed that the growth of *Neurospora crassa* diminished in media containing less than 1 mM Ca^{2+}; extension was more severely impaired than biomass synthesis, resulting in the formation of stubby, bulbous hyphae. Reduced extension and abnormal morphology were correlated with the loss of surface bound Ca^{2+}, probably associated with the cell wall [70]. In the presence of excess nutrients, suitable environmental conditions and the absence of inhibitors, the biomass of unicellular microorganisms increases exponentially. The germ tubes of filamentous fungi extend in a highly polarized manner by inserting new cell wall material exclusively at the extending hyphal tip. Vesicle trafficking is fundamental to tip growth and involves the dynamic and highly organized cycling of vesicles in a multicomponent organelle called the Sptizenkörper and with the involvement of actin and tubulin [71–73].

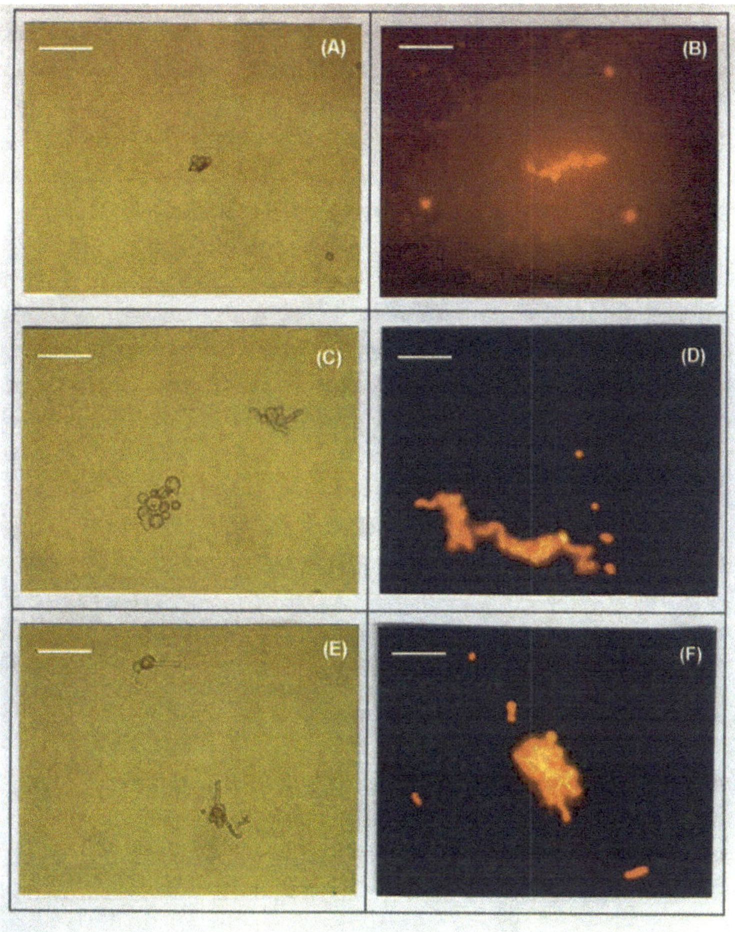

Phase contrast microscope Fluorescence microscope

Fig.3. Spore swelling and germination during the early cultivation period. A and B show non-germinated spores after 5 h of inoculation. C and D show germinated spores and the development of the germ tube after 7 h. E and F show an elongated germ tube after 9 h. (bar: 40 μm). (from [63])

The main hypha also branches, producing further sites of polar growth [74]. A wide range of proteins, Ca^{2+} flux, the COT1 kinase, a mitogen activated protein kinase (MPKA), RhoA genes, and the cAMP-dependent kinase play very important regulatory roles in the polarized growth of hypha [52, 75–77]. A novel hyper branching gene (*hbrB*), which is also required for polarized growth, was recently cloned [72].

As fungal hyphae extend at a linear rate, exponential growth of filamentous fungi is only possible by the formation of an increasing number of growing tips. New tips are formed by branching, often at locations behind the extending tips, and the formation of new branches enables filamentous fungi to increase the total growth surface for better nutrient uptake and product secretion. However, the hyphal diameter, the branching frequency, and the branching pattern are known to change with the specific growth rate [78, 79]. The strong relationship between mitosis and branch formation was recently studied [80]. In wild-type strains of *A. nidulans,* branching intensity was increased when the tip extension was reduced and was reduced when growing on poor substrates. In these situations, the hyphal concentration of nuclei was maintained and it is suggested that branching is correlated to cell cycle progressions in order to maintain a minimum required cytoplasmic volume per nucleus and to avoid the formation of anucleated hyphae in the absence of nuclear divisions.

4.1.2. Macromorphology and microbial pellet classification

Nielsen classified fungal pellets into three types based on the mechanism of pellet formation: coagulative, non-coagulative, and hyphal-element agglomerating [40]. The coagulative type is characterized by the coagulation of spores while germination gives rise to a net of intertwined hyphae. *A. niger* pellets are a good example of this type. In the non-coagulative type, one spore gives rise to one pellet and the number of pellets is directly correlated with the number of spores used as inoculum. Some species of *Streptomyces* belong to this group. In the hyphal-element agglomerating type, the hyphal elements agglomerate and form a clump of hyphal elements that eventually evolve into pellets. *P. chrysogenum* belongs to this group.

However, a new classification of cell morphology into five different classes was proposed by Jin *et al.* [81], as follows:

1- Dispersed mycelia with diffuse filaments
2- Fluffy mycelia with diffuse mycelia
3- Clumpy mycelia with highly aggregated, flock-like mycelia
4- Clumpy pellets with compact nuclei and diffuse boundaries
5- Compact pellets as compact spherules with smooth surfaces

4.1.3. Factors affecting microbial pellet formation

The process of microbial pellet formation and mycelial cell aggregation is influenced by many factors. These factors include the strain used, growth rate, medium composition, surfactants, polymer addition, shear force, aeration, agitation, and many others. The factors influencing micro- and macro-morphology and their relationship to pellet formation have been reviewed by many authors [27, 82–86]. In general, these factors can be divided into three main categories: strain dependent factors, nutrition dependent factors (medium composition), and cultivation conditions, as shown in Fig. 4. Moreover, interactions between these factors must be considered, and it is thus not easy to determine or discuss some of these factors separately. For example, the effect of the cultivation vessel influences the oxygen transfer inside the cultivation medium as well as the shear force. Also, the effect of cultivation conditions such as temperature, oxygen supply, etc. as well as medium components such as the C source, N source, C/N ratio, etc. are reflected in the specific growth rate of the microorganism. Also, the inoculum size, a strain-dependent factor, has an effect on aeration in submerged cultures of fungal cells [87].

Various strain dependent factors, nutrition dependent factors (medium composition), and cultivation conditions that have been shown to affect fungal morphology and pellet formation are listed in Tables 2–4, respectively, and are discussed in the following sections.

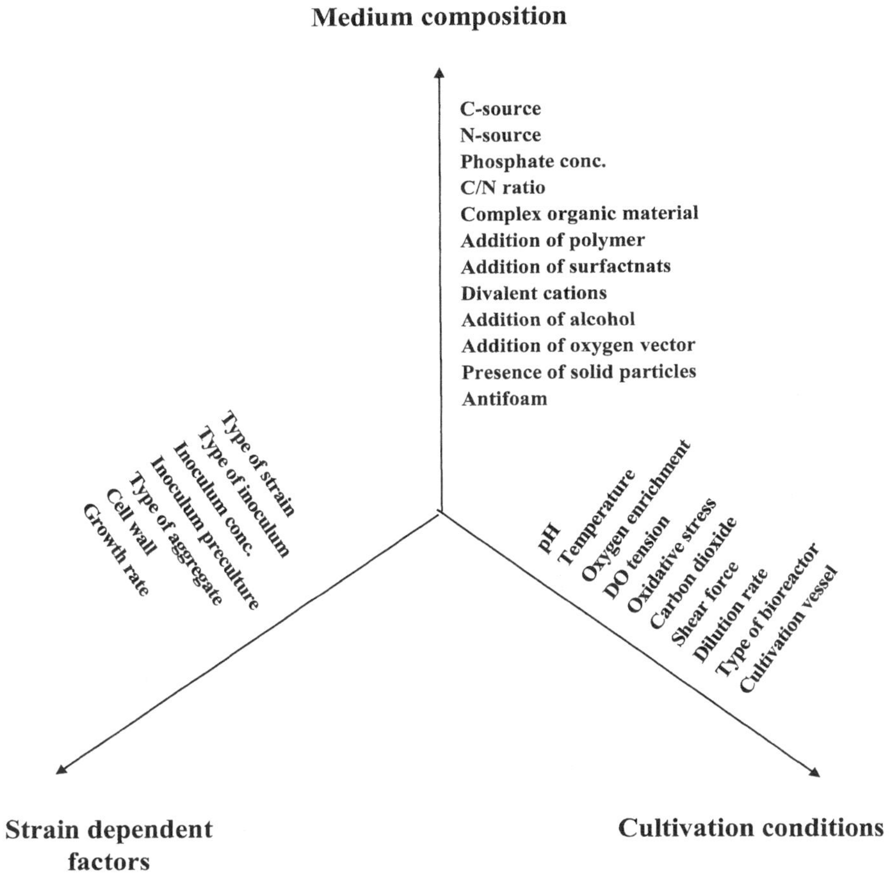

Fig. 4. Different factors affecting the production of microbial biopellets.

4.1.3.1. Strain dependent factors

Strain Type. The strain type has a significant effect on fungal cell morphology. Of the three different strains of *A. oryzae* tested, different growth morphologies were obtained under the same cultivation conditions and ranged from dispersed mycelia to compact pellets [81]. Recently, Molnár *et al.* studied the influence of $fadA^{G203R}$ and $\Delta flbA$ mutations on the morphology of submerged *A. nidulans* cultures [88]. A loss-of-function mutation of the *flbA* gene resulted in an altered germination with unusually thick germination tubes, fluffy pellet morphology, as well as a reduced hyphae fragmentation rate during autolysis. In the $fadA^{G203R}$ mutant strain, conidiophores formed during the stationary phase of growth, and the pellet size shrank considerably.

Table 2
Strain-dependent factors affecting the formation of fungal pellets

Factor	Microorganism	References
Type of strain/ gene regulation	*Aspergillus nidulans*	[88]
	Aspergillus oryzae	[81, 89]
Type of inoculum	*Aspergillus niger*	[34, 90]
	Aspergillus oryzae	[81]
	Penicillium chrysogenum	[84]
Inoculum concentration	Different microorganisms	[39]
	Aspergillus niger	[90]
	Penicillium chrysogenum	[91]
	Phanerochaete chrysosporium	[92]
	Rhizopus chinesis	[93]
	Rhizopus nigricans	[94]
	Rhizopus oryzae	[95]
Inoculum preculture method	*Aspergillus niger*	[34]
	Penicillium chrysogenum	[84, 96]
Type of aggregate	*Aspergillus niger*	[97]
Cell wall composition	*Streptomyces tendae*	[98]
Growth rate	*Aspergillus nidulans*	[99]
	Penicillium chrysogenum	[100, 101]

Inoculum type. Jin *et al.* investigated the influence of inoculum quality on growth morphology using different strains of *A. oryzae* in a stirred tank bioreactor [81]. Three different types of inocula (spore suspension, dispersed mycelia, and compact pellets) were used. For some strains, the inoculum type did not affect the morphological changes in the culture: compact pellets were the sole morphological form. However, larger pellets were obtained upon using compact pellets as inoculum. Hermersdörfer *et al.* [34] reported that the use of *A. niger* spores as inoculum favors the formation of small pellets. Using pre-cultured pellets as inoculum resulted in large pellet formation in the culture.

Inoculum concentration. The effect of inoculum size on pellet formation during the cultivation of *R. nigricans* was studied by Žnidaršič *et al.* [94]. High inoculum concentrations resulted in a homogeneous suspension of mycelium, indicating the non-coagulative type of pellet formation in *R. nigricans*. For the coagulative type of pellet formation, pellets would always be formed regardless of the initial spore concentration [39]. The threshold level, above which filamentous growth occurred and below which pellets were formed, in the case of *R. nigricans*, changed with the agitation speed.

Du *et al.* [93] studied the influence of spore concentration on the cell growth of *Rhizopus chinesis*. It was obvious that a higher initial spore concentration (1×10^9 spores mL^{-1}) led to the production of more hyphae in the early stage of incubation. These hyphae entangled and prevented the formation of pellets. With a lower spore concentration (1×10^5 spores mL^{-1}), pellet size increased and overall biomass production decreased. Another study, done by

Jiménez-Tobon *et al.* [92], revealed that as the initial spore concentration was increased from 1×10^3 to 1×10^6 spores mL^{-1}, a large number of pellets were formed during *Phanerochaete chrysosporium* cultivation. As a result, the average final size of the pellets was much larger for the lower inoculation levels. The initial increase of the average pellet sizes was linear with time and appeared to be independent of the inoculum level.

4.1.3.2. Medium composition

Carbon source. The effect of sucrose concentration on the growth morphology of *A. niger* has been studied by Ryoo [102]. It has been observed that the fractal value of mycelium and the average diameter of pellets decreased with an increase in the sucrose concentration. A study by Sinha *et al.* [104] on the influence of sucrose concentration on the morphology of *Paecilomyces japonicus* showed that the pellet increased in size as the initial sucrose concentration increased from 20 g L^{-1} to 60 g L^{-1}. However, at an initial substrate concentration of 80 g L^{-1}, no pellet formation was observed and the entire fungal population was characterized by filamentous growth. El Enshay *et al.* [103] studied the influence of different carbon sources on growth morphology of *A. niger* in a stirred tank bioreactor. In glucose and xylose media, growth was mainly in filamentous form with small pellets not exceeding 400 μm in diameter. On the other hand, the growth in the fructose culture was mainly in the pellet form. Cho *et al.* [105] studied the influence of the carbon source on the morphology of *Paceilomyces sinclairii.* Cells grown in a sucrose medium were highly branched and showed a longer hyphal length than those grown in a starch medium.

Nitrogen source. Du *et al.* [93] studied the influence of different nitrogen sources on the growth morphology of and antibiotic production by *R. chinesis.* It was observed that the highest antibiotic production, accompanied by pellet growth, was achieved in the medium containing corn steep liqour (CSL). Pellet growth was also observed in the medium using ammonium sulphate, but the compact structure and smooth surface of these pellets with a diameter of about 4 mm were different from the microstructure of those in the medium containing CSL, where fluffy pellets with a compact center, a much looser outer zone and an average diameter of about 3 mm formed. The formation of more pellets was probably due to the high concentration of the organic nitrogen source in CSL, which accelerated spore germination and cell growth. The growth was in the form of dispersed mycelia when the medium was supplemented with peptone and in an entangled filamentous form when the medium was supplemented with yeast extract. However, not only does the type of nitrogen source affect the growth morphology, but its applied concentration does so as well. Bai *et al.* [95] studied the effect of ammonium nitrate on the growth morphology of *R. oryzae.* Increasing the concentration of ammonium nitrate in the preculture medium concomitantly increased the final biomass concentration. An investigation of the fungal morphology revealed a change from a filamentous to a pellet morphology upon the addition of ammonium nitrate to the preculture medium. Further studies showed that increasing the ammonium nitrate concentration caused the fungal pellets to increase in size but to decrease in density and number per unit volume.

Phosphate. A critical relationship between phosphate concentration and the growth morphology of *A. awamori* was reported by Gerlach *et al.* [107]. At the low phosphate concentration of 0.3 g L^{-1}, low amounts of pellets were formed initially but they quickly turned into clumps. At the intermediate phosphate concentration of 1.05 g L^{-1}, pellets were formed. At the high phosphate concentration of 2.1 g L^{-1}, clumps were again formed. Ryoo observed that the addition of phosphate to the medium increased in the mycelial fractal value and the pellet diameter during the cultivation of *A. niger* [102]. Phosphate is a major component of microbial cell walls. With higher phosphate concentrations in the medium, larger pellets were formed by more hydrophobic mycelia. Thus, the effect of phosphate on the growth morphology may be considered strain-specific.

Table 3
Nutrition-dependent factors (medium composition) affecting the pellet formation

Factor	Microorganism	References
Carbon source	*Aspergillus niger*	[34, 102, 103]
	Paecilomyces japonica	[104]
	Paecilomyces sinclairii	[105]
Nitrogen source	*Aspergillus niger*	[34, 106]
	Rhizopus chinesis	[93]
	Rhizopus oryzae	[95]
Phosphate	*Aspergillus awamori*	[107]
	Aspergillus nige	[102]
C/N ratio	*Aspergillus niger*	[108]
	Mortierella alpina	[109]
	Rhizopus arrhizus	[110]
Complex organic materials	*Aspergillus niger*	[111]
	Rhizopus arrhizus	[110, 112]
	Trichoderma reesei	[113]
Polymer	Different Microorganisms	[114]
	Aspergillus niger	[82, 115, 116]
	Mortierella vinacea	[117]
	Penicillium chrysogenum	[118]
	Phanerochaete chrysosporium	[119]
Surfactant	*Aspergillus niger*	[39, 82]
	Rhizopus arrhizus	[111]
	Trichoderma reesei	[112]
Divalent cations	*Aspergillus niger*	[120–122]
Alcohol	*Mucor fragilis*	[123, 124]
Oxygen vector	*Aspergillus niger*	[125]
	Aspergillus terreus	[126]
Solid particles	*Aspergillus awamori*	[127]
	Aspergillus niger	[39, 128]

C/N ratio. The concentration ratio between the carbon and nitrogen source in the media also plays a significant role in affecting growth morphology. A gradual increase in the C/N ratio has been shown to result in increased mycelial growth in *Myrothecium verrucaia* and *A. niger* [108]. In contrast, using a peptone glucose medium, pelleted growth of *R. arrhizus* was found at higher C/N ratios and open mycelial growth at relatively low ratios [110]. The influence of the C/N ratio on the growth morphology of *Mortierella alpine* was studied by Koike *et al.* [109]. The total active biomass volume (pellet diameter times the thickness of the annular region where active cell growth taking place) increased in proportion to the C/N ratio only when this ratio was higher than 20. When the C/N ratio was below 20, the active biomass volume did not change when the C/N ratio was increased.

Complex organic materials. Cells grown in an enriched medium containing peptone and yeast extract usually have higher growth rates due to the presence of growth cofactors, vitamins, polypeptides and other compounds, characteristics of complex nutrients. Thus, oxygen may be depleted and pellet formation is prevented. Domingues *et al.* [113] observed that the cultivation of *T. reesei* Rut C-30 in yeast extract and a peptone supplemented medium prohibited pellet formation and that the growth was mainly in the free mycelial form. However, the addition of yeast extract to the *A. niger* culture changed the growth morphology from the filamentous to the pellet form, with the pellet diameter increasing proportionally with the concentration of yeast extract in the medium [111].

Polymers. Wainwright *et al.* [119] studied the effect of anionic polymers on the aggregation of spores in submerged cultures of *Phanerochaete chrysosporium.* Spore aggregation and swelling are key steps in the development of fungal morphology. Adding polymers such as Junlon (polyacrylic acid) or Hostacerin (sodium polyacrylate) at the beginning of the cultivation process decreased the spore aggregation. The prevention of spore aggregation by polyacrylic acid is a function of polymer concentration, molecular weight and ambient pH. The decrease in spore aggregation resulted in a morphological shift toward smaller pellets and dispersed mycelial forms. Elmayergi *et al.* [82] studied the effect of different types of polymers on the growth morphology of *A. niger.* They observed that the addition of carboxymethycellulose (Carbopol) enhanced the respiration rate of fungal cells. The resulting enhancement of cell growth and activity was largely due to gross morphological differences rather than physiological or biochemical changes in culture. The dispersed growth (which was promoted by the addition of the polymer) showed higher metabolic activity than the discrete pellets because of the increased interface area for nutrient transfer. However, dispersed culture morphology often results in a high bulk fluid viscosity so that the choice of polymer additive depends on an economic balance between increased yields and operating costs.

Rugsaseel *et al.* [116] studied the effects of adding different viscous substances to the fermentation medium of *A. niger* on cell morphology and citric acid production. Adding gelatin at a concentration of 2.0–6.0 mg mL^{-1} as a viscous additive to media containing glucose as the carbon source slightly increased the media viscosity but substantially increased citric acid production. The addition of other viscous substances, including carrageenan, agar, carboxymethylcellulose, and PEG 6000, in low concentrations to the medium also increased citric acid production. The mycelia were thick with stable spherical aggregates consisting of a denser, branched, and partially intertwined network of hyphae in cultures containing viscous additives.

Table 4
Cultivation conditions affecting the formation of fungal pellets

Factor	Microorganism	References
pH	Different microorganisms	[28]
	Aspergillus oryzae	[81]
	Fusarium graminearum	[129]
	Mucor fragilis	[124]
	Penicillium chrysogenum	[27]
Temperature	*Aspergillus niger*	[34]
Oxygen enrichment/ Aeration rate	Different microorganisms	[130]
	Aspergillus awamori	[131]
DO tenstion	*Aspergillus nidulans*	[132]
	Aspergillus niger	[133, 134]
	Paecilomyces sinclairii	[105]
Oxidative stress	Different microorganisms	[135]
	Aspergillus niger	[136, 137]
Carbon dioxide	Different microorganisms	[138]
	Aspergillus niger	[139, 140]
	Penicillium chrysogenum	[141, 142]
Shear force	Different microorganisms	[143]
	Aspergillus awamori	[144]
	Aspergillus niger	[83, 145]
	Aspergillus oryzae	[146, 147]
	Penicillium chrysogenum	[27, 85]
Dilution rate	*Aspergillus niger*	[148]
	Fusarium graminearum	[79]
Type of bioreactor	*Aspergillus awamori*	[107]
	Aspergillus niger	[149, 150]
	Penicillium chrysogenum	[31]
	Phanerochaete chrysosporium	[150]
	Rhizopus oryzae	[151]
Cultivation vessel	*Aspergillus niger*	[111, 152]
	Fusarium moniliforme	[153]
	Rhizopus chinesis	[93]
	Trichoderma hazianum	[154]

Insoluble particles. Cui *et al.* [127] studied the influence of adding wheat bran to the medium on the growth of *A. awamori*. It was found that fungal cells mainly grew on the wheat bran particles (so called adhesion growth) when the initial spore concentration was higher than 1.3×10^5 mL^{-1}. However, when the initial spore concentration was lower than 1.8×10^4 mL^{-1}, cells grew in pellets without adhering to the wheat bran particles, which remained clean without any cells attached to them. Between these two initial spore concentrations, the broth suspension consisted of wheat bran-free pellets, clean wheat bran particles, and adhesion colonies. In general, the presence of wheat bran particles in the fermentation medium seemed to promote the aggregated growth form, probably due to the fact that wheat bran particles damp turbulent flow in the bioreactor; a less turbulent flow favored the aggregated growth form. Papagianni *et al.* [128] also reported that *A. niger* changed from pellets to the filamentous form in the presence of wheat bran during phytase production. This was regarded as an effect of the increased availability of phosphorous in the medium. During the decomposition of wheat bran, phosphorous was liberated by the phytase already produced by the fungal mycelia, and its slow release ensured a continuous presence of phosphorous in the medium.

Alcohol. The pulse addition of ethanol at a low concentration of 0.5% (v/v) to the culture medium did not affect either the morphology or the physiology of *Mucor fragilis*; however, a higher concentration of up to 2.5 % (v/v) resulted in a significant shift in growth morphology from filamentous to yeast-like structures [123]. The induction of dimorphism and morphological change in *M. fragilis* due to alcohol addition has also been reported [124].

Surfactants. In general, the addition of surface-active substances affects pellet growth. The use of Tween 80 does not have the same effect on pellet formation in all species. For example, with *A. niger*, aggregation was increased and larger pellets with a loose structure developed [39]. Domingues *et al.* [112] also showed that the use of Tween 80 in the fermentation medium inhibited pellet formation in *T. reesei* Rut C-30.

Divalent cations. The interaction between different cations in the medium and its effect on growth morphology have been studied. Haq *et al.* [121] reported that the addition of magnesium sulphate to an *A. niger* culture at the low concentration of 2.0×10^{-5} M reduced the Fe^{2+} ion concentration by counter-acting its deleterious effect on mycelial growth. The magnesium ions also induced a loose-pelleted form of growth (0.6 mm in diameter), reduced the biomass concentration, and increased the volumetric production of citric acid. The time when the magnesium ions were added was also a critical factor. If it was added at the beginning of cultivation, cell growth was in the form of mixed pellets. Adding magnesium after 12 or 18 h of cultivation resulted in larger pellets. However, the effect was strain dependent. The counter-ion effect was also reported between copper ions and ferric ions in a molasses-based medium [122]. The addition of copper ions to the medium also induced a loose-pellet form, reduced the biomass concentration, and increased citric acid production.

Oxygen vector. The addition of a non-aqueous liquid phase may provoke a significant increase in oxygen transfer from the gas phase to the cells and result in significant increases in both cell growth and metabolite production [125]. Recently, some oxygen vectors were used to improve growth morphology in submerged fungal cultures. The effect of adding an oxygen vector (oxygen carrier) on *A. terreus* was studied by Lai *et al.* [126]. The addition of n-dodecane (C12) to the medium at a concentration of 0–5% (w/v) resulted in a significant change in the growth form. When C12 was added to the medium at different concentrations of 0%, 2.5% and 5% (w/v), the corresponding fungal pellet diameters varied from 0.75–1.7 μm,

0.65–1.30 μm, to 0.55–2.05 μm and the respective pellet densities also varied (550, 880, and 370 pellets mL^{-1}). Smaller pellet sizes were found to be more favorable for reducing the medium viscosity, leading to easier nutrient transport for cellular utilization, and thus enhanced lovastatin production.

5. EFFECTS OF CULTIVATION CONDITIONS

5.1. Process parameters affecting fungal growth and morphology

5.1.1. pH

In the study of Jin et al. [81] using three different strains of *A. oryzae*, the morphological formation was shown to be critically affected by the growth pH. In the first strain, mycelia were produced at pH 3.0–3.5, whereas compact pellets were observed at pHs between 4.5 and 6.0. Unlike the first one, the second strain produced compact pellets at a low pH from 3.5 to 4.5, and mycelia were formed at pH 3.0. The third strain did not produce any pellets at pHs from 3.0 to 6.0. Braun and Vecht-Lifshitz [28] investigated the influence of pH on morphological changes by cultivating *A. nidulans* in pHs ranging from 3.0 to 6.0. Filamentous and fluffy mycelia occurred at pH from 3.0 to 5.0 and pellet formation took place at pHs above 5.0. However, Van Suijdam and Van Metz [124] stated that no significant difference could be found between morphologies at different pH values in batch and continuous cultures of *P. chrysogenum*. However, it has also been reported that at reduced pH values, *M. fragilis* grew in pellet form [129]. A small pH shift from 2.95 to 2.70 resulted in a significant decrease in pellet growth rate and biomass.

The influence of pH on fungal micromorphology was also studied by others [129]. *F. graminearumi* was cultivated at different pH values between 3.5 and 8.0. The hyphal growth unit length was found to be the longest at pH 6.0 but was only slightly shorter at the other pH values between 4.5 and 8.0 tested. At pH values below 4.5, the hyphal growth unit length decreased with decreases in the pH. On the other hand, hyphal diameters were wider at pHs below 4.0. Between pH 4.0 and 8.0, the hyphal growth unit volume was approximately constant, but at pH 3.0 and 3.5, significant reductions in this value were observed.

5.1.2. Temperature

The effect of temperature on the growth form of *A. niger* in submerged cultures was studied by Hermersdörfer *et al.* [34]. They observed that a temperature shift from 25°C to 30°C shifted the growth from small, short, branched aggregates to large and hairy aggregates. Further increases in temperature up to 35°C resulted in the transformation of fungal growth from the pelleted to the filamentous form. This might be caused by the swelling time of the spores. In general, high temperatures increase spore swelling and germination and thus decrease the spore's tendency to aggregate during the early phase of cultivation and promote filamentous growth.

5.1.3. Aeration (oxygen enrichment) and oxidative stress

Cho *et al.* [105] showed that an increase of the aeration rate from 0.5 vvm to 1.5 vvm resulted in an increase in hyphal branching. Also, a highly vacuolated cell morphology was observed upon increasing the aeration rate to 3.5 vvm. Wongwicharn *et al.* [134] studied the effect on the cell morphology of recombinant *A. niger* by applying different levels of oxygen enrichment to the culture. Two distinct micromorphological states were apparent in these cultures: one, typically seen under oxygen limitation (i.e. 0 to 10% enrichment levels) consisted of long, sparsely branched hyphal element, while a second morphology, typical of oxygen enriched cultures at 30% and 50% oxygen enrichment, was comprised of shorter hyphal elements with more branching. At 50% enrichment, a standard aggregate morphology was observed, possibly as a response to a hyper-oxidant state. Another study showed that when the dissolved oxygen concentration was close to the saturation concentration corresponding to pure oxygen gas, *A. awamori* formed dense pellets and the free filamentous mycelia fraction was almost zero [131]. In the case of very low dissolved oxygen tension, the pellets were rather weak and fluffy, and this had a very different appearance. However, the biomass per pellet volume increased with the dissolved oxygen tension and decreased with the size of the pellets. This means that the smaller pellets formed under a higher dissolved oxygen tension had a higher intrinsic strength. Correspondingly, the porosity of the pellets was a function of the dissolved oxygen tension and the size of the pellets. Oxidative stress in submerged cultures of fungi was also recently reviewed by Bai *et al.* [135]. The culture's response to oxidative stress produced by the addition of exogenous H_2O_2 was recently studied [136]. The addition of exogenous H_2O_2 to *A. niger* cultures resulted in shorter hyphae than those of the control culture and enhanced the formation of large clumps. The hyphal growth unit decreased in oxidatively stressed cultures, with the lowest values when H_2O_2 was present.

5.1.4. Carbon dioxide concentration

The production of CO_2 in large amounts is common in some aerobic fermentations as a consequence of cellular respiration. The presence of CO_2 influenced the morphology of several filamentous fungi, and its effect has been reviewed [138]. Under batch cultivation conditions where the pH was maintained at 6.5, low pCO_2 values of 3% and 5% saturation increased the branching frequency of *P. chrysogenum*. Further increasing in this value up to 15–20% gave rise to swollen, stunted, highly branched, and sometimes pelleted mycelia [141, 142]. McIntyre and McNeil studied the effect of CO_2 on the growth morphology of *A. niger* using an image analysis system [139, 140]. They found that during the initial phase of the batch culture (up to 72 h incubation), pCO_2 levels above 7.5% saturation resulted in larger pellets because of the increased hyphal length of the radially protruding hyphae.

5.1.5. Shear force

Shear force is one of the main critical factors affecting the growth morphology of filamentous microorganisms in submerged cultures. The effect of shear stress due to agitation on morphology has been extensively studied. The primary role of bioreactor agitation is to provide improved mixing, heat transfer, and mass transfer. Good bulk mixing of the fermentation broth is needed to minimize nutrient concentration gradients and to ensure adequate flow rates at heat transfer surfaces. In general, agitation can affect both growth forms (filamentous and pellet), but its effect on pellets is more obvious. The physical effects of agitation on pellets are twofold: it chips hyphal fragments from the pellet surface and ruptures pellets. It is generally accepted that pellet size decreases with an increasing agitation rate [108]. High shear during the beginning of cultivation decreases spore aggregation and prohibits large pellet formation. Agitation under a certain level is necessary to increase the dissolved oxygen tension in culture and results in more hyphae branching. The pellets formed under this condition are usually dense and strong [144]. Higher agitation rates beyond a certain level usually promote hyphal cell growth and induce hyphal fragmentation [146, 147]. After the pellet formation phase, high shear results in the disruption of the pellets into fragments and in erosion of the pellet surface, both of which lead to the release of free mycelia into the medium [143]. However, the rapid breakdown of pellets normally will not occur unless there is a steep change in the agitation conditions or a reduction in the physical strength of hyphae caused by nutrient exhaustion. The morphology of individual mycelia is also significantly affected by agitation. Shorter and wider cells in fungal filaments with many branches have been reported under high agitation conditions [85].

5.1.6 Dilution rate

The effect of dilution rate on the cell morphology of *F. graminearum* was studied by Wiebe and Trinci [79]. They observed that fragment concentration decreased and hyphal diameter increased when the dilution rate was increased. The relationship between the dilution rate and growth morphology was also studied by Schrickx *et al.* [148]. During the cultivation of *A. niger* in chemostat cultures at specific growth rates lower than 0.12 h^{-1}, they observed a change in mycelial morphology: the hyphae were more branched and conidiation took place. At specific growth rates higher than 0.12 h^{-1}, the hyphae were less branched and no conidiophores were formed.

5.1.7. Bioreactor type

Yin *et al.* [151] studied the cultivation of *R. oryzae* in a shake flask and in an air-lift bioreactor. When the inoculated spore concentrations were between 2×10^5 and 2×10^6 spores mL^{-1}, cell morphology appeared as reduced fluffy and pellet mycelia, respectively. However, when these were inoculated into the air-lift bioreactor, separated flocks and pellets were obtained. The morphology observed in the air-lift bioreactor was similar to that in flask cultures, but with a larger size. In another study, cultivating *A. awamori* in an airlift-tower-loop bioreactor, which is characterized by low power input and uniform energy dissipation, produced large, loose, globular pellets with a hairy surface [107]. These pellets were stable because of their low density and high flexibility. The internal mass transfer within the pellets was enhanced by the turbulence. The influence of the pulse frequency on the growth morphology of two fungal strains cultivated in a fluidized-bed bioreactor was studied by Moreira *et al.* [150]. Operating at an optimum pulsation frequency has two effects: a narrow pellet-size distribution, which improves fluidization quality, and enhanced product formation. In the case of *A. niger*, the pellet diameter corresponding to the pulse operated at 0.35 s^{-1} was kept at 3.3 ± 0.1 mm after 22 days of operation; however, in the nonpulsed bioreactor, which was operated for only 11 days, pellets with a diameter of 6.7 ± 0.3 mm were produced. Similar results were obtained with *Phanerochaete chrysosporium*. For the pulsing frequency of 0.0625 s^{-1}, a pellet diameter of 1.6 ± 0.3 mm was maintained after 14 days of operation. On the contrary, the system without pulsation had large conglomerates of mycelia with an average size of 3 cm surrounded by free pellets with a diameter distribution of 2.7 ± 0.5 mm. Another interesting study for the cultivation of *A. niger* in a cyclone reactor was done by Kamilakis and Allen [149]. They observed that increasing the speed of the centrifugal pump resulted in significantly shorter cells with a distinct main hypha and several short branches. Increasing the pump speed significantly increased the length of the lag phase of cell growth.

5.1.8. Shape of the cultivation vessel

The cultivation vessel, either small-scale shake flasks or large-scale bioreactors, significantly influences the growth shape. It has been reported that, during cultivation of *A. niger*, the presence of baffles in shake flasks prevented large spore aggregates from forming and resulted in smaller pellets compared to those obtained in shake flasks without baffles under the same cultivation conditions. The process transfer from shake flasks to a bioreactor also resulted in a significant decrease in pellet diameter concomitant with an increase in pellet density [111]. The decreased pellet size in baffled flasks was also reported during *R. chinesis* cultivation [93]. However, the shape of the internal structure also influenced the growth morphology. The influence of two mixing systems, a turbine mixing system (TMS) and a counterflow mixing system (CMS) on the growth morphology of *Fusarium moniliforme*, was studied by Priede *et al.* [153]. A higher proportion of clumped mycelia with clumps of larger area, perimeter, and roughness were observed in the TMS agitated culture. A correlation between the morphology and productivity was found, and TMS favored the development of more productive mycelia with longer and thinner hyphae. Rocha-Valadez *et al.* [154] also showed that the impeller and sparger configuration have a significant effect on the morphology of *Trichoderma harzianum* in submerged cultures.

5.2. Effect on fungal cell autolysis in submerged cultures

Fungal autolysis, which is regarded as a dynamic phase of cell death, influences numerous biotechnological processes, including secondary metabolite and heterologous protein production. This occurs as a result of hydrolase activity, causing vacuolation and disrupting organelles and the cell wall structure. The process is complicated and usually involves interactions among various cultivation parameters, such as shear stress [155] and nutrient limitation [156], as well as various types of autolytic hydrolases, such as proteases [157], glucanases [158], and chitinases [159]. Autolysis in industrial filamentous fungi was recently reviewed by White *et al.* [160]. In the bioprocessing industry, it is usually desirable to prevent autolysis so that the fermentation can be continued without encountering proteolytic degradation and loss of the protein products (both homologous and heterologous) [161]. As an example, autolysis during the production of penicillins and cephalosporins has been correlated with increased extracellular acylase activity [161, 162]. Although the hydrolytic enzymes degrade these antibiotics, this might also be economically advantageous for the production of 6-aminopenicillinic acid and 7-amino-cephalosporanic acid, which are the precursors for semisynthetic antibiotic production. Autolysis may also be desirable due to its promotion of intracellular product excretion and recovery from the mycelia. Although such lysis may assist downstream processing, McNeil *et al.* [164] reported that greatly increased filtration time during cell removal led to complete filter blockage. Such problems could increase processing times, which is a serious concern in the recovery of susceptible products from the fermentation broth.

However, autolysis is a natural part of filamentous fungal bioprocesses, and its onset can be accelerated or put off by both intrinsic (fungal cell related factors) and extrinsic (cultivation conditions and medium composition) factors. In many instances, autolysis is also a means of survival, with portions of the culture existing by recycling lytic products, freed by hydrolases, which can be utilized by actively growing areas, that is, extending hyphal tips. This type of growth is termed cryptic growth, and can occur as a result of nutrient limitation [160]. It was observed in *P. chrysogenum*, which maintained the growth of bi-cellular fragments after extensive fragmentation in liquid culture and prevented a further reduction in cell dry weight with carbon and nitrogen requirements met by amino acid degradation [165]. Cryptic growth has also been observed in chemostat culture of *A. niger* where protein production was maintained without an increase in biomass [146]. Hyphal fragmentation in bioreactors has been described as a function of shear stress, that is, it occurs when the forces of agitation and turbulent flow exceed the tensile strength of the hyphal cell wall [166]. Direct damage to hyphae from mechanical forces is complicated by the natural process of vacuolation in older hyphal compartments, which may eventually lead to autolysis or an increased susceptibility to shear [167]. The breaking point of each hypha is determined by individual characteristics, such as age, physiological condition, and length. Growth and differentiation occur as a result of cytoplasmic translocation towards the hyphal tip, generating a physiological age gradient along the length of a hypha. The vacuole size in each compartment increases with age and vacuolation generally precedes autolysis in distal regions [168]. This vacuolation occurs as the organism attempts to maintain turgor pressure for cytoplasmic streaming towards the growing apex. In general, extensive vacuolation reduces the growth rate, tip extension rate, and branching frequency due to the reduced availability of cytoplasm for subapical branching [44]. Fragmentation can then occur in aged, vacuolated hyphae, as they have a reduced compartmental turgor pressure and tensile strength [169]. In general, some confusion still surrounds the fragmentation process in filamentous fungi, which is definable by first-order kinetics [170]. Hyphal breakage appears to be spontaneous and unaffected by agitation intensity in chemostat cultures of *P. chrysogenum*. Some authors claim that hyphal cell wall strength is constant, while others state that the composition of hyphal cell walls vary along their length [171]. However, it is generally recognized that the effects of mechanical damage are complicated by the vacuolation, age, and size of the hyphal compartment, as well as by the accumulation of toxic metabolites [160]. In a recent study by Emri *et al.* [172], it was observed that the autolysis of *A. nidulans* hyphae was reduced by the exogenous addition of vitamin E, a powerful antioxidant, to the growth medium. Vitamin E supplementation at a concentration of 1 g/L effectively hindered the intracellular accumulation of reactive oxygen species (ROS), the autolytic loss of biomass, the disintegration of pellets, and the release of chitinase. In recent years, cell autolysis and vacuolation during submerged cultivation of different industrially important fungi have also been studied using image analysis systems [173].

6. EFFECTS OF MORPHOLOGY ON PRODUCTION AND SECRETION

In the case of enzyme production, the macro-morphological features of fungi not only affect enzyme production and excretion but also have a significant effect on productivity. A recent study by Raviraja *et al.* [174] showed a direct relationship between pellet size and ergosterol production in different strains of aquatic hyphomycetes. They observed a significant increase in ergosterol concentration with increased pellet sizes. They assumed that this is because the larger pellets have a larger proportion of more fully differentiated hyphae.

6.1. Effects of micromorphology

Secretory proteins begin their journey to the extracellular medium by entering the endoplasmic reticulum (ER). In the ER, proteins are folded and can undergo distinct modifications such as glycosylation, disulfide bridge formation, phosphorylation, and subunit assembly. Subsequently, proteins leave the ER packed in transport vesicles and head to the Golgi compartment, where additional modifications can take place such as further glycosylation and peptide processing. Finally, again packed in secretory vesicles, proteins are directed to the plasma membrane, from where they are secreted. In some cases, the proteins do not reach the extracellular space, but are targeted to an intracellular compartment, such as the vacuole, either to become resident proteins or to undergo proteolytic degradation [175]. Most recent studies indicate that protein secretion occurs at the apical or subapical regions [176]. Recent work has reinforced this hypothesis [177–179]. Using the novel glucoamylase-green fluorescence fusion protein (GLA-GFP) as a secretion reporter to study protein secretion in *A. niger*, Gordon *et al.* [178] observed that GFP fluorescence was predominant at the hyphal apices and showed that this approach is a promising tool for further research in this field, as it allows *in vivo* monitoring of protein secretion. The apical localization of protein secretion has led to the suggestion of employing morphological mutants displaying an increased apical surface, i.e., hyperbranching mutants, as supersecreting strains [179]. Growth morphology can also affect the final product yield of a heterologous protein by indirectly affecting the secretion of extracellular proteases. Xu *et al.* [180] showed a direct relationship between protease secretion and growth morphology in recombinant *A. niger*. The transformation of growth from filamentous to pellet form increased the heterologous protein production as a result of a significant reduction in native fungal proteases. Thus, this bioprocessing strategy can be effectively used to inhibit protease activity in filamentous fungi and thereby enhance heterologous protein production.

6.2. Effects of macromorphology

The relationship between macroscopic growth morphology (pellet or filamentous) and product formation have been reviewed by many authors [42, 180, 181]. This relationship usually varies among different fungal strains and even among different products from the same strain. For example, pellet growth is preferable for the production of pravastatin precursor by *Penicillium citrinum* [183], citric acid by *A. niger* [39], and antibiotic by *R. chinesis* [93]; while dispersed growth is preferable for production of penicillin by *P. chrysogenum* [85], fumaric acid by *R. arrhizus* [110, 111], and enzymes by *A. niger* [24, 184]. On the other hand, heterologous protein production in *A. niger* was increased by pelleted growth [180]. This growth form inhibits extracellular protease production and thus indirectly increases heterologous protein yield. In addition to the growth morphology, the spatial arrangement of cells inside the pellet, the pellet's surface structure, and the cell density inside the pellet can also influence the product formation. The differentiation of mycelia during pellet formation also has striking effects on enzyme production. The production of polygalacturonidase by *A. niger* is well correlated with the mycelial morphology; the more compact the pellet, the more enzyme is synthesized [34]. A high yield of antibiotic production in *R. chinesis* is also related to the formation of less compact, fluffier, looser pellets [93]. Therefore, there is no general theory for this relation. Some products are highly induced when growth is in filamentous form; others are expressed in high titers when growth is in pelleted form.

The structures of *A. niger* pellets found in shake-flask cultures are illustrated in Fig. 5 and Fig. 6. In general, the pellet structure can be characterized by four distinct layers with a hollow center [111]. The outer layer (layer A) consists of a relatively thin but dense mycelial network. The next layer (layer B) is the thickiest and shows a large decrease in mycelial density. The neighboring inner layer (layer C) appears to be of intermediate density and is composed of hyphal cells and non-germinated spores. The next layer (layer D) contains aggregates of non-germinated spores in addition to germinated spores with short hyphal tips. In large pellets (>3.0 mm in diameter), the inner core of pellets appears to be hollow.

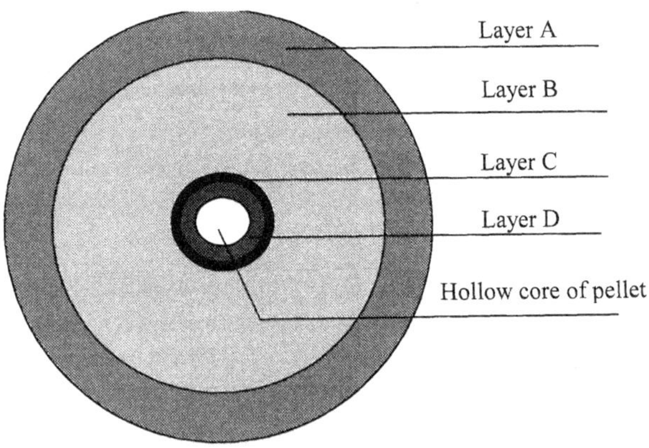

Fig. 5. The general structure of fungal pellet found in shake-flask cultures of *A. niger* [111].

Structural analysis of large fungal pellet and the observation that pellets of larger size and lower density perform less efficiently, clearly demonstrate that the outer denser layer contributes more to metabolite production. The inner, less dense, layer is clearly subjected to substrate limiting conditions resulting in autolysis within the inner part of the pellet. The conclusion that the inner parts of pellet are subjected to substrate limitation or even starvation conditions is additionally corroborated by the presence of fungal spores in the pellet core.

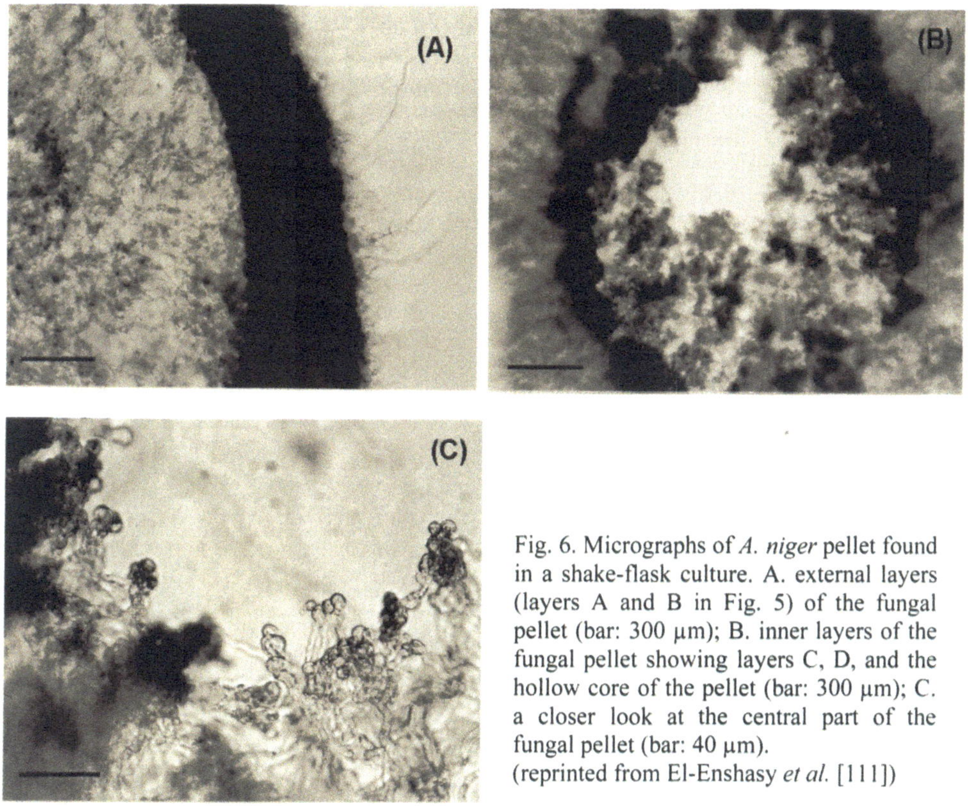

Fig. 6. Micrographs of *A. niger* pellet found in a shake-flask culture. A. external layers (layers A and B in Fig. 5) of the fungal pellet (bar: 300 µm); B. inner layers of the fungal pellet showing layers C, D, and the hollow core of the pellet (bar: 300 µm); C. a closer look at the central part of the fungal pellet (bar: 40 µm).
(reprinted from El-Enshasy *et al.* [111])

Therefore, the relationship between growth morphology, cell physiology and cell productivity is a matter of interest for many researchers. Different studies have been done in recent years to establish the relationship between enzyme production/excretion and cell morphology using new image analysis systems and special staining methods. The physiological status of cells can be determined by a special type of staining. A combination of fluorescence diacetate (FDA) and ethidiumbromide (EB) or propidium iodine (PI) can be used to study the vitality of fungal cell cultures; vital cells fluoresce green and dead cells fluoresce red [65]. Acridine orange (AO) is a very popular dye and was recently used to determine the physiological status of fungal cells (Fig. 7) [63, 65]. It forms green fluorescence-emitting complexes with double stranded RNA (dsRNA) and DNA, and red fluoresce-emitting complexes with single stranded (ss) RNA. This technique was used for the localization of the ssRNA in the fungal mycelium. Based on this method, a direct relationship between cell morphology and productivity in recombinant protein production was determined [63]. The total red fluorescence area, as determined from the image analysis, represents the productive area of the cell mass. This information can be used to estimate the fraction or percentage of the total cell mass that is actively producing the recombinant protein. The fraction of the productive cell mass was also found to be equal to the cell growth yield determined from the changes in the total cell mass during the filamentous growth.

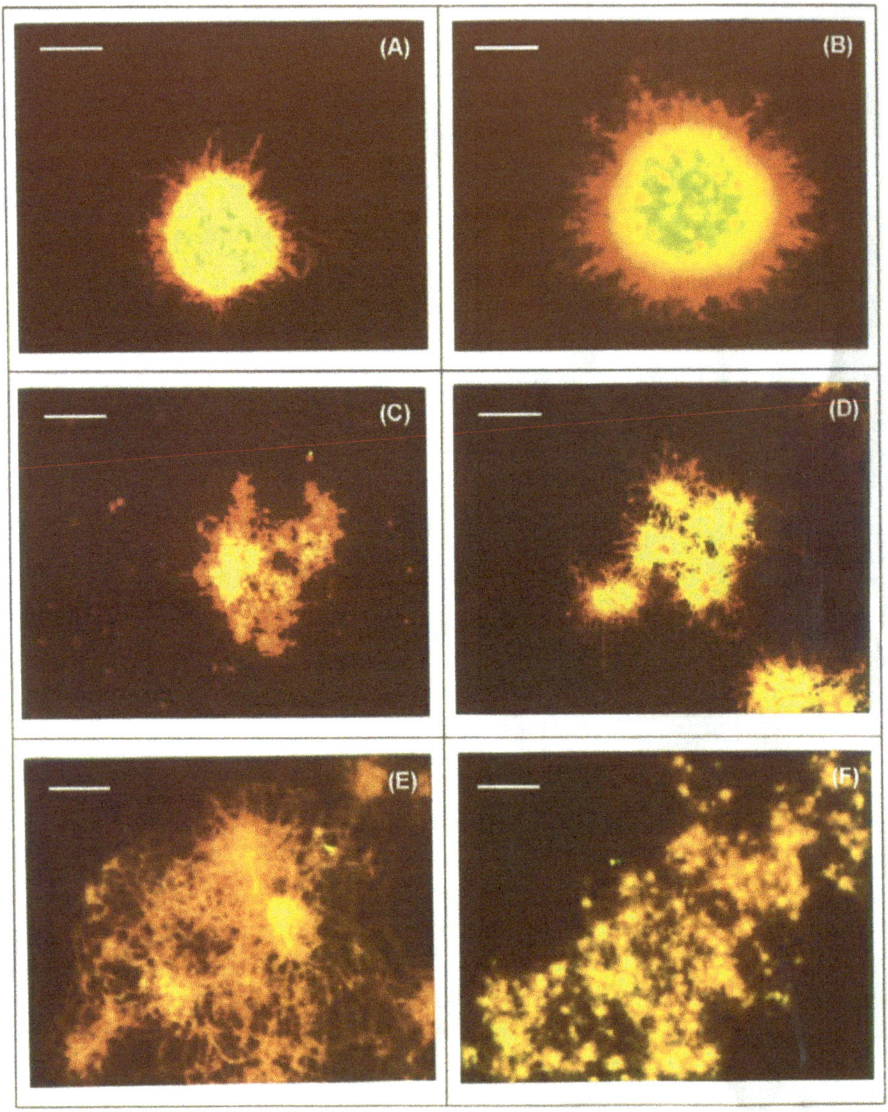

Fig. 7. Fluorescence micrographs of recombinant *A. niger* grown under different agitation speeds at different incubation times. Cells were stained with acridine orange (AO) and seen under a fluorescence microscope. A, C, and E were cells after 19 h growth; and B, D, and F were cells after 25 h growth under agitation speeds of 200, 500 and 800 rpm, respectively. (For A, B, C, D, and F bar = 300 μm and for E bar = 75 μm) (Reprinted from [63] with permission)

7. IMMOBILIZED FUNGAL CELLS

The use of immobilized cells in bioprocesses has many advantages over using them in their free forms. Immobilization allows for easier separation of cells from the fermentation broth, facilitates the isolation and refinement of products, and allows for the repeated use of cells. Moreover, immobilization decreases the cultivation time in repeated batch or continuous operation, which can significantly reduce the production cost. Many papers and reviews on the immobilization of microbial cells have been published [185–188]. In general, there are two methods for immobilization. Entrapment, where microorganisms are entrapped in gel beads such as calcium alginate, and adsorption, where physical or chemical affinity attaches the cells to support materials. Compared to entrapment, adsorption is significantly simpler and cheaper in terms of handling and does not require special apparatus. Adsorption is usually carried out using a porous carrier or a network of woven or non-woven materials. Immobilization using this method can overcome the limitations of entrapment using gel beads. Gel beads are susceptible to shear force and limited by the substrate/air supply to the beads; they are thus considered less effective carriers for large-scale production.

Filamentous microorganisms Immobilized using either a porous carrier, such as SIRAN and celite beads, or adsorption using woven or non-woven materials, such as polyurethane, plant fibers, or glass wool, showed better, more efficient, mass transfer of nutrients and oxygen [189–192]. Different fungal strains were successfully immobilized for the production of a wide range of primary and secondary metabolites; for example, *A. niger* was immobilized on different matrices for the production of glucose oxidase [193]. Immobilized fungal cells are widely used to produce different types of organic acids. In repeated batch culture of *A. niger* immobilized in porous cellulose carriers (Microcube), it was found that citric acid can be semicontinuously produced with a high productivity [194, 195]. For the continuous production of gluconic acid by *A. niger,* immobilization in a glass wool network proved to be a good alternative system to the conventional cultivation method of free cells [196]. Immobilized cells were able to produce a high concentration of gluconic acid in a shorter time compared to free cells. In repeated batch fermentations, using immobilized cells reduced the batch fermentation time from 48 h for free cell fermentations to only 24 h.

Moreover, the immobilization of recombinant fungi is a suitable technique for increasing the production of heterologous proteins. Fungal production of extracellular proteases, which is a significant inhibitor of heterologous products, was found to be significantly reduced after cell immobilization [197], and therefore, the use of recombinant fungi in the immobilized form indirectly increased the production of some heterologous proteins [190, 197]. Furthermore, immobilization can increase strain stability and decrease the possibility of recombinant strain washout from the microbial population.

In the case of filamentous fungal cells, immobilization can also solve the problems of cell morphology to a large extent. *R. oryzae* immobilized in a rotating fibrous-bed bioreactor resulted in good control over the filamentous morphology and allowed for the long-term production of lactic acid from glucose and starch in repeated batch and fed-batch fermentations, achieving a high lactic acid concentration of more than 120 g L^{-1} and a lactic acid yield of more than 90% [198]. By immobilizing fungal mycelia on the fibrous matrix, a cell-free fermentation broth was obtained, which improved oxygen transfer and lactic acid production [199]. Proper immobilization of the fungal mycelia can also increase the total cell surface exposed to the culture medium and thus increase the fraction of productive cells.

8. FUTURE OF FILAMENTOUS FUNGAL CELLS AS BIOFACTORIES

Filamentous fungal cells are major biofactories of the past, the present, and the future. Advances in recent years in the production of many homologous and heterologous products make them a good choice for the manufacturing of high value products in many new application areas. The past 20 years have been a period of progress on several fronts with filamentous fungi as cell factories for food, feed, and pharmaceutical industries. Filamentous fungi offer a very attractive, safe, and cheap expression system for the high-level production of heterologous proteins. Despite their advantages over other cell factories, the use of filamentous fungi as hosts for production of heterologous proteins has not advanced as quickly as expected a few years ago, mainly because the yields and titers of non-fungal proteins have generally been disappointingly low. Various reasons for the low yields of non-fungal products have been suggested, including incorrect folding or processing of the protein, up-regulation of the unfolded protein response (UPR), and proteolytic degradation [161, 200, 201]. In spite of these problems, some antibody fragments and antibody fusion proteins have been produced in filamentous fungi [202]. This opens a new era for antibody production from non-mammalian cells. Nowadays, the therapeutic use of glycoproteins produced in filamentous fungi is very limited, as the glycans formed are dissimilar from mammalian glycans. More knowledge and insight into the glycosylation pathway is needed in order to engineer the glycosylation pathway to mimick the mammalian type of glycosylation [203]. The genomes of several species are being sequenced at some level (expressed sequence tags, genomic libraries, individual chromosomes, and complete genomes). Gene microarrays are also being produced and growing progressively. Thus, the new era will enable a much faster production time for several new heterologous products produced by fungal cell factories. Another field of research with considerable promise is that of -omics technologies. These technologies deal with the overall analysis of gene expression (transcriptomics), protein (proteomics), and metabolite production (metabolomics) at the complete organism level [3]. These techniques require well equipped laboratories with sophisticated techniques and the development of data-analysis and pattern-recognition tools (bioinformatics). On the other hand, the relationship between growth morphology and the expression of homologous and heterologous products using all these -omics technologies is one of the future methods for optimizing the cultivation strategies of filamentous fungi.

REFERENCES

[1] D.J. Jeenes; D.A. MacKenzie; D.A. Roberts, and D.B. Archer, Biotechnol. Gen. Eng. Rev., 9 (1991) 327
[2] R.J. Gouka; P.J. Punt and C.A.M.J.J. Van den Hondel, Appl. Microbiol. Biotechnol., 47 (1997) 1.
[3] P.J. Punt,; N. Van Biezen; A. Conesa; A. Albers; J. Mangnus and C. Van den Hondel, Trends Biotechnol., 20 (2002). 200.

[4] D.B. Finkelstein, D. B, Antonie Van Leeuvenhook J. 53 (1987) 349.
[5] H. Durand; M. Clanet and G. Tiraby, Enz. Microb. Technol. 10 (1988) 341.
[6] M. Devchand and D.I. Gwynne, J. Biotechnol. 17 (1991) 3.
[7] C.A.M.J.J. Van den Hondel; P.J. Punt and R.F.M. Van Gorcom, Antonie Van Leeuwenhoek 61 (1992) 153.
[8] J.C. Verdoes; P.J. Punt and C.A.M.J.J. Van den Hondel, Appl. Microbiol. Biotechnol. 43 (1995) 195.
[9] D.B. Archer, Curr. Opinion Biotechnol. 11 (2000) 478.
[10] J.F. Peberdy; C.E. Caten; J.E. Ogden and J.W. Bennett (eds.), Applied Molecular Genetics of Fungi, Cambridge University Press, 1991, (pp. 103-117).
[11] J. Nicaud; N. Mackman and I. Holland, J. Biotechnol. 3 (1986) 255.
[12] Z. Wang and N.A. Da Silva, Biotechnol. Bioeng. 42 (1993) 95.
[13] N.S. Dunn-Coleman; P. Bloebaum; R.M. Berka; E. Bodie; N. Robinson; G. Armstrong; M. Ward; M. Przetak; G.L. Carter; R. La Cost; L.J. Wilson; K.H. Kodama; E.T. Baliu; B. Bower; M. Lamsa and H. Heinsohn, Bio/Technology 9 (1991) 976.
[14] P.A. Gibbs; R.J. Seviour and F. Schmid, Crit. Rev. Biotechnol. 20 (2000). 17.
[15] J.F. Peberdy, Trends Biotechnol. 12 (1994). 50.
[16] H.A.B. Wösten; S.M. Moukha; J.H. Sietsma and J.G.H. Wessels, J. Gen. Microbiol. 137 (1991) 2017.
[17] S. High, BioEssays 14 (1992) 535.
[18] J.F. Peberdy, J.F. and L. Ferenczy, (eds.), Applications in Biochemistry and Genetics, Marcel Dekker Inc., New York, Basel, 1985 (pp. 3-29)
[19] J.H. Burnett and A.P.J. Trinci, (eds.), Fungal Walls and Hyphal Growth. Cambridge Univ. Press. London, New York, Melbourne, 1979, (pp. 1-25).
[20] C.M. Stagg and M.S. Feather, Biochem. Biophys. Acta. 320 (1973) 64.
[21] J. Trevithick and R. Metzenberg, J. Bacteriol. 92 (1966) 1010.
[22] P.L.Y. Chang and J.R. Trevithick, Arch. Microbiol. 101 (1974) 281.
[23] Heath, I.B. (ed.), Tip Growth in Plant Fungal Walls, Academic Press, San Diego, 1990, (pp. 1-29).
[24] D.A. MacKenzie; L.C.G. Gendron; D.J. Jeenes and D.B. Archer, Enz. Microb. Technol. 16 (1994) 276.
[25] C.H. Bigger; M.R. White and H.D. Braymer, J. Gen. Microbiol. 71 (1972) 159.
[26] G. Martinez-Cadena and J. Ruiz-Herrara, Arch. Microbiol. 148 (1987) 280.
[27] J.C. Van Suijdam and B. Metz, Biotechnol. Bioeng. 23 (1981) 111.
[28] S. Braun and S. Vecht-Lifshitz, Trends Biotechnol. 9 (1991) 63.
[29] J.C. Van Suijdam; N.W. F. Kassen and P.G. Paul, Eur. J. Appl. Microbiol. Biotechnol. 10 (1980) 211.
[30] H. Kobayashi; G. Van Dedem and M. Moo.Young, Biotechnol. Bioeng. 15 (1973) 27.
[31] B. König and K. Schügerl, Biotechnol. Bioeng. 24 (1982) 259.
[32] M. Reuss, M.; H. Chmiel, H.; E.D. Gilles, and H.J. Knackmuss, (eds.), Biochemical Engineering, Gustav Fischer, Stuttgart, New York., 1991, (pp. 393-396).
[33] F. Pottel and K.H. Bellgardt, Proc. DECHEMA Biotechnol. Conf. 5 (1992) 381.
[34] H. Hermersdörfer; A. Leuchtenberger; Ch. Wardsack and H. Ruttloff, J. Basic Microbiol. 27 (1987) 309.
[35] K. Schügerl; R. Wittler and T. Lorenz, Trends Biotechnol. 1 (1983) 120.
[36] R. Wittler; H. Baumgartl; D.W. Lübbers and K. Schügerl, Biotechnol. Bioeng. 28 (1986) 1024.
[37] J.C. Van Suijdam; H. Hold and N.W.F. Kossen, Biotechnol. Bioeng. 24 (1982) 177.
[38] S. Lucatero; C.P. Larralde-Corona; G. Corkidi and E. Galindo, Biotechnol. Prog. 19 (2003) 285.
[39] B. Metz and N.W.F. Kossen, Biotechnol. Bioeng. 19 (1977) 781.
[40] J. Nielsen, Trends Biotechnol. 14 (1996) 438.
[41] M. Pazouki and T. Panda, Bioprocess Eng. 22 (2000) 127.
[42] M. Papagianni, Biotechnol. Adv. 22 (2004) 189.
[43] C.R. Thomas and G.C. Paul, Curr. Opin. in Biotechnol. 7 (1996) 35.
[44] T. Agger; A.B. Spohr; M. Carlsen and J. Nielsen, Biotechnol. Bioeng. 57 (1998) 321.

[45] L.H. Grimm; S. Kelly; J. Hengstler; A. Göbel; R. Krull and D.C. Hempel, Biotechnol. Bioeng. 87 (2004) 213.
[46] E.V. Morozova; M.V. Baranova; V.P. Kozlov; V.M. Tereshina; A.S. Memorskaya and E.P. Feofilova, Mikrobiologiya. 70 (2001) 527.
[47] F. Claverie-Martin; M. Diaz-Torres and M.J. Geoghegan, Curr. Microbiol. 14 (1986) 221.
[48] J.G.H. Wessels, Adv. Microb. Physiol. 38 (1997) 1.
[49] B.C. Van der Aa; M. Asther and Y.F. Dufrêne, Colloids Surf. B: Biointerface. 24 (2002) 277.
[50] U. Fuchs; K.J. Czymmek and J.A. Sweigard, Fungal Gen. and Biol. 41 (2004) 852.
[51] C. d'Enfert; B.M. Bonini; P.D.A. Zapella; T. Fontaine; A.M. da Silva and H.F. Terenzi, Mol. Microbiol. 32 (1999) 471.
[52] H.-J. Bussink and S.A. Osmani, FEMS Microbiol. Lett. 173 (1999) 117.
[53] G.C. Paul; C.A. Kent and C.R. Thomas, Biotechnol. Bioeng. 42 (1993) 11.
[54] A. Bosch; R.A. Maronna and O.M. Yantorno, Process Biochem. 30 (1995) 599.
[55] J. Barrios-González; C. Martínez; A. Aguilera and M. Raimbault, Biotechnol. Lett. 11 (1989) 551.
[56] J.G. Anderson and J.E. Smith, Can. J. Microbiol. 18 (1971) 289.
[57] R.D. Medwid and D.W. Grant, Appl. Env. Microbiol. 48 (1984) 1067.
[58] M. Plascencia-Jatomea; G. Viniegra; R. Olayo; M. Castillo-Ortega and K. Shirai, Macromol. Biosci. 3 (2003) 582.
[59] G.A. Munoz; E. Agosin; M. Cotoras; R. San Martin and D. Volpe, FEMS Microbiol. Lett. 125 (1995) 63.
[60] K.-C. Feng; B.-L. Liu, and Y.-M. Tzeng, World J. Microbiol. Biotechnol. 18 (2002) 217.
[61] E. Zeuthen, E. (ed.) Synchrony in Cell Division and Growth, Interscience Publishers (New York, London, Sydney), 1964, (pp 391-420).
[62] T. Tsukahara, Sabouraudia. 6 (1968) 185.
[63] H.A. El Enshasy, Ph.D. Thesis. Technische Universität Carolo-Wilhelmina, Braunschweig, Germany. (1998). www.biblio.tu-bs.de/ediss/data/19980514a/19980514a.html
[64] K.-B. Oh; Y. Chen; H. Matsuoka; A. Yamamoto and H. Kurata, J. Biotechnol. 45 (1996) 71.
[65] S. Freudenberg; K.I. Fasold; S.R. Müller; D. Siedenberg; G. Kretzmer; K. Schügerl and M. Giuseppin, J. Biotechnol. 46 (1996) 265.
[66] N. Osherov and G. May, Genetics 155 (2000) 647.
[67] S. Khunyoshyeng; S. Cheevadhanarak; S. Rachdawong and M. Tanticharoen, Fungal Gen. Biol. 37 (2002) 13.
[68] K. Esser and P.A. Lemke, (eds.), The Mycota Vol. 1, Springer-Verlag, Berlin, 1994, (pp 89-109).
[69] G.D. Robson; E. Prebble; A. Rickers; S. Hosking; D.W. Denning; A.P.J. Trinci and W. Robertson, Fungal Genet. and Biol. 20 (1996) 289.
[70] J. Schmid and F.M. Harald, J. Gen. Microbiol. 134 (1988) 2623.
[71] S.D. Harris; A.F. Hofmann; H.W. Tedford and M.P. Lee, Genetics. 151 (1999) 1015.
[72] I.M. Gatherar; S. Pollerman; N. Dunn-Coleman and G. Turner, Fungal Gen. and Biol. 41 (2004) 463.
[73] M. Riquelme; R.W. Roberson; D.P. McDaniel and S. Bartnicki-Gracia, Fungal Gen. Biol. 37 (2002) 171.
[74] M. Momany, Curr. Opin. Microbiol. 5 (2002) 580.
[75] S.L. Jackson and I.B. Heath, Microbiol. Rev. 57 (1993) 367.
[76] K.S. Bruno; R. Aramayo; P.F. Minke; R.L. Metzenberg and M. Plamann, EMBO J. 15 (1996) 5772.
[77] G.M. Guest; X. Lin and M. Momany, Fungal Gen. Biol. 41 (2004) 13.
[78] C. Müller; A.B. Spohr and J. Nielsen, Biotechnol. Bioeng. 67 (2000) 390.
[79] M.G. Wiebe and P.J. Trinci, Biotechnol. Bioeng. 38 (1991) 75.
[80] J. Dynesen and J. Nielsen, Fungal Gen. Biol. 40 (2003) 15.
[81] B. Jin; J.H. Van Leeuwen and B. Patel, Processs. Biochem. 34 (1999) 335.
[82] H. Elmayergi; J.M. Scharer and M. Moo-Young, Biotechnol. Bioeng. 15 (1973) 845.
[83] A. Mitard and J.P. Riba, Biotechnol. Bioeng. 32 (1988) 835.
[84] S. Hotop; J. Möller; T. Dullau and K. Schügerl, DECHEMA Biotechnol. Conf. 3 (1989) 597.

[85] J.J. Smith; M.D. Lilly and R.I. Fox, Biotechnol. Bioeng. 35 (1990) 1011.
[86] J. Nielsen; C. Johansen; M. Jacobsen; P. Krabben and J. Villadsen, Biotechnol. Prog. 11 (1995) 93.
[87] D.E. Brown and M.A. Zainudeen, Biotechnol. Bioeng. 20 (1978) 1045.
[88] Z. Molnár; E. Mészáros; Z. Szilágyi; S. Rosén; T. Emri and I. Pócsi, Appl. Biochem. Biotechnol. 118 (2004) 349.
[89] C. Müller; C.M. Hjort; K. Hansen and J. Nielsen, Microbiology. 148 (2002) 4025.
[90] M. Papagianni and M. Moo-Young, Process Biochem. 37 (2002) 1271.
[91] K.G. Tucker and C.R. Thomas, Biotechnol. Lett. 14 (1992) 1071.
[92] G.A. Jiménez-Tobon; M.J. Penninckx and R. Lejeune, Enzyme Microb. Technol. 21 (1997) 537.
[93] L.-X. Du; S.-J. Jia and F.-P. Lu, Process Biochem. 38 (2003) 1643.
[94] P. Žnidaršič; R. Komel and A. Pavko, J. Biotechnol. 60 (1998) 207.
[95] D.-M. Bai; M.-Z. Jia; X.M. Zhao; R. Ban; F. Shen; X.-G. Li and S.-M. Xu, Chem. Eng. Sci. 58 (2003) 785.
[96] S. Hotop; J. Möller; J. Niehoff and K. Schügerl, Process Biochem. 28 (1993) 99.
[97] E. Olsvik; K.G. Tucker; C.R. Thomas and B. Kristiansen, Biotechnol. Bioeng. 42 (1993) 1046.
[98] S.E. Vecht-Lifshitz; S. Magdassi and S. Braun, Biotechnol. Bioeng. 35 (1990) 890.
[99] D. Katz; D. Goldstein and R. Rosenberg, J. Bacteriol. 109 (1972) 1097.
[100] R.C. Righelato; A.P.J. Trinci and S.J. Pirt, J. Gen. Microbiol. 50 (1968) 399.
[101] K.B. Morrison and R.C. Righelato, J. Gen. Microbiol. 81 (1974) 517.
[102] D. Ryoo, Biotechnol. Techniq. 13 (1999) 33.
[103] H. El-Enshasy; K. Hellmuth and U. Rinas, Appl. Biochem. Biotechnol. 90 (2001) 57.
[104] J. Sinha; J.T. Bae; J.P. Park; C.H. Song and J.W. Yun, Enz. Microb. Technol. 29 (2001) 392.
[105] Y.J. Cho; H.J. Hwang; S.W. Kim; C.H. Song and J.W. Yun, J. Biotechnol. 95 (2002) 13.
[106] R.J. Swift; A. Karandikar; A.M. Griffen; P.J. Punt; C.A.M.J.J. Van den Hondel; G.D. Robson; A.P.J. Trinci and M.G. Wiebe, Fungal Gen. Biol. 31 (2000) 125.
[107] S.R. Gerlach; D. Siedenberg; D. Gerlach; K. Schügerl; M.L.F. Giuseppin and J. Hunik, Process. Biochem. 33 (1998) 601.
[108] J. I. Prosser and A.J. Tough, Crit. Rev. Biotechnol. 10 (1991) 253.
[109] Y. Koike; H.J. Cai; K. Higashiyama; S. Fujikawa and E.Y. Park, J. Biosci. Bioeng. 91 (2001) 382.
[110] G.S. Byrne and O.P. Ward, Biotechnol. Bioeng. 33 (1989) 912.
[111] H. El Enshasy; K. Hellmuth and U. Rinas, Appl. Biochem. Biotechnol. 81 (1999) 1.
[112] G.S. Byrne and O.P. Ward, J. Ind. Microbiol. 4 (1989) 155.
[113] F.C. Domingues; J.A. Queiroz; J.M.S. Cabral and L.P. Fonseca, Enz. Microb. Technol. 26 (2000) 394.
[114] P. Jones; D. Moore and A.P.J. Trinci, J. Gen. Microbiol. 134 (1988) 235.
[115] H. Elmayergi, J. Ferm. Bioeng. 53 (1975) 722.
[116] S. Rugsaseel; S. Morikawa; K. Kirimura and S. Usami, Appl. Microbiol. Biotechnol. 42 (1995) 839.
[117] H. Kobayashi and H. Suzuki, J. Ferm. Technol. 50 (1972) 625.
[118] A.G. Pedersen; M.B. Nielsen; J. Nielsen; J. Villadsen and O. Hassager, Biotechnol. Bioeng. 41 (1993) 162.
[119] M.P. Wainwright; A.P.J. Trinci, and D. Moore, Mycol. Res. 97 (1993) 801.
[120] L.M. Pera and D.A. Callieri, Folia Microbiol. 42 (1997) 551.
[121] I.-U. Haq; S. Ali; M.A. Qadeer and J. Iqbal, Process Biochem. 37 (2002) 1085.
[122] I.-U. Haq; S. Ali; M.A. Qadeer and J. Iqbal, Bioresource Technol. 93 (2004) 125.
[123] T.L. Da Silva; H.M. Pinheiro and J.C. Roseiro, Enz. Micorb. Technol. 32 (2003) 880.
[124] I. Serrano; T.L. Silva and J.C. Roseiro, Lett. Appl. Microbiol. 22 (2001) 89.
[125] W. Jianlong, Process Biochem. 35 (2000) 1079.
[126] L-S.T. Lai; T.-H. Tsai and T.C. Wang, J. Biosci. Bioeng. 94 (2002) 453.
[127] Y.Q. Cui; J.N.W. Ouwehand; R.G.J.M. Van der Lans; M.L.F. Giuseppin and K.C.A.M. Luyben, Enz. Microb. Technol. 23 (1998) 168.
[128] M. Papagianni; S.E. Nokes and K. Filer, Process Biochem. 35 (1999) 397.

[129] M.G. Wiebe; G.D. Robson; S.G. Oliver and A.P.J. Trinci, Biotechnol. Bioeng. 51 (1996) 61.
[130] E.S. Olsvik and B. Kristiansen, Biotechnol. Bioeng. 40 (1992) 375.
[131] Y.Q. Cui; R.G.J.M. Van der Lans and K.C.A.M. Luyben, Biotechnol. Bioeng. 57 (1998) 409.
[132] B. L. A. Carter and A. T. Bull, J. Gen. Microbiol. 65 (1971) 265.
[133] K. Zetelaki and K. Vas, Biotechnol. Bioeng. 10 (1968) 45.
[134] A. Wongwicharn; B. McNeil and L.M. Harvey, Biotechnol. Bioeng. 65 (1999) 416.
[135] Z. Bai; L.M. Harvery and B. McNeil, Crit. Rev. Biotechnol. 23 (2003) 267.
[136] M. Kreiner; L.M. Harvey and B. McNeil, J. Biotechnol. 100 (2003) 251.
[137] Z. Bai; L.M. Harvey; S. White and B. McNeil, Enz. Microb. Technol. 34 (2004) 10.
[138] M. McIntyre and B. McNeil, Appl. Microbiol. Biotechnol. 50 (1998) 291.
[139] M. McIntyre and B. McNeil, Enz. Microb. Technol. 20 (1997) 135.
[140] M. McIntyre and B. McNeil, Enz. Microb. Technol. 21 (1997) 479.
[141] C.S. Ho and M.D. Smith, Biotechnol. Bioeng. 28 (1986) 668.
[142] M.D. Smith and C.S. Ho, J. Biotechnol. 2(1985) 347.
[143] A, P.Shamlou; H.Y. Makagiansar; A.P. Ison; M.D. Lilly and C.R. Thomas, Chem. Eng. Sci. 49 (1994) 2621.
[144] Y.Q. Cui; R.G.J.M. Van der Lans and K.C.A.M. Luyben, Biotechnol. Bioeng. 55 (1997) 715.
[145] B. Metz; E.W. Bruijn and J. C. Van Suijdam, Biotechnol. Bioeng. 23 (1981) 149.
[146] A. Amanullah; R. Blair; A.W. Nienow and C.R. Thomas, Biotechnol. Bioeng. 62 (1999) 434.
[147] A.Amanullah; P. Jüsten; A. Davies; G.C. Paul; A.W. Nienow and C.R. Thomas, Biochem. Eng. J. 5 (2000) 109.
[148] J.M. Schrickx; A.S. Krave; J.C. Verdoes; C.A.M.J.J. Van den Hondel; A.H. Stouthamer and H.W. Van Verseveld, J. Gen. Microbiol. 139 (1993) 2801.
[149] E.G. Kamilakis and G.D. Allen, Process. Biochem. 30 (1995) 353.
[150] M.T. Moreira; A. Sanromán; G: Feijoo and J.M. Lema, Enz. Microb. Technol. 19 (1996) 261.
[151] P. Yin; K. Yahiro; T. Ishigaki; Y. Park and M. Okabe, J. Ferm. Bioeng. 85 (1998) 96.
[152] M. Papagianni; M. Mattey and B. Kristiansen, J. Biochem. Eng. 2 (1998) 197.
[153] M.A. Priede; J.J. Vanags; U.E. Viesturs; K.G. Tucker; W. Bujalski and C.R. Thomas, Biotechnol. Bioeng. 48 (1995) 266.
[154] J.A. Rocha-Valadez; E. Galindo and L. Serrano-Carreón, Bioprocess Eng. 23 (2000) 403.
[155] L.M. Harvey; B. McNeil; D.R. Berry and S. White, Enz. Microb. Technol. 22 (1998) 448.
[156] T. Pusztahelyi; I. Pócsi; J. Kozma and A. Szentirmai, Biotechnol. Appl. Biochem. 25 (1997) 81.
[157] M. McIntyre; D.R. Berry and B. McNeil, Appl. Microbiol. Biotechnol. 53 (2000) 235.
[158] S.M. Pitson; R.J. Seviour and B.M. McDougall, Can. J. Microbiol. 43 (1997) 432.
[159] J. Rodriguez; M.J. Santos; J.L. Copa-Patino and M.I. Perez-Leblic, Lett. Appl. Microbiol. 16 (1993) 69.
[160] S. White; M. McIntyre; D. Berry and B. McNeil, Crit. Rev. Biotechnol. 22 (2002) 1.
[161] D.B. Archer; D.A. MacKenzie; D.A. Jeenes and Robert, Biotechnol. Lett. 14 (1992) 357.
[162] F. Reyes; J. Martinez; C. Alfonso; J.L. Copa-Patino and J. Soliveri, J. Pharm. Pharmacol. 42 (1990) 128.
[163] S. White, Ph.D. Thesis, University of Strathcylde, United Kingdom. (1999).
[164] B. McNeil; D.R: Berry; L.M. Harvey; A. Grant and S. White, Biotechnol. Bioeng. 57 (1998) 297.
[165] T. Pusztahelyi; I. Pócsi and A. Szentirmai, Biotechnol. Appl. Biochem. 25 (1997) 87.
[166] P. Kraben and J. Nielsen, Adv. Biochem. Eng. Biotechnol. 60 (1998) 125.
[167] G.C: Paul; C.A. Kent and C.R. Thomas, Biotechnol. Bioeng. 44 (1994) 655.
[168] P.W. Cox; G.C. Paul and C.R. Thomas, Microbiology 144 (1998) 817.
[169] Y.Q. Cui; W.J. Okkerse; R.G.J.M. Van der Lans and K.C.A.M. Luyben, Biotechnol. Bioeng. 60 (1998) 216.
[170] G.C. Paul and C.R. Thomas, Biotechnol. Bioeng. 51 (1996) 558.
[171] C.L. Johansen; L. Coolen and J.H. Hunik, Biotechnol. Prog. 14 (1998) 233.
[172] T. Emri; Z. Molnár; T. Pusztahelyi; S. Rosén and I. Pócsi, Appl. Biochem. Biotechnol. 118 (2004) 337.
[173] P. Jüsten; G.C. Paul and C.R. Thomas, Biotechnol. Tech. 8 (1994) 853.
[174] N.S. Raviraja; L.G. Nikolcheva and F. Bärlocher, Mycologia 96 (2004) 388.

[175] A. Conesa; P.J. Punt; N. Van Luijk and C.A.M.J.J. Van den Hondel, Fungal Gen. Biol. 33 (2001) 155.
[176] D.B. Archer and J.F: Peberdy, Crit. Rev. Biotechnol. 17 (1997) 273.
[177] C.L. Gordon; D.B. Archer; D.J. Jeenes; J.H. Doonan; B. Wells; A.P. Trinci and G.D. Robson, J. Microbiol. Methods. 42 (2000) 39.
[178] C.L. Gordon; V. Khalaj; A.F. Ram; D.B. Archer; J.L. Brookman; A.P. Trinci; D.J. Jeenes; J.H. Doonan; B. Wells; P.J. Punt; C.A. Van den Hondel and G.D. Robson, Microbiology 146 (2000). 415.
[179] I.H. Lee; R.G. Walline and M. Plamann, Mol. Microbiol. 29 (1998) 209.
[180] J. Xu; L. Wang; D. Ridgway; T. Gu and M. Moo-Young, Biotechnol. Prog. 16 (2000) 222.
[181] K. Schügerl; S.R. Gerlach and D. Siedenberg, Adv. Biochem. Eng. 60 (1998) 193.
[182] P. Žnidaršič and A. Pavko, Food Technol. Biotechnol. 39 (2001) 237.
[183] M. Hosobuchi; F. Fukui; H. Matsukawa; T. Suzuki and H. Yoshikawa, J. Ferm. Bioeng. 76 (1993) 476.
[184] A. Wongwicharn; B. McNeil and L.M. Harvey, Enz. Microb. Technol. 24 (1999) 489.
[185] P.K. Walsh and D.M. Malone, Biotechnol. Adv. 13 (1995) 13.
[186] Y. Kourkoutas; A. Bekatorou; I.M. Banat; R. Marchant and A.A. Koutinas, Food Microbiol. 21 (2004) 377.
[187] P. Gerin; M.W. Bellon-Fontaine; M. Asther and P.G. Rouxhet, Biotechnol. Bioeng. 47 (1995) 677.
[188] B. Katzbauer; M. Narodoslawsky and A. Moser, Bioprocess Biosys. Eng. 12 (1995) 173.
[189] Y.-S. Oh; J. Maeng and S.-J. Kim, Appl. Microbiol. Biotechnol. 54 (2000) 418.
[190] M. Gyamerah; G. Merichetti; O. Adedayo; J.M. Scharer and M. Moo-Young, Appl. Microbiol. Biotechnol. 60 (2002) 403.
[191] N.V. Sankpal and B.D. Mulkari, Process Biochem. 37 (2002) 1343.
[192] M.A. Farid; A.I. El-Diwany and H.A. El-Enshasy, Acta Biotechnol. 14 (1994) 67.
[193] J. Fiedurek and Z. Ilczuk, World J. Microbiol. Biotechnol. 7 (1991) 379.
[194] N. Fujii; K. Yasuda and M. Sakakibara, J. Ferm. Bioeng. 78 (1994). 389.
[195] M. Papagianni and M. Mattey, Process. Biochem. 39 (2004) 1963.
[196] H. El-Enshasy, Deutsche Lebensmittel Rundschau. 99 (2003) 409.
[197] F. Liu; W. Li; D. Ridgway; T. Gu and M. Moo-Young, Biotechnol. Lett. 20 (1998) 539.
[198] A. Tay and S.-T. Yang, Biotechnol. Bioeng. 80 (2002) 1.
[199] N. Thongchul and S.-T. Yang, Controlling filamentous fungal morphology by immobilization on a rotating fibrous matrix to enhance oxygen transfer and L(+)-lactic acid production by *Rhizopus oryzae*. B.C. Saha (ed.), ACS Symposium Series 862, Fermentation Biotechnology, Oxford University Press, New York, 2003, pp. 36-51.
[200] J.P.T.W. Van den Hmobergh; P.J.I. Van de Vondervoort; L. Fraissinet-Tachet and J. Visser, Trends Biotechnol. 15 (1997) 256.
[201] M.G. Wiebe; A. Karandikar; G.D. Robson; A.P.J. Trinci; J-L.F. Candia; S. Trappe; G. Wallis; U. Rinas; P.M. Derkx; S.M. Madrid; H. Sisniega; I. Faus; R. Montijn; C.A.M.J.J. Van den Hondel and P.J. Punt, Biotechnol. Bioeng. 76 (2001) 164.
[202] V. Joosten; C. Lokman; C.A.M.M.J. Van den Hondel and P.J. Punt, Microbial Cell Factories 2 (2003). 1.
[203] M. Maras; I. Van Die; R. Contreras and C.A.M.J.J. Van den Hondel, Glycoconjugate J. 16 (1999) 99.

Bioprocessing for Value-Added Products from Renewable Resources
Shang-Tian Yang (Editor)

Chapter 10. Plant Cell and Hairy Root Cultures – Process Characteristics, Products, and Applications

Wei Wen Su[a] and Kung-Ta Lee[b]

[a]Department of Molecular Biosciences and Bioengineering, University of Hawaii, Honolulu, HI 96822, USA

[b]Department of Biochemical Science and Technology, National Taiwan University, Taipei 106, Taiwan

1. INTRODUCTION

The ability to cultivate plant callus cells and organs (such as roots and shoots) in liquid media has laid the foundation for many innovative and crucial technologies in the plant sciences. It has also made an important contribution to modern plant biotechnology. One of the major biotechnological applications of plant cell culture is producing useful compounds, including small molecules (mostly secondary metabolites) as specialty chemicals, as well as macromolecules, including recombinant proteins and polysaccharides. In this context, the two most widely studied culture systems for producing useful compounds are suspension cells and hairy roots. These culture systems may be operated using technologies similar to those employed in conventional industrial fermentation. However, cultured plant cells and hairy roots possess many of their own distinctive properties which require approaches uniquely different from those used for their mammalian or microbial counterparts in developing large-scale industrial culture processes. The upstream bioprocessing (bioreactor design and cultivation strategies) of plant cell and hairy root cultures has been the subject of several comprehensive reviews [1–6]. There are fewer published reports on large-scale downstream processing for purification/recovery of either secondary metabolites or recombinant proteins, and they are mostly on whole plants [7] rather than cultured plant cells. Nonetheless, much can be learned from the ample published data on the analytical-scale purification of plant secondary metabolites or endogenous plant enzymes. In the open literature, process design (integration of both upstream and downstream processing schemes) based on plant cell cultures has been discussed mainly in conceptual terms. Some reports have presented simplified schematic flowcharts of the actual bioprocesses [8, 9], while a handful of reports have addressed the cost analysis issue by conducting detailed design calculations based on hypothetical plant cell culture processes [10, 11].

The purpose of this chapter is to provide a comprehensive review of the current state of knowledge in using *in vitro* plant cultures for producing macromolecules (with an emphasis on using cultured plant cells) and small molecules (with an emphasis on using hairy roots). This review focuses on the latest advances in applied cell physiology and process techniques pertinent to large-scale industrial bioprocess development. In addition, technological innovations are proposed for the improved utilization of renewable resources in industrial plant cell culture processes. The chapter is organized into two sections, one focused on high molecular weight products (mainly recombinant proteins), and the other on low molecular weight products (mainly secondary metabolites). Readers may consult the books/monographs listed in Table 1 for additional reading and for further information on research findings published prior to year 2002.

Table 1
Suggested books/monographs for additional reading

Book/Monograph Title	Comment
Zhong, J.J. (ed.), Advances in biochemical engineering/ biotechnology, vol. 72., Springer, Berlin, 2001	One of the more recent review publications on industrial plant cell cultures
Spier, R.E. (ed.), Encyclopedia of cell technology, Wiley, New York, 2000	A comprehensive resource of basic biological information and process techniques for both plant and animal cell cultures
Doran, P.M. (ed.), Hairy roots, Harwood Academic Publishers, Amsterdam, 1997.	A comprehensive resource for hairy roots, from lab techniques to industrial processing
Misawa, M., Plant tissue culture: an alternative for production of useful metabolites (FAO agricultural services bulletin No. 108), 1994.	A publication from the Food and Agriculture Organization of the United Nations that summarizes key research findings on industrial plant cell culture up to the early 1990s; it contains a comprehensive listing of plant cell products that are of industrial interest.
Endress, R. Plant cell biotechnology, Springer, Berlin, 1994.	A nice reference book that covers both physiological as well as bioprocessing aspects of plant cell cultures
Payne, G.F., Bringi, V., Prince, C., Shuler, M.L. Plant cell and tissue culture in liquid systems, Hanser, Munich, 1992	One of the earliest plant tissue culture books with a process engineering focus

2. PRODUCTION OF MACROMOLECULES

2.1. Products and applications

Plant cell and hairy root cultures have received increasing attention as an alternative large-scale production system for high-value recombinant protein products [12]. Production of plant polysaccharides or gums using plants culture has also been pursued [13], but with much less published data available. Here, we will emphasize on recombinant protein products. Among the different forms of plant cell cultures (including suspension cells, immobilized cells, hairy roots, shoots, and somatic embryos), cultured cells grown in liquid media represent the most practical system for producing recombinant proteins on large scales. Therefore, the focus of discussion is placed on suspension cell cultures, the subject of using hairy root cultures for recombinant protein production is covered elsewhere [12, 14–16].

An important benefit of using plant tissue cultures for recombinant protein production is their capability to perform the complex post-translational modifications necessary for active biological functions of the expressed heterologous proteins [17]. Compared with their mammalian or insect cell counterparts, plant cells are easier and less expensive to culture. In plant cell cultures, the potential human pathogen contamination problem associated with mammalian cell culture is not an issue because simple, chemically defined media are used [18]. Cultured plant cells also possess a number of advantages over transgenic plants. Cultured plant cells generally grow much faster than transgenic plants grown in the field; cell cultures are grown in a confined environment (*i.e.* enclosed bioreactor) and hence devoid of the GMO release problem. Furthermore, cell suspension cultures are composed of dedifferentiated callus cells that lack fully functional plasmodesmata, and hence systemic post-tranacriptional gene silencing (PTGS) may be reduced since PTGS is generally believed to be transmitted *via* plasmodesmata and the vascular system [12, 19].

There are, as for other host systems, some drawbacks in using the plant cell expression systems. Dedifferentiated callus cells are known in some cases to suffer from genetic instabilities due to somaclonal variation. Plant cells generally grow slower than bacterial or yeast cells, and usually have lower recombinant protein expression levels, typically between 0.1–1 mg per liter of culture [18]. The lower protein expression is due in part to the fact that plant cells have a more evolved and more tightly controlled gene/protein regulation machinery; it is hence more difficult to manipulate protein expression in plant cells. That said, with further understanding of gene regulation in plant systems, new findings have emerged reporting very high expression levels. For instance, a product level as high as 129 mg per liter has been reported in the case of recombinant human granulocyte-macrophage colony stimulating factor (hGM-CSF) production in transgenic rice cell suspension cultures [20].

Since the publication of several recent reviews on the subject of recombinant protein production from plant cell or hairy root cultures [12, 18, 21], several new published reports have emerged in the subject area [14–16, 20, 22–34]. Besides studies using reporter/model proteins such as green fluorescent protein (GFP), secreted alkaline phosphatase (SEAP), or β-glucuronidase (GUS), most protein products produced in plant cell cultures are intended for therapeutic or diagnostic applications. Several studies have demonstrated the expression of antibodies or antibody fragments in plant cell suspension cultures and hairy root cultures. Some notable examples are the expression of a secretory anti-phytochrome single-chain Fv (scFv) antibody [35], a TMV-specific recombinant full-size antibody [36], a mouse IgG1 recognizing a cell-surface protein of *Streptococcus mutans* [16], and a mouse scFv [26, 36], all using tobacco suspension cultures. The production of a murine IgG1 using hairy roots derived from transgenic tobacco was investigated by Sharp and Doran [16]. A number of therapeutic proteins have also been expressed in plant tissue cultures, including hepatitis B surface antigen (HBsAg) [25], human α_1-antitrypsin [32, 37], and human cytokines such as interleukin (IL)-2, IL-4 [38], IL-12 [27], and GM-CSF [20, 39].

2.2. Process characteristics

Plant cell culture processes for protein production encompass upstream and downstream processing similar to conventional recombinant fermentation processes. That said, plant cells have distinctive properties that call for unique approaches in bioprocess design and operation. Here we will begin with a discussion on culture characteristics, followed by a review on upstream and downstream processing characteristics, and novel molecular approaches. Culture and upstream processing characteristics of plant cell bioprocesses for recombinant protein production have been extensively discussed in a recent review by Su [34], and hence these aspects will only be briefly reviewed here.

2.2.1. Culture characteristics

From a bioprocess-development perspective, the most relevant plant cell culture characteristics for recombinant-protein production include: 1) cell morphology; 2) degree of cellular aggregation and culture rheology; 3) foaming and wall growth; 4) shear sensitivity; 5) growth rate, oxygen demand, and metabolic heat evolution; and 6) protein biosynthesis characteristics.

2.2.1.1. Growth morphology

Plant cells in suspension cultures generally display semi-spherical or rod (sausage-like) shapes, with cell size ranging from 50–100 μm. The degree of cell aggregation is dependent on the plant species, growth stage, and culture conditions. For recombinant protein production, cell aggregation is generally viewed as undesirable since it complicates bioreactor operation due to problems such as the presence of oxygen/nutrient gradients in cell clumps and sedimentation of large cell aggregates. The formation of large cell clumps also complicates culture broth handling for downstream processing. Cultured cell morphology also depends on the plant species, growth stage, and culture conditions. Elongated, filamentous cells tend to entangle and form a cellular network, resulting in a higher packed cell volume (PCV) for a given number of cells per reactor volume (than spherical cells), and hence higher apparent viscosity. Curtis and Emery [40] reported that elongated cell morphology was responsible for the highly viscous and power-law type rheological properties associated with tobacco suspension cultures. The bioprocess implication is significant in that less biomass can be attained with cultures of elongated cells as opposed to spherical-shaped cells. When cultured in similar high-density perfusion bioreactors, and under comparable growth conditions, we found that tobacco cell culture (mostly elongated, filamentous cells) reached only 10 g/L dry weight but with PCV greater than 60%, whereas *A. officinalis* cell culture (which consists of mostly spherical cells and forms fine suspension with few large aggregates) can reach over 35 g/L cell dry weight with PCV exceeding 60% [41].

2.2.1.2. Rheological properties of culture media

The rheological properties of the culture media have a strong impact on bioreactor mixing, oxygen and heat transfer, and maximum cell concentrations. Factors influencing the rheological properties of suspension plant cell culture include cell concentration (especially in terms of biotic phase volume, as opposed to cell numbers or dry weight), cellular water content, cell size, morphology, and the degree of aggregation. High-density plant cell suspension cultures are generally very viscous. This is due to the fact that plant cell cultures typically attain a very high culture biotic phase volume fraction (PCV over 50%) even in batch cultures. The culture spent media, however, usually is not viscous and behaves as a Newtonian fluid. Cultures that consist of mainly large aggregates are generally shown to be less viscous than those consisting of elongated cells entangled into a filamentous cellular network [40]. Most viscous, high-density, plant suspension cultures exhibit shear-thinning or pseudoplastic characteristics [1, 4]. In this case, the apparent culture viscosity is lower under higher shear. Therefore, mixing and bubble dispersion are expected to be more efficient in the impeller region where high shear exists, whereas the region further away from the impeller may experience a higher apparent viscosity, leading to poor mixing and mass transfer.

2.2.1.3. Aeration and foaming

Bubble aeration is commonly practiced in plant cell bioreactors, but can lead to serious foaming. As a result, a large amount of cells become entrapped in the foam layer, reducing the volumetric biomass concentration in the culture broth. These foam-entrapped cells develop into a thick, meringue-like layer that adheres to the reactor vessel and probes. The accumulated cell crusts may become necrotic and secrete inhibitory substances, such as proteases or superannuated cell organelles. Under severe foaming, foam overflow can clog the air vent filter and make the culture susceptible to contamination. Abdullah *et al.* [42] examined various strategies for overcoming foaming and reactor wall growth in plant cell bioreactors and concluded that bubble-free aeration using thin-walled silicone membrane tubing was the only strategy capable of completely eliminating wall-growth. Bubble-free membrane aeration, however, is not suited for large-scale bioreactors due to a reduced membrane-surface to volume ratio and hence reduced oxygen transfer upon scale-up [43, 44]. We found that, at least in smaller bench-scale bioreactors, silicone-based antifoam and a magnetic scrapper (consisting of two small but strong magnets, one placed on the interior reactor wall and the other on the exterior wall to form a magnetic pair) can reduce the wall growth of transgenic tobacco cells cultured in a sparged stirred-tank bioreactor. Under these circumstances, however, a significant foam layer still built up around the impeller shaft and sensor probes. The foaming problem remains a challenge to overcome in plant cell bioreactor design. Fortunately, as the reactor is geometrically scaled up, the reactor cross-section per volume ratio drops, and wall growth is expected to be less of a problem.

2.2.1.4. Shear sensitivity

Due in part to their large vacuoles and the structure of their primary cell wall, plant cells are generally susceptible to hydrodynamic shear. However, shear sensitivity varies among plant species and can also be affected by the culture age. The cellular response to hydrodynamic shear is affected by the intensity of as well as the duration during which the cells are exposed to shear stress. In this context, cumulative energy dissipation has been suggested as a benchmark for comparing data from shear studies involving a wide range of plant species, hydrodynamic conditions, and physiological indicators [2, 45, 46]. Cumulative energy dissipation serves as a convenient index for estimating hydrodynamic shear damage.

However, it is a global (average) hydrodynamic property, and hence it does not reflect how the energy dissipation rates are distributed within the reactor. Furthermore, under gassing conditions, the impeller power input is reduced, and hence the cumulative energy dissipation resulting from agitation is expected to decrease. While shear damage resulting from the hydrodynamic forces associated with bubble rupture is believed to be insignificant in plant cell cultures [45, 47], there is no evidence indicating overall shear damage is reduced with increasing bubble aeration rates at a fixed stirrer speed. Therefore, the suitability of cumulative energy dissipation as a common index for the extent of hydrodynamic shear in stirred reactors under bubble aeration requires further verification.

2.2.1.5. Growth kinetics

For recombinant protein production, it is often preferred to use plant species that generate fast-growing cell cultures. Topping the list are tobacco and rice cell cultures. Tobacco BY-2 cells are particularly appealing because of their remarkably fast growth rate, as well as their ease of *Agrobacterium*-mediated transformation and cell cycle synchronization. Doubling time as short as 11 hours have been reported for tobacco BY-2 cells [48]. Gao and Lee [49] reported a doubling time of about one day for tobacco NT-1 cells (which are similar to BY-2 cells) expressing GUS. For rice cell cultures, a doubling time of 1.5–1.7 days was reported by Trexler *et al.* [32] for a transgenic rice cell culture expressing human α_1-antitrypsin. Terashima *et al.* [37], on the other hand, reported a very long doubling time of 6–7 days in their α_1-antitrypsin-expressing rice cell cultures. Unlike plasmid-based expression in bacterial cells that lead to a huge amount of over-expression, the metabolic burden resulting from foreign protein expression in plant cells is generally not high enough to substantially impact cell growth or oxygen demand, unless the foreign gene product is toxic or interacts with the plant metabolism to cause altered growth characteristics. Therefore, the cell growth rate, cellular oxygen demand, and metabolic heat evolution are similar in wild-type and transgenic plant cell cultures. Kieran [47] reported that the specific oxygen consumption rate for plant cell cultures is generally on the order of 10^{-6} g O_2/(gdw s) or 0.11 mmol O_2/(gdw h). Maximum specific oxygen uptake rate was 0.78–0.84 mmol O_2/(gdw h) in the transgenic rice cell culture reported by Trexler *et al.* [32]; 0.4–0.5 mmol O_2/(gdw h) for transgenic tobacco NT-1 cells expressing GUS [49]. The metabolic heat evolution rate can be easily estimated from the oxygen demand or the specific oxygen consumption rate since the heat of reaction for aerobic metabolism is approximately -460 J per mmol of oxygen consumed [50]. For instance, for the GUS-expressing tobacco cell culture [49], a metabolic heat evolution rate in the range of 184–230 J/(gdw h) is expected. By assuming comparable heat transfer characteristics between high-density plant cell culture and viscous fungal fermentations, Kieran [47] concluded that efficient heat removal in plant cell bioreactors can be easily achieved even with moderate mixing.

2.2.1.6. Production characteristics of recombinant proteins

In recombinant protein production, the type of promoter used dictates the production pattern. When a constitutive promoter, such as the widely popular cauliflower mosaic virus (CaMV) 35S promoter, is used to drive the transgene expression, the recombinant protein production is considered largely growth associated. If an inducible promoter is used, the transgene is generally induced after the culture reaches a high biomass concentration in the late/post exponential growth phase [51]. In this case, recombinant protein production is decoupled from active cell growth. In order to optimize the efficiency of an inducible gene expression system, it is necessary to examine the inducer dosage and the timing of inducer addition. Depending on the nature of the inducer, repeated inducer feeding may be desirable, and hence, it is necessary to optimize inducer feeding. It is highly desirable to enable the effective secretion of the protein product in order to simplify downstream protein purification. The secretory pathway also provides a better cellular environment for protein folding and assembly than the cytosol, since the endoplasmic reticulum contains a large number of molecular chaperones and is a relatively oxidizing environment with low proteolytic activities, generally allowing higher accumulation of the recombinant proteins [52]. Recombinant proteins could, in principle, be targeted to the ER-Golgi secretion pathway using a proper signal peptide. However, there are exceptions to the rule; factors such as the intrinsic properties of the protein product (large molecular size and other organelle-targeting signals) may dictate the final cellular location. Furthermore, it should be cautioned that the extracellular compartment is not loaded with proteolytic activities that can degrade the proteins of interests. Shin *et al.* [20] observed higher proteolytic activities in the tobacco cell culture than in the rice cell culture. The addition of stabilization agents such as gelatin, polyvinyl pyrrolidone (PVP), and bovine serum albumin (BSA) have met with various degrees of success among the proteins tested for stabilization [18]. Another strategy for stabilizing secreted recombinant proteins in plant suspension cultures is via *in-situ* adsorption. James *et al.* [23] coupled an immobilized protein G and a metal affinity column to a culture flask to recover secreted heavy-chain mouse monoclonal antibody and histidine-tagged hGM-CSF, respectively, by recirculating the culture filtrates through these columns. These researchers noted increased product yields for both proteins as a result of reduced protein degradation.

2.2.2. Upstream processing characteristics

Since steam-sterilizable, chemically defined, nutrient media are commonly used in plant cell cultures, conventional steam sterilization technology currently used in industrial fermentation is expected to be adequate for plant cell processes. Bioreactor design and operation, on the other hand, still present many unique challenges that are yet to be overcome. The culture characteristics described above, coupled with knowledge of cellular stoichiometry, mass/energy balances, reaction kinetics, heat/mass transfer, hydrodynamics and mixing, shear, and process monitoring and control are needed to enable the designing of plant cell bioreactors that not only provide a favorable culture environment for the plant cells to produce a high level of recombinant proteins, but are also cost-effective. General discussions on the topic of plant cell bioreactors can be found in a number of comprehensive reviews. Examples of more recent reviews are those from Doran [2], Kieran [47], and Su [34]. Here we will limit our scope to a general overview of the subject.

2.2.3. Bioreactor design and operation

One common goal in plant cell bioreactor design is to develop a reactor that provides a prolonged, sterile, culture environment with efficient mixing and oxygen transfer without producing excessive foaming and hydrodynamic shears and at a low cost. Using time-constant/regime analysis, Doran [1] concluded that for high-density plant cell cultures (over 30 g dw/L), mixing becomes a limiting factor in airlift bioreactors, leading to poor oxygen transfer and heterogeneous biomass distribution in the reactor. Another problem associated with pneumatically agitated plant cell bioreactors, such as airlift and bubble columns, is foaming. To increase reactor volumetric productivity, it is generally preferable to operate the reactor at high cell densities, and hence stirred tanks remain the reactor of choice.

In designing stirred tank reactors, impeller design is one of the most crucial elements. Doran [2] conducted a detailed theoretical engineering analysis of Rushton turbines (RT) and pitched blade turbines (PBT) for a hypothetical 10 m^3 stirred tank plant cell bioreactor of standard configuration by concurrently considering gas dispersion, solid suspension, oxygen transfer, and shear damage. The analysis results indicated that PBTs operating in the upward-pumping mode were superior to RTs in gas handling and solids suspension under power input setting constrained by shear damage considerations. Unfortunately this analysis was not experimentally verified. Subsequent to this analysis, more recent hydrodynamics studies of upward-pumping axial-flow impellers in two or three-phase systems do support the notion that axial-flow impellers operating at an upward pumping mode exhibit low power-drops upon gassing and high efficiency in solid suspensions. However, there is still no report on using such impeller in plant cell cultures. In addition, as pointed out by Kieran [47], there are also reports indicating the unfavorable mass transfer performance of upward-pumping axial-flow impellers in viscous fermentation broths. One notable study is by Junker *et al.* [53], who reported insufficient oxygen transfer using a Lightnin A315 axial-flow impeller in the up-pumping mode in viscous *Streptomyces* fermentations; while the same impeller operated at the down-pumping mode gave better oxygen transfer under increased broth viscosities. When only physical suspension is required or when solid-liquid reactions are rate limiting, Nienow and Bujalski [54] suggested that wide-blade, axial flow hydrofoils such as the A315 operated in the up-pumping mode should be considered. New impeller designs, such as the low-power number radial flow concave blade disc impellers (*e.g.*, the Chemineer CD-6 and BT-6 impellers), have been shown to provide improved oxygen transfer (over Rushton turbines) in *Streptomyces* fermentations [55]. Unlike the CD-6, which has six symmetrical concave blades, the BT-6 has six vertically asymmetrical blades, with the upper section of the blades longer than the lower section [56]. There is very little power drop upon gassing with the BT-6 impellers, even at very high flow rates, compared with Rushton turbines or high solidity-ratio hydrofoils. As such, the BT-6 is expected to be well suited for dispersing gas in reactors and fermenters where a wide range of gas rates is required [56]. According to Chemineer (Dayton, Ohio) [57], the mass transfer capability of the BT-6 is on the order of 10% higher than that of the CD-6, and the BT-6 is also claimed to be relatively insensitive to viscosity. These new impeller designs hold promise for improving mixing and oxygen transfer in viscous, shear-sensitive high-density plant cell cultures.

In order to reduce the capital costs associated with standard autoclavable, stainless-steel type bioreactors, Curtis and co-workers proposed low-cost, plastic-lined reactors for mass cultures of plant cells [58]. The plastic liners are sterilized by ethylene oxide and mixing/oxygen transfer was provided by simple air sparging. With the reactor operating as a bubble column, the biomass concentration attained in this type of reactors was moderate (about 7 g dw/L) [58]. It may be possible to incorporate mechanical agitation into the plastic-lined reactor and still keep the cost down. Since gas-phase sterilization using ethylene oxide is difficult to implement at large scales due to the high toxicity of this substance, other low-cost sterilization alternatives may be sought. In the meantime, it would be desirable if the plant cells could be modified by metabolic engineering means to resist (or at least reduce) microbial contamination, reducing expensive reactor sterilization procedures. This approach is currently being investigated in our laboratory.

To achieve high biomass densities in plant cell bioreactors, cultures should be operated under fed-batch or perfusion modes. Fed-batch cultures are simple to implement, and when integrated with an effective substrate or inducer feeding strategy, can be valuable systems for improved recombinant protein production [22, 51]. Perfusion cultures offer additional process flexibilities when compared with fed-batch cultures. While fed-batch cultures might be limited by the accumulation of inhibitory substances/metabolites in the medium, such a problem is alleviated by culture perfusion. Another apparent benefit of perfusion cultures is their ability to enable constant harvesting of secretory protein products from the reactor effluent that are pre-clarified by a built-in cell-retention device. Perfusion cultures of *A. officinalis* plant cells have been conducted in uniquely designed air-lift [59] and stirred-tank [41] bioreactors for secreted protein production [60]. A stirred-tank perfusion bioreactor similar to that described in Su and Arias [41] has been used recently to culture transgenic tobacco cells for the production of a constitutively expressed secretory green fluorescent protein (GFP) (Su, W. and Liu, B. unpublished). For more information on plant cell perfusion bioreactors or perfusion reactors in general, readers are referred to reviews by Su [60], Castilho and Medronho [61], and Voisard et al. [62].

2.2.4. Downstream processing characteristics

In designing downstream processing for recovering recombinant protein products from plant cells, one needs to consider the cellular location and application of the products. Regarding the cellular location of the products, it is most important to know whether the product is located extracellularly or intracellularly, and, if it is the latter, in which cellular compartment the product is located. In case the protein product can be used within the dried biomass (*e.g.*, protein products that are intended as edible vaccines [63] or nutraceuticals [29]), downstream processing may simply involve recovering and lyophilizing the biomass from the culture broth without further purification. In most cases, however, recombinant protein products are intended for diagnostic or therapeutic uses and thus require further purification. It is generally preferable for the products to be secreted into the medium to reduce the amount of contaminated endogenous cellular proteins. The physicochemical properties of the native proteins in the extracts or spent media from which the product protein is to be separated will dictate the design of the separation operations [7].

Among the physicochemical properties, solubility, size, and charge characteristics are the most important [7]. It has been suggested that the properties of the native protein components be classified according to the so-called Osborne Method [7]. In the Osborne Method, proteins are classified into four major groups based on their solubility in various solvents: albumins, globulins, glutelins, and prolamins [7]. The classifications of native proteins according to the Osborne Method is available for several major crops. Such information should aid the development of separation schemes for removing contaminating native proteins during the purification of the recombinant protein product based on their differential solubility. Unfortunately, native protein compositions in most cultured plant cells are largely uncharacterized. While information on the native protein composition of whole plants is relevant to the purification of recombinant proteins from plant cell cultures, undifferentiated cell cultures and their whole-plant counterparts are obviously quite different in their protein composition. For instance, the chloroplast enzyme ribulose 1,5-bisphosphate carboxylase-oxygenase (RuBisCo) makes up as much as 50% of the total soluble proteins in tobacco leaves (the main site for harvesting recombinant proteins from transgenic tobacco plants), whereas in cultured non-photosynthetic tobacco cells, RuBisCo is not a major protein. It is therefore necessary to better classify the native proteins of the commonly cultured plant cells with techniques such as the Osborne Method. As for the charge and size characteristics, there is generally a high degree of heterogeneity among the native proteins. Therefore a purification scheme based entirely on these two properties is expected to be ineffective [7]. However, it is still useful to know the isoelectric point (PI) of the recombinant protein. If the PI of the recombinant protein is quite different from that of most native proteins, it may be possible to eliminate the contaminating native proteins based on charge and/or solubility differences by operating the purification at the proper pH. In addition to native proteins, it is also important to know the level of endogenous compounds, such as phenolics, oxalic acid, and phytic acids, in the extracts (for intracellular products) or spent medium (for secreted products), which are known to form complexes with proteins that could interfere with the separation processes [7, 64]. Phenolics have also been reported to cause irreversible protein structural modifications in aqueous extracts [64]. These phytochemicals may also cause resin fouling during adsorption and chromatographic separations [65].

Most reports on recombinant protein purification from plant expression systems deal with whole plants. A recent review has appeared on the recovery of recombinant protein products from transgenic plants [7]. Published data on the recovery of recombinant protein expressed from cultured plant cells are scarce. Fischer et al. [36] reported the purification of a TMV-specific full-size murine IgG-2b/κ antibody expressed in transgenic tobacco cell culture. The N-terminal murine leader peptide was able to target the IgG to the secretion pathway, but the antibody was retained by the cell wall. To purify the IgG, the cell wall was partially digested by enzymatic treatment to release the antibody into the extraction buffer. A three-step procedure was then used to purify the IgG, starting with cross-flow filtration, followed by Protein A affinity chromatography and gel filtration as a final purification step. This procedure recovered more than 80% of the expressed IgG from plant cell extracts. Recently, we have developed a simple three-step separation scheme that enables purification of two

GFP-tagged recombinant proteins (SEAP::GFP and GM-CSF::GFP) from tobacco cell cultures with high purity and yield (Su, W.W. and Peckham, G., unpublished). These GFP-fusion proteins were also tagged with an ER-retention HDEL peptide and were found to accumulate in the ER lumen. To perform the protein recovery, the plant cell extract is first pre-cleaned by being subjected to 30% ammonium sulfate precipitation. The precipitate is removed and the soluble portion is then resolved by hydrophobic interaction chromatography, followed by anion-exchange chromatography.

The ammonium sulfate pre-cleaning step is important since it reduces the phenolics and nucleic acids, which helps the two subsequent purification steps. We also found that it is preferable to use ammonium sulfate precipitation to remove contaminants, rather than using it to concentrate the recombinant proteins, as is typically done. When the GFP fusion proteins end up in the ammonium precipitates (say by using a higher concentrations of ammonium sulfate), the protein appears to form complexes with phenolic compounds, and upon resolubilization of the protein, it becomes more difficult to purify. It has been suggested that the phenolics/protein complex formation is promoted at high ammonium sulfate concentrations [64]. The HIC step is effective considering that GFP is a highly hydrophobic protein. The last step is operated under basic conditions such that the GFP fusion proteins carry negative charges. Since GFP is a widely used reporter tag, which allows for convenient monitoring of recombinant protein production, having a simple and possibly universal separation scheme for purifying GFP-tagged proteins from plant cell cultures should be very useful. Affinity tags such as the hexa-histidine tag has been used for affinity purification of GM-CSF from tobacco cell culture using immobilized metal affinity chromatography [23]. Paramban et al [66] developed a chimeric GFP tag having an internal hexa-histidine sequence. Such a GFP tag allows maximum flexibility for protein or peptide fusions since both the N- and C-terminal ends of the GFP are available. Applications of such a tag in plant cell culture are currently being examined in our lab.

2.3. Molecular approaches

Advances in plant molecular biology enable the development of novel strategies for improving the performance of large-scale plant cell culture processes. Molecular strategies are being used to improve heterologous protein accumulation in plants and plant cells at the transcription, translation, and post-translation levels [67]. Common strategies include the use of strong promoters to increase the transcription levels and the use of appropriate enhancers and leader sequences, such as the tobacco etch virus 5' untranslated region, to improve translation [68]; optimization of codon usage; control of transgene copy number; sub-cellular targeting of gene products (*e.g.,* by using an ER-targeting signal peptide or ER-retention HDEL or KDEL signal); the position in the plant genome at which the genes are integrated [69]; and the removal of mRNA-destabilizing sequences [70]. In some cases, nuclear matrix attachment regions (MARs) have been found to improve transcription efficiency of the transgenes [71]. Viral genes that suppress PTGS, such as the potyvirus hc protease genes, can be used to prevent transgene PTGS [72]. As plants expressing these genes may become more susceptible to viral infection, this approach is not practical for field plants but can work well in suspension cells. Additional ways to increase expression levels include the use of different plant species, integration-independent expression, and enhancing the correct protein folding by co-expressing disulfide isomerases or chaperone proteins [67].

Molecular approaches can also be applied to engineer plant cells for desirable traits that are useful in large-scale plant cell cultures. For instance, plant cells may be engineered to acquire improved tolerance to the physical and biological stresses encountered in large-scale bioprocess. Research in this area is scarce; however, several research groups have tackled the problem to improve the hypoxic stress tolerance of plant cells. As discussed in the proceeding sections, oxygen transfer poses a potential problem to large-scale plant cell/hairy root cultures. If oxygen supply cannot keep up with the cellular oxygen demand, hypoxic or even anoxic conditions may result in the culture. Tolerance to low-oxygen stress by cultured plant cells is expected to be species dependent. While the physiological responses (at the molecular level) of bioreactor-cultured plant cells/hairy roots to extended hypoxic stress is not well documented, it is generally believed that engineering plant cells for improved hypoxic stress tolerance is desirable, or even necessary, to combat the oxygen supply problem in large-scale plant-cell bioreactors, especially for high-density cultures.

Two notable approaches have been taken to engineer cultured plant cells and/or hairy roots for improved tolerance to hypoxic stress. In one approach, it involves the over-expression of bacterial or plant hemoglobin genes. Dordas et al. [73] reported reduced nitric oxide production in maize cell cultures over-expressing a class I barley hemoglobin, and improved tolerance to hypoxic stress as a result. Frey et al. [74] demonstrated that the expression of a bacterial *Vitreoscilla* hemoglobin [74] in tobacco cell cultures relieved nitrosative stress and protected the cells from nitric oxide *in vivo*. In a second approach to improve low-oxygen tolerance, Doran and co-workers [75] found that hairy roots over-expressing *Arabidopsis* pyruvate decarboxylase or alcohol dehydrogenase, the two major enzymes in the fermentation pathway, showed improved growth over control roots under microaerobic conditions. Besides improving culture tolerance to low oxygen, molecular approaches can also be applied to improve the cell vigor under adverse bioreactor culture conditions (such as high shear and over-crowding from high biomass concentrations). Our laboratory is currently examining the expression of an anti-apoptosis gene *Bcl-2* [76] in tobacco cells for its effect on improving culture performance under high cell densities in stirred-tank bioreactors.

2.4. Future outlooks

With the current state of technology, plant tissue culture is generally considered cost effective primarily for producing high value, low to medium volume, products that require stringent quality control [12]. In order to make plant tissue cultures a more competitive expression system (as compared to its mammalian, insect, or yeast counterparts), suitable for producing a broader range of protein products, it is necessary to further improve the protein expression levels and to reduce manufacturing costs. In this context, future breakthroughs are expected to come not only from advances in plant molecular biology and biochemistry, but also from several fronts of technological advances in bioprocessing. Goldstein [77], in his cost analysis of using plant cell and tissue culture for producing food ingredients, pointed out several potential areas of technical advances necessary for bringing down the production costs associated with large-scale plant cell/tissue culture. These technical advances are [77]:

- More productive cells
- Innovations in facility (including bioreactor) design that cut costs without reducing productivity
- Advances which permit more efficient product secretion
- Cell reuse (*e.g.* through culture perfusion)
- Lower-cost nutrients
- Reduced capital costs
- Increased efficiencies in manufacturing and other operations

Most of these technical advances are also necessary for effective recombinant protein production using cultured plant cells. The technological/engineering advances reviewed in the preceding discussions have addressed some of these issues (*e.g.* effective and simple culture perfusion for biomass reuse and secretory product harvesting). Mass culturing plant cells in a physical environment that allows for efficient protein expression is now possible in some cases. To bring the current technology to the next level, innovative approaches are needed to further improve the expression level of recombinant protein products and to enable plant cells to grow in simpler and less expensive culture vessels by a combination of novel reactor design and cellular engineering approaches in order to reduce capital and raw material (nutrient) costs. Limited progress has already been made on addressing this need, *e.g.*, in developing low-cost bioreactors [58] and in engineering plant cells to acquire better tolerance to hypoxic stress [75] (so that plant cells can be cultivated in a simpler bioreactor that is not equipped to provide high oxygen transfer). In addition, since downstream recovery contributes to a large portion of the manufacturing costs, it is crucial to increase recombinant protein secretion efficiency and stability. From a value-added processing perspective, it might be plausible to make plant cells utilize alternative or less expensive carbon sources such as starch hydrolysate (which contains about 95% glucose) [77, 78], refined dextrose, and high fructose corn syrup from corn wet milling [77]. It might also be possible to utilize xylose from lignocellulosic biomass hydrolysate as an alternative carbon source [79] by genetically engineering plant cells to express xylose isomerase [80].

3. PRODUCTION OF SMALL MOLECULES

3.1. Products and applications

Low-molecular weight plant secondary metabolites are an important source of flavors, drugs, colorants, fragrances, and insecticides. Plant cell, tissue, and organ cultures represent attractive alternatives to medicinal plants as sources of these valuable products. *In-vitro* culture of plant cells, tissues, or organs are valuable tools for studying and producing plant secondary metabolites. When the surface of an explant tissue is cut, the cells at the wound site undergo division and form a callus. With proper exogenous growth regulator(s), callus can be cultivated in suspension to produce natural products. The accumulation of high levels of secondary metabolites in suspension cells has been reported for anthocyanins [81], berberine [82], ginseng saponins [83], rosmarinic acid [84], and shikonin [85]. Except for these cases, the biotechnological production of useful secondary metabolites by plant cell culture systems has been largely unsuccessful. One of the major obstacles that still needs to be overcome is the low productivity of most secondary metabolites in dedifferentiated cells grown in the suspension culture. Because many secondary metabolites start to accumulate when organized tissues begin to emerge from callus cells, there seems to be a close link between morphological differentiation and secondary metabolite biosynthesis in plant cells. Unlike suspension cells, hairy roots are differentiated organs. Plant tissues are transformed by the soil-borne bacterium *Agrobacterium rhizogenes* carrying the root inducing (Ri) transfer DNA (t-DNA). Hairy roots are induced after t-DNA integration into the plant genome. Induced hairy roots can be removed from the infected plants/tissues and cultured in liquid media to establish hairy-root cultures. While *A. rhizogenes* infects a wide range of plant hosts, it is difficult to establish hairy roots from some important medicinal plants, such as *Taxus brevifolia* and *Podophyllum peltatum*, after they were co-cultured with *A. rhizogenes*.

Table 2
Valuable secondary metabolites produced by suspension cells or hairy roots

Secondary metabolite	Application	Plant Species	Reference
Suspension cultures			
Catechin	Antiallergic, antimicrobial, and antioxidative	*Camellia sinensis*	[86]
Genistein	Chemopreventive effects (reduce the incident of cancer)	*Glycine max*	[87]
Hypericin	Antidepressant activity	*Hypericum perforatum*	[88]
Paclitaxel	Anticancer	*Taxus brevifolia*	[89]
Podophyllotoxin	Anticancer	*Podophyllum peltatum*	[90]
Hairy root cultures			
Artemisinin	Anti-malaria	*Artemisia annua*	[91]
Camptothecin	Anticancer	*Camptotheca acuminate*	[92]
Coniferin	Precursor of podophyllotoxin	*Linum flavum*	[93]
Tanshinones	Treatment of menstrual disorders	*Salvia miltiorrhiza*	[94]
Polyacetylenes	Cytotoxic activity against leukemia cells	*Panax hybrid*	[95]
Puerarin	Hypothermic, spasmolytic, hypotensive and antiarrhythmic activities	*Pueraria phaseoloides*	[96]

The medicinal applications of plant secondary metabolites have focused on the development of medicines for anticancer, antivirus, antimalarial, anti-inflammation, antidepressant, anti-ischaemia, and immunostimulation activities [97]. Table 2 summarizes plant-derived compounds that have attracted medical and pharmacological interests in the last ten years, which potentially could be produced in plant cells/hairy roots. Among the compounds listed in Table 2, camptothecin, paclitaxel, and podophyllotoxin have attracted considerable interests for anti-tumor application. In addition to the antitumor compounds, hypericin (an antidepressant isolated from St. John's wort) and artemisinin (for malarial therapy) are also considered plant secondary metabolites with pharmaceutical importance. However, because supply of these pharmaceutical compounds is limited to traditional extraction from field cultivated plants, many attempts had been made to develop plant cell cultures to provide an alternative source for these secondary metabolites.

3.2. Process characteristics

3.2.1. Culture characteristics

A. rhizogenes is responsible for hairy root disease in a broad range of dicotyledonous plants and some gymnosperms. Hairy roots can be obtained directly from the cut edges of the petioles of leaf explants or *via* callus two-three weeks after inoculation with *A. rhizogenes* (Fig. 1). Different strains of *A. rhizogenes* showed different hairy root induction efficiency [92, 93]. The strains of *A. rhizogenes* that have usually been applied in hairy root induction of medicinal plants include A4, 15834, LBA9402, MAFF03-01724, R-1601, R-1000, and TR105. Lin et al. [93] reported the *A. rhizogenes* strains differed widely in their ability to induce hairy roots from *Linum flavum* leaf discs, with the LBA9402 strain being the most efficient. The choice of *A. rhizogenes* strains for hairy root induction is host dependent. For instance, although the A4 strain was considered highly virulent and was shown to be highly effective in inducing hairy roots of many plant species, it was not effective in inducing hairy roots from *Linum flavum* leaf discs [93]. Since the natural roots' synthetic capacities are not impaired by the genetic transformation, hairy roots, which can often grow vigorously in hormone-free media and produce secondary metabolites on a level comparable to that of the original plants, have been considered as a potential system for producing important secondary metabolites. Hairy root cultures generally exhibit better genetic and biochemical stability than their cell culture counterparts; for instance, their secondary metabolite production has been reported to remain stable for years. However, the morphology of the root structure also causes problems with inoculation, distribution, and sampling when hairy roots are cultivated in bioreactors, making hairy root cultures less amenable to scale-up.

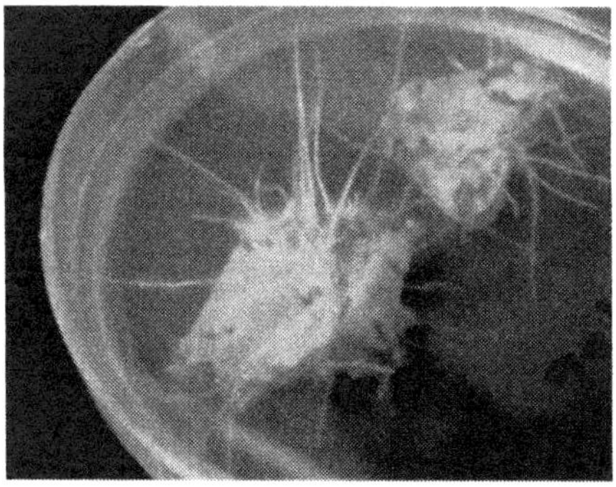

Fig. 1. Induction of *Nicotiana tabacum* hairy roots by the leaf disc method.

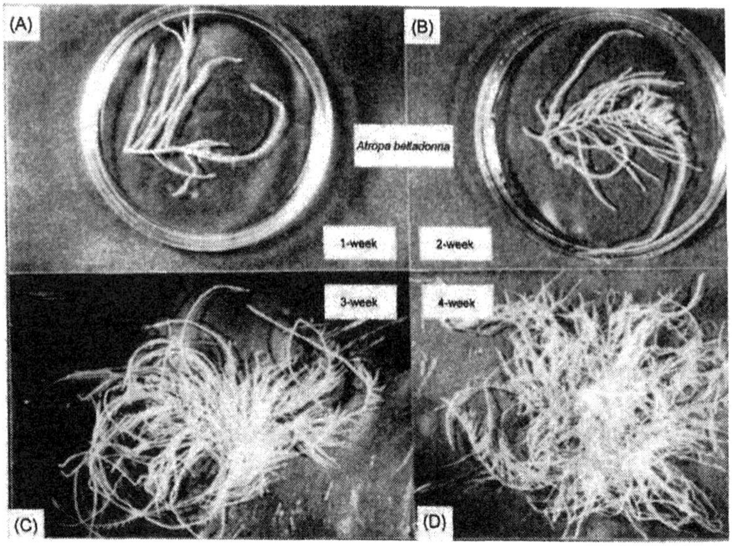

Fig. 2. Morphology of hairy roots (*Atropa belladonna*).

Table 3
Specific growth rates of some medicinal hairy roots

Hairy roots	μ_{max} **(day^{-1})**	**Reference**
Arabidopsis thaliana	0.045–0.101	[98]
Artemisia annua	0.22	[91]
Atropa belladonna	0.3–0.55	[99]
Catharanthus roseus	0.19	[100]
Hyocyamus muticus	0.32	[101]

The growth of *Atropa belladonna* hairy roots in liquid is shown in Fig. 2. One week after being inoculated in flask cultivation, an approximately 1-cm segment of hairy root grew into a root tissue over 8 cm long with main and primary branches (A). With the elongation of the primary branch, the secondary growing tips in the primary branches also appeared (B). With the elongation of the secondary branch, a clump of hairy roots over 13 to 16 cm in diameter was observed after three to four weeks, respectively (C, D). Thus, the growth rate of hairy roots depends on linear extension, the formation of a large number of new growing points on lateral branches, and on a secondary increase in root diameter as the root cells undergo cell expansion and differentiation. Several groups have measured the growth rate of hairy roots cultured in a liquid system. The doubling time was in the range of over 1-day to 1-week, depending on the plant species and culture conditions (Table 3). These data showed that the growth rate of hairy roots is comparable to that of plant suspension cells. Characteristics of suspension cells and hairy roots are compared in Table 4. In summary, suspension cell cultures are relatively easy to establish from a large variety of medicinal plants, and they are amenable to scale-up. Compared with suspension cell cultures, the superior genetic and biochemical stability of biosynthesis of secondary metabolites is the most appealing characteristic of hairy root cultures.

Table 4
Comparison of hairy roots and suspension cells

Characteristics	**Hairy roots**	**Suspension cells**
Size/morphology	highly branched root tissues that can grow into large root biomass	10–200 μm for a single cell; can be aggregated to form cell clusters over 2 mm
Growth rate (doubling time)	generally 1 day–1 week	generally 1 day–1 week
Growth regulator in exogenous supply	not necessary	necessary
Genetic and biochemical stability of biosynthesis of secondary metabolites	differentiated tissues with better genetic and biochemical stability for secondary metabolite production generations	dedifferentiated cells; stability of secondary metabolite production needs to be checked, and constant cell line screening may be necessary
Culture scale up	difficult to scale up	easy to scale up

3.2.2. Upstream processing

The physical structure of the roots poses challenges to inoculation and homogeneous root distribution in a liquid culture. As a result, reduced productivities have often been noted upon culture scale-up [102, 103]. Some attempts had been made to solve the inoculation problem. Ramakrishnan et al. [104] reported an inoculation method that consisted of briefly homogenizing the bulk root cultures of *Hyoscyamus muticus*, *Beta vulgarus*, and *Solanum tuberosum*, then aseptically transferring the slurry to the reactor. The effects of specific excision on root cultures of related species were examined by Falk and Doran [105] and Woo et al. [106]. The effects of the cut treatment on root growth, morphology, and alkaloid content were further investigated in flask cultures. The data showed that hairy roots of *A. belladonna* with a suitable length (longer than 1 cm) retained the ability to grow and produce tropane alkaloids after a cut treatment [107].

After inducing hairy roots and selecting high-producing cell lines, it is necessary to optimize medium components and culture conditions before the culture can be successfully scaled up. Hairy roots can be cultivated without the addition of exogenous hormones, because the t-DNA from *A. rhizogenes* codes for auxin synthesis [108]. However, growth regulators may still affect hairy root growth, organogenesis, and the formation of both primary and secondary metabolites. The accumulation of hyoscyamine and scopolamine could be significantly enhanced in hairy root cultures of *Hyoscyamus muticus* by adding the auxins IAA or NAA [109].

Table 5
Stimulation of plant secondary metabolite production by elicitation

Plant	Elicitor	Function	Reference
Hairy root cultures			
Ammi majus	Benzo(1,2,3)-thiadiazole-7-carbothionic acid *S*-methyl ester	Higher accumulation of coumarins	[110]
Artemisia annua	(22S, 23S)-homobrassinolide; fungal elicitor	Enhancement of artemisinin production	[111]
Beta vulgaris	Micro algal	Enhancement of betalines production	[112]
Catharanthus roseus	$CdCl_2$	Increase in indole alkaloid production	[113]
Cichorium intybus	Fungal elicitor	Production of volatile compounds	[114]
Ocimum basilicum	Fungal cell wall	Enhancement of rosmarinic acid production	[115]
Panax ginseng	Methyl jasmonate	Improving ginsenoside yield	[116]
Salvia miltiorrhiza	Yeast; Ag+	Enhancement of tanshinones production	[117]
Solanum tuberosum	Fungal elicitor	Production of phytoalexins	[118]
Tagetes patula	Micro algal elicitor	Enhancement of thiophenes production	[112]
Suspension cultures			
Glycyrrhiza glabra	Methyl jasmonate	Stimulation of soyasaponin biosynthesis	[119]
Taxus chinensis	2-hydroxyethyl jasmonate/ trifluoroethyl jasmonate	Increase in taxuyunnanine C production	[120]
Taxus canadensis	Methyl jasmonate	Increase in taxoid production	[121]

Elicitors are generally defined as molecules that can stimulate the defense responses of plants, including the formation of phytoalexins. The effects of elicitor on plant secondary metabolite production by hairy roots and suspension cells are summarized in Table 5. Biotic elicitors, such as the cell wall components of filamentous fungi, yeast, and microalgae, have been shown to stimulate the production of antimicrobial compounds in plants. Abiotic elicitors, such as jasmonate (JA) and its methyl ester (MeJA), and salicylic acid, are generally considered to be secondary signals, thus modulating many physiological events in higher plants, including defense responses, flowering, and senescence. They are regarded as a new class of phytohormone. Some secondary metabolites may also be stimulated by heavy metals and synthetic substances [114]. It has been reported that exogenously applying MeJA induced the biosynthesis of terpenoids [122]. MeJA was also reported to stimulate saponin production in cultured ginseng cells [123] and *Bupleurum falcatum* root fragments [124], but the detailed mechanisms responsible for these stimulatory effects remain unevaluated. The elicitation of plant cells and tissues can lead to increased yields, and hence the use of biotic and abiotic elicitors has been considered a viable strategy for improving the yield of plant secondary products. The application of elicitors to plant cell cultures is not only useful for enhancing the biotechnological productivity of valuable secondary metabolites in fermentation systems but also for the study of plant-microbe interactions. An added biotechnological benefit of their use is that they may also promote the liberation of metabolites into the medium [125].

3.2.3. Bioreactors for hairy root culture

The main restriction of hairy root cultures to commercial exploitation is the difficulty in culture scale-up. Several types of reactors have been reported for hairy root cultures, including liquid-phase (rotating drum [126], wave [127], stirred tank [128], bubble column, and air-lift [129]) and gas-phase (trickle bed [130], droplet phase, and mist [131]) reactors. While many of these studies focused only on root growth, examples of hairy root bioreactor studies that also addressed secondary metabolite production are presented in Table 6.

Table 6
Hairy roots cultured in bioreactors for secondary metabolite production

Bioreactor	**Species**	**Culture mode and duration**	**Biomass yield**	**Product content / productivity**	**Ref.**
Mist reactor (1.5L)	*Artemisia annua*	Batch, 28 days	40–105 g fw/L	0.07–0.29 μg artemisinin/g fw	[131]
Air-lift (30L)	*Astragalus membranaceus*	Batch, 20 days	11.5 g dw/L	1.4 mg astragalolide IV/g dw	[129]
Modified stirred tank (3L)	*Ophiorrhiza pumila*	Batch, 8 weeks	87 g fw/L	8.8 mg camptothecin /L	[128]
Wave type (2L)	*Panax ginseng*	fed-batch, 56 days	284.9 g fw/L	145.6 mg ginsenoside/L	[127]

fw: fresh weight, dw: dry weight.

Stirred tanks are commonly used in industrial microbial fermentation, but reactor modifications are necessary to avoid damage to the root tissues from the impellers. By placing a stainless steel net in the bottom of a stirred-tank reactor to prevent direct contact between root tissues and the impeller, hairy root cultures of *A. belladonna* was scaled-up to 30 L for alkaloid production [107]. Compared with flask cultures, no reduction in the alkaloid productivity of the scale-up cultures was observed [107]. Because of the physical characteristics of hairy roots, it is not possible to take homogeneous samples from the cultures during cultivation. To solve this problem, the biomass accumulation of hairy roots can be estimated by detecting the changes in medium conductivity resulting from nutrient consumption [107]. Alternatively, mass balance techniques have been developed to permit accurate aseptic on-line estimation of dry weight, fresh weight, and liquid volume in root cultures [132].

3.2.4. Two-phase culture systems

In order to solve the solubility problem associated with most secondary metabolites, such as diterpenoid taxol, integrated cell culture-separation systems, such as the two-phase culture, were developed. Oleic acid, dibutylphthalate and other organic solvents have been shown to accelerate oxygen transfer in a two-phase culture of simulated plant cells [133]. Tricaprilyn (1,2,3-trioctanoylglycerol) was also shown to be efficient in enhancing taxol production in a *Taxus brevifolia* two-phase cell suspension culture [134]. The circulation of culture medium through an external loop containing a nontoxic organic phase was shown to be efficient for the extraction of secondary metabolites of *Hyoscyasmus muticus* hairy root cultures [135]. In this context, silicon oil was shown to accumulate benzophenanthridine in *Eschscholtzia californica* two-phase cell suspension cultures [136]. The potential to continuously extract specific secondary metabolites of *C. roseus* hairy roots using silicon oil has also been demonstrated. Tikhomiroff et al. [137] showed that the use of silicon oil improved the production of tabersonine and löchnericine but did not affect serpentine and catharanthine yields in a two-phase hairy-root culture.

3.2.5. Permeabilization of plant cells

Plant secondary metabolites are often stored in vacuoles, which makes the continuous production of secondary metabolites by plant tissue cultures difficult. Secondary metabolites are usually extracted from lyophilized plant tissues, their harvest is destructive to the culture and therefore limits the potential productivity of an industrial scale process. Based on the sensitivity of plant cell cultures to alterations to their culture environment, which often leads to permeabilization, the effects of some physical factors – pH, temperature, oxygen starvation and osmotic stress – on secondary metabolite release were investigated. Chemical agents, such as DMSO, Tween-80, Triton X-100, cetyl trimethylammoniumbromide (CTAB), certain monoterpenes, and fatty acids, have also been examined to permeabilize plant cells/roots to release intracellularly stored products. However, these treatments are often too destructive, leading to a loss of hairy root viability. Thus, chemical permeabilization is not favorable for repeatedly harvesting secondary products.

Biological agents, such as live microbial cells, may serve as alternative permabilization agents by releasing hydrolytic enzymes to digest plant cell walls and allow cytosolic contents to seep into the medium or by producing bio-surfactants to alter cell surface activity [138]. These agents hence appear attractive for product recovery in plant hairy root cultures. The effects of food-grade biological agents, including *Lactobacillus helveticus*, *Saccharomyces cereviseae*, *Candida utilis*, and lipid of *L. helveticus*, on the release of batalaines from red beet hairy roots have been examined by Thimmaraju et al [138]. These researchers suggest that lipid of *L. helveticus* is a potentially useful agent for the *in-situ* recovery of betalaines from beet hairy roots [138]. Another cell permeablization strategy involves the use of ultrasound treatments. Brief exposure (1–8 min) to low-energy ultrasound was shown to enhance the release of several secondary metabolites, including ginseng saponins [139], shikonins [140], and paclitaxel [141], from cell cultures. However, the use of ultrasound treatment to promote product release in hairy root cultures has not been published yet.

3.2.6. Supercritical fluid extraction

Medical preparations from medicinal plants are usually based on solvent extraction. Among the available extraction processes, supercritical fluid extraction (SFE) has been used extensively in the food industry for decaffeinating coffee beans and for hop extraction during beer brewing. SFE is a potentially attractive technique for the large-scale extraction of medicinal compounds from plant tissues, due in part to its gas-like mass transfer properties and liquid-like solvating characteristics [142]. Compared with conventional organic solvent extraction, SFE allows for lower solvent consumption and a shorter treating time. Several substances, such as CO_2, N_2O, NH_3 and H_2O, have been used as supercritical fluids. So far, CO_2 has been the most widely used because of its low critical temperature (31.3°C) and because it is non-explosive, safe, and inexpensive, important factors in pharmaceutical applications. It was reported that SFE recovered more podophyllotoxin than conventional 95% ethanol extraction from *Dusosma pleiantha* roots [142]. The enrichment of hyperforin from St. John's Wort extracts by pilot-scale supercritical CO_2 extraction has also been demonstrated [143].

3.2.7. In-situ product recovery

Repeated *in-situ* product recovery is attractive for improving productivity in plant hairy root processes. Extracellular products may be continuously recovered from the medium by *in situ* adsorption either by adding resins (*e.g.,* Amberlite XAD-2; XAD-7) into the medium, or by circulating the spent medium through a resin column external to the reactor. Williams et al. [144] showed that the production level of total sanguinarine was improved using XAD-7 polymeric resins. The addition of polymeric resins to *C. roseus* suspension cell cultures has also been shown to increase the production of catharanthine and ajmalicine [145]. Using the polystyrene resin, Diaion HP-20, camptothencin accumulation in the medium was increased [146]. The amount excreted into the medium increased 5-fold in the presence of Diaion HP-20. Since camptothecin can be absorbed by Diaion HP-20, it was easily recovered from the resin in a fairly pure state after elution with methanol. The timing of the resin addition affects the cell/tissue growth, the production of the secondary metabolite, and the recovery of the secondary metabolite from the culture. Lee-Parsons and Shuler demonstrated that optimized adsorption resin addition resulted in the improvement in ajmalicine production [147].

3.3. Molecular approaches

In recent years, various molecular (metabolic engineering) approaches have been reported for increasing the productivity of valuable plant secondary metabolites in plant cell/hairy root cultures. Specific genes that regulate key steps in biosynthetic pathways could potentially be cloned and expressed in plant cells to modulate cell metabolism. One of the earliest successful examples of metabolic engineering to enhance plant secondary metabolite production is the engineering of *A. belladonna,* a hyoscyamine-rich plant to over-express *Hyoscyamus niger* hyoscyamine 6 β-hydroxylase (H6H), an enzyme that catalyzes the conversion of hyoscyamine to scopolamine, leading to the development of transgenic *A. belladonna* containing a high scopolamine level [148]. Recently, Zhang et al. [117] developed transgenic *H. niger* hairy root cultures overexpressing putrescine N-methyltransferase (PMT) as well as

H6H. Transgenic hairy root lines expressing both PMT and H6H were shown to produce significantly higher levels of scopolamine compared with the wild-type and transgenic lines harboring a single gene (PMT or H6H). The best line was found to produce over nine times more scopolamine than the wild type and more than twice the amount in the highest scopolamine-producing H6H single-gene transgenic line.

The metabolic engineering of shikonin production in *Lithospermum erythrorhizon* has also been attempted. Boehm et al. [149] investigated the effect on shikonin production in *L. erythrorhizon* hairy roots by introducing a bacterial *ubi*A gene which is capable of catalyzing a regulatory reaction in shikonin biosynthesis. In the resulting transgenic root lines, high UbiA enzyme activities could be detected, resulting in an increased accumulation of 3-geranyl-4-hydroxybenzoate. However, no significant correlation between UbiA enzyme activity and shikonin accumulation was observed [149].

Hughes et al. [150] reported the growth of transgenic *C. roseus* hairy roots engineered to express a feedback-resistant *Arabidopsis anthranilate* synthase α subunit under the control of an inducible promoter. According to their results, a large increase in tryptophan and tryptamine was observed, but the levels of most terpenoid indole alkaloids, with the exception of lochnercine, were not significantly altered [150].

So far, metabolic engineering to improve secondary metabolite production has met with mixed success. It is apparently difficult to enhance secondary metabolite productivity by simply up- or down-regulating a single pathway gene. Metabolic engineering by overexpressing transcription factors [151] has shown some promise as a viable approach for increasing secondary metabolite production. Modern integrated approaches based on genomics, proteomics, and metabolomics should accelerate the pace in elucidating metabolic regulation in plant secondary metabolism, which remains central to developing effective metabolic engineering strategies for improving plant secondary metabolite production.

ACKNOWLEDGEMENTS

WWS is grateful to funding support from the United States National Science Foundation (BES97-12916 and BES01-26191), the United States Department of Agriculture (USDA) Tropical & Subtropical Agriculture Research (TSTAR) Program (01-34135-11295), and the USDA Scientific Cooperative Research Program (58-3148-9-080).

REFERENCES

[1] Doran, P.M. (1993); Design of reactors for plant cells and organs. Advances in Biochemical Engineering and Biotechnology 48:115-168.

[2] Doran, P.M. (1999); Design of mixing systems for plant cell suspensions in stirred reactors. Biotechnol Prog 15:319-335.

[3] Chattopadhyay, S.; Farkya, S.; Srivastava, A. and Bisaria, V. (2002); Bioprocess considerations for production of secondary metabolites by plant cell suspension cultures. Biotechnology and Bioprocess Engineering 7:138-149.

[4] Kieran, P.M.; MacLoughlin, P.F. and Malone, D.M. (1997); Plant cell suspension cultures: some engineering considerations. J Biotechnol 59:39-52.

[5] Zhong, J. (2001); Biochemical engineering of the production of plant-specific secondary metabolites by cell suspension cultures. Advances in Biochemical Engineering and Biotechnology 72:1-26.

[6] Su, W.W. (1995); Bioprocessing technology for plant cell suspension cultures. Applied Biochemistry and Biotechnology 50:189-230.

[7] Menkhaus, T.J.; Bai, Y.; Zhang, C.; Nikolov, Z.L. and Glatz, C.E. (2004); Considerations for the recovery of recombinant proteins from plants. Biotechnol Prog 20:1001-1014.

[8] Ulbrich, B.; Weisner, W. and Arens, H. (1985). In: K.H. Neumann and E. Reinhard (Eds.) Primary and Secondary Metabolism of Plant Cell Cultures. Springer-Verlag, Berlin; pp. 293-303.

[9] Tabata, M. and Fujita, Y. (1985); Production of shikonin by plant cell cultures. In: M. Zaitlin, P. Day and A. Hollaender (Eds.) Biotechnology in Plant Science. Academic Press, San Diego, CA; pp. 207-218.

[10] Goldstein, W.; Lasure, L. and Ingle, M. (1980). In: E.J. Staba (Ed.), Plant Tissue Culture as a Source of Biochemicals. C.R.C. Press, Boca Raton, Florida; pp. 191-234.

[11] Drapeau, D.; Blanch, H.W. and Wilke, C.R. (1987); Economic assessment of plant cell culture for the production of ajmalicine. Biotechnology and Bioengineering 30:946-953.

[12] Doran, P.M. (2000); Foreign protein production in plant tissue cultures. Curr Opin Biotechnol 11:199-204.

[13] Xu, J.; Shpak, E.; Gu, T.; Moo-Young, M. and Kieliszewski, M. (2005); Production of recombinant plant gum with tobacco cell culture in bioreactor and gum characterization. Biotechnol Bioeng 90:578-588.

[14] Gaume, A.; Komarnytsky, S.; Borisjuk, N. and Raskin, I. (2003); Rhizosecretion of recombinant proteins from plant hairy roots. Plant Cell Rep 21:1188-1193.

[15] Sharp, J.M. and Doran, P.M. (2001); Strategies for enhancing monoclonal antibody accumulation in plant cell and organ cultures. Biotechnol Prog 17:979-992.

[16] Sharp, J.M. and Doran, P.M. (2001); Characterization of monoclonal antibody fragments produced by plant cells. Biotechnol Bioeng 73:338-346.

[17] Gomord, V. and Faye, L. (2004); Posttranslational modification of therapeutic proteins in plants. Curr Opin Plant Biol 7:171-181.

[18] James, E. and Lee, J.M. (2001); The production of foreign proteins from genetically modified plant cells. Adv Biochem Eng Biotechnol 72:127-156.

[19] Crawford, K.M. and Zambryski, P.C. (1999); Plasmodesmata signaling: many roles, sophisticated statutes. Curr Opin Plant Biol 2:382-387.

[20] Shin, Y.J.; Hong, S.Y.; Kwon, T.H.; Jang, Y.S. and Yang, M.S. (2003); High level of expression of recombinant human granulocyte-macrophage colony stimulating factor in transgenic rice cell suspension culture. Biotechnol Bioeng 82:778-783.

[21] Fischer, R.; Emans, N.; Schuster, F.; Hellwig, S. and Drossard, J. (1999); Towards molecular farming in the future: using plant-cell-suspension cultures as bioreactors. Biotechnol Appl Biochem 30 (Pt 2):109-112.

[22] Liu, S.; Bugos, R.C.; Dharmasiri, N. and Su, W.W. (2001); Green fluorescent protein as a secretory reporter and a tool for process optimization in transgenic plant cell cultures. J Biotechnol 87:1-16.

[23] James, E.; Mills, D. and Lee, J. (2002); Increased production and recovery of secreted foreign proteins from plant cell cultures using an affinity chromatography bioreactor. Biochemical Engineering Journal 12:205-213.

[24] Lee, J.H.; Kim, N.S.; Kwon, T.H.; Jang, Y.S. and Yang, M.S. (2002); Increased production of human granulocyte-macrophage colony stimulating factor (hGM-CSF) by the addition of stabilizing polymer in plant suspension cultures. J Biotechnol 96:205-211.

[25] Smith, M.L.; Mason, H.S. and Shuler, M.L. (2002); Hepatitis B surface antigen (HBsAg) expression in plant cell culture: Kinetics of antigen accumulation in batch culture and its intracellular form. Biotechnol Bioeng 80:812-822.

[26] Xu, H.; Montoya, F.U.; Wang, Z.; Lee, J.M.; Reeves, R.; Linthicum, D.S. and Magnuson, N.S. (2002); Combined use of regulatory elements within the cDNA to increase the production of a soluble mouse single-chain antibody, scFv, from tobacco cell suspension cultures. Protein Expr Purif 24:384-394.

[27] Kwon, T.H.; Seo, J.E.; Kim, J.; Lee, J.H.; Jang, Y.S. and Yang, M.S. (2003); Expression and secretion of the heterodimeric protein interleukin-12 in plant cell suspension culture. Biotechnol Bioeng 81:870-875.

[28] Kwon, T.; Kim, Y.; Lee, J. and Yang, M. (2003); Production and secretion of biologically active human granulocyte-macrophage colony stimulating factor in transgenic tomato suspension cultures. Biotechnology letters 25:1571-1574.

[29] Kwon, S.; Jo, S.; Lee, O.; Choi, S.; Kwak, S. and Lee, H. (2003); Transgenic ginseng cell lines that produce high levels of a human lactoferrin. Planta Medica 69:1005-1008.

[30] Su, W.W.; Guan, P. and Bugos, R.C. (2004); High-level secretion of functional green fluorescent protein from transgenic tobacco cell cultures: characterization and sensing. Biotechnol Bioeng 85:610-619.

[31] Terashima, M.; Ejiri, Y.; Hashikawa, N. and Yoshida, H. (2001); Utilization of an alternative carbon source for efficient production of human alpha(1)-antitrypsin by genetically engineered rice cell culture. Biotechnol Prog 17:403-406.

[32] Trexler, M.M.; McDonald, K.A. and Jackman, A.P. (2002); Bioreactor production of human alpha(1)-antitrypsin using metabolically regulated plant cell cultures. Biotechnol Prog 18:501-508.

[33] Drake, P.M.; Chargelegue, D.M.; Vine, N.D.; van Dolleweerd, C.J.; Obregon, P. and Ma, J.K. (2003); Rhizosecretion of a monoclonal antibody protein complex from transgenic tobacco roots. Plant Mol Biol 52:233-241.

[34] Su, W. (2005); Bioreactor engineering for recombinant protein production using plant cell suspension culture. In: S. Duta Gupta (Ed.), Plant tissue culture engineering. Springer, Berlin.

[35] Firek, S.; Draper, J.; Owen, M.R.; Gandecha, A.; Cockburn, B. and Whitelam, G.C. (1993); Secretion of a functional single-chain Fv protein in transgenic tobacco plants and cell suspension cultures. Plant Mol Biol 23:861-870.

[36] Fischer, R.; Liao, Y.C. and Drossard, J. (1999); Affinity-purification of a TMV-specific recombinant full-size antibody from a transgenic tobacco suspension culture. J Immunol Methods 226:1-10.

[37] Terashima, M.; Murai, Y.; Kawamura, M.; Nakanishi, S.; Stoltz, T.; Chen, L.; Drohan, W.; Rodriguez, R.L. and Katoh, S. (1999); Production of functional human alpha 1-antitrypsin by plant cell culture. Appl Microbiol Biotechnol 52:516-523.

[38] Magnuson, N.S.; Linzmaier, P.M.; Reeves, R.; An, G.; HayGlass, K. and Lee, J.M. (1998); Secretion of biologically active human interleukin-2 and interleukin-4 from genetically modified tobacco cells in suspension culture. Protein Expr Purif 13:45-52.

[39] James, E.A.; Wang, C.; Wang, Z.; Reeves, R.; Shin, J.H.; Magnuson, N.S. and Lee, J.M. (2000); Production and characterization of biologically active human GM-CSF secreted by genetically modified plant cells. Protein Expr Purif 19:131-138.

[40] Curtis, W. and Emery, A. (1993); Plant cell suspension culture rheology. Biotechnology and Bioengineering 42:520-526.

[41] Su, W.W. and Arias, R. (2003); Continuous perfusion plant cell culture: Bioreactor characterization and secreted enzyme production. Journal of Biosciences and Bioengineering 95:13-20.

[42] Abdullah, M.; Ariff, A.; Marziah, M.; Ali, A. and Lajis, N. (2000); Strategies to overcome foaming and wall growth during the cultivation of Morinda elliptica cell suspension culture in a stirred-tank bioreactor. Plant Cell, Tissue and Organ Culture 60:205-212.

[43] Qi, H.N.; Goudar, C.T.; Michaels, J.D.; Henzler, H.J.; Jovanovic, G.N. and Konstantinov, K.B. (2003); Experimental and theoretical analysis of tubular membrane aeration for Mammalian cell bioreactors. Biotechnol Prog 19:1183-1189.

[44] Su, W.W.; Caram, H.S. and Humphrey, A.E. (1992); Optimal design of the tubular microporous membrane aerator for shear-sensitive cell cultures. Biotechnol Prog 8:19-24.
[45] Dunlop, E.; Namdev, P. and Rosenberg, M. (1994); Effect of fluid shear forces on plant cell suspensions. Chemical engineering Science 49:2263-2276.
[46] Kieran, P.M.; Malone, D.M. and MacLoughlin, P.F. (2000); Effects of hydrodynamic and interfacial forces on plant cell suspension systems. Adv Biochem Eng Biotechnol 67:139-177.
[47] Kieran, P.M. (2001); Bioreactor design for plant cell suspension cultures. In: J. Cabral, M. Mota and J. Tramper (Eds.) Multiphase bioreactor design. Taylor and Francis, London; pp. 391-426.
[48] Shinmyo, A.; Shoji, T.; Bando, E.; Nagaya, S.; Nakai, Y.; Kato, K.; Sekine, M. and Yoshida, K. (1998); Metabolic engineering of cultured tobacco cells. Biotechnol Bioeng 58:329-332.
[49] Gao, J. and Lee, J. (1992); Effect of oxygen supply on the suspension culture of genetically modified tobacco cells. Biotechnology Progress 8:285-290.
[50] Cooney, C.; Wang, D. and Mateles, R. (1969); Measurement of heat evolution and correlation with oxygen consumption during microbial growth. Biotechnology and Bioengineering 11:269-281.
[51] Suehara, K.; Takao, A.; Nakamura, K.; Uozumi, N. and Kobayashi, T. (1996); Optimal expression of GUS gene from methyl jasmonate-inducible promoter in high density culture of transformed tobacco cell line BY-2. Journal of Fermentation and Bioengineering 82:51-55.
[52] Fischer, R.; Stoger, E.; Schillberg, S.; Christou, P. and Twyman, R.M. (2004); Plant-based production of biopharmaceuticals. Curr Opin Plant Biol 7:152-158.
[53] Junker, B.; Stanik, M.; Barna, C.; Salmon, P. and Buckland, B. (1998); Influence of impeller type on mass transfer in fermentation vessels. Bioprocess Engineering 19:403-413.
[54] Nienow, A.W. and Bujalski, W. (2002); Recent studies on agitated three-phase (gas-solid-liquid) systems in the turbulent regime. Chemical Engineering Research and Design 80:832-838.
[55] Junker, B.H.; Mann, Z. and Hunt, G. (2000); Retrofit of CD-6 (Smith) impeller in fermentation vessels. Appl Biochem Biotechnol 89:67-83.
[56] Pinelli, D.; Bakker, A.; Myers, K.J.; Reeder, M.F.; Fasano, J. and Magelli, F. (2003); Some features of a novel gas dispersion impeller in a dual-impeller configuration. Chemical Engineering Research and Design 81:448-454.
[57] Anon. (2002); http://www.chemineer.com/impellers.php.
[58] Hsiao, T.Y.; Bacani, F.T.; Carvalho, E.B. and Curtis, W.R. (1999); Development of a low capital investment reactor system: application for plant cell suspension culture. Biotechnol Prog 15:114-122.
[59] Su, W.W.; He, B.J.; Liang, H. and Sun, S. (1996); A perfusion air-lift bioreactor for high density plant cell cultivation and secreted protein production. Journal of Biotechnology 50:225-233.
[60] Su, W.W. (2000); Perfusion bioreactors. In: R. Spier (Ed.), Encyclopedia of Cell Technology. Wiley, New York; pp. 230-242.
[61] Castilho, L. and Medronho, R. (2002); Cell retention devices for suspendedcell perfusion cultures. Adv Biochem Eng Biotechnol 74:129–169.
[62] Voisard, D.; Meuwly, F.; Ruffieux, P.A.; Baer, G. and Kadouri, A. (2003); Potential of cell retention techniques for large-scale high-density perfusion culture of suspended mammalian cells. Biotechnol Bioeng 82:751-765.
[63] Walmsley, A.M. and Arntzen, C.J. (2003); Plant cell factories and mucosal vaccines. Curr Opin Biotechnol 14:145-150.
[64] Jervis, L. and Pierpoint, W. (1989); Purification technologies for plant proteins. J Biotechnol 11:161-198.
[65] Kusnadi, A.R.; Evangelista, R.L.; Hood, E.E.; Howard, J.A. and Nikolov, Z.L. (1998); Processing of transgenic corn seed and its effect on the recovery of recombinant beta-glucuronidase. Biotechnol Bioeng 60:44-52.

[66] Paramban, R.I.; Bugos, R.C. and Su, W.W. (2004); Engineering green fluorescent protein as a dual functional tag. Biotechnol Bioeng 86:687-697.

[67] Goddijn, O. and Pen, J. (1995); Plants as bioreactors. TIBTECH 13:379-387.

[68] Gallie, D. and Walbot, V. (1992); Identification of the motifs within the tobacco mosaic virus 5'-leader responsible for enhancing translation. Nucleic Acids Res 20:4631-4638.

[69] Day, C.; Lee, E.; Kobayashi, J.; Holappa, L.; Albert, H. and Ow, D. (2000); Transgene integration into the same chromosome location can produce alleles that express at a predictable level, or alleles that are differentially silenced. Genes Dev 14:2869-2880.

[70] Hilleren, P. and Parker, R. (1999); Mechanisms of mRNA surveillance in eukaryotes. Annu Rev Genet 33:229-260.

[71] Spiker, S. and William, F. (1996); Nuclear matrix attachment regions and transgene expression in plants. Plant Physiology 110:15-21.

[72] Voinnet, O.; Pinto, Y. and Baulcombe, D. (1999); Suppression of gene silencing: A general strategy used by diverse DNA and RNA viruses of plants. Proc Natl Acad Sci USA 96:14147-14152.

[73] Dordas, C.; Hasinoff, B.; Rivoal, J. and Hill, R. (2004); Class 1 hemoglobins, nitrate and NO levels in hypoxic maize cell suspension cultures. Planta 219:66-72.

[74] Frey, A.D.; Oberle, B.T.; Farrés, J. and Kallio, P.T. (2004); Expression of Vitreoscilla haemoglobin in tobacco cell cultures relieves nitrosative stress in vivo and protects from NO in vivo. Plant Biotechnol. J. 2:221-231.

[75] Shiao, T.L.; Ellis, M.H.; Dolferus, R.; Dennis, E.S. and Doran, P.M. (2002); Overexpression of alcohol dehydrogenase or pyruvate decarboxylase improves growth of hairy roots at reduced oxygen concentrations. Biotechnol Bioeng 77:455-461.

[76] Li, W. and Dickman, M.B. (2004); Abiotic stress induces apoptotic-like features in tobacco that is inhibited by expression of human Bcl-2. Biotechnol Lett 26:87-95.

[77] Goldstein, W. (1999); Economic considerations for food ingredients produced by plant cell and tissue culture. In: T. Fu, G. singh and W. Curtis (Eds.) Plant Cel and Tissue Culture for the Production of Food Ingredients. Kluwer, New York; pp. 195-214.

[78] Yu, H.; Shi, Y.; Yin, J.; Shen, Z. and Yang, S. (2003); Genetic strategy for solving chemical engineering problems in biochemical engineering. J Chem Technol Biotechnol 78:283-286.

[79] Ho, N.W.; Chen, Z.; Brainard, A.P. and Sedlak, M. (1999); Successful design and development of genetically engineered Saccharomyces yeasts for effective cofermentation of glucose and xylose from cellulosic biomass to fuel ethanol. Adv Biochem Eng Biotechnol 65:163-192.

[80] Haldrup, A.; Petersen, S.G. and Okkels, F.T. (1998); The xylose isomerase gene from Thermoanaerobacterium thermosulfurogenes allows effective selection of transgenic plant cells using D-xylose as the selection agent. Plant Mol Biol 37:287-296.

[81] Zhong, J. and Yoshita, T. (1995); High density cultivation of Perilla frutescens cell suspensions for anthocyanin production: effects of sucrose concentration and inoculum size. Biotechnol. Bioeng. 38:653-658.

[82] Matsubaram, K. and Fujita, Y. (1991); Production of berberine. In: A. Komamine, Misawa, A. and DiCosmo, F. (Ed.), Plant Cell Cultures in Japan. CMC, Tokyo; pp. 39-44.

[83] Ushiyama, K. (1991); Large scale cultivation of ginseng. In: A. Komamine, Misawa, A. and DiCosmo, F. (Ed.), Plant Cell Cultures in Japan. CMC, Tokyo; pp. 92-98.

[84] Su, W.W.; Lei, F. and Kao, N.P. (1995); High density cultivation of Anchusa officinalis in a stirred tank bioreactor with in situ filtration. Appl. Microbiol. Biotechnol. 44:293-299.

[85] Takahashi, S. and Fujita, Y. (1985); Production of shikonin. In: A. Komamine, Misawa, A. and DiCosmo, F. (Ed.), Plant Cell Cultures in Japan. CMC, Tokyo; pp. 72-98.

[86] Shibasaki-Kitakawa, N.; Takeishi, J. and Yonemoto, T. (2003); Improvement of catechin productivity in suspension cultures of tea callus cells. Biotechnol. Prog. 19:655-658.

[87] Schmidt, B. and Schuphan, I. (2002); Metabolism of the environmental estrogen bisphenol A by plant cell suspension cultures. Chemosphere 49:51-59.

[88] Walker, T.S.; Bais, H.P. and Vivanco, J.M. (2002); Jasmonic acid-induced hypericin production in cell suspension cultures of Hypericum perforatum L. (St. John's wort). Phytochemistry 60:289-293.
[89] Dong, H.D. and Zhong, J.J. (2001); Significant improvement of taxane production in suspension cultures of Taxus chinensis by combining elicitation with sucrose feed. Biochem. Eng. J. 8:145-150.
[90] Chattopadhyay, S.; Mehra, R.S.; Srivastava, A.K.; Bhojwani, S.S. and Bisaria, V.S. (2003); Effect of major nutrients on podophyllotoxin production in Podophyllum hexandrum suspension cultures. Appl Microbiol Biotechnol. 60:541-546.
[91] Kim, Y.J.; Weathers, P.J. and Wyslouzil, B.E. (2003); Growth dynamics of Artemisia annua hairy roots in three systems. Biotechnol. Bioeng. 83:428-443.
[92] Lorence, A.; Medina-Bolivar, F. and Nessler, C.L. (2004); Camptothecin and 10-hydroxycamptothecin from Camptotheca acuminata hairy roots. Plant Cell Rep. 22:437-441.
[93] Lin, H.W.; Kwok, K.H. and Doran, P.M. (2003); Development of Linum flavum hairy root cultures for production of coniferin. Biotecnol. Lett. 25:521-525.
[94] Chen, H.; Chen, F.; Zhang, Y.L. and Song, J.Y. (1999); Production of lithospermic acid B and rosmarinic acid in hairy root cultures of Salvia miltiorrhiza. J. Ind. Microbiol. Biotechnol. 22:133-138.
[95] Washita, D.; Shimomura, K. and Kitanaka, S. (2003); Polyacetylenes in hairy roots of a Panax hybrid. Planta Med. 69:1163-1165.
[96] Shi, H.P. and Kintzios, S. (2003); Genetic transformation of Pueraria phaseoloides with Agrobacterium rhizogenes and puerarin production in hairy roots. Plant Cell Rep. 21:1103-1107.
[97] Heilmann, J. and Bauer, R. (1999); New medical applications of plant secondary metabolites. In: M. Wink (Ed.), Functions of Plant Secondary Metabolites and Their Exploitation in Biotechnology. Sheffield, London; pp. 274-310.
[98] Shiao, T.L. and Doran, P.M. (2000); Root hairiness: effect on fluid flow and oxygen transfer in hairy root cultures. J. Biotechnol. 83:199-210.
[99] Kanokwaree, K. and Doran, P.M. (1997); The extent to which external oxygen transfer limits growth in shake flask culture of hairy roots. Biotechnol. Bioeng. 55:520-526.
[100] Bhadra, R. and Shanks, J.V. (1997); Transient studies of nutrient uptake, growth, and indole alkaloid accumulation in heterotrophic cultures of hairy roots of Catharanthus roseus. Biotechnol Bioeng. 55:527-534.
[101] Bordonaro, J.L. and Curtis, W.R. (2000); Inhibitory role of root hairs on transport within root culture bioreactors. Biotechnol. Bioeng. 70:176-186.
[102] Kwok, K.H. and Doran, P.M. (1995); Kinetic and stoichiometric analysis of hairy roots in a segmented bubble column reactor. Biotechnol. Prog. 11:429-435.
[103] Woo, S.H.; Park, J.M. and Yang, J. (1996); Root culture using a mist culture system and estimation of scale-up feasibility. J. Chem. Tech. Biotechnol. 66:355-362.
[104] Ramakrishnan, D.; Salim, J. and Curtis, W.R. (1994); Inoculation and tissue distribution in pilot-scale plant root culture bioreactor. Biotechnol. Tech. 8:639-644.
[105] Falk, L.R. and Doran, P.M. (1996)) Influence of inoculum porphology on growth of Atropa belladonna hairy roots and production of tropane alkaloids. Biotechnol. Lett. 18:1099-1104.
[106] Woo, S.H.; Park, J.M. and Yang, J. (1997); Induction of branch roots by cutting methods in Hyoscyamus niger root cultures. Plant Cell Tissue Organ Cul. 48:131-134.
[107] Lee, K.T.; Suzuki, T.; Yamakawa, T.; Kodama, T.; Igarashi, Y. and Shimomura, K. (1999); Production of tropane alkaloids by transformed root cultures of Atropa belladonna in stirred bioreactors with a stainless steel net. Plant Cell Rep. 18:567-4571.
[108] Huffmann, G.A.; White, F.F.; Gordon, M.P. and Nester, G.W. (1984); Hairy root inducing plasmid: physical map and homology to tumor-inducing plasmids. J. Bacteriol 152:269-276.

[109] Vanhala, L.; Eeva, M.; Lapinjoki, S.; Hiltunen, R. and Oksman-Caldentey, K.M. (1998); Effect of growth regulators on transformed root cultures of Hyoscyamus muticus. J. Plant Physiol 153:475-481.

[110] Staniszewska, I.; Krolicka, A.; Malinski, E.; Lojkowska, E. and Szafranek, J. (2003); Elicitation of secondary metabolites in in-vitro cultures of Ammi majus L. Enzyme. Enzyme Microb. Technol. 33:565-568.

[111] Wang, J.W.; Kong, F.X. and Tan, R.X. (2002); Improved artemisinin accumulation in hairy root cultures of Artemisia annua by (22S, 23S)-homobrassinolide. Biotechnol. Lett. 24:1573-1577.

[112] Rao, S.R.; Tripathi, U.; Suresh, B. and Ravishankar, G.A. (2001); Enhancement of secondary metabolite production in hairy root cultures of Beta vulgaris and Tagetes patula under the influence of microalgal elicitors. Food Biotechnol. 15:35-46.

[113] Moreno-Valenzuela, O.A.; Minero-Garcia, Y.; Chan, W.; Mayer-Geraldo, E.; Carbajal, E. and Loyola-Vargas, V.M. (2003); Increase in the indole alkaloid production and its excretion into the culture medium by calcium antagonists in Catharanthus roseus hairy root. Biotechnol. Lett. 25:1345-1349.

[114] Bais, H.P.; Dattatreya, B.S. and Ravishankar, G.A. (2003); Production of volatile compounds by hairy root cultures of Cichorium intybus L under the influence of fungal elicitors and their analysis using solid-phase micro extraction gas chromatography-mass spectrometry. J. Sci Food Agric. 83:769-774.

[115] Bais, P.H.; Walker, T.S.; Schweizer, H.P. and Vivanco, J.M. (2002); Root specific elicitation and antimicrobial activity of rosmarinic acid in hairy root cultures of Ocimum basilicum. Plant Physiol. Biochem. 40:983-995.

[116] Palazon, J.; Cusido, R.M.; Bonfill, M.; Mallol, A.; Moyano, E.; Morales, C. and Pinol, M.T. (2003); Elicitation of different Panax ginseng transformed root phenotypes for an improved ginsenoside production. Plant physiol. Biochem. 41:1019-1025.

[117] Zhang, C.; Yan Q.; Cheuk, W.K. and Wu, J. (2004); enhancement of tanshinone production in Salvia miltiorrhiza hairy root culture by Ag+ elicitation and nutrient feeding. Planta Med. 70:147-151.

[118] Komaraiah, P.; Reddy, G.V.; Reddy, P.S.; Raghavendra, A.S.; Ramakrishna, S.V. and Reddanna, P. (2003); Enhanced production of antimicrobial sesquiterpenes and lipoxygenase metabolites in elicitor-treated hairy root cultures of Solanum tuberosum. Biotechnol. Lett. 25:593-597.

[119] Hayashi, H.; Huang, P. and Inoue, K. (2003); Up-regulation of soyasaponin biosynthesis by methyljasmonate in cultured cells of Glycyrrhiza glabra. Plant Cell Physiol. 44:404-411.

[120] Qian, Z.G.; Zhao, Z.J.; Tian, W.H.; Xu, Y.; Zhong, J.J. and Qian, X. (2004); Novel synthetic jasmonates as highly efficient elicitors for taxoid production by suspension cultures of Taxus chinensis. Biotechnol Bioeng. 86:595-599.

[121] Ketchum, R.E.B.; Gibson, D.M.; Croteau, R.B. and Shuler, M. (1999); The kinetics of taxoid accumulation in cell suspension cultures of Taxus following elicitation with methyl jasmonate. Biotechnol. Bioeng. 62:97-105.

[122] Martin, D.; Throll, D.; Gershenzon, J. and Bohlmann, J. (2002); Methyl jasmonate induces traumatic resin ducts, terpenoid resin biosynthesis, and terpenoid accumulation in developing xylem of Norway spruce stems. Plant physiol. 129:1003-1018.

[123] Lu, M.B.; Wong, H.L. and Teng, W.L. (2001); Effects of elicitations on the production of saponin in cell culture of Panax ginseng. Plant Cell Rep. 20:674-677.

[124] Aoyagi, H.; Kobayashi, Y.; Yokoyama, M.; Kusakari, K. and Tanaka, H. (2001); Efficient production of saikosaponin in Bupleurum falcatum root fragments combined with signal transducers. Appl. Microbiol. Biotechnol. 57:482-488.

[125] Lee, K.T., Yamakawa, t., Kodama, T. and Shimomura K. (1998); Effects of chemicals on alkaloid production by transformed roots of belladonna. Phytochemistry 49:2343-2347.

[126] Kondo, O.; Honda, H.; Taya, M. and Kobayashi, T. (1989); Comparison of growth properties of carrot hairy root in various bioreactors. Applied Microbiology and Biotechnology 32:291-294.

[127] Palazon, J.; Mallol, A.; Eibl, R.; Lettenbauer, C.; Cusido, R.M. and Pinol, M.T. (2003); Growth and ginsenoside productionin hairy root cultures of Panax ginseng using a novel bioreactor. Planta Med. 69:344-349.

[128] Sudo, H.; Yamakawa, T.; Yamazaki, M.; Aimi, N. and Saito, K. (2002); Bioreactor production of camptothecin by hairy root cultures of Ophiorrhiza pumila. Biotechnol. Lett. 24:359-363.

[129] Du, M.; Wu, X.J.; Ding, J.; Hu, Z.B.; White, K.N. and Branford-White, C.J. (2003); Astragaloside IV and polysaccharide production by hairy roots of Astragalus membranaceus in bioreactors. Biotechnol. Lett. 25:1853-1856.

[130] McKelvey, S.A.; Gehrig, J.A.; Hollar, K.A. and Curtis, W.R. (1993); Growth of plant root cultures in liquid- and gas-dispersed reactor environments. Biotechnol Prog 9:317-322.

[131] Wyslouzil, B.E.; Whipple, M.; Chatterjee, C.; Walcerz, D.B.; Weathers, P.J. and Hart, D.P. (1997); Mist deposition onto hairy root cultures: aerosol modeling and experiments. Biotechnol. Prog. 13:185-194.

[132] Ramakrishanan, D.; Luyk, D. and Curtis, W.R. (1999); Monitoring biomass in root culture systems. Biotechnol. Bioeng. 62:711-721.

[133] Wu, Z.L.; Yuan, Y.J. and Hu, Z.D. (2000); Study on volumetric oxygen transfer coefficient of the simulated plant cell two-phase culture system. Chem. Eng. Technol. 23:1111-1114.

[134] Collins-Pavao, M.; Chin, C.K. and Pedersen, H. (1996); Taxol partitioning in two-phase plant cell cultures of Taxus brevifolia. J. Biotechnol. 49:95-100.

[135] Corry, P.J.; Reed, W.L. and Curtis, W.R. (1993); Enhanced recovery of solavetivone from Agrobacterium transformed root cultures of Hyoscyamus muticus using integrated product extraction. Biotechnol. Bioeng. 42:503-508.

[136] Byun, S.Y. and Pedersen, H. (1994); Two-phase airlift fermentor operation with elicitation for enhanced production of benzophenanthridine alkaloids in cell suspensions of E. californica. Biotechnol Bioeng. 44:14-20.

[137] Tikhomiroff, C.; Allais, S.; Klvana, M.; Hisiger, S. and Jolicoeur, M. (2002); Continuous selective extraction of secondary metabolites from Catharanthus roseus hairy roots with silicone oil in a two-liquid-phase bioreactor. Biotechnol. Prog. 18:1003-1009.

[138] Thimmaraju, R.; Bhagyalakshmi, N.; Narayan, M.S. and Ravishankar, G.A. (2003); Food-grade chemical and biological agents permeabilize red beet hairy roots, assisting the release of betalaines. Biotechnol. Prog. 19:1274-1282.

[139] Lin, L.D.; Wu, J.Y.; Ho, K.P. and Qi, S.Y. (2001); Ultrasound-induced physiological effects and secondary metabolite (saponin) production in Panax ginseng cell cultures. Ultrasound Med. Biol. 27:1147-1152.

[140] Lin, L.D. and Wu, J.Y. (2002); Enhancement of shikonin production in single- and two-phase suspension cultures of Lithospermum erythrorhizon cells using low-energy ultrasound. Biotechnol. Bioeng. 78:81-88.

[141] Wu, J.Y. and Ge, X.C. (2004); Oxidative burst, jasmonic acid biosynthesis, and taxol production induced by low-energy ultrasound in Taxus chinensis cell suspension cultures. Biotechnol. Bioeng. 85:714-721.

[142] Choi, Y.H.; Kim, J.Y.; Ryu, J.H.; Yoo, K.P.; Chang, Y.S. and Kim, J.W. (1998); Supercritical carbon dioxide extraction of podophyllotoxin from Dysosma pleiantha roots. Planta Med. 64:482-483.

[143] Rompp, H.; Seger, C.; Kaiser, C.S.; Haslinger, E. and Schmidt, P.C. (2004); Enrichment of hyperforin from St. John's Wort (Hypericum perforatum) by pilot-scale supercritical carbon dioxide extraction. Euro. J. Pharm. Sci. 21:443-451.

[144] William, R.D.; Chauret, N.; Bedard, C. and J., A. (1992); Effect of polymeric adsorbents on the production of sanguinarine by Papaver somniferum cell cultures. Biotechnol. Bioeng. 40:971-977.

[145] Payne, G.F.; Payne, N.N.; Ninomiya, Y. and Shuler, M.L. (1989); Adsorption of nonpolar solutes onto neutral polymeric sorbents. Separation Sci. Technol. 24:457-465.

[146] Saito, K.; Sudo, H.; Yamazaki, M.; Koseki-Nakamura, M.; Kitajima, M.; Takayama, H. and Aimi, N. (2001); Feasible production of camptothecin by hairy root culture of Ophiorrhiza pumila. Plant Cell Rep. 20:267-271.

[147] Lee-Parsons, C.W.T. and Shuler, M. (2002); The effect of ajmalicine and resin addition timing on the production of indole alkaloids from Catharanthus roseus cell cultures. Biotechnol. Bioeng. 79:408-415.

[148] Hashimoto, T.; Yun, D.Z. and Yamada, Y. (1993); Production of tropane alkaloids in genetically engineered root cultures. Phytochemicals 32:713-718.

[149] Boehm, R.; Sommer, S.; Li, S.M. and Heide, L. (2000); Genetic engineering on shikonin biosynthesis: expression of the bacterial ubiA gene in Lithospermum erythrorhizon. Plant Cell Physiol. 41:911-919.

[150] Hughes, E.H.; Hong, S.B.; Gibson, S.I.; Shanks, J.V. and San, K.Y. (2004); Expression of a feedback-resistant anthranilate synthase in Catharanthus roseus hairy roots provides evidence for tight regulation of terpenoid indole alkaloid levels. Biotechnol. Bioeng. 86:718-727.

[151] Yu, O.; Shi, J.; Hession, A.O.; Maxwell, C.A.; McGonigle, B. and Odell, J.T. (2003); Metabolic engineering to increase isoflavone biosynthesis in soybean seed. Phytochemistry 63:753-763.

Bioprocessing for Value-Added Products from Renewable Resources
Shang-Tian Yang (Editor)

Chapter 11. Production of High-Value Products by Marine Microalgae Thraustochytrids

King Wai Fan[a] and Feng Chen[a,b]

[a]Department of Botany, The University of Hong Kong, Pokfulam Road, Hong Kong, China

[b]South China Sea Institute of Oceanology, Chinese Academy of Sciences, Guangzhou, China

1. INTRODUCTION

Man has made use of natural products since ancient times. One example is the use of microalgae as food, dating back to 2000 years ago when *Nostoc flagelliforme*, a cyanobacterium, was utilized as a food source in China [1]. Indeed, microalgae constitute an extremely diverse group of organisms encompassing prokaryotic cyanobacteria and eukaryotic protists. By definition, microalgae are a group of microscopic organisms; the majority possess photosynthetic and accessory pigments for undergoing photosynthesis [2]. Modern applications of microalgal biotechnology have only become the subject of intensive studies in recent decades after the potential of certain microalgae to produce numerous highly useful or high-value products, such as biohydrogen, pigments (e.g., carotenoids, phycobiliproteins, etc.), polyunsaturated fatty acids (PUFAs), sterols, and vitamins was recognized [3–7].

Reviews concerning the benefits and uses of microalgae in biotechnology are numerous [e.g., 8–12]. Unanimously, these reviews concern the potential of using microalgae as producers of high-value products. Specifically, high-value products include those that are extracted for direct human consumption, mainly in the form of functional foods, which are dietary supplements for immune enhancement and disease prevention [13]. In addition, microalgal high-value products also include those that are used as feed additives in the aquaculture industry [14], which are mainly applied as supplements of rare but essential nutrients, such as PUFAs for conditioning zooplankton, which are subsequently employed as food for cultivated aquatic animals [15].

Thraustochytrids have been recently classified as microalgae, especially in a commercial sense [8, 16]. For decades, scientific investigations on thraustochytrids by taxonomists and ecologists mainly focused on their distribution and abundance in natural environments. The biotechnological potential of thraustochytrids was not recognized until the discovery that thraustochytrids contain substantial amounts of docosahexaenoic acid (22:6 DHA ω-3) [17]. Since then, numerous studies have been conducted to investigate the DHA production potential of certain strains of thraustochytrids for marketing as single cell oils [16, 18]. More recently, the potential for using thraustochytrids for the co-production of astaxanthin, squalene, and sterol has also been suggested [7, 19–20] and has led to the concept of using thraustochytrids as cell factories for the production of novel high-value natural products [21].

Although thraustochytrids have drawn more and more attention and have been suggested as cell factories for novel products, up to now, there have been no comprehensive reviews concerning the production of high-value products by thraustochytrids. Singh and Ward [22] provided the only relevant review about the effects of environmental and nutritional factors on DHA production in thraustochytrids. The present review, therefore, intends to focus on some recent advances in the biotechnological applications of thraustochytrids with an emphasis on the production of various high-value products. Moreover, the use of thraustochytrids as feed additives in aquaculture and the poultry industry as well as safety issues that concern their biotechnological applications are also addressed.

2. MODES OF MICROALGAL CULTIVATION

The development of efficient large-scale microalgal cultivation systems is essential for the production of commercially important algal products [8]. In general, microalgal cultivation systems have been broadly categorized into photoautotrophic and heterotrophic systems. The former is light-driven and is suitable for most microalgal species, while the latter requires organic substrates to be used as the sole carbon and energy source and is applicable only to a small number of microalgae [10–11]. Nevertheless, heterotrophic cultivation is considered superior to conventional photoautotrophic cultivation due to the fact that light, the growth-limiting factor, is eliminated [10]. Since there are numerous cultivation modes that can be operated under heterotrophic cultivation, for example, batch, fed-batch, and continuous cultures, the following sections focus on relevant cultivation examples associated with DHA production by thraustochytrids.

Commercially, there are two main types of photoautotrophic systems: open pond and closed photobioreactors; both use natural sunlight or artificial illumination for microalgal cultivation [8]. Open pond is considered as the oldest and simplest system employed for the mass cultivation of microalgae. This system offers culture conditions identical to the external environment and takes advantage of "free" sunlight as its energy source, but the high risk of microbial contamination has restricted the use of such systems to a limited number of microalgae, such as *Dunaliella*, which can tolerate high salinity [8], and *Spirulina*, which can grow well at high pHs [23]. To the best of our knowledge, there is no report on DHA production by microalgae using an open pond system.

Closed photobioreactors, such as tubular, fiber-optical, and helical reactors, use either natural sunlight or artificial illumination. This type of photoautotrophic system has been suggested as the alternative to the disadvantageous open-pond cultivation [24]. Common advantages of using closed photobioreactors include the large illumination surface to volume ratio, the minimization of contamination from the environment, and the efficient control of culture conditions [10]. Close photobioreactors have been employed for DHA production using *Isochrysis galbana* [25–26]. Poisson and Ergan [26] reported a DHA yield of 0.96 mg

L^{-1} by *I. galbana* grown in a conventional bioreactor with cool white fluorescent tubes for continuous illumination. By using optical fiber photobioreactors, the DHA yield in *I. galbana* was further enhanced to 5.4 mg L^{-1}. Vazhappilly and Chen [27] obtained a DHA yield of 1 mg L^{-1} for *Thraustochytrium aureum* ATCC 28211 grown under photoautotrophic conditions in flask cultures (0.1 g L^{-1} biomass after 23 days). The relatively low production of DHA by microalgae using photoautotrophic cultures, coupled with the costly construction and maintenance of closed photobioreactors, has diverted the focus to the heterotrophic mode of production [4].

Heterotrophic systems are a cost-effective alternative for culturing microalgae since light is eliminated for cultivation, and, at the same time, organic substrates are utilized as the sole source of carbon and energy, leading to a high biomass concentration and productivity [10]. Heterotrophic cultivation of microalgae, such as *Crypthecodinium cohnii* and thraustochytrids, for DHA production has been reported using conventional bioreactors or fermenters [28]. Vazhappilly and Chen [29] reported that when *T. aureum* ATCC 28211 was used as the producing organism on 5% glucose, the DHA yield was increased to 4.0 mg L^{-1}, compared to 1.0 mg L^{-1} under photoautotrophic conditions. This illustrates that the heterotrophic mode of growth is superior to the photoautotrophic mode in terms of biomass and DHA production. Bowles *et al.* [30] reported that the high DHA yield of 1.6 g L^{-1} and productivity of 0.49 g L^{-1} day^{-1} were achieved by *Thraustochytrium* strain G13 in bioreactors. More efficiently, a DHA yield of 3 g L^{-1} and productivity of 3.1 g L^{-1} day^{-1} were reported for *Schizochytrium* sp. SR21 when grown in conventional batch bioreactors with 12% glucose [31]. Modifying the batch culture system to become a fed-batch, chemostat or perfusion culture may further enhance the biomass yield and DHA productivity. Thus, heterotrophic systems may provide a cost-effective way for continuously supplying high quality microalgae-produced DHA on a large scale [28].

3. THRAUSTOCHYTRIDS

3.1. The taxonomy of thraustochytrids

Conventionally, thraustochytrids are taxonomically defined as a group of obligate, eukaryotic marine microorganisms characterized by monocentric thalli that can attach to their substrata by means of ectoplasmic net elements that arise from an organelle termed the sagenogenetosome [32]. Reproduction is usually by the asexual formation of biflagellate, heterokont, and zoospores from zoosporangia. Amoeboid stages are also frequently observed. The anteriorly directed flagellum is of the tinsel type, which has a bilateral row of mastigonemes and is the longer of the two, while the posterior flagellum is of the whiplash type.

Thraustochytrids were first described by Sparrow [33] and were placed in the family Thraustochytriaceae and order Saprolegniales (Öomycetes) to include chytrid-like organisms with biflagellate zoospores. However, the presence of L-galactose as the primary monomer of the cell walls [34], the possession of a sagenogenetosome and associated ectoplasmic net elements [35], and the absence of sexual reproduction do not suggest a close relationship with the Oomycota [36]. The inability of *T. aureum* to synthesize lysine via the aminoadipic acid

(AAA) or diaminopimelic (DAP) pathways indicates that they are not related to fungi [37]. Rather, a closer affinity with *Labyrinthula* and *Aplanochytrium* (Labyrinthulids) has been demonstrated [36, 38–39]. Thus, the Thraustochytriaceae and Labyrinthulaceae form a natural grouping [40]. Genera are separated primarily on the morphology of the thallic stage and differences in sporogenesis and spore release, such as sporangium size, and cleavage patterns. However, morphological characters used for species delimitation are sometimes variable and difficult to use for species characterization [41–42].

Advances in molecular techniques have assisted scientists in resolving the problematic phylogenetic position of thraustochytrids, which was impossible to determine with morphological characteristics alone. 5S ribosomal RNA data analysis from *Ulkenia visurgensis* and *Schizochytrium aggregatum* showed a high dissimilarity between thraustochytrids and other protists [43]. Mo and Rinkevich [44] developed a simple, reliable, and fast protocol for DHA extraction in thraustochytrids. The application of such a protocol has expedited data generation for phylogenetic studies of thraustochytrids.

At present, thraustochytrids are members of the phylum Heterokonta of the stramenopiles of the kingdom Straminipila [45]. Stramenopiles organisms are characterized by having tripartite, tubular, flagellar hairs and mitochondria with tubular cristae [45–46]. Sequence analysis of the 18S ribosomal RNA data confirms the unique position of thraustochytrids in the stramenopiles and separates the thraustochytrids into two major groupings: the labyrinthulid phylogenetic group and the thraustochytrid phylogenetic group [47]. Recently, Leander and Porter [48] also indicated that the labyrinthulids and thraustochytrids are closely related when the partial small-subunit ribosomal DNA region was sequenced. The phylum consists of three distinct groupings, namely, the labyrinthulids, the thraustochytrids, and the labyrinthuloids [48]. Alternatively, Huang *et al.* [49] proposed the use of fatty acid profiles of individual species to classify thraustochytrids and have confirmed the groupings derived from PUFAs profile corroborated with the 18S data in resolving the phylogenetic position of the thraustochytrids. Currently, the thraustochytrids contains seven genera: *Althornia, Diplophyhrys, Elina, Japonochytrium, Schizochytrium, Thraustochytrium,* and *Ulkenia* [45, 47, 50].

3.2. Uniqueness of tropical/subtropical thraustochytrids for fermentation

Thraustochytrids are cosmopolitan and have been isolated from a wide range of habitats throughout the world, including some extreme environments, such as hyper-saline lakes and the deep sea [51]. Their roles in the mineralization of substrata from plant and animal origins are well documented [50]. Recently, thraustochytrids from subtropical mangroves have been investigated and suggested as one of the major colonizers of mangrove detritus [52–53]. In mangrove environments, thraustochytrids are found to be associated with decaying mangrove leaves and can also colonize, penetrate, and decompose mangrove leaves with their hydrolytic enzymes [54–55]. Apart from fulfilling their ecological roles, mangrove thraustochytrids are superior to their counterparts in the temperate regions in their biotechnological applications. This is largely because of their abilities to tolerate fluctuating environmental conditions (e.g., growing in various saline environments) and their potential to produce substantial biomass yield [52, 56]. The unique ability of mangrove thraustochytrids to grow at low salinities and to achieve high biomass concentrations offers advantages for the biotechnological production of valuable metabolites through fermentation. Among the mangrove isolates, *Schizochytrium mangrovei* containing numerous oil globules is an ideal candidate for the exploitation of oil production, especially DHA (Fig. 1).

Thraustochytrids can tolerate a range of salinity ranging from as low as 1 ‰ (parts per thousand or g L^{-1}) to full-strength seawater (37 ‰). However, they generally exhibit optimal growth in salinity proximal to the latter [57]. However, from a biotechnological perspective, the high concentration of sodium chloride in full-strength seawater is not suitable for fermentation because of the corrosive effect of chloride ions on conventional bioreactors [58]. Therefore, strains that require minimal sodium chloride for fermentation are ideal candidates because fermentation under low sodium chloride concentrations minimizes the production cost. Tropical and subtropical thraustochytrids, including *Schizochytrium limacinum* and *S. mangrovei*, that only require half-strength seawater for optimal growth are desirable and advantageous from a production point of view [52, 59].

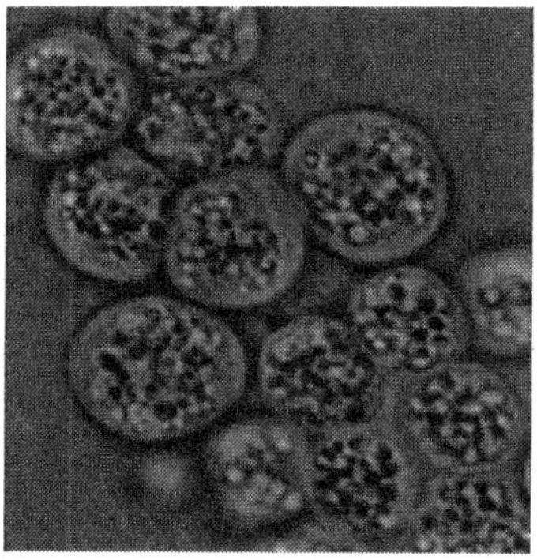

Fig. 1. *Schizochytrium mangrovei* containing numerous oil globules.

Thraustochytrids with the potential to produce high biomass concentrations are beneficial for the production of high-value products. Thraustochytrids isolated from the subtropical regions are high biomass producers. For example, *S. limacinum* SR21 isolated from Yap Island, Japan, can produce a biomass concentration of as high as 48.1 g L^{-1} when grown under optimal conditions [60]. However, selecting the appropriate species is important because biomass production potential does vary from species to species, although they may be isolated from the same locality. Bowles *et al.* [30] conducted an extensive study to investigate the influence of geographical distribution on the biomass concentration of 52 strains on glucose media. Their findings indicated that strains isolated from subtropical locations (25–27°N) tended to produce higher biomass than their counterparts in the cool (59–61°N) and cold (50–51°N) temperate regions. A closer observation of the isolated strains indicated the variability of biomass concentrations among strains within each geographical zone [30]. Fan *et al.* [56] demonstrated that of the 9 strains of thraustochytrids isolated from decaying *Kandelia candel* leaves, only two *S. mangrovei* strains produced biomass concentrations high enough to be considered for potential biotechnological applications. These observations concluded that thraustochytrids isolated from tropical and subtropical regions are robust in terms of biomass concentration but it is also important to select the appropriate species for exploitation (e.g., thraustochytrids with high biomass production as well as DHA content).

4. HIGH-VALUE PRODUCTS FROM THRAUSTOCHYTRIDS

4.1. Astaxanthin

Astaxanthin (3,3'-dihydroxy-ß,-ß,carotene 4,4'-dione), a ketocarotenoid, is widely distributed in nature and is responsible for imparting the distinctive orange-red coloration in animals, particularly in shrimp, crabs, lobster, and salmonids of marine origin [6]. Structurally, astaxanthin exists in three stereoisomers (i.e., 3S, 3′S; 3R, 3′S; and 3R, 3′R) depending on the projection of the hydroxyl group from the two asymmetric carbon atoms located at the C3 and C3′ positions (Fig. 2.) [61]. In red yeast, *Xanthophyllomyces dendrorhous* (previously known as *Phaffia rhodozyma*), the isomer 3R, 3′R is present as the predominant form while synthetic astaxanthin contains a mixture of the three isomers [62]. The provision of 3R, 3′R isomer from *X. dendrorhous*, however, is not a preferred choice for incorporation into feed additives for aquaculture because fish are predominantly pigmented with the 3S, 3′S isomer. The green microalga *Haematococcus pluvialis* is by far the most intensively studied species, owing to the high content of astaxanthin (0.7–3.4% of biomass) within its cell [63]. In *H. pluvialis*, although astaxanthin is produced in the form of 3S, 3′S, the preferred isomer for incorporation in aquaculture feed for salmonids [61], the slow growth rate in heterotrophic cultures and the susceptibility of photoautotrophic cultures to contamination are the major drawbacks for the use of *Haematococcus* for astaxanthin production [64]. Thraustochytrids are capable of heterotrophic growth on an organic substrate in the dark and synthesizing astaxanthin. The potential of astaxanthin production by thraustochytrids is thus of high interest [19, 65].

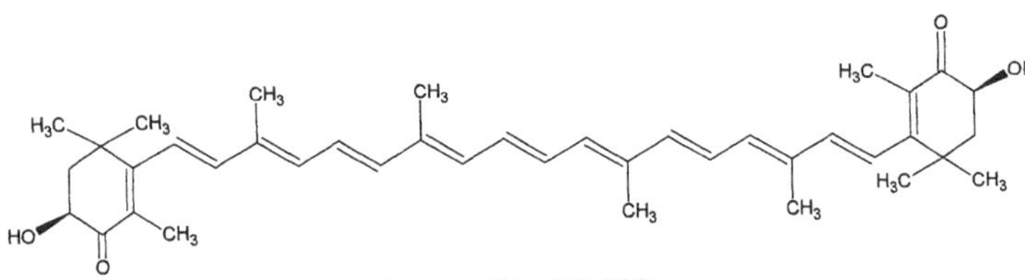

Astaxanthin 3S, 3'S

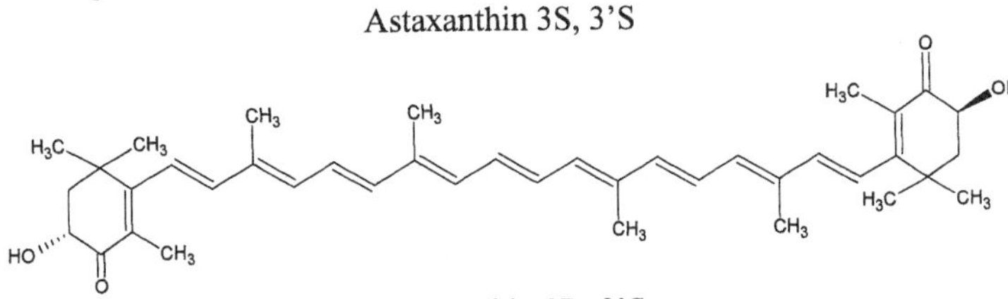

Astaxanthin 3R, 3'S

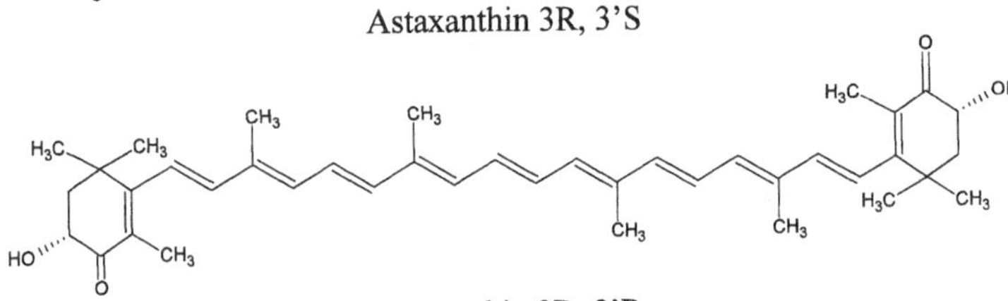

Astaxanthin 3R, 3'R

Fig. 2. Stereoisomers of astaxanthin.

4.1.1. Health benefits and industrial applications

Astaxanthin is a superior antioxidant and is known as "super vitamin E' because its antioxidant activity surpasses that of other carotenoids, such as β-carotene, canthaxanthin, lutein, zeaxanthin, as well as vitamins C and E, especially in quenching singlet oxygens and scavenging free radicals [66]. Increasing evidence has demonstrated the role of astaxanthin in protecting against chemically-induced cancers [67], enhancing the immune system [68], preventing the damaging effects of ultraviolet radiation [69], and treating a number of human diseases, such as atherosclerosis and age-related macular degeneration [70]. Astaxanthin is being marketed as a functional food ingredient in many places of the world [63].

In addition to its role as functional food ingredients or as dietary supplements, the major application of astaxanthin is as feed additive in aquaculture and the poultry industry [61]. In aquaculture, astaxanthin plays an important role in enhancing flesh coloration (i.e., the pink color) of farmed salmonids which is desired by consumers. Since salmonids lack the ability to synthesize astaxanthin *de novo*, the pigment must be incorporated into their diets [61]. In the poultry industry, astaxanthin is effective for increasing yolk coloration in eggs [71].

4.1.2. Production potential

Although Goldstein [72] recognized the formation of bright orange-red pigmentation in *Thraustochytrium roseum*, he did not attempt to characterize or identify the pigments involved in the coloration. It was more than ten years later when the presence of carotenoids, such as β-carotene and canthaxanthin, were confirmed in the thraustochytrid *S. aggregatum* [73]. With the development of modern instrumentation, such as HPLC-MS, the presence of other important carotenoids, astaxanthin and phoenicoxanthin, were recently confirmed to exist in thraustochytrids [66].

A preliminary study of astaxanthin in *Thraustochytrium* sp. CHN-1 indicated that the dominant form of astaxanthin was present as 3S, 3′S, the all trans-isomer, and was synthesized from β-carotene with echinenone and canthaxanthin as intermediates [65, 74]. This study also indicated that content of carotenoids was positively associated with biomass concentration, reaching the highest content of 0.45 mg g^{-1} after 8 days of incubation, and astaxanthin can comprise as much as 50% of the total carotenoids [66]. In addition, Yamaoka *et al.* [74] tested the effect of different light emitting diodes (LEDS) on astaxanthin production in *Thraustochytrium* sp. CHN-1 and indicated that blue LEDS at 470 nm stimulated astaxanthin production in *Thraustochytrium* sp. CHN-1 to a higher cellular content of 0.5 mg g^{-1} as compared to 0.1 mg g^{-1} obtained for the same strain under darkness after 15 days of cultivation. During the same period, a more in-depth study on the carotenoid profile in *Schizochytrium* sp. KH105 was documented [19]. In the study, the effects of carbon and nitrogen concentrations on astaxanthin production were investigated. The result indicated that a high C/N ratio (10% glucose and 0.27% yeast extract and polypeptone) led to the highest astaxanthin yield of 6.1 mg l^{-1} [19]. Long [75] has recently filed for a patent on a formula consisting of corn sugar, processed corn steep liquor, sea salts and thiamine for increasing astaxanthin production in *Thraustochytrium* sp. After 7 days of incubation, an astaxanthin content of 1.5 mg g^{-1} and yield of 8.35 mg L^{-1} were achieved in the *Thraustochytruium* sp.

Due to the ease of heterotrophic cultivation and downstream processing, including the extraction of astaxanthin with acetone, Aki *et al.* [19] suggested the potential of thraustochytrids for the commercial production of astaxanthin. Nevertheless, astaxanthin content in thraustochytrids remains very low as compared to the green microalga *Haematococcus* and thus must be further increased in order to justify commercial production. It is known that a number of environmental and cultural stimulants, such as temperature and metal ions, increase the production of astaxanthin in microorganisms. Further investigations of possible enhancements of astaxanthin production should focus on the effects of these stimulants on astaxanthin synthesis in thraustochytrids [76].

4.2. Docosahexaenoic acid (DHA)

There are two major groups of polyunsaturated fatty acids (PUFAs), namely, ω-3 and ω-6, which are distinguished primarily by the location of the terminal double bond. In ω-3 PUFAs, the terminal double bond is located 3 carbon atoms from the methyl terminus, whereas the ω-6 PUFAs contain a terminal double bond that is 6 carbon atoms from the methyl terminus [22]. Among the PUFAs, DHA, composed of 22 carbon atoms and 6 double bonds in the ω-3 family, has received the most attention because of its importance in the functional development and possible treatment of various diseases in humans [77]. The chemical structure of DHA is shown in Fig. 3.

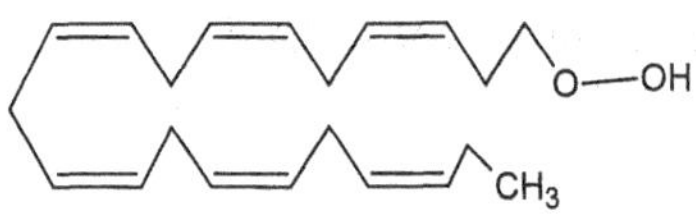

Fig. 3. Chemical structure of docosahexaenoic acid.

Currently, fatty marine fish are the main commercial source of DHA in human diets [78]. With an increasing demand for fish for human consumption and a threatening decrease in fish stocks due to over-fishing, fish and their oils may not be able to sustain the world's future requirements for DHA [79]. Moreover, fish can only synthesize a minute amount of DHA, and most of their PUFAs are acquired from the food chain and originated in marine phytoplankton, including microalgae [79]. Therefore, it is desirable to explore other potential sources for DHA production. Up to now, most research efforts have explored the possibility of producing DHA using marine microalgae [28–29]. The use of microorganisms as a source of DHA is advantageous because there is no seasonal limitation to production, which can be easily achieved using industrial fermentation processes [80]. Moreover, microorganisms usually contain relatively simple fatty acid profiles with a high level of the desired fatty acid, for example, DHA in *Schizochytrium* sp. SR21 [18]. This simplifies purification and reduces unpleasant flavors, which are often associated with impurities.

Among the microalgal groups, thraustochytrids are one of the most promising sources of DHA [21]. Spray-dried *Schizochytrium* sp. biomass has been marketed as Algamac® 2000 and Algamac®3000, and is currently available as a source of DHA to enrich aquaculture species (Aquafauna Bio-Marine, USA). Depending on the strain and growth conditions, the total lipid content of thraustochytrids can be as high as 77% of cell dry weight, as in the case of *S. limacinum* [60]. DHA may represent as much as 50% of the total fatty acids in thraustochytrids [81–82]. Because of the exceptionally high DHA content, the production potential of DHA in thraustochytrids is tremendous.

4.2.1. Health benefits

Because of its various health benefits, DHA has attracted much research and commercial interest [78]. DHA and other ω-3 PUFAs are essential for human development *in utero* and in infancy as well as throughout the lifespan. DHA in biological tissues is mostly present in an esterified form of phospholipids, triglycerides or cholesterol, and is mainly localized in the sn-2 position of the glycerol backbone [83]. DHA is also a major component in many human tissues, including the grey matter of the brain, the membrane of the retina, and spermatozoa [84–86]. Maintaining a sufficient amount of DHA in the brain is essential for normal growth in infants [87]; a deficiency of DHA is shown to be strongly associated with learning disabilities [88]. Therefore, an adequate supply of DHA from mother's milk and/or infant formulae is required to meet the high demand of DHA for normal brain development in pre-terms and infants [86].

Furthermore, evidence from clinical and epidemiological studies has also strongly suggested the beneficial and protective roles of DHA in alleviating atherosclerosis, rheumatoid arthritis, myocardial infarction, and malignant diseases [78, 89]. Thus, a balanced dietary intake of PUFAs that contain sufficient DHA is recommended in order to prevent pathological symptoms related to DHA deficiency. Although the optimal daily intake of PUFAs and the optimal ratio of n-6 to n-3 PUFAs remain unknown, a dietary intake of 1.2–1.6 g day^{-1} of ω-3 PUFAs, as suggested by a panel of nutritional scientists in the US, with a ratio of ω-3 to ω-6 fatty acids of 2.3:1 is recommended for health protection [90].

4.2.2. Distribution of DHA in various lipid classes

Knowing the distribution of DHA in the different lipid classes in thraustochytrids would provide important information for commercial consideration because DHA in the triacylgycerol layer is the only economically feasible form for commercialization and down-stream processing [91]. There are, however, only a few reports investigated in detail the lipid class distribution of thraustochytrids, which are discussed below [18, 92].

Fig. 4. summarizes the distribution of the major lipid classes in thraustochytrids and indicates that the major lipid class in thraustochytrids is neutral lipids, making up from 44.7–95.0% of total lipids. Glycolipids and phospholipids are the minor constituents in thraustochytrids, comprising approximately 0–27.1% and 8.9–28.2% of total lipids, respectively. Nakahara *et al.* [18] found that *Schizochytrium* sp. SR21 contained about 50.0% of biomass as lipids, which was composed of 95.0% neutral and 5.0% polar lipids. Similarly, *Thraustochytrium* sp. ATCC 26185 was found to contain 32.0% of lipids in its biomass with phospholipids accounting for approximately 8.0% of the total lipids [92]. The proportions of different lipid classes in *T. aureum* ATCC 34304 varied [93, 95] when grown in different media, indicating the profound influence of media composition on lipid composition (Fig. 4.).

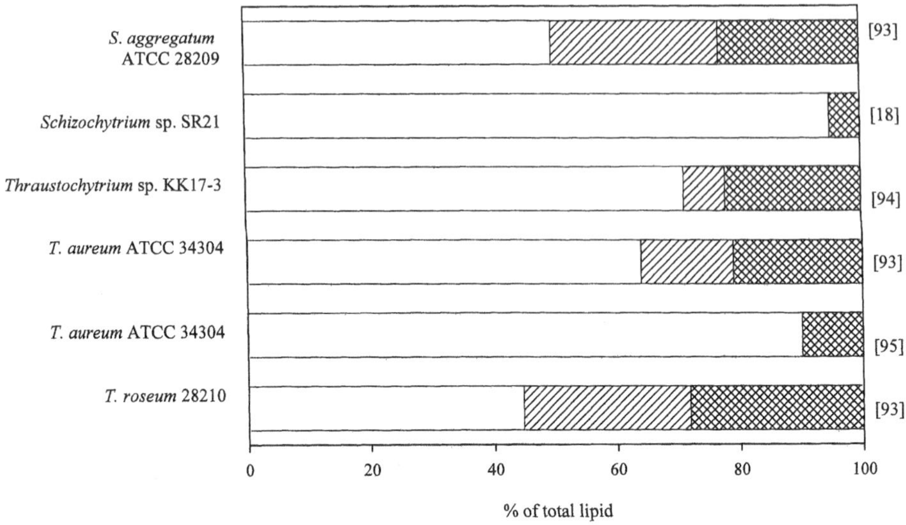

Fig. 4. Distribution of lipid classes in thraustochytrids. (Neutral lipids ▭, Glycolipids ▨, Phospholipids ▩)

In the neutral lipid fractions, triacylglycerols were the major component followed by diacylglycerols and monoacylglycerols (Table 1). Triacylglycerols account for 86.2% of the neutral lipids in *T. aureum* ATCC 34304, but in *Schizochytrium* sp. SR21, it can constitute 98.0% of the neutral lipids. Diacylglycerols and monoacylglycerols, however, were reported to be minor components (< 10.0%) of the neutral lipids of thraustochytrids [93].

In the polar lipid fractions, phospholipids were the major lipid class present in thraustochytrids [93]. Of the different types of phospholipids, phosphatidylcholine was the major component, comprising 49.0–76.2% of the total phospholipids, followed by phosphatidylethanolamine (2.3–11.0%). In *Schizochytrium* SR21, phosphatidylcholine (PC), phosphatidylethanolamine (PE), and phosphatidylinositol (PL) were the major phospholipids, accounting for 71.0%, 11.0% and 5.0%, respectively, of the total phospholipids [18].

PUFAs make up a large proportion of polar lipids, ranging from 36.5% to 72.0%. In general, the proportion of PUFAs in polar lipid is higher than that in neutral lipid (Fig. 5), owning to the regulatory role of PUFAs on membrane fluidity [94]. The major molecules present in the PCs of *Schizochytrium* sp. SR21 are 1-palmitic acid-2-DHA-PC and 1,2-DHA-PC, comprising 70% of the PCs, indicating the abundance of DHA molecules in thraustochytrids [60]. Ashford *et al.* [96] also indicated that 71.8% of the PUFAs in phospholipids was DHA.

Table 1
Composition of neutral lipids and phospholipids from thraustochytrids

	Schizochytrium sp. SR21	*S. aggregatum* ATCC 28209	*Thraustochytrium* sp. ATCC 26185	*T. aureum* ATCC 34304	*T. aureum* ATCC 34304	*T. roseum* ATCC 28210
Composition of neutral lipids (%)						
Triacylglycerols	98.0%	>90.0%	NP	>90.0%	86.2%	>90.0%
Diacylglycerols and monoacylglycerols	NP*	<10.0%	NP	<10.0%	NP	<10.0%
Composition of phospholipids (%)						
Diphosphatidylglycerol	NP	NP	3.7%	NP	NP	NP
Phosphatidylcholine	71.0%	NP	76.2%	NP	49.0%	NP
Phosphatidylethanolamine	11.0%	NP	2.3%	NP	9.0%	NP
Phosphatidylglycerol	NP	NP	1.3	NP	NP	NP
Phosphatidylinositol	5.0%	NP	16.9%+	NP	NP	NP
References	[18]	[93]	[92]	[93]	[95]	[93]

*NP means data not provided
+Data is collectively represented of lysophosphatidylcholine and phosphatidylinositol

PUFAs comprise 25.9–61.5% of neutral lipids (Fig. 5.) and DHA is one of the dominant PUFAs in neutral lipids of thraustochytrids (e.g., 51.2% in *Thraustochytrium* sp. KK17-3 [94] and 69.3% in *T. aureum* [95]). Nakahara *et al.* [18] and Ashford et al. [96] provided evidence to support that DHA was also abundant in triacylglycerols and comprised 27.6% and 56.7%, of total fatty acids in *Schizochytrium* sp. ATCC 20888 and *Schizochytrium* sp. SR1, respectively. Nakahara *et al.* [18] indicated that seven triacylglycerol species were identified in *Schizochytrium* sp. SR21 with PA-PA-DHA as the major species (27.4%) followed by PA-DHA-DHA (16.9%). Ashford *et al.* [96] observed that DHA and DPA were preferentially esterified in the sn-2 position (71.0–75.0%) of the glycerol backbone of triacylglycerols and found the existence of alternating light (11.7 Å) and dark (29 Å) staining bands in triacylglycerols using high pressure freeze substitution. The dark band corresponded to PUFAs while the light band corresponded to saturated and mono-unsaturated fatty acids. Consequently, they proposed a structural model of triacylglycerols and suggested that the light staining band containing saturated and mono-unsaturated fatty acids might be segregated end to end to form a layer, while DHA arranged in sn-2 position was segregated into the dark staining layer [96]. This structural observation explained the abundance of DHA molecules in triacylglycerols of oleaginous thraustochytrids.

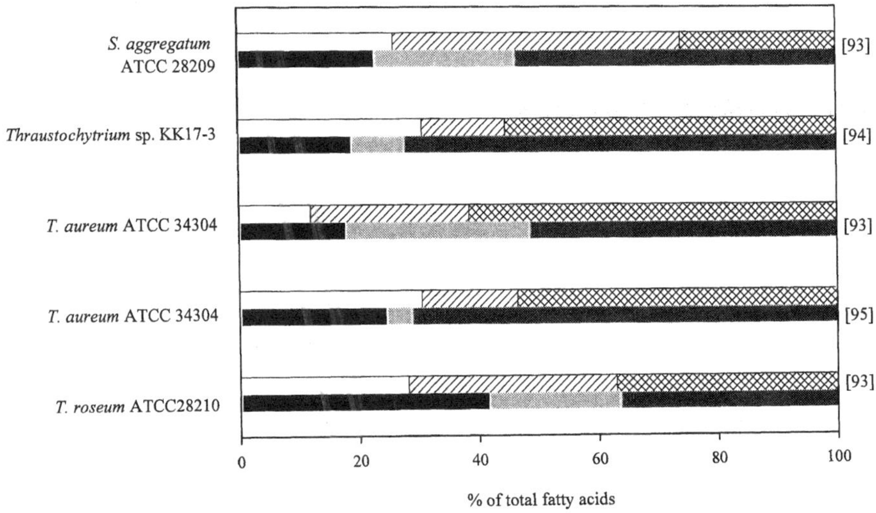

Fig. 5. Percentage of saturated (SFA), monounsaturated (MUFA) and polyunsaturated (PUFA) fatty acids in neutral (NL) and polar (PL) lipid classes in thraustochytrids. NL-SFA ▭, NL-MUFA ▨, NL-PUFA ▩, PL-SFA ■, PL-MUFA ▭, PL-PUFA ■

4.2.3. Production potential

The most important criterion for selecting a microbial source for the commercial production of DHA is the potential of the microorganisms to produce substantial biomass and total lipids [28]. A simple fatty acid profile with a large proportion of DHA in the lipids is another important criterion to be considered when selecting DHA producing microorganisms [97]. Of the various microbial sources, thraustochytrids are the most successful microalgal group in fulfilling these criteria. Table 2 summarizes some of the selected DHA-producing thraustochytrids reported in papers and patents published from 1991 to 2003. Overall, seven species of thraustochytrids from three genera, *Schizochytrium*, *Thraustochytrium* and *Ulkenia*, have been investigated and considered to have DHA production potential. *Thraustochytrium* spp., in general, contained a larger proportion of DHA, ranging from 38.5%–59.5% of the total fatty acids, than *Schizochytrium* spp. and *Ulkenia* sp. SAM 2179, which contained only 30.5%–46.2% of DHA in their total fatty acids. In contrast, however, *Schizochytrium* spp. and *Ulkenia* sp. SAM 2179 contained generally much higher amounts of total lipids than *Thraustochytrium* spp. The total lipid contents of the former might be as high as 77.5% of the biomass while the total lipid content of the latter accounted for in many cases only approximately 20.0% of the biomass. There was, however, an exception: *Thraustochytrium* sp. G13 was reported to contain 78.0% of its biomass as lipids although further investigation might be necessary to confirm the findings [30]. *Schizochytrium* spp. and *Ulkenia* sp. SAM 2179 grew much faster than *Thraustochytrium* spp. Therefore, *Schizochytrium* spp. and *Ulkenia* sp. SAM 2179 have been considered as more promising organisms for the production of DHA. *S. limacinum* is so far the most successful species in terms of biomass, lipid production, and DHA production, followed by *Ulkenia* sp. SAM 2179 and *Thraustochytrium* sp. G13 [30, 60, 101]. The DHA content and productivity of *S. limacinum* were 276.5 mg g^{-1} and 3325.0 mg l^{-1} day^{-1}, respectively. This level of DHA productivity is the highest of any microalgal sources ever reported in the literature.

4.2.4. Factors affecting DHA production

Culture conditions are always important for the optimal production of PUFAs in microalgae, however, they may be species specific [4]. As illustrated in Table 2, the culture conditions for DHA production were different for different species of thraustochytrids, and the production might influenced by the culture mode, growth phase, environment (e.g. medium pH, temperature, etc.), and nutrition (e.g. carbon, nitrogen, phosphorus, sodium, trace elements, growth factors, etc.) [22].

There are many different culture modes available for the cultivation of microalgae in heterotrophic cultures, including batch, fed-batch, continuous, and perfusion cultures [10], but the most widely adopted strategy for DHA production in thraustochytrids is still batch culturing (see Table 2). However, there are obvious limitations associated with this culture mode because of the problem of substrate inhibition with high initial substrate concentrations. A fed-batch strategy developed to overcome the substrate inhibition problem resulted in a two-fold increase in DHA yield in *T. roseum* ATCC 28210 over the batch culture using the same algal strain [100]. Growth phase can also profoundly influence the lipid production of oleaginous microorganisms. Lipid accumulation usually follows a sigmoid pattern, with maximum PUFA production in the culture during the late exponential or early stationary phase [80]. DHA production in thraustochytrids follows the same pattern, with maximum DHA production in most species from day 2 to 4, which corresponds to either their late exponential or early stationary phase.

Table 2
Selected DHA producing thraustochytrids extracted from literatures and patents from 1991-2003.

Thraustochytrids	*S. limacinum* SR21	*S. limacinum* SR21	*S. mangrovei* KF2	*Thraustochytrium* sp. ATCC 20892	*Thraustochytrium* sp. G13	*T. aureum* ATCC 34304	*T. aureum* ATCC 34304	*T. roseum* ATCC 28210	*T. roseum* ATCC 28210	*Ulkenia* sp. SAM 2179
Growth parameters										
Biomass (g l^{-1})	48.1	38.0	13.3	7.1	14.0	3.8	4.9	9.8	17.1	19.5
Total lipid (% of dry weight)	77.5	37.0	59.1	16.9	78.0	16.5	20.3	17.5	24.0	61.0
DHA (% TFA)	35.6	30.5	35.3	59.5	38.5	48.5	51.0	49.8	48.8	46.2
DHA in biomass (mg g^{-1})	276.5	116.6	204.3	100.3	140.4	70.4	103.8	85.9	117.0	282.1
DHA yield (mg l^{-1})	13300.0	4200.0	2762.0	707.4	1965.6	269.6	510.5	841.8	2000.0	5500.0
DHA productivity (mg l^{-1} day^{-1})	3325.0	840.0	1280.6	176.9	1150.6	44.9	85.1	168.4	166.7	1833.3
Environmental and nutritional parameters										
Culture mode	Bioreactor Batch	Shake flask Batch	Shake flask Batch	Shake flask Batch	Bioreactor Batch	Shake flask Batch	Shake flask Batch	Shake flask Batch	Shake flask Fed batch	Bioreactor Batch
Culture period (day)	4	5	2 days 4 h	4	1 day 17 h	6	6	5	12	3
Medium pH	4	NP	6	7	NP	NP	NP	6	6	4
Temp.(°C)	28	25	25	25 followed by 15	24	25	25	25	25	28
Carbon source										
Glucose (g l^{-1})	120.0	90.0	60.0	40.0	40.0	20.0	–	–	–	60
Starch (g l^{-1})	–	–	–	–	–	–	25.0	25.0	25.0 (10.0)	–
Nitrogen source										
Corn steep liquor (g l^{-1})	1.4	20.0	–	–	–	–	–	–	–	0.7
$(NH4)_2SO_4$ (g l^{-1})	4.0	–	–	0.2	–	0.2	0.2	0.2	0.2	2.0
Sodium glutamate (g l^{-1})	–	–	–	2.0	–	2.0	2.0	2.0	2.0 (0.8)	–
Yeast extract (g l^{-1})	–	–	10.0	2.0	5.0	–	–	2.0	2.0	–
C/N ratio	51.9	32.7	24.0	39.2	32.0	33.0	–	–	–	51.9

Other medium components										
Artificial sea salts (g l^{-1})	15.0	15.0	15.0	–	–	–	–	–	–	15.0
$CaCO_3$ (g l^{-1})	–	–	–	0.2	–	0.2	0.2	0.2	0.2	–
KCl (g l^{-1})	–	–	–	1.0	–	1.0	1.0	1.0	1.0	–
KH_2PO_4 (g l^{-1})	3.0	–	–	0.2	–	0.1	0.1	0.1	0.2 (0.08)	3.0
MgSO4 (g l^{-1})	–	–	–	5.0	–	5.0	5.0	5.0	5.0 (2.0)	–
NaCl (g l^{-1})	–	–	–	25.0	–	25.0	25.0	25.0	10.0	–
$NaHCO_3$ (g l^{-1})	–	–	–	–	–	0.1	0.1	0.1	0.1	–
Na_2SO_4(g l^{-1})	–	–	–	–	20.0	–	–	–	–	–
Trace elements										–
Citric acid (mg l^{-1})	–	–	–	5.0	–	–	–	–	–	–
$CoCl_2.6H_2O$ (mg l^{-1})	–	–	–	0.13	–	–	–	–	0.26	–
$CuSO_4.5H_2O$ (mg l^{-1})	–	–	–	0.01	–	–	–	–	0.02	–
$FeCl_3.6H_2O$ (mg l^{-1})	–	–	–	1.45	–	–	–	–	2.9	–
$MnCl_2.4H_2O$ (mg l^{-1})	–	–	–	4.3	–	–	–	–	8.6	–
$ZnCl_2$ (mg l^{-1})	–	–	–	0.3	–	–	–	–	0.6	–
Thiamine (μg l^{-1})	–	–	–	10.0	–	10.0	10.0	–	10.0	–
Vitamin B12 (μg l^{-1})	–	–	–	1.0	–	1.0	1.0	–	1.0	–
References	[60]	[31]	[56]	[98]	[30]	[99]	[82]	[83]	[100]	[101]

Thraustochytrids grow over a wide range of temperatures from 5 to 37°C with an optimum between 20 and 25°C [57]. The optimum temperature for DHA production in thraustochytrids is from 25 to 28°C [60, 83]. Yongmanitchai and Ward [80] found that a lower temperature promotes the biosynthesis of PUFAs in microorganisms because more PUFAs are required in order to regulate the membrane fluidity under a cold environment. For this reason, Nichols *et al.* [102] suggested the biotechnological potential of Antarctic thraustochytrids which require a lower temperature optimum for growth and thus may produce more PUFAs at their optimal temperature, at which both biomass and PUFA yields could be at their maxima. Although lower temperatures might increase DHA biosynthesis, for most other species, it also decreases overall growth. Consequently, the best strategy for DHA production should consider both biomass and DHA production. Singh and Ward [22] suggested that temperature shift might be an effective way of obtaining sufficient biomass with high DHA content in a relatively short time. For example, DHA production was found to increase in shifting temperature conditions (i.e., 3 days at 25°C followed by one day at 15°C) [98].

Thraustochytrids grow at pH ranging from 5.0 to 11.0, with an optimum between 6.0 and 8.0 [103]. Adjusting the initial pH to 6.0–7.0 was the best for DHA production in a number of *Thraustochytrium* spp., such as *T. roseum* ATCC 2810, and *Thraustochytrium* sp. ATCC 20892 [83, 98]. On the contrary, an acidic pH (pH 4.0) was preferred for optimal growth and DHA production in *S. limacinum* SR21 and *Ulkenia* sp. SAM 2179 [31, 101].

In addition to environmental conditions, nutritional conditions are also important. Carbon and nitrogen sources are essential nutrients for cell growth and protein biosynthesis in all organisms including thraustochytrids. Thraustochytrids can utilize a wide variety of carbon source for DHA production, with glucose and starch being the two most efficient and preferred sources [60, 83]. *Schizochytrium* spp. and *Ulkenia* sp. SAM 2179 tend to grow and produce DHA best on glucose, whereas *Thraustochytrium* spp. grow best on starch granules [83]. For instance, a medium with 2.5% starch was found to be the best for biomass production (9.8 g L^{-1}) and DHA production (841.8 mg L^{-1}) in *T. roseum* ATCC 28210 [83]. *Schizochytrium* spp. exhibited the best DHA production in a medium containing 6%–12% glucose, without suffering from substrate inhibition, whereas the optimal glucose concentration in *Thraustochytrium* spp, is less than 4% (Table 2). Although the type and amount of carbon sources used are important for achieving high lipid and DHA production, the amount of carbon in relation to the nitrogen source is also important. A high C/N ratio is generally preferred for lipid accumulation, which can be triggered by nitrogen exhaustion [22]. In oleaginous thraustochytrids, a C/N ratio of 24 or above was reported to be optimal for lipid accumulation (Table 2). For example, *S. limacinum* SR21 accumulated 77.5% of lipids in biomass when grew in a medium with a C/N ratio of 51.9. Likewise, *Thraustochytrium* sp. G13 also attained a high lipid content of 78.0% in biomass when a medium with a C/N ratio of 39.1 was used.

Thraustochytrids utilize undefined organic nitrogen substrates (e.g. corn steep liquor) more efficiently than inorganic nitrogen (e.g., ammonium). However, a combination of corn steep liquor and ammonium sulfate was found to be the best nitrogen source for DHA production in *S. limacinum* SR21 [60]. *Thraustochytrium* spp. prefers sodium glutamate but cannot grow in a medium containing ammonium sulfate [22].

Phosphorous is equally vital to cell growth as well as nucleic acid formation. Its importance also extends to the formation of the phospholipids layer, which, in thraustochytrids, contains a substantial amount of PUFAs [94]. Singh and Ward [100] observed that DHA production significantly increased in *T. roseum* ATCC 28210 when the amount of phosphorous, in the form of KH_2PO_4, increased from 0.1 to 0.2 g L^{-1} in the medium. The optimal concentration of KH_2PO_4 for DHA production in *S. limacinum* SR21 and *Ulkenia* sp. SAM 2179 was even higher (3 g L^{-1}), compared to 0.1–0.2 g L^{-1} in most *Thraustochytrium* spp. [60, 101].

Sodium is another important element for growth in thraustochytrids because sodium ions are essential for phosphate uptake [104]. It can be supplied either as sodium chloride or artificial sea salt. Half-strength seawater was as effective as full-strength seawater; a comparable biomass and DHA yield were obtained for *T. roseum* in the medium containing 25 g L^{-1} and 10 g L^{-1} of NaCl [100]. For DHA production, a medium with half-strength seawater in the form of artificial sea salts is an effective replacement for a defined medium, particularly in *Schizochytrium* spp. (Table 2).

In addition, thraustochytrids require certain trace elements and growth factors, such as vitamins B_{12} (cobalamine) and B_1 (thiamine), for growth [103]. The effect of vitamins on *T. aureum* was tested using a defined medium containing glucose and glutamate [95]. A combination of biotin, cobalamine, nicotinic acid, pantothenic acid, riboflavin, and thiamine produced the highest biomass. *Thraustochytrium globsum, T. roseum* and *S. aggregatum* were all found to require cobalamine for growth [103].

4.2.5. Statistics-based approach for medium optimization

Traditionally, the one-at-a-time strategy is the most popular method employed for optimizing medium components and environmental factors for DHA production in thraustochytrids [31, 98]. The one-at-a-time method is to keep the level of all components constant, with the exception of the one investigated [105]. A major drawback with this method is its inability to account for interactions among factors. Recently, statistics-based experimental design has been employed for the production of PUFAs in microalgae, such as diatoms [106] and thraustochytrids [107]. This method involves three steps, including experimental design (e.g., central composite, factorial, Plackett and Burman designs, etc.), optimization (e.g., response surface methodology, steepest ascent, canonical analysis, etc.), and verification [105].

A statistics-based approach was employed to optimize medium composition and environmental factors for DHA production by *Schizochytrium* sp. ATCC 20888. A fractional factorial design was initially used for screening a range of variables [107]. Concentrations of glucose, yeast extract, and sodium chloride as well as the pH value were the significant variables identified. These variables were further optimized using central composite design and determined using canonical analysis. After optimization, an experimental DHA yield of 498 mg L^{-1} confirmed the validity of the predicated value and was 41% higher than the result from the one-at-a-time experiment (352 mg L^{-1}) [107].

4.2.6. Molecular manipulation of DHA production

Modern biotechnology has witnessed the development of many molecular techniques and their applications for gene manipulation. These molecular techniques have recently been used to discover the pathways leading to DHA production in thraustochytrids [108, 109]. Elucidating the DHA biosynthetic pathway in thraustochytrids is the initial step towards using genetic engineering approaches for DHA enhancement and for genetic manipulation. *Schizochytrium* and *Thraustochytrium* are the two well-recognized genera of thraustochytrids for DHA production [31, 99], but their biosynthetic pathways have only recently been established [108, 109]. Though *Schizochytrium* and *Thraustochytrium* are taxonomically related and are grouped in the same family, it is surprising that two distinct pathways exist for their DHA biosyntheses, namely, the anaerobic pathway, which is also known as a novel polyketide synthase system (PKS) in *Schizochytrium* sp. ATCC 20888 and the aerobic pathway, which involves a series of desaturases and elongases in *Thraustochytrium* sp. ATCC 26185 [110]. Both pathways utilize acetyl-CoA and malonyl-CoA as the principal building blocks for DHA biosynthesis [22, 110].

Schizochytrium sp. ATCC 20888 has been demonstrated to utilize the PKS for DHA biosynthesis [108] (Fig. 6). The exact route for its biosynthesis remains to be determined, but it likely involves condensation and isomerization for introducing double bonds and chain elongation [111]. This pathway differs from the aerobic pathway in that it involves polyketide synthase instead of fatty acid synthase for the repeated full cycle of condensation, ketoreduction, and dehydration reactions. The pathway often omits steps from the full cycle, such as dehydration and reduction, and does not require molecular oxygen and membrane-bound desaturases for introducing double bonds in the developing fatty acid chain [110]. In *Schizochytrium* sp. ATCC 20888, three open reading frames with 11 domains were identified as genes for PUFAs biosynthesis and eight of them had a high homology with the PKS genes of *Shewanella*. The findings suggested the possibility of laterally transferring PKS genes from prokaryotes to eukaryotes [108].

Alternatively, the biosynthesis of DHA in *Thraustochytrium* sp. ATCC 26185 involves the aerobic pathway, which requires two major steps (Fig. 6). Biosynthesis starts with the *de novo* biosynthesis of stearic acid (18:0) from repeated additions of two carbon atoms from acetyl-CoA or malonyl-CoA through a series of condensation, reduction and dehydration reactions [22]. The biosynthetic process proceeds subsequently from 18:0 through a series of desaturation and elongation processes (Fig. 6).

Of particular interest in the acyl chain desaturation and elongation is the controversial debate over the existence of a Δ4 desaturase that introduces a final double bond to docosapentaenoic acid (22:5 DPA ω-3) in order to biosynthesize DHA [112]. Experimental evidence has recently been made available that confirms the presence of Δ4 desaturase in *Thraustochytrium* sp. ATCC 26185 [109, 113]. The Δ4 desaturase is a front-end desaturase, similar to other front-end desaturases, and is capable of desaturating substrata of both ω-3 e.g., DPA and ω-6 e.g., docosatetraenoic acid (22:4 DTA ω-6) [109].

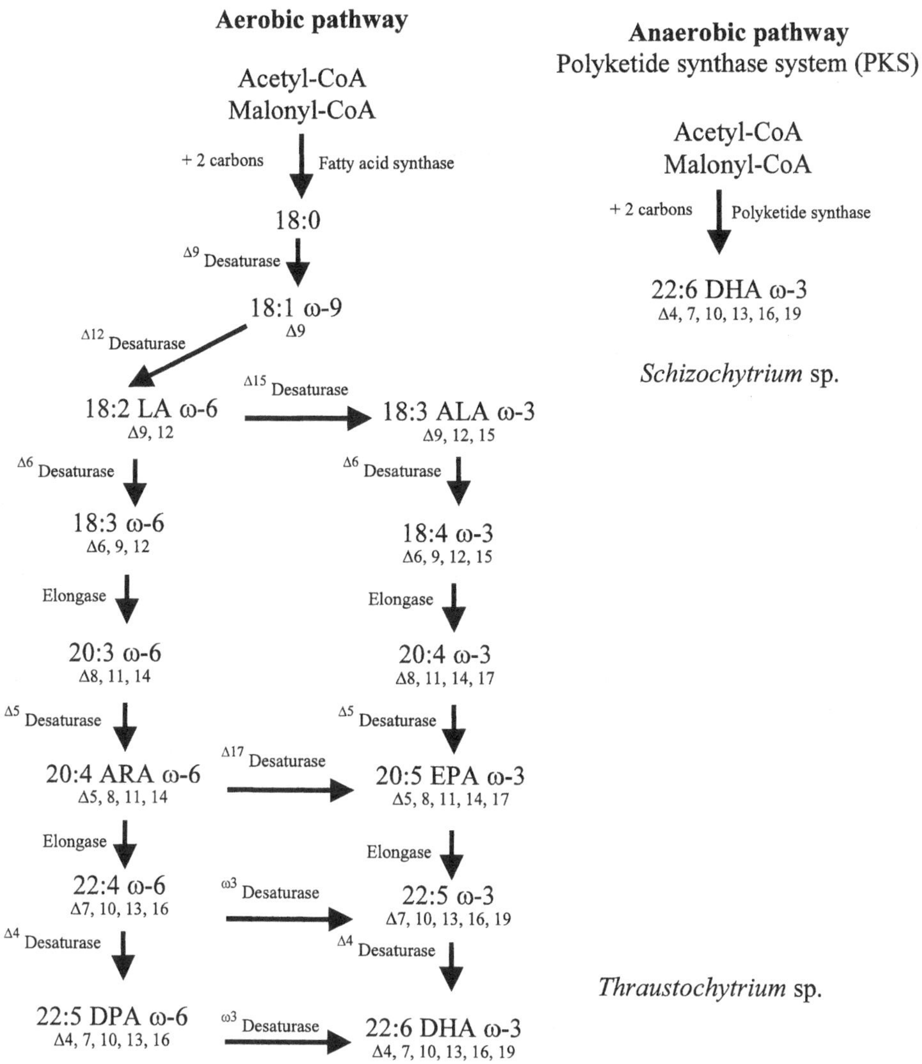

Fig. 6. The aerobic and anaerobic pathways for DHA biosynthesis in thraustochytrids.

The recent successful elucidation of DHA biosynthesis in thraustochytrids has created the possibility of transgenically producing DHA in other systems, especially in oilseed plants [110]. The concept of producing PUFAs by using plant as a 'chemical factories' has been proposed and considered as a potential way to meet future oil demand, owing to the availability of oilseed plant-based production systems [114]. Qiu *et al.* [109] heterologously expressed the Δ4 desaturase gene (FAD4) in the oilseed plant, *Brassica juncea*, which resulted in DHA accumulation (about 6% of total fatty acids). Similarly, Meyer *et al.* [113]

have discovered the existence of ω-3 desaturase gene that channels ω-6 substrata to ω-3 substrata at different stages of the DHA biosynthetic pathway from *Thraustochytrium sp.* ATCC 26185. This gene allows the conversion of linoleic acid (18:2 LA ω-6), which is especially abundant in oilseed plants, and may increase the production of ω-3 PUFAs in transgenic plants [113]. In addition, the discovery of the PKS pathway allows further choices for genetically manipulating DHA production [111].

4.2.7. Downstream processing

Downstream processing involves multi-step procedures for the recovery and purification of products of interest [115]. From a commercial point of view, the most important objective in downstream processing is to maximize product recovery and, at the same time, minimize the cost of production [115]. As for the recovery of biomass and PUFAs from microalgae, the recovery process may well exceed 60% of the production costs [116]. Thus, knowledge on the procedures for purifying PUFAs is required in order to identify the steps that may be omitted or improved in order to reduce the production cost. Robles Medina *et al.* [117] provided an extensive summary of the various useful methods for recovering PUFAs and indicated that the downstream processing of PUFAs includes three major steps: recovering biomass, extracting lipids, and concentrating and purifying the PUFAs [117]. The biomass of microalgae can be harvested and separated from the culture broth by either centrifugation or filtration. The addition of flocculants prior to separation might enhance cell aggregation, which would facilitate the recovery of biomass [116]. Harvested cells are usually subject to lyophilization, as lipids are more readily extracted from freeze-dried biomass than wet biomass. In addition, reports also indicate that spray drying and freeze drying are the two preferred drying methods for thraustochytrids prior to lipid extraction [97, 118]. However, Molina Grima *et al.* [119] showed that lipid is equally and efficiently extracted from both dry and wet biomass of *Phaeodactylum tricornutum*. Thus, the omission of the drying step could significantly reduce the production cost [119].

Finding an appropriate solvent system is important for extracting lipids from algal cells. Lewis *et al.* [120] indicated that the extraction of lipids from thraustochytrids was more efficient when solvents were added in the order of increasing polarity (chloroform > methanol > water). After lipid extraction, PUFAs can be concentrated by chromatographic methods, supercritical fluid extraction, or urea fractionation [117]. Of the various PUFA concentration methods, the HPLC-based method is the most appropriate for the recovery of high purity products. Yamamura and Shimomura [97] described a chromatographic method based on preparative HPLC with reverse phase octadecyl silane and showed that the method was effective for separating DHA with 99% purity from *Schizochytrium sp.* SR21.

4.3. Squalene

Squalene (2,6,10,15,19,23-hexamethyltetracosa-2,6,10,14,18,22 hexaene) is a dehydro-triterpenic hydrocarbon ($C_{30}H_{50}$) with six double bonds (Fig. 7) and is an intermediate for the biosynthesis of phytosterol and cholesterol in plants and animals [121]. Liver oils of deep-sea sharks, currently, represent the richest natural source of squalene [122–123]. The future supply and availability of shark liver oils, however, are questionable because of concerns about the over-exploitation of fishery stock [124]. For this reason, alternative sources, such as those from plant and microorganisms, are being explored for squalene production [123, 125]. Among these potential new sources, microalgae represent one of the best candidates for future exploration because of the well-established fermentation technology for their mass cultivation [20]. Squalene has recently been reported in thraustochytrids when the lipid profiles in the selected thraustochytrids were studied [92, 126]. In addition, a squalene recovery process is recently available and can be handled easily in a simple downstream process based on counter-current chromatography, which has led to the successful purification of squalene with 96% purity [127]. Thus, the production of squalene using thraustochytrids has great potential.

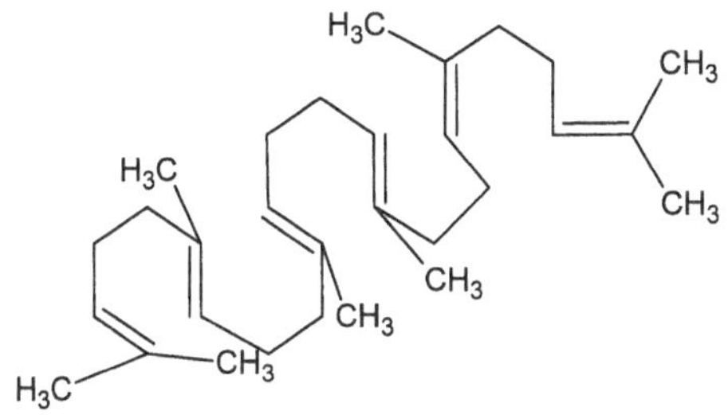

Fig. 7. Chemical structure of squalene.

4.3.1. Health benefits

Squalene has recently been recognized to possess health beneficial properties, owing largely to its strong oxygen scavenging abilities and anti-tumor activities [128–129]. Dessì *et al.* [128] showed that squalene is a strong antioxidant and is capable of protecting PUFAs against temperature and UVA-induced oxidation. Moreover, the antioxidative and protective roles of squalene in alleviating against skin irritation and UVB-induced skin cancer are notable [130]. The high content (10%) of squalene in human skin lipid, therefore, may play an important role in protecting against skin damage [131].

The anti-carcinogenic role of squalene against chemically-induced colon and lung cancers has also been demonstrated in animal models [129]. The mechanism by which squalene may inhibit tumor formation is based on its inhibitory role on ß-hydroxy-ß-methylglutaryl-CoA reductase catalytic activity and subsequent inhibition on farnesylation of Ras oncoproteins [132].

4.3.2. Production potential

Squalene was first reported in *Thraustochytrium* sp. and was noticed to represent 63% of the nonsaponifiable lipid when its lipid profile was analyzed [92]. Weete *et al.* [92], however, did not investigate the effects of nutritional and culture conditions on squalene content [92]. Lewis *et al.* [126] later investigated the influences of culture age, temperature and dissolved oxygen level on the squalene content of thraustochytrid ACEM 6063, a possible member of the genus, *Schizochytrium*. Thraustochytrid ACEM 6063 grown under a low dissolved oxygen level (0–5% saturation) contained a higher squalene content (exceeding 1.0 mg g^{-1}) than that grown at higher dissolved oxygen levels (40%–60%) (0.01 mg g^{-1}) [126]. It was suggested that the high squalene content observed at low oxygen levels might be related to the reduced efficiency of the enzyme, squalene monooxygenase, which requires oxygen molecules to catalyze the oxidation of squalene to 2,3-oxidosqualene for sterol biosynthesis [126].

Squalene content in thraustochytrid ACEM 6063 was found to be inversely associated with culture temperature; for example, a squalene content of 1.2 mg g^{-1} was obtained at 15°C, which decreased to 0.7 mg g^{-1} at 20°C [126]. Lewis *et al.* [126] also showed that squalene content decreased in aged cultures of thraustochytrid ACEM 6063. A similar reduction in squalene content (e.g., 0.16 mg g^{-1} to 0.03 mg g^{-1} in *S. mangrovei* FB1 from day 3 to day 5) was observed in three strains of *S. mangrovei* as the culture aged [20]. A recent study found that a further enhancement of squalene content can be achieved through the application of enzyme inhibitor (K.W. Fan, F. Chen and Y. Jiang, unpublished data). A squalene content of 0.53 mg g^{-1} was observed in *S. mangrovei* FB3 when terbinafine (100 mg L^{-1}), a squalene monooxygenase inhibitor, was applied as compared to the untreated control in which the squalene content was only 0.37 mg g^{-1}. Nonetheless, squalene content in thraustochytrids is still relatively low (0.01–1.8 mg g^{-1}) as compared to conventional sources (e.g., 4.24 mg g^{-1} in olive oil) [20, 126, 123]. Thus, Jiang *et al.* [20] suggested that further optimization of media and other conditions is necessary for enhancing the squalene content in thraustochytrids.

4.4. Other novel bioactive compounds

Over the past decades, thraustochytrids have also been known to produce a number of other bioactive compounds that may be targets for future biotechnological applications [18–19, 133]. The applications of sterols as precursors for steroid production and as natural surfactants are widely recognized in the biotechnology industry [7]. Thraustochytrids are known to produce significant amounts of sterols and thus may be considered as potential candidates for sterols production [92, 126]. Weete et *al.* [92] analyzed the sterol content of *Thraustochytrium* sp. ATCC 26185 and found that cholesterol and 24-ethylcholesta-5,7,22-trien-3β-ol are the major sterols, representing 69% of total sterols. Recently, a thraustochytrid (strain ACEM 6063) isolated from Australian coast was shown to produce 20 sterols, of which 13 were identified. Brassicasterol, cholesterol, stigmasterol, and 24-ethylcholesta-5,7,22-trien-3β-ol were the dominant sterols, representing 50% to 90% of the total sterols [126]. In addition, marine lycosphingolipids are well known to possess excellent bioactivities and are also found in thraustochytrids. Jenkins *et al.* [133] identified glycosphingolipids and thraustochytrosides A-C in *T. globosum* CNK-018, but their bioactivities remained to be determined.

In addition to DHA production, there is evidence to support the possible production of other PUFAs from certain species of thraustochytrids [21]. For example, a substantial amount of DPA was produced with a productivity of 0.44 g $L^{-1}day^{-1}$ in a DHA producing *Schizochytrium* sp. that was isolated from Japanese coasts [18]. In addition, canthaxanthin, a potent antioxidant similar to astaxanthin, was also produced in *Schizochytrium* sp. KH105 at a yield of 10 mg L^{-1} [19]. Thraustochytrids have also been shown to have the ability to penetrate and colonize recalcitrant materials such as cellulose and sporopollenin in the walls of pollen grains [134], suggesting that thraustochytrids may also be a source of degrading enzymes that could be of great industrial importance.

5. OTHER APPLICATIONS OF THRAUSTOCHYTRIDS

5.1. Thraustochytrids in aquaculture

Due to the increasing demand for seafood and the rapid depletion of natural fisheries, 25% of the worldwide seafood production is now from the aquaculture industry [135]. The rapid expansion of aquaculture has led to an increased demand for aquaculture feeds, including microalgae. Indeed, microalgae have played a vital role in modern aquaculture; they are used not only as feeds for adult aquatic animals, but are also essential nutrients for these animals during their larval stages [136]. Of particular interest is the necessity of DHA derived from microalgae for the normal development of juvenile crustaceans, larval fish, and their bloodstock, since most marine animals have a limited ability to synthesize DHA themselves [137–138]. At present, however, fish oil is still the major source of PUFAs for farmed fish, but thraustochytrids have become more and more popular. It has been reported that thraustochytrids are used as feed additives in the aquaculturing of marine mussels, salmonids, and tropical sea cucumbers [139–141].

The first documented application of thraustochytrids in the aquaculture industry was to enrich the DHA content of brine shrimp and rotifers with spray-dried biomass of *Schizochytrium* sp. [118]. Brine shrimp and rotifers are used extensively as live prey for feeding to marine fish larvae, but the PUFA content in these two organisms are insufficient for larval development [142]. Thus, an enrichment procedure with microalgae is commonly required to improve their PUFA content prior to feeding [143]. It was reported that the PUFA content, especially that of DHA in brine shrimp nauplii (100 brine shrimp nauplii per mL) enriched with 400 mg L^{-1} of spray-dried *Schizochytrium* sp., increased from an undetectable level to 0.8% of the biomass [118]. Similarly, DHA content in enriched rotifers (400 rotifers per mL) increased from an undetectable level to 1.38% of the dry biomass after feeding 70 mg L^{-1} of spray-dried cells. The success of *Schizochytrium* sp. used for feeding live prey is mainly due to their high content of DHA (>30% of total fatty acids), suitable size for ingestion, and excellent suspension characteristics in seawater [118].

Spray-dried *Schizochytrium* sp. has also been successfully applied as partial replacement in feeding juvenile mussels, *Mytilus galloprovincialis* [141]. The application of live prey and microalgae as feeds for the larval stages of marine animals is desirable in the aquaculture industry, but the cost is relatively high [144]. Attempts to provide a cost-effective alternative to conventional feed by using non-living algae have been made [144]. When juvenile mussels are fed an inert diet containing 70% spray-dried biomass of *Schizochytrium* sp. and 30% ground biomass of *Spirulina* twice a day at a ration of 20 mg diet per day the juvenile mussels showed a 60% increase in organic dry weight over the control group [141]. This has led to a 50% reduction in feed production cost.

The applications of *Schizochytrium* sp. also extend to the rearing of salmonids [140]. PUFAs are essential for normal development in juvenile fish [138]. Carter *et al.* [140] have recently studied the possibility of incorporating freeze-dried biomass of *Schizochytrium* sp. into feeds for Atlantic salmonids (*Salmo salar*). The results indicated that a *Schizochytrium* sp. based feed can replace fish-oil based feed or canola-oil based feed because of the satisfactory performance in terms of weight gain, whole body chemistry, and immune function [140].

Schizochytrium sp. has been successfully applied as an inducing agent for artificial spawning in *Holothuria scabra*, a commercially important tropical sea cucumber [139] with a high demand worldwide [145]. However, one of the major obstacles to the commercial production of sea cucumbers is the difficulty in artificially achieving reliable spawning and a regular supply of eggs [139]. It is thus possible to use a combination of thermal stimulation and *Schizochytrium* sp. to induce artificial spawning in a number of tropical sea cucumbers [139].

5.2. Enriching foods with thraustochytrids

An insufficient intake of dietary ω-3 PUFAs in western diets has led to increased incidents of cardiovascular diseases and other illnesses [146]. One of the potential methods for increasing the dietary intake of ω-3 PUFAs is to enrich commonly consumed food products with DHA [147]. Some of the ω-3 PUFAs enriched foods on the market include bakery products, dressings, eggs, infant formula, milk, and poultry products [147]. Thraustochytrids have been successfully employed as a source of DHA to enrich food products, including eggs, chicken and lamb meats, as well as milk products from ewes and cows [148–150].

Eggs are one of the first food sources enriched with thraustochytrids [151, 152]. Laying hens that were fed on a diet of *Schizochytrium* sp. equivalent to 165 mg of DHA per hen per day showed a 5-fold increase in egg DHA content (135 mg per egg) compared with eggs produced from hens fed on a control diet with only 28 mg of DHA per egg. Moreover, a further increase in DHA content to 220 mg per egg was achieved when a ration of *Schizochytrium* sp. equivalent to 825 mg of DHA per hen per day was provided [153].

There are a number of benefits claimed for hens fed on an algal-based diet containing *Schizochytrium* sp., including an increased DHA content in eggs and enhanced egg production, feed conversion, and mean body weight [153]. Moreover, eggs enriched with DHA from *Schizochytrium* sp. do not differ in terms of sensory characteristics from eggs produced from hens fed on a control diet and do not negatively impact consumer acceptance. Thus, it is possible to utilize enriched eggs as a food ingredient in other foods that contain eggs, for instance, egg noodles [148, 152].

In addition, research has revealed the promising results of enriching milk fat with PUFAs [154]. The milk fat of dairy cows and ewes can be modified by dietary enrichment with *Schizochytrium* sp., which results in an improved fatty acid profile while retaining acceptable flavor [150, 155]. Franklin *et al.* [155] indicated that cows fed with a diet containing 910 g day^{-1} of *Schizochytrium* sp. produced milk with PUFA and DHA contents of 6.53% and 0.46%, respectively, in milk fat as compared to only 4.03% PUFA and no DHA in milk fat in cows fed with a control diet. Similarly, ewes fed with a diet containing 52 g day^{-1} of *Schizochytrium* sp. had a content of 1.2% DHA in milk fat, but the milk fat of ewes fed with a control diet had an undetectable level of DHA [150].

Manipulating the fatty acid composition of animal meats has also been recently considered as a way to introduce ω-3 PUFAs into diets [156]. The applications of *Schizochytrium* sp. to enrich chicken and lamb meats have been documented [149, 157]. Broiler chicken fed with a diet supplemented with 2.2% of *Schizochytrium* sp. had 2.8-fold and 4.0-fold increases in ω-3 PUFA and DHA content, respectively, and had acceptable sensory characteristics [157]. Similarly, lambs fed with a diet supplemented with 15.5% *Schizochytrium* sp. contained the highest percentage of DHA in the phospholipids and neutral lipid classes of the longissimus muscle, as compared with lambs fed with other diets without algal supplements [149].

6. UTILIZATION OF RENEWABLE RESOURCES

The development of products from the byproducts of agricultural and food industries is one of the major foci for the utilization of renewable resources throughout the globe [158–159]. Vegetable oils contain nutritious components (e.g., fatty acids) that may be used by single-cell microorganisms as organic substrates for the production of high-value products. The three most prominent fatty acids in vegetable oils are oleic acid (OA, 18:1ω-9), LA, and α-linolenic acid (ALA; 18:3 ω-3), comprising well over 70% of the total fatty acids [160]. Lali *et al.* [161] demonstrated that certain strains of thraustochytrids are able to utilize LA and ALA as their sole carbon sources for growth and DHA production. Results indicate that all of these fatty acids are able to sustain growth in the tested thraustochytrids, but DHA accumulation was observed in only three of the tested strains [161]. Of the three fatty acids, ALA was the best source for DHA accumulation, particularly with *Thraustochytrium* F3-1. The ability of thraustochytrids to utilize canola, corn and linseed oils for growth and DHA production further indicates the potential for converting vegetable oils to biomass of oleaginous thraustochytrids [82, 83].

In addition to using vegetable oils as a renewable resource for the production of thraustochytrid biomass, there is an increasing interest in the transformation and conversion of low-value raw materials into higher value products [112, 159]. One such application is the use of thraustochytrids for the transformation of food-processing wastes into valuable oils. Fan *et al.* [162] showed that *S. mangrovei* KF6 is able to utilize industrial food wastes, including bread crusts, soymilk residue (okara powder), and grain husks, for DHA production. Bread crusts fermented with *S. mangrovei* KF6 contained the highest DHA content (12.6 mg g^{-1}), followed by okara powder (7.3 mg g^{-1}) and brewing grain (6.2 mg g^{-1}). Evidence indicated that the 18-carbon fatty acids, particularly LA in the industrial food wastes, are preferentially utilized by thraustochytrids for DHA production [56, 162].

7. SAFETY ISSUES

The increasing number of commercial food products containing DHA derived from thraustochytrids (e.g., DHA enriched eggs) on the market has led to concern over the safe use of thraustochytrids [148]. It seems obvious that DHA has no apparent detrimental effects and is safe for human consumption as observed in the long history of seafood consumption throughout the world. Nevertheless, scientific studies are required to establish beyond doubt that there are no adverse effects associated with the consumption of thraustochytrids-derived oil.

Objective approaches are available to assess the safety of algal-based oils that contains DHA [163–167]. Investigators from Monsanto, Omega Tech (presently known as Martek Biosciences), and Pharmacia have jointly conducted a series of studies specifically to examine the safety of *Schizochytrium* sp. [163–167]. The common algal toxins, such as domoic acid and *Prymnesium* toxin, were not detected in *Schizochytrium* sp. [164, 165]. Furthermore, dietary administration of *Schizochytrium* sp. to Sprague-Dawley rats in subchronic (3 g kg^{-1} body weight day^{-1}), developmental toxicity evaluation (30% of diet), and single generation reproduction (30% of diet) studies indicated that there was no apparent adverse effect on the test rats [164–166]. Similarly, no adverse effects were observed in New Zealand white rabbits and growing swine fed with the biomass of *Schizochytrium* sp. at 1.8 g kg^{-1} body weight day^{-1} and 2.68 kg per pig over 120 days, respectively [163, 165]. *Schizochytrium* sp. was further confirmed not to be mutagenic in the Ames reverse mutation assay [167]. More recently, *Schizochytrium* sp. has been affirmed as Generally Recognized as Safe (GRAS) by the FDA for use in food application [168].

In addition, Kroes *et al.* [169] provided an independent and comprehensive study to test and evaluate the safe use of DHA45-oil, a food-grade oil derived from the marine microalga *Ulkenia* sp. produced in a multi-step fermentation process and containing >95% triacylglycerols and having a DHA content of 45% in the oil. The study affirmed that DHA45-oil is not genotoxic, as demonstrated in the negative results from the Ames and chromosomal aberration assays [169]. Moreover, acute (2 g oil kg^{-1} body weight), subchronic (2 g kg^{-1} body weight day^{-1}), and one-generation reproductive (7.5% of diet day^{-1}) studies also indicated that feeding Sprague-Dawley and Wistar rats with DHA45-oil did not produce significant differences in the evaluated parameters, such as mortality, neurological response, ophthalmology, gross pathology, and histopathology, between the control and DHA45-oil-fed group. This study, along with the studies on *Schizochytrium* sp., conclusively supports the safe use of thraustochytrid-derived oil as a safe dietary source of DHA.

8. CONCLUSIONS

Microalgae represent a largely under-exploited group of natural resources with a tremendous potential to produce high-value natural products. Thraustochytrids not only play an important role in the ecosystem as nutrient recyclers, but also serve as a supreme source of single cell oils with great commercial prospects. We have witnessed the application and commercialization of thraustochytrids for the production of DHA from its initial screening and medium optimization stage to the maturation stage in which the safe consumption of thraustochytrids is evaluated. Through years of dedicated scientific efforts, the commercial production of DHA using thraustochytrids as single-cell factories has now become a reality, and thraustochytrid-derived products have entered the marketplace as aquaculture feed and PUFA-fortified food for the fishery and healthcare sectors.

Thraustochytrids offer great potential for the production of high-value products. Future approaches are to explore means for maximizing DHA production and to further explore the potential of thraustochytrids to produce new products. To maximize DHA production, the use of economical substrates and genetic engineering approach are suggested. Pilot testing using renewable resources, such as plant-based oils and food industry wastes, as cheap substrates for thraustochytrids fermentation has been preliminarily investigated. The use of renewable resources opens up new possibilities for DHA production in thraustochytrids. Nevertheless, the feasibility of these substrates for large-scale production of DHA remains to be assessed. The elucidation of DHA biosynthetic pathways in thraustochytrids using molecular techniques represents an exciting and potentially rewarding subject for future research. The knowledge acquired on DHA biosynthesis in thraustochytrids can be used in the future for genetic manipulation; for example, the construction of thraustochytrids with ultra-high DHA yields.

Thraustochytrids can be further manipulated and manifested as cell factories for the production of other high-value products, such as astaxanthin, squalene, and other valuable metabolites. The isolation and screening of new strains from rare and extreme environments represents a practically sound method for discovering new high-value products. Moreover, the cultivation conditions for the production of each new high-value product require individual optimization.

REFERENCES

[1] K.S. Gao, J. Appl. Phycol., 10 (1998) 37.
[2] M. Olaizola, Biomol. Eng., 20 (2003) 459.
[3] J.R. Benemann, J. Appl. Phycol., 12 (2000) 291.
[4] Z. Cohen, H.A. Norman and Y.M. Heimer, World Rev Nutr. Diet., 77 (1995) 1.
[5] R.D. Hancock and R. Viola, Trends Biotechnol., 20 (2002) 299
[6] E.A. Johnson and G.H. An, Crit. Rev. Biotechnol., 11 (1991) 297.
[7] J.K. Volkman, Appl. Microbiol. Biotechnol., 6 (2003) 495.
[8] K.E. Apt and P.W. Behrens, J. Phycol., 35 (1999) 215.
[9] M.A. Borowitzka, J. Appl. Phycol., 4 (1992) 267.
[10] F. Chen, Trends Biotechnol., 14 (1996) 421.
[11] R.J. Radmer and B.C. Parker, J. Appl. Phycol., 6 (1994) 93.
[12] Z.Y. Wen and F. Chen, Biotechnol. Adv. 21 (2003) 273.
[13] S. López-Varela, M. González-Gross and A. Marcos, Eur. J. Clin. Nutr., 56 **S3** (2002) S29.
[14] M.A. Borowitzka, J. Appl. Phycol., 9 (1997) 393.
[15] G.H. Wikfors and M. Ohno, J. Phycol., 37 (2001) 968.
[16] W.R. Barclay, K.M. Meager and J.R. Abril, J. Appl. Phycol., 6 (1994) 123-129.
[17] B.B. Ellenbogen, S. Aaronson, S. Goldstein and M. Belsky, Comp. Bioch. Physyl., 29 (1969) 805.
[18] T. Nakahara, T. Yokochi, T. Higashihara, S. Tanaka, T. Yaguchi and D. Honda, J. Am. Oil Chem. Soc., 73 (1996) 1421.
[19] T. Aki, K. Hachida, M. Yoshinaga, Y. Katai, T. Yamasaki, S. Kawamoto, T. Kakizono, T. Maoka, S. Shigeta, O. Suzuki and K. Ono, J. Am. Oil Chem. Soc., 80 (2003) 789.
[20] Y. Jiang, K.W. Fan, R.T.Y. Wong and F. Chen, J. Agr. Food Chem., 52 (2004) 1196.
[21] T.E. Lewis, P.D. Nichols and T.A. McMeekin, Mar. Biotechnol., 1 (1999) 580.
[22] A. Singh and O.P. Ward, Adv. Appl. Microbiol., 45 (1997) 271.
[23] F. Chen, Y.M. Zhang and S.Y. Guo, Biotechnol. Lett., 18 (1996) 603.
[24] J.C. Ogbonna and H. Tanaka, J. Appl. Phycol., 12 (2000) 207.
[25] J.G. Burgess, K. Iwamoto, Y. Miura, H. Takano and T. Matsunaga, Appl. Microbiol. Biotechnol., 39 (1993) 456.
[26] L. Poisson, and F. Ergan, J. Biotechnol., 91 (2001) 75.
[27] R. Vazhappilly and F. Chen, Bot. Mar., 41 (1998) 553.

[28] L. Sijtsam and M.E. Swaaf de, Appl. Microbiol. and Biotechnol., 64 (2004) 146.
[29] R. Vazhappilly and F. Chen, J. Am. Oil Chem. Soc., 75 (1998b) 393.
[30] R.D. Bowles, A.E. Hunt, G.B. Bremer, M.G. Duchars and R.A. Eaton, J. Biotechnol., 70 (1999) 193.
[31] T. Yokochi, D. Honda, T. Higashihara and T. Nakahara, Appl. Microbiol. Biotechnol., 49 (1998) 72.
[32] S.T. Moss (ed.), The Biology of Marine Fungi, Cambridge University Press, Cambridge, 1986., pp. 105-129.
[33] F.K. Sparrow, Biol. Bull., 70 (1936) 236.
[34] W.M. Darley, D. Porter and M.S. Fuller, Arch. Mikrobiol., 90 (1973) 89.
[35] F.O. Perkins, Arch. Mikrobiol., 84 (1972) 95.
[36] D. Porter, Veroff. I. Meeresfor. S., 5 (1974) 19.
[37] M. Paton and D.H. Jennings, Mycol. Res., 92 (1989) 470.
[38] C.A. Leander and D. Porter, Mycotaxon., 76 (2000) 439.
[39] F.O. Perkins, Veroff. I. Meeresfor. S., 5 (1974) 45.
[40] D.J. Patterson and J. Larsen (eds.), The Biology of Free-living Heterotrophic Flagellates, Clarendon Press, Oxford, 1991., pp. 415-425.
[41] T. Booth and C.E. Miller, Can. J. Bot.., 47 (1969) 2051.
[42] A. Gaertner, Veroff. I. Meeresfor., 13 (1972) 183.
[43] R.M. MacKay and W.F. Doolittle, Nucleic Acids Res., 10 (1982) 8307.
[44] C. Mo and B. Rinkevih, Mar. Biotechnol., 3 (2001) 100-102.
[45] M.W. Dick (ed.), Straminipilous Fungi: Systematics of the Peronosporomycetes, including accounts of the Marine Straminipilous Protists, the Plasmodiophorids, and Similar Organisms, Kluwer Academic Publishers, Dordrecht, 2001., pp. 276-287.
[46] T. Cavalier-Smith, M.T.E.P. Allsopp and E.E. Chao, Proc. R. Soc. B., 346 (1994) 387.
[47] D. Honda, T. Yokochi, T. Nakahara, S. Raghukumar, A. Nakagiri, K. Schaumann and T. Higashihara, J. Eukaryot. Microbiol., 46 (1999) 637.
[48] C.A. Leander and D. Porter, Mycologia., 93 (2001) 459.
[49] J.Z. Huang, T. Aki, T. Yokochi, T. Nakahara, D. Honda, S. Kawamoto, S. Shigeta, K. Ono and O. Suzuki, Mar. Biotechnol., 5 (2003) 450.
[50] S. Raghukumar, Eur. J. Protistol., 38 (2002) 127.
[51] L. Margulis, J.O. Corliss, M. Melkonian and D.J. Chapman (eds.), Handbook of Protoctista, Jones and Bartlett Publishers, Boston, 1989., pp. 388-398.
[52] K.W. Fan, L.L.P. Vrijmoed and E.B.G. Jones, Bot. Mar., 45 (2002) 50.
[53] K.W. Fan, L.L.P. Vrijmoed and E.B.G. Jones, Mycologia., 94 (2002) 569.
[54] G.B. Bremer and G. Talbot, Bot. Mar., 38 (1995) 37.
[55] S. Raghukumar, V. Sathe-Pathak, S. Sharma and C. Raghukumar, Aquat. Microb. Ecol., 9 (1995) 117.
[56] K.W. Fan, F. Chen, E.B.G. Jones and L.L.P. Vrijmoed, J. Ind. Microbiol. Biotechnol., 27 (2001) 199.
[57] E.B.G. Jones (ed.), Recent Advances in Aquatic Mycology, ELEK Science, London, 1976., pp. 261-278.
[58] W.R. Barclay, Microfloral Biomass Having Omega-3 Highly Unsaturated Fatty Acids, US Patent No. 5518918 (1996).
[59] Honda, T. Yokochi, T. Nakahara, M. Erata and T. Higashihara, 1998. Mycol. Res. 102 (1998) 439.
[60] T. Yaguchi, S. Tanaka, T. Yokochi, T. Nakahara and T. Higashuhara, J. Am. Oil Chem. Soc., 74 (1997) 1431.
[61] R.T. Lorenz and G.R. Cysewski, Trends Biotechnol., 18 (2000) 160.
[62] A.G. Andrewes and M.P. Starr, Phytochemistry., 15 (1976) 1009.
[63] M. Olaizola, J. Appl. Phycol., 12 (2000) 499.

[64] P. Bubrick, Bioresource Technol., 38 (1991) 237.
[65] M.L. Carmona, T. Naganuma and Y. Yamaoka, Biosci. Biotech. Bioch., 67 (2003) 884.
[66] N. Shimidzu, M. Goto and W. Miki, Fisheries Sci., 62 (1996) 134.
[67] T. Tanaka, Y. Morishita, M. Suzui, T. Kojima, A. Okumura and H. Mori, Carcinogenesis., 15 (1994) 15.
[68] H. Jyonouchi, L. Zhang, M. Gross and Y. Tomita, Nutr. Cancer, 21 (1994) 47.
[69] M. Kobayashi and T. Okada, Biotechnol. Lett., 22 (2000) 177-181.
[70] M. Guerin, M.E. Huntley and M. Olaizola, Trends Biotechnol., 21 (2003) 210.
[71] J. Inborr, Feed Mix., 6 (1998) 31.
[72] S. Goldstein, Mycologia., 55 (1963) 799.
[73] L.R.G. Valadon, T. Brit. Mycol. Soc., 67 (1976) 1.
[74] Y. Yamaoka, M.L. Carmona and S. Oota, Biosci. Biotech. Bioch., 68 (2004) 1594.
[75] T.V. Long II, Process for Production of Carotenoid, Xanthophylls and Apo-carotenoids Utilizing Eukaryotic Microorganisms, US Patent No. 0015978 (2002)
[76] P. Bhosale, Appl. Biochem. Biotech., 63 (2004) 351.
[77] L.A. Horrocks and Y.K. Yeo, Pharmacol. Res., 40 (1999) 211.
[78] S.M. Barlow and M.E. Stansby (eds.), Nutritional Evaluation of Long–Chain Fatty Acids in Fish Oil, Academic Press, London, 1982., pp. 25-88.
[79] W. Yongmanitichai and O.P. Ward, Process Biochem., 24 (1989) 117.
[80] I. Gill and R. Valivety, Trends Biotechnol., 15 (1997) 401.
[81] P.K. Bajpai, P. Bajpai and O.P. Ward, J. Am. Oil Chem. Soc., 68 (1991b) 509.
[82] Z.Y. Li and O.P. Ward, J. Ind. Microbiol., 13 (1994) 238.
[83] R.S. Lees and M. Karel (eds.), Omega-3 fatty acids in Health and Disease, Marcel Dekker, INC, New York, 1990., pp. 115-156.
[84] E.C. Jamieson, J. Farquharson, R.W. Logan, A.G. Howatson, W.J.A. Patrick, L.T. Weaver and F. Cockburn, Lipids., 34 (2000) 1065.
[85] A. Lenzi, L. Gandini, V. Maresca, R. Rago, P. Sgrò, F. Dondero and M. Picardo, Mol. Hum. Reprod., 6 (2000) 226.
[86] J.P. San Giovanni, C.S. Berkey, J.T. Dwyer and G.A. Colditz, Early Hum. Dev., 57 (2000) 165.
[87] Y.Y. Linko and K. Hayakawa, Trends Food Sci. Tech., 7 (1996) 59.
[88] S.M. Innis, J. Pediatr., 143 (2003) S1.
[89] H. Tapiero, B.G. Nguyen, P. Couvreur and K.D. Tew, Biomed. Pharmacother., 56 (2002) 215.
[90] P.M. Kris-Etherton, D.S. Taylor, S. Yu-Poth, P. Huth, K. Moriarty, V. Fishell, R.L. Hargrove, G.X. Zhao and T.D. Etherton, Am. J. Clin. Nutr. S., 71 (2000) 179S.
[91] M. Certik and S. Shimizu, J. Biosci. Bioeng., 87 (1999) 1.
[92] J.D. Weete, H. Kim, S.R. Gandhi, Y. Wang and R. Dute, Lipids. 32 (1997) 839.
[93] A. Kendrick and C. Ratledge, Lipids., 27 (1992) 15.
[94] J.Z. Huang, T. Aki, K. Hachida, T. Yokochi, S. Kawamoto, S. Shigeta, K. Ono and O. Suzuki, J. Am. Oil Chem. Soc., 78 (2001) 605.
[95] I. Iida, T. Nakahara, T. Yokochi, Y. Kamisaka, H. Yagi, M. Yamaoka and O. Suzuki, J. Ferment. Bioeng., 81 (1996) 76.
[96] A. Ashford, W.R. Barclay, C.A. Weaver, T.H. Giddings and S. Zeller, Lipids., 35 (2000) 1377.
[97] R. Yamamura and Y. Shimomura, J. Am. Oil Chem. Soc., 74 (1997) 1435.
[98] A. Singh, S. Wilson and O.P. Ward, World J. Microb. Biotechnol., 12 (1996) 76.
[99] P. Bajpai, P.K. Bajpai and O.P. Ward, Appl. Microbiol. Biol., 35 (1991a) 706.
[100] A. Singh and O.P. Ward, J. Ind. Microbiol., 16 (1996) 370.
[101] S. Tanaka, T. Yaguchi, S. Shimizu, T. Sogo and S. Fujikawa, Process for Preparing Docosahexaenoic Acid and Docosapentaenoic Acid with *Ulkenia*, US Patent No. 6509178 B1 (2003).

[102] D. Nichols, J. Bowman, K. Sanderson, C.M. Nichols, T. Lewis, T. McMeekin and P.D. Nichols, Curr. Opin. Biotech., 10 (1999) 240.
[103] G. Bahnweg, Veroff. I. Meeresfor. 17 (1979) 245.
[104] P.A. Siegenthaler, M.M. Belsky and S. Goldstein, Science., 155 (1967) 93.
[105] M. Kennedy and D. Krouse, J. Ind. Microbiol. Biotechnol., 23 (1999) 456.
[106] Z.Y. Wen and F. Chen, Biotechnol. Bioeng., 75 (2001) 159.
[107] S.T. Wu and L.P. Lin, J. Food Biochem., 27 (2003) 127.
[108] J.G. Metz, P. Roessler, D. Facciotti, C. Levering, F. Dittrich, M. Lassner and R. Valentine, Science., 293 (2001) 290.
[109] X. Qiu, H.P. Hong and S.L. MacKenzie, J. Biol. Chem., 276 (2001) 31561.
[110] X. Qiu, Prostag. Leukotr. Ess., 68 (2003) 181.
[111] J.A. Napier, Trends Plant Sci., 7 (2002) 51.
[112] C. Ratledge and J.P. Wynn, Adv. Appl. Microbiol., 51 (2002) 1.
[113] A. Meyer, P. Cirpus, C. Ott, R. Schlecker, U. Zähringer and E. Heinz, Biochemistry., 42 (2003) 9779.
[114] D.L. Alonso and F.G. Maroto, Biotechnol. Adv., 18 (2000) 481.
[115] M.L. Shuler and F. Kargi (eds), Bioprocess Engineering Basic Concepts 2nd edition, Prentice Hall, Upper Saddle River, 2002, pp. 329-384.
[116] E. Molina Grima, E.H. Belarbi, F.G. Acién Fernández, A. Robles Medina and Y. Chisti, Biotechnol. Adv., 20 (2003) 491.
[117] A. Robles Medina, E. Molina Grima, A. Giménez Giménez and M.J. Ibáñez González, Biotechnol. Adv., 16 (1998) 517.
[118] W. Barclay and S. Zeller, J. World Aquacult. Soc., 27 (1996) 314.
[119] E. Molina Grima, A. Robles Medina, A. Giménez Giménez and M.J. Ibáñez González, J. Appl. Phycol., 8 (1996) 359.
[120] T. Lewis, P.D. Nichols and T.A. McMeekin, J. Microbiol. Meth., 43 (2000) 107.
[121] Y. Kohno, Y. Egawa, S. Itoh, S. Nagaoka, M. Takahashi and K. Mukai, BBA. Lipid. Lipid Met., 1256 (1995) 52.
[122] M. Hernández-Pérez, R.M.R. Gallego, P.J.P. Alayón and A.B. Hernández, Mar. Freshwater Res., 48 (1997) 573.
[123] R.W. Owen, W. Mier, A. Giacosa, W.E. Hull, B. Spiegelhalder and H. Bartsch, Food Chem. Toxicol., 38 (2000) 647.
[124] T.F. Ko, Y.M. Weng and R.Y.Y. Chiou, J. Agr. Food Chem., 50 (2002) 5343.
[125] P. Bhattacharjee, V.B. Shukla, R.S. Singhal and P.R. Kulkarni, World J. Microb. Biotechnol, 17 (2001) 811.
[126] T.E. Lewis, P.D. Nichols and T.A. McMeekin, Mar. Biotechnol., 3 (2001) 439.
[127] H.T. Lu, Y. Jiang and F. Chen, J. Chromatogr. A., 994 (2003) 37.
[128] M.A. Dessì, M. Deiana, B.W. Day, A. Rosa, S. Banni and F.P. Corongiu, Eur. J. Lipid. Sci. Tech., 104 (2002) 506.
[129] C.V. Rao, H.L. Newmark and B.S. Reddy, Carcinogenesis., 19 (1998) 287.
[130] A. Budiyanto, N.U. Ahmed, A. Wu, T. Bito, O. Nikaido, T. Osawa, M. Ueda and M. Ichihashi, Carcinogenesis., 21 (2000) 2085.
[131] N. Nicolaides, Science., 186 (1974) 19.
[132] H.L. Newmark, Ann. NY. Acad. Sci., 889 (1999) 193.
[133] K.M. Jenkins, P.R. Jensen and W. Fenical, Tetrahedron Lett., 40 (1999) 7637.
[134] J.J.P. Clokie, Veroff. I. Meeresfor. S., 5 (1974) 159.
[135] A. Frankic and C. Hershner, Aquacult. Int., 11 (2003) 517.
[136] R.M. Gladue and J.E. Maxey, J. Appl. Phycol., 6 (1994) 131.
[137] K.E. Harrsion, J. Shell Fish Res., 9 (1990) 1.
[138] J.R. Rainuzzo, K.I. Reitan and Y. Olsen, Aquaculture., 115 (1997) 103.
[139] S.C. Battaglene, J.E. Seymour, C. Ramofafia and I. Lane, Aquaculture., 207 (2002) 29.

[140] C.G. Carter, M.P. Bransden, T.E. Lewis and P.D. Nichols, Mar. Biotechnol., 5 (2003) 480.
[141] C.A. Langdon and E. Önal, Aquaculture., 180 (1999) 283.
[142] K.I. Reitan, J.R. Rainuzzo, G. Øie and Y. Olsen, Aquaculture., 155 (1997) 207.
[143] P. Sorgeloos, P. Dhert and P. Candreva, Aquaculture., 200 (2001) 147.
[144] R. Robert and P. Trintignac, Aquat. Living Resour., 10 (1997) 315.
[145] C. Conand and M. Byrne, Mar. Fish. Rev., 55 (1993) 1.
[146] J.N. Din, D.E. Newby and A.D. Flapan, Brit. Med. J., 328 (2004) 30.
[147] A.P. Simopoulos, Poultry Sci., 79 (2000) 961.
[148] R. Abril and W. Barclay, World Rev. Nutr. Diet., 83 (1998) 77.
[149] S.L. Cooper, L.A. Sinclair, R.G. Wilkinson, K.G. Hallet, M. Enser and J.D. Wood, J. Anim. Sci., 82 (2004) 1461.
[150] G. Papadopoulos, C. Goulas, E. Apostolaki and R. Abril, J. Dairy Res., 69 (2002) 357.
[151] S.M. Herber and M.E. Van Elswyk, Poultry Sci., 75 (1996) 1501.
[152] S.M. Herber-Mcneill and M.E. Van Elswyk, Poultry Sci., 77 (1998) 493.
[153] J.S. Sims, S. Nakai and W. Guenter (eds.), Egg Nutrition and Biotechnology, CAB international, Wallingford, 2000., pp. 197-202.
[154] T.B. Wright, B. McBride and B. Holub, World Rev. Nutr. Diet., 83 (1998) 160.
[155] S.T. Franklin, K.R. Martin, R.J. Baer, D.J. Schingoethe and A.R. Hippen, J. Nutr., 129 (1999) 2048.
[156] K. Raes, S. De Smet and D. Demeyer, Anim. Feed Sci. Tech., 113 (2004) 199.
[157] J.W. Mooney, E.M. Hirschler, A.K. Kennedy, A.R. Sams and M.E. Van Elswyk, J. Sci. Food Agr., 78 (1998) 134.
[158] F. Karaosmanoĝlu, Energ. Source., 21 (1999) 221.
[159] R. Montgomery, Bioresource Technol., 91 (2004) 1.
[160] F.D. Gunstone, J.L. Hardwood and F.B. Padley (eds.), The Lipid Handbook, Chapman & Hall, London, 1986., pp. 76-78.
[161] S.C. Lali, T. Yokochi and T. Nakahara, J. Oleo. Sci., 50 (2001) 515.
[162] K.W. Fan, F. Chen, E.B.G. Jones and L.L.P. Vrijmoed, 2000. Fungal Divers., 5 (2000) 185.
[163] R. Abril, J. Garrett, S.G. Zeller, W.J. Sander and R.W. Mast, Regul. Toxicol. Pharm., 37 (2003) 73.
[164] B.G. Hammond, D.A. Mayhew, M.W. Naylor, F.A. Ruecker, R.W. Mast and W.J. Sander, Regul. Toxicol. Pharm., 33 (2001) 192.
[165] B.G. Hammond, D.A. Mayhew, J.F. Holson, M.D. Nemec, R.W. Mast and W.J. Sander, Regul. Toxicol. Pharm., 33 (2001) 205.
[166] B.G. Hammond, D.A. Mayhew, K. Robinson, R.W. Mast and W.J. Sander, Regul. Toxicol. Pharm., 33 (2001) 356.
[167] B.G. Hammond, D.A. Mayhew, L.D. Kier, R.W. Mast and W.J. Sanders, Regul. Toxicol. Pharm., 35 (200) 255.
[168] Martek Bioscience Corporation, FDA Gives Favorable Review for Martek DHA Derived from Former OmegaTech Technology, News Release on World Wide Web, (2004)
[169] R. Kroes, E.J. Schaefer, R.A. Squire and G.M. Williams, Food Chem. Toxicol., 41 (2003) 1433.

Bioprocessing for Value-Added Products from Renewable Resources
Shang-Tian Yang (Editor)

Chapter 12. Nonconventional Biocatalysis for Production of Chemicals and Polymers from Biomass

Ping Wang

Department of Chemical Engineering, The University of Akron, Akron, 44325 USA

1. INTRODUCTION

We are probably approaching the peak-high rate in consuming petroleum for energy and materials. Although new reservoirs may continue to emerge, the eventual ending point of natural supplies and the growing concern about environmental quality are driving an unprecedented worldwide search for alternative sources of energy and raw materials. Nature provides us options for alternative energy. In addition to the now industrialized production of fuel ethanol and biodiesel from biorenewable resources, power generation from nuclear reactors, sunlight, or wind is also becoming increasingly efficient and affordable. On the other hand, we are very limited with alternative raw materials. The only alternative to underground fossil carbon sources is materials grown on the surface of the earth.

Various technologies are available to produce chemicals and materials from biorenewable resources. In the processing chain from raw materials and biomass to chemical products, harsh thermochemical treatments and mild biotransformations are complementary and are both necessary, yet in many cases are also competing strategies. This chapter discusses the role and potential of biocatalysis in chemical production from biorenewable resources. Particular focus is on new developments in enzyme technology, i.e., nonconventional biocatalysis. Microbial bioprocessing is discussed in other chapters of this book.

Enzymes are the catalysts that regulate and enable all biological transformations in living organisms. Each reaction in the biological world involves the action of at least one, sometimes up to tens of enzymes. Functioning in controlled *in vitro* environments, enzymes are capable of generating a great variety of valuable products. Global sales of enzymes are over $1 billion per year. The majority of industrial enzymes are consumed in traditional biochemical industries. About $950 million/year is consumed in starch-processing and detergents and $200 million/year by textiles, leather, and pulp and paper industries [1].

We may classify the enzymatic technologies for chemical production from biomass into two categories according to the substrates of the enzymes. The first includes enzymatic treatments of biomass components: cellulose, starch, proteins, oils, and fats that come directly from biological sources with or without simple pretreatments. The second category refers to biotransformation that further processes simple molecules derived from biomass components, such as free sugars, organic acids, and alcohols.

2. ENZYMATIC TREATMENT OF BIOMASS COMPONENTS

Major biomass components, including polysaccharides, proteins, and fats and oils, must be decomposed into small molecules or fractions before further processing. In particular, the hydrolysis of polysaccharides is the first step toward the synthesis of chemicals from biomass carbon sources. Hydrolytic reactions can be carried out either enzymatically or chemically. The path of technology evolution so far favored enzyme technology. For example, since the discovery in the early 1960s that the enzyme amyloglucosidase enables the complete breakdown of starch into glucose, the enzymatic glucose production process quickly replaced the older acidic hydrolysis method. About one decade later, the industrial processes that use α-amylase and amyloglucosidase were able to generate 1.3 billion lbs. of glucose from starch each year. In pursuing cheaper fermentable carbon sources, people are now heavily investigating cellulases, which can break down cellulose into free glucose. The enzymatic processing of fats and oils for the production of chemical products is also an established industrial practice. Since bulk quantities of amino acids are currently produced via microbial processes, protein hydrolysis is not explored as a means for production of amino acids, but is versatile for protein structural studies, food processing [2, 3] and personal care product manipulations [4].

2.1. Enzymatic processing of polysaccharides

A large portion of agricultural products is essentially starch. Corn grains are the main source of industrial starch; they contain about 80% starch by dry weight. Corn production has gone well beyond the need for food in the US. Currently, only about 3% is used for food, while 2% is used for seed, 4% for ethanol, 15% for oil and industrial starch products, and 75% for feed [5]. As mentioned above, the enzymatic production of glucose from starch is quite a sizable industry. Glucose from starch is considered expensive for the production of fuel ethanol. It is, however, affordable for other value-added chemical products, and remains as the major carbon source for various fermentation processes that produce organic acids, alcohols, amino acids, and biopolymers.

Cellulose is an inexpensive alternative source of glucose. The major dry component of all plants is lignocellulosic biomass. Lignocellulosic biomass, such as corn stover, contains 60–70% (dry weight) carbohydrates. The major components of lignocellulosic biomass are cellulose, hemicellulose and lignin. Lignocellulosic biomass may come from wood residues (saw dust, wood waste, and pulp mill wastes), agricultural residues (corn stover, rice hulls, sugarcane bagasse, animal waste) and energy crops (switchgrass, hybrid poplar, willow).

The enzymatic hydrolysis of cellulose has attracted immense research efforts recently, owing to the growing interest in fuel ethanol production from biorenewable resources [6]. The industrial biotransformation of glucose to ethanol is currently a microbial process, whereas the production of fermentable glucose from cellulose prefers enzymatic processes. Cellulase, the enzyme which breaks down cellulose into free glucose, was first applied in an ethanol production process by simply replacing acids used for the hydrolysis of wood. In that process, the hydrolysis of wood and ethanol fermentation were separate steps. People soon realized that replacing harsh inorganic acids with a mild and biocompatible enzyme made it possible to combine the two steps into one. In other words, cellulase and fermenting microbes can be combined in one reactor, and thus simultaneous saccharification and fermentation (SSF) is possible [7]. Over the past two decades, advances in the technology for biomass ethanol production have dramatically improved the overall efficiency of the process, and the biomass ethanol is affordable enough for blending with gasoline. At the same time, production cost must be further reduced if biorenewable ethanol is to be competitive as a pure fuel without government subsidies [6]. One key factor to achieve that is the cost of cellulase enzymes. The US Department of Energy has recently initiated a Biomass Program with the two largest industrial enzyme producers, Genencor International and Novozymes, to seek a ten- to fifty-fold reduction in the cost of cellulase [8].

Directly using of polysaccharides as general polymeric materials, except for simple purposes such as packing materials, adhesives or water absorbents, is usually difficult because of their poor processibility. Their esters with organic acids, however, have versatile applications. Particularly, cellulose acetate and its copolymers [9] have a wide range of applications [10]. The industrial acylation of cellulose is usually realized via a heterogeneous reaction using acetic anhydride in the presence of sulfuric acid. Recently, the use of enzymes for the preparation of cellulose acetate has been examined [11–16]. This is mostly achieved by using a lipase-catalyzed acylation reaction of cellulose with vinyl acetate in organic solvents, such as tert-butanol [14] or mixed dimethyl sulfoxide (DMSO)/para-formaldehyde solvent [12]. The enzymatic process effectively avoids the degradation of cellulose that occurs in the acidic process and provides ease in controlling the degree of substitution. Enzymatic cellulose acylation was observed with both insoluble [13, 14, 16] and solubilized cellulose [11, 12].

2.2. Enzymatic processing of fats and oils

The enzymatic processing of vegetable oils and animal fats has derived a variety of fatty-acid chemicals. Enzymes can tackle lipids through several chemical routes. Enzymatic hydrolysis, esterification and transesterification, and the oxidation of unsaturated lipids have all been extensively pursued. Research in this area is still considerably active. Efforts are not only underway to develop more capable enzymes and reactors, but are also directed towards the exploration of new chemical routes in order to broaden the spectrum of products.

The hydrolysis of triglycerides can be realized at high temperatures and pressures in the presence of steam. Energy consumption is probably the key factor in determining the production cost. Economical industrial practice is feasible only when an inexpensive source of heat is available. Alternatively, lipid hydrolysis can be performed at low temperatures in the presence of lipases. Currently, both enzymatic and thermal processes are practiced industrially [17, 18]. Lipid hydrolysis produces glycerol and fatty acids. Glycerol is an industrial chemical used as a solvent or as an intermediate for the synthesis of other chemicals, such as glyceraldehyde and glyceric acid. Commercial glycerol, however, is mainly produced via various chemical and synthetic routes starting with propene. Valuable fatty acid esters have been developed via an esterification reaction with the free fatty acids produced from the hydrolysis of triglycerides. Esterification is essentially the reverse of the hydrolytic reaction. The same enzymes, lipases, are capable of catalyzing both reactions, depending on the water content and the substrate concentration in the reaction media [19, 20]. Fatty acid esters are important additives in health products. For example, selective enzymatic esterification for the production of γ-linolenic acid esters, which are essential to human health but cannot be synthesized in the human body, has been investigated [21–23]. Compared with chemical catalysts, enzymes can provide desirable selectivities and cause little contamination to the products.

Fatty acid esters can also be produced via lipase-catalyzed transesterification reactions of triglyceride lipids. Transesterification reactions can take place between the triglycerides and an alcohol, carbohydroxylic acid, or another ester. Most commonly, lipases are used in nonaqueous media to catalyze the reactions between tryglycerides with a monohydroxyl alcohol, such as ethanol. The products of the reactions, monoesters of fatty acids, are the key components of biodiesel. Different methods to conduct enzymatic production of biodiesel have been reported recently [24–26].

As mentioned earlier, lipases are key catalysts for both the hydrolysis and esterification reactions of lipids. In the biological world, the major role of lipases is to hydrolyze triacylglycerol lipids. Lipases are quite stable and are widely found in animals, plants, and microorganisms. Industrial lipases are mainly of microbial origin and are used for detergents and paper and food processing in addition to the industrial processing of fats and oils. Lipases are a unique class of enzymes that can assemble and catalyze reactions at the lipid-water interfaces. A number of factors, such as the Coulombic force, Born repulsion, and hydrophobic interactions, can effect such interfacial assemblings of proteins [27–30]. Hydrophobic interactions seem to be the key driver in the interfacial partitioning of lipases. It has been revealed that pancreatic lipase reaches the interfaces via complexation with pancreatic colipase, which provides the necessary hydrophobicity for the interfacial binding of the complex [29, 31–34]. Other lipases, such as *Rhizomucor miehei* lipase, have surfaces hydrophobic enough to achieve interfacial binding [35]. Extensive research results have shown that lipases are also considerably flexible biocatalysts for the acylation or deacylation of a wide range of unnatural substrates, and in most cases with high enantioselectivity [17, 35–37]. Due to their important biological functions, lipases are often the targets of medicinal drugs. For example, lipase inhibitors are being developed for use as antiobesity drugs [27, 36].

In addition to the production of fatty acids and their esters, enzymes are also effective in oxidizing native fats and oils for value-added chemicals. Enzymes from EC subgroups 1.13 and 1.14 are particularly effective catalysts for these reactions [38]. For example, an interesting bienzymatic process for the production of hydroperoxides was reported [39]. The catalytic system consisted of a lipase (EC 3.1.1.3) and a lipoxygenase (EC 1.13.11.12), which were used in a biphasic medium (octane/buffer pH 9, v/v = 1/5). The lipase catalyzed the hydrolysis of triacylglycerol lipids (which were solubilized in octane) at the interface. The liberated unsaturated fatty acids were then oxidized by lipoxygenase in the aqueous phase.

The products of the second reaction, hydroperoxides, were hydrophilic and therefore remained in the aqueous phase. High yields of oxide products were reported.

3. FURTHER PROCESSING OF SIMPLE RENEWABLE MOLECULES FOR VALUE-ADDED PRODUCTS

Simple treatment of polysaccharides, proteins, lipids and other raw materials from biorenewable resources generates chemicals of smaller molecular sizes, such as monosaccharides, amino acids, and fatty acids and their esters. Further processing of simple sugars, either via fermentation or chemical or biochemical transformation, produces an even longer list of chemicals. Combining all of these small molecules provides a solid basis for a great variety of chemical derivation (Table 1) that will eventually produce chemicals and materials that meet a broad range of needs with the strategic goal to replace most of the petroleum-based products with those for biorenewable resources.

The primary focus in sugar processing is currently on fermentation processes. New strains of microbes, either from natural origins or engineered species evolved in research labs via controlled evolution or genetic engineering, continue to emerge. Products so far developed include several important commodity chemicals: ethanol, 1, 3-propanediol, and ascorbic acid, as well as specialty chemicals such as biosurfactants for biomedical applications. These fermentation processes are not the subject of this chapter, since they are reviewed in other chapters; the products evolved from these processes, however, provide many intermediates for further processing. The potentials of using enzyme technology to derive value-added products from these intermediates will be addressed in the following.

3.1. Enzymatic production of chemicals

In order to compete with petroleum-based chemicals in the full range of industrial and societal consumption, there is a need to further process the simple chemicals derived from hydrolyzing biomass and fermentation processes to produce products possessing different physicochemical properties. This is, however, a relatively underdeveloped step. Although there are a great number of chemical and biochemical technologies readily available for this purpose, specific considerations of biomass products may need to be addressed. One critical challenge lies in the fact that biorenewable processing may not always generate the exact chemical structures derived from the petroleum chemistry. Both processing/handling methods and consumption habits must be developed or adjusted to accommodate biomass-based products of similar but not exactly the same structures as petrochemicals. In addition, the environmental compatibility of the processing technologies for biorenewable materials is particularly emphasized. After all, one major reason for exploring biomass comes from the consideration of environmental quality.

As listed in Table 1, the majority of the chemical intermediates available for further processing are alcohols, saccharides and acids, with a few exceptions, such as furan and its derivatives. Chemical derivatives based on these building blocks can be prepared by reacting them with each other. Currently, there is also a great interest in deriving chemicals and materials by reacting bio-based intermediates with petroleum-based chemicals. In either case, due to the distinct chemical properties of bio-based chemicals as compared to petrochemicals, the synthesis of bio-based products often requires slightly different chemical routes from those adapted to pure petrochemical industries. Due to the frequent occurrence of multiple functional groups (such as the hydroxyl groups of sugars) and the chiral property of biomaterials, biocatalysis has unparalleled importance in this endeavor.

Table 1
Biomass-based intermediate chemicals available for further processing

Sugars	**Simple Chemicals**	**Mixtures**
D-Sucrose	Alcohols:	Syngas
D-Galactose	Ethanol	Bio-oil
D-Glucose	D-Sorbitol (D-Glucitol)	
D-Lactose	Dianhydrosorbitol	
D-Fructose	Glycerol	
D-Maltose	Glycols	
D-Leucrose	Xylitol	
D-Xylose	D-Mannitol	
L-Sorbose	1,3-propanediol	
Dextran	2,3-Butanediol	
	Acids:	
	Leuvlinic acid	
	D-Gluconic acid	
	Citric acid	
	Fatty acids	
	Glutamate and other amino acids	
	1,β-ketodiadipic acid	
	Peracetic acid	
	2,2-ketoglutaric acid	
	2-keto-L-gluonic acid	
	Succinic acid	
	Others:	
	Furfural and derivatives	
	Tetrahydrofuran (THF)	
	Levoglucosan	
	Aldehydes	

3.1.1. Glycosylation

Most of the mono- and di-saccharides are available at a relatively low price and in large quantities. Sucrose, for example, is regarded as the most abundantly produced (over 100 million tons per year) low-molecular-weight organic bulk chemical [40]. Accompanying the development of microbial processes, the direct enzymatic processing of simple sugars is currently receiving great interest. Most of the enzymatic processes are one-step transformations of the sugars, mainly providing intermediates that need further processing. One important class of sugar-based chemicals is alkylglycosides [41, 42]. A major application of alkylglycosides is in non-ionic surfactants used in detergents, foods, and pharmaceuticals [43–46]. They are also an important class of drug intermediates. It is estimated that about 70% of lead compounds for drugs are identified from natural products, many of which are glycosylated bacterial metabolites [47]. Chemical glycosylation processes are a well developed industrial practice, yet they are not environmentally friendly and usually produce mixtures of glycosides [48]. The enzymatic route is apparently advantageous from both environmental considerations and product purity standpoint [49, 50]. Most studies on the biosynthesis of glycosides have been dedicated to alkylglucosides. Both glucosylhydrolases (EC 3.2.1) and glucosyltransferases (EC 2.4.1) are able to catalyze the biosynthesis of alkylglucosides. The hydrolases require low-water reaction media to shift the reaction equilibria in favor of synthesis against the hydrolytic reaction [45, 51, 52]. Glucosyltransferases can perform in dilute solutions and have been successfully used for alkylglucoside synthesis in biphasic media, usually with better yields than glucosylhydrolases [50, 53, 54]. Alkylglycosides of other sugars, such as alkyl xylobioside and xyloside were also reported to be possible using enzymatic synthesis [55]. The transglycosylation reaction was also examined in monophasic organic solvents [54]. To increase the solubility of sugars, polar solvents such as acetonitrile, tetrahydrofuran (THF), dimethylformamide (DMF) and pyridine were usually applied. Various lipases have demonstrated good activities for sugar acylation reactions in these solvents [56–59]. For example, the enzymatic synthesis of fatty acid sugar esters in organic media has been conducted in pyridine and acetone [60, 61].

A new development in pursuing sugar derivatives is the use of dextransucrase for the synthesis of oligosaccharide derivatives via glucosylation reactions. The primary use of this enzyme is for the production of dextran from sucrose. In 1986, Ffeifer & Langen developed a biotechnical manufacturing process to produce leucrose using this enzyme [62]. This process also triggered systematic studies of the preparation of different oligosaccharides based on the transglycosylation reaction of sucrose catalyzed by dextransucrase [63]. It was found that the enzyme is able to catalyze reactions which transfer D-glucose from sucrose to different carbohydrate acceptors [64, 65]. A variety of mono-, di-, and oligo-saccharides may act as acceptors, and many different oligosaccharides can then be formed via the acceptor reactions. In addition, it was also found that alditols, aldosuloses, sugar acids, and alkyl saccharides also accept D-glucose from sucrose in the presence of dextransucrase to form interesting, unconventional oligosaccharides.

3.1.2. Esterification

Various esterification or transesterification reactions constitute another major class of enzymatic biomass processing. In general, lipases and various proteases catalyze esterification reactions in organic media with little or no water as a co-solvent. This enables the production of an array of bio-based esters for applications ranging from drug synthesis, cosmetic-product formulation, and industrial intermediates to polymers, surfactants and food additives. Extensive research over the past twenty years on nonaqueous biocatalysis has mostly been centered on this type of reactions. Ideally, such esterification reactions require both substrates have good solubility in an organic solvent, preferably nonpolar, since polar solvents tend to denature enzymes. While this requirement can be easily met with the majority of simple acids and alcohols derived from biomass, it is challenging for sugars or sugar-based acids such as D-gluconic acid. The multiple hydroxyl groups of these substances make them highly polar, and they usually have little solubility in common organic solvents. Currently, there is no particularly efficient method available to overcome this solubility barrier. One common approach is the use of biphasic reaction systems. Another way is to make the sugars hydrophobic enough to gain some organic solubility. This strategy was found effective, for example, for the transesterification reaction between ethyl lactate and *n*-octyl glycoside catalyzed by Novozyme 435 [43, 66] and lipase-catalyzed synthesis of a sugar ester containing arachidonic acid [67].

3.1.3. Oxidation

Enzymatic oxidation of sugars can produce chiral chemical intermediates for the synthesis of drugs and specialty chemicals. It has been demonstrated that the enzyme glucose-2-oxidase catalyzed the efficient conversion of D-glucose to D-glucosone (2-keto-D-glucose) with an optimal conversion of over 99%. Chemical routes of glucose oxidation usually resulted in glucosone yields of less than 50 % and have many byproducts [68]. D-Glucosone can be further transformed into 2-keto-D-gluconic acid via enzymatic oxidation [69] or chemically converted into D-fructose using metal catalysts [70]. Galactose-oxidise (EC 1.1.3.9.) catalyzes the oxidation of D-galactose into 6-aldehydo-D-galactose in the presence of O_2 [71]. The enzymatic oxidation of sucrose to 3-keto-sucrose by glycoside-3-dehydrogenase of *Agrobacterium tumefaciens* was also reported [72]. 3-Keto-sucrose can be further used to derive several value-added chemicals. For example, reductively aminating it produces 3-amino-sucrose, which can be further transformed into biosurfactants and other products by acylation reactions. Various organic acid oxidases, such as amino-acid oxidases [73] and hydroxyl acid oxidases, [74] are also available for the derivation of sugar-based chemical products.

Although researchers have demonstrated the versatility and immense potential of biocatalysis for synthesizing specialty chemicals at the chemistry level, industrial application of new biocatalytic processes is only making initial steps beyond the traditional areas. For the synthesis of bulk and intermediate chemicals that can sell for as little as $1–$5/kg, industrial enzymes, which can cost as much as $10,000–$100,000/kg, may not be economical to use. Another factor hindering the industrialization of biochemical processes is the reluctance of industry to update technologies. Although bioprocessing may be as efficient as traditional processes, the cost of switching processes may exceed the potential economic benefits. Nevertheless, new industrial processes continue to emerge for chemical production from biorenewable resources. DuPont has developed a whole cell process for producing 1,3-propanediol, an important precursor for general-purpose polymers. Genecor has a technology (whole cell system) for biosynthesizing 2-keto-L-gluonic acid (2KLG) – an intermediate in ascorbic acid (vitamin C) production, using glucose as the raw material [1].

3.2. Enzymatic preparation of polymers from biorenewable resources

Interest in biodegradable polymers was probably first triggered by the search for biomedical materials. A few biodegradable aliphatic polyesters are already commercially available for these applications. These include polyglycolic acid (Dexon), glycolide-lactide random copolymer (Vicryl) and trimethylene carbonate block copolymer (Maxon). People have long realized that the use of biodegradable polymers for disposable general purpose materials could greatly benefit the environment. Biodegradable polymers, such as polyesters, polycyanoacrylates, polycarbonates, and polysaccharides and proteins with proper functional properties, can be used for many different applications including packaging materials, industrial chemicals, thickening agents, water absorbents, and personal care products [75, 76].

Often the specific properties and functionalities of biopolymers determine their applications. The most important properties include biodegradability, mechanical strength, hydrophilicity, and the toxicity of the degradation products [75, 76]. The properties of homopolymers, which are constructed with only one type of monomer, are usually inherently limited and do not satisfy a wide range of requirements. Accordingly, much effort has been made to develop biopolymers with tunable properties in order to meet diverse application requirements. A straightforward method is to physically blend two or more homopolymers. These polymer blends are useful for such applications as sustained drug release [77]. The behavior of the polymer blends can be anticipated to be strongly dependent on the degree of mixing of these homopolymers. However, it is usually difficult to mix two immiscible homopolymers well. In the case of PLA, the adjustment of the ratio between the D- and L-stereoisomer represents another mechanism to control the polymer's crystallinity and degradation rate [78, 79]. While these simple approaches are effective to some extent, copolymers that incorporate monomers with different properties can provide more predictable performance, offer much more attractive alternatives, and thus are subject to the most extensive studies. It has been shown that copolymers of lactic acid and glycolic acid could offer properties adjustable depending on their composition [80, 81]. Various copolymers of PLA were developed and examined using glycolide, caprolactone, polyproylene glycol (PPG), and polyethylene glycol (PEG). For example, the incorporation of PPG or PEG significantly altered the properties of PLA [82].

Among the chemicals mentioned above, lactic acid is usually produced by a fermentation process that utilizes agricultural resources; while ethylene glycol and propylene glycol are derived from the petroleum-based chemicals ethylene and propylene, respectively [83]. Polysaccharides, such as cellulose, chitin, and starch, represent another major class of biodegradable and biocompatible materials, and have been used widely as low-cost materials for biomedical and industrial applications [77, 84, 85]. Polysaccharides possess properties that are quite different from those of PLA, PEG or nylons due to their high biodegradability, hydrophilicity, and multiple pendent functional groups. However, like the homopolymers of PLA and nylons, polysaccharides also possess only slightly adjustable physical properties. Accordingly, the current interests in synthesizing polymers from biorenewable sources mostly focus on copolymers that combine the properties of petrochemicals and biomass materials for a wide rang of applications.

Over the last decade or so, there has been a rapidly growing interest in the development of sugar-containing copolymers. Traditional chemical methods, however, have difficulty in controlling the reactions with respect to multiple hydroxyl groups of sugars. The non-selective chemical polymerization processes tend to produce star-branched or crosslinked polymers. In order to produce linear polymers that are required for most applications, complicated blocking and unblocking treatments of the hydroxyl groups of sugars are usually necessary in these chemical processes. For example, the preparation of vinyl-sugar type polymers via a chemical process uses isopropylidene to block the hydroxyl groups of the sugars [86, 87]. In preparing styryl-type polymers with pendent glucose, an acetyl group is used to block the hydroxyl groups of the sugars [88]. The preparation of oligochitin-PPG copolymers involves the blocking of some of the hydroxyl groups of chitin with sodium methoxide followed by hydrolysis to remove the methoxyl group [89]. Other examples include the copolymerization of oligomeric cellulose or amylose with PPG using acetyl as the blocking agent [9, 90]. In addition, most chemical preparations of biodegradable polymers involve the use of heavy metal-based catalysts, which often contaminate the polymer products. For example, metal catalysts, such as the oxides of Zn and Sn, are commonly applied in the preparation of PLA via ring-opening lactide polymerization [91–93]. Extensive purification measures are usually required to remove the metal contaminants from the polymer, especially when the materials are to be used for biomedical applications [94]. In this regard, alternative enzymatic polymerizations are advantageous in that they proceed at mild reaction conditions and involve no metal contaminations.

The unique selectivity of enzymes can provide efficient control over the polymer structure, and often leads to novel polymers that are difficult to derive from chemical processing. As a result, the enzymatic synthesis of polymers has attracted much attention in recent years, particularly for the synthesis of polyesters, polysaccharides, and polyaromatics [95–97]. The first two types are from biorenewable resources, and will later be discussed in further detail. Enzymatic preparations of polyaromatics are an important area in current research of enzymatic processing [98–101]. The major building blocks are petroleum-based, but some special polymers contain sugar [101] or nucleotide [100] side groups. These types of polymers have been reviewed previously [95, 102, 103] and will not be addressed here.

3.2.1. Sugar-containing copolymers

The selectivity of enzymatic catalysis is particularly desirable when preparing sugar-containing polymers. Enzymatic catalysis can provide unparalleled enantio-, stereo-, and regio-selectivity. In the biological world, only L-amino acids are used in the construction of proteins, and D-sugars are usually utilized to build polysaccharide chains. It has been shown that proleather catalyzed the transesterification reaction between sucrose and adipate only at the C-6 and C-1' hydroxyl group of the sugar and leads to a linear copolymer by the enzyme [104, 105]. Similarly, the acylation of a galactose derivative (galactopyranoside) with vinyl acrylate catalyzed by lipase occurred at the C-6 and C-2 hydroxyl groups of the sugar derivative and produced a di-vinyl, hydrophilic crosslinker for vinyl-type polymers [106]. Other examples may include the syntheses of galactoglycerides and glucosides [107, 108], and acylation [57] or deacylation [109] of sugars. These reactions have been reviewed in a recently published paper [20]. Table 2 lists the most representative sugar-containing polymers synthesized by using enzymes.

Vinyl-sugar type polymers have received particular attention. The preparation of this type of polymer includes the use of lipases and other enzymes to catalyze transesterification reactions between sugars and vinyl type esters to produce vinyl monomers with pendent sugar moieties [106, 107, 110, 111]. The vinyl-sugar monomer can form polymers by traditional free radical polymerization of the vinyl moiety. Similar polymers were also prepared using chemical methods [87].

Table 2
Sugar-containing polymers prepared using enzymatic biocatalysis

Type of Polymer	Typical Structure	Reference
Vinyl-sugars	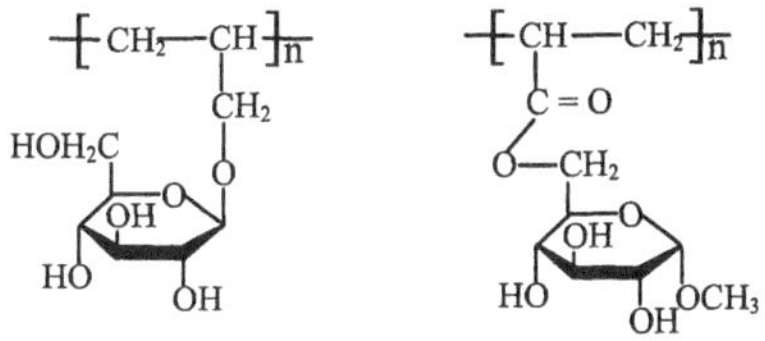	[53, 110, 111]
Linear polyesters	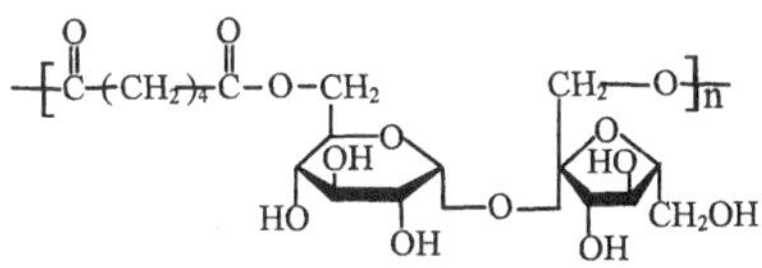	[104, 105]
Other polymers with sugars as side groups.	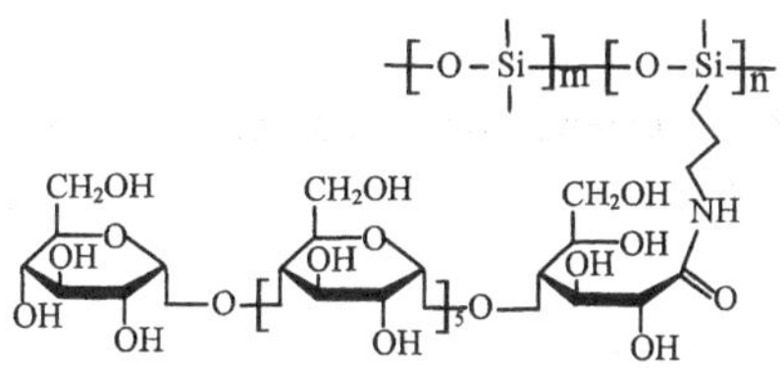	[112]
Homopolymers-artificial cellulose	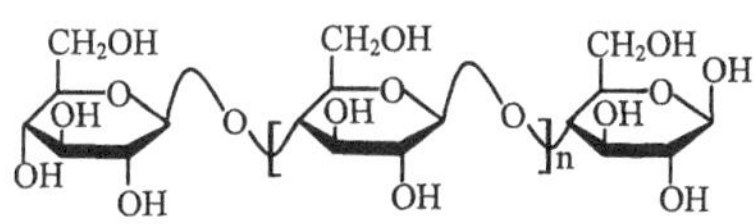	[113]

New sugar-containing polymers continue to appear. A multi-step biocatalytic route for the synthesis of an interesting macromer around a sugar core was recently reported [114]. Lipases were used to catalyze the acylation reaction of 4-C-hydroxymethyl-D-pentofuranose with vinyl methacrylate in dry tetrahydrofuran (THF). One of the sugar acryl derivatives was then used as a multifunctional initiator for ring-opening polymerization of caprolactones (such as ε-CL) catalyzed by Novozyme 435 in toluene, leading to a polymer product with sugar-ending CL chains of number average molecular weight (Mn) of 11,300.

3.2.2. Other polymers

Polyesters are another important class of biomass-based polymers that have been derived using enzymatic biocatalysis. Similar to other enzymatic esterification processes, the enzymatic preparation of polyesters prefers nonaqueous reaction media. The research in this area has mostly focused on the nonaqueous enzymatic preparation of PLA. There are two approaches to the biosynthesis of PLA. One is the enzymatic polycondensation of lactic acid

[115, 116]. The reactions are usually slow; it usually requires several days to reach a conversion higher than 50%. Very interestingly, the rate of enzymatic polycondensation of lactic acid was even lower than that catalyzed by silica gel, which was initially used to absorb the byproduct water of the condensation reaction [115]. The molecular weight of the polymer products were mostly around 2000 Da. In improving enzymes' activities for nanoaqueous polymerization reactions, Ohya et al. examined the use of PEG-modified esterase and lipase in the polycondensation of various α-hydroxy acids, including lactic acid, in organic media [117]. The pegylation only improved the enzymes' activity moderately; the pegylated lipase showed activities that were comparable to or slightly lower than that of the unmodified lipase. The best improvement was observed for pegylated esterase-catalyzed condensation of ethyl glycolate, with the 72-h conversion increased to 65% from the 33% achieved with the unmodified esterase (data read from the reported figures of the paper). In a recent work, enzymatic preparation of PLA using an organic-soluble proleather was reported. Proleather from *Bacillus* sp. was chemically modified with decanoyl chloride. The modified enzyme was highly soluble (up to 44 mg-protein/mL) and active in various organic solvents, including chloroform, THF, pyridine and acetone. The reaction rate observed with the organically soluble proleather was 4 to 22 times that of native proleather, depending upon the solvent applied [118]. However, the improved enzyme activity did not increase the molecular weight of the PLA product, indicating that factors other than enzyme activity control the achievable size of the polymer product.

Enzymatic ring-opening polymerization of lactones has so far achieved more promising results for the production of linear polyesters with higher molecular weights [119–123]. Some of these procedures have been patented [124]. Again, nonaqueous media was the choice. For example, enzyme-catalyzed ring-opening polymerization of lactones has been demonstrated in organic solvents such as toluene and isooctane by heating the reaction mixture to 70~100°C [124–126]. A molecular weight of about 10,000 Da was achieved. Sugar has been identified as a good initiator for such ring-opening polymerization, following the same reaction mechanism as that involved in other simple glyco-acylation reactions [57, 109]. It has been shown that the prime hydroxyl group of ethyl glucoside could initiate the ring-opening polymerization reaction of lactones [127].

4. NEW TRENDS IN ENZYMATIC BIOPROCESSING

Substantial knowledge of protein chemistry and biology has been accumulated in the past century. The beginning of the 21st century has witnessed a great advancement in the science of genomics. All of these bring about a new era in capitalizing the potentials of biological science in industrial practices to benefit the society [128–130]. The use of enzymes in drug synthesis is now well established, while other enzymatic industrial processing is probably still in its infancy. A recent significant development of industrial biocatalysis is the emergence of nonaqueous biocatalysis, which substantially expanded the impact of biocatalysis from traditional aqueous-based bioprocessing to a much broader range of organic synthesis [131]. Subsequently, there has been an upsurge of interest in developing biocatalysts that possess high activities and extended lifetimes at extreme conditions, such as extreme pHs and temperatures, in addition to various nonaqueous environments. However, there seems to be no quick antidote to all the problems encountered, and the challenges will be long-standing. The profusions of genetic engineering, nanoscale science, and high-throughput biocatalysis, on the other hand, are accelerating the pace of development in industrial biocatalysis. The following reviews the most recent advances in this endeavor.

4.1. Biocatalysis in nonaqueous environments

It has been shown that esterification, acylation, and other simple reactions can be conducted using enzymes in nearly anhydrous organic solvents or supercritical carbon dioxide [131–136]. Such monophasic nonaqueous biocatalysis is readily applicable for the preparation of various ester products, as discussed earlier, from biorenewable resources. In addition to fine chemical synthesis, nonaqueous biocatalytic esterification reactions are also suitable for producing biodiesel from vegetable oils [137]. Biodiesel has been produced from biorenewable resources, including sunflower and soybean oils, which are composed of C4–C20 fatty acid triglycerides. These are converted to their respective alkyl esters by substituting triglycerol for the short-chain alcohols, mainly methanol or ethanol [137, 138]. Often, the low activity of enzymes placed in nonaqueous environments is the limiting factors for industrial applications. The observed activities of native enzymes in organic media are usually 2–6 orders of magnitude lower than their aqueous activities [139]. Factors, including structural denaturation and diffusional limitation (due to the insolubility of native enzymes), may contribute to the limited nonaqueous activities of enzymes [140]. Various activation and stabilization methodologies have been examined over the past two decades for biocatalysis in organic solvents. In general, homogeneous reactions with enzymes soluble in organic media showed high enzyme activity. Methods such as attaching hydrophobic chemical groups [118, 141–143], surfactant coating [144, 145], and deglycosylation [146] were all used to prepare organic-soluble biocatalysts. Very interestingly, surfactant-coated protease showed activities that are comparable to those observed in aqueous solutions [145]. Other physical treatments of enzymes include freeze-drying with inorganic salts [135, 147] and complexation polymers [148]; although the enzymes remain insoluble, they also show impressive improvement in their activities. Similar to what has been observed for aqueous biocatalysis, covalent binding of enzymes on solid materials was found very effective in improving the stability of enzymes in organic solvents. It has been demonstrated that the incorporation of enzymes into synthetic polymers, especially via multiple covalent bonds, can significantly improve their activities in nonaqueous environments [149–151]. Particularly, plastic enzymes showed activities that were comparable or even higher than those of enzymes solubilized via ion-pairing with surfactants in organic solvents [149]. Cross-linked enzyme crystals (CLECs) are probably the most powerful approach for stabilizing enzymes against not only organic solvents, but also other factors such as temperature and pH [152–154]. The crystalline structure, however, may introduce great mass transfer resistance and further leads to low specific enzyme activity.

A monophasic nonaqueous approach, however, is not suitable for biotransformations that involve both hydrophobic and hydrophilic substrates and/or cofactors that are insoluble in the same reaction medium. For example, the oxidation of alkenes catalyzed by peroxidases [155,

156], the transglycosylation reactions by galactosidases or glucosidases [157–159], and the degradations of organic pollutants by cofactor-dependent enzymes such as toluene monooxygenases [38] all involve immiscible chemicals/cofactors and cannot be performed efficiently with monophasic reactions. Biphasic reactions have been traditionally applied for these reactions.

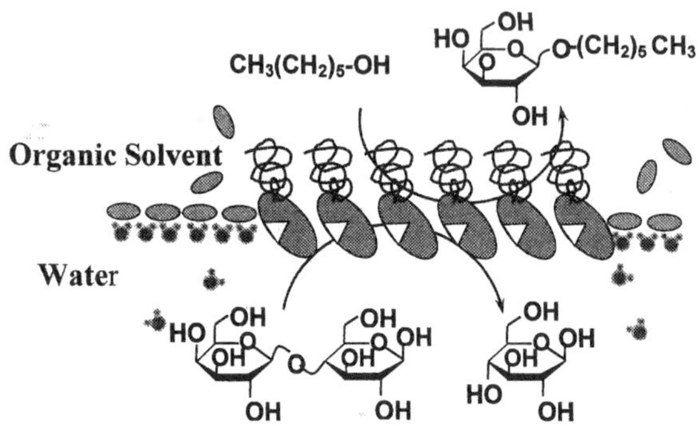

Fig. 1. Novel interface-binding enzymes for biphasic reactions such as transglycosylation (Such a configuration can significantly improve the efficiency of the enzymes [160].)

As discussed earlier, biphasic reactions were applied for the enzymatic synthesis of alkylglucosides using glucosylhydrolases and glucosyltransferases [43]. In contrast to traditional biphasic reactions where the aqueous phase is used as the container for enzymes, the transglycosylation reaction was also reported using an organic-soluble β-galactosidase [46]. An apparent drawback of placing enzymes in either bulk aqueous or organic phase is that the majority of the enzyme is not available for reactions, since only the portion of enzyme exposed to the interface area has a chance to contact substrates hosted in the other bulk phase. In a recent study, a process that applied interface-binding enzymes for biphasic reactions was reported [160]. The transgalactosylation reaction between lactose and 1-hexanol was conducted in a toluene-buffer biphasic system (Fig. 1). The hexanol, whose water solubility is <0.7 wt%, was dissolved in toluene, whereas lactose was contained in an aqueous solution. Polystyrene-conjugated β-galactosidase self-assembled at the interface and catalyzed the reaction. In a control test, native β-galactosidase dissolved in the aqueous phase was used to catalyze the same biphasic reaction. Interestingly, the polymer-conjugated enzyme catalyzed the reaction much more efficiently than the native enzyme, such that the initial reaction rate catalyzed by the former was 145-fold faster. The improvement in enzyme activity for the interfacial biocatalysis is believed to be a result of the simultaneous accessibility of the biocatalyst to substrates contained in both phases.

The use of organic solvents may impact the specificity of the enzymes, mostly due to the structural changes induced by the organic solvents [161, 162]. Native chloroperoxidase (CPO) catalyzes the epoxidation of styrene with an *ee* number of ~47% in pure buffer, but shifted to 36% in the presence of 99% of tert-BuOH [163]. In another study, salt-activated and solubilized subtilisin also produced unexpected alterations in the regioselectivity of the enzyme for nonaqueous acylation reactions [164]. Furthermore, the selectivity of enzymes can be purposely manipulated. Chymotrypsin was modified to accept the D-form of a tryptophane derivative by precipitation with the enzyme inhibitor N-acetyl-D-tryptophane in propanol [165]. The regioselectivity of subtilisins BPN and Carlsberg glyco-ester synthesis was also varied by lyophilizing enzymes with substrates [166].

4.2. High-throughput biocatalysis

Combinatorial synthesis and high-throughput screening of drug candidates has become a key element in R&D for today's pharmaceutical industry. Such technologies usually involve expensive, computerized, automatic processes, and are highly desirable for preparing and testing hundreds and very often thousands of chemical structures in a cost-effective and timely manner. This is especially important for pharmaceutical firms to survive in or lead the intensified worldwide industrial competition. The traditional route usually requires over ten years to screen approximately 10,000 lead compounds for the discovery of a new drug at an overall cost of about $350 million. Viewing the great promise of high-throughput technologies, several researchers have made initial efforts, starting in the late 1990s, to put combinatorial biocatalysis under the spotlight of biotechnology. Promising results have been reported for iterative derivatization of both small and complex molecules, regioselectively controlled libraries, novel one-pot library syntheses, and bioreactor and biocatalyst enhancements [167]

The concept of combinatorial biosynthesis has been demonstrated in several published studies, mainly by Dordick and Clark and their co-workers. An automated, iterative synthesis procedure for the preparation of a 600-compound library was carried out for the derivatization of the flavonoid bergenin [168]. The process used commercially available 96-well filter plates with an automatic liquid handling system for the transport, sampling, and recovery of the reaction products. A chemoenzymatic preparation of a nine-member Ugi condensation library was later reported [169]. The carboxylic acid and amine precursors were first acylated and selectively catalyzed by porcine pancreatic lipase. These derivatives were then subjected to a four-component Ugi condensation reaction in the presence of acetaldehyde and methyl isocyanoacetate. In more recent work reported by Secundo et al., a 39-member library of bile acid derivatives was prepared [170]. A sequential approach that included a hydroxysteroid dehydrogenase-catalyzed oxidation step and a lipase-catalyzed acylation step with a series of different acyl donors led to the modification of the bile acid scaffold. All of the 39 compounds were obtained in high purity and good yield. It is expected that the variety of biocatalytic reactions, including acylation, glycosylation, halogenation, oxidation, and reduction, will eventually afford combinatorial biosynthesis great power in achieving complex libraries similar to and beyond those already illustrated in nature, but on relatively small chips and at a rapid rate [171].

The high-throughput concept is also driving explorations in novel bioreactors and biocatalysts. In addition to the construction of filter-well bioreactors used in combinatorial biosynthesis, microfluidic biochips that enable multienzyme catalysis have been reported [172]. In an interesting study, monolithic porous polymer supports were used to covalently immobilize trypsin [173]. The monolithic reactor maintained good productivity even at a flow velocity of 180 cm/min, which is difficult to reach with packed-bed reactors. A combinatorial experimental technique was also recently reported in the identification of salts and salt mixtures capable of activating penicillin amidase in organic solvents for the transesterification of phenoxyacetate methyl ester with 1-propanol [174]

High-throughput protein synthesis and screening were used as a novel approach in biocatalyst discovery. Taking advantage of *in vitro* protein synthesis, protein function can be screened quickly for catalytic activity or functional biological interactions. It was reported that in vitro synthesis of firefly luciferase on solid-phase template DNA, which was bound to wells in 96-well plates, was achieved using simultaneous transcription and translation in a wheat-germ extract system [175].

4.3. Nanobiocatalysis

Most industrial enzymes are applied in immobilized forms. Cellulase and amylases are exceptional cases, because the substrates of these enzymes are insoluble and require enzymes to be bound to the solid substrates. The structure of the carrier material is vitally important to the performance of immobilized enzymes. Porous materials provide large surface area per unit mass for high enzyme loading, but are subject to great mass-transfer resistance because substrates must diffuse into the materials to reach the enzymes. Nonporous materials, to which enzymes are attached at external surfaces, do not suffer much mass-transfer limitation, but the effective enzyme loading is usually low. In fact, the enzyme loading is frequently below 1% by weight for immobilization with either porous or nonporous materials, i.e., the enzyme is diluted by over 100-fold. Both the mass-transfer resistance and low enzyme loadings directly impinge on the overall efficiency of bioprocessing.

Recent advances in nanoscale science and technology are fueling a new wave of revitalization in the R&D of biocatalysts. Synergizing with materials chemistry, nanostructures have manifested a great potential in enabling biocatalysts with performances well beyond the scope of traditional immobilized enzymes. In addition to enzyme stabilization and activation, nanostructures were also proven powerful in manipulating protein-protein and protein-environment interactions, and thus affect unique biotransformations. Efforts have been made in exploring the use of materials such as nanotubes, nanocapsules, nanofibers, nanoparticles, nanocomposites and nanoporous matrices.

4.3.1. Nanoparticles and nanocapsules

Reducing the size of carrier materials generally improves the efficiency of immobilized enzymes. In the case of surface attachment, smaller particles can provide higher enzyme loading per unit mass; with porous materials, mass-transfer resistance is much reduced in smaller particles owing to the shortened diffusional path of substrates. There have been extensive studies on the use of micrometer-sized materials [152, 154, 176]. Recently, a growing interest has been shown in using nanoparticles as carriers for enzyme immobilization [177–181]. The effective enzyme loading on nanoparticles can be very high (for example, it can reach over 10 wt% with particles smaller than 100 nm [181, 182]), and a large surface area per unit mass is also provided to facilitate the reactions between enzymes and substrates.

Nanoparticles provide an ideal remedy to the usually contradictory issues encountered in the optimization of immobilized enzymes: minimum diffusional limitation, maximum surface area per unit mass, and high effective enzyme loading. In addition to the promising performance features, the unique solution behaviors of the nanoparticles also point to an interesting transitional region between heterogeneous and homogeneous catalysis. Theoretical and experimental studies demonstrated that particle mobility, which is governed by particle size and solution viscosity, can impact the intrinsic activity of the particle-attached enzymes [181].

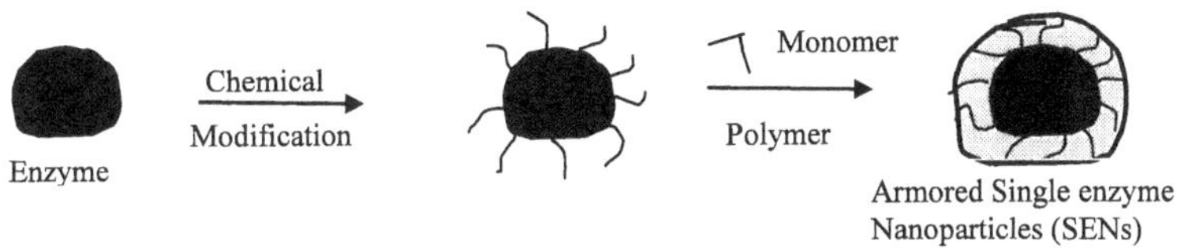

Fig. 2. Armored single enzyme nanoparticles reported by Kim et al. [183]

Nanocapsulated enzymes represent a unique form of nanoparticle biocatalysts. In an early effort, crosslinked polyacrylamide hydrogel nanoparticles with α-chymotrypsin entrapped inside the particle were prepared by forming reversed micelles [184]. The diameter of the particles, measured via quasi-elastic light scattering, was found to be 84 nm, indicating the possibility of multiple enzyme molecules for each particle. The reported enzyme activity appeared to be low though, presumably due to the mass transfer resistance through the hydrogel layer surrounding the enzyme. In a recent study, armored single-enzyme nanoparticles (SENs) (Fig. 2) with a porous composite organic/inorganic network less than a few nanometers thick surrounding each enzyme molecule was reported [183]. This approach significantly stabilized two proteases (α-chymotrypsin and trypsin) with half-lives estimated to be up to 143 days at 30°C. The armor network around α-chymotrypsin is sufficiently thin and porous not to place a large mass-transfer limitation on the substrate. The aqueous activity of the SENs was comparable to that of free enzymes.

4.3.2. Nanoporous media

It is well known that incorporating enzymes into synthetic polymer matrices, especially via multiple covalent bonds, can significantly improve the stability of enzymes [149–151, 185]. It was demonstrated that multiple bonding for enzyme immobilization can also be achieved with mesoporous silica glass, and thus substantially improve the stability of enzymes [186]. A nanoporous sol-gel glass possessing a highly ordered porous structure (with a pore size of 153 Å in diameter) was examined for use as a support material for enzyme immobilization (Fig. 3). A representative enzyme, α-chymotrypsin, was efficiently bounded onto the glass with a bifunctional ligand, trimethoxysilylpropanal. The enzyme loading was measured to be 0.54 wt% via active site titration. The glass-bound chymotrypsin exhibited greatly enhanced stability both in aqueous solution and organic solvents. The half-life of the glass-bound α-chymotrypsin was over 1000-fold higher than that of the native enzyme as measured in anhydrous methanol. The enhanced stability in methanol, which excludes the possibility of enzyme autolysis, particularly reflected the fact that covalent binding provides effective protection against enzyme inactivation caused by structural denaturation.

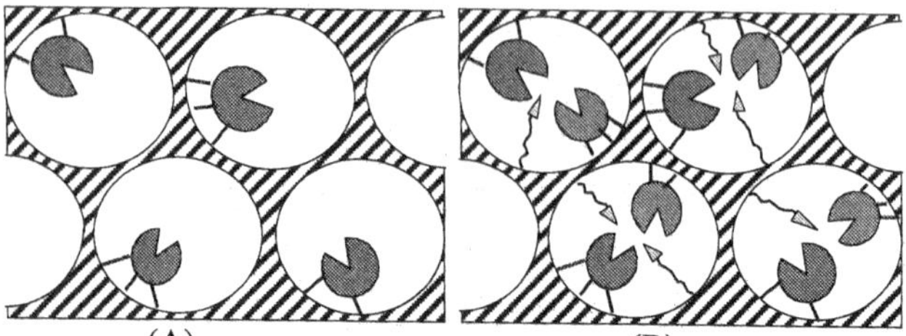

Fig. 3. Configuration of nanoporous biocatalysts. (A) Single-enzyme biocatalyst; (B) Co-immobilized enzyme-cofactor-enzyme biocatalytic system.

Nanoporous structures were also found useful in enabling multienzyme catalysis with co-immobilized enzymes and cofactor [187]. A catalytic system consisting of lactate dehydrogenase (LDH), glucose dehydrogenase (GDH) and cofactor NADH was examined. Commercially available porous spherical glass particles with an average particle diameter of 60 μm and different pore sizes (30 and 100 nm in diameter) were used. The glass was first activated; next, the enzymes and cofactor were covalently attached. The cofactor was immobilized in the form of the reduced state, NADH, which was required to enable the reduction of pyruvate into lactate. NADH can then be regenerated from NAD^+ via the GDH-catalyzed oxidation of glucose. In a test, the glucose concentration initially remained unchanged before the addition of pyruvate; however, the subsequent addition of pyruvate successfully triggered the consumption of glucose. Since the consumption of glucose requires NAD^+, the shuttling of cofactor between the two enzymes appears to be evident. Control tests with activated glass support did not show detectable adsorption of either glucose or pyruvate within three days.

4.3.3. Nanotubes and nanofibers

Nanoparticles provide the upper limits in terms of balancing the contradictory issues including surface area, mass-transfer resistance, and effective enzyme loading; their dispersion in reaction solutions and the subsequent recovery for reuse, however, are often found to be daunting tasks. It appears that the use of nanofibers will overcome this limitation while providing the advantageous features of nanosize materials. Polystyrene nanofibers of 120 nm diameter produced via electrospinning were examined for the immobilization of α-chymotrypsin in a recently reported work [188]. The observed enzyme loading, as determined by active site titration, was up to 1.4 wt%, corresponding to a greater than 27.4% monolayer coverage of the external surface of nanofibers (Fig. 4). The apparent hydrolytic activity of the nanofibrous enzyme in aqueous solutions was over 65% of that of the native enzyme, indicating a high catalytic efficiency as compared to other forms of immobilized enzymes. The nanofibrous α-chymotrypsin also exhibited a much-improved nonaqueous activity that was over three orders of magnitude higher than that of its native counterpart suspended in organic solvents including hexane and isooctane. It appeared that the covalent binding also improved the enzyme's stability against structural denaturation, such that the half-life of the nanofibrous enzyme in methanol was18-fold longer than that of the native enzyme.

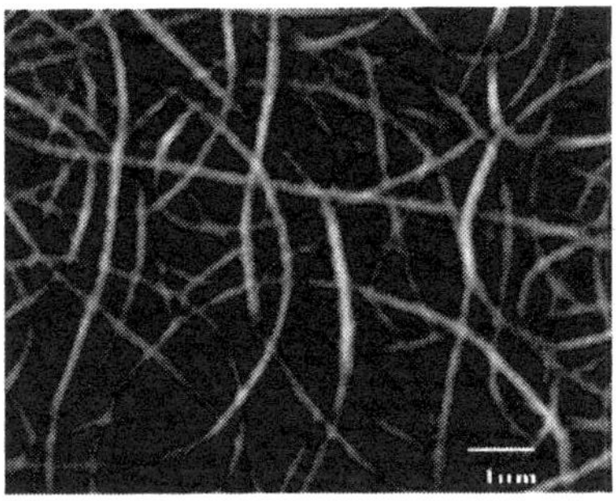

Fig. 4. Nanofibers (130 nm in diameter) of polystyrene with covalently bound enzymes.

Single-walled carbon nanotubes (SWNTs) have a diameter of 1.3 to 2 nm and a length of tens of micrometers. Their unique physicochemical properties have inspired extensive research in a wide spectrum of scientific areas. The interest in using SWNTs has now been extended to biocatalysis. Enzyme-polymer-SWNT composites were prepared and examined for biocatalytic performance [189]. Improved enzyme activity was observed in comparison to similar enzyme-containing composites without using SWNTs. It was suspected that SWNTs, which possess a high specific surface area, may effectively adsorb enzyme molecules and retain the enzyme within the polymer matrix, whereas other forms of enzyme-composites may have suffered from enzyme loss via leaching when they were placed in contact with aqueous solutions.

4.3.4. Genetic engineering for catalyst production

Genetic engineering has advanced rapidly, allowing scientists to design and produce unnatural enzymes with great efficiency. DNA cloning is one major approach. Over expressing enzymes in nonnative microbial hosts makes it possible to produce enzymes in large quantities that would otherwise be impossible. This is extremely important and nowadays an essential step in the large-scale production of enzymes. *E. coli* is probably the most widely used protein expression host. Recent advances in the science and technology of protein transcription, translation and folding are making this bacterium more valuable than ever for the expression of a great variety of proteins from both natural and unnatural amino acid sequences [190]. Particularly pertaining to biorenewable resources, the expression of recombinant lipases in *E. coli* is still subject to substantial investigations [191, 192]. The joint effort between DOE, Genencor, and Novozymes on cellulase is another showcase of the power of genetic engineering to reduce the production cost of industrial enzymes.

Often, the biocatalysts that evolved over millions of years in nature do not offer the desired features for industrial reactors. There is always a need to tailor the properties of a biocatalyst. The current art of molecular biology is so sophisticated that the amino acid sequence of an enzyme may be manipulated at will. The use of site-directed mutagenesis allows investigators to explore unnatural enzymes with desired functionality and efficiency. The rational design of recombinant enzymes, however, is often limited by insufficient knowledge of the structure-function relationships of proteins. Although there are successful examples in developing enzymes using rational design, such as fatty acid unsaturases with specific selectivity [193], cytosine deaminase mutants with altered substrate preference [194], thermlysin-like protease [195] and *Trichoderma reesei* endo-1,4-beta-xylanase II [196] with improved thermal stability, rationally designed recombinant enzymes have so far mostly failed to exhibit the expected properties. However, there is no doubt that, with the immense research effort dedicated to this area, breakthroughs will emerge at a fast pace. One new tool is computational biomolecular simulations, which allow for the quick designing and screening of biocatalysts and are expected to substantially reduce the amount of experimental work. The merger of quickly advancing computational methods with rational genetic design of enzymes will afford us unprecedented speed in discovering and generating efficient biocatalysts [197]. Complementing the rational design approach, directed evolution and gene shuffling have been proven effective and are being practiced widely in both academic and industrial researches [198–201]. However, gene shuffling may not be as potent as rational design in generating novel biocatalysts. Naturally, more and more researchers are combining rational design, combinatorial gene shuffling, and computational biology in the discovery of biocatalysts for desired activities, selectivities and stabilities [197, 202].

5. SUMMARY

Expanding from traditional applications such as food processing, biocatalysis has left a large footprint in modern drug synthesis and discovery, and is now making initial steps toward the production of commodity chemicals. Several examples have shown the promising potential of biocatalysis in the chemical industry. In addition to its environmental compatibility and chemical and structural selectivity, biocatalysis can also enable reactions that are difficult to realize with chemical catalysts. One of the best examples is the production of acrylamide using a nitrilase enzyme. The nitrilase-catalyzed hydrolysis of nitrile in the presence of an ester or an amide is almost impossible to carry out using traditional chemical means. Nitto Chemical has commercialized the enzymatic process in Japan with a production scale of approximately 20,000 metric tons per year. There is no doubt that biomass-based chemicals will rely on biocatalysis to a much higher degree than petrochemicals. It can be expected that, within twenty years or so, the role of biocatalysis in industrial chemical production from biomass will become as critical as it is in today's food processing industry.

REFERENCES

[1] R.L. D'aquino, Chem. Eng., 106(1999) 37.
[2] H.J. Chae, M.-J. In, and M.-H. Kim, Han'guk Nonghwa Hakhoechi, 40(1997) 404.

[3] B.C. Saha and K. Hayashi, Biotechnol. Adv., 19(2001) 355.
[4] A. Gayte-Sorbier, Industries Alimentaires et Agricoles, 94(1977) 291.
[5] J.D. Mcmillan. Biotechnological Routes to Biomass Conversion. in DOE/NASLUGC Biomass and Solar Energy Workshops. Denver, CO, 2004.
[6] M.E. Himmel, M.F. Ruth, and C.E. Wyman, Curr. Opin. Biotechnol., 10(1999) 358.
[7] N.D. Hinman, D.J. Schell, C.J. Riley, P.W. Bergeron, and P.J. Walter, Appl. Biochem. Biotechnol., 34/35(1992) 639.
[8] Website of EERE: http://www.eere.energy.gov/biomass/enzymatic_hydrolysis.html.
[9] S. Kim, V. Stannett, and R. Bilbert, J. Macromol. Sci. - Chem., A10(1976) 671.
[10] L.C. Treece and G.I. Johnson, Chem. Ind., 49(1993) 241.
[11] O.A. El Seoud, G.A. Marson, G.T. Ciacco, and E. Frollini, Macromol. Chem. Phys., 201(2000) 882.
[12] K. Yang, Y.-J. Wang, and M.-I. Kuo, Biotechnol. Prog., 20(2004) 1053.
[13] W. Cui, W.T. Winter, S.W. Tanenbaum, and J.P. Nakas, Enz. Microb. Technol., 24(1999) 200.
[14] V. Sereti, H. Stamatis, C. Pappas, M. Polissiou, and F.N. Kolisis, Biotechnol. Bioeng., 72(2001) 495.
[15] V. Sereti, H. Stamatis, E.G. Koukios, and F.N. Kolisis, J. Biotechnol., 66(1998) 219.
[16] J. Xie and Y.L. Hsieh, J. Polym. Sci., Part A: Polym. Chem., 39(2001) 1931.
[17] F.X. Malcata, H.R. Reyes, H.S. Garcia, and C.G. Hill, Jr., J. Am. Oil Chem. Soc., 67(1990) 890.
[18] L.R. Weatherley, D.W. Rooney, and M.V.U. Niekerk, J. Chem. Tech. Biotechnol., 68(1997) 437.
[19] M.V. Flores, J.J.W. Sewalt, A.E.M. Janssen, and A. Van Der Padt, Biotechnol. Bioeng., 67(2000) 364.
[20] N.N. Gandhi, N.S. Patil, S.B. Sawant, J.B. Joshi, P.P. Wangikar, and D. Mukesh, Catalysis Rev. - Sci. Eng., 42(2000) 439.
[21] C.F. Torres, C.G. Hill, and C. Otero, Biotechnol. Prog., 20(2004) 756.
[22] S.P.J.N. Senanayake and F. Shahidi, J. Food Lipids, 9(2002) 309.
[23] S.P.J.N. Senanayake and F. Shahidi, J. Am. Oil Chem. Soc., 79(2002) 1003.
[24] W. Du, Y. Xu, and D. Liu, Biotechnol. Appl. Biochem., 40(2004) 187.
[25] M. Iso, B. Chen, M. Eguchi, T. Kudo, and S. Shrestha, J. Mol. Catal. B: Enzyme, 16(2001) 53.
[26] H. Noureddini, X. Gao, and R.S. Philkana, Bioresource Technol., 96(2005) 769.
[27] I. Mingarro, G. Heijne, and P. Whitley, Trends Biotechnol., 15(1997) 432.
[28] K. Victor, J. Jacob, and D.S. Cafiso, Biochemistry, 38(1999) 12527.
[29] M. Mosior and S.A. Mclaughlin, Biophys. J., 60(1991) 149.
[30] C.A. Buser, C.T. Sigal, M.S. Resh, and S.A. Mclaughlin, Biochem., 33(1994) 13093.
[31] M.L. Jennens and M.E. Lowe, J. Bio. Chem., 269(1994) 25470.
[32] D. Voet and J.G. Voet, Biochemistry. 2nd ed. John Wiley & Sons, Inc., New York, 1995.
[33] H. Vantilbeurgh, L. Sarda, R. Verger, and C. Cambillau, Nature, 359(1992) 159.
[34] H. Vantilbeurgh, M. Egloff, C. Martinez, N. Rugani, R. Verger, and C. Cambillau, Nature, 362(1993) 814.
[35] A. Louwrier, G.J. Drtina, and A.M. Klibanov, Biotechnol. Bioeng., 50(1996) 1.
[36] R.D. Schmid and R. Verger, Angew. Chem. Int. Ed., 37(1998) 1608.
[37] A. Zaks, Enzymes in organic solvents, in Biocatalysts for Industry, J.S. Dordick, Editor. Plenum: New York, 1991.
[38] V.L. Vilker, V. Reipa, M. Mayhew, and M.J. Holden, J. Am. Oil Chem. Soc., 76(1999) 1283.
[39] M. Gargouri and M.D. Legoy, Enz. Microb. Technol., 21(1997) 79.
[40] F.W. Lichtenthaler, E. Cuny, D. Martin, S. Ronninger, and T. Weber, Practical Routes from Mono- and Disaccharides to Building Blocks with Industrial Application Profiles, in

Carbohydrates as Organic Raw Materials, F.W. Lichtenthaler, Editor. VCH Verlagsgesellschaft: Weinheim, Germany, 1991.
[41] V. Kren and J. Thiem, Chem. Soc. Reviews, 26(1997) 463.
[42] B.M. De Roode, M.C.R. Franssen, A. Van Der Padt, and R.M. Boom, Biotechnol. Prog., 19(2003) 1391.
[43] M.-P. Bousquet, R.M. Willemot, P. Monsan, and E. Boures, Appl. Microbiol. Biotechnol., 50(1998) 167.
[44] M. Bertau and M. Burli, Chimia, 54(2000) 503.
[45] A. Ismail, S. Soultani, and M. Ghoul, J. Biotechnol., 69(1999) 135.
[46] T. Mori, S. Fujita, and Y. Okahata, Carbohyd. Res., 298(1997) 65.
[47] J.S. Thorson, T.J. Hosted, J.Q. Jiang, J.B. Biggins, and A. J, Curr. Org. Chem., 5(2001) 139.
[48] M. Weuthern, R. Kawa, K. Hill, and A. Ansmann, Fat Sci. Technol., 97(1995) 209.
[49] C. Panintrarux, S. Adachi, Y. Araki, Y. Kimura, and R. Matsuno, Enz. Microb. Technol., 17(1995) 32.
[50] P. Monsan, F. Paul, V. Pelenc, and E. Boures, Ann. New York Aced. Sci., 799(1996) 633.
[51] E.N. Vulfson, R. Patel, J.E. Beecher, A.T. Andrews, and B.A. Law, Enz. Microb. Technol., 12(1990) 95.
[52] G. Vic and D. Thomas, Tetrahedron Lett., 33(1992) 4567.
[53] A.M. Blinkovsky and J.S. Dordick, Tetrahedron Symmetry, 4(1993) 1221.
[54] Y. Ooi, T. Hashimoto, and N. Mitsuo, Chem. Parm. Bull., 33(1985) 1808.
[55] S. Matsumura, Y. Kinta, K. Saliyama, and K.K.U. Toshima, Jpn), Biotechnol. Let., 18(1996) 1335.
[56] M. Therisod and A.M. Klibanov, J. Am. Chem. Soc., 108(1986) 5638.
[57] M. Therisod and A.M. Klibanov, J. Am. Chem. Soc., 109(1986) 3977.
[58] A.K. Prasad and J. Wengel, Nucleosides Nucleotides, 15(1996) 1347.
[59] S. Riva, J. Chopineau, A.P.G. Kieboom, and A.M. Klibanov, J. Am. Chem. Soc., 110(1988) 584.
[60] J.A. Arcos, C.G. Hill, and C. Otero, Biotechnol. Bioeng., 73(2001) 104.
[61] J. Chopineau, F. Mccafferty, M. Therisod, and A. Klibanov, Biotechnol. Bioeng., 31(1988) 208.
[62] D. Schwengers, Leucrose, a Ketodisaccharide of Industrial Design, in Carbohydrates as Organic Raw Materials, F.W. Lichtenthaler, Editor. 1991, VCH Verlagsgesellschaft: Weinheim, Germany. p. 183.
[63] J.N. Bemiller, Prospects for Industrial Utilization of Sugar Beet Carbohydrates, in Carbohydrates as Organic Raw Materials, F.W. Lichtenthaler, Editor. 1991, VCH Verlagsgesellschaft: Weinheim, Germany. p. 197.
[64] K. Demuth, H.-J. Jordening, and K. Buchholz, Carbohyd. Res., 337(2002) 1811.
[65] K. Heincke, B. Demuth, H.-J. Jordening, and K. Buchholz, Enz. Microb. Technol., 24(1999) 523.
[66] C. Torres, M. Bernabe, and C. Otero, Biotechnol. Let., 22(2000) 331.
[67] O.P. Ward, J. Fang, and Z. Li, Enz. Microb. Technol., 20(1997) 52.
[68] L. Neidleman, W.F. Amon, and J. Geigert, Process for the Production D-Glucosone and D-Fructose, in Eur. Pat. 4,246,347, 1981.
[69] J. Geigert, S.L. Neidleman, D.S. Hirano, B. Wolf, and B.M. Panschar, Carbohyd. Res., 113(1983) 163.
[70] J. Geigert, S.L. Neidleman, and D.S. Hirano, Carbohyd. Res., 113(1983) 159.
[71] H. Roper, Selective Oxidation of D-Glucose : Chiral Intermediates for Industrial Utilization, in Carbohydrates as Organic Raw Materials, F.W. Lichtenthaler, Editor. 1991, VCH Verlagsgesellschaft: Weinheim, Germany. p. 267.
[72] K. Buchholz, E. Stoppok, K. Matalla, C.-D. Reh, and H.-J. Jordening, Enzymatic Sucrose Modification and Saccharide Synthesis, in Carbohydrates as Organic Raw Materials, F.W. Lichtenthaler, Editor. VCH Verlagsgesellschaft: Weinheim, Germany, 1991.

[73] H.Y. Afeefy, D.J. Buckthal, and G.A. Hamilton, Bioorg. Chem., 8(1990) 41.
[74] S. Gunshore, E.J. Brush, and G.A. Hamilton, Bioorg. Chem., 13(1985) 1.
[75] S.H. Imam, R.V. Greene, and B.R. Zaidi, Biopolymers: utilizing nature's advanced materials. ACS Symposium Series 723. ACS, Washington, DC, 1999.
[76] J.C. Salamone, Polymeric materials encyclopedia. CRC Press, 1996.
[77] P. Sakellariou and R.C. Rowe, Prog. Polym. Sci., 20(1995) 889.
[78] H. Fukuzaki, M. Yoshida, M. Asano, and M. Kumakura, Eur. Polym. J., 25(1989) 1019.
[79] R.T. Macdonald, S.P. Mccarthy, and R.A. Gross, Macromol., 29(1996) 7356.
[80] D. Gilding and A. Reed, Polym., 20(1979) 1459.
[81] B. Jeong, Y.H. Bae, D.S. Lee, and S.W. Kim, Nature, 388(1997) 860.
[82] T. Ymaoka and Y. Kimura, Biodegradable copolymers, in Polymeric materials encyclopedia, J.C.E. Salamone, Editor. 1996.
[83] J. Kroschwitz and M. Howe-Grant, Encyclopedia of chemical technology (4th Ed.). John Wiley & Sons, NY, 1991.
[84] S. Tokura, S.-I. Nishimura, S. Skairi, and N. Nishi, *Macromol. Symp.*, 101(1996) 289.
[85] A.M. Rouhi, Oligosaccharides Coming of Age: Strides in synthesis are steering these biopolymers toward center stage, in C&EN. 6266, 1996.
[86] G. Wulff, J. Schmid, and T. Venhoff, Macromol. Chem. Phys., 197(1996) 259.
[87] G. Wulff, S. Bellmann, J. Schmid, and S. Podzimek, Macromol. Chem. Phys., 198(1997) 763.
[88] M. Okada and K. Aoi, Macromol. Reports, A32(1995) 907.
[89] J. Kadaokawa, K. Ymashita, M. Karasu, H. Tagaya, and K. Chiba, J. Macromol. Sci. - Pure Appl. Chem., A2(1995) 1273.
[90] M. Lynn, V. Stannett, and R. Gilbert, J. Polym. Sci.: Polym. Chem., 18(1980) 1967.
[91] H.R. Kricheldorf, Chemosphere, 43(2001) 49.
[92] S.H. Kim and Y.H. Kim, Macromol. Symp., 144(1999) 277.
[93] A. Albertsson, E. Ulrica, and K. Stridsberg, Macromol. Symp., 157(2000) 39.
[94] S. Hyon, K. Jamshidi, and Y. Ikada, Biomaterials, 18(1997) 1503.
[95] D.L. Kaplan, J.S. Dordick, R.A. Gross, and G. Swift, Enzymes in Polym. Science: An introduction, in Enzymes in Polym. Synthesis, R.A. Gross, D.L. Kaplan, and G. Swift, Editors. ACS: Washington, DC, 1996.
[96] S. Kobayashi, S. Shoda, and H. Uyama, Adv. Polym. Sci., 121(1995) 1.
[97] S. Kobayashi, J. Polym. Sci.: Polym. Chem., 37(1999) 3041.
[98] A.M. Blinkovsky and J.S.U.I. Dordick, J. Polym. Sci.: Polym. Chem., 31(1993) 1839.
[99] H. Uyama, H. Kurioka, I. Komatsu, J. Sugihara, and S. Kobayashi, Macromol. Reports, A32(Suppls. 5&6)(1995) 649.
[100] P. Wang and J.S. Dordick, Macromol., 31(1998) 941.
[101] P. Wang, B.D. Martin, D.G. Rethwisch, and J.S. Dordick, J. Am. Chem. Soc., 117(1995) 12885.
[102] S. Kobayashi and H. Uyama, Enzymatic Polymerization, in Encyclopedia of Polym. science and technology. 2000, John Wiley & Sons.
[103] T. Lalot, M. Brigodiot, and E. Marechal, Enzymic catalysis in organic media for Polymer synthesis, in Compr. Polym. Sci., 2nd Suppl, S.L. Aggarwl. and S. Susso, Editors. Elsevier, Oxford, 1996.
[104] D.R. Patil, D.G. Rethwisch, and J.S. Dordick, Biotechnol. Bioeng., 37(1991) 639.
[105] X. Chen, B. Martin, T. Neubauer, R. Linhardt, J. Dordick, and D. Rethwisch, Carbonhydrate Polymers, 28(1995) 15.
[106] B. Martin, R. Linhardt, and J. Dordick, Biomaterials, 19(1998) 68.
[107] F. Bjorkling and S.E. Godtfredsen, Tetrahedron, 44(1988) 2957.
[108] F. Bjorkling, S.E. Godtfredsen, and O. Kirk, J. Chem. Soc., Chem. Commun., 14(1989) 934.
[109] H.M. Sweers and C.H. Wong, J. Am. Chem. Soc., 108(1986) 6421.
[110] M. Kitagawa and Y. Tokiwa, Biotechnol. Let., 20(1998) 627.
[111] X. Chen, J.S. Dordick, and D.G. Rethwisch, Macromol., 28(1995) 6014.

[112] V. Braunmuhl and R. Stadler, Macromol. Symp., 103(1996) 141.
[113] S. Kobayashi, K. Kashiwa, T. Kawasaki, and S. Shoda, J. Am. Chem. Soc., 113(1991) 3079.
[114] R. Kumar and R.A. Gross, J. Am. Chem. Soc., 124(2002) 1850.
[115] R.D. Sonwalkar, C.C. Chen, and L.-K. Ju, Bioresource Technol., 87(2003) 69.
[116] A. Szakacs-Schmidt, L. Albert, and I. Kelemen-Horvath, Biomedical Chromatography, 13(1999) 252.
[117] Y. Ohya, T. Sugitou, and T. Ouchi, J.M.S. - Pure Appl. Chem., A32(1995) 179.
[118] K.A. Distel, G. Zhu, and P. Wang, Bioresource Technol., 96(2005) 617.
[119] R. Macdonald, S. Pulapura, Y. Svirkin, R. Gross, D. Kaplan, J. Akkara, G. Swift, and S. Wolk, Macromol., 28(1995) 73.
[120] F. Deng and R.A. Gross, Int. J. Biol. Macrom., 25(1999) 153.
[121] Y. Shen, X. Chen, and R. Gross, Macromol., 32(1999) 2799.
[122] F. Deng, K. Bisht, R. Gross, and D. Kaplan, Macromol., 32(2000) 5159.
[123] S. Matsumura, K. Mabuchi, and K. Toshima, Macromol. Rapid Commun., 18(1997) 477.
[124] J.A. Akkara and F.F. Fruno, Enzyme-catalyzed synthesis of maromolecules in organic solvents. USA Patent 5,981,240, 1999.
[125] D. Knani, A.L. Gutman, and D.H. Kohn, J. Polym. Sci.: Polym. Chem., 31(1993) 1221.
[126] A. Kumar, B. Kalra, A. Dekhterman, and R.A. Gross, Macromol., 33(2000) 6303.
[127] K.S. Bisht, F. Deng, R.A. Gross, D.L. Kaplan, and G. Swift, J. Am. Chem. Soc., 120(1998) 1363.
[128] K.M. Koeller and C.H. Wong, Nature, 409(2001) 232.
[129] A. Schmid, J.S. Dordick, B. Hauer, A. Kiener, M. Wubbolts, and B. Witholt, Nature, 409(2001) 258.
[130] J.S. Dordick and D.S. Clark, Curr. Opin. Chem. Biol., 6(2002) 123.
[131] A.M. Klibanov, Nature, 409(2001) 232.
[132] A.J. Mesiano, E.J. Beckman, and A.J. Russell, Chem. Rev., 99(1999) 623.
[133] A. Zaks and A.M. Klibanov, Science, 224(1984) 1249.
[134] J.S. Dordick, Enz. Microb. Technol., 11(1989) 194.
[135] M. Ru, S. Hirokane, A. Lo, J. Dordick, J. Reimer, and D. Clark, J. Am. Chem. Soc., 122(2000) 1565.
[136] B.H. Davison, J.W. Barton, and G.R. Petersen, Biotechnol. Progr., 13(1997) 512.
[137] R.S. Watkins, A.F. Lee, and K. Wilson, Green Chem., 6(2004) 335.
[138] G. Knothe, R.O. Dunn, and M.O. Bagby, Biodiesel: The Use of Vegetable Oils and Their Derivatives as Alternative Diesel Fuels, in Fuels and Chemicals from Biomass, B.C. Saha and J. Woodward, Editors. American Chemical Soc.: Washington, DC, 1997.
[139] J. Dordick, Biotechnol. Progr., 8(1992) 259.
[140] A.M. Klibanov, Trends Biotechnol., 15(1997) 97.
[141] P. Wang, C.A. Woodward, and E.N. Kaufman, Biotechnol. Bioeng., 64(1999) 290.
[142] D.G. Rees and P.J. Halling, Enz. Microb. Technol., 28(2001) 282.
[143] Y. Ito, N. Sugimura, O. Kwon, and Y. Imanishi, Nat. Biotechnol., 17(1999) 73.
[144] M.E. Powers, J. Matsuura, J. Brassell, M.C. Manning, and E. Shefter, Biopolymers, 33(1993) 927.
[145] V.M. Paradkar and J.S. Dordick, J. Am. Chem. Soc., 116(1994) 5009.
[146] R. Vazquez-Duhalt, P.M. Fedorak, and D.W.S. Westlake, Enz. Microb. Technol., 14(1992) 837.
[147] M.T. Ru, J.S. Dordick, J.A. Reimer, and D.S. Clark, Biotechnol. Bioeng., 63(1999) 233
[148] F. Secundo, G. Carrea, G. Vecchio, and F. Zambianchi, Biotechnol. Bioeng., 64(1999) 624.
[149] P. Wang, M.V. Sergeeva, L. Lim, and J.S. Dordick, Nat. Biotechnol., 15(1997) 789.
[150] Z. Yang, A.J. Mesiano, S. Venkatasubramanian, S.H. Gross, J.M. Harris, and A.J. Russell, J. am. Chem. Soc., 117(1995) 4843.
[151] M. Miyanaga, M. Ohmori, K. Imamura, T. Sakiyama, and K. Nakanishi, J. Bioscience and Bioengineering, 87(1999) 463.

[152] D. Haring and P. Schreier, Current Opinion in Chem. Biology, 3(1999) 35.
[153] R.A. Persichetti, N.L. St. Clair, J.P. Griffith, M.A. Navia, and A.L. Margolin, J. Am. Chem. Soc., 117(1995) 2732.
[154] K. Xu and A.M. Klibanov, J. Am. Chem. Soc., 118(1996) 9815.
[155] A.F. Dexter, F.J. Lakner, R.A. Campbell, and L.P. Hager, J. Am. Chem. Soc., 117(1995) 6412.
[156] S. Colonna, N. Gaggero, C. Richelmi, and P. Pasta, Trends Biotechnol., 17(1999) 163.
[157] Y. Okahata and T. Mori, J. Molecular Catalysis B: Enzymatic, 5(1998) 119.
[158] B. Sauerbrei and J. Thiem, Tetrahedron Let., 33(1992) 201.
[159] F. Van Rantwijk, M.W. Oosterom, and R.A. Sheldon, J. Molecul. Catal. B: Enzymatic, 6(1999) 511.
[160] G. Zhu and P. Wang, J. Am. Chem. Soc., 126(2004) 11132.
[161] P.J. Halling, Current Opinion in Chem. Biology, 4(2000) 74.
[162] T. Scheper, H. Meyer, F. Schubert, and F. Kolisis, A comparison of enzymatic reactions in aqueous, organic and multiphase systems, in Biocatalysis in Non-Conventional Media, J. Tramper, M.H. Vermue, and U.V. Stockar, Editors. Elsevier: Amsterdam, 1992.
[163] L. Santhanam and J.S. Dordick, Biocat. Biotransform., 20(2002) 265.
[164] D.H. Altreuter, J.S. Dordick, and D.S. Clark, J. Am. Chem. Soc., 124(2002) 1871.
[165] M. Stahl, M. Mansson, and K. Mosbach, Biotechnol. Let., 12(1990) 161.
[166] J. Rich, B. Bedell, and J. Dordick, Biotechnol. Bioeng., 45(1995) 426.
[167] D.H. Altreuter and D.S. Clark, Curr. Opin. Biotechnol., 10(1999) 130.
[168] P.C. Michels, Y.L. Khmelnitsky, J.S. Dordick, and D.S. Clark, TIBTECH, 16(1998) 210.
[169] X.C. Liu, D.S. Clark, and J.S. Dordick, Biotechnol. Bioeng., 69(2000) 457.
[170] F. Secundo, G. Carrea, M. De Amici, D.V.S. J., and J.S. Dordick, Biotechnol. Bioeng., 81(2003) 391.
[171] J.O. Rich, P.C. Michels, and Y.L. Khmelnitsky, Curr. Opin. Chem. Biol., 6(2002) 161.
[172] M.-Y. Lee, A. Srinivasan, B. Ku, and J.S. Dordick, Biotechnol. Bioeng., 83(2003) 20.
[173] S. Xie, F. Svec, and J.M.J. Frechet, Biotechnol. Bioeng., 62(1999) 30.
[174] J.P. Lindsay, D.S. Clark, and J.S. Dordick, Biotechnol. Bioeng., 85(2004) 553.
[175] M.K.W. Ditursi, J. Cha, M.R. Nevnnan, and J.S. Dordick, Biotechnol. Prog., 20(2004) 1705.
[176] C.P. Govardhan, Curr. Opin. Biotechnol., 10(1999) 331.
[177] F. Caruso and C. Schuler, Langmuir, 16(2000) 9595.
[178] C. Daubresse, C. Grandfils, R. Jerome, and P. Teyssie, Colloid Polym. Sci., 274(1996) 482.
[179] M.B.F. Martins, S.I.D. Simoes, M.E.M. Cruz, and R. Gaspar, J. Mater. Sci.: Mater. Med., 7(1996) 413.
[180] M.-H. Liao and D.-H. Chen, Biotechnol. Lett., 23(2001) 1723.
[181] H. Jia, G. Zhu, and P. Wang, Biotechnol. Bioeng., 84(2003) 406.
[182] J.-P. Chen and D.-R. Su, Biotechnol. Prog., 17(2001) 369.
[183] J. Kim and J.W. Grate, Nano Letters, 3(2003) 1219.
[184] Y.L. Khmelnitsky, I.N. Neverova, A.V. Gedrovich, V.A. Polyakow, A.V. Levashov, and K. Martinek, Eur. J. Biochem., 210(1992) 751.
[185] J.S. Dordick, S.J. Novick, and M.V. Sergeeva, Chem. & Ind., 1998) 14.
[186] P. Wang, S. Dai, S.D. Waezsada, A. Tsao, and B.H. Davison, Biotechnol. Bioeng., 74(2001) 249.
[187] B. El-Zahab, H. Jia, and P. Wang, Biotechnol. Bioeng., 87(2004) 178.
[188] H. Jia, G. Zhu, B. Vugrinovich, W. Kataphinan, D.H. Reneker, and P. Wang, Biotechnol. Progr., 18(2002) 1027.
[189] K. Rege, N.R. Raravikar, D.-Y. Kim, L.S. Schadler, P.M. Ajayan, and J.S. Dordick, Nano Letters, 3(2003) 829.
[190] F. Baneyx, Curr. Opin. Biotechnol., 10(1999) 411.
[191] C. Schmidt-Dannert, J. Pleiss, and R. Schmid, Annals New York Academy of Sciences, 864(1998) 14.

[192] C. Schmidt-Dannert, Bioorganic & Medicinal Chem., 7(1999) 2123.
[193] E.B. Cahoon, Y. Lindqvist, G. Schneider, and J. Shanklin, Proc. Natl. Acad. Sci. U. S. A., 94(1997) 4872.
[194] S.D. Mahan, G.C. Ireton, B.L. Stoddard, and M.E. Black, Biochemistry, 43(2004) 8957.
[195] B. Van Den Burg, G. Vriend, O.R. Veltman, and V.G.H. Eijsink, Proc. Natl. Ac. Sci. USA, 95(1998) 2056.
[196] H.R. Xion, F. F., M. Leisola, and O. Turunen, Extremophiles, 8(2004) 393.
[197] F.H. Arnold, Nature, 409(2001) 253.
[198] J. Shanklin, Curr. Opin. Plant Biol., 3(2000) 243.
[199] V.G.H. Eijsink, A. Bjork, S. Gaseidnes, R. Sirevag, B. Synstad, B. Van Den Burg, and G. Vriend, J. Biotechnol., 113(2004) 105.
[200] F.H. Arnold, P.L. Wintrode, K. Miyazaki, and A. Gershenson, Trends Biochem. Sci., 26(2001) 100.
[201] P.C. Cirino and F.H. Arnold, Curr. Opin. Chem. Biol., 6(2002) 130.
[202] D.N. Bolon, C.A. Voigt, and S.L. Mayo, Curr. Opin. Chem. Biol., 6(2002) 125.

Chapter 13. Biocatalysis for Chiral Synthesis

Hyun-Dong Shin, Xuan Guo, and Rachel R. Chen

School of Chemical & Biomolecular Engineering, Georgia Insitute of Technology, Atlanta, Georgia 30332-0100, USA

1. INTRODUCTION

Global sales of single-enantiomer compounds continue to rise, primarily fueled by the demand from pharmaceuticals but also from at least three other sectors, flavor and aroma chemicals, agricultural chemicals, and specialty materials. It has been estimated that sales will reach nearly $9 billion by the end of 2004, with double digit increases expected over the next few years [1]. The use of biocatalysis in chiral synthesis is increasingly important. This is reflected in the tremendous growth of literature in this research area. In this chapter, we review some of the recent developments of whole-cell and isolated enzymes in chiral synthesis. Due to space restrictions, a comprehensive review is not possible. We apologize for having to omit many interesting papers.

2. CHIRAL SYNTHONS BY FERMENTATION

2.1. Amino acids

Amino acids are important food and feed additives and versatile building blocks for pharmaceutical and agrochemical chemicals. Most of natural L-amino acids are produced by fermentation methods, more commonly with coryneform bacteria. The overall production scale exceeds two million tons per year, with two digit growth expected [2, 3].

Central to the recent improvement of the production process is the use of recombinant DNA technology and bioinformatics tools. Metabolic engineering affords the opportunity to engineer cells to make new products or for better production of existing products [4–6]. Quantitative analysis through metabolic flux analysis and metabolome was used to assist in further strain improvement [7, 8]. In addition, neuronal networks were utilized to develop more reliable multivariate statistical process control [9, 10]. With the help of completely-sequenced *Corynebacterium glutamicum* [11, 12] and powerful metabolic engineering tools, we can expect further and more drastic improvements of the production process in the near future.

2.1.1. L-Cysteine

Despite the fact that a biocatalytic resolution process was implemented in as early as 1982, extraction from human hair, animal feathers and bristles remains the main method for producing L-cystein. Several new developments are noteworthy. *Escherichia coli* over-expressing a major facilitator protein (Orf299) was found to excrete considerable amounts of L-cysteine, especially in the presence of thiosulfate. This strain could be used as the basis for a novel cysteine production bioprocess [13]. Another promising approach was through the metabolic engineering of *C. glutamicum* by introducing a feedback resistant serine acetyltransferase from *E. coli*. The resultant strain produced about 290 mg of L-cysteine plus L-cystine per liter from glucose, about two-fold higher than that of the wild-type strain [14].

2.1.2. Unnatural L-α-amino acids

As with natural L-amino acid, the production of unnatural amino acid from fermentation often requires metabolic engineering. Usually, deregulation by developing feedback resistant enzymes is a necessary step. As illustrated in Fig. 1, the cysteine-biosynthetic pathway of *E. coli* [15] was engineered to produce unnatural L-amino acids. Key to the success of this approach was the relaxed substrate specificity of O-acetylserine sulfhydrylase that allowed incorporating substitutions other than sulfide. One of the unnatural L-amino acids produced with this method, S-phenyl-L-cysteine, an intermediate for a leading HIV-protease inhibitor (Viracept, Agouron Pharmaceuticals), was produced with a final product concentration of 32 g/L and 72% molar yield.

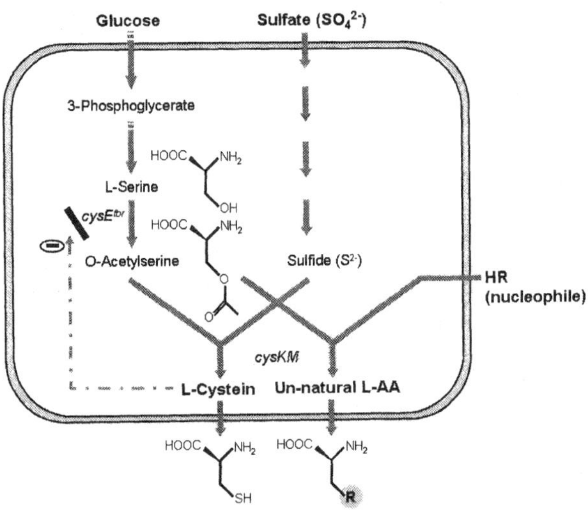

Fig. 1. Unnatural L-amino acid production by the cysteine-biosynthetic pathway [15]. The relaxed substrate specificity of O-acetylserine sulfhydrylases encoded by the *cysK* and *cysM* genes allows the incorporation of externally supplied molecules (HR) instead of sulfide. Moreover, a constant supply of O-acetylserine is ensured by a feedback-resistant variant of the serine acetyltransferase enzyme encoded by *cysE*^{fbr}. AA, amino acid.

2.1.3. D-amino acids

The production of D-*p*-hydroxyphenyl glycine, used for the synthesis of semisynthetic cephalosporins (e.g. cephadroxil) and semisynthetic penicillin (e.g. amoxycillin), is commercialized by a number of companies, including Ajinomoto (Tokyo, JP), Bayer (Leverkusen, GE), DSM, and SNAM Progetti (*Corso Venetia*). These processes are commonly carried out with wild-type strains such as *Agrobacterium radiobacter* and *Pseudomonas putida*, but recently recombinant strains of *E. coli* have been reported [16]. Cloning enzymes from thermophiles and enzyme modifications by both directed and random mutagenesis has led to processes with improved characteristics [17].

For example, the enantioselectivity of a D-hydantoinase from *Arthrobacter* sp. DSM 9771 was reversed through directed evolution, with a simultaneous increase in activity [18]. L-*tert*-Leucine was synthesized with the evolved hydantoinase in combination with two other enzymes. Degussa currently produces this compound on a 100-kg scale using a recombinant microorganism that expresses the aforementioned modified hydantoinase, a carboxylase, and a racemase. In another example, a recombinant D-phenylalanine producing strain was constructed by first deactivating D-amino acid deaminase activity for a D-amino acid used as amino donor (Fig. 2); the mutant was then transformed with a plasmid carrying the genes coding for an L-amino acid deaminase, a D-amino acid aminotransferase, and a racemase [2]. Using this recombinant whole-cell catalyst, D-phenylalanine can be produced from L-phenylalanine and another L-amino acid such as L-alanine, L-glutamine or L-aspartic acid.

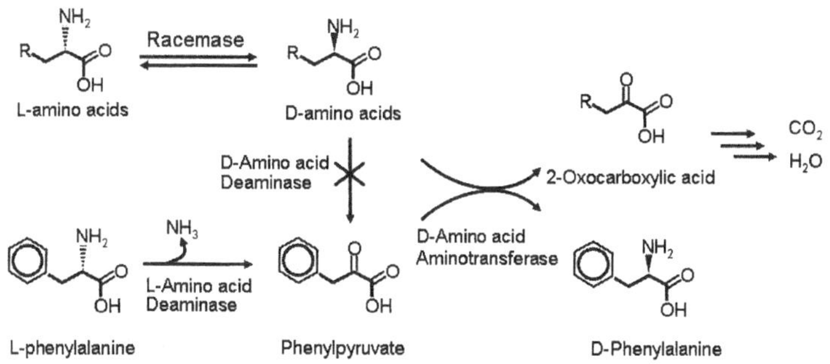

Fig. 2. Production of D-phenylalanine by fermentation with genetically modified microorganisms [2]. L-Alanine, L-glutamine or L-aspartic acids are added as the amino-group donor; L-phenylalanine is formed by endogenous overproduction. The α-keto acid, 2-oxocarboxylic acid is metabolized to CO_2 and H_2O.

2.2. Carboxylic acids

2.2.1. L- and D-lactic acid

Lactic acid has a global market in excess of 100,000 tons per year. Much of the growth is attributable to two emerging products, biodegradable plastic polylactic acid and the environmentally-benign solvent ethyl lactate. Although many microorganisms produce lactic acid, *Lactobacillus* strains are particularly useful due to their high acid tolerance and relative ease of genetic manipulation. However, lactic acid bacteria also have undesirable traits. They are fastidious in terms of nutrient requirements, which often complicates product recovery, and additionally, incomplete or negligible pentose utilization [19, 20] excludes the use of xylose, the second most abundant sugar in nature, in fermentation processes.

Therefore, it is of great interest to develop recombinant *E. coli* strains requiring only a mineral salt medium for D-lactic acid production. Ingram *et al.* has developed such a strain by eliminating several enzymes in the carbon metabolic pathway [21]. The resultant strain produces almost exclusively D-lactic acid, with the chemical purity of D-lactic acid reaching ~98% with respect to soluble organic compounds and the optical purity exceeding 99%. Complementarily, they also engineered a strain that produces only L-lactic acid by functionally replacing the D-lactate dehydrogenase gene with the L-lactate dehyrogenase gene from *Pediococcus acidilactici* [22]. The chemical and optical purity of L-lactic acid were comparable to what was obtained with D-lactic acid.

2.2.2. D-(-)-3-hydroxybutyric acid

D-(-)-3-hydroxybutyric acid (R3HB, Fig. 3A) is an important precursor of 4-acetoxy-zetidinone, an intermediate of carbapenem antibiotics. By utilizing enzymes involved in polyhydroxyalkanoate (PHA) biosynthesis, two metabolic engineering strategies were developed. In the process shown in Fig. 3B where R3HB was made by depolymerizing the accumulated polyhydroxybutyrate (PHB), R3HB homopolymer, R3HB concentration could reach about 7.7 g/L with a yield of 49.5% (85.6% of the maximum theoretical yield) from glucose. The recombinant *E. coli* strain was constructed by chromosome integration of the PHA biosynthesis genes and by further introducing the PHA depolymerase gene on a plasmid [23]. Alternatively, instead of making cells carry the entire PHA biosynthesis pathway, four genes from the PHA were incorporated into the native metabolic pathway of *E. coli* (Fig. 3C), resulting in a hybrid pathway leading to R3HB. With the process shown in Fig. 3C, up to 12 g/L product was accumulated in a 48-h fed-batch process [24].

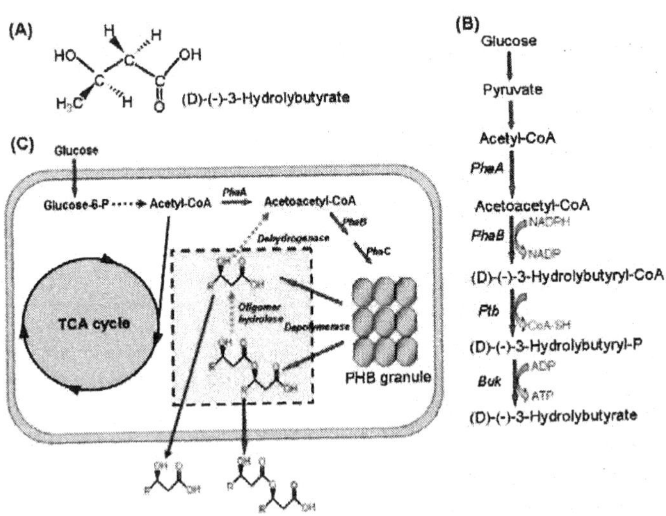

Fig. 3. Poly-(R)-(-)-3-hydroxybutyrate (PHB) metabolism and the metabolic engineering strategies for the production of (R)-(-)-3-hydroxybutyric acid in *E. coli* [23, 24].

3. CHIRAL MOLECULES FROM HYDROLASE

Hydrolases are predominant biocatalysts used for chiral synthesis due to its wide availability, ease of use, and high compatibility with organic solvents.

3.1. Lipase and esterase

Lipase and esterase are the most widely used enzymes, catalyzing the chemo-, regio-, and/or stereo-selective hydrolysis of carboxylic acid esters. They can also catalyze the reverse reaction, esterification, and transesterication under appropriate conditions. Hydrolysis, esterification and transesterification reactions catalyzed by lipases or esterases constitute almost 50% of the reported biocatalysis work worldwide, and some of the processes have been commercialized [25, 26]. Particularly, the lipase-catalyzed process leading to (S)-methoxy-isopropylamine is operated at a volume of greater than 1000 tons of product per year [27]. More often, lipases and esterases are used for the kinetic resolution of chiral secondary alcohols and carboxylic acids/esters, respectively. One drawback of this method is the yield of the product is low, with 50% being the theoretic yield unless the other enantiomer is recycled or is useful itself.

3.1.1. S-Ethyl-2-Ethoxy-3-(4-hydroxyphenyl)propanoate

Novo Nordisk A/S has developed a novel biocatalytic process for the large-scale production of *S*-1 [*S*-2-ethoxy-3-(4-hydroxyphenyl)propanoic acid], a key intermediate in the synthesis of the new anti-diabetic drug Ragaglitazar [NNC61-0029((-)DRF2725)] from its racemic ethylester, *rac*-2 [*rac*-ethyl 2-ethoxy-3-(4-hydroxyphenyl)-propanoate], by enantio-selective hydrolysis (Fig. 4). This chemo-enzymic process was successfully run on a 44-kg pilot scale, with the yield reaching 43–48% and the enantiomeric excess (ee) in the range of 98.4–99.6% [28].

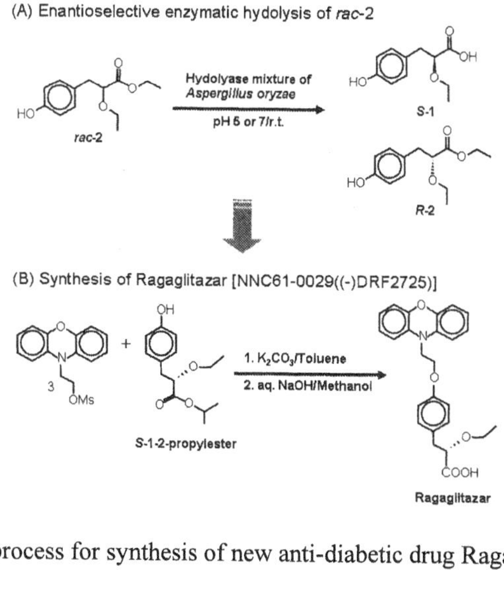

Fig. 4.Chemoenzymatic process for synthesis of new anti-diabetic drug Ragaglitazar [28].

3.1.2. Ethyl-3-hydroxybutyrate

Both enantiomers of ethyl-3-hydroxybutyrate (HEB) are chiral building blocks in the production of numerous biologically active compounds of commercial interest. For example, the (*R*)-enantiomer is an intermediate for an anti-glaucoma drug [29], and (*S*)-HEB is used for the synthesis of pheromones and carbapenem antibiotics [30, 31]. Although the (*R*)-enantiomer can be obtained from hydrolyzing poly-3-hydroxybutyrate in ton quantities, the (*S*)-enantiomer is more difficult to prepare on a large scale.

Recently, a two-step enzymatic resolution process for large-scale production of (S)- and (R)-ethyl-3-hydroxybutyrate using lipase was developed [32]. Both enantiomers were obtained at 99% chemical purity and over 96% enantiomeric excess (ee), with a total process yield of 73%. The first reaction involved a solvent-free acetylation of racemic HEB with vinylacetate for the production of (S)-HEB. In the second reaction, (R)-enriched ethyl-3-acetoxybutyrate (AEB) was subjected to alcoholysis with ethanol to derive optically pure (R)-HEB. Immobilized *Candida antarctica* lipase B (CALB) was employed in both stages, with high productivity and selectivity (Fig. 5).

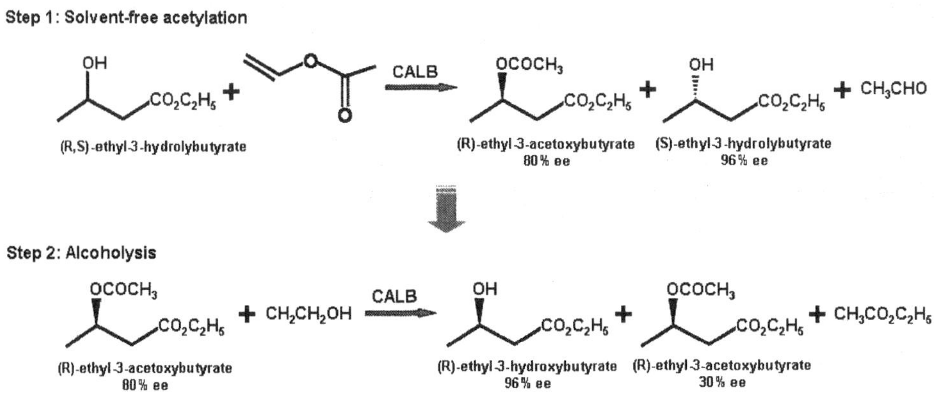

Fig. 5. A two-step enzymatic resolution process for production of (*S*)- and (*R*)- ethyl-3-hydroxybutyrate [32].

3.1.3. Dynamic kinetic resolution of secondary alcohols

Dynamic kinetic resolution (DKR) provides a useful method for the conversion of racemic substrates to single enantiomeric products. Recently, methods based on enzyme-metal catalyst combinations have been developed for DKR, in which a metal complex acts as a racemizing catalyst, an enzyme acts as a resolving catalyst, and together they transform a racemate into an enantiomerically enriched product [33].

Particularly, (S)- as well as (R)-configured alcohols can now be prepared by the use of a commercially available (R)-selective lipase and (S)-selective subtilisin as the enantio-selective acylating catalyst and an aminocyclopentandienylruthenium complex as a racemization catalyst [34]. As shown in Fig. 6, the subtilisin-catalyzed DKR is complementary to its lipase-catalyzed counterpart, and both DKRs were performed with **1i** at 25 °C without an acyl donor, as the substrate itself carries an acyl group. Both DKRs afforded high ee values and excellent yields.

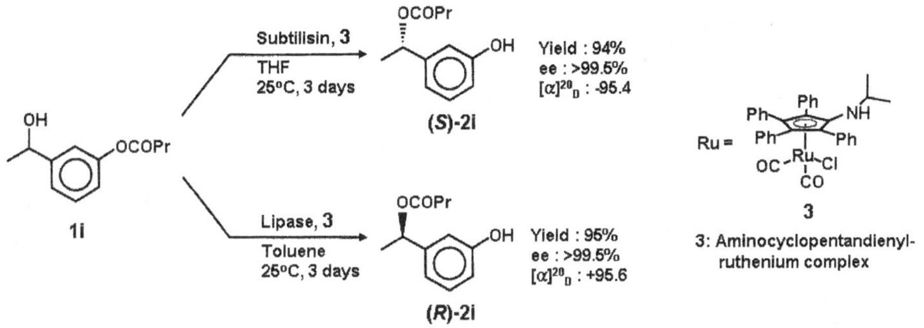

Fig. 6. Dynamic Kinetic Resolution of *m*-Butanoyloxyphenyl-1-ethanol [34].

3.2. Epoxy hydrolase

Racemic mixtures of epoxides can be optically resolved to enantiopure epoxides by kinetic resolution using epoxide hydrolase; these optically active epoxides can be further derivatized, as they readily react with halides, carbon, nitrogen, oxygen, or sulfur nucleophiles [35].

Recently, a new biocatalytic process for the resolution of racemic (*R*)-*para*-chlorostyrene oxide using two complementary enzymes was developed with a one-pot sequential bi-enzymatic strategy, as shown in Fig. 7 [36]. This strategy overcomes the 50% yield limitation inherent to a resolution process. It was demonstrated for the preparation of nearly enantiopure (*R*)-*para*-chlorostyrene diol, the key building block of Eliprodil, a neuroprotective agent. Typically, enzymatic hydrolysis of racemic *para*-chlorostyrene oxide 2 (***rac*-2**) was started using 150 units (*S*)-epoxide specific *St*EH (epoxide hydrolase of *Solanum tuberosum*) in 125 mL buffer solution. After about 50% conversion (18 h at 0°C), 500 units of (*R*)-epoxide specific *An*EH (epoxide hydrolase of *Aspergillus niger*) were added into the reaction vessel and the reaction was completed in about two days. After extraction with ethyl acetate, 4.18 g of diol product (*R*)-*para*-chlorostyrene diol **(*R*)-3** (93% yield) were obtained with an ee of 96%.

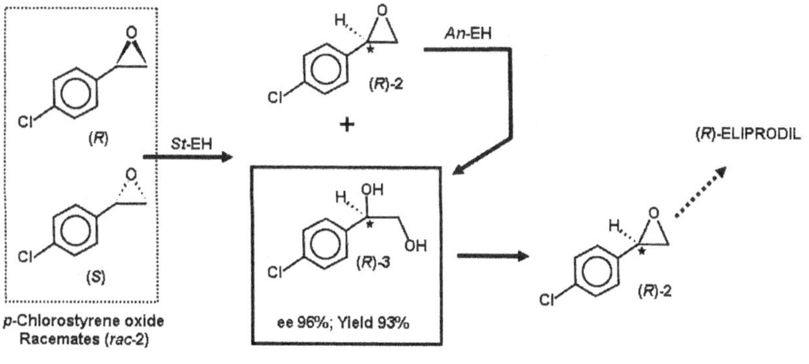

Fig. 7. One-pot kinetic resolution of *para*-chlorostyrene oxide racemate using two regio- and enantio-complementary enzymes [36].

3.3. Nitrile-degrading enzymes

Chiral nitriles, along with optically active amides and carboxylic acids, have wide applications in pharmaceutical and chemical industries. The hydrolysis of nitriles by nitrilases or nitrile hydratase/amidase systems in combination with hydrochloric acid offers useful access to α-hydroxycarboxylic acids [37]. Additionally, hydroxynitrile lyases (HNL) catalyze the stereoselective addition of hydrogen cyanide to aldehydes and ketones yielding cyanohydrin products [38]. α-hydroxycarboxylic acids are then obtained by hydrolyzing the chiral cyanohydrins with hydrochloric acid. In this synthesis, quantitative conversion of an aldehyde into the cyanohydrins product is possible. Recently, HNL-based aromatic α-hydroxycarboxylic acid production processes were implemented on an industrial scale. A number of multi-ton scale nitrilase-based processes were established in Lonza, BASF, and Mitsubishi Rayon for the production of (R)-mandelic acid and its derivatives.

New developments in this area include using Metagenome approach to discover new nitrilase and molecular evolution for improved enzyme [39–41].

4. CHIRAL MOLECULES FROM ENZYMES REQUIRING COFACTORS

4.1. Isolated enzymes

Using oxidoreductases as isolated enzymes necessitate *in–situ* cofactor regeneration. The regeneration of pyridine nucleotides, NAD(H) and NADP(H), is generally carried out by enzymatic methods using a dehydrogenase or a NADH oxidase [42, 43]. Although much progress has been made, cofactor regenerations remain an active area of research. New developments include finding enzymes that are resistant to organic solvent, as most substrates are not readily soluble in aqueous solution, and organic-aqueous biphasic systems are more commonly used.

Novel enzymes are being sought for effectively regenerating both reduced and oxidized cofactors. A recent and notable development is a newly-discovered soluble pyridine nucleotide transhydrolase (STH) from *Pseudomonas fluorescens* useful for the regeneration of NAD and NADP. It was successfully used for hydromorphone synthesis in both cell-free and whole-cell systems [44]. STH can transfer reducing equivalents between NAD(H) and NADP; thus an efficient cofactor recycling in the presence of catalytic amounts of cofactors occurs in a cell-free system, resulting in 84% high yield of hydromorphone. Another promising new enzyme, phosphite dehydrogenase, was discovered from *Pseudomonas stutzeri* WM88 and can be used for NAD regeneration [45]. This enzyme catalyzes the NAD-dependent oxidation of phosphite to phosphate, and is useful for the production of isotopically labeled products with a high isotopic purity. Furthermore, the cofactor specificity of the enzyme was successfully modified by using site-directed mutagenesis; mutants were generated for the regeneration of NADP. Due to the intrinsically large thermodynamic driving force, the innocuous nature of phosphite and phosphate to enzymes, and the low cost of phosphite, this phosphite–phosphite dehydrogenase system may prove complementary to the most widely used formate–formate dehydrogenase system.

Impressive progress has been achieved in recent years in enzyme-coupled regeneration of oxidized cofactors. Particularly, water-forming NADH oxidases were discovered from *Lactobacillus sanfranciscensis* and *Borrelia burgdorferi*.

$$2NADH + O_2 + 2H^+ \rightarrow 2NAD^+ + 2H_2O$$

The enzymes were cloned and heterologously expressed in *E. coli* [46]. The enzyme from *Lactobacillus sanfranciscensis* accepts NADPH with about 30% of the activity detected with NADH, and hence may be used for regenerating both cofactors. The enzyme was applied in the resolution of *rac*-1-phenylethanol with ADH from *L. brevis* yielding enantiomerically pure S-isomer [47]. Unlike regeneration using lactate dehydrogenase or glutamate dehydrogenase, this method does not generate byproducts other than water, but the oxygen sensitivity of the protein remains a disadvantage.

Numerous literatures can be found documenting the combination of an oxido-reductase and an enzyme for cofactor regeneration in chiral synthesis. Some reached preparative scale, as shown in the following example. A biphasic system containing 20% heptan and 80% aqueous buffer (pH 7.0) was used for the asymmetric bioreduction of *p*-chloroacetophenone with (S)-alcohol dehydrogenase from *Rodococcus erythropolis* under *in situ*-cofactor-recycling with a formate dehydrogenase from *Candida boidinii* [48]. Interestingly, with this biphasic system, reductions of poorly water-soluble ketones could be carried out at higher substrate concentrations of 10–200 mM. The resulting (S)-alcohols were formed with moderate to good conversion rates, and with up to 99% ee.

Enantioselective reduction of C-4-substituted 3,5-dixocarboxylates was reported using alcohol dehydrogenase from *Lactobacillus brevis* (LBADH) over-expressed in *E. coli* [49]. LBADH can recycle its cofactor by oxidation of 2-propanol, as shown in Fig. 8. When 2-propanol was used as a co-substrate, this enzyme reduced *tert*-butyl 3,5-dioxohexanoate and *tert*-butyl 3,5-dioxoheptanoate on a preparative scale to the corresponding δ-hydroxy-β-keto esters, *tert*-butyl (*R*)-5-hydroxy-3-oxohexanoate and *tert*-butyl (*R*)-5-hydroxy-3-oxoheptanoate with 99.4% ee and 98.1% ee, respectively. These esters are useful intermediates in the synthesis of pharmaceuticals such as HMG-CoA reductase inhibitor, mevinic acid.

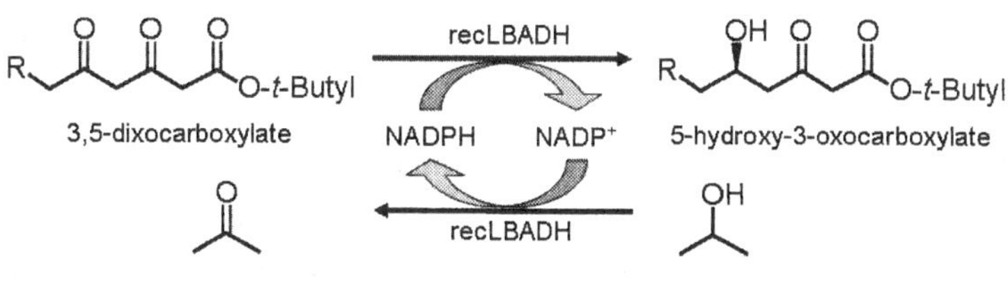

Fig. 8. Enzymatic enantioselective reduction of 3, 5-dioxocarboxylates and substrate-coupling recycling of the cofactor [49]. Cofactor regeneration is achieved with the same enzyme using 2-propanol as hydrogen source.

4.2. Whole-cell systems

An alternative to isolated enzyme systems is whole-cell biocatalysis. Besides the ease of cofactor regenerations, laborious protein purification steps can be avoided. Enzymes are generally more stable within the protection of cell envelopes. Additionally, more complicated reactions requiring multi-enzymes or multi-pathways are possible with whole-cell biocatalysts. However, whole-cell catalyzed reactions are generally much slower than isolated enzyme-catalyzed reactions. A difference of one or two orders of magnitude is not uncommon. This is due to the low rate of substrate entry caused by the permeability barrier of cell envelopes. Fortunately, this problem has begun to receive its due attention and is being addressed using molecular engineering approaches [50]. Another common problem in whole-cell biocatalysis is the presence of interfering enzymes that degrade products or the presence of competing enzymes or pathways that lower the product yield. This problem can be addressed adequately, as detailed in other sections, by metabolic engineering. In this section, we sample a few successful examples of chiral synthesis involving oxidoreductase used in a whole-cell system.

Chiral alcohols are useful intermediates for many pharmaceuticals and chemicals. There have been many attempts to develop enzymatic asymmetrical reduction for the production of chiral alcohols. An impressive example was a biphasic bioreduction using a recombinant *E. coli* strain co-expressing glucose dehydrogenase (GDH) and carbonyl reductase genes [51]. As shown in Fig. 9, when the recombinant *E. coli* cells were incubated in a reaction mixture containing 4-chloro-3-oxobutanoate (COBE), glucose, and a catalytic amount of $NADP^+$, 300 g/L of COBE were almost quantitatively converted to (R)-4-chloro-3-oxobutanoate ethyl ester ((R)-CHBE) in 16 h (molar yield: 94%, optical purity: 92% ee).

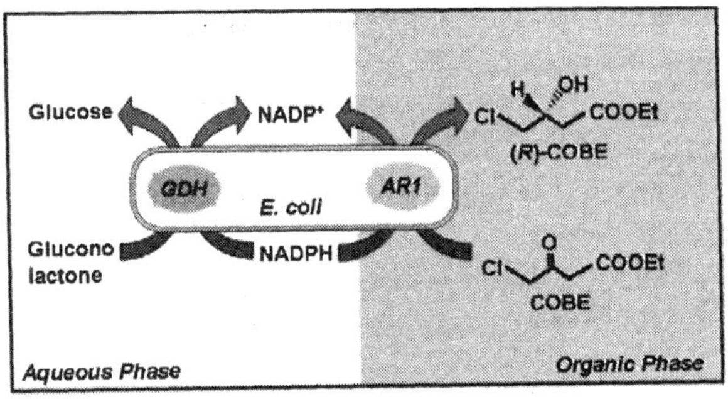

Fig. 9. Stereospecific reduction of 4-chloro-3-oxobutanoate (COBE) in a biphasic system by *E. coli* harbouring glucose dehydrogenase and carbonyl reductase genes [51].

Although several notable studies on the use of monooxygenases in cell-free systems using enzymatic cofactor regeneration have been reported in recent years, the most remarkable results are obtained with the use of either metabolically active or non-growing bacterial or yeast cells.

A combined chemo-enzymic process for the synthesis of several optically pure 4- and 5-substituted lactones was developed using whole cell-catalyzed Baeyer-Villiger oxidations. The substrate is enantiomerically enriched 3-alkyl cyclic ketones [52]. The high regioselective and stereoselective oxidation was achieved using recombinant *E. coli* over-expressing cyclopentanone monooxygenase (CPMO), whereas the substrate for the biocatalysts, chiral ketones, are readily provided by recently developed copper-catalyzed asymmetrical conjugate reductions of the corresponding enones.

Another example is the NADPH-dependent stereoselective reduction of the bicyclic diketone bicyclo[2.2.2]octane-2,6-dione (BCO2,6D) to the chiral ketoalcohol (1R,4S,6S)-6-hydroxybicyclo[2.2.2]octane-2-one (BCO2one6ol) (Fig. 10). A whole-cell batch process for the reduction of carbonyl substrate was developed with a high product yield with respect to co-substrate consumed (formed product/consumed co-substrate), and a high conversion rate under anaerobic conditions with *Saccharomyces cerevisiae* as the biocatalyst and glucose as the co-substrate [53].

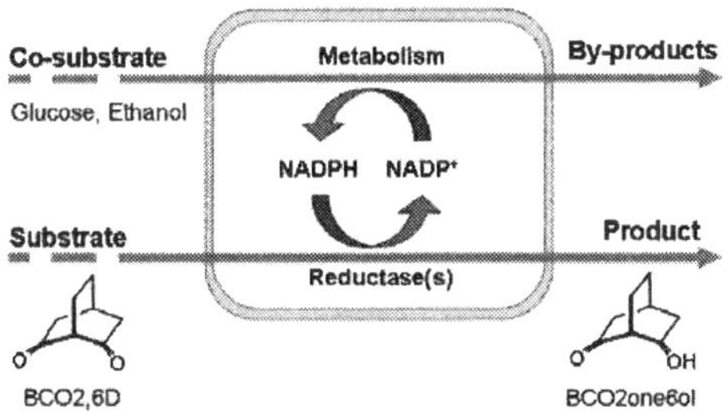

Fig. 10. The stereoselective reduction of the bicyclic diketone substrate bicyclo[2.2.2]octane-2,6-dione (BCO2,6D) to the product (1R,4S,6S)-6-hydroxybicyclo[2.2.2] octane-2one (BCO2 one6ol) [53].

5. IMPROVING ENANTIOSELECTIVITY BY REACTION ENGINEERING

5.1. Reaction engineering using ionic liquids

Ionic liquids are salts that do not crystallize at room temperature. The most common ionic liquids for biocatalysis are imidazolium-based ionic liquids, such as BMIM·BF_4 (1-butyl-3-methyl-imidazolium tetrafluoroborate; Fig. 11). Unlike conventional organic solvents, ionic liquids have no vapor pressure, are able to dissolve many compounds, and can be used to form two-phase systems with many common solvents, such as alcohols, hexane, benzene, *etc.* Biocatalytic reactions in ionic liquids have higher selectivity, faster rates, and greater enzyme stability. Furthermore, unlike organic solvents of comparable polarity, they often do not inactivate enzymes. They can be particularly useful for reactions involving polar substrates, such as sugars. In other words, these solvents permit enzyme-catalyzed reactions on polar substrates in non-aqueous media. These reactions are becoming increasingly important, because natural building blocks (e.g. peptides, sugars, nucleotides, and biochemical intermediates) are important starting materials for pharmaceuticals, fine chemicals and specialty materials, such as hydrogels [54, 55].

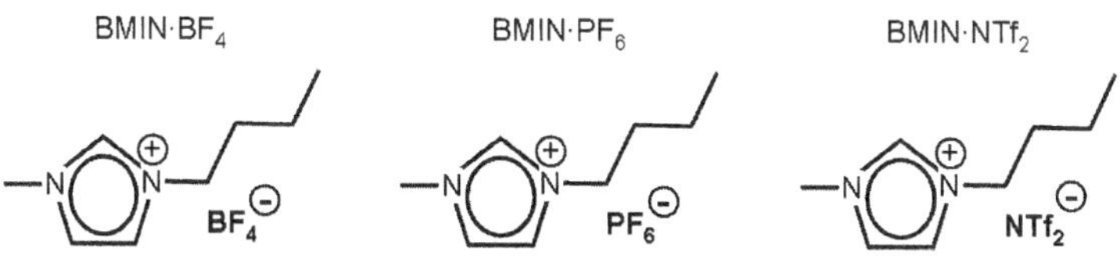

Fig. 11. Common ionic liquids in biocatalysis: the frequently used 1-butyl-3-methyl-imidazolium salts [55].

For example, in the synthesis of N-acetyllactosamine using *Candida antarctica* lipase B (CALB), the conversion yield (83%) when carried out in an ionic liquid was higher than that typical results obtained with organic solvents (50%) (Fig. 12) [56].

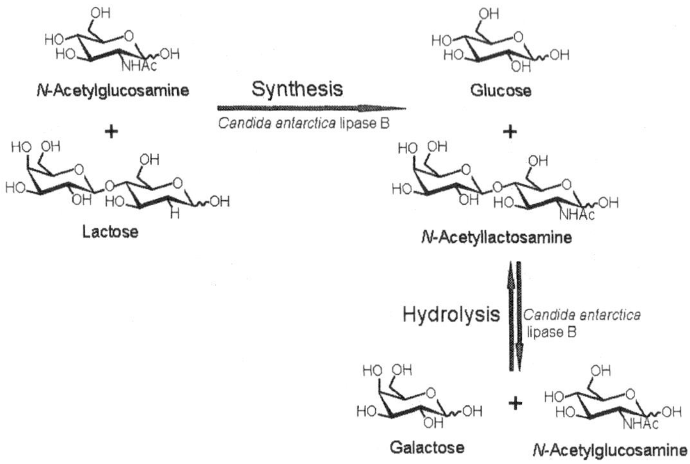

Fig. 12. The synthesis of N-acetyllactosamine using *Candida antarctica* lipase B. Hydrolysis of the product decreases the yield, but adding 25% ionic liquid suppressed hydrolysis and increased yield [56].

Ionic liquid (IL) was recently used with supercritical carbon dioxide ($scCO_2$) in a biphasic reaction system in a *Candida antarctica* lipase B (CALB)-catalyzed ester synthesis (Fig. 13) [57]. Ionic liquids are suitable media for the kinetic resolution of *rac*-phenylethanol. The best enantiomeric excess (94.9%) was obtained with (5-cyanopentyl)-trimethylammonium. Ionic liquids were shown to stabilize enzymes by increasing the free energy of deactivation (to 25 kJ/mol protein), resulting in an approximately 2000-fold increase in the half-life of the enzyme.

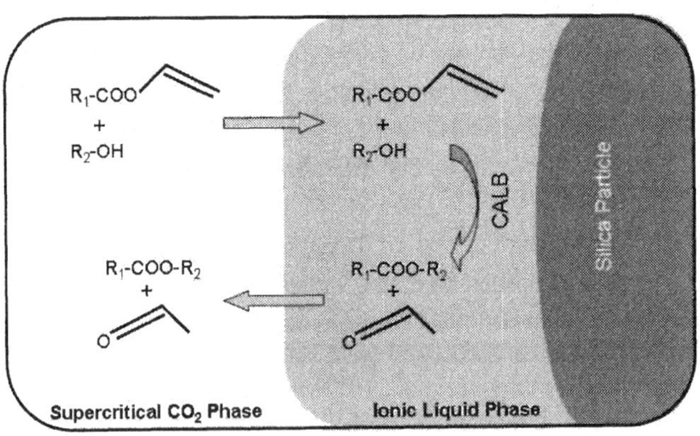

Fig. 13. Scheme of enzyme-catalyzed transesterification in IL/$scCO_2$ biphasic systems [57].

5.2. Reaction engineering using macromolecular compounds

5.2.1. Activation of biocatalysts by crown ethers

Crown ethers are effective complexing agents for alkali metal ions, ammonium ions, and water. Crown ethers generally activate enzymes, such as subtilisin, chymotrypsin, and lipase, in organic media, and consequently enhance the reactivity and enantioselectivity of these enzymes.

Recently, the activation of cytochrome c proteins by crown ether complexation was reported (Fig. 14). This enzyme is of particular interest because it retains its activity at low temperature in organic solvents, such as methanol [58]. (S)-naphthyl methyl sulfoxide was more rapidly oxidized by cytochrome c complexed with crown ether than (R)-isomer, with 73% yield and 47% enentiomeric excess when the reaction was carried out at -40°C.

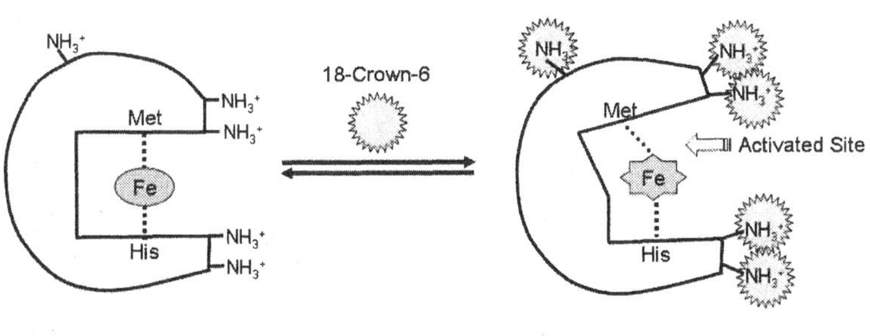

Fig. 14. Activation of cytochrome c via crown ether complexation [58].

5.2.2. Acceleration of enantioselective reaction by cyclodextrins

Cyclodextrins (CD), another class of compound with a macrocyclic structure, have been successfully used to improve enzyme activity and to increase the reaction rate and enantioselectivity. Several kinds of cyclodextrin derivatives were found to enhance the enzyme activity and enantioselectivity of lipases in both aqueous and organic solvents [59, 60]. In one example, the enantioselective hydrolysis of insoluble (*R,S*)-ketoprofen ethyl ester to the optically active (*S*)-ketoprofen was carried out in a dispersed aqueous lipase reaction system induced by the inclusion of chiral cyclodextrins for complexation of the substrate. Hydroxypropyl-*β*-cyclodextrin was the most effective chiral selector and disperser, giving an enantiomeric excess and conversion yield of 99% and 49%, respectively [61].

In another example, *Pseudomonas cepacia* lipase co-lyophilized with peracetylated β-cyclodextrin was immobilized by a sol–gel process, as shown in Fig. 15. The gel-entrapped lipase/cyclodextrin was subsequently used in the kinetic resolution of secondary alcohols using isopropenyl acetate as an innocuous acyl donor and toluene as the organic medium. The (*R*)-alcohol was the faster reacting enantiomer, yielding the (*R*)-ester in high enantiomeric excess and leaving the (*S*)-alcohol in an enantiomerically pure form [62].

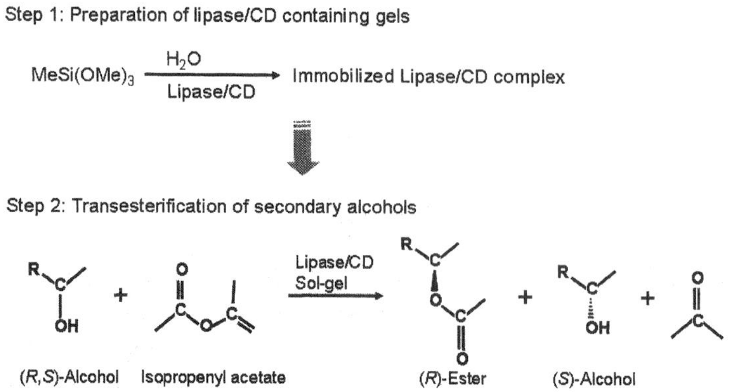

Fig. 15. Lipase/CD entrapped in sol–gel-catalyzed transesterification of secondary alcohols using isopropenyl acetate as an acyl donor in toluene [62].

6. IMPROVING CHIRAL SYNTHESIS BY DIRECTED EVOLUTION AND METABOLIC ENGINEERING

6.1. Directed evolution approach

As revealed in the previous sections, the single most important factor in a successful chiral synthesis process is the enantioselectivity (E) of the enzyme. Although E for any given enzyme can be improved through reaction engineering, modifying catalysts often leads to a more dramatic enhancement of enantioselectivity. Since our current understanding of the molecular basis for enantioselectivity is not complete and does not sufficiently guide the efforts to modify enzymes, directed evolution, which does not require such knowledge, is well suited for this application. Indeed, there have been many successful examples documented [63, 64]. This section will emphasize several key issues in applying directed evolution to improve chiral synthesis. The concept of directed evolution and its applications in other areas of bioprocessing can be found in Chapter 3.

Reetz's group pioneered the use of directed evolution in improving the enantioselectivity of *Pseudomonas aeruginosa* lipase (PAL), a widely used enzyme in chiral synthesis. Initial ep-PCR with a low mutation frequency of 1–2 base substitutions per gene was successful in increasing enantioselectivity from 1.1 to 11.3 after four rounds [65]. "Hot spots" identified from ep-PCR were subjected to saturation mutagenesis, resulting in a variant with five amino acid substitutions with E =25.8. Analyzing the 3-D structure of the wild type and mutant PAL, they concluded that the flexibility of certain loop structures is of prime importance for enantioselectivity [66]. Further improvement came through a combination of ep-PCR at a higher mutation rate, DNA shuffling, and more focused mutagenesis at positions 155 and 162. The improved enzyme showed an E greater than 51, a nearly 50-fold increase. It is remarkable that such enhancement in enantioselectivity was achieved after screening a total of only 40,000 mutants, suggesting that maximum structural diversity is more important than the size of a library [67]. It may also indicate that enantioselectivity is a trait relatively easier to evolve *in vitro*. A summary of the directed evolution efforts is provided by Reetz [68].

As with other applications of directed evolution, if enzymes, wild type and variants, are to be expressed *in vivo*, efficient expression is a requirement. Insufficient expression hinders the directed evolution efforts. For example, Reetz *et al.* [69] found that, although single-round ep-PCR led to a significant improvement in the enantioselectivity of an epoxide hydrolase (from *Aspergillus niger*) from E = 4.6 to E = 10.8 for hydrolytic kinetic resolution of glycidyl phenyl ether, an inefficient expression system prevented more rapid progress in their directed evolution experiments and made it necessary to redirect their efforts to improving *in vivo* expression.

Koga et al [70] reported an alternative strategy to *in vivo* expression in directed evolution. They demonstrated the feasibility of using cell-free expression systems in constructing mutant libraries. Using single-molecule PCR, followed by an *in vitro* coupled transcription/translation system (SIMPLEX), a mutant library was constructed for *Burkholderia cepacia* KWI-56 lipase in an effort to invert its enantioselectivity. Four amino acid positions located close to the catalytic triad were substituted in a combinatory fashion with a limited set of hydrophobic amino acids, resulting in a library with estimated 2401 variants. The new library construction strategy not only circumvented time-consuming cloning and transformation steps, but also helped to overcome common problems associated with the heterologous expression of enzyme variants, such as inclusion body formation [70]. The obvious limitation of this approach is the prerequisite of knowledge about the active sites of the enzymes of interest and the limited size of the library. It thus remains to be seen whether it will be used as a general approach.

Enzyme diversity can also be generated using commercial mutator strains. Using mutator strain *E. coli* XL1-red and subsequent saturation mutagenesis, *Pseudomonas fluorescens* esterase was evolved for the enantioselective hydrolysis of the chiral building blocks (*S*)-3-bromo-2-methyl-propanoate and ethyl-*R*-3-phenylbutanoate [71]. By combining ep-PCR, DNA shuffling, and saturation mutagenesis, an N-acetylneuraminic acid aldolase (Neu5Ac Aldolase) from *E. coli* was evolved to a variant with completely reversed enantioselectivity and producing unnatural N-acetyl-L-neuraminic acid (L-sialic acid) and 3-deoxy-L-manno-oct-2-ulosonic acid (L-KDO). All three beneficial mutations occurred outside the $(\alpha/\beta)_8$-barrel active site [72]. The mutations identified in this case, as with most other directed evolution studies, were located far from the active site. The authors pointed out that, since enzymes contain many more residues far from the active site than close to it, random mutagenesis was biased in favor of distant mutations that have little effect on the active site. It would be more effective to focus mutagenesis experiments on active-site residues. This drawback can be avoided by rational design when enzyme structures are known or active sites are identified [71].

The above examples all started from a non-enantioselective or moderately enantioselective wild-type enzyme. A better starting point for directed evolution could be obtained by harnessing natural diversity using a metagenome approach, i.e. screening enzymes from genomic libraries that are generated by extracting DNA directly from environmental samples [73]. Adopting this strategy, Diversa Corp has discovered over 200 nitrilases that have unique gene sequences. Screening these nitrilases against a panel of chiral substrates revealed a subset of enzymes that exhibited high enantioselectivity [41]. These enzymes were used in directed evolution through Gene Site Saturation Mutagenesis (GSSM) to enhance the enantioselectivity of the enzyme, resulting in an improvement in ee to 98.1% from 87.9% in the synthesis of (*R*)-4-cyano-3-hydroxybutryic acid [40].

6.2. Metabolic engineering approach

Metabolic engineering [74, 75] is a powerful tool for reconstructing cellular metabolic networks for improved production of chiral molecules. Stephanopoulos and coworkers [76] successfully applied the concepts and tools of metabolic engineering in creating a recombinant *Rhodococcus* strain capable of completely converting indene to *trans*-(1*R*, 2*R*) chiral indandiol (Fig. 16), an intermediate in the synthesis of CRIXIVAN®, Merck's HIV-1 protease inhibitor. They first developed a mutant strain with an improved product profile using a chemostat, gradually increasing the selection pressure of the toxic substrate. Using chemostats in metabolic engineering is not a common approach; nevertheless, it was successful in this case. It successfully evolved a mutant producing twice as much desired product without compromising the productivity. Subsequent characterization of the biochemical network and applying systems analysis through flux analysis provided the critical understanding about the interconnectedness of the biochemical reactions producing products and byproducts (Fig. 16). What makes such analysis and understanding useful is the level of details provided and the quantitative nature of the analysis. This allowed them to pinpoint the bottlenecks and correctly identify indan oxide hydrolysis reaction as the target for further genetic modification. By introducing epoxide hydrolase activity on a plasmid, both yield and selectivity were increased to over 90%.

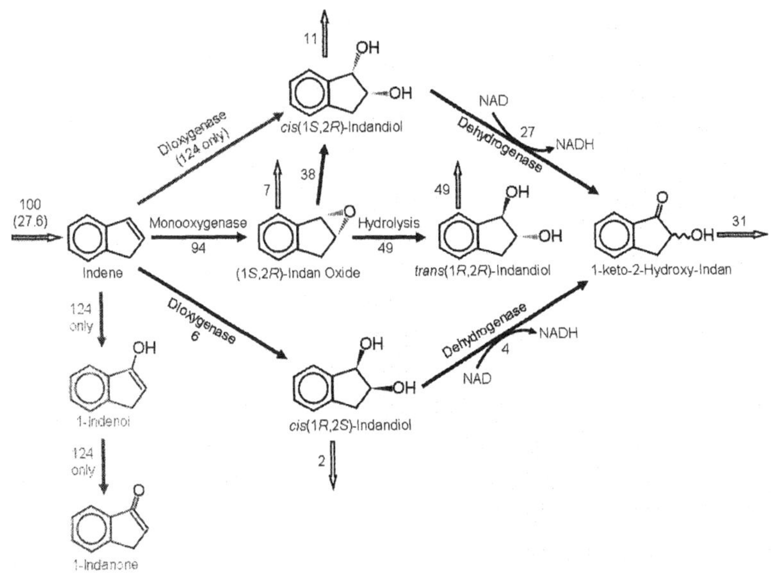

Fig. 16. Indene bioconversion network in *Rhodococcus* spp. I24 and KY1 and steady-state flux distribution for KY1 at 100 ppm indene air-feed concentration and a dilution rate of 0.065 h^{-1}. The oxygenation reactions shaded in gray are observed in I24 only. The fluxes were normalized by the indene uptake rate (in parentheses, $\mu mol \cdot h^{-1}$ per g of dry cell weight). The solid arrows denote intracellular fluxes, and open arrows indicate indene uptake and metabolite excretion fluxes [76].

6.3. Combination of directed evolution with metabolic engineering

The successful use of metabolic engineering in combination with directed evolution was reported by Arnold and coworkers for the production of L-methionine in *E. coli* [77]. The enantioselectivity of a D-selective hydantoinase from *Arthrobacter* sp. DSM 9771 was first inverted to an L-selective enzyme using random and saturation mutagenesis. Co-expressing the evolved enzyme with a hydantoin racemase and an L-*N*-carbamoylase in *E. coli* constructed an effective biocatalyst that produced 91 mM L- methionine from 100 mM D,L-5-(2-methylthioethyl) hydantoin in less than 2 h, and the accumulation of the unwanted intermediate D-carbamoyl-methionine was reduced four fold.

6.4. Whole-genome shuffling

Maxygen Inc. and its subsidiary Codexis Inc. has recently pioneered whole-genome shuffling to accelerate the process of strain improvement. They demonstrated that recursive genomic recombination within a population of bacteria can efficiently generate combinatorial libraries of new strains. The rapid improvement of tylosin production from *Streptomyces fradiae* showed the potential of this non-rationale based method to facilitate cell and metabolic engineering without detailed knowledge of the biochemical networks involved [78]. Whole-genome shuffling, in principle, is applicable to any metabolic network and for any target product. However, practical applications of the method rely on high-throughput screening or a selectable marker. Often, the screening capability and a lack of a selection method limit the use of this method.

6.5. Surface-display whole-cell biocatalysts

Recently, whole-cell biocatalysts were developed by displaying lipases on cell surfaces. Yeast biocatalyst was constructed by surface-displaying a *Rhizopus oryzae* lipase using a novel display system based on Flo1p flocculation functional domain of the yeast cell. The whole-cell catalyst retained its activity in hexane, heptane, cyclohexane and octane, and was found to be very effective in chiral resolution of (*R,S*)-1-phenylethanol by enantioselective trans-esterification with vinyl acetate [79, 80]. The ester product (*R*)-1-phenylethyl acetate reached 39.8 mM (97.3% yield) with a high enantiomeric excess of 93.3%.

Similarly, Lee and co-workers [81] developed an *E. coli* whole-cell biocatalyst by displaying a thermostable *Bacillus* sp. strain TG453 lipase (44.5 kDa) on the cell surface using FadL as an anchoring motif, an outer membrane protein in *E. coli* involved in the binding and transportation of long-chain fatty acids. Stable functional expression of the lipase was achieved with whole cells. The system was demonstrated to be very effective in kinetic resolution of racemic methyl mandelate. Within 36 h, (S)-mandelic acid could be produced with an *ee* of 99% and an enantiomeric ratio of 292, significantly higher than those with crude lipase.

6.6. Development of high-throughput *ee* assays

To support directed evolution and metabolic engineering efforts in the context of improving chiral synthesis as described above, several high-throughput enantiomeric excess (*ee*) screening systems were developed, using a wide range of analytical techniques such as UV-

Vis spectroscopy [65], IR thermography [82], and capillary array electrophoresis [83], allowing 700 to 20,000 *ee* determinations per day with an ee accuracy between ±2% and ±5%.

Notably, Reetz and co-workers developed several systems based on isotopic pseudo-enantiomers that were ^{13}C- or ^{2}H-labeled at appropriate positions, which can be analyzed by several methods with varied capacity and accuracy. In the case of standard mass spectrometry with electrospray ionization (ESI) or matrix-assisted laser-desorption ionization (MALDI), the capacity reached 10,000 samples per day with an accuracy of ±5% [84]. With ^{1}H-NMR as the detection method, the capacity was lower, at 1400 samples per day, but with a higher degree of precision: ±2–3% [85]. With more affordable Fourier transform infrared (FTIR) spectroscopy, a capacity of 10,000 *ee* values per day with a reasonable degree of accuracy, ±7%, can be achieved [86].

Banerjee *et al.* [87] established a colorimetric high-throughput assay using Bromothymol blue as a pH indicator for screening nitrilase-producing microorganisms that hydrolyze mandelonitrile to mandelic acid. A hierarchical screening strategy was adopted. First, nitrilase activity was screened using whole cells in a 96-well microplate format, allowing quick initial identification of organisms with desired enantioselectivity. The positive hits were then subjected to more precise quantitative analysis using HPLC.

Most recently, an automated method was developed using a robotic liquid handler and rapid chiral supercritical fluid chromatography (SFC) analysis. When it was applied to screen hydrolases for enantioselective transesterification of 1-phenyl-ethanol [88], the throughput reached 700 analyses/day with a run time of only 1.5 min under optimal conditions. One of the advantages of chiral SFC as compared to chiral HPLC or GC is its separation efficiency [89]. The low viscosity of carbon dioxide enables high flow rates without losing chromatographic efficiency, which shortens run time and makes it ideal for screening purposes.

Making use of the fact that nitrilases hydrolyze a nitrile substrate and liberate ammonium, which could be the only available nitrogen source to support cell growth, a selection method was developed for positive clones that are active on an optically active nitrile substrate. The successful use of the strategy led to a discovery of 137 unique nitrilases from screening of > 600 biotope-specific environmental DNA (eDNA) with 10^6–10^9 members per eDNA library [39].

7. CONCLUSIONS

Tremendous progress has been made in recent years in using biocatalysis for chiral synthesis. Heterologous expression, directed evolution, and metabolic engineering have become indispensable tools in developing enzyme and whole-cell catalysts. Great strides have been made in solving several long-standing problems, such as cofactor regeneration. An important whole-cell biocatalysis issue, the slower reaction rate rooted in the limited substrate permeability in whole-cell biocatalysis, is being addressed using molecular engineering approaches. We can expect that these laboratory accomplishments will be translated into the successful commercialization of yet more chiral molecules in the future. The challenges ahead are to shorten the time required for biocatalyst development and process development to the level comparable to a typical chemical catalysis process. It is expected that the increasing use of high-throughput technology, embracing various bioinformatics tools made available only recently, and advances in our understanding of enzymatic catalysis mechanisms, protein structure and function relationships, and the fundamentals of metabolic pathways and regulation will bring about more widespread use of biocatalysts in chiral synthesis.

REFERENCES

[1] A.M. Routh, Chemical and Engineering News, June14, (2004), 47-62.
[2] M. Breuer, K. Ditrich, T. Habicher, B. Hauer, M. Keβeler, R. Stürmer and T. Zelinski, Angew. Chem. Int. Ed., 43 (2004) 788-824.
[3] T. Hermann, J. Biotechnol., 104 (2003) 155-172.
[4] S.Y. Lee and E. T. Papoutsakis (eds.), Metabolic Engineering, Marcel Dekker, New York and Basel, 1999, pp. 153-176.
[5] H. Motoyama, H. Yano, Y. Terasaki and H. anazawa, Appl. Environ. Microbiol., 67 (2001) 3064-3070.
[6] E. Radmacher, A. Vaitsikkova, U. Burger, K. Krumbach, H. Sahm and L. Eggeling, Appl. Environ. Microbiol., 68 (2002) 2246-2250.
[7] P. Gourdon and N.D. Lindley, Metab. Eng., 1 (1999) 224-231.
[8] T. Scheper (ed.), Advances in Biochemical Engineering/Biotechnology vol. 79, Springer, Berlin, 2003, pp. 37-58.
[9] S. Albert and R.D. Kinley, Trends Biotechnol., 19 (2001) 53-62.
[10] M.R. Gerigk, R. Bujnicki, E. Ganpo-Nkwenkwa, J. Bongaerts, G. Sprenger and R. Takors, Biotechnol. Bioeng., 80 (2002) 746-754.
[11] M. Pompejus, B. Kroger, H. Schroder, O. Zelder and G. Haberhauer, WO Patent No. 01/00844 (2001).
[12] S. Nakagawa, H. Mizoguchi, S. Ando, M. Hayashi, K. Ochiai, H. Yokoi, N. Tateishi, A Senoh and A. Ozaki, European Patent No. 1108790 (2001).
[13] T. Dassler, T. Maier, C. Winterhalter and A. Bock, Mol. Microbiol., 36 (2000) 1101-1112.
[14] M. Wada, N. Awano, K. Haisa, H. Takagi and S. Nakamori, FEMS Micobiol. Lett., 217 (2002) 103-107.
[15] T.H.P. Maier, Nature Biotechnol., 21(2003) 422-427.
[16] D.C. Lee, S.G. Lee, S.P. Hong, M.H. Sung and H.S. Kim, Ann. New York Acad. Sci., 864 (1999) 401-405.
[17] J. Ogawa, S. Shimizu, Curr Opin. Biotechnol., 13 (2002) 367-375.
[18] A. Berry, R.P. Burlingame and J.R. Millis, PCT WO0004182 (2001).
[19] L. Lapierre, J.E. Germond, A Ott, M. Delley and B. Mollet, Appl. Environ. Microbiol., 65 (1999) 4002-4007.
[20] K. Kyla-Nikkila, M. Hujanen, M. Leisola and A. Palva, Appl. Environ. Microbiol., 66 (2000) 3835-3841.
[21] S. Zhou, T.B. Causey, A. Hasona, K.T. Shanmugam and L.O. Ingram, Appl. Environ. Microbiol., 69 (2003) 399-407.
[22] S. Zhou, K.T. Shanmugam and L.O. Ingram, Appl. Environ. Microbiol., 69 (2003) 2237-2244.
[23] S.Y. Lee and Y. Lee, Appl. Environ. Microbiol., 69 (2003) 3421-3426.
[24] H.J. Gao, Q. Wu and G.Q. Chen, FEMS Micobiol Lett., 213 (2002) 59-65.
[25] S.M. Roberts, Tetrahedron, 60 (2004) 499-500.
[26] A. Zaks, Curr. Opin. Chem. Biol., 5 (2001) 130-136.
[27] A. Schmid, J.S. Dordick, B. Hauer, A. Kiener, M. Wubbolts and B. Witholt, Nature, 409 (2001) 258-268.

[28] H.J. Deussen, M. Zundel, M. Valdois, S.V. Lehmann, V. Weil, C.M. Hjort, P.R. Ostergaard, E. Marcrussen and S. Ebdrup, Org. Proc. Res.Devel., 7 (2003) 82-88.
[29] A.N. Collins, G.N. Sheldrake and J. Crosby (eds.), Chirality in Industry II, Chichester, UK, Wiley, 1997, pp. 245-262.
[30] K. Mori, Tetrahedron, 45 (1989) 3233-3298.
[31] M. Bucciarelli, P. Davoli, A. Forni, I. Moretti, F. Prati, J. Chem. Soc. Perkin Trans., 1 (1999) 2489-2494.
[32] A. Fishman, M. Eroshov, S.S. Dee-Noor, J. van Mil, U. Cogan and R. Effenberger, Biotechnol. Bioeng., 74 (2001) 256-263.
[33] M.J. Kim, Y.S. Ahn and J.W. Park, Curr. Opin. Biotechnol., 13 (2002) 578-587.
[34] M.J. Kim, Y.I. Chung, Y.K. Choi, H.K. Lee, D. Kim and J. Park, J. Am. Chem. Soc., 125 (2003) 11494-11495.
[35] E.J. de Vries and D.B. Janssen, Curr. Opin. Biotechnol., 14 (2003) 414-420.
[36] K.M. Manoj, A. Archelas, J. Baratti and R. Furstoss, Tetrahedron,57 (2001) 695-701.
[37] D. Brady, A. Beeton, J. Zeevaart, C. Kgaje, F. van Rantwijk and R.A. Sheldon, Appl. Microbiol. Biotechnol., 64 (2004) 76-85.
[38] L.M. van Langen, F. van Rantwijk and R.A. Sheldon, Org. Proc. Res. Dev., 7 (2003) 828-831.
[39] D.E. Robertson, J.A. Chaplin, G. DeSantis, M. Podar, M. Madden, E. Chi, T. Richardson, A. Milan, M. Miller, D.P. Weiner, K. Wong, J. McQuaid, B. Farwell, L. A. Preston, X. Tan, M.A. Snead, M. Keller, E. Mathur, P.L. Kretz, M.J. Burk and J.M. Short, Appl. Environ. Microbiol., 70 (2004) 2429-2436.
[40] G. DeSantis, K. Wong, B. Farwell, K. Chatman, Z. Zhu, G. Tomlinson, H. Huang, X. Tan, L. Bibbs, P. Chen, K. Kretz and M.J. Burk, J. Am. Chem. Soc., 125 (2003) 11476-11477.
[41] G. DeSantis, K. Wong, B. Farwell, K. Chatman, Z. Zhu, G. Tomlinson, H. Huang, X. Tan, L. Bibbs, P. Chen, K. Kretz and M.J. Burk, J. Am. Chem. Soc., 124 (2002) 9024-9025.
[42] Z. Li, J.B. van Beilen, W.A. Duetz, A. Schmid, A. de Raadt, H. Griengl and B. Witholt, Curr. Opin. Chem. Biol., 6 (2002) 136-144.
[43] W.A. van der Donk and H. Zhao, Curr, Opin. Biotechnol., 14 (2003) 421-426.
[44] B. Boonstra, D.A. Rathbone, C.E. French, E.H. Walker and N.C. Bruce, Appl. Environ. Microbiol., 66 (2000) 5161-5166.
[45] J.M. Vrtis, A. White, W.W. Metcalf and W.A. van der Donk, Angew. Chem. Int. Ed. Engl., 41 (2002) 3257-3259.
[46] B.R. Riebel, P.R. Gibbs, W.B. Wellborn and A.S. Bommarius, Adv. Synth. Catal., 345 (2003) 707-712.
[47] B. Geueke, B. Riebel and W. Hummel, Enzyme Microb. Technol., 32 (2003) 205-211.
[48] H. Groeger, W. Hummel, S. Buchholz, K. Drauz, T.V. Nguyen, C. Rollmann, H. Husken and K. Abokitse, Org. Lett., 5 (2003) 173-176.
[49] M. Wolberg, W. Hummel and M. Müller, Chem. Eur. J., 7 (2001) 4562-4571.
[50] Y. Ni and R.R. Chen, Biotechnol. Bioeng., 20 (2004) 804-811.
[51] M. Kataoka, K. Kita, M. Wada, Y. Yasohara, J. Hasegawa and S. Shimizu, Appl. Microbiol. Biotechnol., 62 (2003) 437-445.
[52] S. Wang and M. M. Kayser, J. Org. Chem., 68 (2003) 6225-6228.
[53] M. Katz, T. Frejd, B. H. Hägerdal and M. F. Gorwa-Grauslund, Biotechnol. Bioeng., 84 (2003) 573-582.
[54] M.Y. Lee and J.S. Dordick, Curr. Opin. Biotechnol., 13 (2002) 376-384.
[55] S.S. Park and R.J. Kazlauskas, Curr. Opin. Biotechnol., 14 (2003) 432-437.
[56] P. Lozano, T. Diego, S. Gmouh, M. Vaultier and J. L. Iborra, Biotechnol. Prog., 20 (2004) 661-669.
[57] N. Kaftzik , P. Wasserscheid and U. Kragl, Org. Proc. Res. Dev., 6 (2002) 553-557.
[58] D. Paul, A. Suzumura, H. Sugimoto, J. Teraoka, S. Shinoda and H. Tsukube, J. Am. Chem. Soc., 125 (2003) 11478-11479.
[59] A. Ghanem and V. Schurig, Tetrahr. Asymm., 12 (2001) 2761-2766.

[60] Y. Mine, K. Fukunaga, K. Itoh, M. Yoshimoto, K. Nakao and Y. Sugimura, J. Biosci. Bioeng., 95 (2003) 441-447.
[61] S.H. Kim, T.K. Kim, G.S. Shin, K.W. Lee, H.D. Shin and Y.H. Lee, Biotechnol. Lett., 26 (2004) 965-969.
[62] A. Ghanem and V. Schurig, Tetrahr. Asymm., 14 (2003) 2547-2555.
[63] K-E Jaeger and T. Eggert, Curr. Opin. Biotechnol., 15 (2004) 305.
[64] S. G. Burton, D.A. Cowan, and J. M. Woodley, Nat. Biotechnol., 20 (2002) 37.
[65] M. T. Reetz, A. Zonta, K. Schimossek, K, Liebeton, and K.-E. Jaeger, Angew Chem. Int. Ed. Engl. 36 (1997) 2830.
[66] K. Liebeton, A. Zonta, K. Schimossek, M. Nardini, D. Lang, B. W. Dijkstra, M. T. Reetz, and K.-E. Jaeger, Chem. Biol., 7 (2000) 709.
[67] M. T. Reetz, S. Wilensek, D. Zha, and K. -E. Jaeger, Angew. Chem. Int. Ed.Engl., 40(2001)3589.
[68] M. T. Reetz, Proc. Natl. Acad. Sci. USA, 101 (2004) 5716.
[69] M. T. Reetz, C. Torre, A. Eipper, R. Lohmer, M. Hermes, B. Brunner, A. Maichele, M. Bocola, M. Arand, A. Cronin, Y. Genzel, A. Archelas, and R. Furstoss, Org. Lett., 6 (2004) 177.
[70] Y. Koga, K. Kato, H. Nakano, and T. Yamane, J. Mol. Biol., 331 (2003) 585.
[71] G.P. Horsman, A. M. F. Liu, E. Henke, U. T. Bornscheuer, and R. J. Kazlauskas, Chem. Eur. J., 9 (2003) 1933.
[72] M. Wade, C. C. Hsu, D. Franke, M. Mitchell, A. Heine, I. Wilson, and C. H. Wong, Bioorg. Med. Chem., 11 (2003) 2091.
[73] S. Voget, C. Leggewie, A. Uesbeck, C. Raasch, K. E. Jaeger, and W. R. Streit, Appl. Environ. Microbiol., 69 (2003) 6235.
[74] J. E. Bailey, Science, 252 (1991) 1668.
[75] G. Stephanopoulos. Metab. Eng., 1 (1999) 1.
[76] D. E. Stafford, K. S. Yanagimachi, P. A. Lessard, S. K. Rijhwani, A. J. Sinskey, and G. Stephanopoulos, Proc. Natl. Acad. Sci. USA., 99 (2002) 1801.
[77] O. May, P. T. Nguyen, and F. H. Arnold, Nat.Biotechnol., 18 (2000) 317.
[78] Y. Zhang, K. Perry, V. A. Vinci, K. Powell, W. P. C. Stemmer, and S. B. del Cardayré, Nature, 415 (2002) 644.
[79] T. Masumoto, H. Fukuda, M. Ueda, A. Tanaka and A. Kondo, Appl. Environ. Microbiol., 68 (2002) 4517-4522.
[80] T. Masumoto, M. Ito, H. Fukuda and A. Kondo, Appl. Microbiol. Biotechnol., 64 (2004) 481-485.
[81] S. H. Lee, J. Choi, S. J. Park, S. Y. Lee, and B. C. Park, Appl. Environ. Microbiol., 70 (2004) 5074.
[82] M. T. Reetz, M. Hermes, and M. H. Becker, Appl. Microbiol. Biotechnol., 55 (2001) 531.
[83] M. T. Reetz, K. M. Kühling, A. Deege, H. Hinrichs, and D. Belder, Angew. Chem. Int. Ed., 39 (2000) 3891.
[84] M. T. Reetz, M. H. Becker, H.-W. Klein, D. Stöckigt, Angew, Chem. Int. Ed., 38 (1999) 1758.
[85] M. T. Reetz, A. Eipper, P. Tielmann, and R. Mynott, Adv. Synth. Catal., 344 (2002) 1008.
[86] P. Tielmann, M. Boese, M. Luft, and M. Reetz, Chem. Eur. J., 9 (2003) 3882.
[87] A. Banerjee, P. Kaul, R. Sharma, and U. C. Banerjee, J. Biomol.Screen., 8 (2003) 559.
[88] L. Di, O. J. McConnell, E. H. Kerns, and A. G. Sutherland, J. Chromatogr. B., 809 (2004) 231.
[89] S. H. Hoke II, J. D. Pinkston, R. E. Bailey, S. L. Tanguay, and T. H. Eichhold, Anal. Chem., 72 (2000) 4235.
[90] S. Lutz, and W. M. Patrick, Curr. Opin. Biotechnol.,15 (2004) 291.

Bioprocessing for Value-Added Products from Renewable Resources
Shang-Tian Yang (Editor)

Chapter 14. Immobilized Cell Fermentation for Production of Chemicals and Fuels

Ying Zhu

PDL Biopharma, 34801 Campus Drive, Fremont, CA, 94555

1. INTRODUCTION

Cell immobilization was one of the most exciting aspects of biotechnology during the 1970s. It has been used as an effective method to improve the performance and economics of many fermentation processes. Numerous techniques for immobilizing of cells have been developed in the last three decades.

Immobilized cell systems are pertinent to fermentation because of reactor performance advantages that immobilized cells provide over freely suspended cultures, easy separation of the biomass from the liquid and easy product recovery, and specific metabolic improvements or products created upon immobilization. Cell immobilization allows for more efficient operation by reducing the non-productive growth phase. It is well recognized that the high cell density of immobilized cells improves the product yield and the volumetric productivity of bioreactors. Immobilization protects the cells from shear forces and imparts a special stability to the microorganism against environmental stresses (pH, temperature, organic solvents, salts, inhibiting substrates and products, poisons, self-destruction). The activity, viability, and productivity of immobilized cells can be maintained for a long time period, which facilitates continuous cultivation processes and results in better operational stability. Cell wash-out is avoided even at the high dilution rates of the continuous operation mode. Immobilized cells can be handled more easily and recovered from the solution without difficulty; and a cell-free product stream simplifies downstream processing. In addition, because immobilization can influence both diffusion properties of molecules through the support and the physiological behavior of the confined cells, noticeable differences of cell growth, metabolism, and physiology are observed upon immobilization. Higher specific rates of product synthesis or substrate consumption for immobilized cells have been successfully demonstrated. Table 1 shows the comparison between free cell and immobilized cell fermentation. Due to these important advantages of cell immobilization, a variety of immobilized cell bioreactors have been developed to optimize the fermentation processes. Immobilized cells are currently being used industrially for vinegar, organic, and amino acid production, as well as in wastewater treatment [1].

Table 1

Comparison between free cell and immobilized cell fermentations

	Immobilized cell fermentation	Free cell fermentation
Cell growth	Faster fermentation due to the reduction of non-productive growth phase	Slower fermentation because cell growth is required for the production
	High cell density	Low cell density
Production	High substrate concentration	Low substrate concentration to prevent inhibition
	High cell productivity	Low cell productivity
	High product yield	Low product yield
	High final product concentration Improved resistance of cells to inhibitory substrates or products	Product inhibition often results in low final product concentration achieved
Process operation	Batch, fed-batch, or continuous process operation mode in immobilized bioreactor	Batch or fed-batch fermentation using stirred tank reactor
	No cell wash-out in continuous fermentation even at high dilution rate	Difficult to perform continuous process due to cell wash-out
	Smaller fermentor size needed due to the high cell density	Large fermentor required
	Simplified process design due to the separation of products and cells	Effective separation and concentration steps are necessary in downstream processing
	Reuse of cells for prolonged period of time due to cell regeneration	Cells cannot be reused
	Long-term operational stability and constant product quality	Product quality varies lot by lot
	Reduced risk for microbial contamination	More prone to contamination

This chapter will review the immobilization techniques of interest to the fermentation industry, the effects of immobilization on cell physiology, metabolism, genetics, and fermentation behavior, and various types of immobilized cell bioreactor systems and their applications in fermentation.

2. IMMOBILIZATION TECHNIQUES

Cell immobilization is defined as the physical confinement or localization of intact cells to a certain defined region of space with the preservation of some desired activity [2]. The support for cell immobilization permits the exchange of substrates, products, inhibitors, etc., while separating the catalytic cell biomass from the bulk phase containing substrates and products. The support should be inexpensive, stable, reusable, and nontoxic. For fermentation applications, some additional important factors must be considered, such as a simple immobilization procedure, a high immobilization capacity, a high diffusion rate of substance, and excellent mechanical strength for long-term operation.

Immobilization techniques can be classified into two major groups: carrier binding and entrapment [3, 4] (Fig. 1). Carrier binding includes all methods where there is a direct binding of cells to water-insoluble carriers by physical adsorption or by ionic and/or covalent bonds. Entrapment includes the enclosure of cells within a polymeric matrix or behind a semi-permeable membrane. Other techniques such as cross-linking [5] and cell aggregation [6] are less frequently used in fermentation and will not be discussed.

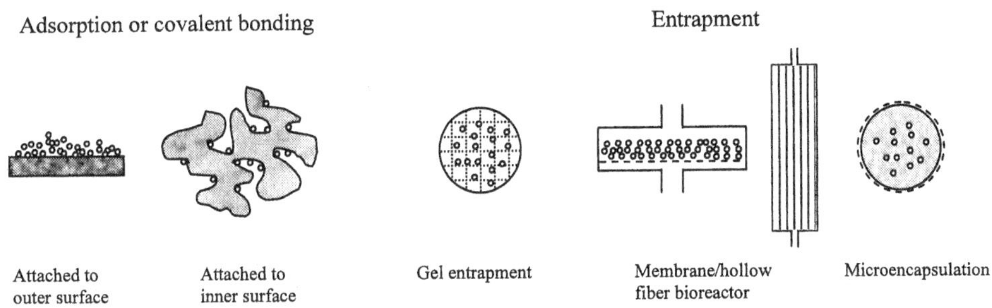

Fig. 1. Cell immobilization techniques.

2.1. Adsorption

Non-specific physical adsorption of cells from the solution on the surface is the simplest and longest known mode of immobilization, because naturally microbial cells spontaneously adsorb to the surfaces of insoluble materials. It has been observed that in their natural biotopes, microorganisms are never found in isolated form in environments such as soil and rumen, but rather in an adsorbed form and in close association with other microorganisms [7].

Adsorption usually requires no special adjustment of the support. It is performed directly in the fermentor by introducing the inoculum and the support to the medium. Cells become immobilized simply as a consequence of their growth. Because of the normally low cost of the support and ease of preparation, adsorption onto porous and/or fibrous materials is very useful in a number of fermentation applications. The supports used include celite [8–10], sand [10], porous brick [10], glass beads [11], clay [12], ceramics [13–15], wood chips [16], ion-exchange resins [17], plastic materials [18–20], organic materials such as cellulosic material [21–24] and activated carbon [25–27], etc. Plastic composite supports (PCS) containing a blend of agricultural material (soybean hulls or oat hulls), complex nutrients, and polypropylene have been developed by high-temperature extrusion and used for cell immobilization [28–31]. Recently, fibrous materials have been developed for cell immobilization because of their high specific surface area, high void volume, low cost, high mechanical strength, and high permeability [32, 33]. Fig. 2 shows the scanning electron micrographs (SEM) of different cells immobilized on cotton fiber surfaces by adsorption. Adsorption is now being successfully applied in industrial processes; vinegar can be produced by acetic acid bacteria (*Acetobacter*) immobilized on wood chips [34].

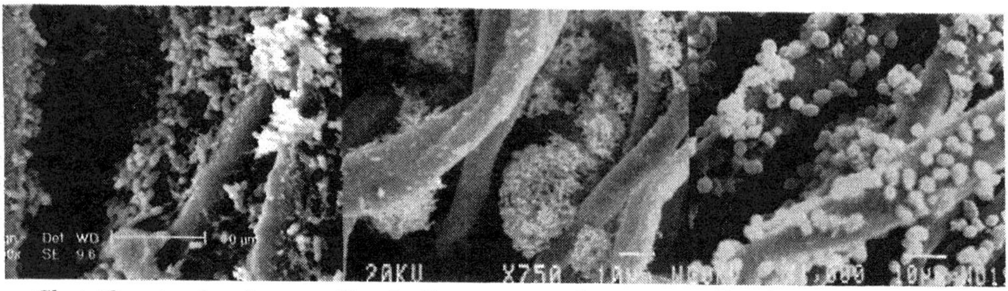

Clostridium tyrobuyricum *Propionibacterium acidipropionici* *Saccharomyces cerevisiae*

Fig. 2. SEM of immobilized cells on cotton fiber (Photos provided by S.T. Yang, The Ohio State University).

The adsorption of cells occurs through interactions such as the van der Waals forces and ionic and hydrogen bonds. Different chemical and physical characteristics of the support surface, including surface roughness, chemical composition, hydrophobicity and its morphology, and cell characteristics such as size, hydrophobicity and surface charge all can significantly affect cell adsorption [35]. Changes in the fermentation conditions, such as media composition, ionic strength, pH, stirring rates, and culture age can also dramatically alter the degree of cell adsorption [36].

Because the attractive forces between the microorganisms and the support are usually low, cells can initially be easily removed by turbulence or shear forces. Once cells are attached to the support, cell growth takes place. They take up and metabolize nutrients, release waste products, synthesize and secrete exopolymers including specific proteins, lectins, or polysaccharides on the outer layer of cells, which is the so-called biofilm formation [37]. Generation of biofilm creates strong binding interactions between cells, results in high cell densities, and improves the productivity of the fermentation processes [4, 28–31].

The adsorption method allows direct contact between the fermentation broth and the cells, and the media are able to flow in and out of the system without restriction, thus minimizing or eliminating mass transfer resistance. It makes the attached cells not suffer from severe nutrient limitation or product inhibition as do those in the entrapment system, thus improving the fermentation efficiency and process stability.

2.2. Covalent bonding

Covalent bonding immobilization is based on covalent bond formation between an activated support and cells in the presence of a binding agent. Compared to physical adsorption, much stronger cell-support binding is formed, which reduces cell loss due to cell detachment from the support. Chemical modification of the support surface is necessary for the specific binding. Some coupling reagents, such as α-amino propyl triethoxy silane, isocyanate, hydroxyethyl acrylate [10, 38, 39], and sodium periodate [40], can be used to activate the support for covalent bonding. Other agents, such as cyanuric chloride [41], can also couple yeasts and bacteria on a cellulose support by a double bond with the hydroxyl groups of the cells and the cellulose. A detailed review of covalent bonding techniques can also be found in the literature [3]. The disadvantage of this method for cell immobilization is the risk of cell damage and loss of activity because the cell membrane is involved in covalent bonding to the support [39]. The toxicity of the chemicals used obviously imposes limitations on the general application of these procedures.

2.3. Entrapment

Entrapment is the most extensively studied method in cell immobilization. It is based on the enclosure of cells within a rigid network. The entrapping itself is tight enough to prevent the release of the cells while still allowing the diffusion of substrates and products. Because it is less likely to cause direct changes in the cell's functions, in principle it is one of the least disruptive methods of cell immobilization. It has been applied widely in biotransformation and fermentation processes producing antibiotics, organic acids, enzymes, and alcohols [38]. Entrapment involves the trapping of microorganisms in the interstices of naturally occurring gel polymers such as agar [42, 43], alginate [44–58], and carrageenan [59–62] or within synthetic polymeric networks formed from monomeric precursors such as acrylamide [63–65], acrylonitrile [63], urethane [66], vinyl alcohol [67–70], and hydroxyethylmethacrylate (HEMA) [71], as well as membrane retention of microorganisms [72] and microencapsulation [73, 74].

2.3.1. Gel entrapment

Entrapment in insoluble calcium alginate beads is recognized as rapid, nontoxic, inexpensive, and versatile and is the most often used method for cell immobilization. Most cell immobilization processes are currently carried out using alginate. Alginate is a family of naturally-occurring unbranched binary copolymers which contain D-mannuronic acid and L-guluronic acids linked by 1,4-glycosidic bonds. Alginate beads are formed by extrusion or emulsification [39]. The extrusion method involves the drop-wise addition of cells suspended in sodium alginate (2–4% w/v) into a calcium chloride (20–100 mM) hardening solution; the emulsification process involves the dispersion of the cell/alginate suspension into an oil phase and gelation then occurs when calcium is added. The factors affecting the gel strength, size, and stability, and subsequently cell activity are: the type of alginate (monomeric composition, block structure, and molecular size), alginate concentration, $CaCl_2$ concentration, cell/alginate ratio, and bead size. [39, 45–47]. Emulsification usually results in smaller diameter beads than the extrusion method. A freeze-dehydration method has been developed to reduce the bead volume and weight by placing the extruded beads at -15 °C for 6–24 h, and the dehydrated beads were successfully used for the production of fructo-oligosaccharides [75].

The main limitation of calcium alginate bead gels is their vulnerability to chelating agents, such as phosphate, lactate, citrate, and EDTA, due to decalcification and the resulting loss of gel strength [39]. The instability of calcium alginate beads and the consequent cell release encountered in filamentous fungi immobilization has become a factor which limits the long-term operation of immobilized cell fermentation [53]. Other ions, such as aluminum, barium, and strontium, can also form strong gels with alginate [76]; however, their toxicity may limit their use in fermentation. Supplementation of the fermentation medium with calcium was applied in order to prevent the dissolving of calcium alginate beads [77]. However, calcium is an important trigger for many metabolic processes and any variation of its concentration in the medium will probably result in altered cellular metabolism.

κ-Carrageenan is another natural polysaccharide which has been proved useful for cell entrapment. It gels in the presence of alkali metal salts, amines, or solvents, and potassium ions are generally used to stabilize the gel. The immobilization procedure is similar to alginate, and it is carried out by adding the warm cell/carrageenan suspension drop-wise (3–4% w/v) to a potassium chloride solution (0.3 M) at room temperature or to an oil phase followed by cooling to room temperature and incubation in potassium chloride solution [39]. Bead size can be controlled by the operating conditions of the gelating process, such as the rate of cooling. Higher mechanical stability of the beads can be obtained by supplementing the polymer with locust bean gum [59, 60, 78] or with hardening treatment using $Al(NO_3)_3$ [79]. Similarly, other natural polymers such as agar have also been employed for cell immobilization. However, the mechanical strength of agar is very weak comparing to alginate or carrageenan, and it may be subject to slow deterioration owing to syneresis [3]. Recently, calcium pectate has been suggested to be a better alternative to calcium alginate due to the enhanced chemical resistance of its matrix [80, 81].

The advantages of natural polymers for entrapment lie in their biocompatibility and acceptability in food applications. However, the synthetic polymers offer higher mechanical strength and better chemical stability over a prolonged period in acidic, alkaline, or saline fermentation broths. Among the synthetic gels, polyacrylamide gel was the first matrix material used for cell immobilization [82]. The monomer used for gel preparation is acrylamide and it is generally crosslinked with N,N-methylene-bis-acrylamide. Ammonium persulfate and N,N,N,N–tetramethylenediamine (TEMED) are added as polymerization reagents. Cell immobilization is carried out by adding these reagents into the cell suspension. The content of the acrylamide and the ratio of cells to acrylamide affect the hardness of the gel formed. The acrylamide to cross-linker ratio determines the pore size of the gel, which significantly affects cell activity and stability, and also the pressure drop when packed in a bioreactor [38]. Polymerization often yields a very stable gel and cells are easily retained inside the gel. The gel is very porous and has high water content. However, due to the toxicity of the reagents used and the radical-dependent polymerization, serious damage to the cell walls and loss of cell viability are often observed [3].

Polyvinyl alcohol (PVA) and HEMA are also widely used for microbial cell immobilization and are less toxic to the cells. These synthetic gels are prepared by irradiation polymerization. N-isopropyl acrylamide (N-IPAAM) can be used to cross-link the PVA/HEMA gel to protect the cells from lysis at 40°C [68]. A new PVA-based immobilization method, LentiKats®, has been developed recently [67, 69, 83]. A cell suspension is mixed into the polymer solution (LentiKat®Liquid) and the gelation is carried out by controlled partial drying at room temperature, resulting in lens-shaped particles containing the entrapped cells. The LentiKat® hydrogel has good mechanical properties, is not biodegradable, and the immobilized cells are not prone to contamination because contaminating cells will not be able to enter the hydrogel and replace the desired biocatalyst.

Comparing to carrier binding, a minor loss of activity of cells is one of the main positive features of the entrapment technique. However, because the nutrients used by the cells as well as the products released from the cells must pass through the gel matrix and the cell wall, mass transfer becomes an important factor to be considered. The nutrient limitation will result in cell death and any enrichment of products may cause inhibition effects and lead to reduced process efficiency. Therefore, diffusion limitation is usually regarded as the most negative feature of this method. To handle and overcome the problems resulting from mass transfer resistances, immobilized cell bioreactors should ensure good mixing characteristics and an optimal mass transfer between gas, liquid, and solid phases.

2.3.2. Membrane retention

Cell immobilization by entrapment in membrane bioreactors is based on the concept of restraining the cells on one side of the membrane while providing nutrients on the other side. Cells can grow and accumulate inside the membrane and biochemical products and metabolic wastes can be easily removed. Microbial cell immobilization in membrane bioreactor has been used in wine production [84] and lactic acid fermentation [85]. Detailed reviews can be found in the literature [86].

Hollow fiber rather than flat sheet membranes have been favored for cell immobilization. It appears attractive in terms of productivity, but high-performance membranes for industrial applications are expensive. Even though high tangential fluid-phase velocity can be assured in most experimental set-ups, it is not possible to prevent membrane fouling due to cell accumulation. The filtration flux would decline with time and bioreactors would gradually lose their productivity. Another severe limitation of this type of immobilization is the supply of gaseous nutrients and the removal of byproduct gases. The narrow flow channels of hollow fibers can be blocked by bubbles, which would result in local regions of nutrient starvation [87].

2.3.3. Microencapsulation

Microencapsulation technology was developed for microbial cell immobilization starting in 1993 to overcome the drawbacks encountered with gel entrapment, such as limited cell loading due to the small void space of polymer matrix and cell leakage [73]. A microcapsule consists of a semi-permeable, spherical, thin, and strong polymer membrane surrounding a liquid core, with a diameter varying from a few microns to 1 mm [39]. Because of the absence of a solid or gelled core and small diameter, mass transfer limitation is reduced in microencapsulation.

The polymer membrane is formed as a result of coacervation or ionic cross-linking between anionic and cationic polysaccharide/polyelectrolytes around a liquid core [39]. There are four techniques used to encapsulate the cells: coacervation, interfacial polymerization, pre-gel dissolving, and liquid droplet forming [73]. Different polymer membranes have been used, including poly-L-lysine/alginate, polyethyleneimine/alginate, calcium alginate, chitosan/terephthaloyl chloride, etc. Wall thickness, pore size, surface charge and mechanical strength can easily be controlled by adjusting the reaction time, concentration, and type of reactant during polymerization of the capsule membrane.

2.4. Modifications of the support surface for cell immobilization

Modifications of the physico-chemical properties of the support surface can be used to promote cell immobilization. Physical methods to increase surface roughness have been applied on PVC and glass beads [88, 89]. The roughness provides shelter for the cells from fluid shear and increases total surface area for immobilization.

Chemical modifications to establish additional interactions, such as ionic and covalent bonds between the cells and the support have also been developed. Glutaraldehyde is a coupling agent usually used for intermolecular cross-linking in enzyme immobilization. In protein chemistry, glutaraldehyde reacts mainly with lysine in the protein structure. Therefore, glutaraldehyde reacts readily with the protein in the lipid bilayer of cell membrane and covalently binds cells [90]. Glutaraldehyde has been successfully applied to immobilization of yeast *Saccharomyces cerevisiae* [90–92], fungus *Aspergillus oryzae* [93], and bacteria including *Escherichia coli* [94], *Propionibacterium acidipropionici* [95] and *Methylosinus trichosporium* [96] on different solid supports, including gelatin, κ-carrageenan, DEAE-cellulose and poly(phenylene oxide).

Hydrophobic and electrostatic interactions play important roles in cell adhesion to the support. Therefore, increased hydrophobicity and increased positive charge of the support surface is favored for cell immobilization. It was shown that adhesion of *Pseudomonas putida* to hydrophobic and positively charged glass beads was substantially greater than to standard beads [89]. Polyethylenimine (PEI) has been used to modify the surface charge of cells and solid supports. PEI is a highly branched polymer, possessing primary, secondary, and tertiary amine groups in a ratio of approximately 1:2:1 [97]. Its high cationic density can increase the positive charge of the cells or support and allows for flocculation of cells and strong cell binding to the support. PEI-coated cotton [98–101], alginate [102], agar [103] and glass [104] have been found to immobilize microbial cells effectively without any loss in activity. PEI and glutaraldehyde are also used together to enhance cell immobilization [36, 103]. The aldehyde groups from glutaraldehyde can cross-link with the amino groups on PEI-coated support and cell surfaces and the cell-support interaction is promoted by chemical bonding. Because of the non-toxicity of both PEI and glutaraldehyde, PEI and/or glutaraldehyde modification will have a good potential for large-scale use in cell immobilization. In addition to PEI, the cationic surface-treating agent chitosan was also used to modify the polymer matrix polypropylene to increase electrostatic interaction between the cells and the support [20].

3. EFFECTS OF CELL IMMOBILIZATION

Immobilization mimics what occurs in nature when cells adhere to and grow on surfaces or within natural structures. Immobilization also provides conditions conducive to cell-to-cell communication. It is natural for cells to modify their pattern of growth and replication as a result of direct contact with a solid surface or with other cells. Immobilization also results in the change of physicochemical properties of the microenvironment, including the presence of ionic charges, reduced water activity, altered osmotic pressure, modified surface tension, and cell confinement [39]. Growth under this situation can induce responses at both biochemistry and genetic levels that cause fundamental changes in the cells. Table 2 lists some of the physiological changes of cells after immobilization.

There have been many studies focusing on the effect of immobilization on fermentation behaviors, including growth rate and cell metabolism [1, 105, 106]. However, contradictory results have been reported. For example, increased growth rate was observed in some immobilized cell systems [107–110], as compared to unchanged [62, 111] or decreased cell growth rate [112–114] found in other examples. Similarly, activation of the cell metabolism with higher specific rates of substrate uptake and product formation was reported in immobilized yeast cells [115–117]. Nevertheless, there are examples where immobilized yeast cells display unchanged or even lower specific productivities [118, 119].

The decrease in cell growth rate and specific productivity implies the mass transfer limitation in immobilized cell systems, which results in lower nutrient and oxygen supplies and an accumulation of inhibitory metabolites, as compared to the suspended cell system. On the other hand, immobilized cells are protected from harsh environmental conditions such as extreme pH and temperature, high shear rate, organic solvents, etc. The mass transfer limitations in this system may exhibit some advantages by reducing the diffusion of an inhibitory substrate. These protective effects from the support contribute to the activation of primary metabolic functions and higher productivity of cells upon immobilization in addition to the high cell density which is usually observed in immobilized cell systems.

Table 2

Effects of cell immobilization on cell physiology

	Physiological responses to immobilization
Cell growth	Increased, static, or decreased growth rate
Cell metabolism	Increased, unchanged, or decreased substrate consumption and product formation rates
	Increase in the specific activity of extracellular or intracellular enzymes
Tolerance to inhibitors	Increased tolerance to toxic compounds
Cellular composition	Alteration in the composition of the cell membrane or cell wall and modified membrane permeability
	High quantities of reserve carbohydrates and structural polysaccharides
Genetics	Increased DNA content and reduced double-stranded RNA content
	Slower total RNA degradation rate
	Different expression of some stress-related genes

Apart from the enhanced specific production rates, it has been reported that specific activities of secreted enzymes from immobilized cells were significantly higher than those of free cells [43]. Differences in the specific activities of some key intracellular enzymes, such as hexokinase [116], alcohol dehydrogenase [117], and acid-forming enzymes [110] have also been highlighted in immobilized cells. The secondary metabolism of immobilized cells is also regulated differently. It was found that immobilized mycelia of *Gibberella fujikuroi* was modulated in a quantitatively different way compared to freely suspended mycelia [49]. It is also reported that nitrogen metabolism was altered in immobilized yeast cells, which resulted in a later onset and lower intensity of cell autolysis in alcohol fermentation [106].

It has been observed that in ethanol [1], acetic acid [107], and butyric acid [110] fermentations, immobilizing cells enhanced their tolerance to toxic products. Doleyres et al. [120] studied the effect of gel immobilization and long-term continuous culturing on the characteristics of lactic acid bacteria and observed that the cells produced from immobilized cell cultures exhibited altered physiologies and had significantly increased tolerances to toxic chemicals such as hydrogen peroxide, simulated gastric and intestinal juices, antibiotics, and nisin as well as physico-chemical stresses such as freeze-drying. Many researchers [110, 121–123] suggested that the increased tolerance was due to a higher level of saturated fatty acids in the cell membrane, which resulted in changes in cell membrane permeability and fluidity [124]. It has been noticed that microorganisms such as *S. cerevisiae* [125] and *E. coli* [126] were able to regulate their membrane lipid composition to increase their tolerance to organic solvents. Because immobilized cells often suffer from a higher concentration of inhibitory products due to their higher productivity and/or mass transfer limitation, it is natural for them to change their membrane properties and increase their resistance to the inhibition in order to survive. Zhu and Yang [110] found that the cell membrane ATPase from immobilized cells was less sensitive to inhibition than the free cells. The environmental conditions in the immobilized systems, such as osmotic pressure [121, 123] and close cell contact [123] have been suggested to be responsible for the increase in tolerance to inhibitors. The different mechanisms of cells against the inhibitors also resulted in morphological differences between alginate-immobilized yeast cells and freely-suspended cells [48]. Immobilized cells had less cell fusion than free cells. Because alginate acted as a cell protector by reducing the attack surface area and decreasing the flow rate towards the entrapped cells, it was not necessary for them to aggregate in order to reduce their contact area with the inhibitor, as the free cells usually did.

Apart from the changes in lipid composition, electron microscopic studies also demonstrated that immobilization induced other changes in cell ultrastructures, such as increased thickness of the cell walls and the surface area of the plasma membrane [106] and modification in the pattern of cell wall mannoproteins [127]. An increase in the structural polysaccharide and storage polysaccharide glycogen was also noticed [115, 128].

Modifications at the genetic level have been reported as a consequence of immobilization. It has been reported that DNA replication in *S. cerevisiae* was affected by immobilization [115]. Immobilized cells had higher DNA content but lower double stranded RNA content than cells in suspension. Another report showed that the total RNA degradation rate of *E. coli* cells immobilized in latex patches was much slower than that of suspended cells over a 16-

day incubation period, which indicated that immobilization stabilized the protein synthesis capacity of the bacteria [129]. Shen et al. [130] studied the expression of several stress-related genes, such as HSP12 and SSA3, in both immobilized and free yeast cells during beer fermentation. They found that cell immobilization exerted certain stresses on the yeast cells, such as nutrient starvation and osmotic stress, and the activated stress responses and nutrient-sensing signaling pathways resulted in different metabolisms, leading to enhanced ester concentrations in the fermentation product. Junter et al. [105] discussed the differences in the global regulation patterns of gene expression in free and immobilized cells in various environmental conditions using a proteomics approach.

4. IMMOBILIZED CELL BIOREACTORS

The choice of bioreactors for immobilized cells is dependent on the immobilization technique, cell metabolism, the shear rate and the mass transfer requirement. The most popular immobilized cell bioreactors employed for various applications include the stirred tank bioreactor, packed-bed bioreactor, fluidized-bed bioreactor, and air-lift bioreactor (Fig. 3).

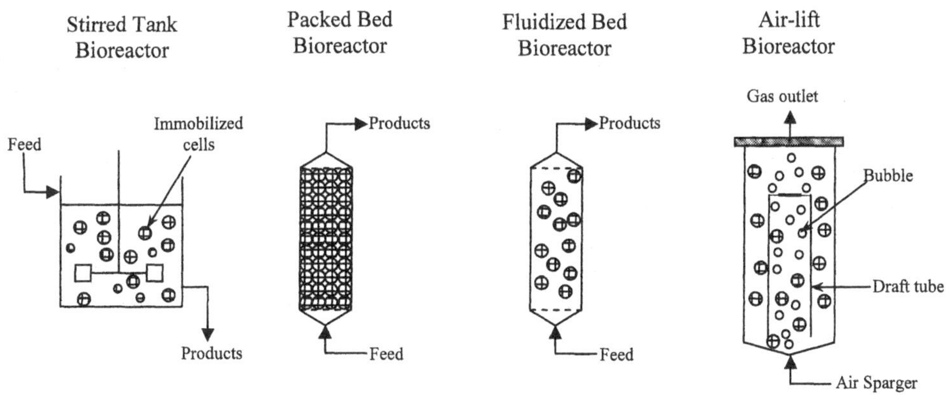

Fig. 3. Various types of immobilized cell bioreactors.

4.1. Continuous stirred tank bioreactor

The most popular bioreactor for continuous fermentation is the continuous stirred tank bioreactor. It provides complete mixing and the fermentation broth is homogenous. For the immobilized cells, high agitation is required to reduce the mass transfer limitation of both substrate and product. However, this may result in high shear forces which can cause damage to both the support and the cells.

Continuous stirred tank bioreactors have been used in ethanol fermentation by calcium alginate immobilized *S. cerevisiae* [54] and propionic fermentation by calcium polygalacturonate immobilized *P. acidipropionici* [131]. A high fermentation stability using a stirred tank bioreactor was demonstrated in the production of lactic starters using carrageenan and locust bean gum immobilized bacteria during a period of 52 days [60].

4.2. Packed bed bioreactor

The packed bed bioreactor consists of a column packed with immobilized cells through which the substrate solution flows. It is operated in the plug-flow mode, with a minimum of back mixing. By selecting an appropriate flow rate, it is possible to achieve a very high reaction rate or even complete conversion of the substrate in a single pass of the column.

Packed bed bioreactors have the advantages of simplicity of operation and low cost. Shear forces are low in packed bed reactors because they are only caused by the fluid-phase velocity. Packed bed bioreactors have been widely used for the production of ethanol [55, 92, 132], acetone-butanol-ethanol [133], lactic acid [20, 134] and enzymes [61, 135] as well as sucrose isomerization [136]. In order to allow gases to flow through, a very low feed flow rate from the reactor top is employed in some packed bed bioreactors called trickle bed reactors [19, 137, 138].

The dimension of the column (length to diameter ratio) can affect the fermentation efficiency greatly due to the effects of gas hold-up and mass transfer [132]. Inhomogeneous concentrations of substrate and products along the packed bed can lead to undesired effects such as inhibition or reduced reaction rates. A high product concentration in the upper portion of the column results in product inhibition whereas cell activities within the lower regions are affected by substrate inhibition. Mass transfer limitations in packed bed typically reduce the reaction rate. Another problem is that a high pressure drop is usually encountered when using particles of relatively small size, which may result in bed compaction. Gas accumulation during the process may induce dead space and reduce the reactor's working volume. Resultant channeling of fluids may affect substrate conversion [39]. Conventional packed bed bioreactors are also difficult to scale-up for commercial applications.

Recently, a novel packed bed bioreactor, the fibrous-bed bioreactor (FBB) has been developed for immobilized cell fermentation. The spiral wound fibrous material was packed loosely inside a column [139]. Its unique packing structure allows for the free flow of gases, liquids, and solids in the reactor bed. A simple *in situ* immobilization of cells was performed by circulating a cell suspension through the FBB. The fibrous-bed bioreactor has been successfully used for several organic acid fermentations, including lactic acid, acetic acid, propionic acid, and butyric acid [32, 33, 107, 140–143], resulting in greatly increased reactor productivity, final product concentration, and product yield. It gave stable, high-rate production of the fermentation product for a long period because of the high density of active cells maintained in the fibrous bed. It can also tolerate a low contaminant level. The most important finding was that the FBB had the ability to quickly adapt to and enrich cultures with a high tolerance to inhibitory fermentation products, resulting in two to three-fold increase in the final product concentration [32, 107, 110]. Cells adapted in the FBB had higher specific growth rates than those of the original culture used to seed the bioreactor. Also, they were less sensitive to inhibitory products than cells in the free form.

4.3. Fluidized bed bioreactor

In the fluidized bed bioreactor, the immobilized cells are maintained in motion by a continuous flow of the feed solution [39]. The fluidized bed bioreactor provides conditions that are intermediate to those of the stirred tank and packed bed bioreactor. Mixing in the fluidized bed is better than packed bed and the shear rate is lower compared to the stirred tank bioreactor. The advantages of this bioreactor compared to stirred tank are lower risk of contamination due to the absence of a shaft seal and higher productivity because of less back-mixing. In contrast to packed bed, it can better accommodate a third phase of gas flow and facilitates multiphase mixing, which reduces the mass transfer limitation and substrate or product inhibition [39].

The fluidized bed bioreactor has been applied in the production of cyclodextrin glucanotransferase by agar immobilized *Bacillus circulans* [43], ethanol production by κ-carrageenan immobilized *Zymomonas mobilis* [144], and acetone-butanol-ethanol fermentation by *Clostridium acetobutylicum* adsorbed onto bone char [145]. The hydrodynamics of fluidized bed bioreactor in wastewater treatment has been well studied using PVA gel immobilized bacteria [146]. The information provided includes the particle density, terminal velocities, drag coefficient, minimum fluidization velocities for beads of various sizes, and bed expansion characteristics. Useful engineering data will be useful for designing fluidized bed bioreactors.

However, fluidized bed bioreactors are subject to unstable bed expansion because of the changes in bead density with time due to cell growth and gas formation, resulting in washout. Axial back mixing is present, resulting in loss of efficiency in the case of product inhibition. Also the fluidized bed bioreactor was shown to exhibit operational and scale-up difficulties [39].

4.4. Air-lift bioreactor

The air-lift bioreactor can be regarded as a special variation of the fluidized bed bioreactor. It usually contains an internal loop gas draft tube inside the column. Gas sparging induces the liquid upflow with the suspended particles in the inner draft tube. Subsequently, gas escapes from the top of the bioreactor and the liquid with the suspended particles is led through the gas-free downcomer [39]. An external loop system may replace the inner draft tube for the recirculation of liquid in some air-lift bioreactors.

Compared to the common fluidized bed bioreactor, the main advantage of the air-lift bioreactor is the improved fluidization characteristics. The particles containing the immobilized cells are more easily fluidized in loop reactor systems and can be kept in suspension due to the circulation of the liquid phase even when working with high liquid-phase and gas-phase velocities. The undesired washout by bed expansion in the fluidized bed bioreactor can be avoided. The gas flow derived mixing in the air-lift bioreactor results in a low shear rate and excellent liquid-solid and gas-liquid mass transfer.

Jovetic et al. [147] carried out the deacylation of antibiotic A40926 in an air-lift bioreactor using *Actinoplanes teichomyceticus* immobilized on calcium alginate beads. The hydrodynamics of the bioreactor, such as information on multiphase flow and distribution of gas and solid phases was studied in ethanol fermentation using alginate immobilized yeast cells [148].

In spite of the promising results obtained on the laboratory and pilot-plant scales with immobilized cells in the new bioreactor configuration, industrial applications have so far been limited to wastewater treatment. One reason for this may be that the installation of new plants and processes requires high costs. In addition, important process parameters can affect the immobilized cell fermentation behavior and the process stability significantly [149]. The physiochemical properties of the support for immobilization, such as size, compressibility and density, are very important when considering the bioreactor design. Immobilized bioreactor configurations must ensure good mixing, minimal mass transfer limitation, good holdback of the cells, and mild operation conditions for shear-sensitive materials. Rational approaches are required for new process design and operations to achieve good process stability as well as a significant reduction of production cost.

5. APPLICATIONS OF IMMOBILIZED CELL TECHNOLOGY

5.1. Alcohols production

Immobilized cell fermentation has been widely used in ethanol production and various methods of immobilization, mostly gel entrapment and adsorption, were tried to immobilize the cells. The best known microorganism for ethanol production is the yeast cell *S. cerevisiae*. Cells were entrapped in calcium alginate beads and continuous ethanol production was carried out using different substrates, such as molasses and forest residual hydrolysate [54, 55, 132, 150, 151]. A high productivity of 60.4 g/L·h was achieved using 1.5% calcium alginate immobilized yeast cells (30-50% w/v) [150]. The effect of alginate concentration on the bead stability in dilute-acid hydrolysate medium was studied and beads containing 3% or 4% alginate were found to be stable in the presence of extra calcium in the medium [131].

Due to the limited substrate range of *S. cerevisiae,* co-immobilization with other microorganisms, such as and *Kluyveromyces fragilis* [152], or enzyme β-galactosidase [52, 153] was used to enhance the ethanol fermentation from other substrates such as lactose. Similarly, *S. pastorianus* was co-immobilized with amylolytic fungus *A. awamori* on cellulose carriers and a simultaneous saccharification and fermentation process was carried out [23]. In addition to alginate, plastic composite support (PCS) was also used as a carrier for *S. cerevisiae* immobilization [30]. Because of the nutrient supply in the PCS blend, it stimulated cell growth and biofilm formation on the support surface. Also, when a low nitrogen-containing ammonium sulfate medium was used for repeated batch fermentation, ethanol production (30 g/L) was 6 times higher in the PCS bioreactor as compared to the bioreactor using only polypropylene support. Cells of *S. cerevisiae* have been efficiently immobilized on loofa sponge in a bubble column bioreactor and this system was scaled up to 50 L for large scale ethanol production [154]. *S. cerevisiae* was also adsorbed on the surface of the mineral Kissiris for continuous ethanol production from molasses [155]. Cell immobilization offers long-term preservation for cells [156]. It was reported that *S. cerevisiae* entrapped in Brazilian chrysotile had an increased life-span of up to three years. In addition to *S. cerevisiae*, another yeast cell, *S. diastaticus* has been immobilized onto beech wood chips for ethanol production in a packed bed bioreactor [157]. The effect of varying particle sizes and pHs was studied, and it was found that pH values in the range 5.0–6.0, and particle size 1.84–1.92 mm had a positive effect on the immobilization process. A high adsorption capacity of 150 mg dry cell mass per gram dry carrier mass was obtained using the 1.84 mm-sized wood chips. With this immobilized system, continuous fermentation was obtained under steady-state conditions for 30 days.

Z. mobilis is a bacterium widely used in ethanol fermentation studies. The polyethylene type of photo-crosslinked resin was used as the carrier for cell immobilization [158]. Cells were mixed with the photo-crosslinkable resin, and polymerization was performed using a light irradiation device. A similar immobilization method was used by Iida et al. [159, 160]. It was found that the molecular structure of the photo-crosslinkable resin could be modulated so as to minimize a disadvantage of this bacterium – poor tolerance to salts in molasses fermentation. Continuous ethanol fermentation was carried out and a high productivity of 60 g/L·h was obtained. In addition, κ-carrageenan was used to immobilize *Z. mobilis* for ethanol production in a synthetic medium containing glucose and xylose [144]. A mutant of *Z. mobilis* was immobilized onto alginate beads and a high ethanol production of 73.5 g/L was achieved in sucrose fermentation [51]. *Z. mobilis* cells were also co-immobilized with the enzyme amyloglucosidase in agar beads for ethanol production from maltose [103]. A mathematic model was developed to study the kinetics of this co-immobilized fermentation system.

Apart from ethanol production, immobilized cell systems have been used in the fermentation of several other alcohols, such as acetone-butanol-ethanol (ABE) and xylitol. Cells of *C. acetobutylicum* were immobilized by adsorption onto bone char and successfully employed in packed-bed and fluidized bed bioreactors for continuous production of ABE from whey permeate [145]. Park et al. [137] used polyester sponge strips for the adsorption of *C. acetobutylicum* cells and utilized them in a trickle bed bioreactor for long term ABE fermentation. A fibrous-bed bioreactor (FBB) was also employed in ABE fermentation using *C. acetobutylicum* [161]. An optimal productivity of 4.6 g/ L·h and a high butanol yield of 0.42 g/g were obtained by using glucose and butyrate as substrates. *C. beijerinckii* has also been used as an ABE producer. After cell adsorption onto clay brick, an immobilized cell biofilm plug-flow bioreactor was established to carry out continuous ABE fermentation [162]. A high reactor productivity of 16.2 g/L·h was achieved with sucrose. In a similar study, production of 6.29 g/L of total ABE was obtained with corn steep liquor as the feed stock [163].

Candida guilliermondii cells were immobilized in calcium alginate beads and used for xylitol production from concentrated sugarcane bagasse hydrolysate [164]. The immobilization conditions were optimized using a statistical design; 20 g/L of sodium alginate, 11 g/L of calcium chloride, and a bead curing time of 24 h were the most appropriate conditions for the immobilization of this yeast cell.

A detailed review of immobilized cell fermentations for alcohols production can be found in the literature [165].

5.2. Organic acids production

Organic acids, such as lactic acid, acetic acid, propionic acid, butyric acid, citric acid, and succinic acid are important microbial products used in a variety of applications in food, pharmaceutical, and chemical industries.

Immobilized cell technology has been used to improve lactic acid fermentation using lactic acid bacteria (LAB), including *Lactobacillus* and *Lactococcus*, or the fungus *Rhizopus oryzae.* The homofermentative LAB *Lactobacillus delbrueckii* subsp. *delbrueckii* was entrapped in calcium alginate gel prepared from 2% sodium alginate, and used for lactic acid production in batch, repeated batch, and continuous fermentation systems [50]. The beads were stable for 14 consecutive batch fermentations without significant activity loss or deformation. A high productivity of 13.92 g/L·h was obtained in continuous immobilized cell fermentation. A mathematical model has been proposed to analyze the mass transfer behavior in lactic acid fermentation using calcium alginate immobilized cells [166]. It was found that the cell density gradient formed in the gel beads was due to the accumulation of inhibitory product lactic acid instead of substrate starvation. To increase the mechanical strength of the gel for cell entrapment, Lamboley et al. [60] used κ-carrageenan and locust bean gum gel beads to immobilize three strains of *Lactococcus*. A high biological and mechanical stability of the beads were demonstrated in the continuous production of mesophilic lactic starters, even with reduced KCl supplementation (0.1 M) or without any KCl addition to the broth medium.

Adsorption was used to immobilize the cells of *Lactobacillus plantarum* [20]. By coupling adsorption with electrostatic interaction, cell adsorption efficiency was enhanced on chitosan-treated polypropylene. A high lactic acid productivity of 7.66 g/L·h was obtained with a yield of 0.97 g/g in a packed-bed bioreactor. Another immobilized upflow packed-bed bioreactor system used in lactic acid production by *Lactobacillus casei* subsp. *casei* demonstrated a four-fold higher volumetric productivity than free cell fermentation [134]. Cells of *Lactobacillus delbrueckii* subsp. *delbrueckii* were adsorbed on activated carbon and simultaneous cell immobilization and lactate adsorption was performed in a repeated batch process [25]. *Lactobacillus brevis* cells were immobilized on delignified cellulosic materials for lactic acid production from whey, and a high operational stability was demonstrated by a series of 10 repeated batch fermentations without any loss in cell activity [22].

Lactic acid was also produced from raw cassava starch using an immobilized co-culture of *Lactococcus lactis* spp. *lactis* and *A. awamori* [167]. *A. awamori* was immobilized directly into cylindrical loofa sponges and the non-flocculating *Lactococcus lactis* was immobilized in sliced loofa sponge with alginate as the polymer support. The simultaneous aerobic starch saccharification and anaerobic lactic acid fermentation was carried out in a circulating loop bioreactor containing both a riser column and a downcomer column to separate different types of immobilized cells.

Fungal fermentation for lactic acid production has been studied using *R. oryzae* immobilized in a rotating fibrous-bed bioreactor (RFBB) [168, 169]. Fungal mycelia were immobilized on a rotating fibrous matrix and the fermentation resulted in a good control of the filamentous morphology. A cell-free fermentation broth was observed, and it improved oxygen transfer and lactic acid production. High lactic acid concentrations of more than 120 g/L were obtained from glucose and starch in repeated batch fermentations.

The immobilization conditions for acetic acid production by *Acetobacter aceti* were studied using three different solid carriers: Siran glass, wood chips, and polyurethane foam [18]. Polyurethane foam was found to be the most successful support for cell adsorption because it allowed a large number of cells immobilized in the shortest time and led to the highest acetification rate. Anaerobic acetic acid production from fructose was studied using *C. formicoaceticum* immobilized in a FBB [32] with a yield of ~1.0 g/g and a final acetate concentration of ~80 g/L. The superior performance of the FBB was attributed to the high cell density of >30 g/L of viable cells immobilized in the fibrous bed, and the lower sensitivity of immobilized cells to acetic acid demonstrated the improved fermentation behaviors of cells after immobilization. The thermophilic heterofermentative bacterium *C. thermolacticum* was entrapped in κ-carrageenan gel for continuous acetic acid production. The maximum productivity obtained was 6.9 g/L·h and a cell concentration of 60 g (dry weight)/L·gel was achieved in the gel at steady state, which was 10–15 times greater than that can be obtained in free-cell fermentations [62]. *C. thermolacticum* cells were also co-immobilized with a homoacetogen *Moorella thermoautotrophica* for acetic acid production from lactose and milk permeate in a FBB [170]. Compared to free cell fermentation, nutrient supplementation was not necessary in the immobilized cell fermentation and plain milk permeate gave a high acetate yield of 0.96 g/g.

Similarly, propionic acid fermentation was carried out using immobilized cells of *P. acidipropionici* in a FBB from whey permeate without nutrient supplementation [142]. A more than ten-fold increase in productivity was obtained compared to conventional batch fermentation using free cells. The highest propionic acid concentration obtained in the FBB fermentation was ~65 g/L [143]. In addition, the FBB offered immobilized cell fermentation with increased resistance to contamination. Non-sterile whey media was used as the feedstock and similar fermentation results were obtained. A mathematical model has been developed on propionic acid fermentation from sweet whey using polygalacturonate entrapped *P. acidipropionici* to study the effect of mass transfer on the fermentation [131].

The FBB was also employed in butyric acid fermentation [33, 171], and a high butyrate production of 57.9 g/L with a high yield of 0.59 g/g xylose and a reactor productivity up to 3.19 g/L·h was achieved. Using cell immobilization in the FBB, a high butyric producer, *C tyrobutyricum,* with enhanced tolerance to acids was obtained [110]. After comparing its growth, fermentation ability, and enzyme kinetics to the wild-type strain used to seed the bioreactor, it was noticed that this immobilization-adapted mutant was physiologically different from the wild type. It demonstrated that immobilization in the FBB was an effective culturing method for in-process strain improvement to achieve mutants with a high tolerance to inhibitory fermentation products.

Immobilized cell fermentations for production of other acids, such as citric acid by *A. niger* [172, 173] and succinic acid by yeast cells [174], have also been studied. Cells of *Saccharomycopsis lipolytica* were immobilized on two polymer carriers based on polyamide and polyacylonitrile, and they exhibited higher resistance to cupric inhibition in citric acid production [63]. Different cell immobilization conditions using PCS blends were evaluated and optimized for succinic acid production from *Actinobacillus succinogenes* [31].

5.3. Antibiotics production

Microbial production of antibiotics by secondary metabolism is one of the key areas in the field of applied microbiology. Because the production of antibiotics is non-growth associated, using cell immobilization to uncouple cell growth and metabolite production is an effective method of improving the process. Different immobilization techniques have been tried for antibiotics production. Cells of *Penicillium chrysogenum* were immobilized in agar beads and used for the production of penicillin in a continuous stirred tank bioreactor [42]. The production phase lasted for 25 days and a 2.5-fold higher productivity was obtained compared to the batch fermentation. Neomycin production was investigated using calcium alginate immobilized *Streptomyces marinensis* [76]. The effects of various immobilization parameters on neomycin production and bead stability were studied, including alginate concentration, different cations and their concentrations, curing time, and bead diameters. Effective neomycin production using repeated batch fermentation was achieved under optimized conditions, and an enhancement of antibiotic productivity by 108% was obtained with immobilized cells as compared to the washed free-cell fermentation. *Streptomyces clavuligerus* cells were immobilized using different materials including sponge, agar, and alginate. Sponge was found to be a better support material than other supports used for immobilization and cephamycin C production [175]. Similarly, the adsorption of *Streptomyces violatus* on sponge tubes yielded better antibiotic production from starch than cells entrapped in alginate beads [176]. The effect of immobilization on antibiotic production was studied using alginate-immobilized *Streptomyces rimosus* [177]. Cell immobilization retarded cell growth rate but increased the duration of growth phase and improved oxytetracycline production. Glass wool with a diameter of 8 μm has also been used for the entrapment of *Amycolatopsis mediterranei* cells for repeated batch production of rifamycin [178].

5.4. Other products

Immobilized cells have also been widely used in other fermentations for production of amino acids such as tryptophan [69] and aspartic acid [94, 179], enzymes such as alkaline protease [11, 45–47], catalase-peroxidase [12], lignin peroxidase [138], β-glucosidase [53], α-amylase [61], cyclodextrin glucanotransferase [43, 70], acid phosphatase [66], and L-glutaminase [135], microbial polysaccharides such as xanthan gum [180, 181], and bacteriocins such as pediocin [182] and nisin [183], as well as biotransformation [8, 48, 57, 65, 68, 71, 184]. These applications share similar performance characteristics as discussed before. A comprehensive review on the applications of immobilized cell fermentation can be found in the literature [38].

6. CONCLUSION

Cell immobilization has now become a mature technology. There is a large potential for the use of immobilized microorganisms in industrial production of fuels and chemicals in continuous bioreactor systems. However, it still requires more engineering studies for process improvements, such as mass transfer, scale-up, and process long-term stability. Future research can also include the closer examination of the impacts of immobilization on cells at both the genomic and proteomic levels as more and more genetically engineered microorganisms are used in the fermentation processes.

REFERENCES

[1] S. Norton and T. D'Amore, Enzyme Microb. Technol., 16 (1994) 365.
[2] R.G. Willaert and G.V. Baron, Introduction. R.G. Willaert, G.V. Baron and L. De Backer (eds.), Immobilized Living Cell Systems: Modelling and Experimental Methods, John Wiley & Sons Ltd., England, 1996.
[3] A. Rosevear, J. Chem. Technol. Biotechnol., 34B (1984) 127.
[4] J.L. Vega, E.C. Clausen and J.L. Gaddy, Enzyme Microb. Technol., 10 (1988) 390.
[5] J. Klein and K.D. Vorlop, Immobilization techniques- cells, M. Moo-Young, C.L. Cooney and A.E. Humphrey (eds.), Comprehensive Biotechnology, Vol. 2, Pergamon Press, NY, 1985.
[6] P. Hasal, A. Cejkova and V. Vojtisek, Enzyme Microb. Technol., 14 (1992) 1007.
[7] P. Monsan, G. Durand and J.M. Navarro, Immobilization of microbial cells by adsorption to solid supports, K. Mosbach (ed.) Methods in Enzymology, Vol. 135, Immobilized Enzymes and Cells, Part B, Academic Press Inc., Orlando, FL, 1987, pp. 307-318.
[8] W.A. Anderson, P. Bay, R.L. Legge and M. Moo-Young, J. Chem. Technol. Biotechnol., 47 (1990) 93.
[9] D. Berk, L.A. Behie, A. Jones, B.H. Lesser and M. Gaucher, Can. J. Chem. Eng., 62 (1984) 120.
[10] P. Brodelius and K. Mosbach, Overview. K. Mosbach (ed.) Methods in Enzymology, Vol. 135, Immobilized Enzymes and Cells, Part B, Academic Press Inc., Orlando, FL, 1987, pp. 173-175.
[11] U. Beshay and A. Moreira, Process Biochem., 38 (2003) 1463
[12] A. Paar, A. Raninger, F. de Sousa, I. Beurer, A. Cavaco-Paulo and G.M. Gubitz, Food Technol. Biotechnol. 41 (2003) 101.
[13] H. Böttcher, U. Soltmann, M. Mertig and W. Pompe, J. Mater. Chem., 14 (2004) 2176.
[14] K. Shimada and K. Shimahara, J. Ferment. Bioeng.,67 (1989) 215.
[15] Y. Zhang, J.F. Kennedy and C.J. Knill, J. Biomater. Sci. Polym. Ed., 7 (1996) 1119.
[16] K. Tapani, P. Soininen-Tengvall, H. Berg, B. Ranta and E. Pajunen, Continuous primary fermentation of beer with immobilised yeast. K.A. Smart (ed.), Brewing Yeast Fermentation Performance (2nd Edition), Blackwell Publishing Ltd., Oxford, UK, 2003, pp. 293-301.
[17] E.N. Kaufman, S.P. Cooper and B.H. Davison, Appl. Microbiol. Biotechnol., 45/46 (1994) 545.
[18] I. de Ory, L.E. Romero and D. Cantero, Process Biochem., 39 (2004) 547.
[19] D. Hekmat, R. Bauer and J. Fricke, Bioprocess Biosyst. Eng., 26 (2003) 109.
[20] S. Krishnan, M.K. Gowthaman, M.C. Misra and N.G. Karanth, J. Chem. Technol. Biotechnol., 76 (2001) 461.
[21] D. Balli, V. Flari, E. Sakellaraki, V. Schoina, M. Iconomopoulou, A. Bekatorou and M. Kanellaki, Process Biochem., 39 (2003) 499.
[22] O. Elezi, Y. Kourkoutas, A.A. Koutinas, M. Kanellaki, E. Bezirtzoglou, Y.A. Barnett and P. Nigam, J. Agri. Food. Chem., 51 (2003) 5285.
[23] N. Fujii, T. Oki, A. Sakurai, S. Suye and M. Sakakibara, J. Ind. Microbiol. Biotechnol., 27 (2001) 52.
[24] A. Sakurai, Y. Nishida, H. Saito and M. Sakakibara, J. Biosci. Bioeng., 90 (2000) 526
[25] C. Chen and L.K. Ju, Appl. Microbiol. Biotechnol., 59 (2002) 170.
[26] H.M. Erhardt and H.J. Rehm, Appl. Microbiol. Biotechnol., 22 (1985) 32.
[27] H.-J. Son, Y.-H. Park and J.-H. Lee, Appl. Biochem. Biotechnol., 112 (2004) 1.
[28] J. Cotton, A. Pometto and J. Gvozdenovic-Jeremic, Appl. Microbiol. Biotechnol., 57 (2001) 626.

[29] A. Demirci, J.C. Cotton, A.L. Pometto III, K.R. Harkins and P.N. Hinz, Biotechnol. Bioeng., 83 (2003) 749.
[30] A. Demirci, A.L. Pometto III and K.L. Ho, J. Ind. Microbiol. Biotechnol., 19 (1997) 299.
[31] S.E. Urbance, A.L. Pometto III, A.A. DiSpirito and A. Demirci, Food Biotechnol., 17 (2003) 53.
[32] Y.L. Huang, K. Mann, J.M. Novak and S.-T. Yang, Biotechnol. Prog., 14 (1998) 800.
[33] Y. Zhu, Z. Wu and S.-T. Yang, Process Biochem., 38 (2002) 657.
[34] J.D. Bryers, Colloids Surf. B: Biointerfaces, 2 (1994) 9.
[35] K. Merritt and Y.H. An, Factors influencing bacterial adhesion. Y.H. An and R.J. Friedman (eds.), Handbook of bacterial adhesion: principles, methods, and applications, Humana Press Inc., Totowa, NJ, 2000, pp. 53-72.
[36] Y. Zhu, Enhanced butyric acid fermentation by Clostridium tyrobutyricum immobilized in a fibrous-bed bioreactor, Ph.D. Dissertation, The Ohio State University, Columbus, OH, 2003.
[37] M.C.M. van Loosdrecht and J.J. Heijnen, Biofilm processes. R.G. Willaert, G.V. Baron, L. DeBacker (eds.), Immobilized Living Cell Systems: Modelling and Experimental Methods, John Wiley & Sons, England, 1996, pp. 255-271.
[38] S.V. Ramakrishna and R.S. Prakasham, Curr. Sci., 77 (1999) 87.
[39] A. Groboillot, D.K. Boadi, D. Poncelet and R.J. Neufeld, Crit. Rev. Biotechnol., 14 (1994) 75.
[40] M. Masek, J. Kas, O. Valentova, K. Demnerova and Z. Vodrazka, Biotech. Lett., 8 (1986) 721.
[41] W.B. Okita, D.B. Bonham and J.L. Gainer, Biotechnol. Bioeng., 27 (1985) 632.
[42] R.A. Swaroopa, A. Jetty and S.V. Ramakrishna, Chem. Biochem. Eng. Quarterly, 17 (2003) 119.
[43] A. Vassileva, N. Burhan, V. Beschkov, D. Spasova, S. Radoevska, V. Ivanova and A. Tonkova, Process Biochem., 38 (2003) 1585.
[44] O. Adachi, S. Tanasupawat, N. Yoshihara, H. Toyama and K. Matsushita., Biosci. Biotechnol. Biochem., 67 (2003) 2124.
[45] K. Adinarayana and P. Ellaiah, J. Sci. Ind. Res., 62 (2003) 589.
[46] U. Beshay, Afr. J. Biotechnol., 2 (2003) 60.
[47] M. Elibol and R. Moreira, Process Biochem., 38 (2003) 1445.
[48] K. Ettayebi, F. Errachidi, L. Jamai, M.A. Tahri-Jouti, K. Sendide and M. Ettayebi, FEMS Microb. Lett., 223 (2003) 215.
[49] I. Garbayo, C. Vilchez, J.E. Nava-Saucedo and J.N. Barbotin., Enzyme Microb. Technol., 33 (2003) 629.
[50] Y. Göksungur and U. Güvenc, J. Chem. Technol. Biotechnol., 74 (1999) 131.
[51] T.R. Kannan, G. Sangiliyandi and P. Gunasekaran, Enzyme Microb. Technol., 22 (1998) 179.
[52] M. Lewandowska, W. Bednarski, M. Adamczak, J. Tomasik and D. Narwojsz, Pol. J. Food Nutr. Sci., 11 (2002) 19.
[53] B.K. McCabe, C. Kuek, G.L.R. Gordon and M.W. Phillips., J. Ind. Microbiol. Biotechnol., 30 (2003) 205.
[54] R. Millati, C. Niklasson and M.J. Taherzadeh, Process Biochem., 38 (2002) 515.
[55] G. Najafpour, H. Younesi and K.S.K. Ismail, Bioresour. Technol., 92 (2004) 251.
[56] S. Silva, F. Ramon-Portugal, P. Andrade, S. Abreu, M.F. Texeira and P. Strehaiano, Am. J. Enol. Vitic., 54 (2003) 50.
[57] H.R. Velankar and M.R. Heble, Electronic J. Biotechnol., 6 (2003) 90.
[58] S.-H. Yoon, R. Mukerjea and J.F. Robyt, Carbohydr. Res., 338 (2003) 1127.
[59] P. Audet, C. Paquin and C. Lacroix, Appl. Microbiol. Biotechnol., 29 (1988) 11.
[60] L. Lamboley, C. Lacroix, J.M. Artignan, C.P. Champagne and J.C. Vuillemard, Biotechnol. Prog., 15 (1999) 646.
[61] C.S. Lin, S.C. Hong, C.Y. Cheng and T.L. Chen, J. Chem. Tech. Biotechnol., 69 (1997) 433.
[62] G. Wang and D.L. Wang, Appl. Biochem. Biotechnol., 8 (1983) 491.
[63] T. Godjevargova and D. Denchev, Biotechnol. Biotechnol. Equip., 16 (2002) 150.

[64] K.A. Lusta, N.G. Starostina, N.B. Gorkina, B.A. Fikhte, V.L. Lozinskii, E.S. Vainerman and S.V. Rogozhin, Biotechnol. Bioeng., 222 (1989) 199.
[65] A.K. Mandwal, C.K.M. Tripathi, P.D. Trivedi, A.K. Joshi, S.C. Agarwal and V. Bihari, Biotechnol. Lett., 26 (2004) 217.
[66] P. Aleksieva, B. Tchorbanov, L. Michailova and L. Nacheva, World J. Microbiol. Biotechnol., 19 (2003) 247.
[67] A. Durieux, X. Nicolay and J.-P. Simon, Landbauforschung Voelkenrode, Sonderheft, 241 (2002), Practical Aspects of Encapsulation Technologies, 131.
[68] A.A. El-Hadi, Process Biochem., 38 (2003) 1659.
[69] B. Klassen, U. Seja, R. Faurie, T. Scheper and K.-D. Vorlop, Landbauforschung Voelkenrode, Sonderheft, 241 (2002), Practical Aspects of Encapsulation Technologies, 141.
[70] R.F. Martins, F.M. Plieva, A. Santos and R. Hatti-Kaul, Biotechnol. Lett., 25 (2003) 1537.
[71] N.Z. Adham, A.A. El-Hady and N. Naim, Acta Pharm. Turcica, 45 (2003) 41.
[72] T. Ikeda, K. Kato, M. Maeda, H. Tatsumi, K. Kano and K. Matsushita, J. Electroanal. Chem., 430 (1997) 197.
[73] J.K. Park and H.N. Chang, Biotechnol. Adv., 18 (2000) 303.
[74] M. Yoshida, E. Mardriyati, D. Tenokuchi, Y. Uemura, Y. Kawano and Y. Hatate, J. Appl. Polym. Sci., 89 (2003) 1966.
[75] H.T. Shin, K.M. Park, K.H. Kang, D.J. Oh, S.W. Lee, S.Y. Baig and J.H. Lee, Lett. Appl. Microbiol., 38 (2004) 176.
[76] B. Srinivasulu, K. Adinarayana and P. Ellaiah, AAPS PharmSciTech., 4 (2003) E57.
[77] C.P. Champagne, F. Girard and N. Rodriguez, Int. Dairy J., 3 (1993) 257.
[78] M.J. Miles, V.J. Morris and V. Carroll, Macromolecules, 17 (1987) 2443.
[79] R. Chamy, M.J. Nunez and J.M. Lema, Enzyme Microb. Technol., 12 (1990) 749.
[80] J. Nahálka, B. Wu, J. Shao, P. Gemeiner and P.G. Wang, Biotechnol. Appl. Biochem., 40 (2004) 101.
[81] M. Navrátil, Z. Dömény, E. Šturdík, D. Šmogrovičová and P. Gemeiner, Biotechnol. Appl. Biochem., 35 (2002) 133.
[82] B. Mattiasson, Immobilization methods. B. Mattiasson (ed.), Immobilized cells and organelles, Vol. 1, CRC Press, Inc., Boca Raton, FL, 1983, pp. 3.
[83] A.G. de Segura, M. Alcalde, F.J. Plou, M. Remaud-simeon, P. Monsan and A. Ballesteros, Biocatal. Biotransform., 21 (2003) 325.
[84] M. Takaya, N. Matsumoto and H. Yanase, J. Biosci. Bioeng., 93 (2002) 240.
[85] S. Tejayadi and M. Cheryan, Appl. Microbiol. Biotechnol., 43 (1995) 242.
[86] T.B.V. Roy, H.W. Blanch and C.R. Wilke, Trends Biotechnol., 1 (1983) 135.
[87] M.A. Mazid, Bio/Technology, 11 (1993) 690.
[88] G.H. Copp and K.J. Kennedy, J. Ferment. Technol. 61 (1983) 333.
[89] A. Gjaltema, N. van der Marel, M.C.M. van Loosdrecht and J.J.Heijnen, Biotechnol. Bioeng., 55 (1997) 880.
[90] O.C. Sitton and J.I. Gaddy, Biotechnol. Bioeng., 22 (1980) 1735.
[91] V. Jirku, J. Turkova, B. Veruovic and V. Kubanek, Biotechnol. Lett., 2 (1980) 451.
[92] J.L. Vega, E.C. Clausen and J.L. Gaddy, Ann. N. Y. Acad. Sci., 506 (1987) 208.
[93] Y.J. Yuan, S.H. Wang, Z.X. Song and R.C. Gao, Biotechnol. Appl. Biochem., 35 (2002) 107.
[94] T. Sato, Y. Nishida, T. Tosa and I. Chibata, Biochim. Biophys. Acta., 570 (1979) 179.
[95] G. Najafpour, J. Sci. Islamic Repub. Iran, 1 (1990) 172.
[96] P.K. Mehta, S. Mishra and T.K. Ghose, Biotechnol. Bioeng., 37 (1991) 551.
[97] R. Bahulekar, N.R. Ayyangar and S. Ponrathnam, Enzyme Microb. Technol., 13 (1991) 858.
[98] S.F. D'Souza and N. Kamath, Appl. Microbiol. Biotechnol., 29 (1988) 136.
[99] S.F. D'Souza and J.S. Melo, Enzyme Microb. Technol., 13 (1991) 508.
[100] N. Kamath and S.F. D'Souza, Enzyme Microb. Technol., 13 (1991) 935.
[101] J.S. Melo and S.F. D'Souza, World J. Microbiol. Biotechnol., 15 (1999) 23.

[102] J.J. Joung, C. Akin and G.P. Royer, Appl. Biochem. Biotechnol., 14 (1987) 259.
[103] L. Hellendoorn, S.P. Ottengraf, J.C. van den Heuvel, J.A. Pennings, V.A. dos Santos and R.H. Wijffels, Biotechnol. Bioeng., 63 (1999) 694.
[104] S.F. D'Souza, J.S. Melo, A. Deshpande and G.B. Nadkarni, Biotechnol. Lett., 8 (1986) 643.
[105] G.A. Junter, L. Coquet, S. Vilain and T. Jouenne, Enzyme Microb. Technol., 31 (2002) 201.
[106] N.N. Martynenko and I.M. Gracheva, Appl. Biochem. Microbiol., 39 (2003) 439.
[107] Y. Huang and S.-T. Yang, Biotechnol. Bioeng., 60 (1998) 499.
[108] G.T. Chun and S.N. Agathos, Biotechnol. Bioeng., 37 (1991) 256.
[109] S. Pashova, L. Slokoska, P. Sheremetska, E. Krumova, L. Vasileva and M. Angelova, Process Biochem., 35 (1999) 15.
[110] Y. Zhu and S.-T. Yang, Biotechnol. Prog., 19 (2003) 365.
[111] P.S. Lau, N.F.Y. Tam and Y.S. Wong, Bioresource Technol., 63 (1998) 115.
[112] E. Inanc, J.E. Miller and D. DiBiasio, Biotechnol. Bioeng., 51 (1996) 697.
[113] G.S. Shreve and T.M. Vogel, Biotechnol. Bioeng., 41 (1993) 370.
[114] T. Yamaguchi, M. Ishida and T. Suzuki, Process Biochem., 34 (1999) 167.
[115] P.M. Doran and J.E. Bailey, Biotechnol. Bioeng., 28 (1986) 73.
[116] B. Hilge-Rotmann and H.J. Rehm, Appl. Microbiol. Biotechnol., 33 (1990) 54.
[117] M.F.M. van Iersel, E. Brouwer-Post, F.M. Rombouts and T. Abee, Enzyme Microb. Technol., 26 (2000) 602.
[118] L. Jamai, K. Sendide, K. Ettayebi, F. Errachidi, O. Hamdouni-Alami, M.A. Tahri-Jouti, T. McDermott and M. Ettayebi, FEMS Microbiology Lett., 204 (2001) 375.
[119] H.-Y. Shen, N. Moonjai, K.J. Verstrepen and F.R. Delvaux, J. Am. Soc. Brew. Chem., 61 (2003) 79.
[120] Y. Doleyres, I. Fliss and C. Lacroix, J. Appl. Microbiol., 97 (2004) 527.
[121] B. Hilge-Rotmann and H.J. Rehm, Appl. Microbiol. Biotechnol., 34 (1991) 502.
[122] V. Jirku, J. Ind. Microbiol. Biotechnol., 22 (1999) 147.
[123] H. Keweloh, H.J. Heipieper and H.J. Rehm, Appl. Microbiol. Biotechnol., 31 (1989) 383.
[124] T. Kawamoto, T. Kanda and A. Tanaka, Appl. Microbiol. Biotechnol., 55 (2001) 476.
[125] G.P. Casey and W.E. Ingledew, Crit. Rev. Microbiol., 13 (1986) 219.
[126] L.O. Ingram, J. Bacterial., 125 (1976) 670.
[127] P. Parascandola, E. de Alteriis, R. Sentandreu and J. Zueco, FEMS Microbiol. Lett., 150 (1997) 121.
[128] M.D. Fumi, G. Trioli, M.G. Colombi and O. Colagrande, Am. J. Enol. Vitic., 39 (1988) 267.
[129] O.K. Lyngberg, V. Thiagarajan, D.J. Stemke, J.L. Schottel,. L.E. Scriven and M.C. Flickinger, Biotechnol. Bioeng., 62 (1999) 44.
[130] H.-Y. Shen, N. Moonjai, K.J. Verstrepen, F. Delvaux and F.R. Delvaux, J. Am. Soc. Brew. Chem., 61 (2003) 175.
[131] C. Coronado, J.E. Botello and F. Herrera, Biotechnol. Prog., 17 (2001) 669.
[132] B.S. Yadav, U. Rani, S.S. Dhamija, P. Nigam and D. Singh, J. Basic Microbiol., 36 (1996) 205.
[133] A Friedl, N. Qureshi and I.S. Maddox, Biotechnol. Bioeng., 38 (1991) 518.
[134] J.M. Bruno-Barcena, A.L. Ragout, P.R. Cordoba and F. Sineriz, Appl. Microbiol. Biotechnol., 51 (1999) 316.
[135] S.R. Kumar and M. Chandrasekaran, Process Biochem., 38 (2003) 1431.
[136] S.-J. Ahn, J.-H. Yoo, H.-C. Lee, S.-Y. Kim, B.-S. Noh, J.-H. Kim and J.-K. Lee, Biotechnol. Lett., 25 (2003) 1179.
[137] C.-H. Park, M.R. Okos and P.C. Wankat, Biotechnol. Bioeng., 36 (1990) 207.
[138] B. Ruggeri and G. Sassi, Process Biochem., 38 (2003) 1669.
[139] S.-T. Yang, Extractive fermentation using convoluted fibrous bed bioreactor, US Patent No. 5 563 069 (1996).
[140] Y.L. Huang, Z. Wu, L.Zhang, C.M. Cheung and S.-T. Yang, Bioresour. Technol., 82 (2002) 51.
[141] E.M. Silva and S.-T. Yang, J. Biotechnol., 41 (1995) 59.

[142] S.-T. Yang, H. Zhu, Y. Li and G. Hong, Biotechnol. Bioeng., 43 (1994) 1124.
[143] S.-T.Yang, Y. Huang and G. Hong, Biotechnol. Bioeng., 45 (1995) 379.
[144] M.S. Krishnan, M. Blanco, C.K. Shattuck, N.P. Nghiem and B.H. Davison, Appl. Biochem Biotechnol., 84–86 (2000), 525.
[145] N. Qureshi and I.S. Maddox, J. Ferment. Bioeng., 80 (1995) 185.
[146] J.-Y. Wu, K.-C. Chen, C.-T. Chen and S.-C.J. Hwang, Biotechnol. Bioeng., 83 (2003) 583.
[147] S. Jovetic, L. de Bresser, J. Tramper and F. Marinelli, Enzyme Microb. Technol., 32 (2003) 546.
[148] J. Klein, A.A. Vicente and J.A. Teixeira, J. Chem. Technol. Biotechnol., 78 (2003) 935.
[149] A. Freeman and M.D. Lilly, Enzyme Microb. Technol., 23 (1998) 335.
[150] A. Sheoran, B.S. Yadav, P. Nigam and D. Singh, J. Basic. Microbiol., 38 (1998) 123.
[151] M.J. Taherzadeh, R. Millati and C. Niklasson, Appl. Biochem. Biotechnol., 95 (2001) 45.
[152] M. Lewandowska and W. Bednarski, Mededelingen - Faculteit Landbouwkundige en Toegepaste Biologische Wetenschappen (Universiteit Gent) 68 (2003) 497.
[153] M. Lewandowska, W. Bednarski and S. Kulesza, Mededelingen - Faculteit Landbouwkundige en Toegepaste Biologische Wetenschappen (Universiteit Gent), 68 (2003) 493.
[154] J.C. Ogbonna, H. Mashima and H. Tanaka, Bioresour Technol., 76 (2001) 1.
[155] A.A. Koutinas, C. Gourdoupis, C. Psarianos, A. Kaliafas and M. Kanellaki, Appl. Biochem. Biotechnol., 30 (1991) 203.
[156] F. Cassiola, H.S. Santos and I. Joekes, Colloids Surf. B: Biointerfaces, 30 (2003) 283.
[157] R. Razmovski and D. Pejin, Folia Microbiol., 41 (1996) 201.
[158] T. Yamada, M.A. Fatigati and M Zhang, Appl. Biochem. Biotechnol., 98-100 (2002) 899.
[159] T. Iida, H. Izumida, Y. Akagi and M. Sakamoto, J. Ferment. Bioeng., 75 (1993) 32.
[160] T. Iida, M. Sakamoto, H. Izumida and Y. Akagi, J. Ferment. Bioeng., 75 (1993) 28.
[161] W.-C. Huang, D.E. Ramey and S.-T. Yang, Appl. Biochem. Biotechnol., 113-116 (2004) 887.
[162] J. Lienhardt, J. Schripsema, N. Qureshi and H.P. Blaschek, Appl. Biochem. Biotechnol., 99 (2002) 591.
[163] N. Qureshi, P. Karcher, M. Cotta and H.P. Blaschek, Appl. Biochem. Biotechnol., 114 (2004) 713.
[164] W. Carvalho, S.S. Silva, A. Converti, M. Vitolo, M.G. Felipe, I.C. Roberto, M.B. Silva and I.M. Mancilha, Appl. Biochem. Biotechnol., 98-100 (2002) 489.
[165] A.M. Lother and M. Oetterer , Rev. Microbiol., 26 (1995) 151.
[166] H. Wang, M. Seki and S. Furusaki, Biotechnol. Prog., 11 (1995) 558.
[167] N.D. Roble, J.C. Ogbonna, H. Tanaka, Biotechnol. Lett., 25 (2003) 1093.
[168] A. Tay and S.-T. Yang, Biotechnol. Bioeng., 80 (2002) 1.
[169] N. Thongchul and S.-T. Yang, Controlling filamentous fungal morphology by immobilization on a rotating fibrous matrix to enhance oxygen transfer and L(+)-lactic acid production by *Rhizopus oryzae*. B.C. Saha (ed.), ACS Symposium Series 862, Fermentation Biotechnology, Oxford University Press, New York, 2003, pp. 36-51.
[170] M. Talabardon, J.P. Schwitzguebel, P. Peringer and S.-T. Yang, Biotechnol. Prog., 16 (2000) 1008.
[171] Y. Zhu and S.-T. Yang, J. Biotechnol., 110 (2004) 143.
[172] B.H. Chung and H.N. Chang, Biotechnol. Bioeng., 32 (1988) 205.
[173] N. Fujii, K. Yasuda and M. Sakakibara, J. Ferment. Bioeng., 78 (1994) 389.
[174] S. Shindo, H. Sahara and S. Koshino, Biotechnol. Lett., 15 (1993) 51.
[175] V. Sridevi and P. Sridhar, Indian J. Exp. Biol., 37 (1999) 274.
[176] M.Y. el-Naggar, M.A. Hassan, W.Y. Said and S.A. el-Aassar, J. Gen. Appl. Microbiol., 49 (2003) 235.
[177] S.S. Yang and C.Y. Yueh, J. Microbiol. Immunol. Infect., 2001 Dec;34(4):235.
[178] M.R. Abu-Shady, A.I. el-Diwany, M.A. Farid and H.A. el-Enshasy, J. Basic Microbiol., 35 (1995) 279.

[179] H. Seko, S. Takeuchi, K. Yajima, M. Senuma and T. Tosa, Ann. N. Y. Acad. Sci., 589 (1990) 540.
[180] S.T. Yang, Y.M. Lo and D.B. Min, Biotechnol. Prog., 12 (1996) 630.
[181] S.T. Yang, Y.M. Lo and D. Chattopadhyay, Biotechnol. Prog., 14 (1998) 259.
[182] H.Y. Cho, A.E. Yousef and S.T. Yang, Appl. Microbiol. Biotechnol., 45 (1996) 589.
[183] X. Liu, Y.-K. Chung, S.T. Yang and A.E. Yousef, Process Biochem., 40 (2004) 13.
[184] S.B. Mahato and S. Garai, Steroids, 62 (1997) 332.

Bioprocessing for Value-Added Products from Renewable Resources
Shang-Tian Yang (Editor)

Chapter 15. Water-in-Oil Cultivation Technology for Viscous Xanthan Gum Fermentation

Lu-Kwang Ju

Department of Chemical Engineering, The University of Akron, Akron, OH 44325-3906

1. INTRODUCTION

Biopolymers are an important source of new polymeric materials. Their production by the microbial fermentation processes, however, suffers from poor mixing, heat transfer, and mass transfer caused by high broth viscosity. Water-in-oil (W/O) cultivation technology has been developed to enhance the performance of highly viscous fermentation processes. In this technology, the aqueous broth is dispersed as fine droplets in a continuous phase of organic solvent (oil). As the biopolymers produced are confined within the dispersed droplets, the overall system viscosity can be kept manageable. Using xanthan fermentation as an example, this new cultivation technology has been studied in both experiments and model simulations. Its feasibility has been demonstrated and the effects of operating factors such as agitation, aeration, aqueous droplet size, and various oil-phase properties have been examined. More recently, "smart," pH-sensitive polymers, with a PMAA backbone as well as MPEG and dodecyl side chains, have been synthesized and characterized. At the fermentation pH (~7), the polymers function as surfactants, promoting the formation of very fine W/O emulsions. The much improved oxygen supply to cells inside the small, aqueous droplets should allow the use of higher cell concentrations to further elevate the xanthan productivity. In addition, the polymers lose surface-activity at acidic pHs (<5.5). The dispersion can be relatively easily phase-separated by acid addition at the end of fermentation for product recovery.

2. XANTHAN FERMENTATIONS

The rheological properties of culture broths strongly affect the fermentation performance, especially if aerobic microorganisms are employed. Very high broth viscosity is encountered in some fermentation systems for the following reasons: the addition of polymeric or solid substrates, as in the fermentations of cellulosic materials [1, 2]; the use of filamentous microorganisms, as in fungal fermentations [3–6]; and the secretion of highly viscous products, particularly biopolymers [7–10]. The high broth viscosity presents serious problems to mixing, heat transfer, and oxygen supply. These problems, in turn, limit the production capacities and efficiencies of the fermentation processes. For example, the volumetric oxygen transfer coefficient, k_La, in penicillin fermentations has been shown to decrease as the broth viscosity increases with cell growth [5].

The viscosity problem in biopolymer-producing fermentations is generally far more serious. Biopolymers produced by microorganisms are rapidly emerging as important sources of new polymeric materials. Among them, extracellular polysaccharides have found a broad range of applications in food, pharmaceutical, and other industries [11–14]. The relatively high production costs of these biopolymers have limited their competitiveness in many potential markets dominated by other plant-derived polysaccharides and synthetic polymers. Economic analysis has shown that the high production costs are often related to the recovery costs, which can be substantially reduced if higher biopolymer concentrations are obtained in the fermentation [15, 16].

Xanthan gum is among the most widely used extracellular polysaccharides [17]. Xanthan is an anionic polysaccharide produced industrially by fermentation using the aerobic bacterium *Xanthomonas campestris*. The xanthan polymer is composed of repeated pentasaccharide units, including two glucopyranose rings and a trisaccharide side chain of mannose-glucuronic acid-mannose. The side chain is attached to the alternate glucose residues in the cellulose-like backbone [18, 19]. Xanthan has a large molecular weight on the order of several millions and a flexible rod-like conformation, which make it powerful in generating high viscosity. This remarkable property makes xanthan uniquely useful; however, it poses major problems to its large-scale production in aerobic fermentation processes. The broth viscosity increases dramatically as the product accumulates along the fermentation. The apparent viscosity of 1% (w/v) xanthan gum solution is higher than 50 poise at the standard shear rate of 1 s^{-1} [20, 21]. In addition, the broth is highly pseudoplastic with yield stress ranging from 2–15 poise, for xanthan concentrations varying from 10 to 50 kg m^{-3} [22]. The rheological property causes the well-known "caverning" phenomenon in typical stir-tank fermentors: the region around the impeller is well mixed while the surrounding fluid is stagnant or nearly so [23]. Consequently, xanthan production with conventional fermentation technology is limited to a final product concentration of about 50 kg m^{-3} and a volumetric productivity of about 0.5 kg m^{-3} h^{-1} [24, 25]. High broth viscosity has long been recognized as the limiting factor in the process.

3. NEW PROCESS DEVELOPMENT

Recognizing the process limitation, various research groups have tried different methods for improving xanthan gum fermentation performance [9, 14, 25–32]. Some of their results are summarized in Table 1. Three important examples are: production in centrifugal fibrous-bed bioreactors [25, 32], the use of immobilized cells in stir-tank fermentors [26, 31], and the development of water-in-oil (W/O) cultivation technology [9, 29, 30, 33–35].

Yang and coworkers [25, 32] studied xanthan production in centrifugal fibrous-bed bioreactors. The cells were immobilized in a rotating, cylindrical fibrous matrix by natural attachment to the fiber surfaces. The circulating medium was introduced along the central axis of the cylinder and forced through the matrix radially by the centrifugal force generated from the rotation. The forced radial flow promoted intimate contact between the medium and the immobilized cells. The centrifugal force also helped to extract the produced xanthan polymer from the fibrous matrix, and produced a relatively cell-free xanthan broth. Also, a higher cell concentration (~7 kg m^{-3}) was achieved. The viability of the immobilized cells was, however, about 40% lower than that of free, suspended cells, leading to a lower specific productivity (0.05–0.075 kg (kg cells) $^{-1}$ h^{-1}, as given in Table 1). The low specific productivity might be an indication of oxygen limitation to cells immobilized in the centrifugal fibrous-bed bioreactors. The system complexities as well as the power consumption required for rotating the fibrous bed and recirculating the medium are some practical issues to be further addressed in the process economics of this technology.

Table 1
Comparison of results obtained with different fermentor designs and technologies as well as from model simulations of W/O fermentation

Fermentor/Technology	Max. Xn (kg m^{-3})	Vol. Productivity (kg m^{-3} h^{-1})	Specific Productivity (kg (kg cells) $^{-1}$ h^{-1})**	Reference
Stirred Tank	22	0.3 – 0.45	0.2 – 0.25	[65]
Bubble Column	25	0.17 – 0.35	0.022 – 0.11	[65]
Air-Lift Fermentor	15	< 0.15	0.02 – 0.05	[66]
Centrifugal Fibrous Bed Fermentor[c]	35	0.7 – 1.0	0.05 – 0.075	[25]
W/O Fermentation	65 - 75	0.5*	0.05	[9]
Model Simulated W/O Fermentation	200 - 300	1.15 – 1.80*	0.07 – 0.08	[33]

* Values based on aqueous-phase volume, roughly half of the total volume.
** The large values (≥ 0.2) for stirred tank were under no oxygen limitation; all others reflected various extents of oxygen limitation in the fermentors/technologies used.

Also using immobilized cells, Robinson and Wang [26, 31] studied xanthan production by immobilized cells of *X. campestris* in small porous Celite particles suspended in shake flasks and stir-tank fermentors. Their rationale was to retain the produced xanthan inside the Celite particles, taking advantage of the extremely slow diffusion of large polymer molecules, so that the viscosity of the medium outside the porous support would not be significantly affected. The approach was successful, to certain extent. For example, the xanthan concentration in the system reached 50 kg m^{-3} within the particles after 140 h, with an additional 20 kg m^{-3} in the liquid medium. In the control with free cells, the xanthan concentration was only 38 kg m^{-3} after 280 h. The study revealed a dilemma in choosing the size of the porous support. With larger particles, the produced xanthan was better retained but oxygen supply by inward diffusion to cells in the inner core of the particles was more severely limited. Accordingly, cell concentration and xanthan productivity achievable was restricted. On the other hand, smaller support allowed better oxygen supply throughout the particles but

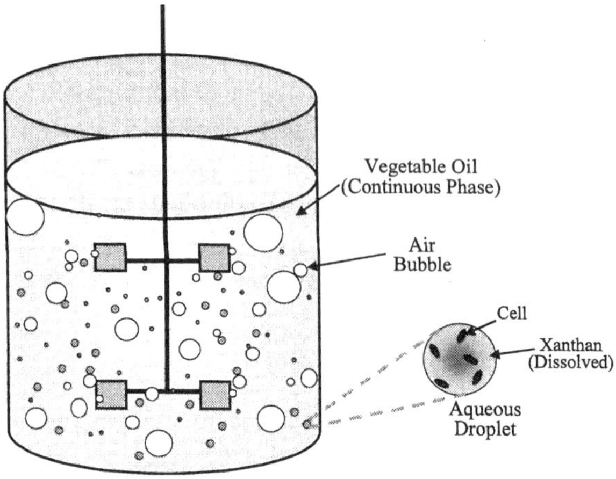

Fig. 1. Schematic diagram of the W/O cultivation system for xanthan production.

poorer cell and xanthan retainment. The xanthan concentration in the suspending medium increased rapidly, after an initial delay period. The system viscosity was high, similar to that in the conventional free-cell fermentation.

The water-in-oil cultivation technology addresses the high viscosity problem of biopolymer fermentations with a similar rationale of retaining cells and produced biopolymers in dispersed droplets/particles. The technology, however, does not suffer from any of the problems of cell and xanthan retainment, and has the potential for using really fine droplets for much improved oxygen supply. The development of W/O cultivation technology is the focus of this work and is described in more detail in the following sections.

4. WATER-IN-OIL CULTIVATION TECHNOLOGY

4.1. Principle

A schematic diagram of the W/O cultivation system is shown in Fig. 1. In this technology, the aqueous culture broth is dispersed as fine droplets in a continuous phase of organic solvent (oil) at an early stage of fermentation when the broth viscosity is still low. The xanthan production and the associated broth-thickening are thus confined within the dispersed aqueous droplets. Consequently, the overall dispersion viscosity can be kept manageable although the viscosity within the droplets increases dramatically during fermentation. The technology has the potential to enhance the performance of viscous fermentations, especially those for production of highly viscous biopolymers such as xanthan gum.

W/O cultivation technology involves complex multiple-phase systems containing gas, oil, an aqueous medium, and particulate cells. The volumetric productivity (Q_p) depends on three parameters: the specific productivity of the microorganisms (q_p), the active cell concentration in the aqueous phase (x), and the aqueous-phase volume fraction (ϕ_w), i.e.,

$$Q_p = q_p \, x \, \phi_w \tag{1}$$

where Q_p is in (kg of xanthan) (m^{-3} of dispersion) h^{-1}, q_p is in (kg of xanthan) (kg^{-1} of dry cells) h^{-1}, x is in (kg of dry cells) (m^{-3} of aqueous phase), and ϕ_w is in (m^3 of aqueous phase) (m^{-3} of dispersion).

Xanthan synthesis by *X. campestris* is partly growth associated [36, 37]. For maximal xanthan production, the fermentation process is generally designed to sustain a reasonable period of xanthan production at the stationary phase before the cells die from the oxygen limitation caused by the increasingly higher viscosity. The maximum cell concentration (x_{max}) is typically determined by the nitrogen source(s) in the medium [37]. Although high x_{max} is desirable for maximum productivity, x_{max} must be designed according to the oxygen supply capability of the cultivation system. In the W/O culture system, the design of x_{max} is expected to depend strongly on the sizes of the aqueous droplets and the oxygen transfer characteristics. The specific productivity, q_p, has also been shown to depend on the oxygen availability to cells [37, 38]. It drops sharply at the onset of oxygen limitation, as DO (dissolved oxygen concentration) reaches zero [39]. At DO = 0, q_p further decreases linearly with a decreasing oxygen transfer rate [39]. In addition, the oxygen transfer rate affects the product quality: the molecular weight of the xanthan produced decreases with the decreasing oxygen transfer rate [39].

Consequently, developing W/O cultivation technology requires careful consideration of the following primary factors: the size of the aqueous droplets, which largely dictates the maximum cell concentration employable for production; the volume fraction of the aqueous phase, where microbial activities actually occur; and the density, viscosity, and oxygen solubility of the oil chosen, which affect the oxygen transfer characteristics and the power consumption required for agitation. In addition, the potential difficulty encountered during phase separation to harvest the product-bearing aqueous phase needs to be addressed. These factors are examined in more detail below.

4.2. Process characteristics

In the multiple-phase W/O culture system, oxygen molecules transfer from gas bubbles to the continuous oil phase then to the aqueous droplets and are consumed by microbial cells in the droplets. With the assumptions of pseudo-steady state and no convection within the highly viscous aqueous droplets, the schematic profile of oxygen partial pressure (p) is shown in Fig. 2. Accordingly, the oxygen transfer can be described by the following equation:

$$(k_La)_{g/o}\, H_o\, (p_g - p_o) = (k_La)_{o/w}\, H_o\, (p_o - p_R) = D_w\, H_w\, a_w \left(\frac{dp}{dr}\right)_{r=R} = \phi_w\, x\, q_{O_2} \tag{2}$$

$(k_La)_{g/o}$ and $(k_La)_{o/w}$ are the volumetric oxygen transfer coefficients associated with the oil films surrounding the gas bubbles and the aqueous droplets, respectively. H_o and H_w are the Henry's law constants of oxygen, defined here as the ratios of dissolved concentration to partial pressure, in oil and aqueous phases. p_g and p_o are the oxygen partial pressures in the gas and bulk oil phase, and p_R is that at the surface of the aqueous droplets ($r = R$). D_w is the oxygen diffusion coefficient in the aqueous phase. q_{O_2} is the specific oxygen uptake rate of the microorganisms.

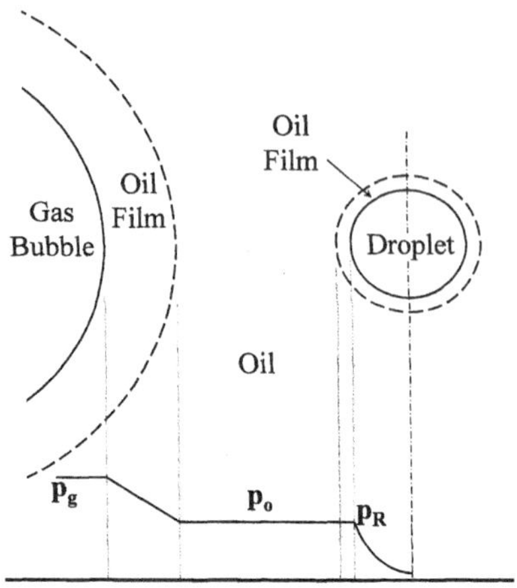

Fig. 2. Schematic diagram of oxygen partial pressure profile in G-O-W multiple-phase systems.

While gas-liquid-solid systems have been under active investigation [40–42], studies on gas-liquid-liquid systems are scarce and mainly in oil-in-water (O/W) systems [43–46]. Studies were therefore performed, in a 6.6×10^{-3} m^3 round-bottomed glass fermentor (BioFlo IIc, New Brunswick Scientific, Edison, NJ), with gassed dispersions of aqueous xanthan solutions in either *n*-hexadecane or vegetable oil, to develop correlations that describe how the droplet size and oxygen transfer characteristics of the W/O xanthan fermentation would be affected by the operating parameters [34]. Various combinations of operating conditions were evaluated: agitation speed (N) – 400, 600, and 775 rpm; aeration rate (G/V) – 0.25, 0.5, and 0.875 vvm; aqueous-phase volume fraction (ϕ_w) – 0.2, 0.3, 0.4 and 0.5; and aqueous-phase xanthan concentration (Xn) – 10, 20, and 40 kg m^{-3}. The results of the study indicated that $(k_La)_{g/o}$ was much smaller than $(k_La)_{o/w}$, thus, $p_R \approx p_o$ and the rate-limiting step was oxygen transfer across the oil films surrounding gas bubbles. In addition, the following correlations were developed for the power input of agitation (P_g, in W), droplet diameter (d_p, in μm), and $(k_La)_{g/o}H_o$ (in kgmol m^{-3} s^{-1} atm^{-1}):

$$P_g = K_1 N^{3.6} G^{-0.27} \phi_w^{-0.23} \tag{3}$$

$$d_p = K_2 (P_g/V)^{-0.23} (G/V)^{-0.04} \phi_w^{0.22} Xn^{0.31} \tag{4}$$

$$(k_La)_{g/o}H_o = K_3 (P_g/V)^{0.36} v_S^{0.59} \phi_w^{0.11} Xn^{-0.26} \tag{5}$$

where v_s is the superficial gas velocity and the dimensional constants K_1, K_2 and K_3 have the numerical values of 7.8×10^{-10}, 29, and 4.1×10^{-4}, respectively. This work was the first for the three-phase G-L-L systems. Nonetheless, the obtained dependencies associated with N, P_g, and G are well supported by the literature on simpler systems, i.e., P_g for two-phase G-L systems [47], d_p correlation in non-aerated L-L dispersions [48], and the k_La correlation in coalescing aqueous systems [49].

4.3. Model simulations

The above process characteristics have been further integrated with the kinetic behaviors of *X. campestris* to simulate the performance of W/O xanthan fermentations [33].

4.3.1. Kinetic behaviors of X. campestris

As mentioned earlier, xanthan synthesis is partly growth-associated. For optimal productivity, fermentation is performed under substrate (commonly the N-source) limitation to regulate the maximum cell concentration [36]. More than half of the product is formed after growth has ceased. Accordingly, cell growth can be described by a Monod-type dependency on N-source concentration (S_N). With additional considerations for the inherent cell death (k_d) and the death caused by oxygen deficiency, the change of cell concentration (x) can be described by the following equation:

$$\frac{dx}{dt} = \left[\mu_{\max} \frac{S_N}{(S_N + K_{S_N})} - k_d - f_d \left(1 - \frac{OUR}{Q_{O_2}} \right) \right] x \tag{6}$$

OUR is the actual oxygen uptake rate while Q_{O_2} is the theoretical oxygen demand of the cells for full aerobic metabolism. Both are based on the volume of the aqueous phase.

The N-source concentration can be calculated accordingly, assuming a constant yield coefficient, Y_{X/S_N}:

$$\frac{dS_N}{dt} = -\frac{x}{Y_{X/S_N}} \mu_{\max} \frac{S_N}{S_N + K_{S_N}} \tag{7}$$

The decrease in glucose (e.g., carbon source) concentration (S_C) is attributed to cell assimilation, maintenance and xanthan production, i.e.,

$$\frac{dS_C}{dt} = -\left[\frac{\mu_{\max}}{Y_{X/S_C}} \cdot \frac{S_N}{(S_N + K_{S_N})} + q_{S_C/X} \frac{S_C}{(S_C + K_{S_C})} \right] x - \frac{1}{Y_{P/S_C}} \left(\frac{dP}{dt} \right) \tag{8}$$

where a Monod-type dependence of glucose consumption is assumed for cell maintenance.

The product (xanthan) formation is described by the Leudeking-Piret relation [50, 51]:

$$\frac{dP}{dt} = \left[A\mu_{\max} \cdot \frac{S_N}{(S_N + K_{S_N})} + B \cdot \frac{S_C}{(S_C + K_{S_C})} \right] x \tag{9}$$

Monod-type dependence is again assumed for the non-growth-associated xanthan production. The coefficient B has been found to be constant when the oxygen availability to cells is not limited [37].

The theoretical oxygen demand (Q_{O_2}) required for fully aerobic metabolisms has been modeled as:

$$Q_{O_2} = \left[\frac{\mu_{\max}}{Y_{X/O_2}} \cdot \frac{S_N}{(S_N + K_{S_N})} + q_{O_2/X} \cdot \frac{S_C}{(S_C + K_{S_C})} \right] x \tag{10}$$

The first term accounts for the demand for cell growth, the second term for the combined cell maintenance and xanthan production. The values for the characteristic parameters used in these equations are summarized in Table 2.

Table 2
Kinetic parameters and diffusion coefficients used in the models [33]

Kinetic Parameters	
K_{S_N} (kg m^{-3})	$5x10^{-4}$
K_{Sc} (kg m^{-3})	1.9
K^*_{Sc} (kg m^{-3})	0.1
$q_{O_2/X}$ [(kg O_2) (kg dry cells)$^{-1}$ h^{-1}]	0.078
B [(kg xanthan) (kg dry cells)$^{-1}$ h^{-1}]	0.107
$q_{Sc/X}$ [(kg glucose) (kg dry cells)$^{-1}$ h^{-1}]	0.069
$Y_{P/Sc}$ [(kg xanthan) (kg glucose)$^{-1}$]	0.86
A [(kg xanthan) (kg dry cells)$^{-1}$]	0.97
Y_{X/O_2} [(kg dry cells) (kg O_2)$^{-1}$]	1.69
$Y_{X/Sc}$ [(kg dry cells) (kg glucose)$^{-1}$]	0.85
Y_{X/S_N} [(kg dry cells) (kg N-source)$^{-1}$]	8.0
μ_{max} (h^{-1})	0.084
k_d (h^{-1})	$3.6x10^{-3}$
f_d (h^{-1})	$6.5x10^{-4}$
Aqueous Phase Diffusion Coefficients (m^2 s^{-1})*	
D_C	$7x10^{-10}$
D_N	$2x10^{-9}$
D_P	$1x10^{-14}$
D_w	$1.5x10^{-9}$

*D_C, D_N, D_P and D_W are diffusion coefficients of C-source (glucose), N-source, product (xanthan), and O_2 in aqueous phase. They are used in the equations describing the transport of these components inside aqueous droplets. These equations are described in detail elsewhere [33].

4.3.2. Model development

For modeling the dynamic behaviors of the W/O xanthan fermentation processes, radial concentration profiles of cells, C- and N-sources, xanthan, and DO have to be determined at every time step. The "local" metabolic activities at that instant can then be calculated according to the afore-mentioned process and kinetic characteristics, i.e., Eqs. (3) - (10). All local metabolic activities are then integrated throughout the average droplet size to obtain the overall aqueous-phase properties at that time step. Ideally, the dynamic frequencies of droplet breakup and coalescence also need to be known at different times of the fermentation process. While oxygen is assumed to be transferred within the droplet only by diffusion, droplet breakup and coalescence can mix and redistribute aqueous-phase components (cells, substrates and product). Breakup and coalescence at high frequencies may lead to a more or less uniform distribution of these components over all of the droplets. On the other hand, at very low frequencies the redistribution of these components is minimal, and concentration gradients may develop inside the droplets.

Xanthan fermentation is characterized by large changes in rheological properties during the course of fermentation. These changes drastically affect droplet breakup and coalescence. At the beginning, aqueous-phase viscosity is low, and frequencies of droplet breakup and coalescence are high. As the fermentation proceeds, the frequencies decrease and the droplet size increases as a result of the increasing aqueous-phase viscosity. This causes the redistribution of the aqueous-phase components to diminish. It is, however, impossible to simulate a fermentation going through such a transition without knowing when and how the transition occurs. W/O xanthan fermentation was thus modeled for two extreme cases [33]. Model 1 assumed uniform redistribution of the aqueous-phase components by frequent droplet breakup and coalescence. Model 2 simulated the condition of viscous aqueous phase with minimal droplet breakup and component redistribution. The two models gave similar qualitative profiles but quite different quantitative results in some cases. Although the exact transition between the two modeled conditions is unknown, the real fermentation should proceed within the bounds set by the two models. Detailed descriptions of the model equations as well as numerical algorithms and procedures for the two models are available elsewhere [33].

4.3.3. Simulation results

Fig. 3 shows the general profiles of cell growth, substrate (glucose and N-source) consumption and product formation simulated for a fed-batch W/O xanthan fermentation. The specific system shown here has an initial N-source concentration (S_{N0}) of 4.5 kg m^{-3} and an aqueous-phase volume fraction (ϕ_w) of 0.6. In the initial stage without oxygen limitation, there is no difference in the various concentration profiles obtained with the two different models. This is correct because under this condition the microbial metabolism is the same throughout the droplet, leading to uniform radial concentrations of all aqueous-phase components (i.e., cells, nutrients and product). It does not matter whether these components are frequently redistributed or not, which is the fundamental difference between the two models.

The simulation results are, however, different once oxygen limitation occurs. In Model 1, cell growth and other metabolic activities inside the oxygen-depleted core are assumed to stop but the cells do not immediately die because they are redistributed to oxygen available zones by frequent droplet breakup and coalescence. The average cell death caused by oxygen limitation is therefore similar to that of conventional all-aqueous systems and is accounted for by the term $f_d \cdot (1 - OUR/Q_{O2})$ in Eq. (6). On the other hand, cells in the oxygen-depleted core are assumed to die immediately in Model 2 because of the negligible redistribution. This is clearly reflected in the different peak cell concentrations obtained: 33.8 kg m^{-3} in Model 1 and only 19.5 kg m^{-3} in Model 2 (Fig. 3). Consequently, the xanthan production rates also differ: after 200 h, 298 kg m^{-3} of xanthan are produced in Model 1 and 201 kg m^{-3} in Model 2. Real process performance can be expected to lie somewhere between the two extreme cases. The above simulation results have taken into account the effects of increasing aqueous-phase viscosity during fermentation, which in turn causes the droplet size to increase and the oxygen transfer coefficient to decrease.

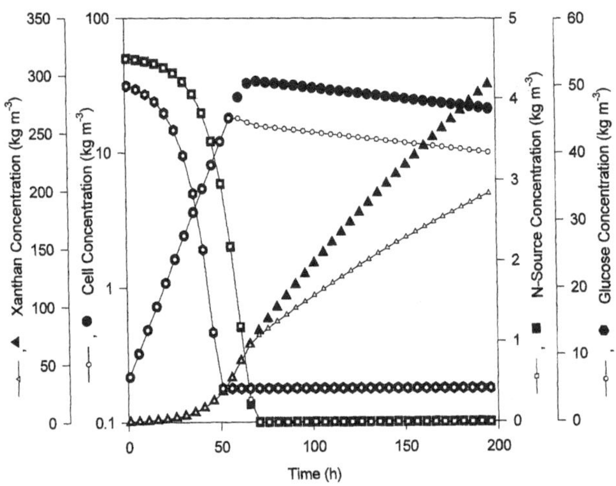

Fig. 3. Simulated profiles of cell, glucose, N-source, and xanthan concentrations ($\phi_w = 0.6$, $S_{N0} = 4.5$ kg m^{-3}) in fed-batch fermentations.

The model simulation has also allowed the effects of various factors such as S_{N0}, ϕ_w, agitation speed, and aeration rate. For example, Fig. 4 is a three-dimensional plot illustrating how S_{N0} affects aqueous-phase xanthan concentrations (Xn) at 200 h for the fermentation of $\phi_w = 0.3$. The corresponding volumetric xanthan productivities (Q_P), based on the total volume of both W and O phases, are shown in Fig. 5. According to Model 1, an increase in S_{N0} (within the simulated range of 0.5 – 7 kg m^{-3}) is always beneficial for both Xn and Q_P although the marginal effect diminishes at high S_{N0}, i.e., ≥ 6 kg m^{-3} in this case. While similar qualitative trends are obtained with Model 2, Xn and Q_P start to decrease slightly with increasing S_{N0} at high concentrations, i.e., $S_{N0} > 4.5$ kg m^{-3}. The latter reflects the rapid expansion of the oxygen-depleted core caused by the overgrowth of cells in the outer shell at these high S_{N0}.

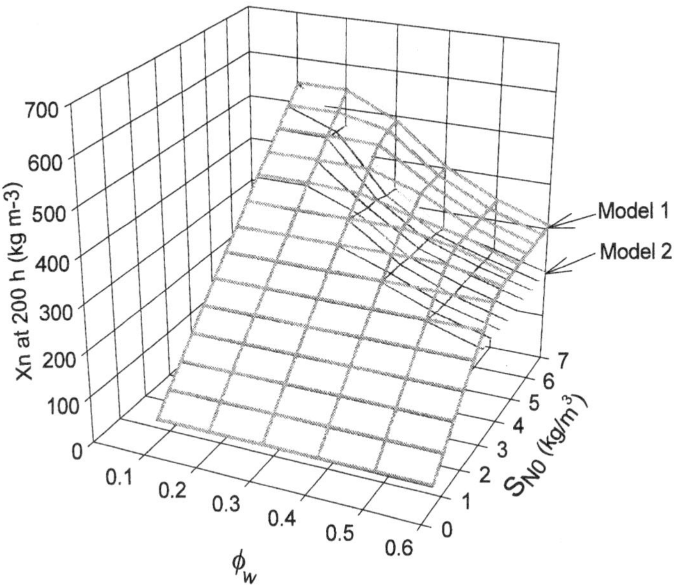

Fig. 4. Effect of S_{N0} and volume fraction (ϕ_w) on xanthan concentration (at 200 h)

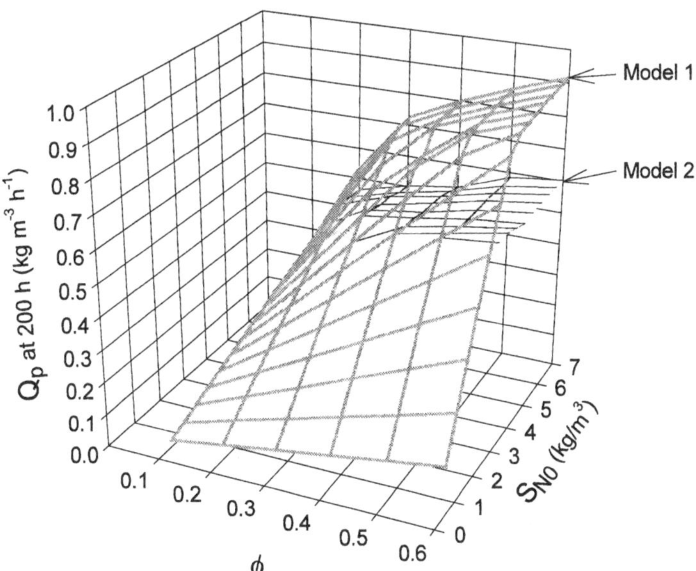

Fig. 5. Effect of S_{N0} and volume fraction (ϕ_w) on volumetric xanthan productivity

The model simulation has indicated the potential of W/O fermentation to produce far higher aqueous-phase xanthan concentrations ($Xn > 200$ kg m^{-3}) and volumetric productivity ($Q_p >$ 0.8 kg m^{-3} h^{-1}) than conventional fermentation (i.e., $Xn \sim 50$ kg m^{-3} and $Q_p \sim 0.5$ kg m^{-3} h^{-1}).

Table 3
Evaluation of different oil phases in W/O fermentation technology

Organic Solvent	*n*-Hexadecane	Vegetable Oil	PFC
	Properties		
Cost	Moderately expensive ($15K m^{-3})	Inexpensive ($1.6K m^{-3})	Expensive ($180K m^{-3})
Oxygen Solubility (% v/v O_2)	22[a]	N/A	37[b]
Oxygen Diffusivity ($\times 10^9$ m^2 s^{-1})	1.8[a]	N/A	4.8[c]
Density ($\times 10^{-3}$ kg m^{-3})	0.77	0.93	2.02[b]
Viscosity (cp)	3.0	58.0	16.2
	W/O Fermentation Performance		
Maximum Xn (kg m^{-3})	65-75 (at 150 h)[d]	>100 (at 300 h)[e]	~ 50 (at 200 h)[e]
Xanthan Yield[f]	~0.5-0.55	~0.67	~0.72
Droplet Size (μm)	~ 500	<120	400-450
Phase-Inversion Volume Fraction	(O/W to W/O with recovered oil)	0.53 (W/O to O/W)	0.56 (W/O to O/W)
Downstream Recovery and Separation	Centrifugation required. Reused oil extracted cell lipids.	Centrifugation required. Reusable, natural surfactants present.	Centrifugation not required. Readily reusable.
Vol. Productivity (kg m^{-3} h^{-1})	~0.5	0.58	0.42
Specific productivity (g Xn) (g^{-1} cells) h^{-1}	~0.05	0.037	0.046

a. Ju and Ho [67]; b. Air Products [68]; c. Ju et al. [69].
d. Xanthan production continuing at termination without apparent limitation.
e. Fermentation terminated due to phase inversion.
f. Yield reported for the period of stationary phase production.

4.4. Effects of different oil phases

In order to be used as the oil phase in W/O cultivation technology, the organic solvent must meet certain basic requirements. The solvent should be practically mutually insoluble with water. It cannot have inhibitory or toxic effects on cells. The solvent also should not dissolve or extract xanthan from the aqueous broth; this keeps its viscosity low and minimizes product loss. Beyond these basic requirements, physical properties such as density, viscosity, and oxygen solubility need to be considered for their effects on the size of aqueous droplets, oxygen transfer characteristics, and the power consumption required for agitation. In addition, the presence of surface-active substances in the oil phase may significantly affect the process performance. On one hand, the surfactants may be beneficial in promoting the formation of fine aqueous droplets minimizing the resistance to radial oxygen transfer (diffusion) within the droplets. On the other hand, at the end of fermentation, the aqueous phase needs to be separated from the oil and removed from the fermentor for downstream product recovery, and the oil phase reused for the next batch. Surfactants may make phase separation more difficult, and may have undesirable effects on product purification and final applications (such as in food).

Three organic solvents have so far been studied. They are *n*-hexadecane, vegetable oil, and perfluorocarbon (PFC; Multifluor APF-215, Air Products) [9, 29]. Some pertinent properties of these oil phases are given in Table 3. The profiles of cell growth, glucose consumption, and xanthan production obtained in W/O fermentation using vegetable oil as the oil phase are shown in Fig. 6 as an example [29]. For comparison, the profiles observed in a control, all-aqueous fermentation conducted under similar operating conditions are shown in Fig. 7 [9]. In the control, the cell concentration reached about 5 kg m^{-3} at 35 h, limited by available N-source in the medium (i.e., 1.5 kg m^{-3} $(NH_4)_2HPO_4$ plus 4 kg m^{-3} peptone). Glucose consumption continued into the stationary phase, while xanthan production stopped at about 60 h, with a concentration of 26 kg m^{-3}. As glucose was still present then, the arrested xanthan synthesis was attributed to oxygen limitation associated with high broth viscosity.

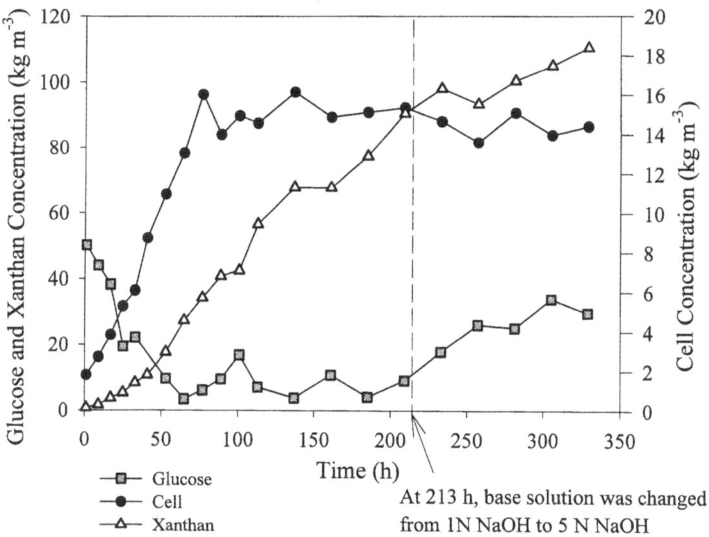

Fig. 6. Concentration profiles of cells, glucose, and xanthan in a W/O fermentation with vegetable oil as the oil phase.

According to the simulation results, the optimal performance of W/O xanthan fermentation in vegetable oil was reached with the medium having an initial N-source concentration of about 4.5 kg m^{-3} of N. The fermentation shown in Fig. 6 was therefore carried out in fresh medium supplemented with 7.15 kg m^{-3} of NH_4NO_3. A much higher cell concentration, 16 kg m^{-3}, was accordingly achieved. In addition, starting from 64 h, glucose and peptone were added at approximately 30 g and 2 g per day, respectively, to sustain continuous xanthan production during the stationary phase. Glucose concentration increased gradually in the later stage but was always within the reasonable range of 3.5–35 kg m^{-3}. The final xanthan concentration obtained was 120 kg m^{-3}.

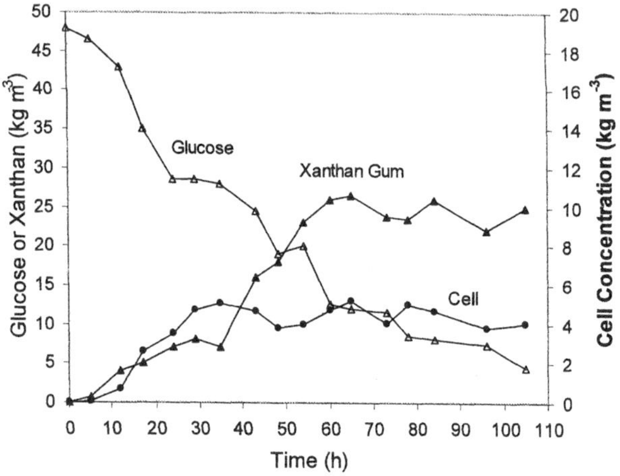

Fig. 7. Concentration profiles of cells, glucose, and xanthan in an all-aqueous control fermentation.

The fermentation was terminated by the occurrence of a phase inversion, from W/O to O/W dispersion. The phase inversion was caused primarily by the continuous increase of the aqueous-phase volume fraction during the fermentation, from 0.30 to 0.54 at the point of phase inversion. The increase in the aqueous-phase volume fraction resulted from the addition of glucose/peptone solution and base (1 N NaOH) for pH control. Accompanying the phase inversion, the highly viscous aqueous broth became the continuous phase. The overall dispersion viscosity increased dramatically, resulting in oxygen limitation. The pH began to rise, indicating cell deterioration and death. The undesirable phase inversion also occurred in the system with PFC as the oil phase.

A comparison of the three oil phases, and their applicability to W/O cultivation technology is summarized in Table 3. Each has its advantages and disadvantages.

For PFC as the oil phase, the main advantages include the following: PFC has a high oxygen solubility and diffusivity, leading to enhanced oxygen transfer rates; being chemically inert, PFC extracted negligible quantities of medium and cell materials, and was shown to be readily recoverable and reusable; downstream processing for phase separation and product recovery is easier due to the high density and chemical inertness of PFC. Accompanying these advantages are some disadvantages as described below: high power consumption due to high density ($\propto \rho^{0.9}$), leads to increased operating costs; the phase-inversion occurs at low aqueous-phase volume fractions (~0.56); large droplet sizes, 400–450 μm in diameter, limits the oxygen supply to cells in the inner core of the droplets; the material is expensive, at about $200 per liter.

The use of vegetable oil as the oil phase also has some potential advantages over the other oils studied. The advantages are: easy availability and low cost (~$1.6 per liter); nontoxicity and edibility; the use of natural surfactants in the promotion of W/O dispersion; and the formation of finer emulsions with droplet sizes smaller than 120 μm, enabling higher xanthan productivity and concentrations (up to about 110 kg m^{-3}). Nonetheless, phase inversion also occurred at relatively low aqueous-phase volume fractions (~0.53) despite the presence of natural surfactants in the oil. More seriously, downstream processing may be more complicated if the complete removal of vegetable oil is required because of the more stable emulsions formed with vegetable oil.

At the current stage of W/O fermentation technology, vegetable oil appears to be the most suitable oil phase among those investigated. Its low cost, smaller droplets in emulsion, and higher xanthan concentrations confirm the economic benefits attainable by W/O fermentation technology.

5. PH-SENSITIVE SURFACTANTS FOR WATER-IN-OIL CULTIVATION

5.1. Principle

The previous observations indicated that it is desirable to develop smart surfactants which allow for reversible emulsification and de-emulsification (phase separation) with changes in simple operational factors (such as temperature, pH, or ionic strength). Ideally, these surfactants will generate very fine aqueous droplets under fermentation conditions so that more cells can be supported by the enhanced oxygen supply and yield very high xanthan productivity. At the end of the fermentation, the operation conditions change, causing the surfactants to lose activity and allow ready phase separation so that the product-bearing aqueous phase can be harvested and the oil phase, containing the retained surfactants, can be reused in the next batch of fermentation.

With pH being the easiest factor to adjust (by the simple addition of an acid or base), it is desirable to develop pH-sensitive surfactants for the above purposes. Because most microorganisms can tolerate acidic conditions much better than basic conditions, the final pH is preferably adjusted to a low value (e.g., 3–5) so that a small fraction of the aqueous broth can be retained in the fermentor as "seeds" for the next batch of fermentation. The desired surfactants should therefore have good surface-activity at pH ~7 in order to promote the formation of W/O dispersions, and lose the activity at pH <5.

Terpolymers formed by methacrylic acid (MAA), methoxy poly(ethylene glycol) methacrylate (MPEGMA), and lauryl methacrylate (C_{12}, LMA) represent one such potential surfactant system [30, 35]. As shown in Fig. 8, these polymers are expected to have a PMAA backbone with MPEG and dodecyl side chains. Their pH-sensitivity results from the ability to form complexes between a segment of the PMAA backbone and the MPEG side chain [52–58]. The complexation occurs under acidic conditions (pH <pKa of PMAA), due to extensive H-bonding between the protonated (undissociated) carboxylic acid groups of PMAA and the other oxygens of MPEG. At neutral pH (and higher), the complexation dissolves as the H-bonds break due to the proton dissociation of the carboxylic acid groups. The terpolymers are therefore surface-active, with PMAA and MPEG as the hydrophilic moiety and the dodecyl chains as the hydrophobic moiety. Ideally, at low pHs, complete complexation between PMAA and MPEG may occur if the terpolymers have a 1:1 molar ratio between the MAA and EG repeat units. The complexation occupies the hydrophilic functional

groups of both PMAA and MPEG and, thus, renders the entire terpolymer molecule hydrophobic and surface-inactive. The water-insoluble hydrophobic terpolymers also tend to partition into the oil phase and are retained in the fermentor for reuse in the next batch of fermentation without appreciable polymer loss [30].

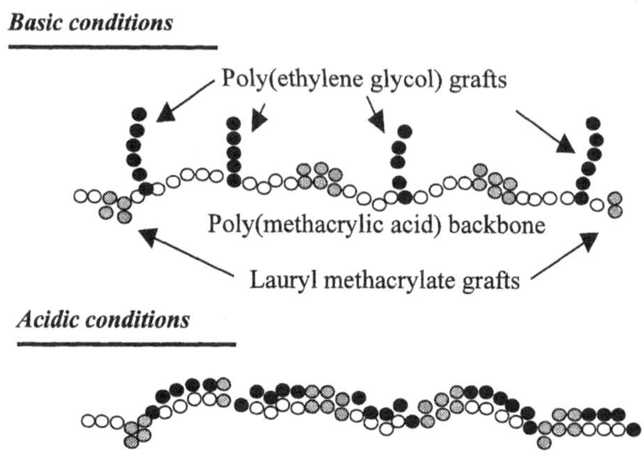

Fig. 8. Schematic depiction of conformations of the terpolymer containing stoichiometric MAA:EG repeat unit ratio under basic and acidic conditions.

5.2. Synthesis

Terpolymers were synthesized by free radical polymerization at 70°C under a nitrogen (N_2) atmosphere with various ratios of the three monomers MAA, MPEGMA, and LMA [30, 35]. The macromonomer, MPEGMA, used had an average molecular weight of 1,000. At different concentrations, hydrogen peroxide (H_2O_2) and azobisisobutyronitrile (AIBN) were evaluated as initiators. Tetrahydrofuran (THF) or a 50/50 water-ethanol mixture was used as the reaction solvent. The study results (some of which were later presented) showed that the terpolymer best for the intended application had the molar ratio of MAA:MPEGMA:LMA = 1:0.04:0.76 (equivalent to the weight ratio of 1:0.52:2.24), which corresponded to a 1:1 molar ratio between the MAA and EG repeat units, to allow more complete complexation between the PMAA backbone and the MPEG grafts at low pHs. The most suitable synthesis condition was with 0.45% of AIBN (wt of AIBN to total wt of initial monomers) as the initiator, THF as the reaction solvent, and a reaction time of 10 h. For example, the time profiles of monomer conversion are compared in Fig. 9 for polymerization with different initiators. The more parallel conversions of all monomers in the system with 0.45% AIBN ensured better control of the desired terpolymer composition.

After 5–10 h of reaction, the product was dialyzed in water for 3 days and then in isopropanol for 3 days. (LMA is practically insoluble in water. The second dialysis in isopropanol was necessary for its complete removal.) The dialysis membrane had a molecular weight cut-off (MWCO) of 3,500 Daltons, chosen to allow all the remaining reactants to pass through while retaining the vast majority, if not all, of the products. The dialyzed polymer was dried under 0.22-μm filtered air at room temperature for at least 1 week, ground to a fine powder, and then kept at room temperature in a vacuumed desiccator for later studies.

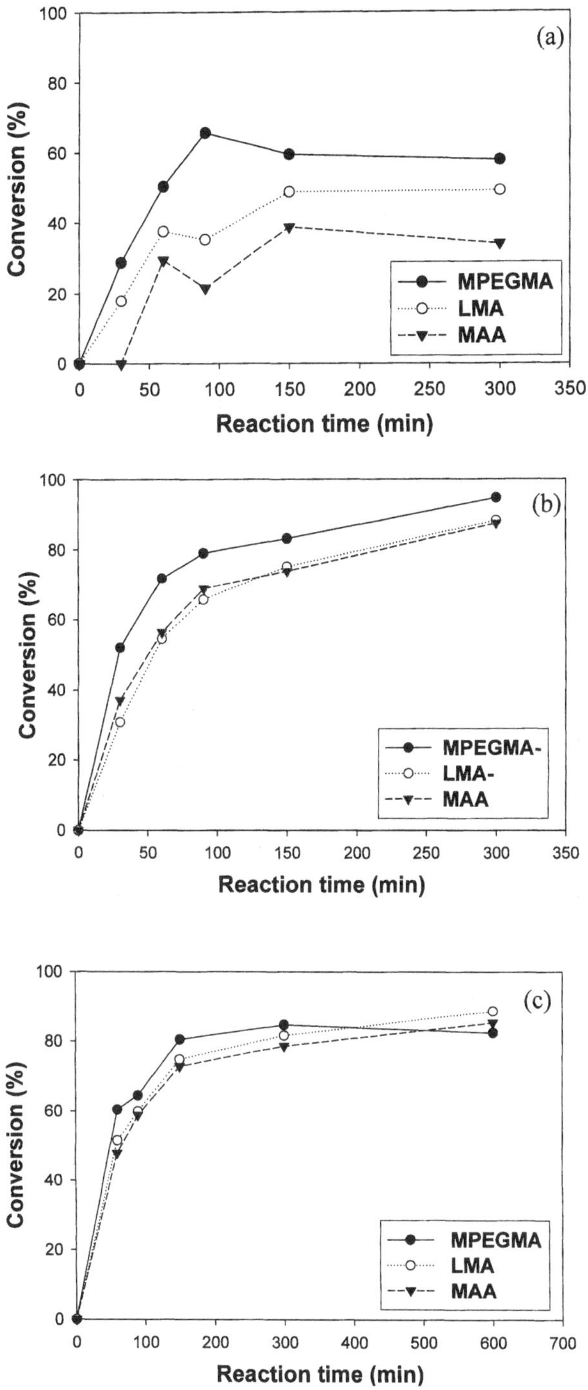

Fig. 9. Monomer conversions in polymerization with different initiators: (a) hydrogen peroxide (10% v/v), (b) 1% AIBN, and (c) 0.45% AIBN.

5.3. Characterization

The terpolymers were analyzed by typical equipment such as GPC, size-exclusion HPLC, NMR, and TGA. The polymer synthesized under the desired conditions, as described above, had a weight-averaged molecular weight of 300,000 with a polydispersity index of ~2.5. Both ^{1}H NMR and TGA analyses offered evidences of complexation in the terpolymers [59]. However, these analyses were not done with polymers in the W/O environment for the intended application. The results of more pertinent characterization are described below.

5.3.1. Polymer structure by reactivity ratio study

It is particularly important to assess the structural composition of the MAA-MPEGMA-LMA terpolymers because it dictates the extent to which the pH-sensitive complexation essential to the intended application in W/O fermentation occurs. The structural composition of co- and ter-polymers, in turn, depends on the reactivity ratios of the monomers involved [60-63]. The reactivity-ratio study was therefore performed with the most promising polymer system, as described above. The results suggested a polymer structure of alternating blocks of MAA and LMA, with isolated PEG grafts, as shown in Fig. 10 [35]. The structure is compatible with the desired pH-dependent complexation between the PMAA blocks and the MPEG grafts, but does not promise complete complexation.

In order to achieve complete complexation, each PMAA block must have the same number of MAA units as that of EG units in the adjacent PEG graft. For random terpolymerization, this is statistically impossible. It would be more probable if ideal "template polymerization" (TP) takes place between MAA and MPEGMA, without interruption by LMA blocks. The effect of MPEGMA on MAA conversion observed in the reactivity-ratio study was somewhat consistent with the well-known Type-II template polymerization of PEG-MAA systems [52, 54, 57, 64], where the polymerization starts in the solution but upon reaching a critical length, the oligoradicals complex with the template and the subsequent propagation proceeds along the template by the addition of monomer molecules from the surrounding solution. Nonetheless, the occurrence of template polymerization is still inconclusive and the effect of the presence of LMA on template polymerization is unknown. The pH-sensitivity of the polymer needed to be empirically evaluated, as described in the next section.

5.3.2. pH-dependent emulsification and phase separation

Standard bottle tests were performed. Emulsions with specific W/O phase volume ratios were prepared by vortex-mixing water and vegetable oil with 0.1% (w/v) polymer for 3 min. The emulsion pH was then adjusted to the desired value with 1 N HCl or 1 N NaOH. The emulsions were mixed for an additional 3 min and transferred into graduated cylinders. The phase separation with time was then tracked by determining the fraction of water remaining emulsified. For example, the results were compared in Fig. 11 for polymers with 1:1 MAA:EG ratio and different LMA contents at two different pHs (7 and 4) in test systems with a W/O phase ratio of 50/50 [30]. The polymer with 16% LMA appeared to have a weaker emulsification ability at pH 7, as indicated by the faster phase separation. The other two showed good emulsification in the neutral range and fast phase separation at pH 4. Complete separation was observed for the polymer with 47% LMA, while separation was less complete for that with 76% LMA. Incomplete phase separation in the polymer with very high LMA content is not surprising, considering the possible interference of the LMA grafts to the complexation between PEG and PMAA.

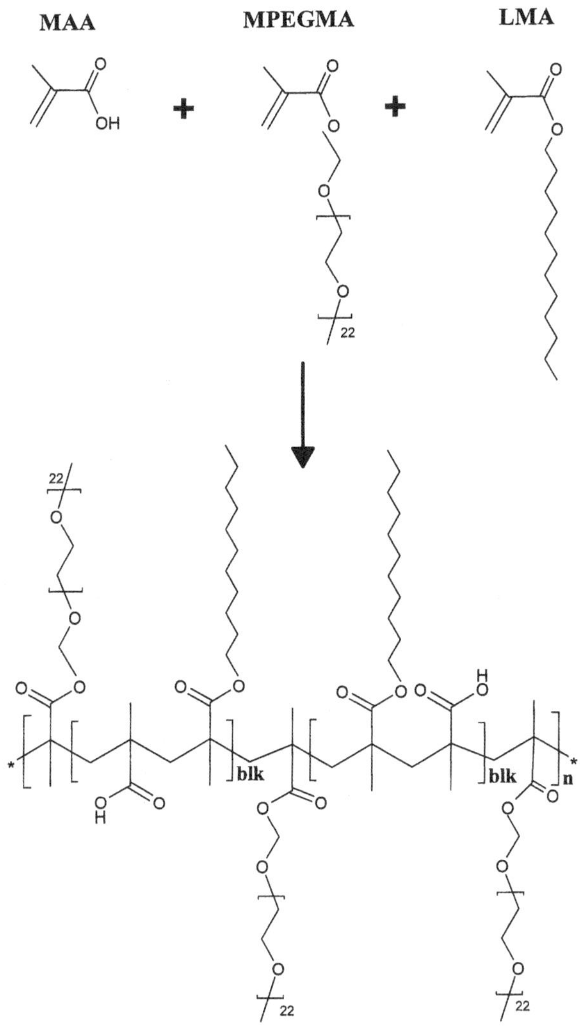

Fig. 10. Terpolymer structure proposed according to results from reactivity ratio study.

5.3.3. Emulsion droplet sizes

A standard procedure for evaluating the ability of polymers in emulsifying xanthan solution in oil was established [30]. The test involved forming the emulsion using a Waring blender, fixing the aqueous xanthan droplets with 2 M $FeCl_3$, and then measuring the droplet sizes under a microscope. The results are summarized in Table 4 for polymeric surfactants having a 1:1 MAA:EG ratio and different LMA contents. The polymers with higher LMA contents produced finer emulsions, which are desirable for supporting higher cell concentrations in fermentation, as described earlier.

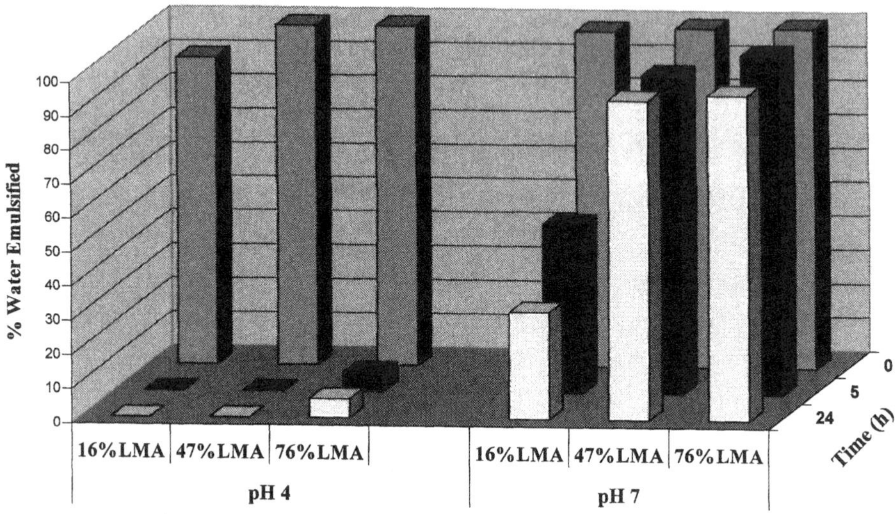

Fig. 11. Comparison of phase separation behaviors for emulsions (1/1 W/O) formed with terpolymers of 1:1 MAA:EG ratio and different LMA contents.

Table 4
Droplet sizes and size distributions obtained with polymeric surfactants having 1:1 MAA:EG ratio and different LMA contents

LMA (mol% to MAA)	Droplet Size/Distribution (μm)	
	Average	Standard Deviation
16	148.2	84.4
47	82.3	50.1
76	63.6	37.2

5.4. W/O xanthan fermentation with pH-sensitive surfactants

For comparison, a preliminary W/O fermentation was carried out under the same conditions as before, except that 0.15% of the polymeric surfactant (1.5 g of polymer per liter of total dispersion volume) was dissolved in the vegetable oil prior to autoclaving [59]. The profiles of cell growth, xanthan production and glucose concentration are shown in Fig. 12.

The maximal cell and xanthan concentrations obtained, ~14 and 105 kg m^{-3}, respectively, were similar to those in Fig. 6 for the W/O fermentation without the polymeric surfactant. However, cell growth was slow and active xanthan production started late (after ~5 days). The causes to these rather unusual behaviors are unknown. As expected, the polymer made the dispersion much finer, with the average size of aqueous droplets in the range of 30–40 (± 40) μm, as compared to ~120 μm in the systems without polymer addition. A study is underway with the goal of optimizing the operation of polymer-supplemented W/O xanthan fermentations.

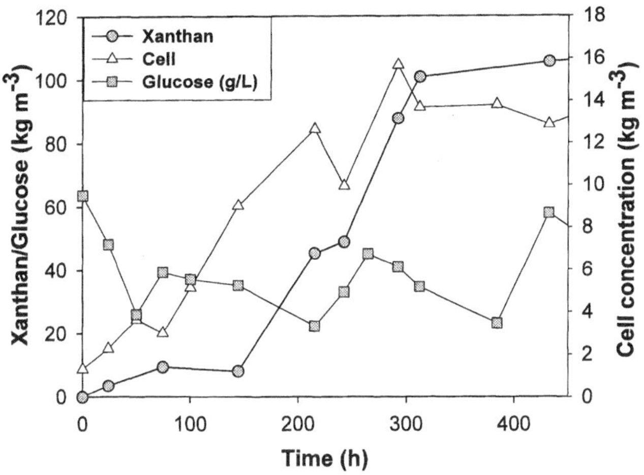

Fig. 12. Time profiles for W/O fermentation with 0.15 wt% polymeric surfactant.

6. CONCLUSIONS

W/O cultivation is a potentially advantageous process for producing highly viscous biopolymers in microbial fermentation. Using xanthan fermentation as an industrially important example, studies have been made with both experiments and model simulations to demonstrate its feasibility and evaluate the effects of operating factors such as agitation, aeration, aqueous droplet size, and various oil-phase properties. "Smart," pH-sensitive polymeric surfactants, with PMAA backbones as well as MPEG and dodecyl side chains, have also been synthesized and characterized. When applied to W/O fermentation, the surfactants are shown to reduce the sizes of aqueous droplets during fermentation so that a better oxygen supply to cells inside the aqueous droplets and a higher xanthan productivity can be achieved. The surfactants have also been shown to lose surface-activity at pH <5.5 and, thus, potentially allow easy phase separation for product recovery at the end of fermentation. Although the studies of W/O cultivation technology have so far been focused on xanthan fermentation, the technology is expected to be applicable to other highly viscous fermentation processes.

REFERENCES

[1] D.D. McLean, K. Abear, and M.F. Podruzny, Can. J. Chem. Eng. 64 (1986) 588.
[2] L.-K. Ju and O.A. Afolabi, Biotechnol. Prog. 15 (1999) 91.
[3] H. Furuse, I. Yabe, T. Asakura, O. Miyawaki, and K. Toda, J. Biosci. Bioeng. 95 (2003) 544.
[4] S.W. Kim, H.J. Hwang, C.P. Xu, J.W. Choi, and J.W. Yun, Let. Appl. Microbiol. 36 (2003) 321.
[5] L.-K. Ju, C.S. Ho, and J.F. Shanahan, Biotechnol. Bioeng. 38 (1991) 1223.
[6] J.T. Bae, J.P. Park, C.H. Song, C.B. Yu, M.K. Park, and J.W. Yun, J. Biosci. Bioeng. 91 (2001) 522.

[7] F.P. Duta, F.P. De Franca, E.F.C. Servulo, L.M.D.A. Lopes, A.C.A. Da Costa, and A. Barros, Appl. Biochem. Biotechnol. 113-116 (2004) 639.
[8] A.M. Dlamini and P.S. Peiris, Appl. Microbiol. Biotechnol. 47 (1997) 52.
[9] L.-K. Ju and S. Zhao, Biotechnol. Techn. 7 (1993) 463.
[10] M.M. Francis, T.R. Jack, and S.K. Cheema, Biopolymer Production Process, Canadian Patent No. 2058355, (1993).
[11] N.M. Bikales (ed.), Water-Soluble Polymers, Plenum Press, New York, 1973.
[12] A. Margaritis and G.W. Pace, in Comprehensive Biotechnology: The Principles, Applications and Regulations of Biotechnology in Industry, Agriculture and Medicine, Blanch, H.W., Drew, S., and Wang, D.I.C. (eds.), Pergamon Press, New York, 1985, 1005.
[13] R.A. Moraine and P. Rogovin, Biotechnol. Bioeng. 13 (1971) 381.
[14] A. Mulchandani, J.H.T. Luong, and A. LeDuy, Biotechnol. Bioeng. 33 (1989) 306.
[15] D.C. MacWilliams, J.H. Rogers, and T.J. West, Polymer Sci. Technol. 2 (1973) 105.
[16] A. Vincent, Topics Enz. Ferment. Biotechnol. 10 (1985) 109.
[17] I.W. Sutherland, Biotechnology of Microbial Exopolysaccharides, Cambridge University Press, New York, 1990.
[18] P.E. Janson, L. Kenne, and B. Lindberg, Carbohydrate Res. 45 (1975) 275.
[19] L.D. Melton, L. Mindt, D.A. Rees, and G.R. Sanderson, Carbohydrate Res. 46 (1976) 245.
[20] A. Jeanes, J.E. Pittsley, and F.R. Senti, J. Appl. Polymer Sci. 5 (1961) 519.
[21] Kelco, Kelco Xanthan Gum Book, Second ed., San Diego, California, 1978.
[22] E. Galindo, B. Torrestiana, and A. Garcia-Rejon, Bioprocess Eng. 4 (1989) 113.
[23] J. Solomon, T.P. Elson, A.W. Nienow, and G.W. Pace, Chem. Eng. Commun. 11 (1981) 143.
[24] B. Atkinson and F. Mavituna, Biochemical Engineering and Biotechnology Handbook, 2nd ed., The Nature Press, New York, 1983.
[25] S.-T. Yang, Y.-M. Lo, and D.B. Min, Biotechnol. Progress, 12 (1996) 630.
[26] D.K. Robinson and D.I.C. Wang, Ann. N. Y. Acad. Sci. 506 (1987) 229.
[27] L.G. Maury, Production of Xanthan Gum by Emulsion Fermentation, US Patent No. 4,352,882, (1982).
[28] K. Engelskirchen, W. Stein, M. Bahn, L. Schieferstein, J. Schindler, and R. Schmid, Process for Producing *Xanthomonas* Biopolymers, US Patent No. 4,618,582, (1986).
[29] S.G. Kuttuva, A.S. Restrepo, and L.-K. Ju, Appl. Microbiol. Biotechnol. 64 (2004) 340.
[30] A.S. Restrepo and L.-K. Ju, Ind. Eng. Chem. Res. 42 (2003) 4296.
[31] D.K. Robinson and D.I.C. Wang, Biotechnol. Progress, 4 (1988) 231.
[32] Y.M. Lo, C.H. Hsu, S.T. Yang, and D.B. Min, Bioprocess Biosyst. Eng. 24 (2001) 187.
[33] S.G. Kuttuva, A. Sundararajan, and L.-K. Ju, J. Disp. Sci. Technol. 19 (1998) 1003.
[34] S. Zhao, S.G. Kuttuva, and L.-K. Ju, Bioprocess Eng. 20 (1999) 313.
[35] A.S. Restrepo, N.M. Pinzon, and L.-K. Ju, J. Polymer Sci., A: Polymer Chem. 42 (2004) 2950.
[36] A.V. Quinlan, Ann. N. Y. Acad. Sci. 469 (1986) 259.
[37] H.U. Peters, I.S. Suh, A. Schumpe, and W.D. Deckwer, Can. J. Chem. Eng. 70 (1992) 742.
[38] I.S. Suh, H. Herbst, A. Schumpe, and W.D. Deckwer, Biotechnol. Let. 12 (1990) 201.
[39] I.S. Suh, A. Schumpe, and W.D. Deckwer, Biotechnol. Bioeng. 39 (1992) 85.
[40] C.M. Chapman, A.W. Nienow, M. Cooke, and J.C. Middleton, Chem. Eng. Res. Des. 61 (1983) 71.
[41] J. Cieszkowski and M. Dylag, Proceed. Eur. Conf. Mixing, 6th (1988) 421.
[42] M. Greaves and V.Y. Loh, Proceed. Eur. Conf. Mixing, 5th (1985) 451.
[43] T.R. Das, A. Bandopadhyay, R. Parthasarathy, and R. Kumar, Chem. Eng. Sci. 40 (1985) 209.
[44] C.S. Ho, L.-K. Ju, and R.F. Baddour, Biotechnol. Bioeng. 36 (1990) 1110.
[45] L.-K. Ju, J.F. Lee, and W.B. Armiger, Biotechnol. Progress, 7 (1991) 323.
[46] J.D. McMillan and D.I. Wang, Ann. N. Y. Acad. Sci. 589 (1990) 283.
[47] B.J. Michel and S.A. Miller, AIChE J. 8 (1962) 2626.
[48] R.V. Calabrese, C.Y. Wang, and N.P. Bryner, AIChE J. 32 (1986) 677.
[49] K. Van't Riet, Ind. Eng. Chem. Process Des. Dev. 3 (1979) 357.

[50] R.M. Weiss and D.F. Ollis, Biotechnol. Bioeng. 22 (1980) 859.
[51] A. Pinches and L.J. Pallent, Biotechnol. Bioeng. 28 (1986) 1484.
[52] V.Y. Baranovsky, I.V. Kotlyarsky, V.S. Etlis, and V.A. Kabanov, Eur. Polymer J. 28 (1992) 1427.
[53] V.Y. Baranovsky, A.A. Litmanovich, I.M. Papisov, and V.A. Kabanov, Eur. Polymer J. 17 (1981) 969.
[54] J. Ferguson and S.A.O. Shah, Eur. Polymer J. 4 (1968) 611.
[55] Y.Y. Tan and G. Challa, Macromol. Chem., Macromol. Symp., 10/11 (1987) 215.
[56] J. Matuszewska-Czerwik and S. Polowinski, Eur. Polymer J. 24 (1988) 791.
[57] S. Polowinski, Progress Polymer Sci. 27 (2002) 537.
[58] A.M. Mathur, B. Drescher, and A.B. Scranton, Nature, 392 (1998) 367.
[59] A.S. Restrepo, Smart pH-Sensitive Polymeric Surfactants for Water-in-Oil Xanthan Fermentation, PhD Dissertation, University of Akron, 2004.
[60] H.F. Naguib, S.M. Mokhtar, R.O. Ali, and M.Z. Elsabee, J. Polymer Sci., A: Polymer Chem. 41 (2003) 3180.
[61] L.G. Stanek, S.M. Heilmann, and W.B. Gleason, J. Polymer Sci., A: Polymer Chem. 41 (2003) 3027.
[62] L. Martin-Gomis, R. Cuervo-Rodriguez, M.C. Fernandez-Monreal, E.L. Madruga, and M. Fernandez-Garcia, J. Polymer Sci., A: Polymer Chem. 41 (2003) 2659.
[63] H.S. Bisht, S.S. Ray, and A.K. Chatterjee, J. Polymer Sci., A: Polymer Chem. 41 (2003) 1864.
[64] J. Matuszewska-Czerwik and S. Polwinski, Eur. Polymer J. 8 (1988) 791.
[65] A. Pons, C.G. Dussap, and J.B. Gros, Bioprocess Eng. 5 (1990) 107.
[66] I.S. Suh, A. Schumpe, and W.D. Deckwer, Biotechnol. Bioeng. 39 (1992) 85.
[67] L.-K. Ju and C.S. Ho, Biotechnol. Bioeng. 34 (1989) 1221.
[68] Air-Products, Product Manual on Multifluor Inert Fluid APF-215, Allentown, PA, 1987.
[69] L.-K. Ju, J.F. Lee, and W.B. Armiger, Biotechnol. Bioeng. 37 (1991) 505.

Bioprocessing for Value-Added Products from Renewable Resources
Shang-Tian Yang (Editor)

Chapter 16. Extractive Fermentation for the Production of Carboxylic Acids

Shang-Tian Yang,[a] Hanjing Huang,[a] Abdullatif Tay,[a] Wei Qin,[b] Lucita De Guzman,[c] and Ellen C. San Nicolas[d]

[a]Department of Chemical and Biomolecular Engineering, The Ohio State University, 140 W. 19th Avenue, Columbus, Ohio 43210, USA

[b]Department of Chemical Engineering, Tsinghua University, Beijing, China

[c]Food Technology Program, School of Technology, University of the Philippines in the Visayas, Miag-ao, Iloilo, Philippines

[d]College of Engineering, Central Colleges of the Philippines, 52 Aurora Blvd., Quezon City, Philippines

1. INTRODUCTION

Carboxylic acids are small organic acids with one or more carboxylic acid groups. Their properties vary significantly with their carbon-chain length, molecular structure, and the presence of additional functional groups. Figure 1 shows some important carboxylic acids and their chemical structures. These carboxylic acids are currently produced either from petroleum-based feedstocks through chemical synthesis or from carbohydrates via fermentation. Originally, all industrial carboxylic acids were produced by biochemical processes. By the mid-twentieth century, petrochemical processes had begun to replace biochemical routes, and they are now the primary industrial methods for manufacturing many carboxylic acids, including acetic, propionic, butyric, fumaric, malic, and acrylic acids. However, several carboxylic acids, such as citric, gluconic, and itaconic acids, continue to be produced solely by fermentation, largely because their complex chemical structures are difficult to produce via chemical synthesis.

Citric, fumaric, malic, succinic, and itaconic acids are multifunctional carboxylic acids [1]. Their current industrial applications include uses as food acidulents and in the manufacture of polyester resins. They can also be used as building blocks for the synthesis of esters and biodegradable polymers [2]. Commercially, maleic anhydride is the precursor for the production of fumaric (*trans*-butenedioic acid), malic (hydroxysuccinic acid), and succinic acids. Annual US production of maleic anhydride, mainly by vapor-phase oxidation of nbutane, is ~260,000 metric tons, not including maleic anhydride produced as an intermediate in the synthesis of 1,4-butanediol.

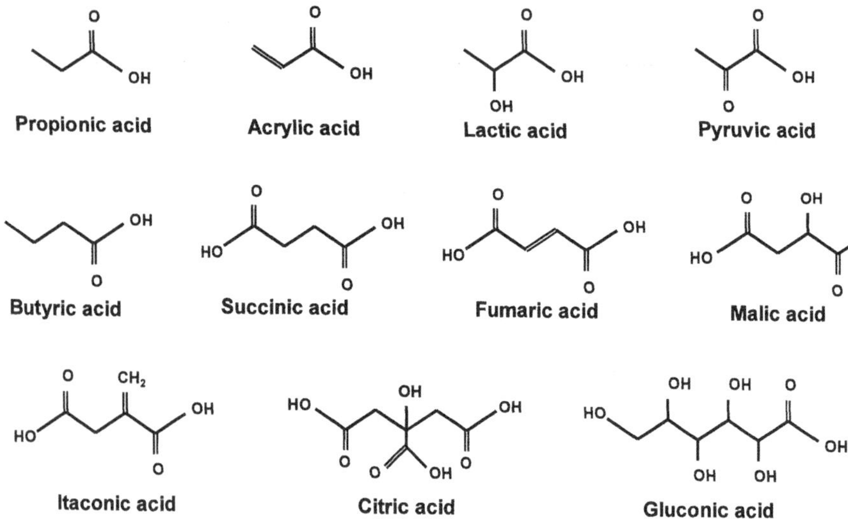

Fig. 1. Chemical structures of some carboxylic acids.

Acetic, propionic, and butyric acids are short-chain fatty acids. Acetic acid is a commodity chemical and an important chemical feedstock. It can also be used to produce non-corrosive deicers, such as calcium magnesium acetate (CMA) and potassium acetate [3]. Propionic and butyric acid are specialty chemicals widely used in the food, pharmaceutical, and chemical industries [4]. Acrylic acid (2-propenoic acid) is a commodity chemical with an estimated annual production capacity of 4.2 million metric tons [5]. Lactic acid is an important chemical with wide applications in food, pharmaceutical and cosmetic industries. Optically pure lactic acid is used to synthesize polylactic acid (PLA), a biodegradable polymer that has a large potential market of over one million metric tons per year [6]. Lactic acid can also be used to synthesize esters, such as ethyl lactate, that can replace toxic industrial solvents. Gluconic acid is produced by oxidizing the aldehyde group of glucose with either bacteria or filamentous fungi [7]. It is a strong chelating agent widely used in foods and various industrial products. Table 1 summarizes the industrial applications and market sizes of these carboxylic acids.

With the increasing oil price, concerns about oil supplies and environmental pollution caused by petrochemical processes, and consumer demand for natural food ingredients, there has been a high level of interest in producing carboxylic acids from renewable resources using bioprocesses [8–11]. The production of industrially important carboxylic acids from cheap and abundant recycled biomaterials can reduce wastes and our dependency on imported oils. In addition, fermentation can produce a pure isomer of the carboxylic acid (such as either D(–)- or L(+)-lactic acid) for better processing and/or application purposes, an advantage over chemical synthesis, which usually produces a mixture of optical isomers that are difficult to separate from each other.

However, fermentation also has many disadvantages, mainly low productivity and product concentration that can limit its industrial applications. In this chapter, current fermentation and separation methods for the production of several carboxylic acids will be discussed first, followed by a review of extractive fermentation technologies that can alleviate product inhibition and improve fermentation for economical production of many industrially important carboxylic acids that are currently produced mainly by petrochemical routes.

Table 1
Important carboxylic acids that are or can be produced from sugars by fermentation

Organic acid	$pK_{a,\,25^{\circ}C}$	Market size (tons/year)	Industrial production method	Use / Application
Acetic acid	4.76	3,500,000	Chemical synthesis; some fermentation	Commodity chemical, vinegar
Acrylic acid*	4.26	4,200,000	Partial oxidation of propene	Commodity chemical
Butyric acid	4.81	30,000	Chemical oxidation of butyraldehyde	Specialty chemical, food flavoring, pharmaceuticals
Citric acid	3.13 4.76 6.40	1,200,000	Fungal fermentation of glucose	Food, beverages, pharmaceuticals, detergents
Fumaric acid	3.5 4.5	30,000	Isomerization of maleic acid	Polymer intermediate, food
Gluconic acid	3.62	50,000	Fermentation; biological oxidation of glucose	Food, pharmaceuticals, chelating agent
Itaconic acid	3.65 5.13	20,000	Fungal fermentation of glucose	Polymer intermediate
Lactic acid	3.86	120,000	Fermentation; some chemical synthesis	Food, pharmaceuticals, polylactic acid
Malic acid	3.04 5.05	–	Chemical hydration of maleic or fumaric acid	Polymer intermediate, food, beverages
Propionic acid	4.88	180,000	Chemical synthesis; some fermentation	Specialty chemical, food preservative
Pyruvic acid	2.49	>1,000	Fermentation; Dehydration and decarboxylation of tartaric acid	Pharmaceuticals, specialty chemical, food
Succinic acid	4.21 5.64	25,000	Chemical conversion from maleic anhydride	Polymer intermediate

*There is no known biological process that can produce acrylic acid, but it is possible to metabolically engineer microorganisms to produce acrylic acid from glucose [5].

2. CARBOXYLIC ACID FERMENTATION

2.1. Fermentation

The fermentation process and its performance for carboxylic acid production depend on the type of microorganisms and the biochemical pathways through which the acid is produced. Table 2 lists some carboxylic acid fermentations that are either currently used in industry or have good potential for industrial applications. Citric, fumaric, and malic acids are the intermediate metabolites in the tricarboxylic acid (TCA) cycle and are over-produced by some filamentous fungi under aerobic conditions. Itaconic, gluconic, and pyruvic acids are also produced under aerobic conditions by some fungi. Acetic and lactic acids can be produced under either aerobic or anaerobic conditions by different groups of microorganisms. On the other hand, butyric, propionic, and succinic acids are produced mainly by anaerobic bacteria.

Table 2
Microbial production of carboxylic acids from sugars in aerobic and anaerobic fermentations

Organic acid	Microorganism	Yield (g/g sugar)	Productivity (g/L·h)	Final Conc. (g/L)	Reference
Aerobic fermentation					
Acetic acid	*Acetobacter aceti* *Ga. europaeus*	~0.6*	6.4	>100	[12–14]
Citric acid	*Aspergillus niger*	0.7–0.9	0.5–1.0	140	[15, 16]
Fumaric acid	*Rhizopus oryzae*	0.75	3.78	74	[17]
	Rhizopus arrhizus	0.36	0.2	63	[18]
Gluconic acid	*Aspergillus niger*	0.9	3–8	140–230	[19]
	Aureobasidium pullulans	0.93	10–19.3	208–260	[20]
Itaconic acid	*Aspergillus terreus*	0.68	0.25–0.64	65	[21–23]
Lactic acid	*Rhizopus oryzae*	0.7–0.9	1–6	120	[24]
Malic acid	*Aspergillus flavus*	0.94	0.59	113	[25]
	Schizophyllum commune	0.86	0.39	50	[26]
Pyruvic acid	*Torulopsis glabrata*	0.68	1.28	60.4	[27, 28]
	Recombinant *S. cerevisiae*	0.54	1.35	135	[29]
Anaerobic fermentation					
Acetic acid	*C. thermoaceticum*	0.8–0.95	0.66	~100	[30]
	C. formicoaceticum	0.9–0.98	1.23	80	[31, 32]
Butyric acid	*C. tyrobutyricum*	~0.45	0.8–6	80	[33, 34]
Lactic acid	*Lactobacillus* spp.	0.85–0.95	3–15	~130	[35, 36]
Propionic acid	*P. acidipropionici*	0.4–0.65	0.2–1	~75	[37, 38]
Succinic acid	*A. succinogenes*	0.8	1.34	106	[39]
	Recombinant *E. coli*	1.10	1.30	99	[40]

*Overall acetic acid yield from glucose, which is first converted to ethanol by yeast.

A fermentation must have high productivity, yield, and final product concentration in order to be economical, as in the cases of citric and gluconic acid fermentations. Specialty chemicals with a relatively high market price, such as itaconic (~$4/kg) and pyruvic (~$8/kg) acids, can also be produced by fermentations with moderate productivities and yields. In general, economical production of carboxylic acids requires a productivity greater than 1 g/L·h and a final product concentration greater than 50 g/L. Although almost all carboxylic acid fermentations listed in Table 2 meet these criteria, many of them are still not used in commercial production because their high performance was achieved through cell immobilization in non-conventional bioreactors [17, 24, 41], which have not been proven or accepted by the fermentation industry. Free-cell fermentations in either stirred-tank or air-lift fermentors are the standard large-scale fermentation processes.

In general, conventional fermentation processes are not economical to compete with petrochemical processes for the production of carboxylic acids with market prices of ~$1/kg, such as acetic, butyric, and propionic acids, because fermentation usually suffers from low productivity and difficulties in product recovery and purification. Therefore, for fermentation-based carboxylic acids to penetrate the organic chemical market, substantial improvements in fermentation and recovery technology are needed. This is especially true for those carboxylic acids for which more economically competitive synthetic sources are available. Current and novel separation methods for recovering and purifying carboxylic acids from fermentation broth will be discussed in the next section. The integration of product separation with fermentation, so called extractive fermentation, to reduce end product inhibition and improve fermentation productivity will be discussed thereafter.

2.2. Separation of carboxylic acids from fermentation broth

Fermentatively produced carboxylic acids are usually recovered by distillation (for volatile products), precipitation (for non-volatile products), or solvent extraction. Adsorption with ion-exchange resins and electrodialysis with bipolar membranes also have been used for some carboxylic acids. Table 3 compares various methods used in carboxylic acid separation and recovery from fermentation broth. The choice of the separation method depends on the type of carboxylic acid and its concentration and purity in the fermentation broth. In general, a product concentration of ~10% (wt/vol) is needed in order for the recovery process to be economical.

Distillation can be used for the separation of volatile compounds, such as acetic acid. Non-volatile compounds, such as lactic acid, can be converted to esters first before being separated by distillation. However, distillation is energy intensive and is not the preferred choice for separating carboxylic acids. The calcium salts of many carboxylic acids, such as citric, fumaric, and lactic acids, have low solubilities in water. Precipitation thus has been the separation method most widely used in the production of these carboxylic acids. However, the process requires the acidification of the calcium salt in order to produce the free acid as the final product, which generates a large amount of the undesirable solid waste $CaSO_4$, which is expensive to dispose of by landfill. Consequently, current industrial production processes for citric and lactic acids in the US use solvent extraction, which also has good potential for use as an *in situ* recovery method for an integrated fermentation-separation process. Adsorption and electrodialysis also have been widely studied for use in separating fermentation-produced carboxylic acids, although they are not yet widely used in industry.

Table 3
Separation methods for recovery of carboxylic acids from fermentation broth

Method	Description	Advantages	Disadvantages
Precipitation	$CaCO_3$ or CaO is added in the medium to neutralize acid. The calcium carboxylate solution is concentrated by evaporation, then crystallized and separated from the mother liquor	low impurities in product; low capital costs; high yield	requires the use of H_2SO_4 to release carboxylic acid, which generates $CaSO_4$, a solid waste requiring landfill disposal
Distillation	NH_3 is used to neutralize acid. Ammonia carboxylate then reacts with alcohol to form ester, which is separated by distillation	high product purity; the byproduct $(NH_4)_2SO_4$ can be used as a fertilizer	requires hydrolysis of ester and distillation to separate the alcohol from carboxylic acid. High capital and energy costs associated with distillation require economy of scale
Extraction	Use organic solvents to extract carboxylic acid from the broth.	low costs, high yield, best for carboxylate salt production	the solution needs to be acidified to allow efficient extraction of the free carboxylic acid. Extractant needs to be regenerated by distillation or back-extraction (stripping).
Adsorption	Usually using ion exchange resins to adsorb carboxylate ions from the broth	easy to operate	low adsorption capacities, high resin costs, requires energy-intensive resin regeneration, separation is not highly selective
Electrodialysis	electric current is applied to move negatively charged carboxylate ions through an anion-exchange membrane towards the anode in the electrodialyzer	Carboxylate is concentrated in aqueous solution, does not require acid addition to adjust the solution pH	product purity is low and may require further purification, high energy input; membrane fouling; difficult to scale up

2.2.1. Adsorption

Anion exchange resins have been widely studied for use as adsorbents of carboxylic acids, including lactic, citric, fumaric, and acetic acids [42–45]. In general, the undissociated acid is adsorbed onto weak or strong base polymer resins containing tertiary or quaternary amine groups, and the adsorbed acid molecules are later eluted or desorbed with methanol, ammonia, or H_2SO_4. Desorption with steam is also possible, but the recovery yield is usually lower than 70%. The main disadvantages of the adsorption process for carboxylic acid separation are its low adsorption capacity (usually less than 100 mg/g resin) and the requirement of additional chemicals for acidifying the feed broth and for eluting/desorbing the acid molecules from the adsorbents. The presence of other anions, such as SO_4^{2-} and Cl^-, in the fermentation broth can significantly reduce the adsorption efficiency due to competition for active sites on the ion exchange resin. The adsorption process usually uses packed columns operated under a batch mode, although continuous processes using fluidized and simulated moving beds have also been studied [46–48]. Other materials, including activated carbon, polyvinyl pyridine, and Silicalite (Zeolite) molecular sieves, have also been studied for the adsorption of lactic acid [49, 50]. Compared to solvent extraction, adsorption is more expensive and thus has not been used in industry for carboxylic acid separation from fermentation broth.

2.2.2. *Electrodialysis*

Electrodialysis (ED) is a membrane-based separation process in which ions are driven through an ion-selective membrane and separated and concentrated under the influence of an electric field [51, 52]. Two types of ED are available for separating carboxylic acids from fermentation broth. Conventional ED, which consists of cation and anion exchange membranes stacked between anode and cathod, can be used to concentrate and partially purify carboxylates (see Figure 2). A three-compartment bipolar membrane eletrodialysis (BMED) consisting of a stack of cation exchange membranes (CAM), anion exchange membranes (AEM), and bipolar membranes (BM), which split water into H^+ and OH^-, can be used to produce concentrated free acid and base from salt [53, 54]. Figure 3 illustrates the process concept of a three-compartment BMED. In practice, a two-compartment BMED with either CAM or AEM can be used to convert carboxylates to carboxylic acids if the feed is relatively free from impurities. The recovered base can be recycled for use in controlling fermentation pH.

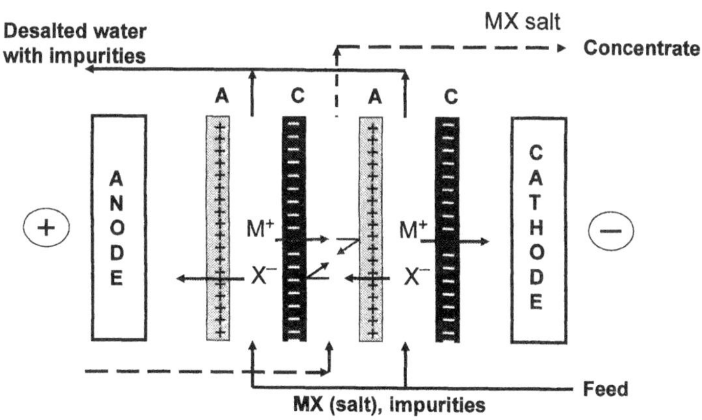

Fig. 2. Principle of desalting electrodialysis. A: anion exchange membrane; C: cation exchange membrane; M^+: cation; X^-: anion.

Industrially, ED has been used in desalination and separating organic and inorganic acids from waste water [51–53]. Compared to adsorption and solvent extraction, ED has the advantage of producing free acid from salt without using acidifying chemicals. It has been widely studied for the separation of many carboxylic acids, including acetic, butyric, citric, formic, gluconic, lactic, propionic, and pyruvic acids [55–68]. Recent studies have shown that ED can be an economical method for recovering lactic acid from fermentation broth [6, 53].

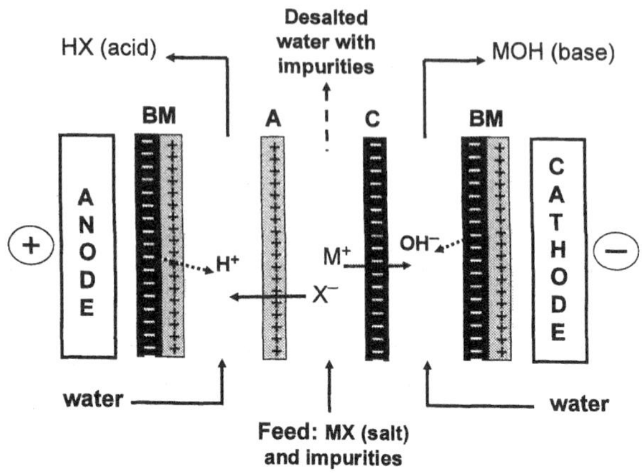

Fig. 3. Principle of three-compartment water-splitting electrodialysis using bipolar membranes (BM). A: anion exchange membrane; C: cation exchange membrane; M^+: cation; X^-: anion.

Fermentation broths usually contain significant amounts of Ca^{2+} and Mg^{2+}, which form insoluble hydroxides at the interface of the bipolar membrane where the ions separate. These divalent cations must be reduced to about 1–5 ppm in order to prevent the fouling of bipolar membranes. Conventional or desalting ED can be used to concentrate lactate salt by over 2-fold, from 8–10 wt% in the feed broth to ~20 wt%, purify the lactate salt from non-ionic impurities (proteins and sugars) by a factor of 10 to 20, and reject ~99% of divalent ions, from ~1000 ppm to ~10 ppm, using monovalent cation selective membranes [6]. The recovery yield of lactate is usually greater than 95%, and the power consumption is relatively low: ~0.33 kWh/kg lactate [6]. However, the divalent ions must be further reduced to ~1 ppm by chelation before the lactate salt is converted to lactic acid using a two-compartment water splitting ED, which can achieve 99% acidification of the lactate salt with a low power consumption (0.55 kWh/kg lactic acid) at a high current density (1000 A/m^2) [6]. Since CEM, which have a much lower fouling tendency than AEM, are used, the process is relatively stable, with a membrane life exceeding 15,000 hours. The two-stage ED process has been demonstrated at the pilot scale and recently scaled up for commercial production of organic acids [53]. One-stage electrodialysis using three-compartment BMED has also been shown to be feasible for lactic acid recovery from fermentation broth [69]. However, to produce high-purity (>99.9%) polymer-grade lactic acid, further polishing and purification, including treatments with ion exchange resins, esterification with ethanol, and distillation and hydrolysis of the ester to lactic acid, are required [6, 53].

For ED, current efficiency has been reported to vary from 96% to 25%. In general, current efficiency increases with increasing acid (salt) concentration and current density [64]. To achieve a high current density requires a high fluid conductivity. However, for ED conversion of carboxylic salt to carboxylic acid, the conductivity and thus current efficiency of carboxylic acid are low and decrease with increasing molecular mass or size [65]. High concentrations of uncharged sugars and amino acids in the broth can also lower current efficiency [56]. To increase the conductivity and reduce the energy consumption, the weak acid compartment can be filled with granular cation exchange resins (e.g., Amberlite IR 120 Plus) in hydrogen form that conduct current via hydrogen ions.

The recovery yield and purity of the final product are adversely affected by diffusional leakage of carboxylic acid and sugar molecules through membranes. In general, CEM is less leaky than AEM [70]. Back diffusion of acid molecules and electro-osmosis and osmosis also reduce the current efficiency and final acid concentration [63]. Water transport (osmotic dilution) from the diluted compartment to the concentrated compartment can be minimized by maintaining an overpressure in the concentrated compartment of the electrodialysis stack, which also improves glucose rejection. However, overpressure can cause fluid flow resistance that hinders ion transfer, thus requiring more energy [58].

2.2.3. Liquid-liquid extraction

Fermentation-produced citric and lactic acids are separated by solvent extraction, which is a well established process widely used in chemical, food, and pharmaceutical industries. In liquid-liquid extraction, the product (extract) is partitioned between two immiscible liquid phases, an aqueous phase and the other usually an organic solvent, which can be categorized into three types: conventional oxygen-bearing and hydrocarbon solvents, phosphorus-bonded oxygen-bearing extractants (e.g., trioctylphosphine oxide or TOPO), and high-molecular weight aliphatic amines [71]. More recently, ionic liquids and aqueous two-phase systems have also been studied for extracting carboxylic acids from fermentation broth [72, 73]. Unlike organic solvents, ionic liquids and aqueous two-phase systems are environmentally benign. However, they have low partition coefficients (~1 or less) and are usually more expensive.

Three main criteria in selecting a solvent for extraction are high distribution coefficient (K_d), selectivity, and ability to cause phase separation. In addition, biocompatibility, which relates to the biological effect of a solvent on living cells, is also an important consideration for extractive fermentation and the environment [74]. Conventional solvents, including most alcohols, ketones, ethers, and aliphatic hydrocarbons, are not efficient for extracting hydrophilic carboxylic acids, which are usually present at low concentrations in fermentation broths and wastewater. Therefore, reactive extraction with long-chain aliphatic amines has been more commonly used for the recovery of carboxylic acids from dilute solutions [75–80]. Among the long-chain aliphatic amines, secondary (e.g., ditridecyl amine or Adogen 283) and tertiary amines (e.g., tricaprylyl amine or Alamine 336) are the most widely used ones because of their low solubility in water and high distribution coefficients for carboxylic acids (see Table 4). In general, the extractability of carboxylic acids by amines increases with chain length, as the longer-chain carboxylic acid is more hydrophobic.

Unlike physical extraction by conventional solvents, the extraction of carboxylic acids by amines involves a complexation reaction between undissociated acids (HA) and amines (R_3N), as follows [71, 76]:

$$m[R_3N]_{org} + n[HA]_{aqu} \Leftrightarrow [(R_3N)_m(HA)_n]_{org}$$

The complexation reaction results in a high distribution coefficient. In general, secondary amines are easier to complex with acids and therefore possess higher distribution coefficients than tertiary amines. A diluent is usually required to improve amine solvation power, which has a significant influence on the distribution coefficient because a proper diluent can solvate the amine-acid complex to avoid its precipitation and the formation of a separate phase. The effects of diluent on the extractability of amine solvents vary with the type of diluent (polar vs. nonpolar) and carboxylic acid [81–84]. In general, the extractability of tertiary amines increases with their chain length when a polar diluent such as 1-octanol is used [82]. However, the opposite trend was observed in the nonpolar diluent n-heptane. Diluent is also used to lower the viscosity of the solvent phase.

The extraction of carboxylic acids by aliphatic amines is greatly affected by the pH [78, 79]. Since the complexation reaction takes place between the undissociated acid and amine molecules, the extraction generally works better at a lower pH (<<pK_a). However, it has been reported that quaternary amines, such as Aliquat 336, are able to extract both dissociated and undissociated acids [78], and therefore can also work well at a high pH (>pK_a). In general, the K_d value increases with decreasing pH except at extremely high or low pHs where K_d does not change with the pH. K_d goes to almost zero when the pH is greater than 7, no matter how high K_d was at low pH. The effect of pH on K_d can be modeled by the following equation [78]:

$$K_d = \frac{K_1 + K_2 K_a/[H^+]}{1 + K_a/[H^+]}$$

where K_a is the equilibrium or dissociation constant of the acid, and K_1 and K_2 are two intrinsic distribution coefficients at extremely low and high pH values, respectively. At low pH values or $[H^+] >> K_a$, $K_d = K_1$. At high pH values or $[H^+] << K_a$, $K_d = K_2$ (which is zero for most amine extractants, except for Aliquat 336). Therefore, the solvent can be easily stripped and regenerated by back extraction with a strong base [78].

Table 4
Distribution coefficients for several carboxylic acids at 25°C and pH <<pK_a

	Distribution Coefficient, K_d		$pK_{a,25\,C}$
	Alamine 336	Adogen 283	
Acetic acid	2.8	16.1	4.76
Lactic acid	3.5	8.4	3.86
Propionic acid	8.4	~20	4.88
Butyric acid	16.5	~50	4.81

The complexation reaction between acids and amines is exothermic, and therefore, K_d decreases when the temperature increases. Back extraction can also be done by stripping with hot water or temperature swing and diluent swing [76, 77]. For volatile acids, amine extractants can be regenerated by distillation if the diluent is relatively non-volatile. In some cases, salt solutions and strong acid solutions are used to strip amine solvents, where ion exchange actually plays the role of regeneration. If pure acid instead of acid salt needs to be produced, the acid could be back extracted into aqueous ammonia, forming ammonium salt which is then concentrated and heated so as to decompose into the acid and ammonia.

2.2.4. Extraction with supported liquid membrane in hollow fibers

One concern in solvent extraction is the difficulty of controlling phase contact and separation. Conventional phase contact is created by mixing two phases to increase their mass transfer interface and then letting them settle into two separate phases, usually done in two steps, but poor control of the water-solvent interface can cause emulsion and incomplete phase separation. To overcome this problem, membrane-immobilized phase contact via a supported liquid membrane (SLM) was first introduced in 1984 [85]. In SLM, the static phase interface is immobilized in the micro-pores of polymer membranes in a hollow-fiber module, providing a large specific surface area for efficient mass transfer. The hollow-fiber membrane-based extraction process makes solvent extraction less dependent on the physical properties of the solvent, reduces entrainment of the solvent, and reduces emulsification. Compared to conventional solvent extraction, membrane-based extraction also has the advantages of no flooding, no limitations on the density of the two phases, and higher volumetric mass transfer efficiency. It also greatly reduces the toxic effects of organic solvents on microorganisms in an extractive fermentation. The membrane-based extraction process has been widely studied for the separation of carboxylic acids [86–90].

A supported liquid membrane (SLM) also can combine extraction and stripping, which are generally carried out in two separate steps in conventional solvent extraction processes, into one step, providing the maximum driving force for the separation of a target species and thus leading to the best recovery of the species. In SLMs, the liquid membrane phase is the organic solvent imbedded in the pores of a microporous support, e.g., microporous polypropylene hollow fibres. When the organic liquid contacts the microporous support, it readily wets the pores of the support, and the SLM is formed. For the extraction of a target species from an aqueous feed solution, the organic-based SLM is placed between two aqueous solutions, the feed solution and the strip solution, where the SLM acts as a semi-permeable membrane for the transport of the target species from the feed solution to the strip solution. As illustrated in Figure 4, the SLM containing the secondary amine (R_2NH) facilitates the transport of lactic acid (R'COOH) across the SLM from the feed solution to the strip solution containing NaOH. However, SLM suffers from a gradual loss of the organic membrane phase to the aqueous feed and strip solutions, due to emulsification (e.g., resulting from lateral shear forces) at the membrane-aqueous interfaces, and to the osmotic pressure difference across the membrane. The osmotic pressure difference displaces the organic membrane phase from the micropores of the support. Displacement of the

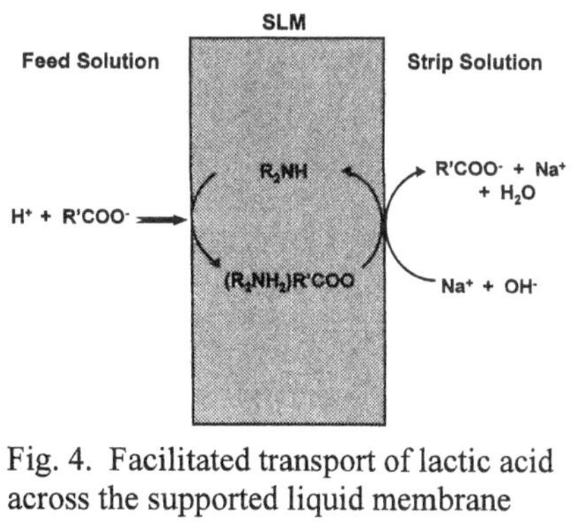

Fig. 4. Facilitated transport of lactic acid across the supported liquid membrane (SLM).

organic membrane phase from the pores can ultimately allow feed and strip solutions to mix, leading to complete failure of the separation unit.

The poor stability of supported liquid membranes can be overcome by continuously replenishing the solvent phase in the liquid membrane. One way to do this is by dispersing the strip solution in the organic solvent, which forms the continuous phase and is circulated through one side of the hollow-fiber membrane [91]. This approach, however, requires the separation of the two phases present as the water-in-oil emulsion, which can be problematic, as in the conventional solvent extraction process. Alternatively, extraction and stripping (back extraction) can be physically separated into two hollow-fiber (HF) membrane contactors with solvent continuously circulating through them (see Figure 5). This process requires only a small amount of solvent that is continuously regenerated and used to extract carboxylic acid from the fermentation broth with a good long-term stability [90]. When a strong base, such as NaOH, is used in the stripping solution, lactic acid can be separated and recovered from the fermentation broth as a concentrated sodium lactate solution at a high concentration close to its solubility in water.

Compared to adsorption and electrodialysis, membrane-based solvent extraction consumes less energy and is easier to scale up. However, the extraction efficiency is highly dependent on the broth pH, which is usually not at the optimum for extraction. Acidification of the fermentation broth with a weak acid (e.g., CO_2 or carbonic acid) before extraction is thus necessary. A strong inorganic acid (e.g., HCl and H_2SO_4) cannot be used as it competes more favourably than the weak carboxylic acid for complexation with amines. Broth acidification can be eliminated when the extraction process is integrated with fermentation in an extractive fermentation process, which is discussed in the next section.

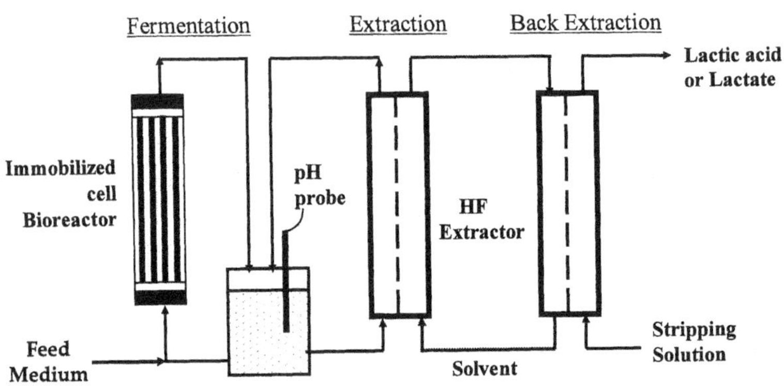

Fig. 5. An extractive fermentation process using immobilized cell bioreactor for fermentation and hollow-fiber membrane extractors for continuous product separation. The fermentation product, e.g., lactic acid, is extracted in the first HF unit and then stripped (back extracted) in the second HF unit. The final product is a concentrated sodium lactate solution when NaOH is used, or lactic acid when a strong acid or hot water is used, as the stripping solution.

3. INTEGRATED FERMENTATION-SEPARATION PROCESSES

Conventional carboxylic acid fermentations are subject to end-product inhibition, which significantly reduces cell growth and metabolic activities and limits acid production, decreasing fermentation productivity and product yield, limiting final product concentration, and requiring the use of dilute substrates in the fermentor. In addition, overproduction of a major fermentation product may prompt cells to shift to produce other byproducts. For example, in conventional propionic acid fermentation, the accumulation of propionic acid in the fermentation broth alters the bacterial metabolism to produce more byproducts – acetic, lactic and succinic acids. Therefore, removing the fermentation end product can have a major effect on cell metabolism and increase the final product yield. *In situ* product removal from the fermentation broth in a carboxylic acid fermentation thus can have many advantages, including:

- ability to use high substrate concentrations, reducing the total liquid volume handled in the process
- ease in reactor pH control without addition of a neutralizing base, as the product acid is continuously removed instead of accumulating in the broth
- higher productivity due to reduced end product inhibition
- higher product yield and purity due to a metabolic shift favorable to the formation of the major fermentation end product
- higher final product concentrations
- reduction of downstream processing load and recovery cost

The application of *in situ* product recovery techniques and integrated fermentation-separation processes, often synonymously called extractive fermentation, have been widely studied and published works before 2003 have been reviewed in several articles [92–94]. The separation of the product in an extractive fermentation process can be performed either within the bioreactor or by circulating the fermentation broth through an external separation unit (see Figure 5). In the latter case, the external separation unit can be an adsorption column, extractor, or electrodialysis unit. For volatile products, pervaporation also can be used. Here we will discuss only those involving adsorption, electrodialysis, and solvent extraction as they are most relevant to carboxylic acids and likely to have industrial applications in the near future.

Table 5 lists some integrated fermentation-separation processes that produce several carboxylic acids from sugars. In general, *in situ* product separation can increase the final product concentration and reactor volumetric productivity by up to several fold, depending on the fermentation conditions, as compared to the same fermentation without product removal. In most cases, the product yield is also improved significantly, mainly due to reduced production of byproducts. It is noted, however, that *in situ* product separation does not always increase reactor productivity because the fermentation may be complicated by problems such as solvent toxicity and nutrient removal, which can lower cell productivity. Also, to facilitate product extraction, extractive fermentation may have to be operated at a suboptimal pH, lowering reactor productivity.

Table 5
Comparison of integrated fermentation-separation processes for carboxylic acid production

Product	Microorganism / fermentation conditions	Yield (g/g sugar)	Productivity (g/L·h)	Final Conc. (g/L)	Reference
Product removal by adsorption with ion exchange resins					
Lactic acid	*L. delbrueckii* Batch fermentation	1.17 (0.95)	0.9 (1.3)	1.84 (105)	[95]
	L. casei / Continuous fermentation with cell recycle	(0.98)	(138)	(80)	[96]
	L. delbrueckii Batch fermentation	1.12 (0.93)	5.3 (1.7)	1.25 (100)	[97]
Fumaric acid	*Rhizopus oryzae* Rotary biofilm reactor	1.13 (0.85)	1.12 (4.25)	1.15 (85)	[44, 17]
Citric acid	*Aspergillus niger* Batch fermentation	1.15 (0.95)	1.6 (0.54)	1.2 (78)	[43]
Electrodialysis fermentation					
Lactic acid	*L. lactis* / batch fermentation with periodic electrodialysis	~1.0 (0.78)	4.7 (3.4)	~1.0 (62)	[99]
	R. oryzae / Ca-alginate beads fluidized bed bioreactor	0.96 (0.71)	1.7 (14.8)	5.2 (52)	[100]
Propionic acid	*P. freudenreichii* Batch fermentation	1.2 (0.14)	1.8 (0.22)	2.0 (38)	[101]
Extractive fermentation					
Citric acid	*A. niger* / batch fermentation	-	1.7 (0.15)	5.2 (232)	[102]
Butyric acid	*C. tyrobutyricum* / batch fermentation with immobilized cells	1.1 (0.45)	1.1 (7.37)	6.9 (300)	[103]
Lactic acid	*R. oryzae* / rotating fibrous bed bioreactor, fed-batch fermentation	1.02 (0.92)	0.3 (0.73)	2.4 (293)	[104]
Propionic acid	*P. acidipropionici* / batch fermentation	1.3 (0.66)	8.3 (1.0)	3.9 (75)	[105]

Note: The numbers in the table indicate the relative performance as compared with the same fermentation without simultaneous product removal (= 1). The numbers in the parenthesis show the actual performance data for the integrated fermentation-separation processes.

3.1. Product removal by adsorption

Extractive fermentations with adsorption using anion-exchange resins have been extensively studied for the production of lactic acid and several other carboxylic acids [43, 44, 95–98]. In general, anion-exchange resins are not toxic to cells and the removal of the acid product can provide good pH control and increase reactor productivity significantly (see

Table 5). However, the fermentation is limited by the capacity of the ion exchange resins, and final product concentrations are not much higher than those from conventional fermentations. The process is generally stable, provided that the adsorption column is properly regenerated and not fouled or clogged by cells. The adsorption of cells onto the resins can be prevented by cell immobilization, microfiltration of the fermentation broth before it passes through the adsorption column, or shielding the adsorbent with some coating [98]. Anion exchange resins can also adsorb other acids, such as H_2PO_4. However, the loss of nutrients did not affect citric acid fermentation [43].

Adsorption with packed bed column is usually operated in a batch mode. To achieve continuous operation, Davison and Scott (1992) proposed a biparticle fluidized bed for lactic acid fermentation with simultaneous adsorption [48]. Particles are stratified in the fluidized bed based on size and density. High density adsorptive particles (activated carbon) are continuously added to the top of the bed. Cells are immobilized in low density beads, and the high density particles are continuously removed at the bottom as they settle out, without removing any of the cell-containing beads. They demonstrated the feasibility of the fermentation with κ-carrageenan beads seeded with inorganic oxides to vary the densities. The addition of activated carbon for the adsorption of lactic acid controlled the pH and improved the fermentation. This process bypasses the usual cell separation steps necessary for adsorption from fermentation broths. Gailliot et al. (1990) demonstrated that fluidized bed adsorption using high density synthetic resins was effective in removing dilute products from whole fermentation broths on an industrial scale [47].

3.2. Electrodialysis fermentation

Electrodialysis fermentation (EDF) uses an electrodialyzer with ion-exchange membranes to remove ionized product carboxylates. EDF can provide good pH control without requiring base, reducing chemical use and waste generation. EDF has been extensively studied for lactic acid production from glucose [99–101, 106–111]. In general, EDF can improve reactor productivity and produce a concentrated and relatively pure product stream, simplifying further downstream processing. However, EDF usually gives lower product yields due to product loss and low conversion rates for the fermentation substrate, glucose. At a higher glucose conversion rate, productivity would be significantly reduced. In addition, the final product concentrations from EDF are lower than those of other fermentation processes (see Table 5). These drawbacks and its complexity greatly limit EDF for the production of carboxylic acids.

One main concern in EDF is membrane fouling by cells and proteins. Cell immobilization or removal by filtration is necessary to reduce membrane fouling and to allow electrodialysis to operate for a longer period. Boyaval *et al.* (1987) combined tangential flow filtration (TFF) with electrodialysis in a continuous *L. helveticus* fermentation to produce a concentrated product stream of lactic acid (85 g/L) [106]. With TFF cell recycle alone, they could increase productivity to as much as 22 g/L·h, but at a low conversion rate (~65%) and product concentration. For near complete conversion, the maximum productivity achieved was only ~6 g/L·h. When electrodialysis was incorporated into the scheme, product concentration rose, but productivity dropped to 3.2 g/L·h. Total cell recycle is often correlated with decreasing specific productivity, and therefore, a bleed stream is required to remove dead cells and maintain a reasonable cell concentration. Furthermore, they found that a high cell concentration resulted in a fouled filtration membrane, dramatically restricting the permeate flow. However, cell fouling of the electrodialysis membranes did not seem to be a problem. Yao and Toda (1990) used a hollow-fiber filter module to remove bacterial cells from broth entering the electrodialysis unit and observed improved productivity for EDF as compared to control fermentations [107]. The average productivity for their best electrodialysis run was 8.67 g/L·h. Siebold et al. (1995) used ultrafiltration to provide a cell-free permeate to the electrodialysis unit [108]. Without ultrafiltration to remove the cells and most of the protein that foul the membrane, long-term operation was not possible. The ED membrane required cleaning every 24 hours with a strong alkali surfactant. With ultrafiltration, cleaning was unnecessary. Current yield was higher for ultrafiltered broth than for non-filtered broth.

Another problem for EDF is the removal of nutrients (e.g., amino acids) and inorganic ions (phosphates, calcium, etc.) by electrodialysis. Amino acids have different ionic forms at different pHs; thus the loss of amino acids in fermentation broth can occur in EDF. A bipolar ion exchange membrane was found to reduce amino acid loss significantly as compared to conventional electrodialysis. EDF also may perform better when electrodialysis is operated intermittently instead of continuously [99, 110]. It was found that hydrogen gas produced during electrodialysis had a negative effect by reducing the redox potential of the fermentation medium [99].

3.3. Extractive fermentation

Extractive fermentation using an organic solvent to remove carboxylic acids from the fermentation broth has been extensively studied [102–105, 112–114]. Most studies have used long-chain aliphatic amines because of their high distribution coefficients. In general, extractive fermentation allows continuous production and recovery of the desired fermentation product in one continuous step, significantly improving reactor productivity and final product concentration by reducing end-product inhibition. Membrane-based extractive fermentation with product stripping by a strong base can produce a carboxylate product at a high concentration and purity [103]. Nevertheless, extractive fermentation usually suffers from solvent toxicity and is limited by suboptimal operating pHs, both of which can greatly limit reactor productivity. Here we will briefly discuss methods that can be used to reduce or avoid solvent toxicity and the effects of pH on extractive fermentation, followed with a few examples of membrane-based extractive fermentation from our own studies.

3.3.1 Solvent toxicity

Most organic solvents are toxic to bacteria: they either inhibit or stop microbial growth. Solvent toxicity can be exerted on microorganisms at the molecular level and at the phase level [115]. Toxicity at the phase level comes from direct contact between the solvent phase and the cells due to emulsions, which may block nutrient diffusion from the medium to cells due to solvent coating and may disrupt the cell wall due to increased surface tension. Phase toxicity can be eliminated by immobilizing cells or separating cells from the fermentation broth to prevent direct contact with the solvent [103]. Immobilizing the solvent in hydrophobic membranes to avoid the formation of emulsions can also protect cells from direct contact with the solvent [103].

Molecular toxicity comes from dissolved solvents, which can inhibit enzymes or modify cell membrane permeability. The toxicity of a solvent is closely related to its hydrophobicity or log P value, which is the octanol/water partition coefficient. In general, solvents with a log P higher than 4 are biocompatible, and those with a value lower than 2 are not biocompatible [115]. The molecular toxicity can be reduced to a minimal level by reducing the "dissolved" solvent in water. This can be done by mixing/diluting toxic solvents with a nontoxic solvent, such as oleyl alcohol, entrapping toxic solvents in non-toxic oils, such as corn and sunflower oils, encapsulating the solvent in microcapsules [114], and passing the solvent-laden fermentation broth through an activated-carbon filter [113]. Immobilized cells were also found to be more resistant to solvent toxicity than free cells [103].

Solvent toxicity also can be avoided by using aqueous two-phase systems [116]. Two water soluble polymers immiscible to each other can be used to generate two separate aqueous phases [117]. As around 80% of the two phases are composed of water, aqueous two-phase extraction provides a biocompatible environment [118]. Cell partition between the two phases is dependent on polymer concentration, types of polymers, phosphate buffering, and pH. The uneven partition of cells in an aqueous two-phase system allows product removal from the cell-free phase. However, the partition coefficient for carboxylic acids in aqueous two-phase systems is only 1, since the acid is evenly distributed in both phases. Therefore, the product is produced at a relatively low concentration, as in the conventional fermentation process. Furthermore, maintaining the two phases for long-term operation can be complicated since the distribution of phase-forming polymers may change during the extractive fermentation. In general, aqueous two-phase systems have very limited beneficial effects and are not considered a viable option for carboxylic acid fermentation.

3.3.2. Effects of pH

The most critical process factor affecting extractive fermentation is pH, which has significant effects on both carboxylic acid fermentation and extraction. In general, efficient solvent extraction requires a pH value below the pK_a value of the carboxylic acid, which is less than 5 for most carboxylic acids (see Table 1). On the other hand, most carboxylic acid fermentations have an optimal pH between 5 and 7. In an extractive fermentation, the fermentor pH will be self-regulated by balancing between the fermentation and extraction. Therefore, the fermentor pH does not have to be externally controlled with base addition; instead, it can be kept at a stable level via the removal of acid products by extraction. Thus, the extractive fermentation will be operated at a pseudo-steady-state pH at which the rate of acid production from fermentation is equal to the rate of acid removal by extraction. This has been illustrated in extractive fermentation studies of butyric, lactic, and propionic acids [103–105] and will be discussed below.

3.3.3. Membrane-based extractive fermentation

The hollow-fiber membrane-based extractive fermentation process shown in Figure 5 has been demonstrated with several carboxylic acid fermentations. Figure 6 shows the kinetics of

L-lactic acid production by *Rhizopus oryzae* in a fed-batch extractive fermentation. As can be seen in this figure, the fermentation was maintained at ~pH 4, and lactic acid was produced continuously at a stable rate, reaching a high concentration of ~293 g/L in the stripping solution. The overall lactic acid yield was higher than 90% based on glucose consumption, and almost no byproduct (ethanol) was produced in the fermentation. It is noted, however, that the reactor productivity (~0.73 g/L·h) was lower than that obtained from a comparable fermentation without product extraction but operated at the optimal fermentation pH of 6.5, at which the microorganism was able to produce lactate to a final concentration of ~120 g/L [24]. In this extractive fermentation, the reactor productivity was greatly limited by the lower operating pH because the extraction capacity of the hollow-fiber units, which is proportional to the total membrane surface area, used in the study was relatively small. Increasing the extractor capacity should allow the fermentation to operate at a higher pH and thus will increase the reactor productivity.

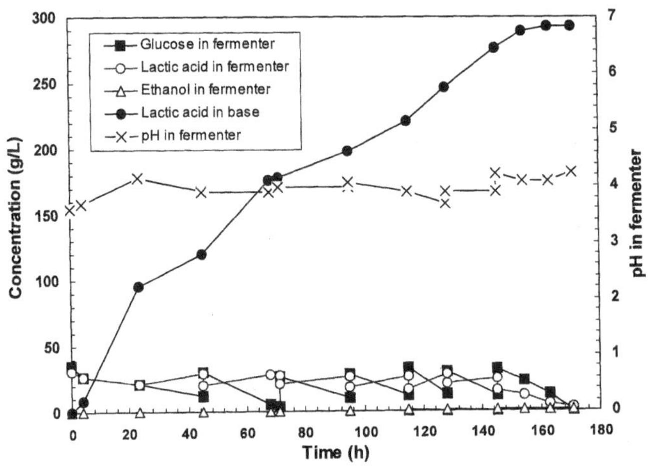

Fig. 6. Kinetics of extractive fermentation for lactic acid production from glucose by *Rhizopus oryzae* immobilized in a rotating fibrous-bed bioreactor with product extraction by Alamine 336 (30% in oleyl alchol) and back extraction by *6N* NaOH in hollow-fiber membrane extractors (see Figure 5 for the process scheme).

Figure 7 shows the kinetics of an extractive fermentation for propionic acid production by *Propionibacterium acidipropionici* immobilized in a fibrous-bed bioreactor. In this fermentation, the fermentation pH was maintained at ~4.8, and the final propionate concentration reached ~170 g/L, which is 2.4 times higher than that produced in a comparable fermentation at pH 7.0 (see Table 6). It is noted that the extractive fermentation not only gave significantly higher reactor productivity, but also much higher propionate yield and product purity because the production of acetate and succinate were dramatically reduced. The increased product purity also can be attributed to the higher selectivity of amine extraction for propionic acid than for acetic and succinic acids (see distribution coefficients in Table 4). Similar results have also been obtained with extractive fermentation for butyric acid production from glucose by *C. tyrobutyricum*, which reached a final butyrate concentration of

~300 g/L with a 10% increase in product yield and reactor productivity even though the fermentation was maintained at a suboptimal pH of ~5.5 [103].

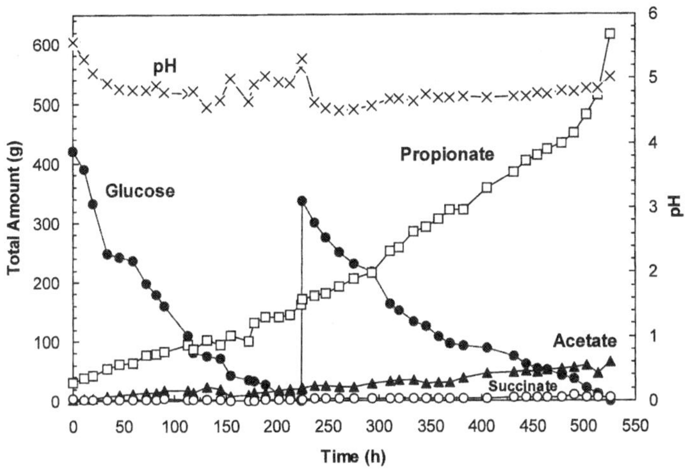

Fig. 7. Kinetics of fed-batch extractive fermentation for propionic acid production from glucose by *P. acidipropionici* immobilized in a fibrous-bed bioreactor with product extraction by Adogen 283 (5% in oleyl alchol) and back extraction by *6N* NaOH in hollow-fiber membrane extractors (see Figure 5 for the process scheme).

Table 6
Comparisons of propionic acid production in extractive and conventional fermentations

	Batch Fermentation			Extractive Fermentation	
	pH 7.1	pH 5.0	pH 7.0*	pH 5.3	pH 4.8*
Productivity (g/L·h)	0.2	0.12	0.09 / 0.26*	0.98	0.4 / 2.5*
Product Yield (g/g)					
Propionic acid	0.31	0.54	0.4–0.65	0.66	0.78
Acetic acid	0.12	0.13	0.10	0.07	0.11
Succinic acid	0.10	0.09	0.09	0.02	0.01
P/A Ratio	2.58	4.15	4.0	9.8	7.1
Product Purity	58%	71%	69%	88%	88%
Final Propionate Concentration (g/L)	12.5	18.5	71.7	75	170

*Fermentation with cells immobilized in a fibrous-bed bioreactor (FBB). The higher productivity value is based on the FBB volume, whereas the lower value is based on the total liquid volume in the system. Some data are from [38, 105].

It is clear that extractive fermentation can only be efficiently carried out at pHs significantly higher than the pK_a value of the carboxylic acid produced in the fermentation. However, most anaerobic bacteria grow better at pHs near 7, at which solvent extraction does not work well. To overcome this pH limitation, a novel two-step extractive fermentation, shown in Figure 8, has been proposed for the production of acetic acid from lactose using a coculture of homoacetogen (*C. formicoaceticum*) and homolactic acid bacteria [31]. In this process, lactic acid produced from lactose by homolactic acid bacteria immobilized in the first bioreactor is used as the substrate for acetic acid production by the homoacetogen immobilized in the second bioreactor operated at its optimal pH of 6.5–7.0. The high pH is not good for the extraction of acetic acid. Therefore, the acetate-laden fermentation broth is recycled back to the first bioreactor, where lactic acid produced from lactose lowers the pH to ~5 and thus acidifies the acetate to acetic acid. Then, the fermentation broth containing lactate and acetic acid enters the extractor, where acetic acid is selectively extracted, leaving the unextracted lactate for further conversion to acetate in the second bioreactor. This two-step fermentation and simultaneous separation process concept has been demonstrated in a simulated extractive fermentation with results shown in Figure 9. The concentrations of acetic and lactic acid were maintained at 10 and 1 g/L, respectively, in the simulated fermentation broth (pH 5.5), which was continuously extracted by passing through a hollow-fiber extractor. As can be seen in Figure 9, acetic acid was selectively extracted at a stable rate with a high product purity of ~90% in this system.

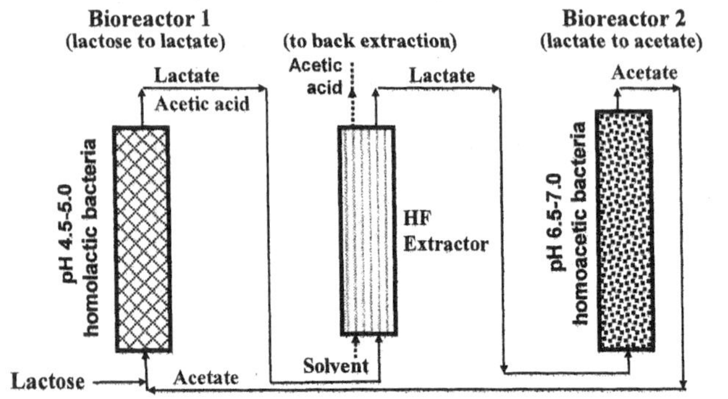

Fig. 8. A two-step extractive fermentation for acetic acid production from lactose using lactic acid produced by lactic acid bacteria to acidify the acetate produced by homoacetogens to allow efficient extraction of acetic acid by solvent in the hollow-fiber extractor.

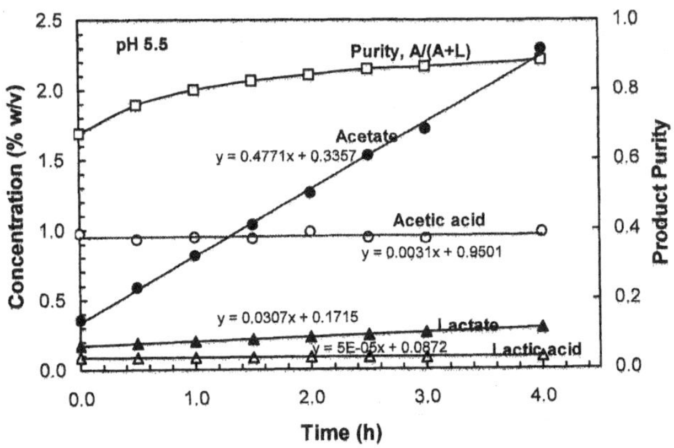

Fig. 9. Continuous extraction of acetic acid from a simulated fermentation broth containing 1% acetic acid and 0.1% lactic acid at pH 5.5. This experiment demonstrated the feasibility of the two-step extractive fermentation process concept shown in Fig. 8.

4. SUMMARY AND OUTLOOK

Although many important carboxylic acids can be produced by microorganisms from biomass, with only a few exceptions, the fermentative routes usually cannot compete with petroleum-based chemical synthesis because fermentative production of chemicals usually suffers from a number of serious limitations, mainly low product yield, reactor productivity, and final product concentration, which are caused by severe end-product inhibition. Extractive fermentation to selectively separate the desirable product *in situ* has the advantages of reducing product inhibition and increasing the fermentation rate and product yield. Also, extractive fermentation with simultaneous back extraction with a base can produce a highly concentrated and relatively pure carboxylate that can be efficiently converted to carboxylic acid by water-splitting electrodialysis. With the oil price exceeding $70 per barrel, it is attractive to produce acetic, butyric, propionic, fumaric, and other carboxylic acids that are currently produced from petroleum-based feedstocks from biomass. Extractive fermentation provides a promising way to produce these carboxylic acids from sugars.

REFERENCES

[1] G.T. Tsao, N.J. Cao, J. Du, C.S. Gong, Production of multifunctional organic acids from renewable resources. Adv Biochem Eng/Biotechnol. 65 (1999) 242–280.

[2] H. Song and S.Y. Lee, Production of succinic acid by bacterial fermentation. Enz. Microb. Technol. 39 (2006) 352–361.

[3] S.T. Yang, Y.L. Huang, Z. Jin, Yan Huang, H. Zhu, and W. Qin, *Calcium magnesium acetate at lower-production cost: production of CMA deicer from cheese whey*. Final report to Federal Highway Administration and New York State Energy and Development Authority; Publication No. FHWA-RD-98-174; April 1999.

[4] M.J. Playne, Propionic and butyric acids. In M. Moo-Young, ed., *Comprehensive Biotechnology*, Pergamon: New York, 1985, Vol 3, p. 731.

[5] A.J.J. Straathof, S. Sie, T.T. Franco, L.A.M. van der Wielen, Feasibility of acrylic acid production by fermentation. Appl Microbiol Biotechnol. 67 (2005) 727–734.

[6] R. Datta and M. Henry, Lactic acid: recent advances in products, processes and technologies – a review. J Chem Technol Biotechnol. 81(2006) 1119–1129.

[7] C.P. Kubicek and L. Karaffa, Organic acids, In C. Ratledge and B. Kristiansen (eds.) *Basic Biotechnology*, 3rd edition, Cambridge University Press, Chapter 15, 2006.

[8] H. Ohara, Biorefinery. Appl Microbiol Biotechnol 62 (2003) 474–477.

[9] B. Kamm and M. Kamm, Principles of biorefineries. Appl Microbiol Biotechnol 64 (2004) 137–145.

[10] Th. Willke and K.-D. Vorlop, Industrial bioconversion of renewable resources as an alternative to conventional chemistry. Appl Microbiol Biotechnol 66 (2004) 131–142.

[11] M. Gavrilescu and Y. Chisti, Biotechnology – a sustainable alternative for chemical industry. Biotechnol Adv 23 (2005) 471–499.

[12] H. Ebner and H. Follmann, Vinegar. In: G. Reed (ed.) *Biotechnology*, vol 5. *Food and feed production with microorganisms*. Verlag Chemie, Weinheim, pp. 425–446 (1983)

[13] T.K. Ghose and A. Bhadra, Acetic acid. In M. Moo-Young (ed.), *Comprehensive Biotechnology*, Vol. 3, Pergamon Press, New York, 1985. Chapter 36.

[14] J. Trcek, H. Toyama, J. Czuba, A. Misiewicz, K. Matsushita, Correlation between acetic acid resistance and characteristics of PQQ-dependent ADH in acetic acid bacteria. Appl Microbiol Biotechnol 70 (2006) 366–373.

[15] M. Roehr, C.P. Kubicek, and J. Kominek, Citric acid. In M. Roehr (ed.) *Biotechnology*, 2nd edition, vol. 6. *Products of Primary Metabolism*, Verlag Chemie, Weinheim. pp. 308–345, 1996.

[16] L. Karaffa and C.P. Kubicek, *Aspergillus niger* citric acid accumulation: do we understand this well working black box? Appl Microbiol Biotechnol 61 (2003)189–196.

[17] N. Cao, J. Du, C. Chen, C.S. Gong, G.T. Tsao, Production of fumaric acid by immobilized *Rhizopus* using rotary biofilm contactor. Appl Biochem Biotechnol. 63/65 (1997) 387–394.

[18] E. Riscaldati, M. Moresi, F. Federici, M. Petruccioli, Ammonium fumarate production by free or immobilised *Rhizopus arrhizus* in bench- and laboratory-scale bioreactors. J Chem Technol Biotechnol. 77 (2002) 1013–1024.

[19] M. Roehr, C.P. Kubicek, and J. Kominek, Gluconic acid. In M. Roehr (ed.) *Biotechnology*, 2nd edition, vol. 6. *Products of Primary Metabolism*, Verlag Chemie, Weinheim. pp. 347–362, 1996.

[20] S. Anastassiadis, A. Aivasidis, C. Wandrey, Continuous gluconic acid production by isolated yeast-like mould strains of *Aureobasidium pullulans*, Appl Microbiol Biotechnol. 61 (2003) 110–117.

[21] T. Willke and K.D. Vorlop, Biotechnological production of itaconic acid. Appl Microbiol Biotechnol. 56 (2001) 289–295.

[22] T. Willke, K. Welter and K.D. Vorlop, Biotechnological production of itaconic acid from sugar. Zuckerindustrie 126 (2001) 444–447.

[23] K. Yahirol, T. Takahama, S.-R. Jai, Y. Park, M. Okabe, Comparison of air-lift and stirred tank reactors for itaconic acid production by *Aspergillus terreus*. Biotechnol Lett. 19 (1997) 619–621.

[24] A. Tay and S.T. Yang. Production of L(+)-lactic acid from glucose and starch by immobilized cells of *Rhizopus oryzae* in a rotating fibrous bed bioreactor. Biotechnol. Bioeng., 80 (2002) 1–12.

[25] E. Battat, Y. Peleg, A. Bercovitz, J.S. Rokem, I. Goldberg, Optimization of L-malic acid production by *Aspergillus flavus* in a stirred fermenter. Biotechnol Bioeng. 37 (1991) 1108–1116.

[26] M. Kawagoe, K. Hyakumura, S.-I. Suye, K. Miki, K. Naoe, Application of bubble column fermentors to submerged culture of *Schizophyllum commune* for production of L-malic acid. J Ferment Bioeng. 84 (1997) 333–336.

[27] Y. Li, J. Chen, S.-Y. Lun, Biotechnological production of pyruvic acid. Appl Microbiol Biotechnol. 57 (2001) 451–459.

[28] R. Miyata and T. Yonehara, Breeding of high-pyruvate-producing *Torulopsis glabrata* with acquired reduced pyruvate decarboxylase. J Biosci Bioeng. 88 (1999) 173–178.

[29] A.J.A. van Maris, J.M.A. Geertman, A. Vermeulen, M.K. Groothuizen, A.A. Winkler, M.D.W. Piper, J.P. Van Dijken, J.T. Pronk, Directed evolution of pyruvate decarboxylase-negative *Saccharomyces cerevisiae*, yielding a C-2-independent, glucose tolerant, and pyruvate-hyperproducing yeast. Appl Environ Microbiol. 70 (2004) 159–166.

[30] S.R. Parekh and M. Cheryan, High concentration of acetate with a mutant strain of *C. thermoaceticum*. Biotechnol Lett. 16 (1994) 139–142.

[31] Y. Huang and S.T. Yang, Acetate production from whey lactose using co-immobilized cells of homolactic and homoacetic bacteria in a fibrous-bed bioreactor. Biotechnol Bioeng. 60 (1998) 498–507.

[32] Y.L. Huang, K. Mann, J.M. Novak, and S.T. Yang, Acetic acid production from fructose by *Clostridium formicoaceticum* in a fibrous-bed bioreactor, Biotechnol Prog., 14 (1998) 800–806.

[33] Y. Zhu, Z. Wu and S.T. Yang, Butyric acid production from acid hydrolysate of corn fiber by *Clostridium tyrobutyricum* in a fibrous bed bioreactor, Process Biochem. 38 (2002) 657–666.

[34] Y. Zhu and S.T. Yang, Adaptation of *Clostridium tyrobutyricum* for enhanced tolerance to butyric acid in a fibrous-bed bioreactor. Biotechnol Prog. 19 (2003) 365–372.

[35] S.K. Singh, S.U. Ahmed, and A. Pandey, Metabolic engineering approaches for lactic acid production. Process Biochem. 41 (2006) 991–1000.

[36] D.M. Bai, Q. Wei, Z.H. Yan, X.M. Zhao, XG Li, S.M. Xu, Fedbatch fermentation of *Lactobacillus lactis* for hyper-production of L-lactic acid. Biotechnol Lett. 25 (2003)1833–1835.

[37] S.T. Yang, Y. Huang, and G. Hong, A novel recycle batch immobilized cell bioreactor for propionate production from whey lactose. Biotechnol Bioeng., 45 (1995) 379–386.

[38] S. Suwannakham and S.T. Yang. Enhanced propionic acid fermentation by *Propionibacterium acidipropionici* mutant obtained by adaptation in a fibrous-bed bioreactor, Biotechnol Bioeng. 91 (2005) 325–337.

[39] J.G. Zeikus, M.K. Jain, P. Elankovan. Biotechnology of succinic acid production and markets for derived industrial products. Appl Microbiol Biotechnol 51 (1999) 545–552.

[40] G.N. Vemuri, M.A. Eiteman, and E. Altman. Succinate production in dual-phase *Escherichia coli* fermentations depends on the time of transition from aerobic to anaerobic conditions. J Ind Microbiol Biotechnol. 28 (2002) 325–332.

[41] Y.L. Huang, Z. Wu, L. Zhang, C.M. Cheung, and S.T. Yang, Production of carboxylic acids from corn meal by immobilized cell fermentation in a fibrous-bed bioreactor, Bioresource Technol. 82 (2002) 51–59.

[42] X. Cao, H.S. Yun, and Y.-M. Koo, Recovery of L-(+)-lactic acid by anion exchange resin Amberlite IRA-400. Biochem Eng J. 11 (2002) 189–196.

[43] J.L. Wang, X.H. Wen, D. Zhou, Production of citric acid from malasses integrated with in-situ product separation by ion-exchange resin adsorption. Bioresource Technol. 75 (2000) 231–234.

[44] N. Cao, J. Du, C. Chen, C.S. Gong, G.T. Tsao, Simultaneous production and recovery of fumaric acid from immobilized *Rhizopus* with a rotary biofilm contactor and an adsorption column. Appl Environ Microbiol. 62 (1996) 2926–2931.

[45] H.M. Anasthas and V.G. Gaikar, Adsorption of acetic acid on ion-exchange resins in non-aqueous conditions, Reactive Functional Polymers 47 (2001) 23–35.

[46] P. Gluszcz, T. Jamroz, B. Sencio, S. Ledakowicz, Equilibrium and dynamic investigations of organic acids adsorption onto ion-exchange resins. Bioprocess Biosyst Eng. 26 (2004) 185–190.

[47] F.P. Gailliot, C. Gleason, J.J. Wilson, Fluidized bed adsorption for whole broth extraction. Biotechnol. Progr. 6 (1990) 370–375.

[48] B.H. Davison and C.D. Scott, A proposed biparticle fluidized-bed for lactic acid fermentation and simultaneous adsorption. Biotechnol. Bioeng. 39 (1992) 365–368.

[49] L. Ju and C. Chen, Adsorption characteristics of polyvinylpyridine and activated carbon for lactic acid recovery from fermentation of *Lactobacillus delbrueckii*. Sep Sci Technol. 33 (1998), 1423–1437.

[50] I.H. Aljundi,., J.M. Belovich, O. Talu. Adsorption of lactic acid from fermentation broth and aqueous solutions on Zeolite molecular sieves. Chem Eng Sci. 60 (2005) 5004–5009.

[51] R.K. Nagarale, G.S. Gohil, V.K. Shahi, Recent developments on ion-exchange membranes and electro-membrane processes. Adv Colloid Interface Sci. 119 (2006) 97–130.

[52] T. Xu, Ion exchange membranes: State of their development and perspective. J Membrane Sci. 263 (2005) 1–29.

[53] M. Bailly, Production of organic acids by bipolar electrodialysis: realizations and perspectives. Desalination 144 (2002) 1.57–162.

[54] J. Wiśniewski, G. Wiśgniewska, T. Winnicki, Application of bipolar electrodialysis to the recovery of acids and bases from water solutions. Desalination 169 (2004) 11–20.

[55] Z. Wang, Y. Luo, P. Yu, Recovery of organic acids from waste salt solutions derived from the manufacture of cyclohexanone by electrodialysis. J Membrane Sci. 280 (2006) 134–137.

[56] V.H. Thang, W. Koschuh, K.D. Kulbe, S. Novalin, Detailed investigation of an electrodialytic process during the separation of lactic acid from a complex mixture. J Membrane Sci. 249 (2005) 173–182.

[57] E.G. Lee, S.-H. Moon, Y.K. Chang, I.-K. Yoo, H.N. Chang, Lactic acid recovery using two-stage electrodialysis and its modelling. J Membrane Sci. 145 (1998) 53-66.

[58] L. Madzingaidzo, H. Danner, R. Braun, Process development and optimisation of lactic acid purification using electrodialysis. J Biotechnol 96 (2002) 223–239.

[59] V. Habova, K. Melzoch, M. Rychtera, B. Sekavova, Electrodialysis as a useful technique for lactic acid separation from a model solution and a fermentation broth. Desalination 163 (2004) 361–372.
[60] Y.-J. Wee, J.-S. Yun, Y.Y. Lee, A.-P. Zeng, H.-W. Ryu, Recovery of lactic acid by repeated batch electrodialysis and lactic acid production using electrodialysis wastewater. J Biosci Bioeng. 99 (2005) 104–108.
[61] L. Yu, T. Lin, Q. Guo, J. Hao, Relation between mass transfer and operation parameters in the electrodialysis recovery of acetic acid. Desalination 154 (2003) 147–152.
[62] M. Fidaleo, M. Moresi, Modeling of sodium acetate recovery from aqueous solutions by electrodialysis. Biotechnol Bioeng. 91 (2005) 556–568.
[63] G.S. Luo., X.Y. Shan, X. Qi, Y.C. Lu, Two-phase electro-electrodialysis for recovery and concentration of citric acid. Separation Purification Technol. 38 (2004) 265–271
[64] J.S.J. Ferrer, S. Laborie, G. Durand, M. Rakib, Formic acid regeneration by electromembrane processes. Membrane Sci. 280 (2006) 509–516.
[65] R.K. Nagarale, G.S. Gohil, V.K. Shahi, G.S. Trivedi, S.K. Thampy, R. Rangarajan, Studies on transport properties of short chain aliphatic carboxylic acids in electrodialytic separation. Desalination 171 (2004) 195–204.
[66] T. Godjevargova, R. Dayal, S. Turmanova, Gluconic acid production in bioreactor with immobilized glucose oxidase plus catalase on polymer membrane adjacent to anion-exchange membrane. Macromol. Biosci. 4 (2004) 950–956.
[67] M. Fidaleo and M. Moresi, Assessment of the main engineering parameters controlling the electrodialytic recovery of sodium propionate from aqueous solutions. J Food Eng. 76 (2006) 218–231.
[68] B. Zelić and D. Vasić-Rački, Process development and modeling of pyruvate recovery from a model solution and fermentation broth. Desalination 174 (2005) 267–276.
[69] Y.H. Kim, S.H. Moon, Lactic acid recovery from fermentation broth using one-stage electrodialysis J. Chem. Technol. Biotechnol., 76 (2001) 169–178.
[70] A. Narebska, M. Kurantowicz, M. Staniszewski, Separation of fermentation products by membrane techniques. IV. Electrodialytic conversion of carboxylates to carboxylic acids. Sep. Sci. Technol., 36 (2001) 443–455.
[71] A.S. Kertes and C.J. King, Extraction chemistry of fermentation product carboxylic acids. Biotechnol Bioeng. 28 (1986) 269–282.
[72] M. Matsumoto, K. Mochiduki, K. Fukunishi, K Kondo, Extraction of organic acids using imidazolium-based ionic liquids and their toxicity to *Lactobacillus rhamnosus*. Sep Purification Technol. 40 (2004) 97–101.
[73] J. Planas, Kozlowski, J.M. Harris, F. Tjerneld, B. Hahn-Hagerda, Novel polymer-polymer conjugates for recovery of lactic acid by aqueous two-phase extraction. Biotechnol. Bioeng., 66 (1999) 211-218.
[74] Y. Wang and L.E.K. Achenie, Computer aided solvent design for extractive fermentation. Fluid Phase Equilibria 201 (2002) 1–18.
[75] C.J. King, Amine-based systems for carboxylic acid recovery. CHEMTECH, May (1992) 285–291.
[76] J. Prochazka, A. Heyberger, V. Bizek, M. Kousova, E. Volantova, Amine extraction of hydroxycarboxylic acids. 2. Comparison of equilibria for lactic, malic, and citric acids. nd Eng Chem Res. 33 (1994) 1565–1573.
[77] J.A. Tamada and C.J. King, Extraction of carboxylic acids with amine extractants. 3. Effect of temperature, water coextraction, and process considerations. Ind Eng Chem Res. 29 (1990) 1333–1338.
[78] S.T. Yang, S.A. White, S.T. Hsu, Extraction of carboxylic acids with tertiary and quaternary amines: effect of pH. Ind Eng Chem Res. 30 (1991) 1355–1342.

[79] A. M. Eyal and R. Canari, PH dependence of carboxylic and mineral acid extraction by amine-based extractants: effects of pK_a, amine basicity, and diluent properties, Ind Eng Chem Res. 34 (1995) 1789–1798.
[80] N. Tik, E. Bayraktar, Ü. Mehmetoglu, In situ reactive extraction of lactic acid from fermentation media. J Chem Technol Biotechnol. 76 (2001) 764–768.
[81] K.L. Wasewar, V.G. Pangarkar, A.B.M. Heesink, G.F. Versteeg, Intensifcation of enzymatic conversion of glucose to lactic acid by reactive extraction. Chem Eng Sci. 58 (2003) 3385–3393.
[82] Y.K. Hong and W.H. Hong, Equilibrium studies on reactive extraction of succinic acid from aqueous solutions with tertiary amines. Bioprocess Eng. 22 (2000) 477–481.
[83] A. Senol, Influence of diluent on amine extraction of pyruvic acid using Alamine system. Chem Eng Processing 45 (2006) 755–763.
[84] A. Senol, Effect of diluent on Amine extraction of acetic acid: modeling considerations. Ind Eng Chem Res. 43 (2004) 6496–6506.
[85] A. Kiani, R.R. Bhave, K.K. Sirkar, Solvent extraction with immobilized interfaces in a microporous hydrophobic membrane. J Membrane Sci. 20 (1984) 125–145.
[86] A.Sengupta, R. Basu, K.K. Sirkar, Separation of solutes from aqueous solutions by contained liquid membranes. AIChE J. 34 (1988) 1698–1708.
[87] R. Basu and K.K. Sirkar, Hollow fiber contained liquid membrane separation of citric acid. AIChE J. 37 (1991) 383–393.
[88] A.M. Eyal and E. Bressler, Mini-review industrial separation of carboxylic acid and amino acids by liquid membranes: applicability, process considerations and potential advantages. Biotechnol Bioeng. 41 (1993) 287–295.
[89] M.S. Solichien, D. O'Brien, E.G. Hammond, C.E. Glatz, Membrane-based extractive fermentation to produce propionic and acetic acids: toxicity and mass transfer considerations. Enz Microb Technol. 17 (1995) 23–31.
[90] H. Huang, S.T. Yang, D.E. Ramey, A hollow-fiber membrane extraction process for recovery and separation of lactic acid from aqueous solution. Appl Biochem Biotechnol. 114 (2004) 671–688.
[91] W.S.W. Ho, Combined supported liquid membrane/strip dispersion process for the removal and recovery of penicillin and organic acids, U. S. Patent 6,433,163 (2002)
[92] D. Stark and U. von Stockar, In Situ Product Removal (ISPR) in Whole Cell Biotechnology During the Last Twenty Years. Adv Biochem Eng/Biotechnol. 80 (2003) 149-175.
[93] P. Fernandes, D.M.F. Prazeres, J.M.S. Cabral, Membrane-Assisted Extractive Bioconversions. Adv Biochem Eng/Biotechnol. 80 (2003) 115-148.
[94] J.J. Malinowski, Two-phase partitioning bioreactors in fermentation technology. Biotechnol Adv. 19 (2001) 525–538.
[95] C.C. Chen and L.-K. Ju, Coupled lactic acid fermentation and adsorption. Appl Microbiol Biotechnol. 59 (2002) 170–174.
[96] A. Gonzalez-Vara y R., G. Vaccari, E. Dosi, A. Trilli, M. Rossi, D. Matteuzzi, Enhanced production of L-(+)-lactic acid in chemostat by *Lactobacillus casei* SM 20011 using ion-exchange resins and cross-flow filtration in a fully automated pilot plant controlled via NIR. Biotechnol Bioeng. 67 (2000) 147–156.
[97] A. Srivastava, P.K. Roychoudhury, V. Sahai, Extractive lactic acid fermentation using ion-exchange resin. Biotechnol. Bioeng., 39 (1992) 607–613.
[98] A. Senthuran, V. Senthuran, R. Hatti-Kaul, B. Mattiasson, Lactate production in an integrated process configuration: reducing cell adsorption by shielding of adsorbent. Appl Microbiol Biotechnol. 65 (2004) 658–663.
[99] P. Vonktaveesuk, M.Tonokawa, A. Ishizaki, Simulation of the rate of L-lactate fermentation using *Lactococcus lactis* IO-1 by periodic electrodialysis. J. Ferment. Bioeng., 77 (1994) 508-512.

[100] L. Xuemei, L. Jianping, L. Mo'e, C. Peilin, L-lactic acid production using immobilized *Rhizopus oryzae* in a three-phase fluidized-bed with simultaneous product separation by electrodialysis. Bioprocess Eng., 20 (1999) 231-237.

[101] J.P.L.C. da Costa, C. Schorm, A. Quesada-Chanto, K.W. Boddeker, R. Jonas, On-line dialysis of organic acids from a *Propionibacterium freudenreichii* fermentation: evaluation of a new pH control strategy. Appl. Biochem. Biotechnol. 76 (1999) 99-106.

[102] S. Wieczorek and H. Brauer, Continuous production of citric acid with recirculation of the fermentation broth after product recovery. Bioprocess Eng. 18 (1998) 75-77.

[103] Z. Wu and S.T. Yang, Extractive fermentation for butyric acid production from glucose by *Clostridium tyrobutyricum*, Biotechnol Bioeng. 82 (2003) 93-102.

[104] A.Tay, Production of L(+)-lactic acid from glucose and starch by fermentations with immobilied cells of *Rhizopus oryzae*. Ph.D. thesis, Ohio State University, Columbus, Ohio, 2002

[105] Z. Jin and S.T. Yang, Extractive fermentation for enhanced propionic acid production from lactose by *Propionibacterium acidipropionici.* Biotechnol Progress, 14 (1998) 457–465.

[106] P. Boyaval, C. Corre, S. Terre, Continuous lactic acid fermentation with concentrated product recovery by ultrafiltration and electrodialysis. Biotechnol. Lett., 9 (1987) 207-212.

[107] P.X. Yao, K. Toda, Lactic acid production in electrodialysis culture. J. Gen. Appl. Microbial., 36 (1990) 111-125.

[108] M. Siebold, P. Vonfrieling, R. Joppien, D. Rindfleisch, K. Schugerl, H. Roper, Comparison of the production of lactic acid by three different Lactobacilli and its recovery by extraction and electrodialysis. Process Biochem., 30 (1995) 81-95.

[109] H. Li, R. Mustacchi, C.J. Knowles, W. Skibar, G. Sunderland, I. Dalrymple, S.A. Jackman, An electrokinetic bioreactor: using direct electric current for enhanced lactic acid fermentation and product recovery. Tetrahedron, 60 (2004) 655-661.

[110] M.-t. Gao, M. Hirata, M. Koide, H. Takanashi, T. Hano, Production of L-lactic acid by electrodialysis fermentation (EDF). Process Biochem. 39 (2004) 1903–1907.

[111] M.-t. Gao, M. Koide, R. Gotou, H. Takanashi, M. Hirata, T. Hano, Development of a continuous electrodialysis fermentation system for production of lactic acid by *Lactobacillus rhamnosus*. Process Biochem. 40 (2005) 1033–1036.

[112] M. Kilic, E. Bayrakatar, S. Ates, U. Mehmetoglu, Investigation of extractive citric acid fermentation using response-surface methodology. Process Biochem. 37 (2002) 759-767.

[113] K. Ye, S. Jin, K. Shimizu, Performance improvement of lactic acid fermentation by multistage extractive fermentation. J. Ferment. Bioeng., 81 (1996) 240-246.

[114] K. Kondo, T. Otono, M. Matsumoto, Preparation of microcapsules containing extractants and the application of the microcapsules to the extractive fermentation of lactic acid. J. Chem. Eng. Jpn., 37 (2004) 1-6.

[115] M. Matsumoto, K. Mochiduki, K. Kondo, Toxicity of ionic liquids and organic solvents to lactic acid-producing bacteria. J Biosci Bioeng. 98 (2004) 344–347.

[116] J. Planas, P. Radstrom, F. Tjerneld, B. HahnHagerdal, Enhanced production of lactic acid through the use of a novel aqueous two-phase system as an extractive fermentation system. Appl. Microbiol. Biotechnol., 45 (1996) 737-743.

[117] G. M Zijlstra, C.D de Gooijert, J. Tramper, Extractive bioconversions in aqueous two-phase systems. Curr Opin Biotechnol. 9 (1998) 171–176.

[118] J. Sinha, P.K. Dey, and T. Panda, Aqueous two-phase: the system of choice for extractive fermentation. Appl Microbiol Biotechnol. 54 (2000) 476–486.

Bioprocessing for Value-Added Products from Renewable Resources
Shang-Tian Yang (Editor)

Chapter 17. Fungal Fermentation for Medicinal Products

Chin-Han Shu

Department of Chemical and Materials Engineering, National Central University, #300, Jungda Road, Jungli 32054, Taiwan, R.O.C.

1. INTRODUCTION

Fungi have played important roles as foods and medicines in both ancient and modern biotechnological processes. Fungi range from microscopic yeasts and molds to macroscopic mushrooms. Their applications include production of antibiotics, alcohols, enzymes, organic acids, and numerous pharmaceuticals. The advent of recombinant DNA technology enables fungi to utilize novel carbon sources and to be hosts for the production of heterogonous proteins. Although several reviews of fungi as microbial cell factories for food use [1–3] and enzyme production [4, 5] have been published recently, reviews of fungi as cell factories for medicinal products are relatively limited and scattered.

Recently, new drug candidates from fungi have been found with anti-tumor, anti-hypertensive, immunosuppressant, anti-diarrheal, or anti-mutagenic properties. Increasing scientific evidence from animal tests and clinical studies has supported the idea that some fungi could be used as adjuvant cancer treatments. Thus, fungi play important roles in the booming nutraceutical and functional food markets. Although more scientific evidence is required to substantiate the therapeutical effects of medicinal fungi, possible healing mechanisms and their key compounds have been proposed. This review highlights fungal medicinal products, their therapeutical potential and engineering aspects in manufacturing these products.

2. FUNGAL PHYSIOLOGY

Fungi are the most accomplished chemists in nature due to their abilities in synthesizing various metabolites. The compounds they synthesize range from primary metabolites such as citric acid and enzymes, to secondary metabolites such as ergot alkaloids and antibiotics. In general, the primary metabolites behave as the building blocks for the secondary metabolites, and their simplified interrelationships are shown in Fig. 1. Most of the secondary metabolites from fungi are classified among five major metabolic sources: amino acids, the shikimic acid pathway, the polyketide pathway, the mevalonic acid pathway, and polysaccharides [6]. Most of the therapeutic effects of fungal metabolites on humans are limited to the secondary metabolites.

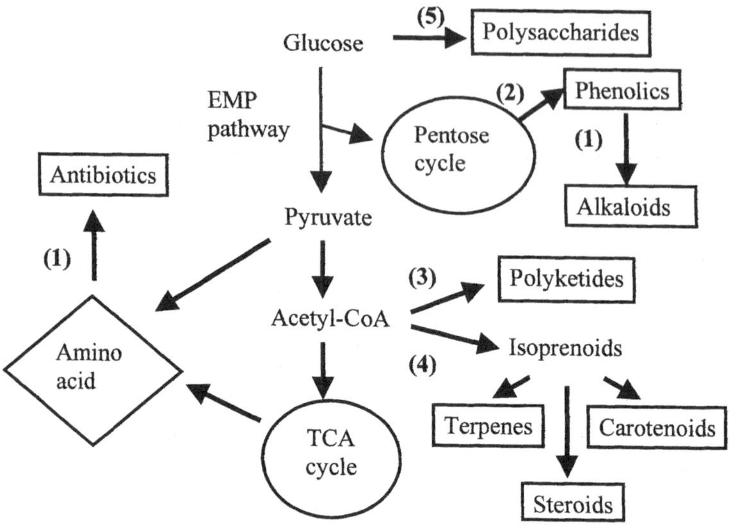

Fig. 1. Simplified interrelationships of metabolic pathways in primary and secondary metabolism. The major pathways of secondary metabolism are numbered as follows: (1) amino acid-derived pathway; (2) shikimic acid pathway; (3) the polyketide pathway; (4) mevalonic acid pathway; and (5) glucose-derived metabolites.◇: primary metabolites; □: secondary metabolites.

3. FUNGAL METABOLITES

3.1. Current fungal therapeutics

3.1.1. Amino acid-derived products and enzymes in the biosynthesis of therapeutics or their intermediates

From amino acids fungi produce a wide spectrum of therapeutic products, which include antibiotics such as penicillin and cephalosporins and peptides such as cyclosporins (Table 1).

Table 1
Current therapeutics from fungal amino acid-derived metabolites

Drug name	Structure	Microorganism	Description	Reference
Penicillins	β-lactam antibiotics	*Penicillium chrysogenum*	Inhibitor of cell wall synthesis in gram-positive bacteria	[11]
Cephalosporins	β-lactam antibiotics	*Cephalosporium acremonium*	Inhibitor of cell wall synthesis in both gram-positive and gram-negative bacteria	[12]
Cyclosporins	cyclic undecapeptide	*Tolypocladium inflatum*	Immunosuppressive, anti-parasitic, anti-inflammatory, and antifungal activity	[13]
Caspofungin	lipopeptide	*Glarea lozoyensis*	glucan synthesis inhibitors	[10]

Penicillin, the first β-lactam antibiotic, discovered by Fleming [7], is considered to be one of the most important antibiotics in terms of annual production and therapeutic use [8]. Due to significant strain improvement and process development in industry, penicillin G and penicillin V are the natural penicillins produced by fermentation. Only a relatively small amount of these penicillins is used directly as therapeutic agents while the majority is used for the production of the β-lactam nucleus, 6-amino penicillanic acid (6-APA). A wide range of semi-synthetic penicillins are derived from 6-APA. These newer antibiotics are claimed to have better properties in terms of increased stability, easier absorption and fewer side effects. However, the excessive use of penicillins has caused the development of drug resistance in pathogens.

Cephalosporins are β-lactam antibiotics but show broader spectrum antibiotic activity against gram-negative as well as gram-positive bacteria as compared to penicillin. Cyclosporin A is a cyclic undecapeptide with a broad spectrum of pharmacological properties, including immunosuppressive, anti-parasitic, anti-inflammatory, and antifungal activity. Cyclosporin A can be produced by both submerged and solid-state cultures, and is a potent immunosuppressant drug for human transplant patients [9].

Caspofungin, derived from pneumocandin, is an antifungal peptide which inhibits fungal β-1,3-glucan synthesis [10]. This novel antifungal agent is the first echinocandin licensed in the USA for the treatment of refractory invasive aspergillosis.

Although most of the enzymes from fungi cannot be used as therapeutics directly, some of them are used as biological catalysts in the synthesis of pharmaceutical intermediates and drugs. Table 2 lists some fungal enzymes that are or can be used to produce drugs and drug intermediates.

Table 2
Fungal enzymes for production of pharmaceutical intermediates and drugs

Enzyme name	Drug name	Source of enzyme/microorganism	Description	Reference
Penicillin G amidohydrolase	6-APA	*Penicillium chrysogenum*	β-lactam antibiotics intermediate	[14]
Penicillin V amidohydrolase	6-APA	*Fusarium* sp. SKF 235	β-lactam antibiotics intermediate	[15]
Protease	(S)-Ibuprofen	*Aspergillus oryzae*	anti-inflammatory and pain relief medicine,	[16]
Lipase B	506U78	*Candida antarctica*	anti-leukemia agent	[17]
Lipase B	(S)-2-Pentanol	*Candida antarctica*	anti-Alzheimer's	[18]
Epoxide hydrolase	(S) epoxide	*Aspergillus niger* SC 16310	melatonin receptor agonist	[19]
Formate dehydrogenase	Omapatrilat	*Pichia pastoris*	antihypertensive	[20]

3.1.2. Shikimic acid derived products

The shikimic acid pathway provides an alternative route to aromatic compounds, particularly the aromatic amino acids. Various metabolites, including alkaloids, phenolics and mycotoxins, are derived from aromatic acids. Several bioactive compounds from this pathway are listed in Table 3.

Psilocybin acts as a product and its hydroxy metabolite psilocin represents the true pharmacologically active agent [25]. Although these compounds are Schedule I drugs in the United States, they are important drugs in psychotherapy.

Asperlicin is a potent antagonist of cholecystokinin, which is synthesized from tryptophan and leucine by *Aspergillus alliaceus* [21]. Cholecystokinin is a recognized hormone neurotransmitter involved in the control of pancreatic secretion, gallbladder contraction, and gut motility.

Ergot comprises a group of indole alkaloids, which are predominantly found in various species of *Claviceps*. The ergot alkaloids and derivatives have a broad spectrum of pharmacological action that includes central, neurohumoral and peripheral effects [22]. More than 50 proprietary and generic drugs are available in the USA for the treatment of migraine headache and to modulate uterine contraction [23]. Psilocin and psilocybin, which are structurally related to the neurotransmitter serotonin, are the compounds responsible for the hallucinogenic properties of "magic mushrooms" [24].

Table 3.
Current fungal pharmaceuticals from shikimic acid-derived metabolites.

Drug name	Structure	Microorganism	Description	Reference
Asperlicin	benzodiazepin	*Aspergillus alliaceus*	cholecystokinin antagonist	[21]
Ergot alkaloids	indole alkaloids	*Claviceps purpurea*	dopamine antagonists, serotonin, 5-hydroxytryptamine and α-adrenoreceptors modulator	[23]
Psilocin	indole alkaloids	*Psilocybe cubensis*	binding to 5-hydroxytryptamine receptor	[24]
Chlorflavonin	flavonoid	*Aspergillus candidus*	antifungal agents	[27]

3.1.3. Polyketides and acetate pathway derived products

Most of the phenolics are major natural antioxidants used in functional foods. Likewise, *Aspergillus* can produce not only natural antioxidants [26] but also a flavonoid antibiotic, chlorflavonin [27].

In spite of the fact that almost two-thirds of the known bioactive polyketide natural products are isolated from the actinomycetes, several important mycotoxins such as aflatoxins and therapeutic statins are produced by filamentous fungi via the polyketide pathway. Natural statins including lovastatin and mevastatin are produced by *Aspergillus* and *Penicillium*, respectively. Semisynthetic statins include chemically modified simvastatin and pravastatin and synthetically derived atorvastatin, fluvastatin and cerivastatin. Both natural and semisynthetic statins are potent competitive inhibitors of 3-hydroxy-3-methylglutaryl-coenzyme A (HMG-CoA) reductase, a key enzyme that catalyzes the rate-limiting step within the cholesterol biosynthetic pathway. Lovastatin, known as mevinolin and produced by *Aspergillus terreus*, is commercially available as MEVACOR for the treatment of hypercholesterolemia [28]. Mevastatin, known as compactin, was isolated from *Penicillium citrinum* [29].

Griseofulvin, isolated from *Penicillin griseofulvum,* is a fungistatic compound active against microtubule formation during fungal mitosis [30]. It is used in the treatment of ringworm and other fungal infections of the skin or nails.

Mycophenolic acid (MPA), produced by *Penicillium brevicompactum*, is an antibiotic that exhibits antiproliferative activity on a variety of tumors in mice and rats by functioning as a potent uncompetitive, reversible inosine monophosphate dehydrogenase (IMPDH) inhibitor [31]. IMPDH is the rate-limiting enzyme in *de novo* synthesis of guaosine nucleotides. Immune cells, such as T- and B-lymphocytes, are specifically dependent on this pathway. Thus, it has been introduced into mycophenolate mofetil (MMF) as an immunosuppressant drug for kidney transplant patients [32]. Table 4 summarizes these therapeutics from fungal polyketides and acetate pathways.

Table 4.
Current therapeutics from fungal polyketides and acetate pathways.

Drug name	Structure	Microorganism	Description	Reference
Lovastatin	naphthalene β-hydroxylactone	*Aspergillus terreus*	HMG-CoA reductase inhibitor	[28]
Mevastatin	naphthalene β-hydroxylactone	*Penicillium citrinum*	HMG-CoA reductase inhibitor	[29]
Griseofulvin	chlorine containing antibiotics	*Penicillium griseofulvum*	Inhibitor of microtubule assembly in dermatophytic fungi	[30]
Mycophenolic acid	aromatic polyketide	*Penicillium brevicompactum*	IMPDH enzyme inhibitor	[32]

3.1.4. Mevalonate pathway derived products

Terpenoids and steroids are derived from the mevalonate pathway. Currently, the known bioactive compounds from the mevalonate pathway are isolated from plants. However, some therapeutic compounds, such as taxol and fusidic acid, also could be produced by fungi (Table 5).

Taxol, a diterpene ester isolated from the Pacific yew (*Taxus brevifolia*) is being used clinically in the treatment of ovarian and metastatic breast cancers [33]. Chemical synthesis of the complete taxol molecule is a tedious and complex process. Thus, alternative production methods are needed. *Taxomyces andreanae* [34] and *Pestalotiopsis microspora* [35] have been discovered as the taxol-producing endophytic fungi, and further optimization in both strain and process might provide an alternative path to commercial production of taxol.

Fusidic acid, derived from the fungus *Acremonium fusidioides*, has a narrow spectrum of activity against staphylococci, including strains resistant to other classes of antibiotics. This makes the drug of increasing importance given the continued emergence of resistances in this genus [36].

Table 5.
Current and potential therapeutics from fungal mevalonate pathway.

Drug name	Structure	Microorganism	Description	Reference
Taxol	diterpenoid	*Pestalotiopsis microspora*	Anticancer agent; inhibits tubulin formation during cell division	[33]
Fusidic acid	steroidal antibiotic	*Acremonium fusidioides*	Narrow-spectrum antibiotics against drug-resistant gram positive bacteria; inhibits protein biosynthesis	[36]

3.1.5. Glucose-derived products

Polysaccharides, β-glucan, from several edible and medicinal mushrooms have been demonstrated to have anti-tumor and anticancer activities. Several commercial polysaccharides, such as lentinan, polysaccharides-Krestin (PSK), polysaccharide-peptide (PSP), and sonifilan, have passed clinical trials and are recognized as therapeutics in Japan and China [37]. These fungal polysaccharides and their therapeutic applications are listed in Table 6.

Table 6.
Current carbohydrate-based therapeutics from fungi.

Drug name	Structure	Microorganism	Description	Reference
Lentinan	Polysaccharides	*Lentinus edodes*	Antitumor agent	[38]
Sonifilan	Schizophyllan	*Schizophyllum commune*	Antitumor agent	[39]
Krestin; PSK	Protein bound polysaccharides	*Coriolus versicolor*	Antitumor agent	[40]
PSP	Protein bound polysaccharides	*Coriolus versicolor*	Antitumor agent	[41]

Lentinan isolated from *Lentinus edodes* is a cell wall glucan with β-1,3 linkage backbone and 1,6 linkage branch [38]. Sonifilan is isolated from *Schizophyllum commun* [39]. PSK [40] and PSP [41] are isolated from *Coriolus versicoler.*

PSP, 100 kDa protein bound polysaccharide, is composed of a polypeptide abundant with glutamic and aspartic acids and a polysaccharide chain composed of monosaccharides with α-1,4 and β-1,3 glucosidic linkages. Protein-bound polysaccharides vary in their monosaccharide compositions; for instance, fucose is present in PSK only and rhamnose and arabinose in PSP [42]. The antitumor activity of these polysaccharides results from several mechanisms, including the stimulation of host defense mechanism, cytotoxicity against endothelial cells, decreased IL-6 production, and enhanced apoptosis [43, 44].

3.2. Potential therapeutics

Although many therapeutics from fungi have been developed, as discussed above, many potential therapeutics from the secondary metabolites of medicinal fungi, listed in Table 7, show interesting bioactivities, including antitumor, antimicrobial, antiviral, hypolipidemic, and hypoglycemic activity.

Some fungal polysaccharides derived from the glucose pathway, not yet recognized as therapeutics, are able to stimulate cytokine production in both *in vitro* and *in vivo* studies [45–47]. Orally administered beta glucan could enhance cytotoxicity and synergized with a specific anti-tumor monoclonal antibody in killing tumor cells [48]. Several potential therapeutics from fungal glucans are listed in Table 7.

Some antimicrobial metabolites derived from the polyketide and shikimic acid pathways of fungi are listed in Table 7. Merlinic acids derived from the polyketide pathway show antimicrobial activity with minimum inhibitory concentration (MIC) values of 0.4–10 μg/mL [50]. Strobilurins, 9-methoxystrobilurin and oudemansin A derived from the shikimic acid pathway, are aromatic antifungal compounds with MIC values of 0.1–1 μg/mL [51, 52].

Antiviral metabolites from fungi can be classified into two categories according to the action of their mechanisms: metabolites that act as biological response modifiers (BRM); and metabolites that act directly as viral inhibitors. Most of the fungal polysaccharides act as BRM; however, PSK inhibits reverse transcriptase of avian myeloblastosis virus *in vitro* [53]. Antiviral triterpenoids and phenolic components are listed in Table 7. For example, ganoderic acid β, isolated from *Ganoderma lucidium*, showed significant anti-HIV-1 protease activity with an inhibitory concentration 50% (IC_{50}) value of 20 μM [54]. Ganodermadiol, lucidadiol, and applanoxidic acid G isolated from *Ganoderma pfeiferri* showed antiviral activity against influenza virus type A and HSV-1 [55]. Hispidin and hispolon, isolated from *Inonotus hispidus*, showed antiviral activity against influenza viruses type A and B [56].

Some cytotoxic and antineoplastic metabolites from fungi are potential anti-cancer therapeutics, as listed in Table 7. Montadial A isolated from *Bondarzewia montana* showed cytotoxicity against lymphocytic leukemia L1210 cells in mice with an IC_{50} value of 10 μg/mL [57]. Zhankuic acids, isolated from *Antrodia camphorata*, are new steroids with cytotoxic activity against P-388 murine leukemia cells with an IC_{50} value of 1.8 μg/mL [58]. Lucidenic acid N, a triterpenoid isolated from *G. lucidum*, showed significant cytotoxic effect on Hep G2 cells with an IC_{50} value of 2.06×10^{-4} μM [59]. Poricoic acid G, isolated form *Poria cocos*, showed significant cytotoxicity to leukemia HL-60 cells with an IC_{50} value of 3.93×10^{-2} μM [60].

Some fungal polysaccharides [61-64] and steroids [65] showed hypoglycemic activity. Dehydrotrametenolic acid, a novel steroid isolated from *Wolfiporia cocos*, reduces hyperglycemia in mice with noninsulin-dependent diabetes mellitus (NIDDM) and acts as an insulin sensitizer in glucose tolerance tests [65]. Some fungal polysaccharides also showed hypocholesterolemic activity. Besides the fungi listed in Table 7, many other medicinal fungi are sources for potential therapeutics.

Table 7.
Potential therapeutics from fungi.

Bioactivity	Key compounds	Microorganism	Pathway	Reference
Antitumor	PGG-glucan	*Saccharomyces cerevisiae*	Glucose-derived	[45]
	SSG-glucan	*Sclerotinia sclerotiorum*		[46]
	Polysaccharides	*Ganoderma lucidum*		[47]
	grifolan	*Grifola frondosa*		[49]
Antimicrobial	Merlinic acids	*Merulius tremellosus*	Polyketide	[50]
		Phlebia radiata		[50]
	Strobilurins	*Agaricus* sp.	Shikimic acid	[51]
	9-methoxystrobilurin	*Favolaschia pustulosa*		[52]
	oudemansin A	*Favolaschia pustulosa*		[52]
Antiviral	PSK	*Coriolus versicolor*	Glucose-derived	[53]
	Ganoderic acid β	*Ganoderma lucidum*	Mevalonate	[54]
	Ganodermadiol, lucidadiol, applanoxidic acid G	*Ganoderma pfeiferri*		[55]
			Shikimic acid	
	Hispidin, hispolon	*Inonotus hispidus*		[56]
Antineoplastic	Montadial A	*Bondarzewia montana*	Mevalonate	[57]
	Zhankuic acid A	*Antrodia camphorata*		[58]
	Lucidenic acid N	*Ganoderma lucidum*		[59]
	Poricoic acid G	*Poria cocos*		[60]
Hypoglycemic	Acidic polysaccharides	*Tremella aurantia*	Glucose-derived	[61]
	Glucuronoxylomannan	*Tremella fuciformis*		[62]
	Galactomannan	*Pestalotiopsis* sp.		[63]
	Polysaccharides	*Cordyceps sinensis*	Mevalonate	[64]
	Dehydrotrametenolic acid	*Walfiporia cocos*		[65]
Hypolipidemic	Acidic polysaccharides	*Tremella aurantia*	Glucose-derived	[61]
	Protein bound polysaccharides	*Cordyceps sinensis*		[64]

4. PATHWAY MANIPULATION

Advanced engineering approaches, including genetic engineering and bioprocess engineering, can manipulate fungal pathways to enhance metabolite yields.

4.1. Genetic Engineering

Advances in genetic engineering, such as metabolic engineering, enable the development of better strains with modified pathways which produce high yields of desired metabolites [66, 67].

4.1.1. Pathway engineering

Advanced metabolic engineering technologies and tools have made the pathway engineering of fungi possible. In general, pathway engineering includes four major principles [68]: optimization of the primary metabolic pathway to the targeted metabolites, including removal of the rate-limiting steps, transcriptional and allosteric regulation; blockage of competing pathways; enhancement of carbon flux toward the primary metabolic pathway from the central metabolism; and modification of secondary metabolite metabolic pathways as necessary to improve energy metabolism and availability of required enzymatic cofactors.

Advanced technologies in this field are listed as follows:

1. Efficient gene manipulation system
   a. directed evolution
   b. DNA shuffling
   c. combinational biosynthesis
   d. stable expression system
   e. transcriptional control strategy— modification mRNA stability
2. Metabolic flux quantification methodology and analytical tools

Useful metabolic flux analysis required the dilute concentrations of various types and fluxes of metabolites inside the cell. The reactants, metabolites, nucleotides and cofactors inside the cell could be accurately identified and quantified by the following analytical tools:

   a. ^{13}C-labeling techniques
   b. two-dimensional nuclear magnetic resonance (2D NMR), *in situ* NMR
   c. liquid chromatography (LC)-double mass spectrometry (MSMS), LC-MS
   d. gas chromatography-mass spectrometry (GC-MS)

3. Metabolic flux analysis

Computer models have been developed to accommodate the tremendous amounts of data and integrate the genetic and biochemical pieces. Metabolic fluxes of the biosynthetic pathway can be calculated by computer models. Metabolic flux analysis has occupied a central place in metabolic engineering, which provides the metabolic state of a cell and its physiology under a set of environmental conditions. The changes of metabolic fluxes in response to various types of genetic and environmental perturbations can reveal the control sites of the metabolic fluxes. Metabolic control analysis (MCA) [69, 70] can therefore be used to identify the flux-controlling enzyme(s) in the biosynthesis of desired metabolites. MCA has generally revealed that no single enzyme limits the flux through a pathway; hence, it is necessary to express more than one gene simultaneously in the pathway as demonstrated in the following examples including homologous and heterologous gene expression systems.

Homologous gene expression system for the production of naturally occurring β-lactams has been demonstrated in *Penicillium* species. Most penicillin V or penicillin G is produced by fermentation of the high-yield production strains of *Penicillium chrysogenum*. The production strains are developed by repeated rounds of mutation and selection. However, the strain improvement of *P. chrysogenum* could further be accomplished by pathway engineering after elucidation of the penicillin biosynthetic pathway. The penicillin biosynthetic pathway consists of three enzymatic steps as indicated in Fig. 2, and the genes *pcb*AB, *pcb*C and *pen*DE responsible for the three enzymes α-aminoadipylcysteinyl-valine

(ACV) synthase, isopenicillin-N (IPN) synthase, and acyltransferase, respectively, are located in a single gene cluster in chromosome I [71]. These enzymes have been considered as the rate-limiting steps by metabolic control analysis. Amplification of only one gene is not as effective as overexpression of the whole gene cluster. Also, a 40% increase in penicillin production was observed in a transformant strain of *P. chrysogenum* Wisconsin 54-1255 with extra copies of a DNA fragment containing two genes *pcb*C and *pen*DE [72]. In addition, several tandem repeats of the genes have been found in penicillin overproduction strains obtained from the classic mutation [73].

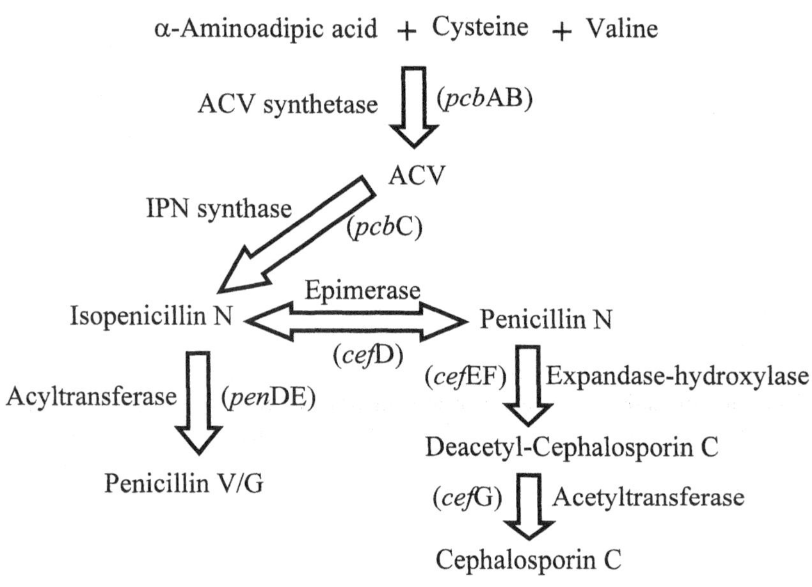

Fig. 2. Biosynthetic pathways of naturally occurring penicillin V/G and cephalosporin C.

Likewise, the limiting enzymes in cephalosporin biosynthesis have been reported to be the bifunctional enzymes expandase-hydroxylase and acetyltransferase in *Cephalosporium acremonium* (see Fig. 2), and cloning extra copies of its corresponding genes *cef*EF and *cef*G has resulted in a significant increase in the production of cephalosporin [74, 75].

Heterologous gene expression systems allow strain improvement to produce new metabolites by direct alteration of the biosynthetic pathway. Most of the current therapeutic cephalosporins are semi-synthetic, being chemically synthesized from 7-aminocephalosporanic acid (7-ACA) or 7-aminodeacetoxycephalosporanic acid (7-ADCA). However, the chemical steps involved are expensive and environmental unfriendly. Thus, searching for proper biological fermentation is promising. A heterologous gene expression system for production of β-lactams has been achieved in species of *Cephalosporium* [76] and *Penicillium* [77], as shown in Fig. 3.

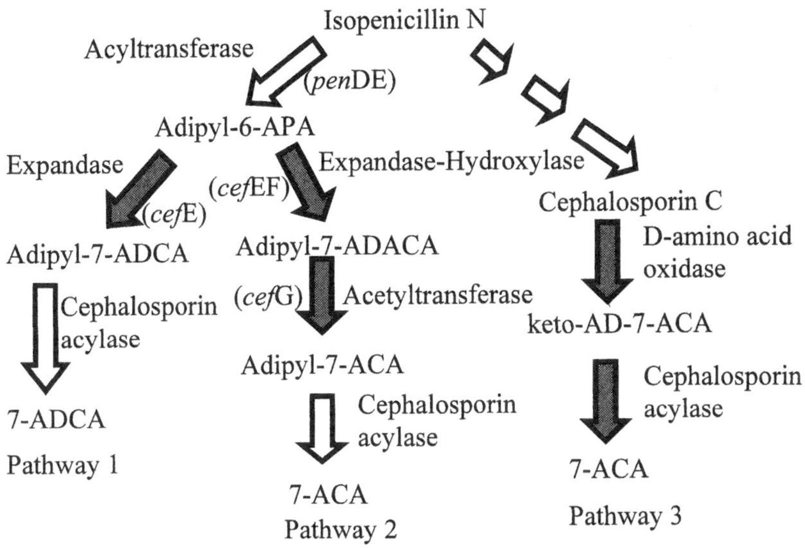

Fig. 3. Modified β-lactam biosynthetic pathways for *in vivo* production of 7-aminocephalosporic acid (7-ACA) and 7-aminodeacetoxycephalosporanic acid (7-ADCA) in fungi. Pathway 1 includes the *cef*E gene (encoding expandase) from *Streptomyces calavuligerus* in *Penicillium chrysogenum*. Pathway 2 includes the *cef*EF (encoding expandase-hydroxylase) and *cef*G (encoding acetyltransferase) genes from *Cephalosporium acremonium* in *P. chrysogenum*. Pathway 3 includes the cDNA encoding D-amino acid oxidase from *Fusarium solani* and genomic DNA encoding cephalosporin acylase from *Pseudomonas diminuta* in *C. acremonium*. (Dark arrow indicates the constructed metabolic flux; white arrow indicates the original metabolic flux.)

4.1.2. Heterologous protein production

With advances in recombinant DNA techniques, overproduction of homologous and heterologous proteins in filamentous fungi has recently been achieved. Although many fungal expression systems are available for heterologous proteins, their expression levels are usually much lower as compared to homologous proteins. Among heterologous proteins, only limited cases of human origin with pharmaceutical applications have been developed; for instance, α-interferon [78], interleukin-6 [79] and insulin [80]. However, the complex post-translational protein modification in filamentous fungi remains a challenge for commercial development of these microorganisms as hosts for pharmaceutical proteins production.

4.2. Bioprocess Regulation

In the competitive environment, microorganisms are under a strict control mechanism to avoid overproducing metabolites. Deregulation of the desired metabolic flux in microorganisms is mainly obtained from mutations through classical and molecular genetic manipulations. Most therapeutic metabolites from fungi are considered to be secondary metabolites. The regulation mechanisms of secondary metabolites are closely related to the status of the nutrients in the broth, which includes substrate induction, carbon catabolite repression, nitrogen or phosphate regulation, and product feedback regulation. Bioprocess operation provides optimal environmental conditions for the production of fungal therapeutic metabolites. Nevertheless, the optimal performance of the overproducing mutants is governed by a proper coordination of chemical factors including nutritional status and physical factors including environmental conditions.

The fungal morphology is very complex and ranges from dispersed mycelial filaments to densely interwoven mycelial masses referred as pellets, which are controlled by both physical and chemical factors in the submerged culture [81]. The morphology of filamentous fungi significantly affects their product formation, and therefore, is the key feature in optimizing the filamentous fungal fermentation. The major process parameters affecting fungal morphology and fermentation performance are briefly discussed below.

4.2.1. Substrate induction

Supplementation of amino acids can trigger the production of several amino-acid derived metabolites. For example, methionine supplementation could induce ACV synthetase in the cephalosporin pathway of *A. chrysogenum* [82] and glutamate addition enhanced penicillin G formation in *P. chrysogenum* [83].

4.2.2. Carbon catabolite repression

Carbon sources are mainly used for cell growth and product formation. However, the production of the secondary metabolites usually occurs when carbon sources in the media are limited; that is, the presence of carbon sources would repress the formation of the secondary metabolites. For instance, the β-lactam antibiotic cephalosporin C is repressed by glucose in *A. chrysogenum*. The *pcb*C and *cef*EF biosynthetic genes are repressed by the carbon source repressor due to the presence of glucose [84]. Fed-batch operation could be used to increase the production of metabolites by keeping the repressive carbon source at a very low level to avoid repressive effects as demonstrated in penicillin production [85].

4.2.3. Nitrogen regulation

The type of nitrogen sources is another key affecting cell growth and product formation. In general, the biosynthesis of fungal secondary metabolites is limited by the nitrogen sources favoring cell growth, such as ammonium salts and certain types of amino acids. However, other nitrogen sources, such as proline and urea, might be poor for cell growth but are good to stimulate the production of secondary metabolites. For instance, cephalosporin C and penicillin production are limited by using NH_4^+ and lysine as the nitrogen source, but are stimulated by using asparagine and glutamate, respectively [86].

4.2.4. Phosphate regulation

Inorganic phosphorus is not only a growth-limiting nutrient but also a regulator of secondary metabolites. Production of cephalosporin and ergot alkaloids are stimulated by phosphate limitation [86]. Polyketide metabolites in *Monascus* are also regulated by the phosphate [87].

4.2.5. Product feedback regulation

Product feedback inhibition is commonly present in microorganisms. Likewise, the secondary metabolites inhibit their own biosynthesis. For instance, dimethylallyltrytophan synthase, the first enzyme of the ergot alkaloid synthetic pathway, is inhibited by the end product [23]. End-product analogs, such as 5-methyltryptophan, can be used to select feedback insensitive strains with a high ergot alkaloids production capability [88].

4.2.6. Culture pH

Culture pH affects both fungal morphology and productivity. The hyphal length of *P. chrysogenum* decreased as the culture pH increased from 6.0 to 7.4. However, pellet size increased as the culture pH was higher than 7.4 in a continuous culture. The optimal pH for penicillin production was around 7.4 [89]. Likewise, the optimum pH for exopolysaccharide formation was 4.5, but that for biomass was 3.5 [90]. A two-stage strategy is thus commonly proposed to produce the metabolite.

4.2.7. Inoculum density

The inoculum density plays a critical role in the development of fungal morphology, as demonstrated in penicillin production [91, 92]. *P. chrysogenum* forms pellets only when the inoculum size is below 10^4 spores/ml. As the inoculum size increased from 10^2 to 10^4 spores/ml, penicillin production in the shaker flask increased significantly from 500 to 5000 U/ml.

4.2.8. Temperature

Pellet formation could be induced by reducing the culture temperature from the optimal value [93, 94]. In the fermentation with *Aspergillus*, only pellets were observed at a low temperature of 25 °C; at 30 °C, the pellets initially formed but decomposed later; at 35 °C, filamentous mycelium was the major morphology [94].

4.2.9. Shear rate

Agitation is crucial for good mixing and mass transfer in submerged cultures; however, it might induce pellet formation and cause mechanical damage to fungal mycelia. High agitation speeds ranging from 1250 to 1500 rpm would favor cell growth but inhibit penicillin formation as compared to those at 900 rpm in a 24-liter fermenter [95]. Likewise, a decrease in the pellet size and the specific penicillin production rate with increasing power input per unit mass were observed in three different scales of fermentation, 5 dm^3, 100 dm^3 and 1000 dm^3 [96].

4.2.10. Dissolved oxygen tension (DOT)

Production of various metabolites in filamentous fungi can be greatly affected by the dissolved oxygen tension in the culture media. A common feature of oxygen metabolism of fungi indicates that the critical DOT for growth and the critical DOT for product biosynthesis are distinct. For example, the critical DOT of penicillin was 30% air saturation but the critical

DOT of cell growth was 7% air saturation [97]. The critical DOT of cephalosporin C was 20% air saturation and below this value a repression occurred [98].

4.2.11. Dissolved carbon dioxide tension

Product formation and fungal morphology are also significantly associated with dissolved carbon dioxide tension. Low dissolved carbon dioxide tension might induce fungal growth in mycelial form [99]. On the other hand, a high dissolved carbon dioxide tension resulted from gassing the fermentation with 20% carbon dioxide would inhibit cell growth and penicillin production [100].

5. CONCLUSIONS

Fungi are the sources of various known medicinal products and potential therapeutics. The understanding of their biosynthetic pathways has allowed scientists to further improve the strains through pathway manipulation. A homologous gene expression system has successfully optimized the naturally occurring metabolites, such as β-lactam antibiotics in *Penicillium* species. A heterologous gene expression system allows strain improvement to produce new metabolites by direct alteration of the biosynthetic pathway.

Advances in screening methodologies and pharmacology would speed up the process of discovering novel fungal medicinal products. Advances in fungal fermentation techniques should continue to eliminate the limitations of current production techniques. Optimization of fungal fermentation through engineering aspects should also continue to play an important role in the production of medicinal products despite the advances in the understanding of fungal physiology and biosynthetic pathways of desired metabolites.

REFERENCES

[1] D.B. Archer, Curr. Opin. Biotechnol., 11 (2000) 478.
[2] P.R. Anupama, Biotechnol. Adv., 18 (2000) 459.
[3] A. Pandey (ed.), Solid-state Fermentation, Wiley Eastern, New Delhi, 1994.
[4] P.J. Punt, N. van Biezen, a. Conesa, A. Albers, J. Mangnus and V. van den Hondel, Trends Biotechnol., 20 (2002) 200.
[5] T. Godfrey and S. West (eds.), Industrial Enzymology, Macmillan Press Ltd, London, 1996.
[6] W.B. Turner and D.C. Aldridge (eds.), Fungal Metabolites II. Academic Press, London, 1983.
[7] A. Fleming, Br. J. Exp. Pathol., 10 (1929) 226.
[8] Miguel A. Peñalva, Robert T. Rowlands and Geoffrey Turner, Trends Biotechnol., 16 (1998) 483.
[9] J.F. Borel, Progr. Allergy, 38 (1986) 9.
[10] S. Tawara, F. Ikeda, K. Maki, Y. Morishita, K. Otomo, N. Teratani, T. Goto, M. Tomishima, H. Ohki, A. Yamada *et al.*, Antimicrob. Agents Chemother., 44 (2000) 57.
[11] R.T. Rowlands, 50 Years of Penicillin Application—History and Trends (H. Kleinauf and H. von Döhren, eds.), Technische Universität Berlin, Berlin, Germany, 1991.
[12] A. Smith, Current Commercial Products (M. Moo-Young, ed.), Pergamon Press, New York, 1985, 163.
[13] Z. Rahacek and Z. De-Xiu, Process Biochem., 26 (1991) 157.
[14] P.A. Whiteman and E. P. Abraham, FEBS Lett., 394 (1996) 31.

[15] V.K. Sudhakaran J.G. Shewale, Hind Antibiot. Bull., 37 (1995) 9.
[16] Sepracor, International patent WO 88/07582.
[17] M. Mahmoudian, J. Eaddy and M. Dawson, Biotechnol. Appl. Biochem., 29 (1999) 229.
[18] J.E. Audia, T.C. Britton, J. Droste, B.K. Folmer, G.W. Huffman, V.L. John, Preparation of *N*-(phenylacetyl)di- and tripeptide derivatives for inhibiting β-amyloid peptide release. PCT International Application: WO 9822494 A2 19980528; Can. 129:41414 (1998).
[19] A. Goswami, M.J. Totleben, A.K. Singh and R.N. Patel, Tetrahedron: Asymm., 10 (1999) 3167.
[20] C.T. Hou, R.N. Patel, A.I. Laskin and N. Barnabe, Arch. Biochem. Biophys., 216 (1982) 296.
[21] M.A. Goetz, M. Lopez, R.L. Monaghan, R.S. Chang, V.J. Lotti and T.B. Chen, J. Antibiot. (Tokyo) 38 (1985) 1633.
[22] L. Ögren, F. Agnati, K. Andersson, H. Kall, C. Köhler and B. Fredholm, Advances in Biochemicals and Psychopharmacology 23 (M. Goldstein, D.B. Caine, A. Lieberman and M.O. Thorner, eds.), Raven Press, New York, 1980.
[23] I. Ninomiya and T. Kiguchi (eds.), The Alkaloids: Chemistry and Pharmacology 38, Academic Press, San Diego, 1990.
[24] A. Hofmann, A. Frey, H. Ott, T. Petrzilka and F. Troxler, Experienta., 15 (1958) 397.
[25] F. Hasler, D. Bourquin, R. Brenneisen, T. Bär and F.X. Vollenweider, Pharm. Acta Helv., 72 (1997) 175.
[26] G.C. Yen, Y.C. Chang, F. Sheu and H.C. Chiang, J. Agric. Food Chem., 49 (2001) 1426.
[27] J. E. Munden, D. Butterworth, G. Hanscomb and M. S. Verrall, Appl. Microbiol., 19 (1970) 718.
[28] W. Alberts, Proc. Natl. Acad. Sci. USA, 77 (1980) 3957.
[29] A. Endo, M. Kuroda and Y. Tsujita. J. Antibiot., 29 (1976) 1346.
[30] K. Gull and A. P.J. Trinci, Nature, 244 (1973) 292.
[31] P.W. Clutterbuck, A.E. Oxford, H. Raistrick and G. Smith. Biochem. J., 26 (1932) 1441.
[32] A. Jekkel, I. Barta, A. Kónya, J. Sütõ, S. Boros, Gy. Horváth and G. Ambrus, J. Mol. Catal. B: Enzymatic, 11 (2001) 423.
[33] M.C. Wani, H.L. Taylor, M.E. Wall, P. Coggen and A.T. McPhail, J. Am. Chem. Soc., 93 (1971) 2325.
[34] A. Stierle, G. Strobel and D. Stierle, Science, 260 (1993) 214.
[35] G. Strobel, X.S. Yang, J. Sears, R. Kramer, R.S. Sidhu and W.M. Hess, Microbiology, 142 (1996) 435.
[36] M.J. Perry, A. Hendricks-Gittins, L.M. Stacey, M.W. Adlard and W.C. Noble, J. Antibiot. (Tokyo), 36 (1983) 1659.
[37] P. Miles and S.-T. Chang, (eds.), Mushroom Biology. Concise Basics and Current Developments, World Scientific, Singapore, New Jersey, London, Hong Kong, 1997.
[38] T. Ikekawa, N. Uehara, Y. Maeda, M. Nakanishi and F. Fukuoka, Cancer Res., 29 (1969) 734.
[39] N. Komatsu, S. Okubo, S. Kikumoto, K. Kimura, G. Saito and S. Sakai, Gann., 60 (1969) 137.
[40] S. Tsukagoshi, Y. Hashimoto, G. Fujii, H. Kobayashi, K. Nomoto and K. Orita, Cancer Treatment Rev., 11 (1984) 131.
[41] C.N. Chen, M.M.P. Yang and P. Zhu, Proceedings of PSP International Symposium (Q.Y. Yang and C.Y. Kwok, eds.), Fudan University Press, Shanghai, China, 1993.
[42] T.B. Ng, Gen. Pharm., 30 (1998) 1.
[43] B. Sano, Y. Sugiyama, K. Kunieda, J. Sano and S. Saji, Surgery Today, 32 (2002) 503.
[44] M. Fisher and L.X. Yang, Anticancer Res., 22 (2002) 1737.
[45] M.L. Patchen, T. Vaudrain, H. Correira, T. Martin and D. Reese, Exp. Hematol., 26 (1998) 1247.
[46] K. Hashimoto, I. Suzuki, M. Ohsawa, S. Oikawa and T. Yadomae, J. Pharmacobio-dyn., 13 (1990) 512.
[47] N.K. Cheung, S. Modak, A. Vickers and B. Knuckles, Cancer Immunol. Immunother., 51 (2002) 557.
[48] Y. Sone, R. Okuda, N. Wada, E. Kishida and A. Misaki , Agric. Biol. Chem., 49 (1985) 2641.

[49] H. Nanba, A. Hamaguchi and H. Kuroda, Chem. Pharm. Bull. (Tokyo), 35 (1987) 1162.
[50] B.M. Giannetti, W. Steglich, W. Quack, T. Anke, F.Z. Oberwinkler and C. Naturforsch, J. Biosci., 33 (1978) 807.
[51] W. Weber, T. Anke, M. Bross and W. Steglich. Planta Med., 56 (1990) 446.
[52] K.A. Wood, D.A. Kau, S.K. Wrigley, R. Beneyto, D.V. Renno, A.M. Ainsworth, J. Penn, D. Hill, J. Killacky and P. Depledge, J Nat Prod., 59 (1996) 646.
[53] K. Hirose, M. Hakozaki, J. Kakuchi, K. Matsunaga, C. Yoshikumi, M. Takahashi, T.S. Tochikura and N. Yamamoto, Biochem. Biophys. Res. Commun., 149 (1987) 562.
[54] B.S. Min, N. Nakamura, H. Miyashiro, K.W. Bae, M. Hattori, Chem. Pharm. Bull., 46 (1998) 1607.
[55] R.A.A. Mothana, N.A. Awadh Ali, R. Jansen, U. Wegner, R. Mentel and U. Lindequist, Fitoterapia, 74 (2003) 177.
[56] N.A. Awadh Ali, R.A.A. Mothana, A. Lesnau, H. Pilgrim and U.Lindequist, Fitoterapia, 74 (2003) 483.
[57] B. Sontag, N. Arnold, W. Steglich, T. Anke, J. Nat. Prod., 62 (1999) 1425.
[58] C.H. Chen, S.W. Yang and Y.C. Shen, J. Nat. Prod., 58 (1995) 1655.
[59] T.S. Wu, L.S. Shi and S.C. Kuo, J. Nat. Prod., 64 (2001) 1121.
[60] M. Ukiya, T. Akihisa, H. Tokuda, M. Hirano, M. Oshikubo, Y. Nobokuni, Y. Kimura, T. Tai, S. Kondo and H. Nishino, J. Nat. Prod., 65 (2002) 462.
[61] T. Kiho, H. Morimoto, M. Sakushima, S. Usui and S. Ukai, Biol. Pharmaceut. Bull., 18 (1995) 1627.
[62] T. Kiho, Y. Tsujimura, M. Sakushima, S. Usui and S. Ukai, J. Pharmaceut. Soc. Jap., 114 (1994) 308.
[63] T. Kiho, S. Itahashi, M. Sakushima, T. Matsunaga, S. Usui, S. Ukai, H. Mori, H. Sakamoto and Y. Ishiguro, Biol. Pharmaceut. Bull., 20 (1997) 118.
[64] T. Kiho, A. Yamane, J. Hui, S. Usui and S. Ukai, Biol. Pharmaceut. Bull., 19 (1996) 294.
[65] M. Sato, T. Tai, Y. Nunoura, Y. Yajima, S. Kawashima and K. Tanaka, Biol. Pharm. Bull., 25 (2002) 81.
[66] G.N. Stephanopoulos, A.A. Aristidou and J. Nielsen (eds), Metabolic Engineering: Principles and Methodologies, Academic Press, San Diego, 1998.
[67] J.E. Bailey , Science, 252 (1991) 1668.
[68] G. Chotani, T. Dodge, A. Hsu, M. Kumar, R. LaDuca, D. Trimbur, W. Weyler and K. Sanford. Biochim. Biophys. Acta, 1543 (2000) 434.
[69] H. Kacser and J.A. Urns, Symp. Soc. Exp. Biol., 27 (1973) 65.
[70] D.A. Fell, Biotech. Bioeng., 58 (1998) 121.
[71] J.F.S. Martín, S. Gutiérrez and A.L. Demain, Fungal Biotechnology (T. Anke, ed.), Chapman & Hall, Weinheim, Germany, 1997.
[72] A.E. Veenstra, P. van Solingen, R.A.L. Bovenberg and L.H.M. van der Voort. J. Biotechnol., 17 (1991) 81.
[73] F. Fierro, J.L. Barredo, B. Díez, S. Gutiérrez, F. J. Fernández and J.F. Martín, Proc. Natl. Acad. Sci. USA, 92 (1995) 6200.
[74] P.L. Statrud, A.J. Tietz, T.D. Ingolia, C.A. Cantwell, D.L. Fisher, J.L.Chapman and S.W. Queener, Bio/Technol., 7 (1989) 477.
[75] L. Mathison, C. Soliday, T. Stephan, T. Aldrich and J. Rambosek, J. Curr. Genet. 23 (1993) 33.
[76] T. Isogai, M. Fukagawa, I. Aramori, M. Iwami, H. Kojo, T. Ono, Y. Ueda, M. Kohsaka and H. Imanaka, Bio/Technol., 9 (1991) 188.
[77] L. Crawford, A.M. Stepan, P.C. McAda, J.A. Rambosek, M.J. Conder, V.A. Vinci and C.D. Reeves, Bio/Technol., 13 (1995) 58.
[78] D.I. Gwynne, F.P. Buxton, S.A. Williams, S. Garven and R.W. Davies, Bio/Technol., 5 (1989) 713.

[79] D. Carrez, W. Janssens, P. Degrave, C.A.M.J.J. van den Hondel, J.R. Kinghorn, W. Fiers and R. Contreras, Gene, 94 (1990) 147.
[80] S. Mestric, P.J. Punt, R. Valinger and C.A.M.J.J. van den Hondel, Fungal Genetics Newslett., 43B (1996) 25.
[81] M. Papagianni, Biotechnol. Adv., 22 (2004) 189.
[82] J. Zhang, G. Banko, S. Wolfe and A.L. Demain , J. Indust. Microbiol., 2 (1987) 251.
[83] F. Lara, R.C. Mateos, G. Vázquez and S. Sanchez , Biochem. Biophys. Res. Commun., 105 (1982) 172.
[84] K. Jekosch and U. Kück , Appl. Microbiol. Biotechnol., 54 (2000) 556.
[85] R. Bajpai and M. Reuss, Biotechnol. and Bioeng., 23 (1981) 717.
[86] S. Sanchez and A.L. Demain, Enzyme Microb. Technol., 31 (2002) 895.
[87] T.F. Lin and A.L. Demain , J. Indust. Microbiol., 12 (1993) 361.
[88] A.I. Scott, H. Mizukami and S.L. Lee, Phytochem., 18 (1979) 795.
[89] S.J. Pirt and D.S. Callow, Nature, 184 (1959) 307.
[90] Y. Wang and B. McNeil, Enzyme Microb. Technol., 17 (1995) 124.
[91] C.T. Calam (ed.), Cambridge Studies in Biotechnology vol. 4, Cambridge Univ. Press, Cambridge, 1987.
[92] J. Nielsen, C.L. Johansen, M. Jacobsen, P. Krabben and J. Villadsen, Biotechnol. Prog., 11 (1995) 93.
[93] S. Braun and S.E. Vecht-Lifshitz, Trends Biotechnol., 9 (1991) 63.
[94] K. Schügerl, S.R. Gerlach and D. Siedenberg, Adv. Biochem. Eng. Biotechnol., 60 (1998) 195.
[95] B. König, C. Seewald and K. Schügerl, Eur. J. Appl. Microbiol. Biotechnol., 12 (1981) 205.
[96] H.Y. Makagiansar, A.P. Shamlou, C.R. Thomas and M.D. Lilly, Bioprocess Eng., 9 (1993) 83.
[97] F. Vardar and M.D. Lilly, Eur. J. Microbiol. Biotechnol., 14 (1982) 203.
[98] W. Zhou, K. Holzhauer-Rieger, M. Dors and K. Schugerl, Enzyme Microb. Technol., 14 (1992) 848.
[99] M.D. Smith and C.S. Ho, J. Biotechnol., 2 (1985) 347.
[100] C.S. Ho and M.D. Smith, Biotechnol. Bioeng., 28 (1986) 668.

Chapter 18. Solid State Fermentation and Its Applications

Liping Wang and Shang-Tian Yang

Department of Chemical and Biomolecular Engineering, The Ohio State University,
140 West 19th Avenue, Columbus, Ohio 43210, USA

1. INTRODUCTION

Solid-state fermentation has long been applied to the food industry. SSFs are processes carried out with microbes growing on nutrient impregnated solid substrate with little or no free water. The growth of koji, an enzyme-rich mold grown on shallow trays of steamed rice, is a classical example of SSF. Solid state fermentation (SSF) can be directly carried out with abundant low-cost biomaterials (starch, cellulose, lignin, hemicellulose, chitin, etc.) with minimal or no pretreatment, and thus is relatively simple, uses less energy than submerged fermentation (SmF), and can provide unique microenvironments conducive to microbial growth and metabolic activities. Currently, SSF is undergoing a renewed surge of interest, primarily because of the opportunities that SSF affords for increased productivity and product concentration as compared to SmF [1, 2]; new product possibilities, cheaper product recovery, and the prospect of using a wide range of agri-industry commodities and waste streams as substrates. Large amounts of excess plant biomass are produced by the agri-industry. It is desirable to use this as a renewable resource for sustainable chemical production via microbial cultivation. If not used to generate a value-added product, the biomass would remain in the waste stream and require expensive disposal or treatments. The major reason that Western industry is reluctant to use SSF is a lack of knowledge and of scalable bioreactor technologies. There are very few data on growth and product formation kinetics, reactor design, or process control in SSF available in the literature. However, with the increased interest in SSF with the goal of developing industrially applicable SSF systems, progress is being made. This chapter will review the recent advances in SSF system research and development.

2. PRODUCTS FROM SOLID STATE FERMENTATION (SSF)

SSF has been applied in the preparation of traditional foods since ancient times, especially in the Orient. Koji processing is a well-known example. In the koji process, steamed rice is inoculated with the spores of *Aspergillus oryzae* or *Aspergillus sojae* and kept in a temperature and humidity controlled room for a number of days. The mold germinates and produces hydrolytic enzymes which act on the rice starch, and when the fermentation is finished, the koji is used for the enzyme treatment of other fermentation substrates or is itself used as a substrate for a subsequent fermentation. Soy sauce production, for example, begins with koji fermentation. With the development of modern biotechnology, solid state fermentation can be applied to produce various products, including enzymes, organic acid, secondary metabolites (e.g., antibiotics), biofuels (e.g., ethanol), aroma compounds, and even many bioactive products like mycotoxins, plant growth factors, immuno-suppressive drugs, etc. [3–5]. Enzymes produced by SSF have many industrial applications, such as enzyme assisted ensiling, bioprocessing of crops and crop residues, fiber processing, feed supplement, biopulping, soil bioremediation, biopesticide, etc. [6]. Tables 1–3 list some important products from SSF, both on a laboratory and an industrial scale.

2.1. Organic acids from SSF

Some organic acids, such as citric acid, have long been produced from SSF. Others, lactic acid, fumaric acid, oxalic acid, gluconic acid, etc., were reported to be produced by SSF only in recent years (Table 1).

2.1.1. Citric acid

There is a large and still growing market for citric acid in the food, pharmaceutical, and other industries all over the world. It is one of the world's largest fermentation products with an estimated annual production of 1,000,000 tons [7]. Currently, it is produced mainly by SmF using *Aspergillus niger* or *Candida sp.* from different sources of carbohydrates, such as molasses and starch based media [8].

With the renewed interest in SSF, production of citric acid in SSF with low-cost agricultural products or residues has attracted attention (Table 1). Cost reduction can be expected when using these less expensive substrates, such as cassava bagasse [9], sugar cane bagasse [10, 11], fruits waste [12], sugar beet pulp [13, 14], apple pomace [15], coffee husk, and cassava bagasse [10].

In solid state fermentation, citric acid production is affected by several critical factors including carbon, phosphorous and nitrogen sources, trace elements, alcohols, pH value, oxygen availability, and CO_2 accumulation [9, 16]. The fermentation yield of citric acid (g/kg dry substrate) varies depending on the substrates added; e.g., it was higher with tapioca and cane molasses than with wheat bran or potato waste pulp [17]. Several researchers have reported that the addition of methanol (3–6% w/w) into the solid substrates stimulates the production of citric acid [18–20]. It was also found that the stimulation of citric acid production by methanol was affected by moisture content of the substrates and of the strains. Addition of methanol increased production of citric acid only when using low-active strains of *A. niger* or highly moist substrate. However, the addition of methanol always eliminated the sporulation of *A. niger* strains [12, 13].

With *A. niger*, citric acid production is aided by a low aeration rate and a high concentration of CO_2 because a low-oxygen environment limits the respiration activity of the cells, which consequently turn to citric acid synthesis [9, 21]. Various kinds of SSF reactors for citric acid production have been tested. Packed-bed reactors were found to be superior to flasks, trays and drum reactors [3, 9, 22]. Improved heat and mass transfer are thought to be the reason. However, reactor design and scale up are complicated tasks. A detailed study will be necessary to test all the factors that may affect the fermentation process and to evaluate the efficiency of various kinds of reactors for the production of citric acid.

Table 1
Some organic acids produced in solid state fermentations

Product	Organism	Substrate	Reference
Citric acid	*Aspergillus niger*	Kumara (starch)	[23]
		Sweet potato	[24]
		Cassava bagasse	[9]
		Sugarcane bagasse	[11, 12]
Lactic acid	*Lactobacillus amylophilus*	Wheat bran (starch)	[25, 26]
	Lactobacillus paracasei	Sweet sorghum	[27]
	Lactobacillus casei	Sugar-cane press mud	[28]
	Rhizopus sp. and Acremonium thermophilus	Corncob (cellulose)	[29]
	Rhizopus oryzae and Aspergillus niger	Carrot processing waste	[30]
	Rhizopus oryzae	Sugarcane bagasse	[31, 32]
Oxalic acid	*Aspergillus niger*	Sweet potato	[24]
	White rot fungi	Spruce sapwood chips	[33]
Fumaric acid	*Rhizopus arrhizus*	Orange peels	[34]
	Rhizopus sp.	Raw cassava starch	[31, 32]
Gluconic acid	*Aspergillus niger*	Fig extract	[35]
		Glucose	[36]
		Fig	[37]
Gallic acid	*Aspergillus foetidus and Rhizopus oryzae*	Mixed tannin-rich agro-products	[38]

2.1.2. Lactic acid

Lactic acid can be produced by the fermentation of bacterial or fungal strains. *Lactobacillus sp.* (bacteria) and *Rhizopus sp.* (fungi) are the commonly used strains. Although industrial lactic acid fermentation is currently mainly carried out with homolactic acid bacteria in SmF, there has been increasing interest in fungal fermentation with *Rhizopus oryzae* for lactic acid production because of its unique ability to produce optically pure L(+)-lactic acid from glucose, xylose, and starch. However, controlling the fungal mycelial morphology and broth rheology is important to fungal fermentation. The highly branched fungal mycelia usually cause complex (viscous) broth rheology and difficulty in mixing and aeration in SmF with conventional agitated tank fermenters. Various cell immobilization methods to control the cell morphology and to achieve high cell density and high reaction rate have been studied. Tay and Yang [39] have developed a rotating fibrous bed bioreactor for immobilizing fungal cells and achieved excellent lactic acid production from glucose and soluble starch, with high lactic acid yield (>90% from glucose) and productivity (2.5 g/L·h or 467 g/h·m^2). However, their good fermentation results were only obtained with increased oxygen tension (90%), and the results were not as good at a lower oxygen tension (25%–50%), indicating that oxygen transfer limitation was severe in the immobilized cell culture. This disadvantage has also greatly limited the industrial application of immobilized fungal cell fermentation. Furthermore, the reactor productivity is limited by the available surface area for cell attachment, and such a surface-dependent immobilized cell bioreactor is difficult to scale up. There is another disadvantage with SmF when starch is used as the substrate: starch has a relatively low solubility in water, and insoluble starch granules are difficult to handle (e.g., pumping and mixing) in conventional SmF. It is very difficult to use crude (insoluble) starch directly in SmF, especially with immobilized cells, as there would be increased mass transfer barriers. Consequently, the fermentation is much slower. In recent years, production of lactic acid from SSF has been reported [3]. Different crops or crop residues, such as sweet sorghum, corncob, sugarcane press mud, and carrot-processing waste, were used as substrates (Table 1). Soccol et al. [31, 32] evaluated lactic acid production in SmF and SSF (with inert support) by using a strain of *R. oryzae*, and found that both production level and productivity were higher in SSF. This indicates that SSF also has the good potential to economically produce lactic acid and other organic acids and alcohols from starch present in solid substrates.

2.2. Enzymes from SSF

Historically, enzymes have long been produced from SSF. Several reviews of the production of enzymes from SSF have been published in recent years [3, 5, 40]. Recordings of SSF can be found in Asia starting from thousands of years ago, e.g., the Koji process mentioned above. Evidently, the SSF process originated from food fermentation and production of enzymes. In theory, all the enzymes that are presently known and produced by any means are able to be produced under SSF. Some examples are listed in Table 2. The microorganisms involved can be filamentous fungi, yeasts, or bacteria.

2.2.1. Amylases

Amylases are among the most important industrial enzymes in commercial biotechnology [3]. Generally, hydrolytic enzymes and amylolytic enzymes are commonly produced by filamentous fungi; the preferred strains include *Trichoderma spp.*, *Aspergillus spp.*, and *Rhizopus spp.* [2]. Four of the amylases are of special interests: α-amylase, β-amylase, glucoamylase, and pullulanase. Each of them attacks the starch molecule at different sites, resulting in products of various chain lengths. Amylase production has been reported in the species of the genera *Aspergillus, Penicillium, Cephalosporium, Mucor, Candida, Neurospora, Rhizopus,* etc. [40, 41].

In industry amylases are mainly produced by fungi and bacteria through SmF. In recent years, the feasibility of applying SSF to the production of amylases has been intensively investigated. All kinds of agro-products or residues can be used as substrates in SSF, such as oil cakes (coconut oil cake, sesame oil cake, groundnut oil cake, palm kernel cake, olive oil cake, etc.), wheat bran, rice husks, spent brewing grains, banana peels, potato peels, etc. [42–48]. The most attractive feature of SSF for the production of enzymes is that it is much more economical than SmF. It is generally recognized that production of many industrial enzymes by SSF is much less expensive than by submerged fermentation [2]. In submerged fermentations, the product enzyme is often produced at a low concentration and with a low yield and productivity. Recovery of the enzyme product from the dilute fermentation broth is also costly. Current methods for using either starch or cellulose as fermentation feedstock usually include a hydrolysis step followed by fermentation. The hydrolysis of these polymeric biomaterials for fermentation use is, however, costly. Of special interest to industry are processes which combine microbial cultivation simultaneously with the hydrolysis of the biomass, as found in many SSF systems. Simultaneous saccharification and fermentation eliminates the hydrolysis step, thereby reducing production costs. Furthermore, many fungal enzymes can only be produced at high quantities under SSF conditions.

Table 2
Some enzymes produced in solid state fermentations

Product	Organism	Substrate	Reference
Amylases	*Aspergillus oryzae*	Brown rice	[49]
Glucoamylase	*A. niger*	Tea waste	[50]
Lipase	*A. niger*	Gingelly oil cake	[51]
	Penicillium simplicissimum	Soy cake	[52]
Acid protease	*A. niger*	Wheat bran	[53]
Cellulase, Glucosidase	*A. ellipticus, A. fumigatus;* *Fusarium oxysporum*	Lignocellulosic waste Corn stover	[54] [55]
Cellobiase	*A. niger*	Waste pulp	[56]
Xylanase	*A. niger*	Wheat bran	[57]
	Fusarium oxysporum	Corn stover	[55]
	Trichoderma longibrachiatum	Wheat bran-malt sprouts mixture	[58]
Hemicellulase	*Thermomonospora strain* 29	Coffee waste	[59]
Phytase	*A. niger*	Wheat bran	[60]
	A. ficuum, Mucor racemosus, Rhizopus oligosporus	Canola, soybean meal, cracked corn, wheat bran	[61, 62]
Mannanase	*A. ochraceus*	Coeffee waste	[63]
Pectinase	*A. niger*	Soy and wheat bran	[64]
Pectin lyase	*A. niger*	Misc.	[65]
Chitinase	*Trichoderma harzianum*	Wheat bran + chitin	[66]
Tannase	*A. niger; A. foetidus, R.. oryzus*	Tannin-rich powdered fruits	[39]

2.2.2. Phytase

Phytase is an enzyme that makes the phosphorus from phytin available for animal digestion. Up to now, phytase has been mainly used as a dietary supplement for swine and poultry. As phytase is increasingly used in animal feeds, science and technology related to this enzyme are rapidly evolved. Several detailed reviews have been published in recent years [67, 68].

The benefits of phytase are its double effects on reducing the use of expensive inorganic phosphorus in animal diets and the environment pollution from excessive manure phosphorus runoff [68]. In grains used for animal feed, about 60–75% of phosphorus is in an organically bound form known as phytate, which cannot be utilized by monogastric animals. Thus, the majority of this phytin phosphorus will pass directly through the digestive track of the animals and winds up in manure and liquid effluent. In the U.S., large-scale animal production (pig and poultry farms and cattle feedlots) generates enormous quantities of potentially hazardous wastes, with phosphorus being a major pollutant. It has been shown that adding 500 to 1000 units of phytase to monogastric animal feeds can replace approximately 1 gram inorganic phosphorus supplementation and reduce phosphorus in the manure by about 30–50% [68]. The rate of phytase inclusion in animal diets is depending on both the desired degree of phytin reduction and economical considerations. Phytase currently produced by submerged fermentation is relatively expensive, and may add US $2–3 per metric ton to the feed cost [61]. SSF would be an economical alternative for the production of the enzyme. Phytase produced in SSF can be easily extracted with water and the production cost is expected to be much lower than that in submerged fermentation because of higher enzyme concentrations and activities. Phytase produced by filamentous fungi on selected feed ingredients contains also accessory enzymes, fungal protein and organic acids that increase feed digestibility and access to phytin in plant cells [61]. The product can be directly mixed in feed rations as a value-added supplement, further reducing the cost for use in animal feed.

2.2.3. Chitinases

Chitinases are hydrolytic enzymes responsible for the degradation of chitin, a high molecular weight linear polymer of N-acetyl-D-glucosamine units. As chitin degrading enzymes, chitinases have received increasing attentions in applications in the biocontrol of fungal pathogens, because chitin is the major structural component of fungal cell walls. Biological control is an attractive alternative for fungal pathogen control, since it neither causes environmental pollution nor induces pathogen resistance, which is usually a side effect of synthetic antifungal agents [69, 70]. Chitinases can be found in a wide range of organisms, including fungi, viruses, bacteria, insects, plant, and animal. However, chitinases of filamentous fungi have been shown to have higher activity levels and a wider antifungal spectrum than those of plants and bacteria [71]. Li [72] has reviewed fungal chitinases in details. The potential of industrial production of chitinases by either bacteria or fungi has been studied in both submerged and solid state fermentations [66, 71, 73–79]. *Trichoderma harzianum* is one of the most widely studied microorganisms for the production of chitinases. Mass production of *Trichoderma* spores for biocontrol purposes is of a high interest. Unfortunately, large-scale production of chitinases at present is still too expensive and uneconomical to make this enzyme available in sufficient quantities. SSF production of chitinase and *Trichoderma* spores could greatly reduce the production costs and thus should have a good potential for commercial applications.

In spite of the high productivity and low cost of SSF process, currently the application of SSF in the production of industrial enzymes is still mostly limited to lab scales. Most of the research is focusing on strain screening, process parameter optimization, small scale reactor design, etc. The main problem is that the development of a simple, practical, and automated SSF fermenter has not yet been achieved.

2.3. Biological control agents (BCA) from SSF

Chemical pesticides have been extensively used for many years in agriculture all over the world. It is a general concern that the use of these chemical pesticides poses adverse effects on human health and the environment. For this reason, there has been interest in researching biological control agents (BCA) since the 1960s [90] (Table 3). A biocontrol agent may be a living microorganism, a natural product of microbial origin, or a chemically modified natural product of microbial origin. Several steps are required to develop a microbial pesticide, including isolation of the microorganism, identification and characterization, and pilot trials under real conditions [91]. Both submerged and solid state fermentation processes have been used for BCA production. In this part, we will discuss only the production of BCAs by solid state fermentation.

Table 3
Some biocontrol agents and other compounds produced in solid state fermentations

Product	Organism	Substrate	Reference
Biocontrol agents (aerial spores or conidia)	*Trichoderma harzianum*	Solid substrates;	[80]
	Epicoccum nigrum	Peat/vermiculite/lentile meal	[81]
	Coniothyrium minitans		[82]
Biofungicide	*Collectotrichum truncatum*	Perlite + corn meal	[83]
Bioinsecticide	*Beauveria bassioana*	Clay granules + liquid medium	[84]
Neomycin	*Streptomyces marinensis*	Various solid substrates	[85]
Penicillin	*Penicillium chrysogenum*	Sugar cane bagasse	[86]
Griseofulvin	*Penicillium griseofulvum*	Rice bran	[87]
Red pigment	*Monascus sp. EBF1*	Barley	[88]
Biopulping	*Phanerochaete chrysosporium* *Ceriporiopsis subvermispora*	Wood chips	[89]

In the development of BCAs, it is important to develop methods for mass production. Usually, bacteria and yeast are produced in submerged fermentation, while many fungi are fermented in solid state fermentation. It was found that SSF has some advantages over SmF in producing high quality BCAs inocula. SSF can produce abundant, robust, and healthy conidia because SSF is more like fungal natural environmental conditions and because habitat aeration is better, while many filamentous fungi sporulate poorly in SmF. *Coniothyrium minitans* is a promising fungal BCA of the plant pathogen *Sclerotinia sclerotiorum*. It sporulates relatively poorly in liquid media, but sporulates well on solid substrates [82, 92–94]. *C. minitans* spores produced by SSF are of better quality, with greater resistance to UV-irradiation and desiccation during recovery procedures, and are more viable after storage [90]. De Cal et al. [95] reported that the conidial production of *Penicillium frequentans* was higher in SSF than in SmF. Fresh conidia produced by solid-state fermentation reduced the incidence of brown rot on plums by 75%. Agosin et al. [80] compared the aerial spores of *T. harzianum* P1, a potential biocontrol agent, produced in SSF against those produced in SmF. They found that spores from SSF had higher productivity, greater UV-resistance, and a longer shelf life. The SSF spores had a thicker outer wall and fewer organelles. The SmF spores, however, were usually mostly collapsed, inhibited many cytoplasmic organelles, and had a much thinner outer wall. Even some bacteria-originated BCAs had higher productivities in SSF than in SmF. *Bacillus subtilis* isolated by Shoda [96] exhibited a broad suppression spectrum to various plant pathogens mainly by producing lipopeptide antibiotic, iturin A. When this bacterium was grown on soybean curd residue in SSF, the productivity of iturin A was about 10 times higher than when grown in SmF [96].

However, the commercialization of these microorganisms as BCAs has been hampered by the lack of a cost-effective way of producing sufficient amounts of biomass for use. Design and scale-up of an appropriate SSF reactor with control and automation is a complicated task and a breakthrough has not yet been achieved.

3. ADVANTAGES AND UNSOLVED PROBLEMS

As listed in Table 4, SSF has many unique process characteristics that differ from and often provide advantages over SmF [1], especially for filamentous fungi that are usually more difficult to operate in SmF than bacteria and yeasts. Filamentous fungi have long been employed in the fermentation industry and continue to be the principal source of antibiotics and enzymes. As saprophytes, filamentous fungi naturally secrete large amounts of enzymes and metabolic products. As an additional advantage, secretion assures simplified and inexpensive product recovery because the product is purified from the spent medium rather than from the host biomass [97]. Currently, about 40% of industrial enzymes are produced by filamentous fungi [98], mainly by SmF processes. Compared to SmF, SSF is a process with low cost and high productivity. Many industrial enzymes have been shown to be better produced by SSF than by SmF [65]. It has been reported that the production cost of cellulase in SmF is ~$20/kg; while the cost is only ~$0.2/kg in SSF [3]. The production of pectin lyase by *A. niger* was 3 times higher in SSF than in SmF, and the produced enzyme in SSF constituted 65% of total extracellular proteins produced by the fungus [65]. Maldonado and De Saad [99] reported that enzyme production was four to six times higher in SSF than in SmF. In general, SSF is more suitable than SmF for the growth of filamentous fungi under conditions where catabolite repression applies [100].

Table 4
Some unique characteristics and advantages of solid state fermentation

Characteristics	Advantages/Comments
Low moisture content	lower reactor volume required for a given productivity
	lower purification costs because of higher product concentrations
	lower costs for treatment of liquid effluent
	inhibition of contaminants
High interfacial surface area to liquid volume ratio	aeration is easily achieved
	generally has lower energy requirements
Simulates the natural environment for microbial growth	allows more complete genetic expression in the microbe
	some products are produced at higher rates in SSF
	some products are produced only in SSF
	yields are reliable and reproducible
Simple media	often no more than unprocessed grains with minimal mineral supplementation or no supplementation at all
	may consist of agri-industry wastes such as corn fiber and bagasse
Substrate availability	may increase during fermentation (or decrease or remain constant, as well) rather than always decrease as it does in SmF

In addition to the higher production titers and yields, phytase produced in SSF showed better heat resistance and retained most of its activity after agglomeration and palletizing in the process of producing animal feeds [60]. Also, fungal spores or conidia produced in SSF had a thicker outer wall and showed greater resistance to UV light and heat, and also had a longer shelf-life when used as a biocontrol agent [80].

Recent advances in the molecular genetics of filamentous fungi have also allowed development of commercially promising recombinant fungal strains for production of enzymes, heterologous proteins, and other biochemicals [98, 101]. However, most filamentous fungi, including *Aspergillus*, secrete a diversity of extracellular proteases [102] which may cause a major problem for heterologous protein production because these extracellular proteases degrade heterologous proteins. The construction and use of protease-deficient host strains may alleviate this problem, but often at the expense of reduced expression of the protein product. Development of an optimal production medium and classical strain improvement programs (random mutagenesis and selection) have also been studied with limited success in improving protein yields. Although gene expression dynamics of filamentous fungi under SSF conditions remain largely unknown, there is a possibility that the attack of fungal proteases on heterologous proteins may be reduced in SSF, thus allowing filamentous fungi to be used as a more efficient producer for secreted recombinant protein (enzyme) production.

However, there are several major problems in the development of SSF on an industrial scale, including the mass and heat transfer limitations and difficulty in handling solids in existing reactors and the lack of kinetic and design data on various fermentations [103, 104].

Heat and mass transfer are particularly critical issues for scaling-up SSF processes. Several workers have suggested forced air convection in packed bed reactors as at least a partial solution to the mass and heat transfer difficulties [105–109] noted in even shallow tray fermenters [110]. Although packed bed reactors successfully increase protein productivity, a temperature gradient exists in the reactor, because the lack of free-flowing water causes inefficient heat removal [111]. It is estimated that even with an airflow as high as one volume per reactor volume, up to 85% of the enzyme produced by the microorganism can be denatured because of the high heat accumulation by the end of the fermentation [112]. It also has been shown that the growth of biomass in the void space of packed beds is responsible for much of the decrease in effective diffusion coefficients observed in these reactors and is therefore much of the cause of the formation of gaseous concentration gradients [113]. Furthermore, it has been suggested that steric hindrance in packed beds prevents the continued growth of biomass [114]. Another common problem in SSF is bed caking caused by substrate matrix change during the fermentation process, which in turn causes difficulties in process control and downstream processing. Difficult solid handling in the process as run in existing reactors also presents a major problem in the development of SSF on an industrial scale [107]. In general, packed bed bioreactors are difficult to scale up and pose problems in solid handling [107]. They are also difficult to operate as continuous reactors.

4. SSF REACTORS

To commercialize SSF processes for the production of enzymes, chemicals, and biologically active compounds (including enzymes and therapeutic proteins) from polymeric biomaterials produced in large quantities as low-value commodities or byproducts in the agricultural industry, it is necessary to develop efficient bioreactor systems [115]. The most commonly used SSF bioreactors are tray reactors, drum reactors, and packed bed reactors [3, 116]; they vary in their agitation and aeration conditions. It is recognized that continuously-mixed beds with forced aeration have the potential to perform better than other bioreactors due to their good heat and mass transfer characteristics. In recent years, many mixed SSF reactors have also been studied [3], such as rotating drum reactors and fluidized beds reactors. Durand [117] gave a detailed review of SSF reactor designs. Table 5 lists and compares some important SSF reactors.

4.1. Tray reactors

Tray reactors, based on traditional Koji fermentation, are the simplest SSF reactors. They are typically stacks of shallow trays loaded with fermentation substrates, which are inoculated with microorganisms, in an aerated and controlled environment room (Fig. 1). The top surface of trays is exposed to the air. The bottom surface is usually perforated but without forced air circulation through the tray. The temperature is controlled by controlling the temperature of the gas stream into the room. Moisture saturation is maintained by spraying sterilized water into the room, thus keeping an optimal moisture content level for the fermentation. The current industrial applications of SSF usually use this kind of reactor, as it is the most traditional form of SSF reactor and a lot of knowledge and experience have been accumulated. However, there are unsolved problems, among which heat removal is the most important one. The heat removal in trays depends mainly on conduction through the tray walls to air. In general, conductive cooling is insufficient to remove the metabolic heat for a large scale SSF system, even if it is continuously mixed [118]. Other problems include a high contamination risk and low substrate utilization efficiency resulting from low heat and mass transfer rates.

Table 5
Features and problems of commonly used SSF reactors

Reactor type	Features / problems	References
Tray reactors	Simple structure, easy to operate Non-aeration, non-mixed Heat accumulation Temperature and moisture gradients generated Bed caking	[110, 118, 119]
Packed-bed reactors	Forced-aeration, non-mixed Axial temperature and gas concentration gradients exist Difficult to scale-up Bed caking	[103, 107, 112, 120, 121]
Rotating-drum reactors	Forced aeration, continuously or intermittently mixed Improved mass and heat transfer Shear effect may cause damage to organisms Slumping flow may cause little mixing Complicated reactor construction Difficult operation	[122, 123 – 129]
Fluidized-bed reactors	Continuously mixed High mass and heat transfer rate No bed caking High power requirements High shear damage to microorganisms Difficult to fluidize large, coarse and sticky particles	[130 – 134]
Spouted-bed reactors	Continuously or intermittently mixed High mass and heat transfer rate Lower power requirements than fluidized-beds Good in handling large, coarse, non-uniformly sized, and sticky particles usually used in SSF Need further investigations on characterization and scale-up	[49, 135–141]

When heat is not sufficiently removed, temperature and moisture gradients will develop in the matrix. Rathbun and Shuler [110] have experimentally studied temperature and gas concentration gradients in tray fermenters. They established a temperature gradient in a tray containing various depths of medium, and then measured temperature and concentration as cultivation proceeded over time. The vertical temperature gradient was larger than the lateral gradient as a result of the heating. When thin (~0.5 cm) beds were compared to relatively thick beds (~3.2 cm), little or no transport resistance was found for the shallow beds, allowing intrinsic growth parameters to be determined. The thicker beds had noticeable transport limitation effects, developing considerable gradients in both temperature and oxygen concentration.

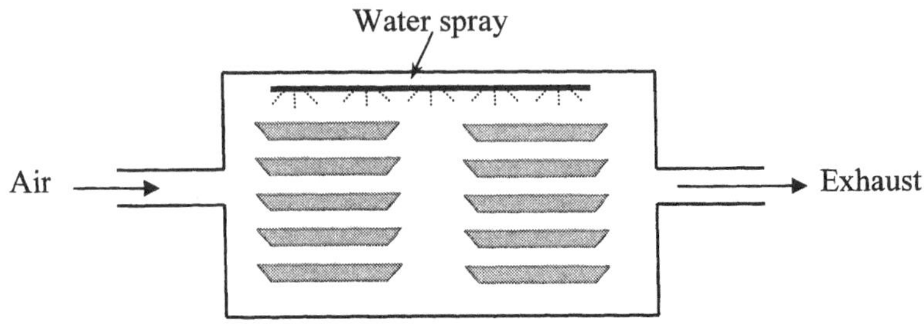

Fig. 1. Schematic of a tray reactor.

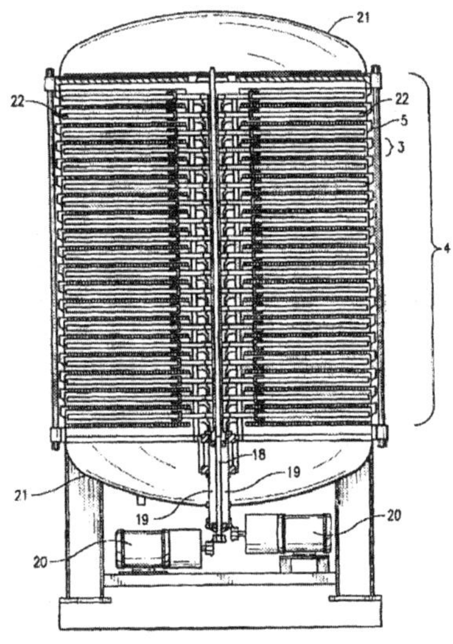

Fig. 2. Schematic of a PlaFractor™ stacks fermenter (adapted from [119]).

The evaporating of water from the matrix removes a little of the metabolic heat, but still cannot solve the problem because of the poor mass and heat transfer inside the fermentation matrix. Thus, the limitation is that only a thin layer in the tray can be effectively fermented. Bed caking, which is caused by the agglomeration of mycelium, makes the problem even worse. The caking can be a major resistance to the air flow. Some of these systems incorporate a "mixing" arrangement. However, mixing can cause shear damage to the mycelium and loss of productivity, so this has to be balanced against the productivity loss resulting from improper temperature control or bed drying. In many cases, the productivity of unmixed tray culture systems is better than that of mixed systems [119].

There have been many efforts in improving this traditional fermentation process, e.g., Biocon India Ltd. developed a technology called PlaFractor™ in 2003 (Fig. 2) [119]. A PlaFractor™ stack is made of several tray modules arranged upon each other and sealed by gaskets in between. Each tray module has a mixing arm with blades, which rotate around the axis formed by the arms. It seems to be featured of contained environment, superior temperature control, automation and in-situ operations, and space and energy saving.

4.2. Packed bed reactors

A packed bed bioreactor for SSF is a column with a perforated base for forced aeration. It may have a jacket for water circulation to control the temperature during fermentation. Packed bed reactors allow forced convective mass transfer as air is pumped through the bed. Smaller gradients than those found in trays are found in packed bed reactors, but reduction in bed porosity with time can still be a problem [103]. Axial temperature and gas concentration gradients have been documented in packed columns [120], although they are much smaller than those observed in tray reactors [121]. Compared to tray reactors, packed bed reactors successfully increased the enzyme productivities. However, they are difficult to scale up, and with bed caking in the columns, solid handling becomes difficult [107]. They are also difficult to operate as continuous reactors. Despite these problems, efforts have been made in order to improve the performance of packed beds and to scale up it.

In order to scale up the reactor, a mathematical model predicting temperature, moisture, biomass, and substrate profile is a valuable tool. Several models have been published. Earlier models usually concerned only energy balance [122, 142–146], which gave information on radial and axial temperature gradients, because heat removal is the main problem for large scale SSF reactors. The axial temperature gradients promote evaporation even if saturated air is used to aerate the column because the water carrying capacity of the air increases as it heats up. This evaporation not only removes 65% of the waste metabolic heat [147] but also dries the substrate, thus inhibiting cell growth. Water balance is therefore also very important. Oxygen supply has been reasonably ignored in most models if the bed is forced aerated. In reactors with forced aeration, the supply of oxygen to particle surfaces is not a limiting factor [120]. Several recent models have included the water balance. Weber et al. [142] proposed a model with enthalpy and water balances.

Enthalpy balance:

$$\frac{\partial}{\partial t}\left((1-\varepsilon)\cdot\sum_i (C_i \cdot h_i)+\varepsilon\cdot\sum_j (C_j \cdot h_j)\right)=-r_0''' \cdot \Delta H_0 - F_a'' \cdot \frac{\partial}{\partial z} h_a$$

C_i and C_j indicate concentrations of all components in the moist solid matrix and the gas phase, respectively.

Water balance:

$$\frac{\partial}{\partial t}\left[(1-\varepsilon)\cdot(\varphi_s \cdot C_s + x_{wx} \cdot C_x)+\varepsilon \cdot C_{wg}\right]=r_w''' - F_a'' \cdot \frac{\partial}{\partial z}\varphi_g$$

These equations can be simplified based on the following assumptions [142]: first, in the accumulation terms of the balances, all contributions of gases and all mass accumulation terms are negligible; second, there is a pseudo-steady state with respect to temperature and oxygen consumption rate; third, the air is at equilibrium with the solid matrix at any point in the bed; and last, the oxygen consumption rate is independent of temperature, provided that the air inlet and outlet temperatures are chosen within the optimal range found by McQuilken et al. [148]. The simplified equations are:

$$0 = -r_0''' \cdot \Delta H_0 - F_a'' \cdot \frac{d}{dz}(h_a)$$

$$\frac{\partial}{\partial t}\left[(1-\varepsilon)\cdot(\varphi_s \cdot C_s + x_{wx} \cdot C_x)\right] = -r_0''' \cdot Y_{w0} - F_a'' \cdot \frac{\partial}{\partial z}\varphi_g$$

The model was validated with the fermentation of the fungi *Coniothyrium minitans* and *Aspergillus oryzae*. The model gave accurate temperature predictions when online oxygen measurement was used as input.

Von Meien and Mitchell [149] developed a two-phase model for water and heat transfer within an intermittently-mixed solid-state fermentation bioreactor with forced aeration.

Water balance in gas phase:

$$\varepsilon \cdot \rho_g \frac{\partial \varphi_g}{\partial t} + G\frac{\partial \varphi_g}{\partial z} = K'a(\varphi_s - \varphi_s^*)$$

Water balance in solid phase:

$$\frac{\partial (S \cdot \varphi_s)}{\partial t} = -K'a(\varphi_s - \varphi_s^*) + Y_{WB}(S\frac{\partial b}{\partial t} + C_x \frac{\partial S}{\partial t})$$

Energy balance in gas phase:

$$\varepsilon \cdot \rho_g (C_{Pg} + \varphi_g \cdot C_{Pv})\frac{\partial T_g}{\partial t} + (C_{Pg} + \varphi_g \cdot C_{Pv}) \cdot G\frac{\partial T_g}{\partial z} = -ha(T_g - T_s)$$

Energy balance in solid phase:

$$S \cdot (C_{Ps} + \varphi_s \cdot C_{Pw})\frac{\partial T_s}{\partial t} = ha(T_g - T_s) - \lambda \cdot K'a(\varphi_s - \varphi_s^*) + Y_Q\left(S\frac{\partial b}{\partial t} + C_x \frac{\partial S}{\partial t}\right)$$

This model predicts that it is impossible to prevent the bed from drying out when it is forced aerated unless the bed is intermittently mixed.

Even with heat removal by evaporation, some extra care may still be needed when scaling-up solid-state fermentation processes for the production of thermo-labile products. Muller dos Santos et al. [112] studied thermal denaturation in packed beds. Based on the mathematical model of a well-mixed bioreactor, it is suggested that even with airflows as high as one vvm, up to 85% of the enzyme produced by the microorganism can be denatured by the end of the fermentation.

4.3. Rotating drum reactors

In a rotating drum reactor, the substrate bed is held within a horizontal or near horizontal drum, which can be continuously or intermittently mixed (Fig. 3). The effect of operating conditions, particularly the rotating speed, have been studied by many researchers. Stuart et al. [123] revealed that the extent of the effect of shear caused by drum rotation is controversial. In some studies, high rotating speed reduced productivity, presumably because of the deleterious effects of shear. However, in some other studies, highest productivity was obtained at the highest speed. Based on their own experiments, Stuart et al. [123] concluded that the effects of operational variables on the performance of the reactors were mediated by their effects on transport phenomena, such as mixing, heat and mass transfer, and shear within the fermentation bed.

Another critical issue concerned with the performance of the rotating drum reactors is the slumping flow of the fermentation bed. At the low rotational speeds (a few rpm) typically used for rotating drum bioreactors, the bed in an unbaffled drum undergoes a slumping flow in which the bed as a whole slides down the internal surface of the drum, causing relatively little mixing within the substrate bed. Under such conditions, the rotating drum does not perform any better than a tray bioreactor. In order to avoid the slumping flow, two strategies have been considered: high rotating speed (10-50 rpm) [123] and the addition of baffles inside the drums [124–126]. However, high rotation rates can adversely affect growth due to shear damage [123]. Thus, the use of baffles is preferred. Schutyser et al. [127] studied straight baffles and curved baffles in rotating drums. It was found that, in a drum with curved baffles, complete mixing in the radial and axial direction was achieved much faster than in the other designs. Problems arise in that as construction becomes more complicated, manual tasks become more difficult, and loading and unloading of the solid may also become more difficult.

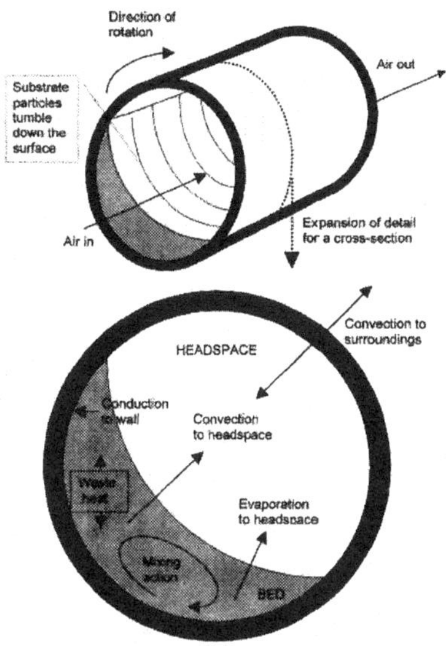

Fig. 3. Schematic of a rotating drum reactor. The reactor can be continuously or intermittently rotated. (adapted from [120]).

The gas-flow pattern in rotating drum reactors is another issue for concern. The gas flow in the headspace associated with the end-to-end aeration typically follows the pattern of plug-flow with axial dispersion [128], which causes significant axial temperature gradients in the substrate bed [123, 129]. This suggests that axial mixing, such as the use of angled lifters, is necessary [129]. However, this, possibly together with the replacing the end-to-end air flow with introduction and removal air at several points along the reactor axis, will greatly increase the complexity of the reactor construction. The scale up of rotating drum reactors is also limited by vessel size and mechanical gear design.

4.4. Fluidized bed reactors

Gas-solid fluidized beds have also previously been used to improve mass and heat transfer in SSF [130, 131]. Yeasts have been successfully grown in fluidized beds [132]. The fluidized beds were intensively studied in the 1980s, but have attracted relatively little attention in recent years [3]. In a fluidized bed, the gas-flow rate required for fluidization is high enough to confirm a very good rate of heat and mass transfer. However, as a bioreactor for SSF, its shortcoming is obvious. Fluidization requires the use of fine particles and the minimum air velocity for fluidization is high [133], resulting in high power requirements; also, the high air flow rate causes high shear damage to the microorganisms, especially filamentous fungi which grow in the form of delicate hyphae. Large, coarse, and sometimes sticky particles, which usually appear in SSF, are difficult to fluidize. A spouted bed may provide good heat and mass transfer as well, while requiring a lower gas-flow rate and can easily handle the coarse particles. In the late 1980s, a spouted bed for the production of cellulase using immobilized *Trichoderma viride* cells was reported [134]; however it was a gas-solid-liquid reactor used for submerged fermentation.

4.5. Spouted Bed Bioreactor (SBB)

Recently, a novel gas-solid spouted bed bioreactor has been proposed to overcome the difficulties in SSF [49]. Gas-solid spouting, a variant of fluidization initially developed in the early 1950 for grain drying, permits good gas-solid contact for solid materials that are too coarse or dense for stable fluidization [135]. Spouted beds now have many applications in granulations, combustion, drying, and coating. Spouted bed reactors have been found to have many advantages over conventional fluidized beds [136, 137]. They are used when homogeneous and stable flow region is not attained in the fluidized bed, as in the cases of non-spherical particles and polydisperse and fine disperse systems [138]. Spouting is particularly appropriate for handling the large, coarse, non-uniformly sized, and sticky particles that are often used as SSF substrates. In a spouted bed, heat and mass transfer rates within the reactor are high, and solids are well mixed [139, 140]. The minimum air velocity required for spouting is lower than that required for fluidization. Gas-solid spouting in SSF also can prevent mycelial caking during the fermentation [49]. Because the particles in a spouted bed do not become packed together, problems associated with packed bed reactors will not arise in a spouted bed bioreactor.

In a gas-solid spouted bed (see Fig. 4), a stream of high velocity gas is injected into a bed of coarse particles; it jets through the middle of the bed, carrying a stream of particles with it, which then rain back onto the bed [136]. The rise of the particles in the middle of the bed is accompanied by the sinking of particles in the annular region. Agglomerations of particles are broken apart by high velocity impacts in the core region. Large particles, which can not be fluidized because they are too large, can often be spouted. Somewhat finer particles, which can be fluidized but only at high gas velocities, can be spouted at lower gas velocities. Therefore, the use of spouting rather than fluidization has the advantages of being applicable to larger substrate particles (such as grains) without requiring grinding and requiring less energy for air compression during fermentation.

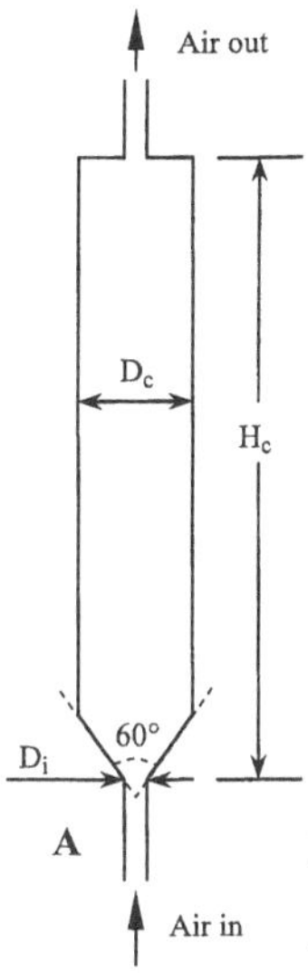

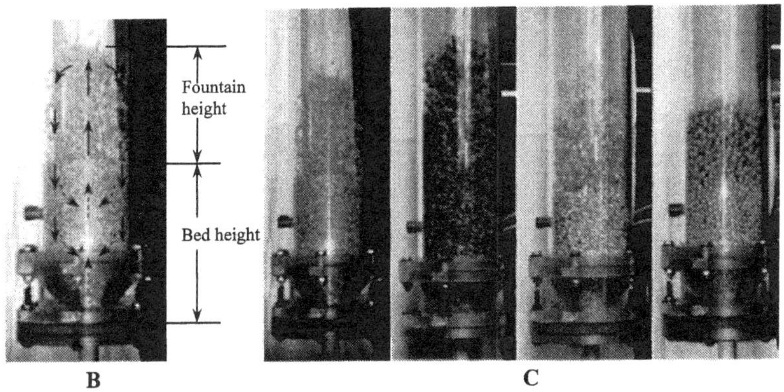

Fig. 4. A gas-solid spouted bed reactor – Gas flow upward while particles flow downward except in the central jet area where both gas and particles flow upward. A. Dimensions of the bioreactor. B. Spouted-bed structure. Arrows represent the grain flow direction. C. Spouting of various grains: brown rice, lentils, cracked corn, and soy bean (from left to right).

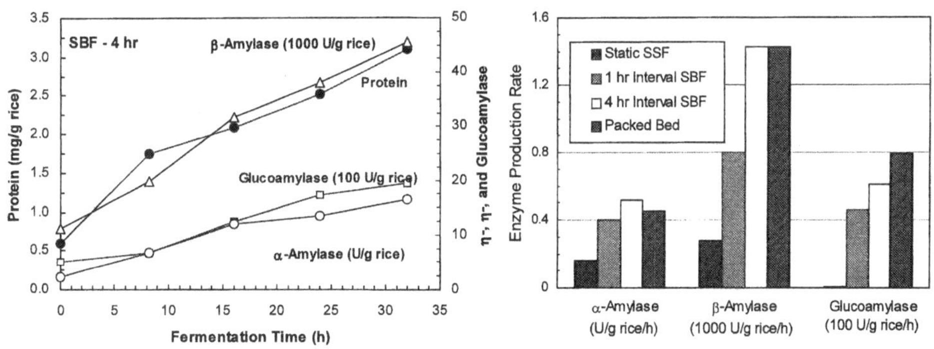

Fig. 5. A. Kinetics of enzyme production from brown rice in a spouted bed bioreactor. B. Comparison of enzyme productivities in SBB and other types of reactors (adapted from [49]).

SSF have been studied for the production of α-amylase, β-amylase, and glucoamylase from rice by *Aspergillus oryzae* in spouted bed, packed bed, and tray reactors [49]. The results showed that tray reactors with surface aeration had poor mass and heat transfer. The packed bed reactor with continuous aeration through the rice bed produced high protein and enzymes, but the fermented rice was difficult to remove and process due to the formation of large chunks of rice aggregates knitted together with fungal mycelia. Also, the fermentation in the packed bed was not uniform. The spouted bed bioreactor with intermittent air spouting achieved high production levels in both total protein and enzymes that were comparable to those found in the packed bed bioreactor, but without the non-uniformity and solids handling problems. Fig. 5A shows typical protein and amylases production kinetics in SSF with rice as the substrate. Fig. 5B shows that SSF with intermittent spouting had high enzyme production comparable to that of packed bed bioreactors with forced aeration and much higher than that of the static tray reactor. Fig. 6 shows the mycelium growth on grain particles in SBB.

Fig. 6. Morphology of filamentous fungi in SSF. Fungal mycelia with spores grown on solid substrate with starch granules. The right photo shows that the fungal mycelia penetrate through the grain's skin and reach starch granules.

As a SSF bioreactor, SBB demonstrates high potential in obtaining better mass and heat transfer, and thus a higher productivity of enzymes and other products. Further investigation will be needed in order to better characterize and scale up the spouted bed bioreactor for solid state fermentation. In order to scale up the spouted-bed bioreactor, an understanding of reactor hydrodynamics is necessary. We have studied the minimum spouting velocity (U_{ms}), particle mixing time, and the effects of column diameter (D_c), inlet diameter (D_i), bed height (H), particle density (ρ_p) and fluid density (ρ_f) [141]. In general, the U_{ms} increases when the static bed height, cone angle, or gas inlet diameter are increased. Numerous correlations have been suggested for the prediction of minimum spouting velocity, a critical design parameter in reactor scale-ups [141, 150, 151]. Many of these are empirical in nature, though a few have a dimensional analysis or theoretical basis. One of the earliest proposed, and still one of the most popular, is that of Mathur and Gishler [152]. Several improved or situation-specific correlations have since been proposed, although most of this work was done in the early 1970's. Table 6 lists several of these correlations.

Of the existing correlations, most were developed for roughly spherical particles. Many SSF substrates are not spherical, for example, rice grains have a length to width ratio of 3 or more. Furthermore, although grain in general and rice in particular have been studied in spouting beds before, these studies were all concerned with drying rice to fairly low moisture contents (generally less than 30%) [153], not with the somewhat higher moisture contents (30% to 45%) usually used in solid state fermentation. The higher moisture content can cause significant changes in particle size, density, and surface characteristics which can greatly affect the spoutability of the solid particles and hydrodynamics in the spouted bed reactor. A critical question that must be answered is whether the solid substrates suitable for SSF can be properly spouted in a spouted bed bioreactor. A comparison of experimental results with existing correlations is also necessary in order to determine the correlations best suited to the design of a spouted bed bioreactor for SSF.

Table 6
Correlations for the minimum spouting velocity [141]

Correlation equation	Reference
$U_{ms}=\left(\frac{d_p}{D_c}\right)\left(\frac{D_i}{D_c}\right)^{1/3}\left[2gH\frac{(\rho_p-\rho_f)}{\rho_f}\right]^{1/2}$ SI units	[154]
$U_{ms} = kV^{0.6}\rho_p^{0.6}d_p D_i^{0.2}$ CGS units	[155]
$U_{ms} = d_p\left(\frac{g(\rho_p-\rho_f)}{\rho_f D_c}\right)^{1/2}\left[0.64+26.8\left(\frac{D_i}{D_c}\right)^2\right]\left(\frac{H}{D_c}\right)^{0.5-1.76 D_i/D_c}$ SI units	[156]
$U_{ms} = 0.0143 d_p^{0.741} H^{0.592}\left[2g\frac{(\rho_p-\rho_f)}{\rho_f}\right]^{1/2}$ CGS units	[157]
$U_{ms}=0.977\left(\frac{d_p}{D_c}\right)^{0.615}\left(\frac{D_i}{D_c}\right)^{0.274}\left[2gH\frac{(\rho_p-\rho_f)}{\rho_f}\right]^{0.324}$ SI units	[158]
$U_{ms} = a\left(\frac{d_p}{D_c}\right)^{0.714}\left(\frac{D_i}{D_c}\right)^{0.697}H^{0.677}\left[2g\frac{(\rho_p-\rho_f)}{\rho_f}\right]^{0.530}$ SI units; $a = 0.832$ for Reactor I and 0.493 for Reactor II	[141]

5. CONCLUSIONS

Solid state fermentation has been applied to food production for ages. With a new surge of interest in SSF since the 1980s, its applications are much expanded across many fields of science and industry. Organic acids and industrial enzymes are the two most common categories of products from SSF. With little or no free water in the fermentation bed, SSF generates higher volumetric productivity and less waste water. The extraction of products from the fermentation bed can provide more concentrated solutions than submerged fermentation for the downstream processes. All the advantages of SSF suggest that SSF has a high potential as a much more economical industrial process. However, its industrial applications are very limited, especially in western countries. The main problem is that the development of a simple and practical automated fermenter for SSF processes has not yet been achieved. Traditionally, SSF is carried out in trays or packed-bed bioreactors. These conventional reactors are simple to construct and widely used, but cannot provide enough mass and heat transfer, which are very important for fermentation processes. In recent years, many new designs of SSF reactors have been proposed in laboratories: rotating drums, fluidized beds, etc. However, each has its shortcomings, and is not yet feasible for scale-up. Recently, a novel gas-solid spouted bed bioreactor has been proposed to overcome the difficulties in SSF [49]. This reactor can provide good mass and heat transfer, easier solid handling, and experiments showed that higher enzyme production was obtained compared to static and packed bed reactors. It is demonstrated to have a high potential for industrial application, but further investigations are needed to better characterize it and to scale it up.

ACKNOWLEDGEMENTS

This work was supported in part by the National Research Initiative of the USDA Cooperative State Research, Education and Extension Service (grant number 00-01797) and the Consortium for Plant Biotechnology Research, Inc. (grant number GO12026-229).

NOMENCLATURE

a, b, c, d, and *e*	constants
C_s	concentration of support in bed (kg dry support m^{-3} support)
C_{Pg}	dry gas heat capacity (J kg^{-1} K^{-1})
C_{Ps}	heat capacity of dry solid (J kg^{-1} K^{-1})
C_{Pv}	water vapor heat capacity (J kg^{-1} K^{-1})
C_{Pw}	heat capacity of liquid water (J kg^{-1} K^{-1})
C_{wg}	concentration of water in air (kg water m^{-3} air)
C_x	concentration of biomass on support (kg dry biomass m^{-3} support)
d_p	particle diameter; m
d_{ave}	particle average diameter; m
d_{eqv}	particle equivalent diameter; m
D_c	reactor body diameter; m
D_i	air inlet diameter; m
F''_a	superficial aeration rate (kg dry air m^{-2} s^{-1})
h_a	enthalpy of (moist) air (J [kg dry air]$^{-1}$)
ha	heat transfer coefficient for heat transfer between the solid and gas phases (J s^{-1} m^{-3} K)
H	Bed height; m
ΔH_o	reaction enthalpy (J [kg O_2]$^{-1}$)
k	constant in Charlton correlation
$K'a$	mass transfer coefficient for water transfer between the solid and gas phases (kg s^{-1} m^{-3})
r'''_o	oxygen production biomass (kg O_2 m^{-3} reactor s^{-1})
r'''_w	water production biomass (kg H_2O m^{-3} reactor s^{-1})
S	volumetric concentration of total dry solid (kg m^{-3})
t	time (s)
T_g	temperature of gas phase (K)
T_s	temperature of solid phase (K)
U_{ms}	reactor superficial minimum spouting velocity; m/s
V	volume of spouting bed in Charlton correlation; cm^3

x_{wx} water content of biomass (kg water [kg dry biomass]$^{-1}$)
Y_Q heat yield from growth (J kg^{-1})
Y_{WB} stoichiometric coefficient relating water production to growth (kg kg^{-1})
z axial position in bed (m)
ρ_g gas phase density (kg m^{-3})
ρ_p particle density; kg/m^3
ρ_f spouting gas density; kg/m^3
φ_g water content of gas phase (kg kg^{-1} dry air)
φ_s water content of solid phase (kg kg^{-1} dry solid)
φ_s^* solid phase Water content for equilibrium with the gas phase at T_g (kg kg^{-1})
λ enthalpy of evaporation of water (J kg^{-1})
ε void fraction (m^3 air m^{-3} reactor)

REFERENCES

[1] G. Viniegra-Gonzalez, E. Favela-Torres, C.N. Aguilar, S.D. Romero-Gomez, G. Diaz-Godinez, C. Augur, Biochem. Eng. J., 13 (2003) 157.
[2] A. Pandey, P. Selvakumar, C.R. Soccol, P. Nigam., Curr. Sic., 77 (1999) 149.
[3] A. Pandey, C.R. Soccol, D. Mitchell, Process Biochem., 35 (2000) 1153.
[4] D.K. Arora, P.D. Bridge, D. Bhatnagar (eds.), Handbook of fungal biotechnology, Marcel Dekker, New York, 2004. pp 267-273.
[5] D.K. Arora, P.D. Bridge, D. Bhatnagar (eds.), Handbook of fungal biotechnology, Marcel Dekker, New York, 2004. pp 275-286.
[6] A. Pandey (ed.), Advances in Biotechnology, Educational publishers & Distrbutors, New Delhi, 1998. pp 13-16.
[7] C.R. Soccol, F.C. Prado, L.P.S. Vandenberghe, A. Pandey, in: Encyclopedia on Bioresource Technology, Haworth Press, New York, 2003. pp. 652-664.
[8] L.P.S. Vandenberghe, C.R. Soccol, A. Pandey, J.-M. Lebeault, Braz. Arch. Biol. Techn., 42 (1999) 263.
[9] F.C. Prado, L.P.S. Vandenberghe, C. Lisboa, J. Paca, A. Pandey, C.R. Soccol,. Eng. Life Sci., 4 (2004) 179.
[10] L.P.S. Vandenberghe, C.R. Soccol, A. Pandey, J.-M. Lebeault, Bioresource Technol., 74 (2000) 175.
[11] D. Kumar, V.K. Jain, G. Shanker, A. Srivastava, Process Biochem., 38 (2003) 1731.
[12] D. Kumar, V.K. Jain, G. Shanker, A. Srivastava, Process Biochem., 38 (2003) 1725.
[13] E. Gasiorek and W. Lesniak, Prace Naukowe Akademii Ekonomicznej imienia Oskara Langego we Wroclawiu, 949 (2002) 67.
[14] E. Gasiorek and W. Lesniak, Pol. J. Food Nutr. Sci., 11 (2002) 31-37.
[15] S.A. Shojaosadati and V. Babaeipour, Process Biochem., 37 (2002) 909.
[16] S.S.A. Fatemi and S.A. Shojaosadati, Amirkabir, 11 (2000) 314.
[17] E. Gasiorek and W. Lesniak, Zywnosc, 10 (2003) 48.
[18] T. Roukas, Enzyme Microb. Tech., 24 (1998) 54.
[19] C.T. Tran, D.A. Mitchell, Biotechnol. Lett., 17 (1995) 1107.
[20] Y.D. Hang and E.E. Woodams,. Biotechnol. Lett., 9 (1987) 183.
[21] J. Pintado, A. Torrado, M.P. Gonzalez, M.A. Murado,. Enzyme Microb. Technol., 23 (1998) 149.
[22] L.P.S. Vandenberghe, C.R. Soccol, F.C. Prado, A. Pandey, Appl. Biochem. Biot., 118 (2004), 293.
[23] M.Y. Lu, I.S. Maddox, J.D. Brooks, Process Biochem., 33 (1998) 117.
[24] S. Leangon, I.S. Maddox, J.D. Brooks, World J. Microb. Biot., 15 (1999) 493.

[25] B.J. Naveena, M. Altaf, K. Bhadrayya, G. Reddy, Food Technol. Biotech., 42 (2004) 147.
[26] B.J. Naveena, C. Vishnu, M.D. Altaf, G. Reddy, J. Sci. Ind. Res. 62 (2003) 453.
[27] K. Richter, and A. Traeger, Acta Biotechnol., 14 (1994) 367.
[28] S. Xavier and B.K. Lonsane, Appl. Microbiol. Biot., 41 (1994) 291.
[29] S. Miura, T. Arimura, N. Itoda, L. Dwiarti, J.B. Feng, C.H. Bin, M. Okabe, J. Biosci. Bioeng., 97 (2004) 153.
[30] N. Garg, Y.D. Hang, J. Food Sci. Tech. Mys., 32 (1995) 119.
[31] C.R. Soccol, B. Matin, M. Raimbault, J.M. Lebeault, Appl. Microbiol. Biot., 41 (1994a) 286.
[32] C.R. Soccol, B. Marin, M. Raimbault, J.M. Lebeault, Appl. Microbiol. Biot., 41 (1994b), 330.
[33] M. Makela, S. Galkin, A. Hatakka, T. Lundell, Enzyme Microb. Tech., 30 (2002) 542.
[34] P. Buzzini, M. Gobbetti, J. Rossi, Annali della Facolta di Agraria, Universita degli Studi di Perugia, 44 (1993) 661-73.
[35] T. Roukas and M. Liakopoulou-Kyriakides, Food Biotechnol., 16 (2002) 17.
[36] O.V. Singh, R.K. Jain, R.P. Singh, J. Chem. Technol. Biot., 78 (2003) 208.
[37] T. Roukas, J. Ind. Microbiol. Biot., 25 (2000) 298.
[38] G. Mukherjee and R. Banerjee, J. Basic Microb., 44 (2004) 42.
[39] A. Tay and S.T. Yang, Biotechnol. Bioeng., 80 (2002) 1.
[40] A. Pandey, C.R. Soccol, J.A. Rodriguez-Leon, P. Nigam, Solid-state fermentation in biotechnology: fundamentals and applications. New Delhi: Asiatech Publishers, Inc., 2001, pp. 98-110.
[41] R. Bigelis, Food Technol-Chicago, 46 (1992) 151.
[42] S. Ramachandran, A.K. Patel, K.M. Nampoothiri, F. Francis, V. Nagy, G. Szakacs, A. Pandey, Bioresource Technol., 93 (2004a) 169.
[43] S. Ramachandran, A.K. Patel, K.M. Nampoothiri, S. Chandran, G. Szakacs, C.R. Soccol, A. Pandey,. Braz. Arch. Biol. Techn., 47 (2004b) 309.
[44] J. Zhang, K. Wu, J. Cai, R. Pan, M. Marinus, Shipin Yu Fajiao Gongye, 28 (2002) 1.
[45] Z. Baysal, F. Uyar, C. Aytekin, Process Biochem., 38 (2003) 1665.
[46] K.M. Nampoothiri, R.R. Singhania, C. Sabarinath, A. Pandey, Process Biochem., 38 (2003) 1513.
[47] S. Kokab, M. Asghar, K. Rehman, M.J. Asad, O. Adedyo, Int. J. Agric. Biol., 5 (2003) 36-39.
[48] A.P. Goes and J.D. Sheppard, J. Chem. Technol. Biot., 74 (1999) 709.
[49] E.M. Silva, and S.T. Yang. Biotechnol. Progress, 14 (1998) 580.
[50] P. Selvakumar, L. Ashakumary, A. Pandey, Bioresource Technol., 65 (1998) 83.
[51] N.R. Kamini, J.G.S. Mala, R. Puvanakrishnan,. Process Biochem., 33 (1998) 505.
[52] M. Di Luccio, F. Capra, N.P. Ribeiro, G.D.L.P. Vargas, D.M.G. Freire, D. De Oliveira, Appl. Biochem. Biotech., 113-116 (2004) 173.
[53] E. Villegas, S. Aubague, L. Alcantara, R. Auria, and S. Revah, Biotech. Adv., 11 (1993) 387.
[54] A. Gupte and D. Madamwar.. Biotechnol. Lett., 14 (1997) 623.
[55] G. Panagiotou, D. Kekos, B.J. Macris, P. Christakopoulos, Ind. Crop Prod., 18 (2003) 37.
[56] L. Xia, Shipin Yu Fajiao Gongye, 25 (1999) 1.
[57] C. Wang, H. Ye, H. Hong, H. Gao, 1998. International Conference on Biotechnology in the Pulp and Paper Industry, 7th, Vancouver, B. C., June 16-19, 1998, C65-C67.
[58] S. Sargin and G. Ongen, Ege Universitesi Ziraat Fakultesi Dergisi, 40 (2003) 145.
[59] K.C. Srivastava,. Biotech. Adv., 11 (1993) 441.
[60] M. Papagianni, S.E. Nokes, K. Filer,. Process Biochem., 35 (1999) 397.
[61] B. Bogar, G. Szakacs, J.C. Linden, A. Pandey, R.P. Tengerdy, J. Ind. Microbiol. Biot., 30 (2003) 183.
[62] B. Bogar, G. Szakacs, A. Pandey, S. Abdulhameed, J.C. Linden, R.P. Tengerdy, Biotechnol. Progress 19 (2003) 312.

[63] V. Alcaraz-Sandoval, G. Saucedo-Castaneda, S. Huerta-Ochoa, 1997. Mannanase production by filamentous fungi in solid state fermentation, Adv. Solid State Fermentation, Proc. Int. Symp. Solid State Ferment., 311-319.
[64] L.R. Castilho, R.A. Medronho, T.L.M. Alves, Bioresource Technol., 71 (1999) 45.
[65] V.M. Taragano and A.M.R. Pilosof, Enzyme Microb. Technol., 25 (1999) 411.
[66] K.M. Nampoothiri, T.V. Baiju, C. Sandhya, A. Sabu, G. Szakacs, A. Pandey, Process Biochem., 39 (2004) 1583.
[67] P. H. Selle, V. Ravindran, R. A. Caldwell, W. L. Bryden. Nutr. Res. Rev., 13 (2000) 255.
[68] X. G. Lei, J. M. Porres. Biotechnol. Lett., 25 (2003) 1787.
[69] W.T. Chang, C.S. Chen, S.L. Wang. Curr. Microbiol. 47 (2003) 102.
[70] N. Dahiya, R. Tewari, R.P. Tiwari, G.S. Hoondal. World J. Microb. Biot. 21 (2005) 1611.
[71] C. Sandhya, L. K. Adapa, K. M. Nampoothiri, P. Binod, G. Szakacs, A. Pandey. J. Basic Microbiol. 44 (2004) 49.
[72] D.-C. Li. Mycopathologia. 161 (2006) 345.
[73] D. Neetu, T. Rupinder, P. T. Ram, S. H. Gurinder. Curr. Microbiol. 51 (2005) 222.
[74] C. Sandhya, P. Binod, K. M. Nampoothiri, G. Szakacs, A. Pandey. Appl. Biochem. Biotech. 127 (2005) 1.
[75] A.S. Gad, New Egyptian J. Microbiol., 5 (2003) 112.
[76] N. Rattanakit, S. Yano, M. Wakayama, A. Plikomol, T. Tachiki, J. Biosci. Bioeng., 95 (2003) 391.
[77] B. Bhushan, J. Appl. Microbiol., 88 (2000) 800.
[78] C. Wiwat, P. Siwayaprahm, A. Bhumiratana, Curr. Microbiol., 39 (1999) 134.
[79] P.A. Felse and T. Panda, Process Biochem., 34 (1999) 563.
[80] E. Agosin, M. Cotoras, G. Munoz, R. San Martin, D. Volpe, 1997. Comparative properties of Trichoderma harzianum spore produced under solid state and submerged culture conditions. In: S. Roussos (ed), Advances in Solid State Fermentation, Proceedings of the International Symposium on Solid State Fermentation, 2nd, Montpellier, Fr., Feb. 27-28, 1997. pp 463-473.
[81] I. Larena, A. De Cal, P. Melgarejo, Int. J. food microbiol., 94 (2004) 161.
[82] T. De Vrije, N. Antoine, R.M. Buitelaar, S. Bruckner, M. Dissevelt, A. Durand, M. Gerlagh, E.E. Jones, P. Luth, J. Oostra, W.J. Ravensberg, R. Renaud, A. Rinzema, F.J. Weber, J.M. Whipps, Appl. Microbiol. Biot., 56 (2001) 58.
[83] R.W. Silman and T.C. Nelsen, FEMS Microbiol. Lett., 107 (1993) 273.
[84] C. Desgranges, C. Vergoignan, A. Lereec, G. Riba, and A. Durand, Biotech. Adv., 11 (1993) 577.
[85] P. Ellaiah, B. Srinivasulu, K. Adinarayana, Process Biochem., 39 (2004) 529.
[86] J. Barrios-Gonzalez, H. Gonzalez, A. Mejia, Biotechnol. Adv., 11 (1993) 539.
[87] S.S. Saykhedkar, R.S. Singhal, Biotechnol. Progress, 20 (2004) 1280.
[88] T.H. Cho, D.J. Seo, G.J. Woo, D.K. Kang, Han'guk Misaengmul-Saengmyongkong Hakhoechi, 30 (2002) 253.
[89] M.B. Wall, D.C. Cameron, E.N. Lightfoot, Biopulpling process design and kinetics. Biotech. Adv., 11 (1993) 645.
[90] D.P. Roberts and S.M. Lohrke,. Pest Manag. Sci., 59 (2003) 654.
[91] E. Montesinos, Int. Microbiol., 6 (2003) 245.
[92] M.P. McQuilken and J.M. Whipps, Eur. J. Plant Pathol., 101 (1995) 101.
[93] M.P. McQuilken, S.P. Budge, J.M. Whipps, Biocontrol. Sci. Technol., 7 (1997) 23.
[94] J. Oostra, J. Tramper, A. Rinzema, Enzyme Microb. Tech., 27 (2000) 652.
[95] A. De Cal, I. Larena, B. Guijarro, and P. Melgarejo, Biocontrol Sci. Techn., 12 (2002) 715-725.
[96] M. Shoda, Mededelingen - Faculteit Landbouwkundige en Toegepaste Biologische Wetenschappen (Universiteit Gent), 64 (1999) 275.
[97] D.B. Archer, D.J. Jeenes, and D.A. MacKenzie, Antonie Van Leeuwenhoek, 65 (1994) 245.

[98] D.B. Archer and J.F. Peberdy, Crit. Rev. Biotechnol., 17 (1997) 273.
[99] M.C. Maldonado and A.M.S. De Saad, J. Ind. Microbiol. Biotechnol., 20 (1998) 34.
[100] E. Favela-Torres, J. Cordova-Lopez, M. Garcia-Rivero, M. Gutierrez-Rojas, Process Biochem., 33 (1998) 103.
[101] S.F. Covert, and D.J. Cullen, In Frontiers in Mycology, G.F. Leatham, editor, Chapman and Hall, New York, 1992, pp. 66-77.
[102] S. Bartling, J.P.T.W. van den Hombergh, O. Olsen, D. von Wettstein, and J. Visser, Curr. Genet. 29 (1996) 474.
[103] M.V. Ramana Murthy, N.G. Karanth, K.S.M.S. Raghava Rao. Adv. Appl. Microbiol., 38 (1993) 99.
[104] R.E. Mudgett, Enzyme Microb. Technol., 2 (1980) 273.
[105] A. Durand and D. Chereau, Biotechnol. Bioeng., 31(1988) 476.
[106] A. Durand, R. Renaud, S. Almanza, J. Maratray, M. Diez, C. Desgranges, Biotech. Adv., 11 (1993) 591.
[107] E. Gumbira-Sa'id, D.A. Mitchell, P.F. Greenfield, and H.W. Doelle.. Biotechnol. Lett., 14 (1992) 623.
[108] G. Saucedo-Castañeda, M. Gutiérrez-rojas, G. Bacquet, M. Raimbault. Biotechnol. Bioeng., 35 (1990) 802.
[109] G. Saucedo-Castañeda, B.K. Lonsane, M.M. Krishnaiah, J.M. Navarro, S. Roussos, M. Raimbault'. Proc. Biochem., 27 (1992) 97.
[110] B.L. Rathbun and M.L. Shuler, Biotechnol. Bioeng., 38 (1983) 353.
[111] K.S.M.S. Raghavarao, T.V. Ranganathan, N.G. Karanth, Biochem. Eng. J., 13 (2003) 127.
[112] M. Muller dos Santos, A. Souza da Rosa, S. Dal'Boit, D.A. Mitchell, N. Krieger, Bioresource Technol., 93 (2004) 261.
[113] R. Auria, J. Palacios, S. Revah, Biotechnol. Bioeng., 39 (1992) 898.
[114] J.J. Laukevics, A.F. Apsite, U.E. Viesturs, R.P. Tengerdy,. Biotechnol. Bioeng., 27 (1985) 1687.
[115] L.R. Lynd, C.E. Wyman, T.U. Gerngross, Biotechnol. Prog., 15 (1999) 777.
[116] V. Bellon-Maurel, O. Orliac, P. Christen,. Process Biochem., 38 (2003) 881.
[117] A. Durand, Biochem. Eng. J., 13 (2003) 113.
[118] M.A.I. Schutyser, F.J. Weber, W.J. Briels, A. Rinzema, R.M. Boom, Biotechnol. Bioeng., 82 (2003) 552.
[119] S. Suryanarayan, Biochem. Eng. J., 13 (2003) 189.
[120] M.K. Gowthaman, N.P. Ghildyal, M.S. Raghava Rao, N.G. Karanth., J. Chem. Technol. Biot., 56 (1993) 233.
[121] N.P. Ghildyal, M. Ramakrishna, B.K. Lonsane, N.G. Karanth,. Bioprocess Eng., 8 (1992) 67.
[122] D.A. Mitchell, N. Krieger, D.M. Stuart, A. Pandey, Process Biochem., 35 (2000) 1211.
[123] D.M. Stuart, D.A. Mitchell, M.R. Johns, J.D. Litster, Biotechnol. Bioeng., 63 (1999) 383.
[124] A.J. Marsh, D.A. Mitchell, D.M. Stuart, T. Howes, Biotechnol. Lett., 20 (1998) 607.
[125] C.J. Fung and D.A. Mitchell, Biotechnol. Tech., 9 (1995) 295.
[126] A. J. Marsh, D. A. Mitchell, D. M. Stuart, and T. Howes, Biotechnol. Lett., 20 (1998) 607.
[127] M.A.I. Schutyser, F.J. Weber, W.J. Briels, R.M. Boom, A. Rinzema, Biotechnol. Bioeng., 79 (2002) 284.
[128] D. Stuart, D. Mitchell, T. Howes, Proceedings of the 4th Pacific Rim Biotechnology Conference, Melbourne, Victoria, 6-9 Feb, 1995, p. 262.
[129] D.A. Mitchell, A. Tongta, D.M. Stuart, N. Krieger,. Biotechnol. Bioeng., 80 (2002) 114.
[130] P.B. Kokitkar, R.D. Tanner,. Enzyme Microb. Technol., 12 (1990) 552.
[131] M. Tanaka, A. Kawaide, R. Matsuno, Biotechnol. Bioeng., 28 (1986) 1924.
[132] O. Moebus and M. Teuber, Chem. Mikrobiol. Technol. Lebensm., 10 (1986) 99.
[133] D. Kunii and O. Levenspiel, Fluidization Engineering, 2d edition. Butterworth-Heinemann, Stoneham, Massachusetts, 1991, pp. 57-58.

[134] C. Webb, H. Fukuda, B. Atkinson, Biotechnol. Bioeng., 28 (1986) 41.
[135] K. Viswanathan, Can. J. Chem. Eng., 64 (1986) 223.
[136] J. F. Davidson, R. Clift, and D. Harrison (eds.), Fluidization, (2nd ed.), Academic Press, London. pp. 201-224.
[137] M. Olazar, S. Alvarez, R. Aguado, M.J. San Jose,. Chem. Eng. Technol., 26 (2003) 845.
[138] E. Piskova, K. Kojouharov, D. Mitev, G. Krueger, L. Moerl, Journal of the University of Chemical Technology and Metallurgy, 38 (2003) 623.
[139] F. Larachi, B.P.A. Grandjean, J. Chaouki, Chem. Eng. Sci., 58 (2003) 1497.
[140] R.G. Szafran and A. Kmiec, Ind. Eng. Chem. Res., 43 (2004) 1113.
[141] L.P. Wang, E. Silva, S.T. Yang, J. Chin. Inst. Chem. Eng., 36 (2005) 285.
[142] F.J. Weber, J. Tramper, A. Rinzema, Biotechnol. Bioeng., 65 (1999) 447.
[143] S.D.M Hasan, J.A.V. Costa, A.V.L. Sanzo, Biotechnol. Tech., 12 (1998) 787.
[144] P. Sangsurasak, D.A. Mitchell, Biotechnol. Bioeng., 60 (1998) 739.
[145] P. Sangsurasak, T. Srinophakhun, D.A. Mitchell, Kasetsart Journal: Natural Sciences, 31 (1997) 452.
[146] S. Rajagopalan and J.M. Modak, Bioprocess Eng., 13 (1995) 161.
[147] M. Gutierrez-Rojas, S. Amar Aboul Hosn, R. Auria, S. Revah, E. Favela-Torres,. Proc Biochem, 31 (1996) 363.
[148] M.P. McQuilken, S.P. Budge, J.M. Whipps. Mycol. Res., 101 (1997) 11.
[149] O.F. Von Meien and D.A. Mitchell, Biotechnol. Bioeng., 79 (2002) 416.
[150] M. Olazar, M.J. San Jose, S. Alvarez, A. Morales, J. Bilbao,. Ind. Eng. Chem. Res., 43 (2004) 655.
[151] S. Aravinth and T. Murugesan, Bioprocess Eng., 16 (1997) 289.
[152] K.B. Mathur and N. Epstein, Spouted Beds. Academic Press. New York, NY. pp.1-5.
[153] K. Viswanathan, M.S. Lyall, B.C. Raychaudhuri. Can. J. Chem. Eng., 64 (1986) 223.
[154] K.B. Mathur and P.E. Gishler, AIChE J., 1 (1955) 157.
[155] B.G., Charlton, J. B. Morris, Williams, 1965. "An experimental study of spouting beds of spheres," Rep. No. AERE-R4852. UK. At. Energy Authority, Harwell, cited in Mathur and Epstein, 1974.
[156] J.E. Smith and K. V. S. Reddy, Can. J. Chem. Eng., 42 (1964) 206.
[157] G. Brunello, G. Della Nina, F. C. S. Nunes, C. A. O. Nascimento, Can. J. Chem. Eng., 52 (1974) 170.
[158] O. Uemaki, R. Yamada, M. Kugo, Can. J. Chem. Eng., 61 (1983) 303.

Bioprocessing for Value-Added Products from Renewable Resources
Shang-Tian Yang (Editor)

Chapter 19. Algal Photobioreactor for Production of Lutein and Zeaxanthin

Attaya Wasanasathian[a] and Ching-An Peng[a,b]

[a]Department of Chemical Engineering and [b]Department of Materials Science, University of Southern California, Los Angeles, CA 90089, USA

1. INTRODUCTION

1.1. Prevalence of age-related macular degeneration

Damage to the photoreceptors in the macula can have devastating effects on visual function as the macula is the region of the retina responsible for the keenest visual acuity. Moreover, cone photoreceptors, the only photoreceptors in the macula, appear to be more susceptible to damage from visible light than rod photoreceptors. Accumulation of photodamage to the macula with age can have clinical manifestations related to age-related macular degeneration (AMD). In the USA and other developed countries, AMD is the leading cause of severe vision loss and blindness [1]. Unlike cataracts, there is no treatment available for most cases of AMD. Estimates of prevalence of age-related macular changes in population-based samples of Americans ranged from approximately 8% for men and women aged 43 to 54 years to 30% for those aged 75 years or older [2].

1.2. Evidence for lutein/zeaxanthin in AMD

Carotenoids are essential components of the photosynthetic apparatus and play a vital role in the protection of photosynthetic organisms against photo-oxidative damage caused by excessive light. They are able to quench chlorophyll triplet states and highly reactive singlet oxygen species [3]. In addition, they serve the function of a light-harvesting antenna, i.e., absorbed light energy can be either transferred to chlorophyll or be dissipated in the case of excess radiant energy [4]. One of the most abundant carotenoids in higher plants and green algae is the β,β-carotenoid violaxanthin. These are associated with the light-harvesting complexes (LHC I and II) of photosystems (i.e., PS I and PS II). Under physiological conditions, the polar carotenoids – leutin and zeaxanthin – occur only in trace amounts within the LHCs [5, 6]. However, under high light exposure, there is a rapid change in the carotenoid composition of the LHCs through operation of the xanthophylls cycle. That is, the di-epoxide xanthophylls violaxanthin is rapidly converted via the intermediate antheraxanthin (one epoxide group) to the epoxide-free lutein and zeaxanthin [7, 8]. Lutein and zeaxanthin may be important antioxidants in the thylakoid membrane bilayer itself, where it could scavenge reactive oxygen species and/or terminate lipid peroxidation chain reactions [9]. In addition to the antioxidant roles leutein and zeaxanthin play on thylakoids, they are known as the primary carotenoids found in the macular region of the eyes. High concentration of the xanthophylls such as leutein and zeaxanthin is responsible for the yellowish color of this region of the retina as well as its designation as the macula lutea or "yellow spot." Lutein and zeaxanthin exist there primarily to protect against photodamage of the retina by filtering out damaging short-wavelength visible light, thus minimizing free radical damage. Recent studies have found that diets rich in lutein and zeaxanthin may play a role in reducing the risk of AMD and cataracts – two diseases that are usually a result of the aging process [10]. Quite a few publications on the production of lutein and zeaxanthin, including a biosynthetic approach and genetic manipulation, have been reported [11–15]. Here, we focus on the overproduction of lutein and zeaxanthin in unicellular green microalgae (i.e., *Dunaliella salina*) cultivated in a photobioreactor exposed to high-intensity light.

2. CLASSIFICATION OF CLOSED ALGAL PHOTOBIOREACTORS

2.1. Experimental photobioreactors

Microalgae are traditionally cultivated in open ponds where the culture environment (e.g., salinity and pH) has been selected to favor algal growth relative to that of contaminating microorganisms. However, there are tens of thousands of species which cannot reliably be cultured in open systems. The obvious drawback of open systems is the potential for contamination from competitors, pathogens, and predators. Open cultures have also suffered from a lack of control, such as inadequate control of physical and chemical conditions (e.g. turbulence, nutrient concentrations, dissolved gases) which greatly affect cell growth and product formation of the microbial culture. Also, production of high-value algal products from strains that cannot be maintained in open ponds requires closed photobioreactors. The advantages of closed systems over conventional open ponds are that they can be erected over any open space, can operate at high biomass concentrations, prevent the loss of water by evaporation, and avoid atmospheric contamination.

Several types of photobioreactors have been designed and experimented with since the work of Pirt and coworkers sparked renewed interest in closed systems in the early 1980s [16]. Most of these are small-scale systems; some are at the pilot or pre-pilot stage of development. For scale-up photobioreactors to commercial size, only a few attempts have been made. For convenience, these systems are described herein by categories: (1) tubular photobioreactor and (2) flat-panel photobioreactor.

2.1.1. Tubular type

Among closed systems, tubular photobioreactors are ideal when a high degree of control over the culture conditions is required, like in large-scale culture of microalgae. Tubular solar receivers effectively harvest much of the sunlight impinging on the reactor surface. Unlike the open-culture systems, the design of closed tubular photobioreactors is more complex and the basic knowledge required for predictable performance is just beginning to emerge [17, 18]. The tubular photobioreactors can be arranged in a meandering pattern with parallel tubes connected by U-shaped fittings at their ends or in a helical arrangement wrapped around a tower. The main design considerations have been: (i) to reduce the light path so that the algae absorb the maximum amount of light, (ii) to maintain homogeneous mixing, (iii) to remove excess photosynthetically produced O_2, (iv) to add CO_2 efficiently, and (v) to control the culture temperature. Small-scale experiments with such systems have shown higher productivity and higher cell densities compared with conventional outdoor open ponds [19].

2.1.2. Flat-panel type

Flat-panel photobioreactors have often been used to grow phototrophic microorganisms in the laboratory because they provide a simple geometry and greatly reduce the light penetration depth through the culture surface. The first panel reactors built and experimented with in Florence by Tredici and coworkers were placed vertically, with the channels running parallel to the ground. A pump was used to circulate the suspension. The flat panels used are constructed from 16-mm-thick Plexiglas® alveolar sheets commonly available on the market. The main features that characterize these photobioreactors are the high surface to volume (S/V) ratio, the vertical or tilted inclination from the horizontal of the channels, and no mechanical device used to mix cell suspension. Moreover, circulation and degassing of the culture is operated by bubbling air at the base of each channel. Alveolar plates, due to their high S/V ratio, achieve high volumetric productivities and maintain operation at high cell concentrations. However, these systems typically give lower areal yields compared to tubular photobioreactors [20]. The lower performance achieved by cultures in flat-panel photobioreactors has been attributed to the fact that these systems, unlike tubular photobioreactors, have very short light penetration depths and do not offer light dilution (unless they are placed at a high inclination with the horizontal), thus leading to photoinhibition of microalgal growth [20]. Alveolar plates, like other small flat-panel systems, may be used profitably for research or in small production plants; however, due to the fact that the surface area of the cultivation unit is limited to a few square meters, they are not suitable for commercial-scale settings.

2.2. Commercial-scale photobioreactors

Only a few large commercial-scale photobioreactors have been built and operated. Some plants are continuously running and others suffered from mistakes with respect to fundamental technical aspects. Below are some examples.

2.2.1. Photo Bioreactors Ltd. (PBL) photobioreactor

A commercial-scale tubular photobioreactor fully activated by PBL in 1990 for *Dunaliella* production was established in Cartagena (Murcia, Spain). Circulation was provided by an airlift, and temperature was controlled by shading the reactors with nets or by water spraying. Several major technical errors are apparent in this system: a) oxygen degassing was inadequate, b) the tube material degraded quickly under sunlight, c) wall growth was almost unavoidable, and d) control of temperature was problematic. The very high S/V ratio with respect to the length of the tubes together with the insufficient circulation and the improper management of the culture, led to poor growth of the algae, biofouling, and serious contamination. After an almost complete crash of the culture, PBL never got into mass production and was closed at the end of 1991 [21].

2.2.2. Algatechnologies Ltd. photobioreactor

Vertical tubular photobioreactors are currently used by Algatechnologies Ltd. (Kibbutz Kentura, Israel) to develop and commercialize advanced agrobiologic products utilizing microalgae. The company's technologies are purchased and licensed from the Microalgal Biotechnology Laboratory of Ben Gurion University. The initial product is natural astaxanthin derived from *Haematococcum pluvialis*. Since oxygen concentrations can build up to detrimental levels that inhibit the growth of *H. pluvialis*, removal of oxygen is done by shortening the length of tubes or adding gas outlet devices along long tubes [22].

2.2.3. Mera Pharmaceuticals' photobioreactor

Mera Growth Module (MGM), a wave-type photobioreactor developed by Mera Pharmaceuticals, provides closed and controlled environment needed to optimize the growth of a large variety of microalgae species and produce high value products. The MGM utilizes high-level computer controls to monitor, maintain and adjust the growing environment for all critical conditions, such as temperature, light and nutrient levels. This level of control allows maintenance of conditions that promote the most desirable rate of growth. The unique requirements of each microalgal species (except, perhaps, for a few extremophile species) can be met simply by programming the controls on the platform technology - MGM. Mera Pharmaceuticals claims the MGM is the largest cost-effective photobioreactor ever operated, and that it will lead to the development of multi-billion dollar markets for the products derived from microalgae.

3. DESIGN CHALLENGES FOR CLOSED ALGAL PHOTOBIOREACTORS

To attain efficient and reliable large-scale culture of microalgae, the fundamental design criteria needed to be considered for the operation of closed photobioreactors include reactor configuration, S/V ratio, mixing and degassing devices, temperature regulation, orientation, and the material used to build the solar receiver. Some of these aspects are briefly discussed in the following sections.

3.1. Efficient gas transfer

Accumulation of photosynthetically generated oxygen is one of the main factors that limit scale-up of photobioreactors. Dissolving oxygen concentrations over 35 mg/L may be easily reached in outdoor dense algal cultures at mid-day. Such high oxygen levels are toxic to many phototrophs and, if coupled with prolonged exposure to full sunlight, may lead to the photo-oxidative death of the culture [23, 24]. Since volumetric oxygen production is directly related to volumetric biomass productivity, oxygen build-up becomes an especially serious problem in high S/V ratio photobioreactors. At maximal rates of photosynthesis, a 1-cm-diameter tubular photobioreactor would accumulate about 8–10 mg of oxygen L^{-1} min^{-1} [19]. In such systems, the maintenance of dissolved oxygen below the toxic concentration requires adoption of very short loops and this fact makes the serpentine design difficult to scale up. The situation is less dramatic in tubes of greater diameter; since the volumetric rate of oxygen evolution falls in proportion to the decrease of the S/V ratio, but it has been shown that even in 14-cm-diameter tubular photobioreactors dissolved oxygen may reach levels that significantly reduce productivity [25]. In serpentine tubular photobioreactors, the time cycle, the speed of circulation, and the length of the loop must be very carefully chosen based on the expected volumetric productivity to prevent build-up of inhibitory oxygen concentrations. For example, in a tubular photobioreactor with a 2.6-cm ID and a culture flow speed of 0.3 m/s, the time cycle should not exceed 6 min, hence the length of the reactor must be less than 100 m [23]. To extend the length of a photobioreactor it is necessary to increase the culture circulation speed or reduce the S/V ratio. Internal gas-exchange systems (e.g., bubble columns) offer some advantage in this respect. Unlike serpentine tubular photobioreactors, in which oxygen accumulates during the loop cycle and is removed only at the degasser, bubble columns have the merit that the whole culture is subjected simultaneously to the degassing air.

In the flat-panel photobioreactors, photosynthetically produced oxygen is removed directly from the culture by the degassing functionality of the air used to circulate the culture. However, in tubular photoreactors the excess oxygen is removed in a degasser or in the gas riser located at the end of the tubes. This means that the oxygen concentration in the culture will build up during the passage of the microalgae through the tubular part of the reactor, rapidly reaching supersaturation. If an airlift is used to circulate the culture, this problem is exacerbated since the airlift virtually ensures that the medium always will be at least oxygen saturated. While the practical feasibility of tubular photobioreactors for microalgal culture has already been established, little work has been done to determine to what degree the high O_2 concentration reduces productivity. At this time there are very limited design options available to provide photobioreactors with good degassing capabilities. The only solution is to keep the tube length as short as economically possible so that the degree of oxygen build-up can be minimized.

3.2. Light utilization efficiency

When considering commercial-scale microalgal culture, the source of light is usually sunlight since artificial light is too expensive. The amount of light available to the microalgal cells is the critical factor affecting the productivity of microalgal cultures. In general, shallower or thinner cultures can reach a greater cell density and, ultimately, a greater productivity since the effects of self-shading are minimized. Closed algal systems can be designed so that the light penetration path to the microalgae is very short, thus optimizing the light utilization efficiency. This has been achieved mainly by using flat transparent panels [26], tubes in various configurations [27, 28], and introducing light via fiber optics [29, 30]. It is interesting to note that most early tubular photobioreactors used tubes 10–30 cm in diameter [25], but almost all tubular reactors in use now have a tube diameter less than 4 cm. The narrower tube diameter not only improves the light utilization efficiency, but also provides more mixing which enhances growth. For example, It has been reported that reducing the tube diameter of a tubular photobioreactor from 5 to 3 cm increased the productivity of *Anabaena siamensis* from 310 to 370 mg L^{-1} day^{-1} [31].

The orientation of photobioreactors with respect to the sun is also important so that the algae receive the maximum amount of light throughout the day when operated outdoors. The flat-panel photobioreactors can be mounted so that their orientation can be changed during the day; however, tubular photobioreactor orientation is fixed. Most of the tubular photobioreactors consist of serpentine tubes lying on the ground, and therefore the lower parts of the tubes receive less light than the upper parts. Torzillo et al. have attempted to optimize light availability by arranging the tubes in two planes with the tubes in the upper plane placed in the vacant space between the ones in the lower plane [32]. An alternative arrangement is to arrange the tubes vertically and the most efficient way to do this is to have the tubes wrapped around a vertical tower. If the diameter of the tower is sufficiently great, then little shading takes place and this also makes the most efficient use of the available land area. This is the Biocoil design used by Biotechna Ltd. [33]. Although all of these systems are quite effective on the laboratory scale, several of these systems cannot be scaled up easily or cost-effectively. Thus, they may be adequate for the production of very high value products such as biopharmaceuticals, but would be too expensive for low-value products such as microalgae for aquaculture feeds. Three basic designs, however, are promising for large-scale production: 1) the array of hanging bags [34], 2) the horizontal tubular photobioreactor [28, 32, 35], and 3) the helical tubular photobioreactor [33]. The flat panel reactors, although very efficient, are probably too expensive for very large-scale commercial operations.

3.3. Homogeneous mixing

The type of device used to mix and circulate the culture suspension is essential in the design of a successful photobioreactor. Both the productivity of a photobioreactor and the cost of its construction and operation are affected to a great extent by the type of device used for mixing [36]. The beneficial effects that mixing has on growth and productivity of mass cultures of phototrophs are well known [24, 35]. Mixing is necessary to prevent cells from settling, to distribute nutrients and break down diffusion gradients at the cell surface, to remove photosynthetically generated oxygen, and above all, to ensure that all the cells experience alternating periods of light and darkness. The method of providing adequate mixing is critical since many algae are quite shear sensitive [37, 38]. There is a general consensus that bubble columns and airlift systems are the mixing systems with low shear stress and therefore are recommended for shear-sensitive microorganisms. It has been reported that productivity may increase up to 75% when pumps are replaced by airlift systems [38]. Other noticeable advantages of bubble columns and airlifts, from the industrial point of view, are their simplicity of operation and low maintenance costs due to the absence of moving parts. In addition, narrow tube diameters also have the advantage of allowing faster flow rates that not only increase the algal productivity but also reduce biofouling on tube walls

3.4. Temperature control

Although photobioreactors offer a better process control than open ponds do, the maintenance of optimal temperatures in closed photobioreactors is not an easy task, especially when the systems are operated outdoors. Photobioreactors behave as solar receivers. Any light absorbed by the photobioreactor walls or by the culture and not used in photosynthesis is converted into thermal energy, and may cause an increase of the culture temperature, resulting in culture failure even with tolerant species [26]. Among the solutions experimented with to control overheating in outdoor photobioreactors are water-bath immersion and water spraying. Cooling by water spraying has been shown to be reliable and cost-effective when used in dry climates [20, 23]. The tubular photobioreactor operated at Cadarache, France used another method, which involved floating the photobioreactor tubes in a large water-filled tank. In order to heat the culture, the tubes were floated in the surface of the tank; and if cooling was required, the tubes were submerged [28].

4. OVERCOMING CHALLENGES FOR LUTEIN/ZEAXANTHIN PRODUCTION

4.1. Oxygen removal

Excess O_2 has to be removed in order to guarantee a successful cultivation of microalgae. It has been reported that oxygen levels above air saturation (0.2247 mole O_2/m^3 at 20°C) could inhibit photosynthesis in many microalgal species, even if carbon dioxide concentration is maintained at elevated levels [39]. Therefore, in order to enhance mass production of microalgae, it is necessary to remove oxygen, which is deleterious to microalgae if its concentration becomes too high.

In the first detailed study of tubular reactors, Pirt et al. designed a serpentine tubular system using 52 glass tubes, 1 m length and 1 cm bore, connected in series by silicone rubber U-bends. Also, an air-lift system was included that generated an effective surface/volume ratio of 127 m^{-1} [16]. However, this design disregarded the oxygen accumulation problem. That work became the basis upon which an industrial culture facility was built in 1990 using 200 km of polyethylene tubing (1.2 cm in diameter) by PBL mentioned above. However, the operation failed, principally because a great proportion of the tube length was unproductive due to oxygen inhibition [17]. Weissman et al. also pointed out that a hypercritical O_2 concentration is one of the dominating factors limiting the mass culture of microalgae in tubular systems [19]. At this time, there seem to be few alternatives to the reactor design. Theoretically, the accumulated O_2 can be removed by increasing the medium flow rate or shortening the length of tubes. It has been previously reported that excessive shear in the fluid must be avoided to prevent destruction of the microbial mass [38]. Thus, the hydrodynamic stress must be controlled in order to conserve cell integrity. Furthermore, if cell residence time is shortened by decreasing the hydraulic residence time of the culture liquid, the cell concentration tends to decline due to dilution, decreasing the reactor productivity.

In order to tackle the problem of a high concentration of oxygen being a strong inhibitor to the algal biomass production, we have developed an innovative methodology which involves gradually supplying carbon dioxide as well as removing the accumulated oxygen through the use of perfluorocarbons (PFCs) [40]. In clinical studies PFC emulsions, due to their high solubility of respiratory gases (50 mL O_2/dL and 160-210 mL CO_2/dL, respectively) and inert chemical characteristics, have been used to transport and deliver fresh oxygen to animal tissues and remove metabolically accumulated carbon dioxide [41]. For photosynthetic microalgae, the carbon in CO_2 is their nutrient and oxygen, if accumulated, is their harmful waste. Therefore, the function of PFCs used with microalgae will be the reverse to that with animals. That is, PFCs act as the CO_2-releasing and O_2-stripping vehicles (see Fig. 1).

4.1.1. PFC emulsion-based tubular photobioreactor

Prior to the operation of the tubular photobioreactor, we carried out a series of algae culturing studies in the presence of PFC emulsions in Erlenmeyer flasks using an orbital shaker. It has been determined that the growth of microalgae is not affected by the addition of PFC emulsions under 5% v/v. Apparently, the light penetration efficiency was not decreased by the small amounts of PFC emulsions used. The microalgae used for this study were *Dunaliella salina* and *Chlorella vulgaris*. The PFC liquid used for this investigation was perfluorooctyl bromide (PFOB) which has been widely used to carry oxygen for the purpose of oxygen supply and regulation in the field of animal cell cultures. The PFC emulsions were mediated by a surfactant (Pluronic F-68) and a co-surfactant (perfluoroalkylated PEG-1450).

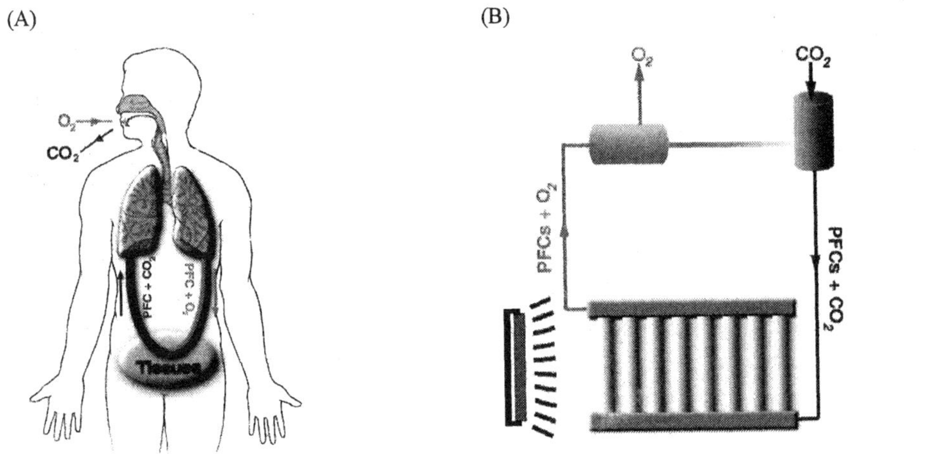

Fig. 1. The application of PFCs on gas exchange systems for (A) human and (B) microalgal photobioreactor. For human, PFCs carry the oxygen from lung to tissues. In the meantime, the carbon dioxide released from cells is transported to the lung for ventilation by PFCs. The function of PFCs used in the microalgal photosynthetic system is opposite to its function in human system. (Adopted from [40])

Fig. 2(A) gives the schematic view of the vertical airlift tubular photobioreactor used to test the ability of PFC emulsions to purge out high concentrations of oxygen generated within the microalgal culture medium. The bench-top tubular photobioreactor (total volume of 700 cm^3) given in Fig. 2(B) consists of a riser (length of 70 cm and inside diameter of 0.8 cm) and a degasser (volume of 60 cm^3) for low-shear propulsion of culture fluid through a tubular loop that constitutes the solar receiver. At the bottom end of the solar receiver tube is connected to the gas-injected riser of the airlift device, and the top end is connected to the degasser. The solar receiver comprises 14 transparent Pyrex glass tubes (with the dimension of 55 cm in length and 1 cm in outer diameter) jointed into a loop by an extremely low gas permeability PharMed tube with a T-connector for sampling. The flow velocity of the culture medium was selected to be 20 cm/s for preventing potential sedimentation of microalgae. The flow regime is well-controlled by regulating the gas supply to the riser section of the airlift via a flow meter.

The reactor was sterilized by autoclave at 121°C for 30 min and then assembled as shown in Fig. 2(B). The microalgal culture media were sterilized separately and pre-fed into the reactor by a peristaltic pump before the airlift process started. The carbon source for microalgae consisted of 5% CO_2 balanced with N_2. The cell inoculation density was 1×10^5 cells/mL. The reactor was operated under a 50 μE $m^{-2}\,s^{-1}$ light intensity. Samples were taken daily from 5 sampling ports to determine cell concentrations and oxygen concentrations. Cell concentrations were determined using a hemacytometer while the oxygen concentrations were measured using an oxygen microelectrode sensor. PFC emulsions were added at day one of the culture period. The concentration of PFC emulsions in the photobioreactor was 5% v/v.

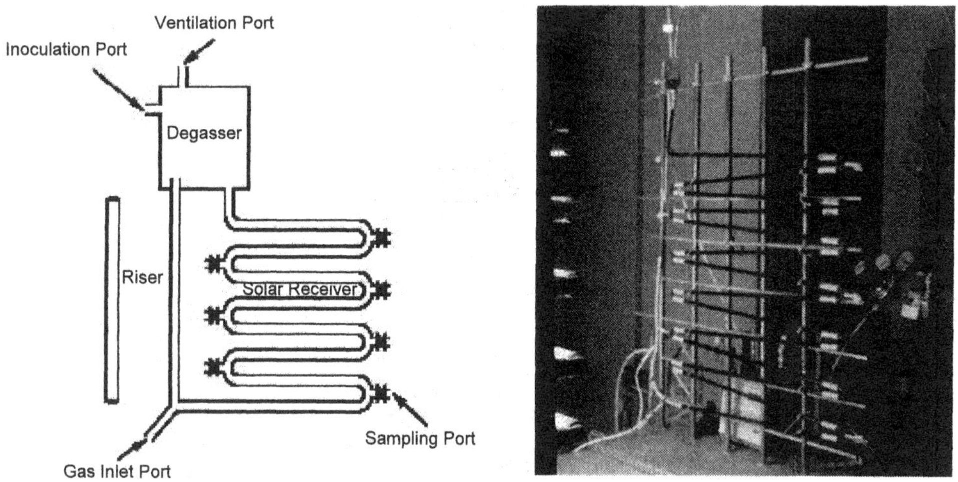

(A) (B)

Fig. 2. (A) Schematic diagram of an airlift tubular photobioreactor; (B) Bench-top tubular photobioreactor

The growth kinetics of *Chlorella vulgaris* without adding PFC emulsions is shown in Fig. 3(A). The maximum cell concentration obtained was 10.2×10^6 cells/mL. Also, the cell concentrations were decreased gradually to 4.4×10^6 cells/mL along the length of reactor (from port 1 to 5) due to the accumulation of oxygen within the reactor. Fig. 3(B) shows the evolution of oxygen concentrations at different ports of the reactor. The oxygen concentration was accumulated to 47 % within 9 days at port 5 (the sampling port farthest from the degasser). Obviously, the oxygen concentration gradient was built up in the reactor. However, with the presence of PFC emulsions, the oxygen concentration gradient was eliminated along the reactor (Figure 4(B)). In addition, oxygen concentrations in the reactor were dropped from 18 % to 4 % within 8 days due to the circulation of O_2-stripping PFC emulsions. The growth kinetics of *Chlorella vulgaris* in the reactor with the addition of PFC emulsions is presented in Fig. 4(A). As expected, there were no cell concentration gradients established at three sample ports, in comparison with the cell concentration gradients shown in Figure 3(A). Moreover, the maximum cell concentration was 40.0×10^6 cells/mL which is 4-fold higher than the one without using PFC emulsions as the internal de-oxygenators.

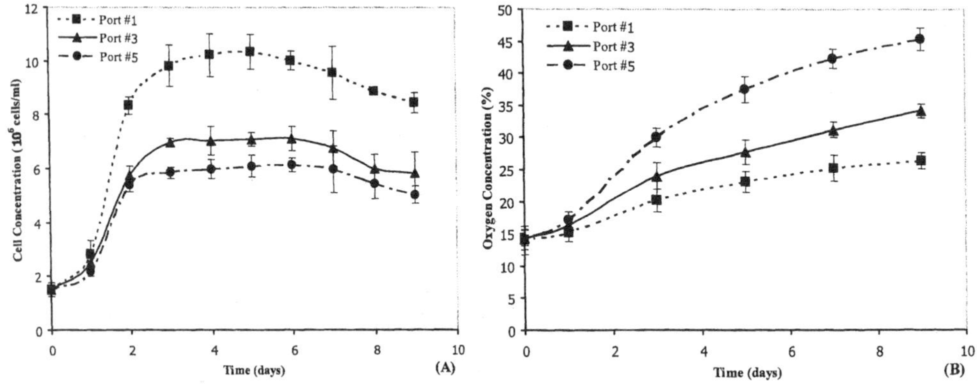

Fig. 3. (A) Growth kinetics of *Chlorella vulgaris* and (B) oxygen evolution at different sampling ports of the photobioreactor without the addition of PFC emulsions. Error bars represent mean ± standard deviation (n = 3).

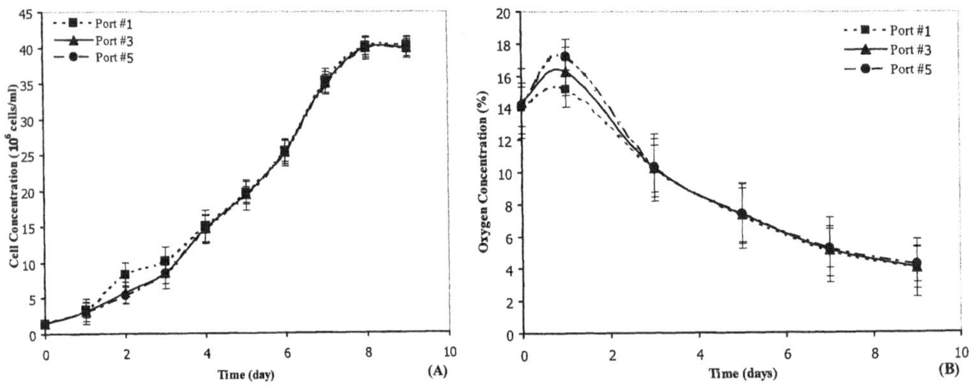

Fig. 4. (A) Growth kinetics of *Chlorella vulgaris* and (B) oxygen evolution at different sampling ports of the photobioreactor with the addition of PFC emulsions. Error bars represent mean ± standard deviation (n = 3).

4.2. Photoinhibition reduction

Even though light is the ultimate substrate for photosynthetic energy conversion, it can also harm green microalgae if exposed to high levels of it. If a tubular photobioreactor is operated under direct sunlight (photon flux density over 2,000 μE m^{-2} s^{-1} at noon), the microalgae will eventually experience a phenomenon known as photoinhibition, which results in a depression of photosynthetic activity. Moreover, high levels of irradiance combined with the elevated oxygen concentration built in the photobioreactor can lead to severely photo-oxidative death of the culture [23, 24]. It is well-accepted that the primary target for light damage is the water-splitting photosystem II (PS II). This photodamage leads to the degradation of 32-KDa D_1-polypeptide in the reaction center. The degradation of this very important protein appears to be a direct consequence of PS II chemistry involving highly oxidizing radicals and toxic oxygen species. Under high light stress, there seem to be several self-protecting mechanisms in the PS II reaction center itself. Among them, there is the xanthophylls cycle which will be initiated within the D_1/D_2 heterodimer to be able to quench chlorophyll triplets and scavenge for toxic oxygen species [3]. Also, in many green algae, xanthophylls undergo cyclic changes (through the addition or removal of oxygen), which help photosynthetic cells disperse excess energy under conditions of high illumination, thus protecting them from destructive photo-oxidation [42]. These reactions, known as epoxidation (oxygen addition to form epoxide groups) and de-epoxidation (removal of oxygen), result in the interconversion of xanthophylls located in the plastids, and constitute what is termed the xanthophyll cycle.

Typical products of the protective xanthophyll cycle are lutein and zeaxanthin which are synthesized from β-carotene by hydroxylation and epioxidation reactions [43]. We have cultivated *Dunaliella salina* under high levels of irradiance (2,000 μE m^{-2} s^{-1}) to induce lutein and zeaxanthin production. However, the microalgae only grew continuously for 3–4 days. To mitigate the photodamage on cells, a high concentration of CO_2 (40% balanced with nitrogen) was introduced to prolong the growth rate of *Dunaliella salina* over one week. To our knowledge, this observation has not been reported anywhere else. It is conceivable that the high CO_2 concentration alleviated the photodamage probably because of the rate increment of forward electron transport through PS II. This speculation is based on the hypothesis that photodamage occurs when there is an imbalance between light energy adsorption and utilization in PS II [44].

4.2.1. Effects of high light intensity and high CO_2 concentration on cell growth

We have performed a series of experiments measuring the growth of *Dunaliella salina* under different irradiance levels and CO_2 concentrations. Fig. 5(A) shows cells cultivated under four different sets of culture conditions. For cell growth at a high irradiance (2,000 μE m^{-2} s^{-1}, designated as HL) and under a limiting amount of inorganic carbon (i.e., in the presence of low concentrations of $NaHCO_3$ in the growth medium and atmospheric CO_2, designated as LC), the green algae *D. salina* developed signs of light stress. It can be clearly seen that the receding green chlorophyll unmasked yellow accessory xanthophyll pigments which normally function as antenna to funnel light energy to photosynthetic reaction centers, and to draw off excess energy that could damage the system. The cell concentration under the HLLC culture condition peaked at day 4 and started to show photobleach at day 9. However, this photodamage can be reduced by continuously pumping 40% CO_2 (designated as HC) balanced with N_2 gas. It can be seen that the color of the culture returned back to green from day 7 to 9. However, the cells grown under the conditions of HC and low light (25 μE m^{-2} s^{-1}, designated as LL) showed a total growth-arrest. Fig. 5(B) gives the growth kinetics of these four sets of cultures. The growth temperature was maintained at 27°C by using a water-containing fish tank as a heat absorber or filter.

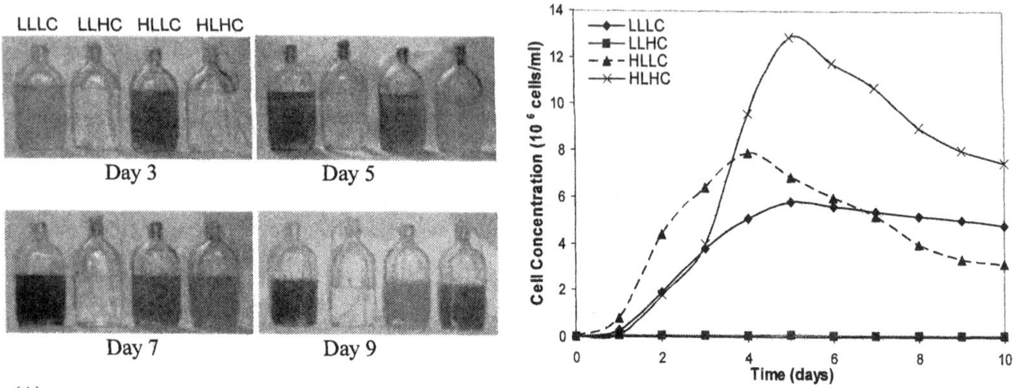

(A) (B)

Fig. 5. (A) Photographs of *Dunaliella salina* cultured under different irradiance levels and CO_2 concentrations (LLLC: low light, low CO_2; LLHC: low light, high CO_2; HLLC: high light, low CO_2; HLHC: high light, high CO_2); (B) Growth kinetics of *D. salina* under four different sets of culture conditions.

The levels of lutein and zeaxanthin produced within *Dunaliella salina,* cultivated in an airlift tubular photobioreactor (shown in Fig. 2B) under four different culture conditions (i.e., LLLC, LLHC, HLLC, and HLHC) with the aid of 5% v/v PFC emulsions, were examined. Three milliliters of algal suspensions were collected from the degasser of the photobioreactor (see Fig. 2A) every 48 h starting from day one and centrifuged at 3,750 rpm for 15 min. Pigments were extracted from algal cells by adding 0.5 mL 100% acetone to the pellet and vortexing at maximum speed for 2 min. The extract was centrifuged in an Eppendorf centrifuge at 3,750 rpm and 20 μL of supernatant subjected to high-performance liquid chromatography (HPLC). Carotenoid pigments were separated by HPLC on a Spherisorb ODS2 4.6 × 250 mm C_{18} column eluted by acetonitrile-tetrahydrofuran-methanol-water (80:10:5:5) at a flow rate of 1.0 mL/min for 25 min. Absorption spectra for individual peaks were obtained with a photodiode array detector set at 445 nm. Carotenoid pigments were identified by their absorption spectra and by their typical retention times. Fig. 6 shows the chromatograms of pigments extracted from the cells collected on day 5. It is clear that under all culture conditions the amount of zeaxanthin produced was much higher than the amount of lutein produced. Under light stress (2,000 μE m^{-2} s^{-1}) culture condition, the amount of zeaxanthin can be extracted from cells is very low if the concentration of CO_2 imposed is low (i.e., LC). However, with 40% CO_2 supplement (i.e., HC) to the light stress culture, it led to an approximately 10-fold increase in zeaxanthin accumulation.

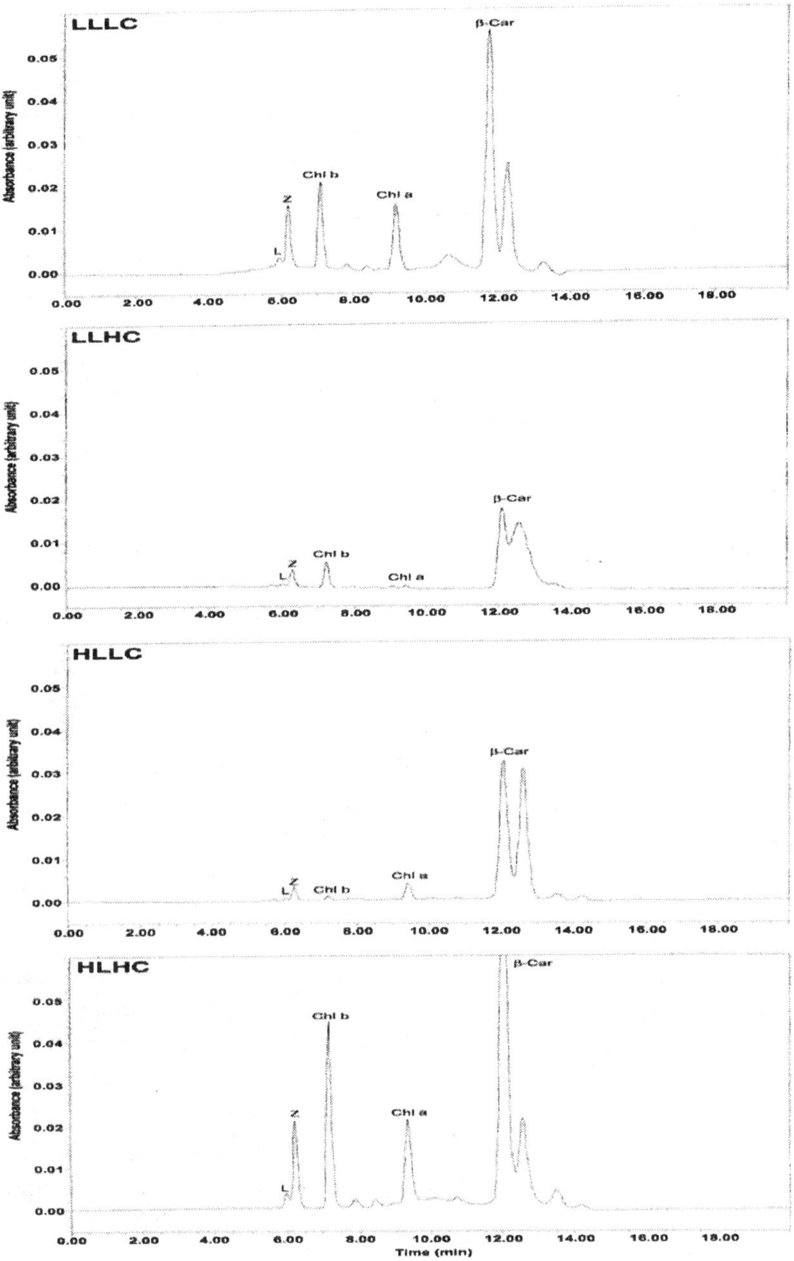

Fig. 6. HPLC elution profiles of pigments extracted from *Dunaliella salina* after culture for 5 days at four different conditions (LLLC, LLHC, HLLC, and HLHC). Chromatograms were recorded as a function of the absorbance at 445 nm. Identification of lettered peaks is as follows: L, lutein; Z, zeaxanthin; Chl *b*, Chlorophyll b; Chl a, Chlorophyll a; β-Car, β-carotene.

5. CONCLUSIONS

Growth and productivity of photosynthetic cells within the photobioreactor are affected by many physicochemical and biological factors. Certain requirements of photobioreactors, including the need for strict control of oxygen accumulation, efficient light intensity, and cooling make these systems more expensive to build and operate than open ponds. Thus, despite their advantages, the use of photobioreactors must be limited to production of very high value compounds from phototrophs that cannot be cultured in open ponds. In this study, O_2, CO_2, and light were considered as the pivotal design and scale-up parameters. Perfluorocarbon emulsions were used to internally purge out the accumulated inhibitor – oxygen – within the airlift tubular photobioreactor for the enhancement of microalgal production. Stressful culture conditions – high levels of irradiance and high concentrations of CO_2 – were harnessed to induce the mass production of high-value commercial xanthophyll products (i.e., lutein and zeaxanthin), which play a role in reducing the risk of age-related macular degeneration.

REFERENCES

[1] N.M. Bressler, S.B. Bressler and S.L. Fine, Surv. Ophthalmol. Vis. Sci., 32 (1988) 375.
[2] B.E.K. Klein, R. Klein and K.L.P. Linton, Ophthalmol, 99 (1992) 933.
[3] T.W. Goodwin, The Biochemistry of Carotenoids, Vol I. Chapman & Hall, New York, NY, 1980.
[4] K.K. Niyogi, O. Bjorkman and A.R. Grossman, Proc. Natl. Acad. Sci. USA, 94 (1997) 14162.
[5] A.V. Ruban, A.J. Young, A.A. Pascal and P. Horton, Plant Physiol., 104 (1994) 227.
[6] A.S. Verhoeven, W.W. Adams III, B. Demmig-Adams, R. Croce and R. Bassi, Plant Physiol., 120 (1999) 727.
[7] H. Levy, T. Tal, A. Shaish and A. Zamir, J. Biol. Chem., 268 (1993) 20892.
[8] H.Y. Yamamoto, Methods Enzymol., 110 (1985) 303.
[9] P. Muller, X.-P. Li and K.K. Niyogi, Plant Physiol., 125 (2001) 1558.
[10] J.T. Landrum and R.A. Bone, Arch. Biochem. Biophys., 385 (2001) 28.
[11] G. Banet, U. Pick and A. Zamir, Planta, 210 (2000) 947.
[12] D. Lagarde, L. Beuf and W. Vermaas, Appl. Environ. Microbiol., 66 (2000) 64.
[13] J.C.B. McDermott, D.J. Brown, G. Britton and T.W. Goodwin, Biochem. J., 144 (1974) 231.
[14] A. Ruther, N. Misawa, P. Boger and G. Sandmann, Appl. Microbiol. Biotechnol., 48 (1997) 162.
[15] H. Schubert, B.M.A. Kroon and H.C.P. Matthijs, J. Biol. Chem., 269 (1994) 7267.
[16] S.J. Pirt, Y.K. Lee, M.R. Walach, M.W. Pirt, H.H. Balyuzi and M.J. Bazin, J. Chem. Tech. Biotechnol., 33B (1983) 35.
[17] F. Camacho Rubio, F.G. Acien Fermaadez, J.A. Sanchez Perez, F. Garcia Camacho and E. Molina Grima, Biotechnol. Bioeng., 62 (1999) 71.
[18] E. Molina Grima, J.A. Sanchez Perez, F. Garcia Camacho, J.M. Fernandez Sevilla, F.G. Acien Fernandez and J. Urda Cardona, Appl. Microbiol. Biotechnol., 62 (1999) 71.
[19] J.C. Weissman, R.P. Goebel and J.R. Benemann, Biotechnol. Bioeng., 31 (1988) 336.
[20] M.R. Tredici and G. Chini Zittelli, Biotechnol. Bioeng., 57 (1998) 187.
[21] P. Kretschmer, O. Pulz, C. Gudin and V. Semenenko (eds.), Proc. 2nd Eur. Workshop Biotechnology of Microalgae, Institut fur Getreideverarbeitung, Bergholz-Rehbrucke, Germany, 1995, pp. 21-24.
[22] Personal Communication with Dr. Harold Wiener of Algatechnologies Ltd.
[23] A. Vanshak (eds.), *Spirulina platensis (Arthrospira):* Physiology, Cell Biology, and Biotechnology, Taylor & Francis, London, 1997, pp. 101-115.

[24] F.E. Round and D.J. Chapman (eds), Progress in Phycological Research, Vol. 7, Biopress, Bristol, UK, 1990, pp 269-330.
[25] G. Torzillo, B. Purshparaj, F. Bocci, W. Balloni, R. Materassi and G. Florenzano, Biomass, 11 (1986) 61.
[26] M.R. Tredici and R. Materassi, J. Appl. Phycol., 4 (1992) 221.
[27] T. Stadler, J. Mollion, M.C. Verdus, Y. Karamanos, H. Morvan and D. Christiaen (eds.), Algal Biotechnology, Elsevier Applied Science, London, 1988, pp. 219-228.
[28] T. Stadler, J. Mollion, M.C. Verdus, Y. Karamanos, H. Morvan and D. Christiaen (eds.), Algal Biotechnology, Elsevier Applied Science, London, 1988, pp. 199-208.
[29] J.G. Burgess, K. Iwamoto, Y. Miura, H. Takano and T. Matsunaga, Appl. Microbiol. Biotechnol., 39 (1993) 456.
[30] R.D. MacElroy and D.T. Smernoff (eds.), Controlled Ecological Life Support System, NASA, Moffett Field, CA, 1987, pp. 45-50.
[31] S. Boussiba, Microb. Releases, 2 (1993) 35.
[32] G. Torzillo, P. Carlozzi, B. Purshparaj, E.Montaini and R. Materassi, Biotechnol. Bioeng., 42 (1993) 891.
[33] R.C. Cresswell, T.A.V. Rees and N. Shah (eds.), Algal and Cyanobacterial Biotechnology, Longman Science, London, 1989, pp. 294-316.
[34] E. Cohen and S. Arad, Biomass, 18 (1989) 59.
[35] A. Richmond, S. Boussiba, A. Vonshak and R. Kopel, J. Appl. Phycol., 5 (1993) 327.
[36] K.L. Terry and L.P. Raymond, Enzyme Microbiol. Technol., 7 (1985) 474.
[37] H.J. Silva, T. Cortinas and R.J. Ertola, J. Chem. Tech. Biotechnol., 40 (1987) 41.
[38] C. Gudin and D. Chaumont, Bioresour. Technol., 38 (1991) 145.
[39] S. Aiba, Adv. Biochem. Eng., 23 (1982) 85.
[40] A. Wasanasathian and C.-A. Peng, Art. Cells, Blood Subs. Immob. Biotechnol., 29 (2001), 47.
[41] K.C. Lowe, Science Progress, 80 (1997) 169.
[42] B. Demmig-Adams and W.W. Adams, Annu. Rev. Plant. Physiol., 43 (1993) 599.
[43] G.E. Bartley and P.A. Scolnick, Plant Cell, 7 (1995) 1027.
[44] S. Powles, Annu. Rev. Plant Physiol., 35 (1984) 15.

Bioprocessing for Value-Added Products from Renewable Resources
Shang-Tian Yang (Editor)

Chapter 20. Power-Generation from Biorenewable Resources: Biocatalysis in Biofuel Cells

Ping Wang and Hongfei Jia

Department of Chemical Engineering, The University of Akron, 200 E. Buchtel Commons, Akron, 44325-3906, USA

1. INTRODUCTION

Biofuel cells refer to a class of fuel cells that apply microbial or enzymatic biocatalysts. This definition includes fuel cells that have biocatalysts such as laccase on the cathode to reduce oxygen, although the fuel is not necessary a biofuel. Biofuel cells are sometimes called biological, biochemical, biocatalytic or bioelectrochemical fuel cells. In this chapter, they will be called biofuel cells. Biocatalysts in biofuel cells may function in a way similar to inorganic catalysts such as Pt to catalyze redox reactions in the vicinity of the electrodes. In this case, the biofuel cells are called primary or direct biofuel cells. In another form, the biocatalysts' role is to produce simple fuels such as hydrogen or methane from biochemicals such as sugars. The simple fuels can then be oxidized *in situ* by other catalysts, usually inorganic, at the surface of the electrodes for the generation of electricity. Such fuel cells are classified as secondary or indirect biofuel cells. Early observations of electricity generation from glucose and other organic compounds in the presence of microbes dates back to the early 1900's [1, 2]. In the 1960's, NASA showed much interest in generating power from human wastes on space shuttles, inspiring many R&D efforts for biofuel cells. Biofuel cells capable of generating power from urea [3] and methane [4] were built and tested during this period of time. The first enzymatic biofuel cell was also reported in 1964, using glucose oxidase (GOx) as the anodic catalyst [5]. Although many other potential applications of biofuel cells have been proposed since then, there is currently no biofuel cell in commercial operation for power generation. This is mostly due to the fact that the performance of biofuel cells, in terms of power density, lifetime, and operational stability, falls far below that of chemical fuel cells. That narrows the potential of biofuel cells to areas where chemical fuel cells are not suited. The growing interest in searching for bio-based alternative energy resources, biocompatible batteries for biomedical electronics and environmentally friendly portable energy devices are spurring a new wave of biofuel cell development. Challenges still lie in overcoming their weakness in comparison with chemical fuel cells. On the other hand, recent advances in materials science, protein chemistry and genetic engineering provide new opportunities for more capable biocatalysts for use in fuel cells.

Several reviews published earlier provide detailed historical accounts of the evolution of biofuel cells [6–9]. More recent progress in biofuel cell technologies has also been reviewed in several papers. These include reviews on enzymatic biofuel cells based on monolayer-functionalized electrodes [10], miniature biofuel cells [11], and enzymatic biofuel cells for implantable devices [12], as well as microbial biofuel cells used for energy generation from wastewater [13, 14]. While the previous reviews mostly focused on either cell construction or applications, this review attempts to examine the function of and challenges for biocatalysis in biofuel cells.

2. ROLE OF BIOCATALYSTS IN BIOFUEL CELLS

In nature, biocatalysis regulates the biotransformations that help organisms survive and grow in different environments. To date, sophisticated technologies are available to conduct biotransformations under controlled conditions in order to meet our seemingly endless need for materials and energy. Among other developments, biosynthesis of fuels, including methane, methanol, ethanol, and more recently, hydrogen, have been realized, in some cases with remarkable efficiency. Production of fuel ethanol and biodiesel from biomass has already formed a sizable industrial sector. Table 1 lists several fuel chemicals for which bioprocessing technologies have been developed.

Table 1
Biocatalysis for production of fuels

Fuel	Feed Biochemicals	Biocatalysts	Indirect biofuel cell studied
Ethanol	Cane sugar, corn starch	Yeast/fungi	No investigation
Methane	Acetate, methanol, methylated amines, pyruvate, H_2+CO_2	Methanogenic archaea	No investigation
Hydrogen	Organic polymer, carbohydrates, fats, amino acids	Bacteria (such as *Clostridium butyricum*)	[16–20]
Biodiesel	Vegetable oil, animal fat	Thermal cracking, transesterification (lipase)	No investigation
Methanol	CO_2 and glucose or amino acid	Enzymatic [21]	No investigation
	Biomass	Microbial: Biomethanol is usually produced by hydrogenation of CO_2	

Theoretically, a combination of a fuel biosynthesis reactor with a chemical fuel cell makes a biofuel cell, usually called an indirect biofuel cell (Fig. 1). One critical challenge in developing indirect biofuel cells is how to effectively interface the electrical cell and the bioreactor. In other words, how to balance the operating conditions in order to optimize overall cell efficiency. A typical factor of consideration is temperature. Most bioreactors require ambient conditions, and subsequently, low-temperature fuel cells are preferred. Compared to other fuel cells using CO, CH_4 and CH_3OH, H_2 fuel cell is a more established technology for low-temperature operations. In fact, only H_2-based indirect biofuel cells so far have been studied. The efficiency of H_2 production remains as a critical challenge, not only in terms of biotransformation yield, but also with respect to space-time productivity. In most cases, the available fuel product concentration is the factor limiting the performance of indirect biofuel cells. Some special requirements of biocatalysis, such as light for algae, may complicate the interfacing even further.

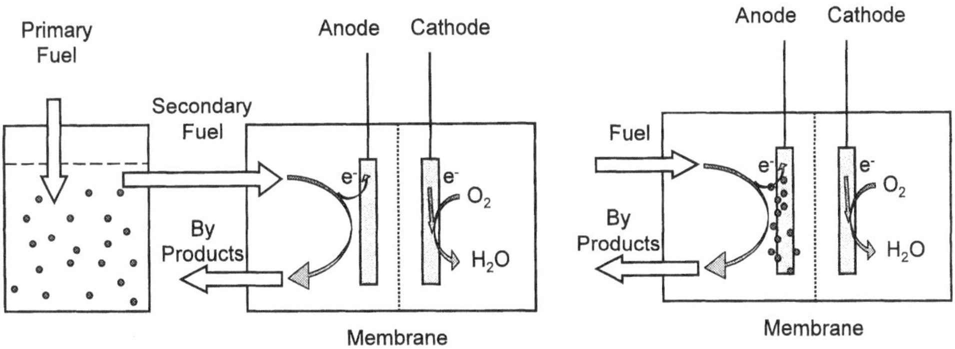

Fig. 1. Configurations of two types of biofuel cells: indirect (left) and direct (right) biofuel cells.

The development of direct fuel cells, in which biocatalysts are directly involved in the redox reactions or reaction chains for electricity generation, has generated much more interest. The efficiency of electron conduction between the biocatalysts and electrodes is a critical factor in the design of direct biofuel cells. For living microbes, electron conduction may be achieved through complicated biochemical processes [22]. In the case of isolated enzymes, direct electronic communication with electrodes usually suffers from the nonconductivity of protein hulls that host their active sites. Although knowledge of electron conduction through protein hulls of native enzymes is a subject yet to be fully explored, direct electron transfer was observed for cytochrome c, laccase, hydrogenase and several peroxidases [23–26]. Methods of chemical modifications that make enzymes more conductive have also been developed. Enhanced electron transfer to electrodes has been achieved for glucose oxidase via either the modification of the protein hull or wiring the active site/cofactor to the electrodes [27–30]. An alternative way to overcome the biocatalysts-electrode electron conduction resistance is the use of mediators that shuttle between the two. Even though this introduces additional steps in the redox reaction chains from fuel to electricity, much higher cell efficiencies were usually observed. The challenge in using mediators, which are usually expensive, is how to maintain them in the cells if continuous feeding of fuels is required. The use of mediators has been reported for both microbial and enzymatic biofuel cells [7, 9, 15, 31]. The literature has indicated confusion regarding whether to define mediator-facilitated biofuel cells as direct or indirect biofuel cells. We classify these as direct fuel cells since the biocatalysts initiate chains of redox reactions for power generation. In other words, biocatalysts in direct biofuel cells catalyze redox reactions for electricity generation, while indirect biofuel cells have biocatalysts for fuel synthesis.

3. FUEL OPTIONS IN BIOFUEL CELLS

Hydrogen and methanol are the most popular fuels examined for general-purpose fuel cells. Other chemicals such as ammonia, hydrocarbons, ethanol, propanol, ethylene glycol, glycerol, cyclic alcohols, formic acid, and formate have also been investigated [32]. Most of these fuels, however, have less electrochemical activity than hydrogen and methanol. There was a great interest in using hydrazine (N_2H_4) for fuel cells during the 1960's, because of its high electrochemical activity and ease in transportation and storage. The drawbacks of using hydrazine are its toxicity and high production cost.

Biofuel cells afford a much broader range of fuel options. Biocatalysts are capable of catalyzing the oxidation of most of the conventional fuels as mentioned above. On the other hand, biocatalysts are also efficient in oxidizing more complicated chemicals such as carbohydrates and various organic wastes for electricity generation. Table 2 summarizes fuels that have been used to biofuel cells.

For conventional fuels, such as hydrogen and methanol, biofuel cells have to compete with chemical fuel cells. Although the performance of most biofuel cells, in terms of power density and lifetime, must be improved significantly in order to become comparable with the chemical fuel cells (this will be discussed in further detail later), biofuel cells have several advantages. Compared to noble metal-catalyzed fuel cells, biofuel cells offer ambient operation conditions and can eventually be cheaper as the production cost of the key biocatalysts continues to drop as a result of developments in genetic engineering. Enzymes also possess better tolerance against impurities in fuels. For example, Pt can be sensitive to trace amounts of CO, which is often present as an impurity in hydrogen produced from reforming gas. CO can also be generated as a product of incomplete oxidation of methanol. Biocatalysts are usually inert to CO. Karyakin et al. [33] examined the performance of an anode using a hydrogenase from *Thiocapsa roseopersicina*. The electrode showed no CO toxicity interference, and generated potentials and current density similar to those of chemical electrodes using Pt. This impurity tolerance can significantly reduce the costly requirement of fuel purification.

Carbohydrates are the most abundant renewable resource available for energy production. Unfortunately, it is difficult, if not impossible, to use carbohydrates in traditional chemical fuel cells, since their catalysts are usually not capable of catalyzing this type of reactions. Biocatalysts can overcome this limitation. Carbohydrates, such as sugars and starch, can be directly oxidized in biofuel cells to generate electricity. In some cases, pretreatment of the biomass may be required. For example, cellulose can first be hydrolyzed by cellulase to produce glucose. In microbial biofuel cells, utilization of glucose for energy generation achieved a high efficiency. Theoretically, complete oxidation of one mole of glucose generates 24 moles of electrons:

$$C_6H_{12}O_6 + 6\ H_2O \rightarrow 6\ CO_2 + 24\ H^+ + 24\ e^- \tag{1}$$

Table 2
Fuels used in biofuel cells

Fuels	Biocatalysts	References
Hydrogen	Hydrogenase	[33, 34]
Methane	Methanogenic archaea	[4]
Methanol	Dehydrogenases	[35]
Ethanol	Dehydrogenases	[36, 37]
Propanol	Dehydrogenases	[36]
Glucose	Glucose Oxidase	[38–47]
	Glucose Dehydrogenase	[36, 48]
	Microorganisms	[49–51]
Galactose, maltose, sucrose, trehalose	*Proteus vulgaris*	[51]
Starch	*Clostridium butyricum* or *Clostridium beijerinckii*	[52]
Lactate	Lactate dehydrogenase	[53]
Acetate	*E. coli*	[54]
Mysteric acid	Co-Enzyme A	[36]
Urea	Ureanase	[3]
Marine sediments	Bacteria	[55]
Wastes containing sulfates	*Desulfovibrio desulfuricans*	[56]
Wastewater	Microorganisms	[57–60]
Sewage sludge	*E. coli* K12	[61]
Light/H_2O	Cyanobacteria	[62–65]

A recent study reported that 83% coulombic yield was achieved in a microbial biofuel cell constructed with *Rhodoferax ferrireducens* [66], while yields of up to 89% were also reported using a mixed and enriched culture [49]. An earlier study reported that coulombic yield of sucrose was close to 100% [67]. These results are impressive considering that microorganisms also consume energy to stay alive. In indirect biofuel cells, carbohydrates were frequently used as substrates for the production of simple fuels, especially hydrogen, which then feed a chemical fuel cell [7]. The coulombic yield of such indirect biofuel cells tends to be low. Even under ideal conditions, one mole of glucose can only produce four moles of hydrogen corresponding to eight moles of electrons (Fig. 2) [14]. The actual yield was usually as low as 25% of the theoretical estimation [19]. This means that the biofuel cell only converted 1/12 of the energy stored in glucose to electricity. Glucose was also a popular fuel for enzymatic biofuel cells. Oxidation of glucose is often catalyzed by either glucose oxidase or glucose dehydrogenase. The inherent efficiency of the enzymatic glucose fuel cells is even lower. One mole of glucose only produces two moles of electrons (Eqs. 2 and 3). However, enzymatic glucose biofuel cells may be valuable for applications such as implantable devices, where living organisms are not desirable.

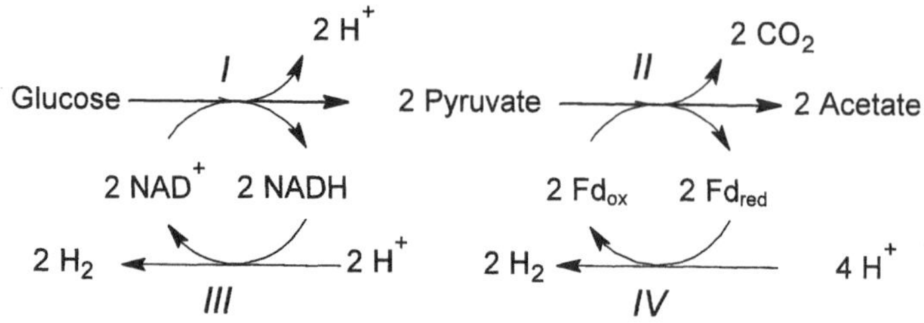

Fig. 2. Biological production of hydrogen from glucose. ***I***: glucose metabolism or glycolysis via the Entner-Doudoroff pathway; ***II***: oxidative decarboxylation of pyruvate by ferredoxin oxidoreductase (Fd); ***III*** and ***IV***: formation of hydrogen by hydrogenase.

$$glucose + FAD \xrightarrow{\text{Glucose Oxidase}} gluconolactone + FADH_2 \tag{2}$$

$$glucose + 2\ NAD \xrightarrow{\text{GDH}} gluconolactone + 2\ NADH \tag{3}$$

Industrial and agricultural organic wastes can be a low-cost resource for generating electricity. An early study reported a microbial fuel cell system that used rice husks generated 40 mA at 6 V [68]. Another microbial fuel cell continuously generated power for five years by digesting wastewater-containing sulfates [69]. Efforts to optimize microbial biofuel cells based on organic wastes are continuing. The focus is mostly on cell design [57, 58], electrode materials [61], new bacterial strains [70, 71], and operational conditions [59]. One interesting idea was reported to generate electricity from a marine sediment-water interface by simply placing the anode inside the sediment and the cathode in the seawater [55, 72, 73].

Human and animal wastes are another cheap source of fuels for biofuel cells. NASA has shown much interest in developing microbial biofuel cells that treat human waste and generate power. A trip to Mars may accumulate tons of organic waste on the vehicle, which can be burdensome if have to be brought back to Earth for treatment. Earlier efforts studied indirect biofuel cells that consumed the ammonia generated from urea [3]. Recent reports show that NASA continues to fund the development of biofuel cells utilizing human wastes (http://www.space.com).

A unique form of biofuel cells is the use of biophotosynthesis to facilitate the conversion of light to electricity. Under illumination, cyanobacteria are able to catalyze the oxidation of water to oxygen (Eq. 4). An energy conversion efficiency of 2–2.5% was recently achieved using a photosynthetic anode combined with an enzymatic oxygen cathode [65].

$$H_2O + 2\ h\upsilon \xrightarrow{\text{Cyanobacteria}} 1/2\ O_2 + 2\ H^+ + 2\ e^- \tag{4}$$

4. BIOCATALYSIS IN CATHODE REACTIONS

Oxygen is the primary oxidant for fuel cells. It is cheaper and safer than any other chemical oxidant. Bio-cathodes developed in the early days of biofuel cells involved several reaction steps. In one report [74], oxygen was first reduced to hydrogen peroxide on the surface of gold, and was then converted to water in the presence of an enzyme, chloroperoxidase (CPO) (Eqs. 5 and 6):

$$O_2 + 2\,H^+ + 2\,e^- \xrightarrow{Au} H_2O_2 \tag{5}$$

$$H_2O_2 + barbituric\ acid + HCl \xrightarrow{CPO} 5-chlorobarbituric\ acid + H_2O \tag{6}$$

This was probably the first report of a biocatalytic cathode for fuel cells. Although the chemistry could be done, the overall design is problematic since it used another fuel, barbituric acid in this case, at the cathode in addition to the fuel used at the anode. A few years later, a bacteria-catalyzed cathode was tested by Bennetto et al. [75] with an iron-oxidizing bacterium, *Thiobacillus ferroxidans*. However, the same concern remains, since the bacteria may require certain nutrition amendment.

Table 3
Enzymatic cathodes for biofuel cells

Oxidant	Biocatalyst	References
O_2	Laccase	[34, 45, 76–78]
	Cytochrome oxidase	[41, 42, 79]
	Bilirubin oxidase	[46, 80]
H_2O_2	Horseradish peroxidase (HRP)	[43]
	Microperoxidase-11	[44, 47, 81]

One-step reduction of oxygen to water is certainly more desirable (Eq. 7). The last decade has witnessed a significant progress in developing enzymatic electroreduction of oxygen (Table 3). That is usually achieved with a high overpotential, i.e. the potential loss caused by the irreversibility of the process. High overpotential also exists for chemical catalysts. Katz et al. reported an oxygen cathode consisting of cytochrome c and cytochrome oxidase on a gold electrode [10]. The enzymatic cathode was able to catalyze the four-electron reduction of oxygen to water without the generation of hydrogen peroxide. Compared to a bare Au electrode, the overpotential was reduced by ~0.3 V. Laccase and bilirubin oxidase (BOD) were also used to catalyze the reduction of oxygen [34, 45, 46, 76–78, 80]. Impressively, the fastest electroreduction of oxygen was demonstrated with a BOD cathode, which only was polarized by 0.07 V at a current density of 0.5 mA/cm^2, much less than the polarization of 0.37 V at the same current density observed for a Pt electrode [77].

$$O_2 + 4\,H^+ + 4\,e^- \xrightarrow{\text{laccase / bilirubin oxidase}} 2\,H_2O \tag{7}$$

The use of hydrogen peroxide as an oxidant for an enzyme-catalyzed cathode has also been explored. Horseradish peroxidases adsorbed on carbon black [43] and microperoxidase-11 (MP-11) covalently attached to a cystamine [10] were examined (Table 3).

5. ELECTRON TRANSFER IN BIOFUEL CELLS

The power density of biofuel cells achieved so far is generally 2–3 orders of magnitude lower than that of chemical fuel cells. Among other factors, the internal electron transfer of the fuel cells is critical. In chemical fuel cells, catalyst metals are usually deposited on electrodes and the generated electrons thus have an easy path from the reaction sites to the electrodes. In biofuel cells, especially direct biofuel cells, the use of biocatalysts introduces additional resistance to electron conduction from the reaction sites to the electrodes. When the active sites of the catalysts are close to the surface of the electrode, either through the use of small-size catalysts or some other manipulation that brings them closer [24, 25, 82], direct electron transfer (DET) from the active sites to the electrode is possible. Alternatively, an electron-carrier chemical, i.e., a redox mediator, can be used to facilitate electron transfer from the reaction sites to the electrode. In either case, there is much higher resistance than that of the electron conduction process in chemical fuel cells.

5.1. Biofuel cells based on direct electron transfer (DET)

Since the late 1970s, studies in bioelectrocatalysis have found quite a few oxidoreductases capable of performing DET [23]. Among others, several enzymes have been examined for DET biofuel cells. These include hydrogenase, glucose oxidase, laccase, bilirubin oxidase and peroxidase. The substrate specificity of the enzymes determines their potential for biofuel cells.

Laccase and bilirubin oxidase, which catalyze the electrochemical reduction of oxygen as discussed earlier, show excellent electron conductivity. The electron flow in this case is from the electrode to the active sites of the enzymes. It was reported that the redox potential of an electrode prepared by physical adsorption of laccase on carbon black was ~1.2 V vs. standard hydrogen electrode (SHE) in the presence of molecular oxygen. That potential is about the same as the standard thermodynamic potential of O_2/H_2O, 1.223 V [83]. A glucose-air biofuel cell has been built with a mediated glucose anode and a DET cathode using laccase or bilirubin oxidase [84]. The system is able to deliver 100 $\mu A/cm^2$ of electricity at 0.6 V, and 250 $\mu A/cm^2$ at 0.5 V. The overall power output of the biofuel cell was in the range of 50–250 $\mu W/cm^2$.

A biohydrogen anode was also developed by physically adsorbing hydrogenase from *Thiocapsa roseopersicina* on carbon filament materials (CFM) [33, 85]. At room temperature and pH 7.0, the electrocatalytical activity of the hydrogenase, measured by the rate of electron exchange on cyclic voltammetry (CV) generated per molecule of the catalyst, is two orders of magnitude higher than that of a platinum-based electrode. A prototype biofuel cell with a

DET anode using hydrogenase and a cathode using laccase generated a current density of 400 μA/cm^2 at 0.8 V, corresponding to a power density of 320 μW/cm^2 [34]. Interestingly, an electromotive force (EMF) of 1.35 V was also reported for that unit. That potential is slightly higher than the standard O_2/H_2O potential.

DET is also feasible with microbes. Kim and coworkers reported a series of studies on mediator-free microbial biofuel cells [59, 60, 70, 71, 86–93]. Anaerobically-grown ferric ion [Fe(III)] reducing bacterium, *Shewanella putrefaciens*, was found to be electrochemically active using CV [94]. A redox potential of 0.2 V vs. Ag/AgCl reference electrode was observed with cell suspension and glassy carbon as the working electrode. According to the report, membrane cytochromes with oriented heme groups toward the cell surface may essentially have enabled DET. A lactate-O_2 microbial biofuel cell was constructed using *Shewanella putrefaciens*. Both the anode and cathode were built with carbon felt with a dimension of 5×5×0.3 cm. Good current generation was observed with three strains of the bacteria on the anode upon the addition of lactate, while no current was obtained in the control test with *E. coli*. An open circuit voltage (OCV) of ~0.6 V was obtained, and a current output of 40 μA, corresponding to a density of 1.6 μA/cm^2, was generated with a 1000 Ω external load.

DET was also achieved with microbes digesting wastewater [59, 70, 71, 90, 91, 95]. For example, a current of ~210 μA (8.4 μA/cm^2) for a load of 1000 Ω was observed for such fuel cells [94]. The corresponding power density of these microbial fuel cell systems is generally low, mostly less than ~0.07 μW/cm^2. The efficiency of the O_2 cathode is one of the limiting factors. Many reported microbial fuel cells did not use any catalyst at the cathode. A recent study showed that a substantial power drop (~80%) occurred when Pt was not used on the carbon-based cathode [96]. A novel single chamber microbial biofuel cell was developed by Liu et al. using a gas diffusion air cathode and an anode with bacteria from domestic wastewater [57]. A maximum power of 2.6 μW/cm^2 was achieved. When glucose was used as the fuel, the power density improved 10-fold to 26 μW/cm^2. The highest power density from mediator-free microbial biofuel cells so far reported is 1.2 mW/cm^2, also using glucose as the fuel [97].

5.2. Biofuel cells using redox mediators

Redox mediators are chemicals with electrochemical activity. In a bioelectrocatalysis process, mediators may exchange electrons with fuels or oxidants at the reaction sites of the biocatalysts, and then diffuse to the surface of electrode and exchange electrons there. This process is repeated, and the mediator functions as an electron shuttle between the biocatalyst and electrode. Many studies on mediators in biofuel cells have been reported [79, 15, 98, 99]; more references to mediators may be found in studies of biosensors [8, 100–109].

Mediators should have appropriate redox potentials to coordinate with the biocatalysts. Their potentials should be within the range of the thermodynamic potentials of the anode and cathode. The use of a mediator will enable and accelerate the internal electron conduction of the biofuel cell. The trade-off is a drop in OCV. The OCV of most mediated biofuel cell systems reported so far varied between 0.2 V and 1 V [7, 15]. A mediated H_2/O_2 biofuel cell was reported recently to have an OCV of 1.17 V [110], which is very close to the standard thermodynamic potential of the fuel cell (1.22 V). Methyl viologen was used to mediate the oxidation of hydrogen catalyzed by *Desulfovibrio vulgaris*, while 2,2'-azinobis(3-ethylbenzothiazoline-6-sulfonate) (ABTS) was used to mediate the reduction of O_2 with bilirubin oxidase.

To achieve highly efficient electron transfer, mediators should have good activity and diffusivity. In some cases, a combination of two mediators may provide better performance. For example, the mediating effects of mixed Fe(III) EDTA and thionine were studied in a microbial biofuel cell with *E. coli* in the anode [111]. Thionine could be reduced much faster than Fe(III) EDTA by the bacteria, but its oxidation rate was much slower than Fe(III) EDTA on the electrode. Because Fe(III) EDTA can oxidize the reduced form of thionine at a relatively fast reaction rate ($k_{et} = 4.8\times10^4\ M^{-1}s^{-1}$), the combination of these two mediators greatly improved the performance of the biofuel cell.

Free cofactors can be viewed as mediators in many ways despite their catalytic role in the biotransformation reactions. Many dehydrogenases are NAD(H)-dependent. The enzymatic reduction of NAD^+ is usually vigorous, whereas the electrochemical oxidation of NADH, on the other hand, is much less dynamic. The thermodynamic redox potential of NAD(H) is –0.59 V vs. SCE, but direct electrochemical oxidation of NADH usually occurs at a high overpotential. At pH 7.0, the reported overpotential is about 1.1 V at carbon [112] and 1.3 V at the platinum electrode [113]. With such an overpotential, it is impossible to build a biofuel cell using NAD(H)-dependent enzymes with direct electrochemical oxidation of NADH on the electrode. One way to improve the oxidation of NADH is to use a mediator. Palmore et al. reported a methanol-oxygen biofuel cell using such a strategy [35]. In the anodic chamber of the biofuel cell, complete oxidation of methanol was achieved with three NAD(H)-dependent enzymes (ADH, AldDH and FDH). NADH was oxidized by benzylviologen (redox potential: 0.55 V vs. SCE), and the reduced benzylviologen was oxidized electrochemically at the electrode. The reaction between NADH and benzylviologen was catalyzed by another enzyme, diaphorase. The biofuel cell equipped with a chemical oxygen cathode achieved an OCV of 0.8 V and a power output of 680 $\mu W/cm^2$ at 0.49 V.

6. IMMOBILIZATION OF BIOCATALYSTS AND MEDIATORS

Like any bioprocessing, immobilization of catalysts is generally preferred for biofuel cells when continuous feeding of fuels is required. Often immobilization stabilizes enzymes or the catalysts against denaturation. Accordingly, enzyme immobilization was also sought as an important method of extending the lifetime of biofuel cells.

An easy method for immobilizing bacteria on an electrode is to form biofilms: it requires no special treatment [91, 95]. It was shown that microbial cells attached to the electrode surface were primarily responsible for the generation of current [66]. Co-immobilization of bacteria with mediators was also reported. *Desulfovibrio desulfuricans* modified with mediators, polyviologen and N,N,N',N'-tetracyano-p-quinodimethane, was immobilized onto electrodes via physical adsorption [114]. The microbial fuel cell was able to generate electricity, but with a relatively low current density, 5 $\mu A/cm^2$.

The immobilization of enzymes can be achieved either chemically or physically. Physical entrapment has been extensively applied for the immobilization of DET enzymes on electrodes. For example, hydrogenase and laccase have been physically adsorbed on carbon black particles to construct composite electrodes [34]. A similar method was also employed in the study of glucose/air biofuel cells [84, 115]. Keeping the active sites of the enzyme close to the surface of the electrode is critical for DET. For laccase-catalyzed direct electroreduction of oxygen, a critical distance of 20 Å between the enzyme's active site and the electrode surface was proposed [116]. A greater distance slowed the overall reaction; while a shorter distance made the enzymatic reaction the limiting step. A similar phenomenon was also reported for HRP, for which the critical distance was 18 Å [117].

Pizzariello et al. [43] reported a glucose/H_2O_2 biofuel cell using ferrocene-modified composite electrodes. GOx or HRP was first adsorbed on synthetic graphite particles and used to form a suspension solution with 2-hexadecanone and ferrocene in chloroform. The composite electrodes were then prepared by spray-sprinting the suspension on a polyester substrate. The biofuel cell continuously worked for 30 days with a negligible voltage drop; however, the power density was low.

Polymer matrices are often used to entrap enzymes and mediators. Minteer et al. reported a method of entrapping enzymes in Nafion membrane [36]. NAD^+-dependent dehydrogenases (such as alcohol dehydrogenase, aldehyde dehydrogenase, formaldehyde dehydrogenase, glucose dehydrogenase and lactate dehydrogenase) were physically mixed with tetralkylammonium bromide-modified Nafion solution, which was then casted on methylene green-modified glassy carbon electrodes. The immobilized enzymes were subjected to a 25 min annealing treatment at 140°C. Cofactor NAD^+ was immobilized via ion exchange with the Nafion membrane. No statistical difference in enzyme activity was observed before and after heat treatment. Ethanol/O_2 biofuel cells constructed using this method generated a power density as high as 2.04 mW/cm^2 [36]. Cyclic voltammetry measurements indicated that mass transfer, not the reaction kinetics, is the limiting factor in such fuel cells.

Heller and co-workers extensively explored the use of redox polymers for the construction of miniature biofuel cells [11, 39, 40, 45, 46, 77]. Two types of redox polymers, both containing Os redox centers but with different redox potentials, were developed. The polymer of higher redox potential, 0.58–0.79 V vs. SHE, was used for the cathode; while the one with lower redox potential, 0.02–0.32 V vs. SHE, was used for the anode. Enzymes were mixed with the redox polymers along with a crosslinker, poly(ethylene glycol) (Mw 400) diglycidyl ether (PEGDGE). The electrodes were built by casting the enzyme-polymer solution onto 7-μm carbon fibers. A recent report showed that a glucose-oxygen biofuel cell was able to deliver a power density of up to 0.35 mW/cm^2 at 0.88 V [77].

Efficient covalent binding of enzymes and mediators has been demonstrated. Katz et al. reported several studies on biofuel cells using co-immobilized enzyme-cofactor-mediator complexes on metal electrodes [10, 15, 41, 42, 44, 47, 79, 81]. They modified the electrode surface with a monolayer of redox mediator-cofactor arrays and then integrated the enzymes via bioaffinity. For example, a redox monolayer was formed by covalently grafting pyrroloquinoline quinone (PQQ) to a cystamine-modified Au-electrode, followed by attaching N6-(2-aminoethyl)-NAD^+ to the PQQ monolayer. Lactate dehydrogenase (LDH) was then adsorbed to the PQQ-NAD^+ monolayer via bioaffinity, and was further stabilized by cross-linking using glutaraldehyde [118, 119]. Following the same approach, GOx-FAD was assembled on an Au electrode with mediators such as PQQ [30], nitrospiropyran [120], rotaxane [121], C-60 [122] and Au nanoparticles [123]. Since the affinity between FAD and GOx was strong, no crosslinking was required for this type of electrode. Interestingly, such immobilized GOx showed higher activities than did native enzymes with the traditional electron acceptor, oxygen. This is attributed to the enhanced efficiency of electron conduction [123]. The monolayer assembly method was also used for the construction of biocathodes, such as an H_2O_2 electrode using microperoxidase-11 and O_2 electrode with cytochrome C/cytochrome oxidase [10].

7. ENGINEERING ASPECTS OF BIOFUEL CELLS

7.1. Electrode materials

An ideal electrode should offer several features: low-cost, electron-conductivity, physical and chemical stability, and a large specific surface area. Noble metals, especially platinum and gold, have been widely used in electrochemical studies. However, they are too expensive for large-scale fuel cell applications. Cheaper electrodes have thus been constructed using carbon-based materials. Table 4 summarizes the application of carbon-based electrodes in previous studies of biofuel cells.

Table 4
Carbon-based electrodes in biofuel cells

Electrodes	Configuration	Surface area	References
Carbon fiber (single)	1-D	Medium	[39, 45, 46, 77]
Graphite plates or rods	2-D	Low	[35, 36, 43, 49, 54, 57, 59, 61, 66, 69, 72, 91, 93, 124-126]
Carbon felt, paper and cloth	3-D	Medium	[57, 58, 96, 110]
Composite electrode using carbon black (composite)	3-D	High	[34, 84]

Carbon fibers were employed for the construction of miniature enzymatic fuel cells targeting implantable power supplies [11, 39, 40, 45, 46, 77]. Different forms of 2-D electrodes, such as graphite plates and sheets, glassy carbon rods, and platinum and gold foils have also been used in biofuel cells.

In general, nonporous electrodes can provide only very limited power density for biofuel cells. The size of biocatalysts is much larger than that of metal atoms. A GOx molecule, for example, has a cross section of 58 nm^2. Assuming the enzyme is 100% active, a monolayer coverage of GOx on a flat electrode surface will generate a current density of 200 $\mu A/cm^2$ [41]. Due to the restriction of cell voltage (usually less than 1 V), biofuel cells based on 2-dimentional electrodes can only achieve a power density of less than 1 mW/cm^2. Higher power density is possible with porous electrodes if the internal surface area is not accounted in the power density calculation. Porous electrodes such as carbon felt, paper, cloth and various composite electrodes allow 3-D distribution of the biocatalysts throughout the matrix for high catalyst loadings.

7.2. Mass transfer

Three processes are subject to mass transfer limitations in biofuel cells: diffusion of the fuel or oxidant to the active sites of the biocatalysts, proton transfer through the membrane, and diffusion of redox mediators between electrodes and biocatalysts, or alternatively, the electron transfer between the active sites of the catalysts to the electrodes. As in conventional fuel cells, high resistance to the mass transfer process of fuels tends to build a concentration difference between the supply end and the reaction sites, thus slowing down the reactions and leading to the polarization of the electrodes. The performance of porous composite electrodes is usually greatly limited by the mass transfer of fuels. Typical engineering methods, such as patterned electrode design and the introduction of convective transport by forced flow or mechanical stirring, may alleviate the problem.

Polymeric mediators also suffer from mass transfer limitations. The apparent diffusion coefficient of these redox polymers is 10^{-9}–10^{-8} cm^2/s, much less than the typical value of small mediators (10^{-6}–10^{-5} cm^2/s). However, a comb-like structure with redox centers grafted to polymer backbones may afford much-improved apparent diffusion rate. A 1000-fold increase of the diffusion coefficient was recently reported by increasing the length of the spacers between the polymer backbone and the redox center [39, 46, 127].

7.3. Membrane-less biofuel cells

The majority of the reported biofuel cells consist of two chambers: the anodic and cathodic chambers. Separating the two chambers is usually an ion-exchange membrane, mostly Nafion membrane. The function of the membrane is to prevent direct contact between the fuel and oxidant and at the same time conduct protons. When the separation of the fuel and the oxidant is not necessary, the membrane is not needed. Liu et al. explored the possibility of operating microbial fuel cells without a proton-exchange membrane [58]. A special single-chamber fuel cell was constructed using a piece of plastic tube with the anode and cathode mounted at opposite ends. Compared to biofuel cells with an ion-exchange membrane, the membrane-less fuel cell showed a power density that was about twice as high, but the coulombic efficiency dropped by ~80%. It was postulated that the low efficiency was a result of the diffusion of oxygen into the anode area. However, for cheap fuel sources such as wastewater, this coulombic efficiency loss is probably acceptable.

The other type of membrane-less biofuel cells reported was glucose-O_2 enzymatic fuel cells. The major concern is again the interference of oxygen to the anodic reaction. Persson et al. examined the use glucose dehydrogenase (GDH) for a membrane-less glucose biofuel cell [128, 129]. As shown in Equation 3, GDH catalyzes the oxidation of glucose with the action of cofactor NAD, which has a good tolerance to O_2. An alternative enzyme is GOx, which is more stable and active than GDH. However, GOx tends to take in oxygen and generate hydrogen peroxide. Special treatment is needed to ensure that the electrons are not accepted by the oxygen but rather by redox mediators in GOx membrane-less biofuel cells. An effective approach is to enhance DET to compete with the reaction with oxygen by wiring the enzyme molecules to the anode with redox polymers [11, 39, 40, 45, 46, 77, 127]. Similarly, Katz et al. wired the active centers of GOx with PQQ to the surface of gold electrode [41, 79]. Both methods effectively reduced the interference caused by oxygen.

8. OVERALL PERFORMANCE OF BIOFUEL CELLS

Table 5 summarizes the general performance of typical biofuel cells reported so far. Similar to other energy generation devices, biofuel cells are expected to function over a reasonably long period of time with a certain level of power output. This appears to be a long-standing challenging goal to achieve for most types of biofuel cells.

The lifetime of biofuel cells has always been a concern. There are several factors to be considered. For mediated biofuel cells, the loss or degradation of redox mediators limits the lifetime of cells [12, 126]. In most cases, the stability of biocatalysts is largely the determining factor. Thus, living microbes are advantageous since they have the ability to reproduce. As a result, a lifetime of months or years is typically expected of microbial fuel cells. The longest-lasting microbial fuel cell was reported by Habermann et al. It worked for more than five years without malfunction or maintenance [69]. In contrast, most enzymatic fuel cells usually survive only a few days. Closely related to lifetime, operational stability of biofuel cells is also affected by the stability of biocatalysts. Microbial fuel cells can maintain stable power generation for up to months [55, 66]. Enzymatic biofuel cells often last from hours to days [44, 47, 130].

An enzyme's lifetime can be extended upon immobilization. A miniature biofuel cell with GOx and BOD immobilized in Os-containing redox polymer has the potential to last 20 days at 37°C (estimated by extrapolating the power decay curve reported in reference [39]). More promising results were reported by Moore et al. using tetrabutylammonium bromide modified Nafion membranes to entrap dehydrogenases [131]. The half lifetimes of the native parent enzymes are only 7–8 h in solution. After immobilization, active lifetimes of more than 45 days were achieved. Furthermore, biofuel cells built with this technique showed no significant power decay during several weeks of continuous operation [132].

Another major issue is power density, which is usually measured by power generation per surface area of electrode, or per weight or volume of the cell. The achievable power density of microbial biofuel cells is generally much lower than that of an enzymatic biofuel cells. Accordingly, microbial biofuel cells are preferred for the applications where the volume and weight of cells are not of concern; while enzymatic fuel cells can be designed to supply power for compact devices. Compared to the performance of biofuel cells two decades ago, the power density of newly developed biofuel cells was about 1–2 orders of magnitude higher [52, 97, 124, 132]. These advances have pushed biofuel cell technology one more step closer to commercial applications. In fact, biofuel cells with a power density greater than 1 mW/cm^2 may already be powerful enough for cellular phone chargers [133].

Table 5
Performance of typical biofuel cells reported recently

Fuel/oxidant	Anode/cathode catalysts	Performance[a]	Lifetime[c]	Reference
Glucose/O_2	GOx/ laccase or BOD	OCV: up to ~1.0 V I_{SC}: up to ~1 mA/cm^2 PD: Up to 0.35 mW/cm^2	>1 wk @ 37°C	[39, 40, 45, 46, 77, 127]
Glucose/H_2O_2	GOx/HRP	OCV: 0.22 V I_{SC}: ~7.2 μA/cm^2 PD: 0.15 μW/cm^2	30 days	[43]
Glucose/O_2	GOx/ Cytochrome oxidase	OCV: up to 0.12 V I_{SC} [b]: ~5.5 (0.55) mA/cm^2 PD [b]: ~43 (4.3) μW	No loss for 2 days	[41, 42]
Glucose/H_2O_2	GOx/ microperoxidase	OCV: 0.31 V I_{SC} [b]: ~1.7 (0.114) mA/cm^2 PD [b]: ~2.4 (0.16) mW/cm^2	50% loss in 3 h	[44]
Glucose/ cumene peroxide	GOx/ microperoxidase	OCV: 0.99 V I_{SC} [b]: ~13 (0.83) mA /cm^2 PD [b]: ~4.1 (0.26) mW/cm^2	N/A	[47]
Alcohols, glucose mysteric acid/O_2	Dehydrogenases/ chemical	OCV: up to 0.82 V PD: up to 2.04 mW/cm^2	>45 days	[36, 131, 134]
Methanol/O_2	Dehydrogenases/ chemical	OCV: 0.8 V I_{SC}: ~1.3 mA/cm^2 PD: 0.67 mW/cm^2	N/A	[35]
H_2/O_2	Hydrogenase/ laccase	PD: 0.32 mW/cm^2	N/A	[34]
H_2/O_2	Bacteria (hydrogenase)/ BOD	OCV: 1.17 V I_{SC}: ~0.2 mA/cm^2 PD: 0.18 mW/cm^2	N/A	[110]
Light, H_2O/O_2	Cyanobacteria/ BOD	OCV: 0.6 V I_{SC}: up to ~0.23 mA/cm^2 PD: 30–40 μW/cm^2	N/A	[65]
Glucose/O_2	*E. coli* and sewage sludge/Chemical	PD: 78.7 μW/cm^2	N/A	[61]
Glucose/O_2	Mixed bacterial culture/Chemical	PD: 360 μW/cm^2	N/A	[49]
Glucose/O_2	*R. ferrireducens*/ carbon	PD: 3.3 μW/cm^2	No loss in ~ 40 days	[66]
Marine sediment/O_2	Bacteria/ carbon	OCV: up to ~0.75 V PD: up to 4 μW/cm^2	>250 days	[55, 73]
Glucose/O_2	Bacteria/ chemical	OCV: up to ~0.7 V PD: up to 49 μW/cm^2	> 40 days	[57, 58, 96]
Glucose/ ferricyanide	*E. coli* K12/ chemical	OCV: 0.895 V I_{SC}: up to ~4.0 mA/cm^2 PD: 1.2 mW/cm^2	N/A	[97]

a. OCV: open-circuit voltage; I_{SC}: short-circuit current density; PD: power density. Data of power and current densities are based on apparent electrode area

b. Data of power and current densities estimated based on the surface area. Data in parentheses were data originally reported in the reference based on the absolute surface area of the electrodes considering the roughness factor.

c. At room temperature if no temperature was specified.

9. CONCLUSION

Exciting technical advances in the field of biofuel cells have recently emerged. Compared with chemical fuel cells, biofuel cells afford more fuel options, bio-compatibility and mild operation conditions. Although the power density of biofuel cells is usually 2–3 orders of magnitude lower than that of chemical fuel cells, they are attractive for special applications such as implantable devices, sensors, drug delivery, microchips, and portable power supplies. Microbial biofuel cells also have great potential in digesting organic wastes and biomass for power generation. It is important to note that interest in developing biofuel cells is rapidly growing. As new enabling technologies in materials science, nanotechnology and genetic engineering continue to evolve, high performance biofuel cells may soon take a significant role in the dynamic energy market.

REFERENCES

[1] M.C. Potter, Proc. R. Soc. Lond. B Biol. Sci., 84(1912), 260.
[2] B. Cohen, J. Bacteriol., 21(1931), 18.
[3] M.G.D. Duca, J.M. Fuscoe and R.W. Zurilla, Develop. Ind. Microbiol., 4(1963), 81.
[4] W.V. Hees, J. Electrochem. Soc., 112(1965), 258.
[5] A.T. Yahiro, S.M. Lee and D.O. Kimble, Biochim. Biophys. Acta, 88(1964), 375.
[6] J.O.M. Bockris and S. Srinivasan, Fuel Cells: Their Electrochemistry, McGraw-Hill, New York, 1969,612.
[7] G.T.R. Palmore and G.M. Whitesides, in ACS Symposium Series, edited by, 1994, 271.
[8] W.J. Aston and A.P.F. Turner, Biotechnology & Genetic Engineering Reviews, 1(1984), 89.
[9] G. Govil and A. Saran, J. Indian Chem. Soc., 59(1982), 1226.
[10] E. Katz and I. Willner, in Advanced Macromolecular and Supramolecular Materials and Processes, edited by K.E. Geckeler, Kluwer Academic/Plenum Publishers, New York, 2003, 175.
[11] A. Heller, Phys. Chem. Chem. Phys., 6(2004), 209.
[12] S.C. Barton, J. Gallaway and P. Atanassov, Chem. Rev., 104(2004), 4867.
[13] B.E. Logan, Environ. Sci. Technol., 38(2004), 160A.
[14] L.T. Angenent, K. Karim, M.H. Al-Dahhan, et al., Trends Biotechnol. , 22(2004), 477.
[15] E. Katz, N.A. Shipway and I. Willne, in Handbook of fuel cells - Fundamentals technology and applications, edited by W. Vielstich, A. Lamm, and H.A. Gasteiger, John Wiley and Sons Ltd, Chichester, 2003, 355.
[16] I. Karube, S. Suzuki, T. Matsunaga, et al., Ann. N.Y. Acad. Sci., 369(1981), 91.
[17] S. Suzuki, I. Karube, T. Matsunaga, et al., Biochimie, 62(1980), 353.
[18] S. Suzuki and I. Karube, Applied Biochemistry and Bioengineering, 4(1983), 281.
[19] S. Suzuki, I. Karube, H. Matsuoka, et al., Ann. N.Y. Acad. Sci., 413(1983), 133.
[20] S. Dawar, B.K. Behera and P. Mohanty, Int. J. Energ. Res., 22(1998), 1019.
[21] R. Obert and B.C. Dave, J. Am. Chem. Soc., 121(1999), 12192.
[22] G.T.R. Palmore, Trends Biotechnol. , 22(2004), 99.
[23] A.L. Ghindilis, P. Atanasov and E. Wilkins, Electroanal., 9(1997), 661.
[24] S.D. Varfolomeev, I.N. Kurochkin and A.I. Yaropolov, Biosens. Bioelectron., 11(1996), 863.
[25] R.S. Freire, C.A. Pessoa, L.D. Mello, et al., J. Brazil. Chem. Soc. , 14(2003), 230.
[26] W. Schuhmann, Rev. Mol. Biotechnol., 82(2002), 425.

[27] A. Guiseppi-Elie, C. Lei and R.H. Baughman, Nanotechnology, 13(2002), 559.
[28] Y.-D. Zhao, W.-D. Zhang, H. Chen, et al., Anal. Sci., 18(2002), 939.
[29] C. Cai and J. Chen, Anal. Biochem., 332(2004), 75.
[30] I. Willner, V. Heleg-Shabtai, R. Blonder, et al., J. Am. Chem. Soc., 118(1996), 10321.
[31] K. Lewis, Bacteriological Reviews, 30(1966), 101.
[32] G. Sandstede, E.J. Cairns, V.S. Bagotsky, et al., in Handbook of fuel cells - Fundamentals technology and applications, edited by W. Vielstich, A. Lamm, and H.A. Gasteiger, John Wiley and Sons Ltd, Chichester, 2003, 145.
[33] A.A. Karyakin, S.V. Morozov, E.E. Karyakina, et al., Electrochem. Commun., 4(2002), 417.
[34] M.R. Tarasevich, V.A. Bogdanovskaya, N.M. Zagudaeva, et al., Russ. J. Electrochem., 38(2002), 335.
[35] G.T.R. Palmore, H. Bertschy, S.H. Bergens, et al., J. Electroanal. Chem., 443(1998), 155.
[36] S.D. Minteer, N.L. Akers and C.M. Moore, Enzyme immobilization for use in biofuel cells and sensors, US Patent No. 2004101741.
[37] M.D. Arning, B.L. Treu and S.D. Minteer, Polym. Mater. Sci. Eng., 90(2004), 566.
[38] L. De La Garza, G. Jeong, P.A. Liddell, et al., J. Phys. Chem. B, 107(2003), 10252.
[39] N. Mano, F. Mao and A. Heller, J. Am. Chem. Soc., 124(2002), 12962.
[40] N. Mano, F. Mao, W. Shin, et al., Chem. Commun. (Cambridge, United Kingdom), 2003), 518.
[41] E. Katz, I. Willner and A.B. Kotlyar, J. Electroanal. Chem., 479(1999), 64.
[42] E. Katz and I. Willner, J. Am. Chem. Soc., 125(2003), 6803.
[43] A. Pizzariello, M. Stred'ansky and S. Miertus, Bioelectrochemistry, 56(2002), 99.
[44] I. Willner, E. Katz, F. Patolsky, et al., J. Chem. Soc., Perkin Trans. 2 1998. 1817.
[45] T. Chen, S.C. Barton, G. Binyamin, et al., J. Am. Chem. Soc., 123(2001), 8630.
[46] N. Mano, F. Mao and A. Heller, J. Am. Chem. Soc., 125(2003), 6588.
[47] E. Katz, B. Filanovsky and I. Willner, New J. Chem., 23(1999), 481.
[48] M.K. Islam, T. Ohashi, T. Yasukawa, et al., Chemical Sensors, 20(2004), 744.
[49] K. Rabaey, G. Lissens, S.D. Siciliano, et al., Biotechnol. Lett., 25(2003), 1531.
[50] T. Yagishita, S. Sawayama, K.-I. Tsukahara, et al., Bioelectrochem. Bioenerg., 43(1997), 177.
[51] N. Kim, Y. Choi, S. Jung, et al., Biotechnol. Bioeng., 70(2000), 109.
[52] J. Niessen, U. Schroder and F. Scholz, Electrochem. Commun., 6(2004), 955.
[53] E. Simon, C.M. Halliwell, C.S. Toh, et al., Bioelectrochemistry, 55(2002), 13.
[54] D.H. Park, S.K. Kim, I.H. Shin, et al., Biotechnol. Lett., 22(2000), 1301.
[55] C.E. Reimers, L.M. Tender, S. Fertig, et al., Environ. Sci. Technol., 35(2001), 192.
[56] M.J. Cooney, E. Roschi, I.W. Marison, et al., Enzyme Microb. Technol., 18(1996), 358.
[57] H. Liu, R. Ramnarayanan and B.E. Logan, Environ. Sci. Technol., 38(2004), 2281.
[58] H. Liu and B.E. Logan, Environ. Sci. Technol., 38(2004), 4040.
[59] G.-C. Gil, I.-S. Chang, B.H. Kim, et al., Biosens. Bioelectron., 18(2003), 327.
[60] B. Kim, I. Chang, M. Hyun, et al., A biofuel cell using wastewater and active sludge for wastewater treatment, WO Patent No. 0104061.
[61] D.H. Park and J.G. Zeikus, Biotechnol. Bioeng., 81(2003), 348.
[62] T. Yagishita, S. Sawayama, K. Tsukahara, et al., Renew. Energ. , 9(1996), 958.
[63] T. Yagishita, S. Sawayama, K.I. Tsukahara, et al., Sol. Energ. , 61(1997), 347.
[64] T. Yagishita, S. Sawayama, K. Tsukahara, et al., J. Biosci. Bioeng. , 88(1999), 210.
[65] S. Tsujimura, A. Wadano, K. Kano, et al., Enzyme Microb. Technol., 29(2001), 225.
[66] S.K. Chaudhuri and D.R. Lovley, Nat. Biotechnol., 21(2003), 1229.
[67] H.P. Bennetto, G.M. Delaney, J.R. Mason, et al., Biotechnol. Lett., 7(1985), 699.
[68] K.R. Williams, An Introduction to Fuel Cells, Elsevier Pub. Co., New York, 1966,248.
[69] W. Habermann and E.H. Pommer, Appl. Microbiol. Biotechnol., 35(1991), 128.
[70] H.S. Park, B.H. Kim, H.S. Kim, et al., Anaerobe, 7(2001), 297.
[71] C.A. Pham, S.J. Jung, N.T. Phung, et al., FEMS Microbiol. Lett., 223(2003), 129.
[72] D.R. Bond, D.E. Holmes, L.M. Tender, et al., Science, 295(2002), 483.

[73] L.M. Tender, C.E. Reimers, H.A. Stecher, et al., Nat. Biotechnol., 20(2002), 821.
[74] C. Laane, W. Pronk, M. Franssen, et al., Enzyme Microb. Technol., 6(1984), 165.
[75] H.P. Bennetto, D.K. Ewart, A.M. Nobar, et al., in Charge and Field Effects in Biosystems, edited by M.J. Allen, S.F. Cleary, and F.M. Hawkridge, Plenum, New York, 1989, 339.
[76] G.T.R. Palmore and H.-H. Kim, J. Electroanal. Chem., 464(1999), 110.
[77] V. Soukharev, N. Mano and A. Heller, J. Am. Chem. Soc., 126(2004), 8368
[78] G. Gupta, V. Rajendran and P. Atanassov, Electroanal., 16(2004), 1182.
[79] E. Katz, A.F. Bueckmann and I. Willner, J. Am. Chem. Soc., 123(2001), 10752.
[80] S. Tsujimura, K. Kano and T. Ikeda, Electrochemistry, 70(2002), 940.
[81] I. Willner, G. Arad and E. Katz, Bioelectrochem. Bioenerg., 44(1998), 209.
[82] A. Ghindilis, Biochem. Soc. Trans., 28(2000), 84.
[83] P. Atanassov, Abstracts of Papers, 223rd ACS National Meeting, Orlando, FL, United States, April 7-11,2002
[84] P. Atanassov, F. Colon and V. Rajendran, Abstracts of Papers, 228th ACS National Meeting, Philadelphia, PA, United States, August 22-26, 2004
[85] S.V. Morozov, E.E. Karyakina, N.A. Zorin, et al., Bioelectrochemistry, 55(2002), 169.
[86] B.H. Kim, D.H. Park, P.K. Shin, et al., Biofuel cell without electron transfer mediator, EP Patent No. 827229.
[87] H.J. Kim, H.S. Park, M.S. Hyun, et al., Enzyme Microb. Technol., 30(2002), 145.
[88] B.H. Kim, I.S. Chang, G.C. Gil, et al., Biotechnol. Lett., 25(2003), 541.
[89] K.H. Kang, J.K. Jang, T.H. Pham, et al., Biotechnol. Lett., 25(2003), 1357.
[90] N.T. Phung, J. Lee, K.H. Kang, et al., FEMS Microbiol. Lett., 233(2004), 77.
[91] B.H. Kim, H.S. Park, H.J. Kim, et al., Appl. Microbiol. Biotechnol., 63(2004), 672.
[92] I.S. Chang, J.K. Jang, G.C. Gil, et al., Biosens. Bioelectron., 19(2004), 607.
[93] B.H. Kim, I.S. Chang, J.K. Jang, et al., Membraneless and mediatorless microbial fuel cell for treating wastewater, WO Patent No. 2003096467.
[94] B.H. Kim, T. Ikeda, H.S. Park, et al., Biotechnol. Tech., 13(1999), 475.
[95] J.Y. Lee, N.T. Phung, I.S. Chang, et al., FEMS Microbiol. Lett., 223(2003), 185.
[96] S. Oh, B. Min and B.E. Logan, Environ. Sci. Technol., 38(2004), 4900.
[97] U. Schroder, J. Niessen and F. Scholz, Angew. Chem. Int. Ed., 42(2003), 2880.
[98] T. Ikeda and K. Kano, Biochim. Biophys. Acta, 1647(2003), 121.
[99] A.K. Shukla, P. Suresh, S. Berchmans, et al., Curr. Sci. (India), 87(2004), 455.
[100] T.S. Wong and U. Schwaneberg, Curr. Opin. Biotechnol. , 14(2003), 590.
[101] K. Habermuller, M. Mosbach and W. Schuhmann, Fresenius, J. Anal. Chem. , 366(2000), 560.
[102] M.D. Leonida, Curr. Med. Chem., 8(2001), 345.
[103] P.U. Abel and T. Von Woedtke, Biosens. Bioelectron., 17(2002), 1059.
[104] T. Ikeda and K. Kano, J. Biosci. Bioeng. , 92(2001), 9.
[105] K. Kano and T. Ikeda, Anal. Sci., 16(2000), 1013.
[106] P.W. Stoecker and A.M. Yacynych, Bull. Electrochem., 8(1992), 181.
[107] E.J. Calvo and C. Danilowicz, J. Brazil. Chem. Soc. , 8(1997), 563.
[108] S.F. White and A.P.F. Turner, in Handbook of Biosensors and Electronic Noses:Medicine, Food, and the Environment, edited by E. Kress-Rogers, CRC Press, Boca Raton, 1997, 43.
[109] J.A. Duine, Biosens. Bioelectron., 10(1995), 17.
[110] S. Tsujimura, M. Fujita, H. Tatsumi, et al., Phys. Chem. Chem. Phys., 3(2001), 1331.
[111] K. Tanaka, C.A. Vega and R. Tamamushi, Bioelectrochem. Bioenerg., 11(1983), 289.
[112] J. Moiroux and P.J. Elving, Anal. Chem., 50(1978), 1056.
[113] H. Jaegfeldt, J. Electroanal. Chem. and Interfacial Electrochem., 110(1980), 295.
[114] D.H. Park, B.H. Kim, B. Moore, et al., Biotechnol. Tech., 11(1997), 145.
[115] F. Colon, V. Rajendran and P. Atanassov, 206th ECS meeting, Honolulu, Hawaii, United States, October 3-8, 2004

[116] A.I. Yaropolov, T.K. Sukhomlin, A.A. Karyakin, et al., Dokl. Akad. Nauk SSSR, 260(1981), 1192.
[117] J.J. Kulys and A.S. Samalius, Bioelectrochem. Bioenerg., 13(1984), 163.
[118] E. Katz, V. Heleg-Shabtai, A. Bardea, et al., Biosens. Bioelectron., 13(1998), 741.
[119] A. Bardea, E. Katz, A.F. Bueckmann, et al., J. Am. Chem. Soc., 119(1997), 9114.
[120] R. Blonder, I. Willner and A.F. Bueckmann, J. Am. Chem. Soc., 120(1998), 9335.
[121] E. Katz, L. Sheeney-Haj-Ichia and I. Willner, Angew. Chem. Int. Ed., 43(2004), 3292.
[122] F. Patolsky, G. Tao, E. Katz, et al., J. Electroanal. Chem., 454(1998), 9.
[123] Y. Xiao, F. Patolsky, E. Katz, et al., Science, 299(2003), 1877.
[124] J. Niessen, U. Schroder, M. Rosenbaum, et al., Electrochem. Commun., 6(2004), 571.
[125] K. Jones Anne, E. Lamle Sophie, R. Pershad Harsh, et al., J. Am. Chem. Soc., 125(2003), 8505.
[126] R.M. Allen and H.P. Bennetto, Appl. Biochem. Biotechnol., 39-40(1993), 27.
[127] H.H. Kim, N. Mano, X.C. Zhang, et al., J. Electrochem. Soc., 150(2003), A209.
[128] B. Persson, L. Gorton and G. Johansson, Bioelectrochem. Bioenerg., 16(1986), 479.
[129] B. Persson, L. Gorton, G. Johansson, et al., Enzyme Microb. Technol., 7(1985), 549.
[130] C. Kang, H. Shin, Y. Zhang, et al., Bioelectrochemistry, 65(2004), 83.
[131] C.M. Moore, N.L. Akers, A.D. Hill, et al., Biomacromolecules, 5(2004), 1241.
[132] R. Winder, Chem. Ind.-London, 17(2003), 15
[133] S.R. Narayanan and T.I. Valdez, in Handbook of Fuel Cells - Fundamentals, Technology and Applications, edited by W. Vielstich, A. Lamm, and H.A. Gasteiger, John Wiley and Sons Ltd, Chichester, 2003, 1133.
[134] N.L. Akers and S.D. Minteer, in The second international conference on fuel cell science, engineering and technology. 2004. Rochester, New York USA: The American Society of Mechanical Engineering. 497.

Bioprocessing for Value-Added Products from Renewable Resources
Shang-Tian Yang (Editor)

Chapter 21. Biological Production of Hydrogen from Renewable Resources

Zhinan Xu

Institute of Bioengineering, Department of Chemical and Biochemical Engineering, School of Material Science and Chemical Engineering, Zhejiang University, Hangzhou 310027, P. R. China

1. INTRODUCTION

Energy is vital to global prosperity. At present, 90% of the world's energy requirements are fulfilled by fossil fuels, which are often regarded as endless and cheap. However, we now know that the Earth possesses a finite amount of fossil fuels [1–2], and that their indiscriminate use will eventually lead to the foreseeable depletion of limited fossil energy resources [3]. Presently, the utilization of fossil fuels is causing global climate change, mainly due to the emission of pollutants like COx, NOx, SOx, CxHx, soot, ash, droplets of tar and other organic compounds, which are released into the atmosphere as a result of combustion. In addition, fossil fuel-based industry contributes to extensive damage to the environment and to human health. There is now a global effort focused on the development of non-polluting and sustainable energy sources that will replace fossil fuels. Among the future alternative fuels (such as butanol, ethanol, methanol, methane, biodiesel, and hydrogen), hydrogen is widely recognized as the most promising fuel [4]. It has the highest energy content per unit weight of any known fuel (143 GJ $(\text{tonne})^{-1}$) and is the only fuel that is not chemically bonded to carbon. Therefore, burning hydrogen does not contribute to the greenhouse effect, ozone depletion, and acid rain. When hydrogen burns in air, it gives off nothing more than water vapour and heat energy.

Hydrogen is already an industrial gas which has gained some limited applications in industry: as a reactant in hydrogenation processes, as an O_2 scavenger to prevent oxidation and corrosion, as a fuel in rocket engines, and as a coolant in electrical generators, etc [5–6]. However, there are many obstacles to the large-scale production of hydrogen as a clean and renewable fuel to supplement or substitute fossil fuel, from the cost-effective production of sufficient quantities of hydrogen to its storage, transmission, and distribution [7]. Hydrogen may be produced by a number of processes, including by electrolysis of water, the thermocatalytic reformation of hydrogen-rich organic compounds, and biological processes. Currently, nearly 90% of hydrogen is produced by the reactions of natural gas or light oil fractions with steam at high temperatures [8]. Coal gasification and the electrolysis of water are other industrial methods for hydrogen production. These industrial methods mainly consume fossil fuels as energy sources, and sometimes hydroelectricity. However, these processes are highly energy-intensive and not always environmentally benign. Moreover, the petroleum reserves of the world are depleting at an alarming rate. Thus, biological hydrogen production assumes paramount importance as an alternative energy resource.

Biological processes are carried out largely at ambient temperature and pressure, and hence are less energy-intensive than chemical or electrochemical ones. These processes are not only environmentally friendly, but they also lead to a new avenue for the inexhaustible utilization of renewable energy resources [9]. In addition, they can also consume various waste materials, which facilitate waste recycling. Various organizations have performed research in this area, and several national and international programs have been initiated (e.g. European Union programs COST 818 'Hydrogenases and their biotechnological applications' and COST 841 'Chemical and biological diversity of hydrogen metabolism'). Over the past quarter century, many hundreds of publications have appeared on biological H_2 production, and advances towards practical applications are pushing the transition from a fossil fuel-based economy to a hydrogen-based economy. In this chapter, the principles of biohydrogen syntheses are first introduced, then various efforts are reviewed on how to improve the availability of biohydrogen process for practical applications, and some new concepts and strategies are also included to fundamentally reform biohydrogen production. Finally, some outlooks for future biohydrogen production are presented.

2. PRINCIPLES OF BIOHYDROGEN PRODUCTION SYSTEMS

A large number of microorganisms, including significantly different taxonomic and physiological types, can produce molecular hydrogen. Biological hydrogen production processes can be classified as follows: (1) direct biophotolysis; (2) indirect biophotolysis; (3) photo-fermentation; (4) dark-fermentation [10–11].

2.1. Direct biophotolysis

The process of biophotolysis was first demonstrated in the early 1940s by Hans Gaffron, who observed hydrogen metabolism in the green algae *Scenedesmus obliquus* and *Chlamydomonas reinhardtii* [12–13]. Hydrogen is produced by direct biophotolysis, which is composed of light reaction and dark reaction. As shown in Fig.1, in the light reaction, radiation energy is captured by chlorophyll (Chl) molecules and then used to split water and to generate chemical energy in the form of ATP [14–15]. The electrons withdrawn from water are used to lift the redox potential of ferredoxin (Fd). In the dark reaction, the chemical energy from ATP and the reducing power from Fd are used to fix CO_2 into carbohydrates. At the same time, Fd is an efficient mediator in these cells for the hydrogen evolving enzyme, [Fe]-hydrogenase, and links the soluble [Fe]-hydrogenase to the electron transport chain in the green algal chloroplast [16–17]. The absence of CO_2 enhances the light-driven H_2-production, suggesting a competition for electrons between the CO_2-fixation and the H_2-production processes. Normally, hydrogen production by green algae requires several minutes to a few hours of anaerobic incubation in darkness to induce and/or activate enzymes, including hydrogenase. Hydrogenase combines protons (H^+) in the medium with electrons (donated by reduced ferredoxin) to form and release H_2 gas. This process results in the simultaneous production of H_2 and O_2 at a 2:1 ratio [18–19], and can be expressed in a general reaction:

$$2H_2O \xrightarrow{\text{Light energy}} 2H_2 + O_2 \quad (1)$$

Apparently, this mechanism holds the promise of generating hydrogen continuously and efficiently through the solar conversion ability of the photosynthetic apparatus.

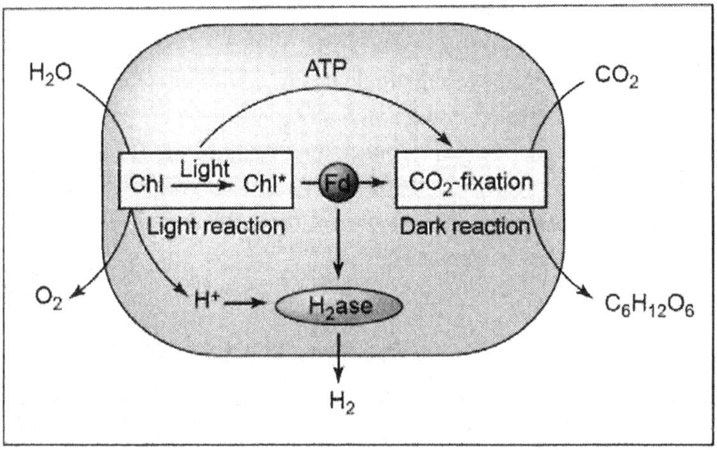

Fig.1 Schematic drawing of the light and dark reactions that occur within a green algae chloroplast. Chl, chlorophyll molecule; Fd, ferredoxin; H_2ase, hydrogenase. (Adapted from [20])

2.2. Indirect biophotolysis

In direct biophotolysis, one major problem is the high sensitivity of the hydrogen evolving process to oxygen which is produced simultaneously during water photolysis [21–22]. This problem can be potentially circumvented by temporally and/or spatially separating oxygen evolution and hydrogen evolution, i.e., indirect biophotolysis. Cyanobacteria (also known as blue-green algae) are a large and diverse group of photoautotrophic microorganisms [23–24], and can also synthesize and evolve H_2 through photosynthesis via the following processes:

$$6H_2O + 6CO_2 \xrightarrow{\text{Light energy}} C_6H_{12}O_6 + 6O_2 \quad (2)$$

$$C_6H_{12}O_6 + 6H_2O \xrightarrow{\text{Light energy}} 12H_2 + 6CO_2 \quad (3)$$

Species of cyanobacteria are morphological diverse, and several enzymes are directly involved in hydrogen metabolism and the synthesis of molecular H_2. These include nitrogenases, which catalyze the production of H_2 as a by-product of nitrogen reduction to ammonia, uptake hydrogenases which catalyze the oxidation of H_2 synthesized by the nitrogenase, and bidirectional hydrogenases [24]. Within the filamentous cyanobacteria, vegetative cells may develop into structurally modified and functionally specialized cells, such as heterocysts, the specialized cells that perform nitrogen-fixation [4, 24]. The localization of nitrogenase in heterocysts provides an oxygen-free environment and enables the heterocystous cyanobacteria to fix nitrogen from air [25–26]. Further studies show that filament integrity is important because filament breakage leads to a loss of nitrogenase activity and hydrogen evolution [27].

In addition, one elaboration of the indirect biophotolysis concept is suggested by separating the H_2 and O_2 evolution reactions into separate stages [28]. The whole bioprocess involves four distinct steps: 1) production in open ponds at 10% solar efficiency of a biomass high in storage carbohydrates; 2) concentration of the biomass from the ponds in a settling pond; 3) anaerobic dark fermentation to yield H_2 and acetate using glucose stored in the algal cells; 4) conversion of acetate to H_2 using algal cells in a photobioreactor. After this last step, the algal biomass would be returned to the ponds to repeat the cycle. Support systems include the anaerobic digestion of any wasted biomass, an inoculum production system to provide make-up biomass and a gas system to separate H_2 and recycle CO_2. Actually, this is an integrated system which couples biophotolysis and dark fermentation through CO_2 fixation, and a very high yield of H_2 can be expected.

2.3. Photo-fermentation

Photosynthetic bacteria have long been studied for their capacity to produce hydrogen through the action of their nitrogenase systems. The photosynthetic device of purple bacteria is simple and has only one photosystem (PS), which is fixed in the intracellular membrane and not powerful enough to split water [29]. Under anaerobic conditions, however, these bacteria are able to use simple organic acids or hydrogen disulfide as electron donor. The electrons that are liberated from the organic carbon or H_2S are pumped around through a large number of electron carriers. During electron transport, protons are pumped through the membrane, and a proton gradient is developed and then used to generate ATP by ATP synthase. The extra energy in the form of ATP can be used to transport the electrons further to the electron acceptor ferredoxin (Fd). When molecular nitrogen is not present, the electrons that are placed on ferrodoxin can be used by nitrogenase to reduce protons to hydrogen. This whole process can be expressed by eq (3) given before.

Carbon monoxide can also be used for hydrogen production via the water-gas shift reaction by some photosynthetic bacteria as follows [30–32]:

$$CO(g) + H_2O(l) \longrightarrow CO_2(g) + H_2(g) \tag{4}$$

This CO can be generated from thermally gasified wood chips. Apparently, the CO-linked hydrogenase is most suited for practical applications, and oxygen-resist enzymes have been identified. The enzyme mediates hydrogen production from CO at rates up to 96 mmol H_2/(L·h) [10].

2.4 Dark-fermentation

Dark hydrogen fermentation is a ubiquitous phenomenon under anoxic or anaerobic conditions (i.e., no oxygen present as an electron receptor). When bacteria grow on organic substrates (heterotrophic growth), these substrates are degraded by oxidation to provide building blocks and metabolic energy for growth. This oxidation generates electrons which need to be disposed of to maintain electronic neutrality. In aerobic or oxic environments, oxygen is reduced and water is the product. In anaerobic or anoxic environments, other compounds e.g., protons, which are reduced to molecular hydrogen (H_2), need to act as electron acceptors.

In the fermentation process of glucose to hydrogen, pyruvate is a key anaerobic metabolite formed by glucose catabolism. The breakdown of pyruvate is catalyzed by one of two enzyme systems:

(1) Pyruvate: formate lyase (PFL)

$$Pyruvate + CoA \longrightarrow Acetyl - CoA + Formate \qquad (5)$$

(2) Pyruvate: ferredoxin oxidoreductase

$$Pyruvate + CoA + 2Fd(ox) \longrightarrow Acetyl - CoA + CO_2 + 2Fd(red) \qquad (6)$$

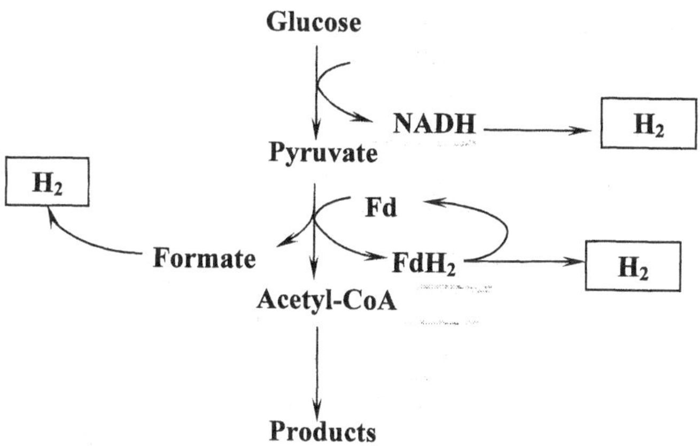

Fig. 2. Representative hydrogen production pathways by anaerobic bacteria.

As illustrated in Fig. 2, in the absence of oxygen, the pyruvate is used to produce acetyl-CoA, from which ATP can be derived, and either formate or reduced ferredoxin, from which hydrogen can be derived by hydrogenase [28]. The enteric bacteria derive hydrogen from formate by formate lyase and strict anaerobes derive hydrogen from *Fd* (*red*) by hydrogenase (see Fig. 2). Depending on the fermentation conditions and bacteria used in the process, acetic and butyric acid are the main anaerobic metabolites along with hydrogen gas [33–34].

$$C_6H_{12}O_6 + 2H_2O \longrightarrow 2CH_3COOH + CO_2 + 4H_2 \qquad (7)$$

$$C_6H_{12}O_6 + 2H_2O \longrightarrow CH_3CH_2CH_2COOH + 2CO_2 + 2H_2 \quad (8)$$

In addition, there is another pathway for hydrogen evolution called the NADH pathway. NADH-ferredoxin oxidoreductase also oxidizes NADH and reduces ferredoxin, which then derives the generation of hydrogen [35].

3. MICROORGANISMS AND ENZYMES FOR HYDROGEN PRODUCTION

3.1. Microorganisms

3.1.1. Green algae

About 50 years ago, Gaffron et al. discovered that the eukaryotic unicellular green algae *Scenedesmus obliquus* is able to evolve molecular hydrogen by means of a hydrogenase in light under anaerobic conditions. H_2 production by green algae was achieved with *Scenedesmus obliquus*, *Chlamydomonas reinhardtii* and *C. moewusii* [36–37]. Among them, *C. reinhardtii* showed a relatively higher ability for producing hydrogen. In order to reduce the high sensitivity of hydrogenase to O_2, *C. reinhardtii* was employed to carry out a two-stage process by incubating the microalgae in the medium that does not contain sulfur-containing nutrients at the second stage [38]. In this two-phase process, CO_2 is first fixed into H_2-rich substrates during normal photosynthesis (Phase I), this is followed by the light-mediated generation of molecular H_2 when the microalgae are incubated under anaerobic conditions (Phase II). Using this sulfur-deprived medium, the rate of O_2 synthesis and CO_2 fixation decline significantly, after about 22 h, *C. reinhardtii* cultures become anaerobic and begin to synthesize H_2. The attainable rate of H_2 production is ca. 0.07 mmol $H_2/(L \cdot h)$ [39–40].

3.1.2. Cyanobacteria

Indirect biophotolysis processes are the paths followed by cyanobacteria. In this system, photosynthesis (O_2 evolution and CO_2 fixation) and N_2-fixation (H_2 production) are either spatially or temporally separated from each other. Cyanobacteria contain photosynthetic pigments, such as Chl a, carotenoids, and phycobiliprotein, and can perform oxygenic photosynthesis. They are a morphologically diverse group that includes unicellular, filamentous and colonial species, and can be further divided into two types: heterocystous and nonheterocystous. Initial work by several authors focused on the heterocystous filamentous cyanobacterium *Anabaena cyclindrica* B-624 [41]. Its vegetative cells may develop into structurally modified and functionally specialized cells, such as heterocysts. In the heterocyst, nitrogenase is protected from O_2 by a heavy cell wall so that nitrogen-fixation and hydrogen generation can be effectively performed there. Another advantage of cyanobacteria is its simple nutritional requirements: air (N_2 and O_2), water, mineral salts, and light. Because of the high rates of H_2 production, *Anabaena* species and strains have been subjected to intense study for the past several years. In addition, hydrogen production has also been explored with other species, including *Nostoc muscorum*, *N. spongiaeforme*, *Westiellopsis prolifica*, *Oscillotoria Miami* BG7, *Aphanothece halophytico* [41–43].

3.1.3. Photosynthetic bacteria

It is well known that purple non-sulfur bacteria can evolve molecular H_2 catalyzed by nitrogenase under nitrogen-deficient conditions using light energy and reduced compounds. Many photoheterotrophic bacteria are found to generate H_2, including *Rhodobater*, *Rhodopseudomonas* [44–45], *Rhodospirillum* [46], *Chromatium* [47], *Chlorobium* [48], and *Halobacterium* [49], etc. Recently, a few mutants of the existing photosynthetic bacteria were isolated to improve the production of hydrogen. Some uptake hydrogenase-negative mutants of *Rhodobacter capsulatus* and *Rhodospirillum rubrum* showed increased H_2 photoproduction, depending on the nitrogen and the carbon sources employed. Similarly, several mutants of *Rhodobacter sphaeroides* with the inactivated PHA synthase have been shown to have enhanced hydrogen productivity because of polyhydroxyalkanoate (PHA) accumulation was abolished and no longer in competition with H_2 photoproduction [50].

3.1.4. Dark-fermentation bacteria

A large number of microbes living in anaerobic conditions are known to produce H_2 as a fermentative means of disposing excess reducing equivalents. *Clostridium* and *Enterobacter* are the most studied fermentative microorganisms for hydrogen production from carbohydrates [51–52]. In *Clostridium* sp., *C. butyricum*, *C. beijerinckii* and *C. acetobytylicum* are often used to evolve H_2, but produce different end metabolites [53]. Because of having some tolerance to oxygen, a variety of *Enterobacter* strains have attracted intensive study. *Enterobacter aerogenes* is the first species in this genus reported for its fermentative H_2 production, and several other groups searching for H_2-producing microbes have also independently isolated various strains of *E. aerogenes*. A newly isolated strain, *E. aerogenes* III-BT 08, was shown to have a high H_2-producing potential [54]. More recently, some thermophilic bacteria were discovered to have the ability to produce fermentative H_2. Such organisms include *Thermotoga neapolitana*, *Thermotoga elfii*, and *Caldicellulosiruptor saccharolyticus* [55]. In addition, many unidentified mixed anaerobic bacteria have been used to produce hydrogen from waste water and some renewable raw materials.

3.2. Major enzymes for hydrogen production

All the processes of biological hydrogen production are fundamentally dependent upon the presence of a hydrogen-producing enzyme. This enzyme catalyzes what is arguably the simplest chemical reaction: $2H^+ + 2e^- \rightarrow H_2$. However, a survey of all presently known enzymes capable of hydrogen evolution shows that they contain complex metallo-centers as active sites and that the active enzyme units are synthesized in complex process involving auxiliary enzymes and protein maturation steps. At present, three groups of enzymes performing this reaction are known: nitrogenase, [NiFe] hydrogenase, and [Fe] hydrogenase.

Nitrogenase is a two-component protein system that uses ATP (2ATP/ e^-) and low-potential electrons derived from reduced ferredoxin or flavodoxin to reduce a variety of substrates. This enzyme is very oxygen labile, and evolves hydrogen concomitantly with the fixation of N_2 to NH_3.

$$N_2 + 8H^+ + 8e^- + 16ATP \longrightarrow 2NH_3 + H_2 + 16ADP + 16P_i \quad (9)$$

Among algae, only the blue-green algae (cyanobacteria) have nitrogenase. Photosynthetic bacteria also use this enzyme to produce H_2. In the absence of other substrates, nitrogenase continues to turn over, reducing protons to hydrogen. The turnover of this enzyme complex is extremely low (6.4 s^{-1}), and extra ATP is consumed [28]. Actually, this enzyme system is very complex, and the products of at least 20 genes are necessary for co-factor synthesis and insertion as well as metal metabolism. Considering the low turnover number, the considerable energy inputs necessary for biosynthesis and the requirement of ATP for catalysis, nitrogenase is not very metabolically active to produce H_2.

Many microorganisms contain [NiFe] hydrogenase, which is usually thought of as functioning as an "uptake" hydrogenase because its normal metabolic function is to derive reductants from H_2. The [NiFe] hydrogenases are heterodimeric proteins consisting of both small (S) and large (L) subunits. The small subunit contains three iron-sulfur clusters, two [4Fe-4S] and one [3Fe-4S]. The large subunit contains a unique, complex nickel iron center with co-ordination to 2 CN and one CO, forming a biologically unique metallo-center [56]. Activities in uptake direction are usually in the order of 300-400 μmol/min mg, and the rates of H_2 evolution are ca. 65 μmol/min mg [57], which corresponds to a turnover rate of 98 s^{-1}. Thus, even working in reverse of its normal function, this class of hydrogenase appears to be a better catalyst for hydrogen evolution than nitrogenase.

Many algae and fermentative H_2 producers contain [Fe] hydrogenase to produce H_2. It contains a unique complex, Fe-S center, in which one of the Fe atoms is complexed with CO and CN. The highly reactive nature of this cluster together with the proposed formation of an iron-hydride intermediate during proton reduction may make searching for an oxygen stable hydrogenase a rather elusive goal. [Fe] hydrogenase has extremely high turnover number: 6000 s^{-1} for *C. pasteurianum* and 9000 s^{-1} for *Desulfovibrio* spp. [28]. This is a thousand times faster than that of nitrogenase. This is a classical and reversible hydrogenase which attracts wide studies on its enzymatic reaction mechanism.

The above three types of hydrogenases should be important for hydrogen evolution because the quantity or inherent activity of these enzymes could limit the performance of the overall process. However, the production of hydrogen involves many metabolic pathways, including electronic transportation, energy metabolism and redox balance, etc. Thus, it is necessary to apply some global research tools to examine and regulate the bioactivities of related enzymes for enhanced H_2 evolution in some model H_2-producing microorganisms.

4. COMPARATIVE STUDIES ON BIOHYDROGEN PRODUCTION PROCESSES

Biohydrogen can be produced by four different types of bioprocesses with highly diverse microorganisms. It has been found that most of the biological processes are operated at an ambient temperature and normal pressure. Therefore, these processes are not energy intensive. The relative advantages and disadvantages of different biological processes are presented in Table 1. Each biohydrogen production process has its advantages and disadvantages, but some common challenges exist that have to be overcome in the future. In the photobiological hydrogen production process, the main drawbacks are oxygen inhibition to the photoautotrophic hydrogen production process, a low H_2 evolving rate and low light conversion efficiency in both the photoautotrophic and photoheterotrophic processes. In the dark-fermentation process, relatively lower hydrogen yield is the main drawback; however, it is a promising process due to its higher production rate of H_2 as well as the versatility of the substrates used.

Table 1
Comparison of different biological hydrogen production processes

Process	Type of microorganism	Advantages	Disadvantages
Direct biophotolysis	Green algae	Can produce H_2 directly from water Solar conversion energy increased by 10 fold as compared to trees, crops	Require high intensity of light O_2 can be poisonous to the system
Indirect biophotolysis	Cyanobacteria	Can produce H_2 from water Has the ability to fix N_2 from atmosphere	Low photochemical efficiency Uptake hydrogenase enzymes can degrade H_2 ~30% O_2 in the gas mixture O_2 is inhibitory to nitrogenase
Photo-fermentation	Photosynthetic bacteria	Wide-spectrum light energy can be used Can use different waste materials such as distillery effluents, whey, etc	Low light conversion efficiency Low light intensity for the saturation of H_2-production
Dark fermentation	Fermentative bacteria	Can produce H_2 all day long without light A variety of carbon sources can be used as substrates Produces valuable metabolites such as butyric and acetic acids as by-products It is an anaerobic process, so there is no O_2 limitation	Lower achievable yields of H_2 As yields increase, H_2 fermentation becomes thermodynamically unfavorable Product gas mixture contains CO_2 that has to be separated

5. IMPROVEMENTS OF PHOTOBIOLOGICAL HYDROGEN PRODUCTION

5.1. Overcoming the O_2 sensitivity of key enzymes (nitrogenase and hydrogenase)

The most critical problem to the biophotolysis hydrogen process arises from the fact that the H_2 evolution system is strongly inhibited by oxygen, while during the hydrogen production process oxygen is simultaneously emitted. Several research studies concentrated on ways to overcome this problem. O_2 tensions could be reduced by increased gas transfer. Greenbaum et al. sustained a photosynthetic $2H_2O \rightarrow H_2 + O_2$ process continuously for days by sparging the reaction mixture with helium, thus removing the product gas (O_2 and H_2) from the vicinity of the cells. Some regenerable or irreversible oxygen absorbers were suggested to be used for this purpose [58], but this approach is not considered practical for scale-up. Indirect biophotolysis processes have been developed to overcome the O_2 sensitivity problem, such as the two-stage bioprocess by S deprivation, as described in section 2.2. In order to increase the likelihood of successful commercial exploitation, the continuity of this two-stage process needs to be addressed, because H_2-production by S-deprivation of the algae (*C. reinhardtii*) is time-limited [59]. After about 100 h of S-deprivation culture, the algae need to go back to normal photosynthesis in order to be rejuvenated by replenishing endogenous substrate [60]. Moreover, the productivity of H_2 gas accumulation (~2 mL/L·h) represents about 15% of the photosynthetic capacity of the cells when the latter is based on the capability for O_2 evolution under physiological conditions. Recently, some efforts have been made to mutagenize the H_2-producing enzymes (hydrogenase and nitrogenase) with the objective of altering or removing the oxygen sensitivity of the enzyme, thereby permitting light-driven O_2 and H_2 co-production in green algae [61]. Some powerful molecular tools, such as DNA shuffling, have been introduced to rapidly evolve these enzyme molecules for reduced sensitivity to molecular oxygen.

5.2. Maximizing solar conversion efficiency under mass culture conditions

It is well agreed that, in theory, photosynthesis in general and microalgae cultures in particular can achieve as much as 10% total light energy conversion into a primary product, such as CO_2 fixed into biomass or even H_2 [28]. However, such extrapolations are based on theoretical considerations or data obtained under low-light conditions. When the cultures grow under full sunlight, the conversion efficiency is disappointingly low, typically well below 1%. The reason for this inefficiency is that the rate of the dark reactions is roughly ten-times lower than the rate of light capture by photosynthetic pigments (e.g. chlorophyll). This results in up to 90% of the photons captured by the photosynthetic apparatus under full sunlight not being used in photosynthesis but rather decaying as heat or fluorescence. A similar situation also exists for photosynthetic bacteria under mass culture conditions. This so-called "light-saturation" is a major reason that algae productivities are not nearly as high as those projected from extrapolations of laboratory data at low light intensities.

Various solutions to this problem had already been proposed some 50 years ago: rapid mixing, dilution of light incident on the surface of algae cultures, and algal mutants with reduced chlorophyll contents. Rapid mixing may be expected to create in the eddies of turbulence surrounding the algal cells the "flashing light effect", and some profound effect on productivity was observed in mass cultures [62]. The use of light attenuation devices that transfer sunlight into the depths of a dense algal culture is another approach to overcoming the light saturation effect. The simplest approach is to arrange photobioreactors in vertical arrays to reduce direct sunlight. One attracting alternative is the use of optical fiber photobioreactors, in which light energy is collected by large concentrating mirrors and piped into small photobioreactors with optical fibers. However, this approach still presents some large technical and economical challenges.

A more practical approach is to find or develop mutants of algal cells with reduced pigment content, that is, with smaller or truncated Chl antenna size. A truncated Chl antenna will diminish the over-absorption and wasteful dissipation of excitation energy by the cells, and it will also diminish the photoinhibition of photosynthesis at the surface of the culture. Moreover, a truncated Chl antenna will alleviate the rather serious light gradient and mutual cell shading and permit more uniform illumination of the cells in mass cultures. Such altered optical properties of cells will result in much greater photosynthetic productivity and better solar conversion efficiency in the culture. One molecular mechanism has been provided to explain the regulation of the size and composition of the light-harvesting Chl antenna during chloroplast development. Excitation pressure was used as a tool to generate green algae (*Dunaliella salina*) with a truncated Chl antenna size. The photon use efficiency as a function of incident irradiance was measured in fully pigmented and truncated Chl antenna cells. At low intensities (100 μmol photons/m^2), both cell types performed with a relatively high photon use efficiency. At increasing incident intensities, however, photon use efficiencies for the fully pigmented cells declined sharply, reaching a value of ca. 5% at an irradiance corresponding to full sunlight (2500 μmol photons/m^2). The cells with the truncated Chl antenna size exhibited a smaller decline in photon use efficiency with irradiance, still reaching a value of ca. 0.45 at the intensity of full sunlight. By isolating microalgal mutants with truncated Chl antenna size, a 50% increase in H_2 productivity was achieved in continuous laboratory cultures operating at high light intensities, compared with the wild type [63–64]. In order to further construct genetically engineered green algae with these characteristics, it is important to identify genes that confer a truncated Chl antenna size using *C. reinhardtii* [65]. This research direction is very active, and several related genes in *C. reinhardtii* have recently been identified [66]. Once a library of such genes is on hand, the overexpression or down-regulation of their expression, as needed, can be applied to *C. reinhardtii* and other green algae that might be suitable for commercial exploitation and H_2 production.

5.3. Enhancing H_2 production with metabolic engineering

The production of biohydrogen is a complex process mediated by different metabolic pathways in cyanobacteria and photosynthetic bacteria. One of the major obstacles to efficient solar energy generation of H_2 in heterocystous cyanobacteria might be the presence of hydrogenase in the heterocysts. Many heterocystous cyanobacteria contain both uptake hydrogenase (Hup) and bidirectional (or reversible) hydrogenase (Hox), although a few have only Hup [67]. In hetercystous cyanobacteria, Hup occurs predominantly in the hetercysts and recovers some of the H_2 produced by the nitrogenase reaction. Hox occurs in both vegetative cells and hetercysts, and is also considered to absorb H_2 due to its low Km for H_2 [68]. By mutation breeding, several mutants of the cyanobacterium *Anabaena variabilis* were obtained, in which one or both hydrogenase activities were greatly reduced and which produced significantly higher amounts of H_2 than the wild type [69]. Happe et al. created a *hup*-deletion mutant of *Anabaena variabilis* which produced H_2 at four to six times the rate of the wild type, and one of these mutants (PK84) evolved 167.6 μmol H_2/h·(mg chl), which was demonstrated to have the possibility of outdoor hydrogen production [70]. Recently, Masukawa et al. applied molecular biology tools to construct three genetically defined hydrogenase mutants from *Anabaena* sp.: hupL$^-$(deficient in uptake hydrognase), hox H$^-$(deficient in the bidirectional hydrogenase), and hupL$^-$/hoxH$^-$(deficient in both genes) [71]. The results showed that the hupL$^-$ mutant produced H_2 at a rate four to seven times that of the wild type under optimal conditions. The hoxH$^-$ mutant produced significantly lower amounts of H_2 and had slightly lower nitrogenase activity than the wild type. H_2 production by the hupL$^-$/hoxH$^-$ mutant was slightly lower than but almost equal to that of the hupL$^-$mutant. These results show that mutants deficient in hydrogen uptake are favourable and need to be used for effective photobiological hydrogen production.

In photosynthetic bacteria, the amount of H_2 that evolves anaerobically from organic substrates is determined by the interaction of several metabolic pathways: H_2 evolution, mediated by the enzyme nitrogenase; H_2 uptake (recycling), by a membrane-bound uptake hydrogenase that reduces the net amount of gas evolved; and biosynthesis of alternative electron sinks for reductants, in particular, poly-3-hydroxybutyrate (PHB) in the form of cytoplasmic granules. In batch cultures with synthetic growth media, uptake hydrogenase-negative mutants of *Rhodobacter capsulatus* and *Rhodospirillum rubrum* showed increased H_2 photo-production [50, 72–73]. Similarly, the biosynthesis of storage energy reserves, specifically PHB, reduced nitrogenase-mediated H_2 evolution by photosynthetic bacteria. Franchi et al. constructed three differently metabolically engineered strains, single PHA$^-$ and Hup$^-$ mutants and one double PHA$^-$/Hup$^-$ mutant, of the purple nonsulfur photosynthetic bacterium *Phodobacter sphaeroides* RV [74]. With the lactic-acid-based synthetic medium, the single Hup$^-$ and double PHA$^-$/Hup$^-$ mutants exhibited increased rates of H_2 photoproduction about one third higher than that of the wild-type strain. The PHA$^-$ mutant did not obviously increase the rate of H_2 evolution because the amount of the produced PHB is very low in this synthetic medium. With the food-waste-derived growth medium, only the single Hup$^-$ mutant showed higher rates of H_2 production, but all the mutants sustained a longer-term H_2 photo-production phase than the wild-type strain, with the double mutant exhibiting overall the largest amount of H_2 evolved. In another study, some similar results were obtained using the same genetic improvement strategy in the recently isolated strain *Rhodobacter Sphaeroids* KD 131 [75]. It was shown that the rate of hydrogen production in the wide-type strain was improved from 1.62 ml H_2/ ml broth to 2.2 ml H_2/ ml broth by using a double-deficient (Phb$^-$/Hup$^-$) strain in 48 h of culturing, and that the amount of hydrogen produced was, in the increasing order, the wild type strain of *Rb Sphaeroids* KD 131, Phb$^-$, Hup$^-$, Phb$^-$/Hup$^-$ mutants. These studies demonstrated the feasibility of single and multiple gene engineering of microorganisms to redirect their metabolisms for improving H_2 photoproduction using actual waste-derived substrates.

Very few reports are available on the recombinant expression of hydrogenase. Actually, direct evolution of hydrogenase molecule against O_2 sensitivity by DNA shuffling and its recombinant expression in suitable hydrogen-evolving microorganisms should be attractive for enhanced H_2 evolution and simpler bioprocess development. Because the potential and the challenges of applying metabolic engineering tools to improve biohydrogen production have been demonstrated, further research advances, from one gene to multiple genes and multi pathway engineering, from synthetic media to the media derived from raw materials, and practical bioreactor bioprocesses can be expected.

5.4. Immobilized cultures

Immobilized cells are widely used for both practical and academic purposes. Immobilized culture technique has been developed for cell stabilization, biomass increase and easy operation. One of the key limitations for practical H_2 production by photobiological systems is that the rates of H_2 evolution are very low. Typically, all light-dependent biohydrogen systems (direct photosynthesis, indirect photosynthesis and photo-fermentation systems) have rates of H_2 synthesis well below 1 mmol/L·h [76]. However, the rates of H_2 syntheses are well above 1 mmol/L·h in the dark-fermentation systems. In general, rates of H_2 production by phototrophic bacteria are higher when the cells are immobilized in or on a solid matrix. Phototrophic bacteria were suitable to be entrapped into translucent gels like agar [77], carrageena [78], poly(vinyl alcohol) and alginate [79]. *Rhodobacter sphaeroides* O.U. 001 was immobilized in calcium alginate beads, and this was used for continuous hydrogen production. However, the increase of H_2 production rate was limited by substrate diffusion through gel matrix. Immobilization on a porous transparent matrix eliminates this problem. In addition, the transparency of a matrix allows light to be delivered at any point due to multiple refraction of light beams inside the matrix. Porous glass has the best transparency and provides a high surface-to-volume ratio for effective medium exchange. One disadvantage with this material is the difficulty of immobilizing bacteria because the negative charges on the glass surface decrease bacteria adsorption. A method was proposed to make the positive charge modification of the glass surface by the treatment of 3-(2-aminoethyl-aminopropyl)-trimethoxysilane [80]. Various species and strains of phototrophic microorganisms, including green microalge, cyanobacteria and anoxygenic photosynthetic bacteria, were shown to be able to bind to the activated glass surface. Over the freely suspended fermentation, the rate of H_2 synthesis was improved about ten-fold by immobilizing one *Rhodobacter sphaeroides* strain on porous activated glass, and the highest rate attained was 3.6–4.0 L/L·h [76].

Quick immobilization was suggested for cyanobacteria. It was possible to polymerize polyurethane foam together with microalgae or cyanobacteria without essential loss of activity. Immobilization of microorganisms on activated surfaces or polymerization of foam with microorganisms is quick but needs careful handling and expensive reagents. Autoimmobilization is a cheaper alternative approach. The time of autoimmobilization depends on the strain origin and matrix properties, varies from days to several weeks and starts from biofilm formation. After biofilm formation, the biomass growth depends mostly on the medium content and might be as quick as the growth rate of microorganisms. Application of immobilization technique and designing matrix with thin layer is a favourable strategy to reach high density culture for phototrophic bacteria. Thin layer allows keeping the optimal concentration of cells per unit of illuminated surface even with high density of culture.

In general, bacteria immobilized on thin matrixes provided highest volumetric rates of hydrogen photoproduction. In addition, immobilized cultures showed remarkably higher stability than batch cultures or resting cells, and stable H_2 production could be maintained up to 3600 h by the application of continuous medium flow. However, immobilization process is not always cost-effective for H_2 production, and both the technology and process economics require significant improvements.

5.5. Photobioreactors

Algae culture biotechnology has evolved recently into a commercially viable sector, with many companies utilizing both open culture systems and controlled closed photobioreactors. For the purpose of biological hydrogen production, it is essential to use enclosed photobioreactors in which monocultures can be maintained for an extended time period, preferably with sunlight as the energy source. The productivity of photobioreactors is light limited, and a high surface-to-volume ratio is a prerequisite for a photobioreactor. Light energy falling on the light-exposed surface, however, is not always used efficiently. Even under low-intensity sunlight, the photochemical efficiencies are low in most photosynthetic organisms, and tend to decrease under high-intensity sunlight. In addition to the truncated Chl antenna size of the photosystems, many engineering tools have been introduced to create an efficient biological process, including rapidly mixing the culture, diluting light and reasonably distributing light. Thus, it is important to meet the above requirements through rational photobioreactor design.

A number of photobioreactors have been developed. Three of the most noteworthy are pneumatically agitated vertical column reactors, tubular reactors, and flat panel reactors. Depending on the reactor type and the operation mode, cells are exposed to different light/dark cycles. When the cycles are in the range of micro or milli seconds, the photosynthetic efficiency (PE) increases and approaches that at low light intensities [81]. However, when they are from several seconds to tens of seconds, there is no improvement and even a decrease in PE has even been reported in comparison to the efficiency under continuous light. The depth of the photic zone depends on the dimensions and operations of the reactor, biomass concentration, and the specific absorption coefficient of the biomass. On the basis of model calculation and empirical data, flat panel reactors and tubular reactors show the highest efficiencies with rational light regimes in these reactors [82]. In addition, gas accumulation and shear stress should be considered carefully in these reactor designs to overcome their limitations to the productivity. Considering the highest efficiencies attained by flat panel reactors and tubular reactors, these two types of photobioreactors are worthwhile to be further discussed.

Flat panel reactors consist of a rectangular transparent box with a depth of only 1-5 cm. The height and width can be varied to some extent, but in practice only panels with a height and width both smaller than 1 m have been studied. The photobioreactor are mixed with air introduced via a perforated tube at the bottom of the reactor. In order to create a high degree of turbulence, 2.8–4.2 L of air per liter of reactor volume per minute has to be provided. Usually the panels are illuminated from one side by direct sunlight and the panels are placed vertically, or inclined versus the sun. Light /dark cycles are short in these reactors, and this is probably the key factor leading to the high PE. A disadvantage of these bioreactors is that the power consumption of aeration is high, although mixing is always necessary in any reactor. As shown in Fig.3, a flat panel airlift photobioreactor was designed for the cultivation of *Chlorella vulgaris* [83]. This new design uses flat panels to reduce light path and baffles to induce a regular light cycling of microalgae. The large-scale flat-plate reactor is a rectangular air-lift photobioreactor with a large number of light re-distributing plates fixed a few centimeters from each other. Many scaled-up versions of photobioreactors consist of repeating many of the smaller photobioreactor units, with its practical implications. Since the scaled-up reactor consists of only one unit, it is still practical to sterilize it and only one regulatory unit is needed.

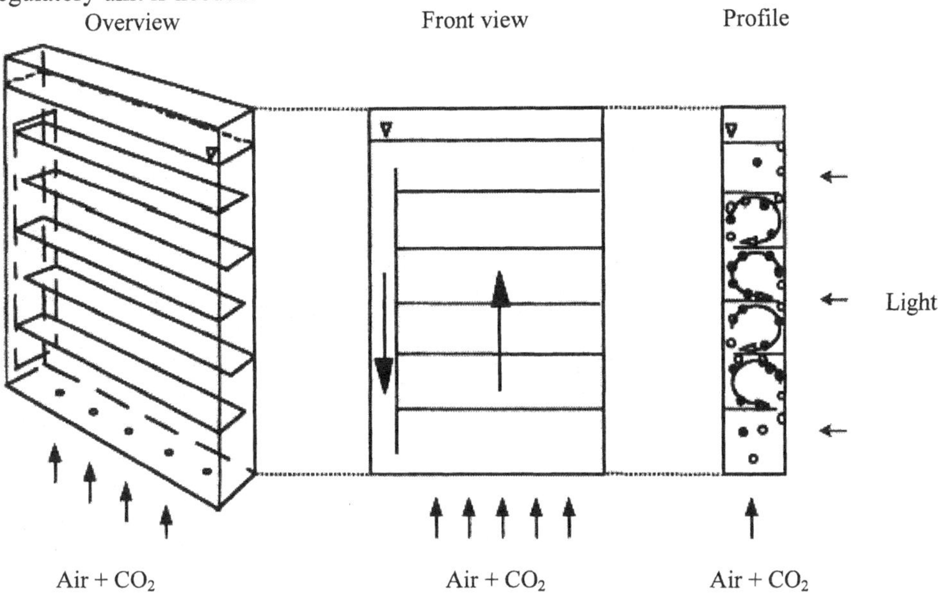

Fig.3. Flat panel airlift photobioreactor. (Adapted from [83])

Tubular photobioreactors consist of long transparent tubes with diameters ranging from 3 to 6 cm, and lengths ranging from 10 to 100 m. The culture liquid is pumped through these tubes by means of mechanical or air-lift pumps. The tubes can be positioned on many different ways: in a horizontal plane as straight tubes with a small or large number of U-bends; vertical, coiled as a cylinder or a cone; in a vertical plane, positioned in a fence-like structure using U-bends or connected by manifolds; horizontal or inclined, parallel tubes connected by manifolds; in addition, horizontal tubes can be placed on different reflective surfaces with a certain distance between the tubes. A 0.2-m^3 tubular airlift photobioreactor was designed for continuous outdoor culture of the microalge *P. tricornutum* (Fig. 4) [84]. This design method effectively combines the relevant aspects of external irradiance-dependent cell growth, oxygen accumulation in the solar loop, oxygen removal in the airlift device, and hydrodynamics of the airlift system that determine the flow velocity through the solar receiver.

Although tubular reactor design is very diverse, the predominant effect of the specific designs on the light regime is a difference in the photon flux density incident on the reactor surface. In most designs, the shape of light gradient and the cycling of dark/light are similar. The length of the tubes is limited because of accumulation of gas. The way to scale up is to connect a number of tubes via manifolds. One big photobioreactor, which consisted of 25,000 glass tubes with a total surface area of 12,000 m^2, was designed and used for the production of *Chlorella* sp.

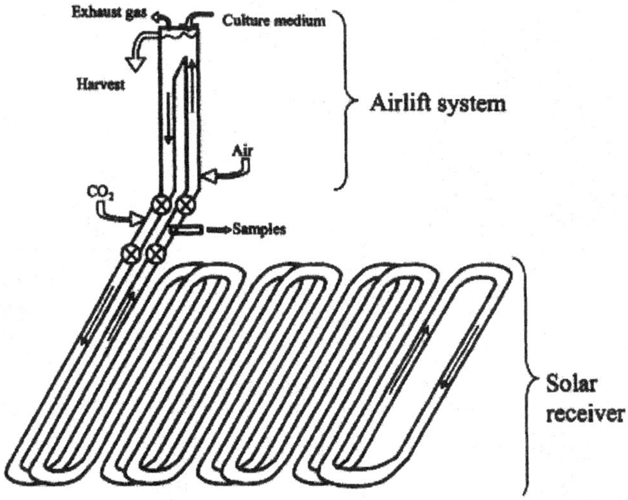

Fig. 4. Tubular airlift photobioreactor. (Adapted from [84])

A favorable design strategy for the photobioreactor is to separate light collection from biological cultivation [29]. Solar beam irradiation in 'clear sky' areas can be collected and concentrated into optical fibres with lenses or parabolic mirrors. Via the fibres, light can be guided into a large-scale photobioreactor. The design of a photobioreactor with a light redistributing system is a great challenge for process engineers. Various types of bioreactors (stirred-tank reactor, vertical bubble column) were integrated with a large number of glass fibers or a few solid transparent bars (glass or quartz). Recently, one more promising integrated system has been proposed [82]. As shown in Fig. 5, a large number of light redistributing plates are fixed a few centimeters from one another within a rectangular airlift photobioreactor. And these light redistributing plates can be connected to the optical fibers. The predicted problem is how to design light-redistributing plates with uniform radiation across the entire surface. In this system, mixing is provided by air injected between adjacent plates and the culture liquid rises in between. Only the space between the two center plates is not aerated, acting as a downcomer. In this system, the liquid culture volume as a whole is mixed, and this bioreactor is scalable. With the decrease of the production costs of lenses, mirrors, solar tracking devices and optical fibres, this new cultivation strategy is generally applicable.

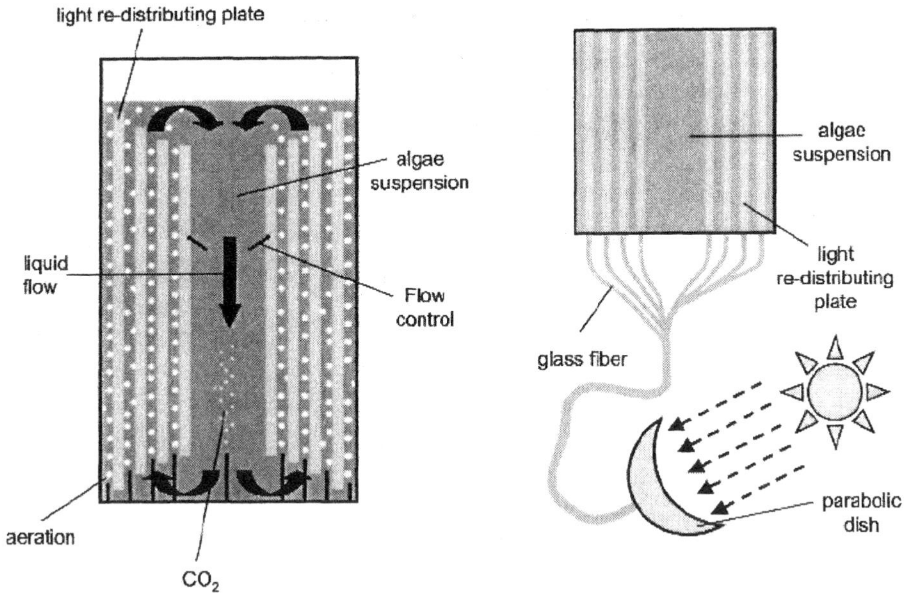

Fig. 5. A rectangular air-lift photobioreactor with light redistributing plates and external light collection. (A) Cross section vertical plane; (B) Cross section horizontal plane. (Adapted from [82])

6. IMPROVEMENTS OF FERMENTATIVE HYDROGEN PRODUCTION

6.1. Factors affecting hydrogen yields in dark fermentation

Dark hydrogen fermentation is a ubiquitous phenomenon. Actually, one type of indirect photolysis has one dark fermentation process that can produce hydrogen by utilizing carbohydrates, which are photosynthesized in the first stage. For the development of a practical H_2 production bioprocess, one of the main constraints of this fermentative process is its low hydrogen yield. Depending on the fermentation conditions and bacteria used in the process, acetic and butyric acid are the main anaerobic metabolites produced with hydrogen. Theoretically, 2 to 4 moles of hydrogen can be produced from each mole of glucose fermented with acetic and /or butyric acid as the co-products. The actual hydrogen yield is often lower than the theoretical yield, however. Therefore, it is important to identify the factors which affect hydrogen yields in dark-fermentation.

The relatively low yield of hydrogen during fermentation is a natural consequence of the fact that fermentation has been optimized by evolution to produce cell biomass and not hydrogen. Thus, a portion of the substrate is used to produce ATP and other metabolites, which can be used to maintain cell metabolism and increase biomass. Moreover, the actual yields of hydrogen are reduced in many microorganisms by the presence of one or more uptake hydrogenases, which consume a part of the hydrogen produced [27]. Also, different anaerobic strains have different H_2 yields. The yields of hydrogen among various microorganisms are listed in Table 2. In general, the hydrogen yield from glucose was ca. 2 mol H_2/mol glucose with *Clostridium* sp. and only ca. 1 mol H_2/mol glucose with *Enterobacter* sp.. *Clostridium* sp., *C. pasteurianum*, *C. butyricum*, and *C. beijerinkii* are strong hydrogen producers, while *C. propionicum* is a poor hydrogen producer.

Table 2
Hydrogen production from carbohydrates by various fermentative bacteria

Microorganism	Substrate	H_2 yield (mol/mol)	Productivity (L/L·h)	Reference
Escherichia coli	glucose	0.6	0.001	[85]
E. aerogenes E.82005	glucose	1.1	0.52	[52]
E. aerogenes AY-2	glucose	0.35	0.03	[86]
E. aerogenes	molasses	0.52, 1.58	0.08	[52]
E. aerogenes HO-39	glucose	0.8-1.0	0.12	[87]
E. aerogenes AY-2	glucose	1.17	0.047	[88]
E. cloacae III-BT08	sucrose	2.2	0.79	[89]
Clostridium sp.	sucrose	1.8	2.14	[90]
C. acetobutylicum	glucose	2.0	0.36	[91]
C. butyricum	glucose	1.8-2.0	0.74-1.52	[92]
C. beijerinckii AM21B	glucose	1.97-2.2	0.52-0.53	[93, 94]
Citrobacter sp. Y19	glucose	2.49	0.093	[95]
Clostridium sp. No.2	glucose	2.36	0.45	[96]
C. butyricum, E. aerogenes &	potato starch	2.4, 7.0	0.006, 0.024	[97]
Rhodobacter sp. M-19		4.5, 7.2	0.17	[98]
C. butyricum & *E. aerogenes*	starch	2.6	1.3	[99]
Rhodopseudomonas palustris P4	glucose	2.76	1.33	[100]
Thermotoga neapolitana	glucose	8.5	1.94×10^{-7}	[101]
C. thermolacticum	lactose	3.0	0.063	[102]
Mixed culture (thermphilic)	glucose	1.11	1.05	[103]
Mixed culture (*Clostridium* sp.)	wheat starch	1.3, 1.9	0.075	[104]
Mixed culture (*Clostridium* sp.)	glucose	1.43	0.2	[105]
Mixed culture	sucrose	3.03	7.3	[106]
Mixed culture (*Clostridium* sp.)	glucose	1.0	0.32	[107]

Even for the same strain, the H_2 yield is affected significantly by many physiological conditions, and metabolic pathway shifts determine the productivity of hydrogen. In addition to volatile fatty acids (VFAs), anaerobic fermentation also leads to the formation of alcohols. These reduced end-products, such as ethanol, butanol and lactate, contain additional H atoms that are not liberated as gas [10]. Therefore, alcohol production results in a lower hydrogen yield. In order to maximize the yield of hydrogen, bacterial metabolism must be directed away from alcohols and reduced acids towards VFAs [10]. On the other hand, the conversion of pyruvate to ethanol, butanediol, lactic acid, and butyric acids will be involved in the oxidation of NADH. It will decrease the yield of H_2 through the reduced oxidation of NADH. Some proton suicide techniques and allyl alcohol were employed to block the formation of alcoholic and acidic metabolites, resulting in high yields (3.8 mol H_2/mol glucose) in *E. cloacae* and *E. aerogenes* [108].

Some other fermentation conditions affect the H_2 yield via metabolic pathway shifts. *C. pasteurianum* is a classic VFA and hydrogen producer, but its metabolism of glucose can be shifted away from hydrogen production towards solvent production by maintaining high glucose concentration (12.5% w/v), by introducing CO (one inhibitor of hydrogenase), and by iron limitation [10, 53]. *Clostridia* sp. produces VFAs and hydrogen during the exponential growth phase, and rapid alcohol production occurs in the late growth phase. When mixed anaerobic bacterial cultures are used for waste water disposal and hydrogen production, the transition from the production of hydrogen and VFAs to alcohol production still exists. Hydraulic retention time (HRT) also has a pronounced effect on metabolic balance [109]. Actually, if we knew the actual metabolic pattern, it would be possible to drive the pathways toward enhanced hydrogen production by controlling environmental conditions, such as pH, HRT, nutrition, C/N ratio, organic loading rate, etc.

6.2. Integration with photosynthetic hydrogen production

Photosynthetic bacteria can use short-chain organic acids as electron donors for the production of hydrogen at the expense of light energy. These bacteria have several advantages over their fermentative counterparts, such as high theoretical conversion yield and the utilization of wide spectral light energy to decompose organic acids into hydrogen and CO_2. This positive free-energy reaction is impossible to be accomplished by anaerobic digestion. In addition, photosynthetic bacteria lack oxygen-evolving activity, which otherwise poses oxygen inactivation problems in different biological systems. The combination of photosynthetic and anaerobic bacteria can provide an integrated system for the maximization of hydrogen yield. Miyake et al. first reported that high-yield hydrogen production of 7 mol H_2/mol glucose was attained from glucose by immobilized cells of *C. butyricum* and *Rhodobacter spheroids* [110]. Yokoi et al. reported that a mixed culture of *C. butyricum* and *Rhodobacter* sp. M-19 produced H_2 from starch with a yield of 6.6 mol H_2/mol glucose in a fed-batch culture [99]. In 2001, they further reported that two-step repeated batch cultures by the above mixed culture produced a high yield of 7.0 mol H_2/mol glucose from the starch remaining in sweet potato starch residue [111]. Kim et al. combined dark fermentation with photofermentation to improve hydrogen productivity from food-processing wastewater and sewage sludge [75]. In a recent study, Lee et al. described hydrogen production using a two-phase fermentation system in which *Rhodopseudomonas palustris* produced hydrogen from effluents of dark fermentation [112]. In this study, an anaerobic sequencing batch reactor (ASBR), upflow anaerobic sludge blanket (UASB) and continuous stirred tank reactor (CSTR) were used for dark fermentation experiments. The effluents from these carbohydrate-fed reactors were then tested for the second-phase hydrogen production, and the results showed that, among these different effluents, CSTR effluent was the most suitable for photohydrogen production.

The high hydrogen yield from glucose was achieved by using the above integrated systems, but there are still some limiting factors that need to be overcome. The main products of anaerobic fermentation are acetic and butyric acids, but the components of the resulted effluent are complicated when practical wastewater or raw material is used. The conversion of these acids into hydrogen by photofermentation should be further investigated for the synergy of the integrated process. Moreover, the concentration of ammonia in the effluents from dark fermentation should be controlled at a low level; it otherwise inhibits the H_2 elution ability of nitrogenase in photosynthetic bacteria. One more disadvantage is that these integrated systems reduce the overall production rate of hydrogen as compared to single dark fermentation. Therefore, special attention should be paid to the compatibility between dark fermentation and photo-fermentation before using this novel H_2 production system.

6.3. Enhancing fermentative H_2 production by metabolic engineering

Metabolic engineering is a powerful tool to improve the genetics of microorganisms for the enhanced synthesis of the targeted metabolite via the redirection of metabolic fluxes. Fermentative H_2-producing bacteria can be metabolically engineered in several ways to enhance hydrogen productivity. Some of these include: (1) overexpression of cellulases, hemicellulases, and lignases that can maximize substrate availability; (2) elimination of uptake hydrogenase; (3) overexpression of H_2-evolving hydrogenases that have themselves been modified to be hydrogen tolerant; (4) elimination of metabolic pathways that compete for reducing equivalents required for H_2 synthesis.

Although many anaerobic H_2-evolving bacteria have a relatively strong ability to utilize a broader range of substrates, such as starch, xylose, fructose, glucose, etc., the development of practical H_2 production based on complex industrial wastewater needs *Clostridia* sp. to have stronger or special enzyme bioactivity for both H_2 elution and wastewater treatment. As shown before, mutants deficient in hydrogen uptake are favourable and have to be used for effective photobiological hydrogen production [20]. Thus, mutants deficient in uptake hydrogenase might also improve H_2 productivity. Anaerobic bacteria generally have the ability to produce hydrogen gas during catabolism of carbohydrates and [Fe]-hydrogenase (EC1.12.7.2) is known to release hydrogen gas from the reduced form of ferredoxin in *Clostridium* and *Desulfovibrio* species [113]. [Fe]-hydrogenase is highly sensitive towards oxygen and possesses 100-fold more activity than [NiFe]-hydrogenase [114]. *Clostridium paraputrificum* M-21 was isolated and characterized as a chitin-degrading hydrogen-producing anaerobe [115–116]. One recombinant *Clostridium paraputrificum* carrying multiple copies of hydA (encoding [NiFe]-hydrogenase gene) was constructed which showed a 1.7-fold increase in H_2 production as compared with the wild type [117]. It was found in this recombinant strain that overexpression of hydA abolished lactic acid production, and increased acetic acid production by over-oxidation of NADH, which is required for the reduction of pyruvic acid to lactic acid in the wild type. Another similar work was targeted at cloning [Fe]-hydrogenase from *Enterobacter cloacae* IIT-BT 08 because this facultative anaerobe showed a high hydrogen-production yield of 6 mol H_2/mol sucrose [118]. This hydA ORF gene was expressed in non-hydrogen producing *E. coli* BL-21, and its expressed protein showed *in vivo* and *in vitro* bioactivities. Hydrogen production by *E. coli* is mediated by the formate hydrogenlyase (FHL) system. *E. coli* strain HD701, which cannot synthesize the FHL repressor (HycA) and is, therefore, upregulated with respect to FHL expression, has been constructed [119]. Further studies showed that *E. coli* HD701 evolved ca. 2 times more hydrogen than *E. coli* MC4100 (host strain), and similar H_2 productivity was also achieved when industrial wastes with a high sugar content were applied [120].

It is also an important strategy for H_2-producing anaerobes to redirect metabolic fluxes or pathways for enhanced production or reduced consumption of reducing equivalents required for H_2 synthesis. *Clostridium tyrobutyricum* is a typical H_2-producing anaerobe with many beneficial characteristics. It can use various types of crude feedstock as the substrate to produce mixed organic acids, mainly butyric and acetic acids, and simultaneously evolve a significant amount of H_2. In a recent attempt to decrease acetate formation and improve butyrate production by *C. tyrobutyricum*, some metabolic shift phenomena appeared [121–122]. The fermentation pathways leading to butyrate and H_2 production are shown in Fig. 6.

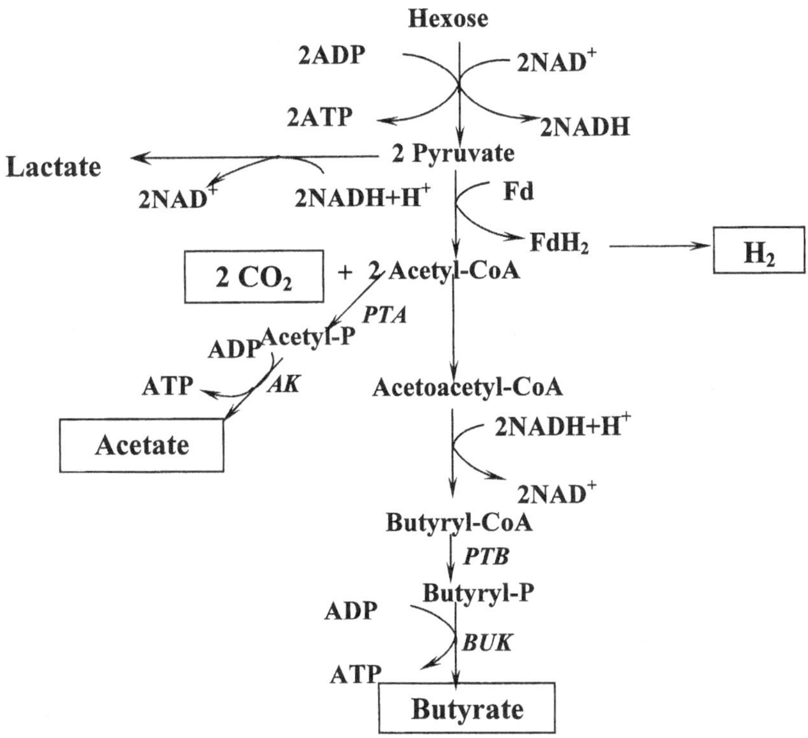

Fig. 6. Possible metabolic pathways for H_2 and butyrate production in *Clostridium tyrobutyricum*. (Adapted from [123])

Acetate kinase (AK) is a key enzyme for acetate formation, whereas butyrate kinase (BUK) and phosphotransbutyrylase (PTB) are two key enzymes for butyric acid biosynthesis. The *ack*-knock out mutant (PAK-Em) had twice the hydrogen yield than the wild type while gas production in the mutant (pTHBUT) overexpressing BUK and PTB was reduced by 67% [124–125]. Overexpressing BK and PTB would increase carbon flow through butyrate formation and also would reduce the amount of NADH available, thus reducing hydrogen production via NADH-ferredoxin oxidoreductase, which oxidizes NADH and reduces ferredoxin to FdH_2. As expected, inactivating *ack* increase butyrate productivity ca. 50% over the wild type, however, hydrogen yield was also improved from 1.35 mol H_2/mol glucose to 2.88 mol H_2/mol glucose rather than the expected decrease [123]. These results suggested that the metabolic pathways in *C. tyrobutyricum* can be manipulated by gene inactivation and /or gene overexpression to increase or decrease the NADH pool, which in turn affects hydrogen production. Moreover, the observed effects of H_2 production by inactivating *ack* suggested that both local and global effects of gene inactivation and expression should be systematically investigated using some novel genomic and proteomic tools in order to enhance H_2 production.

6.4. Hydrogen production by low-cost substrates with mixed bacteria

Research in dark fermentative hydrogen production has often focused on pure glucose or sucrose as the substrate and pure cultures of bacteria, such as the metabolically diverse spore-forming *Clostridia* and *Enterobacter* spp. In order to increase the economical potential of H_2 production, many efforts have been made to utilize low-cost substrates in fermentative H_2 bioprocesses in the past decade. Especially, it is desirable to produce H_2 continuously by utilizing waste materials containing high concentrations of organics, such as municipal solid waste, industrial wastewater, and agricultural wastes, because this type of bioprocesses may simultaneously provide economic and environmental benefits.

However, the production of hydrogen from waste materials creates new challenges because the waste materials are not sterile and it is too costly to sterilize them and maintain aseptic conditions. In addition, waste materials usually are composed of a variety of substrates that can be most efficiently utilized by mixed species of bacteria. Unfortunately, some of the bacteria present in microbial inocula or wastewater will consume hydrogen, lowering the overall efficiency of hydrogen production. This is the so-called hydrogen interspecies transfer phenomena. Particularly, methanogens can convert hydrogen to methane, a gas that has only 42% of the energy content of hydrogen (mass basis) and seriously increase the difficulty of H_2 separation. Strategies for controlling the growth of methanogens include maintaining a low pH in the bioreactor (in the range of 5.0–6.0) [126–128], using an inoculum that is heat-treated to kill non-spore-forming methanogens [129–130], and using short hydraulic retention times (HRT) [131]. For example, by keeping the hydraulic retention time below 1 day (a typical maximum growth rate of methanogens) in a completely mixed reactor, methanogenic bacteria can be excluded from a continuous flow reactor. Maintaining a low pH and using a heat-treated inoculum do not necessarily prohibit the existence of methanogens after long-term H_2 fermentation, and short retention times also reduce the efficiency of substrate utilization by the bacterial and thus the overall process efficiency. Recently, some molecular biological procedures were introduced to rapidly analyze microbial communities, such as RISA analysis (Ribosomal Intergenic Spacer Analysis) and denaturing gradient gel electrophoresis (DGGE) [107, 132], which will facilitate the detection and control of methanogens during fermentative H_2 processes. Besides the H_2-consuming methanogens, many non-H_2 producing bacteria, such as lactate-, ethanol-, and propionate-producing bacteria, coexist in the reactor [53]. Noike et al. reported that a substantial decrease or even a complete cessation of H_2 production by *Clostridium* at 35°C was observed when lactate-producing bacteria are added [133]. Recently, some thermophilic biohydrogen fermentation bioprocesses were developed because most lactate-forming bacteria are suppressed efficiently by high temperature (60 °C) [90, 134]. Furthermore, the H_2 production rate and yield increased to a great extent compared to those at mesophilic conditions. A higher temperature (70°C) is used to carry out fermentative H_2 production using *Thermotoga neapolitana* and *Caldicellulosiruptor saccarolyticus* [135–136]. *T. neapolitana* is tested to be a microaerophile and can produce a very high yield of 8.4 mol H_2/mol glucose. With many good characteristics, the order *Thermotogales* attracts more and more attention in fermentative H_2 production development.

It is known that the anaerobic digestion processes of wastewater treatment are well established on an industrial scale in many countries. It will greatly facilitate the development of biohydrogen dark-fermentation bioprocesses by using a variety of substrates (glucose, xylose, sucrose, soluble starch, starch, microcrystalline cellulose, etc.) and different-source wastewater containing organic matters [137–141]. In the past five years, increasing interests and great progress have been made in this area, such as biomass immobilization, bioreactor design, and microbial community control; however, long-term and stable H_2 production with practical wastewater treatment has not been satisfactorily achieved and needs to be studied further.

6.5. Gas sparging and bioreactor optimization

Hydrogen evolution pathways are sensitive to H_2 concentrations and are subject to the inhibition of end-products. As the hydrogen concentration increases, H_2 synthesis decreases and metabolic pathways shift towards the production of more reduced metabolites, such as lactate, ethanol, acetone, butanol, or alanine. Continuous H_2 synthesis requires a pH_2 of <50 kPa at 60°C, <20 kPa at 70°C, <2 kPa at 98°C [136]. Gas sparging has become a common method to reduce hydrogen partial pressure in the liquid phase for the increase in its yield. It was reported that the specific hydrogen production rate would be improved from 1.446 ml H_2/min·(g biomass) to 3.131 ml H_2/min·(g biomass) by nitrogen sparging [105]. Also, Ar and fuel cell exhaust gas are tested or suggested as effective sparging gases to lower the H_2 partial pressure. Membrane bioreactors can also be used to reduce biogas partial pressure in anaerobic fermentative H_2 production. A hollow fiber/silicone rubber membrane efficiently reduced biogas partial pressure in a dark fermentation system, resulting in a 10% improvement in the hydrogen production rate and a 15% increase in H_2 yield [142]. Pd-Ag membrane and synthetic polyvinyltrimethyl silane membrane are also employed to selectively remove H_2 [143]; however, the economics of these bioreactors and their operations are far from satisfactory. Stripping is also an efficient tool to reduce H_2 partial pressure to enhance H_2 production, and three different types of stripping (stripping by boiling, recirculating gas, and evaporation) are compared [144]. In order to improve the economics, the heat consumption of stripping may be compensated from fuel-cell hot exhaust gas. The H_2 production is also affected by the availability of CO_2 because anaerobes can synthesize succinate and formate with CO_2, pyruvate and NADH via the hexose monophosphate pathways. Thus, the efficient CO_2 removal from the culture medium prevents the consumption of NADH and increases the hydrogen yield. With many H_2 gas sparging operations, CO_2 can also be removed simultaneously. For economical success, gas sparging technology still needs to be improved further.

For the purpose of low-price H_2 production at a large scale, the optimization of bioreactor design has a pronounced effect on hydrogen yield and bioprocess development. In many cases, the conventional CSTR (continuous flow stirred tank reactor) process was used for hydrogen fermentation, but its hydrogen performance was considerably restricted by low dilution rates due to the low specific growth rates of hydrogen producers, and the washout of the biomass usually occurs at low HRTs. To solve this problem, the immobilized culture was introduced to enhance biomass retention. The results showed that, with physical or biological immobilization of cells, hydrogen production rates could be improved to a high range of 0.25–1.85 L/L·h, which was much higher than that of the CSTR [87, 145–146]. In one study, immobilized fixed-bed bioreactors obtained a high hydrogen production rate of 2.71 L/L·h [147]. It is well established that trickling biofilter reactors (TBR) are very suitable for treating high-strength wastewater. Van Groenestijin et al. first applied it as a high-rate bioreactor to produce hydrogen in the presence of hyperthermophilic bacteria, which form a biofilm on the surface of packing materials [148]. Oh et al. further studied the long-term performance of TBR to produce H_2 under various conditions, and a very high H_2 production rate of 23,52 ± 1,41 L/L·h was achieved and maintained for nearly one hundred days [149]. This favorable bioprocess was attributed to the high biomass density (18–24g VSS/L), a high percentage of H_2-evolving bacteria, and low gas hold-up. Nevertheless, the matrices used for cell immobilization inevitably occupy significant space in the reactor, limiting cell density and possibly creating mass transfer barriers to substrates and products. To avoid the problems caused by using immobilization matrices, granular sludge was generated to simultaneously enhance cell retention and biomass concentration [149–150]. Recently, a carrier-induced granular sludge bed bioreactor (CIGSB) was developed to produce H_2 from sucrose. After the optimization of operation conditions, this bioreactor achieved the optimal volumetric hydrogen production rate of 7.3 L/L·h and a maximum hydrogen yield of 3.03 mol H_2/mol sucrose when it was operated at a 0.5-h HRT [106]. In addition to the above bioreactors, three-phase fluidized-bed bioreactors and anaerobic sequencing batch reactors have also been used for H_2 production with improved H_2 productivity [151–152].

Recently, a fibrous-bed bioreactor (FBB) was developed for high-density immobilized-cell fermentation to produce hydrogen with high reactor productivity. The fibrous bed bioreactor, originally developed for multiphase fermentation at the Ohio State University, has been successfully used for several organic acid and solvent fermentations, including butyric acid [121–122, 125]. The FBB is a column vessel packed with spiral wound fibrous matrix with built-in flow channels between layers of fibrous matrices to allow fluid and particles flowing through the fibrous bed in the axial direction. The fibrous bed bioreactor is novel in its packing design and advantageous in its ability to immobilize a high density of producing cells (up to 100 g dcw per liter) while circumvents clogging and fouling problems commonly occurring to conventional immobilized cell bioreactors (packed bed and membrane bioreactors). Consequently, the fibrous bed bioreactor has stable long-term performance over its entire operation period even with a feed stream containing suspended solid (e.g., starch granules and corn fibers). This attribute is particularly important when "dirty" byproduct streams from corn milling plants are used as the feedstock for fermentation. Because of the high cell density in the FBB, contamination is not a problem after reactor start-up and even non-sterile cheese whey permeate can be used as a continuous feed to the reactor as has been shown in propionic acid and lactic acid fermentations [153]. Also, cells immobilized in the fibrous matrix are protected from trace oxygen and thus have stable hydrogen production. With the above advantages, the FBB was employed to produce H_2 and butyric acid using a metabolically engineering *C. tyrobutyricum*. The highest butyric acid concentration (81 g/L) was achieved in this bioreactor, whereas the rate of H_2 production reached ca. 0.8 L/L·h with a high yield of 2.36 mol H_2/mol glucose [123, 125]. In the view of process economics, the above H_2 production process may be desirable because butyric acid (ca. $0.75/lb) can be a main revenue-generating product, in addition to hydrogen as a clean fuel.

7. NEW CONCEPTS AND STRATEGIES FOR BIOHYDROGEN PRODUCTION

In view of the importance of biohydrogen energy and the lack of the availability of commercial biohydrogen production using currently developed technologies, some new concepts and strategies have been suggested to fundamentally solve the bottleneck of this bioprocess. Although some of these strategies are merely speculative or not proven yet by experiments, they are interesting and encouraging.

In the past decade, very few new organisms have been reported to produce H_2 (e.g. *Caldicellulosiruptor saccharolyticus, Gloeocapsa alpicola, Rubrivivax gelatinosus* and *Thermotoga elfii*), and there has not been much significant improvement in H_2 production [55, 154–155]. New methods to screen for hydrogenase or H_2-evolving organisms are imperative. A chemochromic screening method for agar plates has been developed to detect nanomolar quantities of hydrogen via a thin film sensor containing tungsten oxide and palladium, which changes color when exposed to H_2 [59]. Alternatively, a fluorescence technique for on-line monitoring has also been developed [156]. These high-throughput or enabling screening methods will speed up the discovery of new or mutagenesized hydrogenase or microbes from the natural or artificial mutant banks. Recently, more and more efforts have been made to create new hydrogenases or nitrogenases with good characteristics using directed molecular evolution via DNA shuffling, and many metabolic engineering strategies have been suggested to improve H_2 yield and productivity.

With the advent of genomics, more and more microorganisms have been fully sequenced and others have been partially sequenced by the rapidly developing genome projects, and genes are recognized and annotated from the sequence information using bioinformatics. The identification of potential hydrogen producers can be achieved by sequence analysis and pathway alignment of hydrogen metabolism in complete and incomplete genome data [157]. Genome data and systems biology tool can also be used to provide extensive information about an organism's physiology and metabolic fluxes. *Synechocystis* sp. PCC 6803 is a useful model cyanobacteria for future metabolic engineering exploitation to improve its hydrogen production, and this organism's genome has been sequenced. Recently, Reardon et al. apply cDNA microarrays and 2-D gel electrophoresis to achieve much information about the dynamics of *Synechocystis* sp. PCC 6803 during light-dark cycling [158]. These proteomic and transcriptomic assessments provide one global exploration of H_2 biosynthesis in this model organism. Therefore, experimental systems biology tool will become an important approach to improve the potential of H_2 production in the future.

Although in theory the amount of hydrogen that can be generated from renewable sources such as cellulose is vast, only 16-24% of the maximum stoichiometric yield of hydrogen from glucose (about 12 mol H_2 per mol glucose) is typically achieved by biological methods. The stoichiometric yield of H_2 from glucose was recently demonstrated by Wood et al. in a cell-free system using bacterial pentose phosphate pathway (PPP) enzymes [159]. But this yield is only attainable under near-equilibrium conditions, which implies very slow rates and or very low partial pressures of H_2, and this enzymatic bioprocess is cost-prohibitive. Some new efforts are trying to construct a biomolecular device to produce H_2 by combining with light-driven water-splitting as it occurs in the natural process of photosynthesis in plants [160]. Such a semi-artificial device should combine the best suited components found in various native systems, and arrange them on the surface of electrode materials to finally achieve light-derived hydrogen production. To avoid the recombination of oxygen with hydrogen and the inhibition of the hydrogenase by oxygen, separate reaction chambers should be connected by a salt bridge and a conducting wire. Advantages of such a system are: (1) it is modular, i.e. all components can be easily exchanged; (2) separated development and optimization of all components is possible; (3) the regeneration of the key components in native systems is possible; (4) the highest possible stability of all components can be realized. Some efficient procedures have been developed to isolate photosystem components, and the immobilization of hydrogenase on electrode surfaces has also been achieved. However, this semi-artificial H_2 system is primarily a concept that needs further investigation.

Another direction of research aiming to maximize hydrogen productivity involves the exploration of an accessible and rich source of electron and biochemical electron pump, together with an active hydrogenase. The evolution of hydrogen through NADH pathways is driven by the necessity of reoxidizing the residual NADH of metabolic reactions, as follows:

$$NADH + H^{+} \longrightarrow NAD^{+} + H_{2} \tag{10}$$

Therefore, the yield of hydrogen will be improved if metabolic fluxes can be adjusted and /or redirected to increase the amount of NADH available in cells. In anaerobic bacteria, 2 mol NADH are produced from glucose glycolysis; however, if pyruvate can be further metabolized by TCA cycle, 8 mol NADH and 2 mol $FADH_2$ will be obtained. Therefore, it is very intriguing to couple the TCA cycle and H_2 synthesis by NADH, and 10 mol H_2 per mol of glucose can be produced, theoretically [161]. This respiration-driven H_2 evolution, acting through a so-called "reverse electron flow", could replace photosynthetic reactions as the driving force for maximal H_2 production. However, this conceptual strategy meets many challenges; for example, TCA works only under aerobic conditions, but hydrogenase is an oxygen-sensitive enzyme. Moreover, given that NADH is reoxidized by oxygen under aerobic conditions, reoxidation of NADH by electron transport chain would be inhibited for hydrogen production. There have been some experiments in which H_2 production can be enhanced to some extent by respiration. Of course, the metabolism of H_2-evolving microorganisms would need to be fundamentally re-engineered to couple a respiratory process (TCA) to the low redox electron transport pathway required for support of hydrogenase-mediated H_2 production. In principle and perhaps in the future, this concept and strategy may offer the best approach to overcome the bottleneck of biological hydrogen production.

8. CONCLUSION

Biological hydrogen production is one of the most challenging areas in biotechnology, with respect to environmental and energy-source problems. In the past decade, hydrogen energy has progressed on all fronts, making in road into all areas of energy. Existing technologies offer high potential for the development of practical H_2 production bioprocesses. Further research and development aimed at increasing rates of synthesis and final yields of H_2 are essential. Bioprocess integration, optimization of bioreactor design, rapid removal and purification of hydrogen, and especially, directed evolution of hydrogenase and metabolic engineering of the H_2-evolving microorganism offer exciting prospects for biohydrogen systems, and some novel strategies will also be very encouraging and exciting in the future. The rapid advances of biological and engineering sciences will greatly facilitate the overcoming of existing bottlenecks as well as new challenges and create new opportunities for economical hydrogen production in the near future.

REFERENCES

[1] T.N. Veziroglu and F. Barbir, Int. J. Hydrogen Energy, 17 (1992) 391.
[2] J.O.M. Bockris, Int. J. Hydrogen Energy, 27 (2002) 731.
[3] T.N. Veziroglu, Int. J. Hydrogen Energy, 20 (1995) 1.
[4] A. Hansel and P. Lindblad, Appl. Environ. Microbiol., 50 (1998) 153.
[5] T.A. Veziroglu, S.A. Knez and D.S. Newsome (eds), Othemer encyclopedia of Chemical technology, Wiley, New York, 1996, P.884.
[6] R. Ramachandran, R.K. Menon. Int. J. Hydrogen Energy, 23 (1998) 593.
[7] S. Dunn, Int. J. Hydrogen Energy, 27 (2002) 235.
[8] D. Das and T.N. Veziroglu, Int. J. Hydrogen Energy, 26 (2001) 13
[9] J.R. Benemann, Int. J. Hydrogen Energy, 22 (1997) 979.
[10] D.B. Levin, L. Pitt and M. Love, Int. J. Hydrogen Energy, 29 (2004) 173.
[11] R. Nandi and S. Sengupta, Crit. Rev. Microbiol., 24 (1998) 61.
[12] H. Gaffron, Nature, 143 (1939) 204.
[13] H. Gaffron, J. Gen. Physiol., 26 (1942) 219.
[14] R. Wuenschiers, Handbook of Photochemistry Photobiology, 4 (2003) 353.
[15] R. Wuenschiers, Handbook of Photochemistry Photobiology, 4 (2003) 295
[16] Y. Miura, Process Biochem., 30 (1995) 1.
[17] J.R. Benemann, Nature Biotechnol., 14 (1996) 1101.
[18] E. Greenbaum, R.R.L. Guillard, W.G. Sunda, Photochem. Photobiol., 37 (1983) 649.
[19] L. Florin, A. Tsokoglou and T. Happe. J. Biol. Chem., 276 (2001) 6125.

[20] R. Mertens, A. Liese, Curr. Opin. Biotechnol. 15 (2004) 343
[21] J.W. Lee, E. Greenbaum, Appl. Biochem. Biotechnol., 105 (2003) 303.
[22] E. Greenbaum, Biophys J., 54 (1988) 365.
[23] B.A. Witton and M. Potts (eds.), The Ecology of Cyanobacteria, Kluwer, Dordrecht, 2000.
[24] P. Tamagnini, R. Axelsson, P. Lindberg, F. Oxelfelt, R. Wunschiers and P. Lindblad, Microbiol. Mol. Biol. Rev., 66 (2002) 1.
[25] P. Fay, Microbiol. Rev., 56 (1992) 340.
[26] J. R. Gallon, New Phytol., 122 (1992) 571.
[27] J.C. Weissman and J.R. Benemann, Appl. Environ. Microbiol., 33 (1977) 478.
[28] P.C. Hallenbeck and J.R. Benemann, Int. J. Hydrogen Energy, 27 (2002) 1185.
[29] I. Akkerman, M. Janssen, J. Rocha and R. H. Wijffels, Int. J. Hydrogen Energy, 27 (2002) 1195.
[30] R.L. Uffen, Proc. Natl. Acad. Sci. USA, 76 (1976) 3298.
[31] J.E. Champine and R.L. Uffen, J. Bacteriol., 169 (1987) 4784.
[32] R.L. Kerby, P.W. Ludden and G.P. Robert, J. Bacteriol., 177 (1995) 2241.
[33] L.T. Angenent, K. Karim, M.H. Al-Dahhan, B.A. Wrenn and R. Domiguez-Espinosa, Trends Biotechnol., 22 (2004) 477.R.
[34] P.A.M. Classen, J.B. van Lier, A.M. Lopez, E.W.J. van Niel, L., Sijtsma, A.J.M. Stams, S.S. Viries and R.A. Weusthuis, Appl. Microbiol. Biotechnol., 52 (1999) 741.
[35] K. Nath and D. Das, Appl. Microbiol. Biotecnol., 65 (2004) 520.
[36] T.N. Veziroglu and P.K. Takashashi (eds.), Proceedings 8th WHEC, Pergamon Press, New York, 1990.
[37] J. Schnackenberg, R. Schulz and H. Sanger, FFB lett., 327 (1993) 21.
[38] D.D. Wykoff, J.P. Davies, A. Melis and A.R. Grossman, Plant Physiol., 117 (1998) 129.
[39] S. Kosourov, A. Tsygankov, M. Seibert and M.L. Ghirardi, Biotechnol. Bioeng., 78 (2002) 731.
[40] A. Melis, L. Zhang, M. Forestier, M.L. Ghirardi and M. Seibert, Plant Physiol., 122 (2000) 127.
[41] F.A.L. Pinto, O. Troshina and P. Lindblad, Int. J. Hydrogen Energy, 27 (2002) 1209.
[42] S. Kumazawa and A. Mitsui, Int. J. Hydrogen Energy, 6 (1981) 339.
[43] M. Banerjee, A. Kumar and H.D. Kumar, Int. J. Hydrogen Energy, 12 (1989) 871.
[44] E. Krahn, K. Schnerder and K. Muller, Appl. Microbiol. Biotechnol., 46 (1996) 285.
[45] C.H. Sasikala, C.V. Ramana and P.R. Rao, Int. J. Hydrogen Energy, 20 (1995) 123.
[46] S.P. Singh and S.C. Srivastava, Int. J. Hydrogen Energy, 16 (1991) 404.
[47] Y. Ohta, J. Frank and A. Mitsui, Int. J. Hydrogen Energy, 6 (1981) 451.
[48] I.N. Gogotov, N.A. Zorin and L.T. Serebriakova, Int. J. Hydrogen Energy, 16 (1991) 393.
[49] M.M.T. Khan and J. Bhatt, Int. J. Hydrogen Energy, 16 (1991) 683.
[50] N.A. Zorin, T. Lissolo, A. Colbeau and P.M. Vignais, J. Mar. Biotechnol., 4 (1996) 28.
[51] S. Tanisho, N. Wakao and Y. Kokako, J. Chem. Eng. Jpn., 16 (1983) 529.
[52] S. Tanisho, Y. Suzuki and N. Wakoo, Int. J. Hydrogen Energy, 12 (1987) 623.
[53] F.R. Hawkes, R. Dinsdale, D.L.Hawkes and I. Hussy, Int. J. Hydrogen Energy, 27 (2002) 1339.
[54] N. Kumar and D. Das, Process Biochem., 35 (1999) 589.
[55] E.W.J. van Niel, M.A.W. Budde, G.G. de Hass, F.J. Wal, P.A.M. Claassen and A.J.M. Stams, Int. J. Hydrogen Energy, 27 (2002) 1391.
[56] A.J. Pierik, W. Roseboom, R.P. Happe, K.A. Bagley and S.P.J. Albracht, J. Bio. Chem., 274 (1999) 3331.
[57] M. Rousset, Y. Montet, B. Guigliarelli, N. Forget, M. Asso and P. Bertrand, Proc. Natl. Acad. Sci. USA, 95 (1998) 1625.
[58] A. Rosenkranz and A.J. Krasna, Biotechnol. Bioeng., 26 (1984) 1334.
[59] M.L. Ghirardi, L. Zhang, J.W. Lee, T. Flynn, M. Seibert, E. Greenbaum and A. Melis, Trends Biotechnol., 18 (2000) 506.
[60] D.D. Wykoff, J.P. Davies, A. Melis and A.R. Grossman, Plant Physiol., 117 (1998) 129.
[61] M.L. Ghirardi, R.K. Togasaki and M. Seibert, Appl. Biochem. Biotechnol., 63 (1997) 141.
[62] A. Richmond, J. Appl. Phycol., 8 (1996) 381.
[63] Y. Nakajima and R. Ueda, J. Appl. Phycol., 11 (1999) 195.

[64] Y. Nakajima, M. Tsuzuki and R. Ueda, J. Appl. Phycol., 13 (2001) 95.
[65] J.E.W. Polle, S. Kanakagiri, E. Jin, T. Masuda and A. Melis, Int. J. Hydrogen Energy, 27 (2002) 1257.
[66] J.E.W. Polle, J.R. Benemann, A. Tanaka and A. Melis, Planta, 21 (2000) 335.
[67] P. Tamagnini, J.L. Costa, L. Almeida, M.J. Oliveira, R. Salema and P. Lindblad, Curr. Microbiol., 40 (2000) 356.
[68] O. Schmitz and H. Bothe, FEMS Microbiol. Lett., 135 (1996) 97.
[69] L.E. Mikheeva, O. Schmitz, S.V. Shestakov and H. Bothe, Z Naturforsch, 50c (1995) 505.
[70] T. Happe, K. Schutz and H. Bohme, J. Bacteriol., 182 (2000) 1624.
[71] H. Masukawa, M. Mochimaru and H. Sakurai, Appl. Microbiol. Biotechnol., 58 (2002) 618.
[72] A. Jahn, B. Keuntje, M. Dorffler, W. Klipp and J. Oelze, Appl. Microbiol. Biotechnol., 40 (1994) 687.
[73] M. Kern, W. Klipp and J.H. Klemme, Appl. Envirn. Microbiol., 60 (1994) 1768.
[74] E. Frachi, C. Tosi, G. Scolla, G. D. Penna, F. Rodriguez and P.M. Pedroni, Marine Biotechnol., 6 (2004) 552.
[75] J. Miyake, Y. Igarashi and M. Rogner (eds), Biohydrogen III, Elsevier, Kidlington, 2004.
[76] D.B. Levin, L. Pitt and M. Love, Int. J. Hydrogen Energy, 29 (2004) 173.
[77] M. Vincenzini, W. Balloni, D. Mannelly and G. Florenzano, Experientia, 37 (1981) 710
[78] N. Francou and P. Vignais, Biotechnol. Lett., 6 (1984) 639.
[79] P.C. Hallenbeck, Enzyme Microb. Technol., 5 (1983) 171.
[80] A.A. Tsgankov, Y. Hirata, Y. Asada and J. Myake, Biotechnol. Tech., 7 (1993) 283.
[81] M. Janssen, P. Slenders, J. Tramper, L.R. Mur and R.H. Wijffels, Enzyme Microb. Tech., 29 (2001) 298.
[82] M. Janssen, J. Tramper, L.R. Mur and R.H. Wijffels, Biotechnol. Bioeng, 81 (2003) 193.

[83] D. Jorg, U. Andrea, R. Axel, S. S. Ulrike, T. Walter, J. Biotechnol., 92 (2001) 89.
[84] E. Molina, J. Fernandez, F.G. Acien, Y. Chisti, J. Biotechnol. 92 (2001) 113
[85] R. Nandi, S. Dey and S. Sengupta, Biotechnol. Bioeng., 75 (2001) 492.
[86] Y. Nakashimada, M.A. Rachman, T. Kakizono and N. Nishio, Int. J. Hydrogen Energy, 27 (2002) 1399.
[87] H. Yokoi, Y. Maeda, J. Hirose, S. Hayashi and Y. Takasaki, J. Ferment. Bioeng., 80 (1995) 571.
[88] M.A. Rachman, Y. Furutani, Y. Nakashimada, T. Kakizono and N. Nishio, J. Ferment. Bioeng., 83 (1997) 358.
[89] N. Kumar and D. Das, Process Biochem., 35 (2000) 589.
[90] S.K. Khanal, W.H. Chen, L. Li and S. Sung, Int. J. Hydrogen Energy, 29 (2004) 1123.
[91] H.L. Chin, Z. Chen and C.P. Chen, Biotechnol. Prog., 19 (2003) 383.
[92] M. Heyndrick, P. D. Vos, B. Thibau, P. Stevens and J. D. Ley, Sys. Appl. Microbiol., 9 (1987) 163.
[93] F. Taguchi, J.D. Chang, N. Mizukami, T. Saito, K. Hasegawa and M. Morimoto, Can. J. Microbiol., 39(1993) 726.
[94] F. Taguchi, N. Mizukami, K. Hasegawa and T. Saito, Can. J. Microbiol., 40(1994) 228.
[95] Y.K. Oh, E.H. Seol, E.Y. Lee and S. Park, Int. J. Hydrogen Energy, 28 (2003) 1353.
[96] F. Taguchi, N. Mizukami, T. Saito and K. Hasegawa, Can. J. Microbiol., 41(1995) 536.
[97] H. Yokoi, A. Saitsu, H. Uchida, J. Hirose, S. Hayashi and Y. Takasaki, J. Biosci. Bioeng., 91 (2001) 58.
[98] H. Yokoi, R. Maki, J. Hirose and S. Hayashi, Biomass Bioenergy, 22 (2002) 389.
[99] H. Yokoi, T. Tokushige, J. Hirose, S. Hayashi and Y. Takasaki, Biotechnol. Lett., 20 (1998) 143.
[100] Y.K. Oh, E.H. Seol, E.Y. Lee and S. Park, Int. J. Hydrogen Energy, 27 (2002) 1373.
[101] A. Suellen, V. Ooteghem, S.K. Beer and P.C. Yue, Appl. Biochem. Biotechnol., 98 (2002) 177.
[102] C.N. Collect, N. Adler, J.P. Schwitzguebel and P. Peringer, Int. J. Hydrogen Energy, 29 (2004)1479.

[103] Y.K. Oh, S.H. Kim, M.S. Kim and S. Park, Biotechnol. Bioeng., 88 (2004) 690

[104] I. Hussy, F.R. Hawkees, R. Dinsdale and D.L. Hawkes, Biotechnol. Bioeng., 84 (2003) 619.

[105] O. Mizuno, R. Dinsdale, F.R. Hawkes, D.L. Hawkes and T. Noike, Bioresource Technol., 73 (2000) 59.

[106] K. S. Lee, J.F. Wu, Y.S. Lo, Y.C. Lo, P.J. Lin and J.S. Chang, Biotechnol. Bioeng., 87 (2004) 648.

[107] S.E. Oh, P. Lyer, M. A. Bruns and B. Logan, Biotechnol. Bioeng., 87 (2004) 119.

[108] N. Kumar and D. Das, Enzyme Micro Tech, 29 (2001) 280.

[109] J.J. Lay, Biotechnol. Bioeng., 68 (2000) 269.

[110] J. Miyake, X.Y. Mao and S. Kawamura, J. Ferment. Technol., 62 (1984) 531

[111] H. Haruhiko, A. Saitsu, H. Uchida, J. Hirose, S. Hayashi and Y. Takasaki, J. Biosci. Bioeng., 91 (2001) 58.

[112] C.M. Lee, P.C. Chen, C.C. Wang, Y.C. Tung, Int. J. Hydrogen Energy, 27 (2002) 1308.

[113] J. Meyer, and J. Gagnon, Biochemistry, 30 (1991) 9697.

[114] M.W.W. Adams, Biocheim. Biophs. Acta., 1020 (1990) 115.

[115] D. Evvyernie, S. Yamazaki, K. Morimoto, S. Karita, T. Kimura, S. Sakka and K. Ohmiya, J. Biosci. Bioeng., 89 (2000) 89.

[116] D. Evvyernie, K. Morimoto, S. Karita, T. Kimura, K. Sakka and K. Ohmiya, J. Biosci. Bioeng., 91 (2001) 339.

[117] K. Morimoto, T. Kimura, K. Sakka and K. Ohmiya, FEMS Microbiol. Lett., 246 (2005) 229.

[118] J. Mishra, S. Khurana, N. Kumar, A.K. Ghosh and D. Das, Biochem. Biophys. Res. Commun., 324 (2004) 679.

[119] M. Sauter, R. Bohm, A. Bock, Mol. Microbiol., 6 (1992) 1523.

[120] D.W. Penfold, C.F. Forster, L.E., Enzyme Micro Tech, 33 (2003)185.

[121] Y. Zhu and S. T. Yang, Biotechnol Prog., 19 (2003) 365

[122] Y. Zhu, Z. Wu and S.T. Yang, Process Biochem., 38 (2002) 657.

[123] X. Liu, Production of butyric acid and hydrogen by metabolically engineered mutants of *Clostridium tyrobutyricium*, Ph.D. dissertation, Ohio State University, Columbus, Ohio.

[124] Y. Zhu, Metabolic engineering of *Clostridium tyrobutyricum* for enhanced butyric acid fermentation, Ph.D. dissertation, Ohio State University, Columbus, Ohio.

[125] Y. Zhu, X. Liu and S.T. Yang, Biotechnol. Bioeng., 90 (2004) 154.

[126] C.Y. Lin and R.C. Chang, J. Chem. Technol. Biotechnol., 74 (1999) 498.

[127] H.H.Fang and H. Liu, Bioresour. Technol., 82 (2002) 87.

[128] S.E. Oh, S.V. Ginkel and B.E. Logan, Environ. Sci. Technol., 37 (2003) 5186

[129] R. Robinson, C. Batt and P. Patel (eds), Encyclopedia of food microbiology, Academic Press, New York, 1999.

[130] M. Nakamura, H. Kanbe and J. Matsumoto, Water Sci. Technol., 28 (1993) 81.

[131] J.J. Zhang, X.Y. Li, S.E. Oh and B.E. Logan, Biotechnol. Bioeng., 88 (2004) 854.

[132] D.F. Xing, N.Q. Ren, M.L. Gong, J.Z. Li and Q.B. Li, Sci. China Ser. C-Life Sci, 48 (2005) 155.

[133] T. Noike, H. Takabatake, O., Mizuno and M. Ohba, Int. J. Hydrogen Energy, 27 (2002) 1367.

[134] H.Q. Yu, H.H. Fang and G.W. Gu, Process Biochem., 38 (2002) 447-454.

[135] S.A.V. Ooteghem, S.K. Beer and P.C. Yue, Appl. Biochem. Biotechnol., 98 (2002) 177.

[136] E.W.J. vas Neil, P.A.M. Claassen and A.J.M. Stams., Biotechnol. Bioeng., 81 (2003) 255.

[137] G.Z. Liu and J.Q. Shen JQ, J. Biosci. Bioeng., 98 (2004) 251.

[138] J.J. Lay, Biotechnol. Bioeng., 74 (2001) 280.

[139] J.H. Wu and C.Y. Lin, Water Sci. Technol., 49 (2004) 223.

[140] C.C. Wang and C.W. Chang, C.P. Chu, D.J. Lee, B.V. Chang and C.S. Liao, J. Envirn. Sci. Health, 38 (2003) 1867.

[141] J.J. Lay, K.S. Fan, J.I. Hwang, J.I. Chang and P.C. Hsu, J. Envirn. Eng., 131 (2005) 595.

[142] T.M. Liang, S.S. Cheng and K.L. Wu, Int. J. Hydrogen Energy, 27 (2002) 1157.

[143] A.T. Nielsen, H. Amandusson, R. Bjorklund, H. Dannetun, J. Ejlertsson, L.G. Ekedahl , I. Lundstrom and B. Svensson B, Int. J. Hydrogen Energy 26 (2001) 547.
[144] J.W. van Groenestjn, J.H.O. Hazewinkel, M. Nienoord and P.J.T. Bussmann, Int. J. Hydrogen Energy, 27 (2002) 1141.
[145] S.Y. Wu, C.N. Lin, J.S. Chang, K.S. Lee and P.J. Lin, Biotechnol Prog., 18 (2002) 921.
[146] N. Kumar and D. Das, Enzyme Micro Tech 29 (2001) 280.
[147] J.S. Chang, K.S. Lee and P.J. Lin, Int. J. Hydrogen Energy 27(2002) 1167.
[148] J.W. van Groenestijn, J.H.O. Hazewinkel, M. Nienoord and P.J.T. Bussmann, Int. J. Hydrogen Energy, 27 (2002) 1141.
[149] H.H.P. Fang, H. Liu and T. Zhang, Biotechnol. Bioeng., 78 (2002) 44.
[150] M.A. Rachman, Y. Nkashimada, T. Kakizono and N. Nishio, Appl Microbiol Biotechnol 49 (1998) 450.
[151] S.Y. Wu, C.N. Lin and J.S. Chang, Biotechnol. Prog., 19 (2003) 828.
[152] C.Y. Lin and C.H. Chou CH, Eng. Life Sci., 4 (2004) 66.
[153] E.M. Silva and S.T. Yang, J. Biotechnol., 41 (1995) 59.
[154] T. Olga, S. Larissa, S. Marina and L. Peter, Int. J. Hydrogen Energy, 27 (2002) 1283.
[155] P.C. Maness and P.F. Weaver, Int. J. Hydrogen Energy, 27 (2002) 1407.
[156] M. Seibert, V. Makarova, A.A. Tsygankov and A.B. Rubin AB, US Patent No. 125762003088736 (2003).
[157] V.C. Kalia, S. Lal, R. Ghai, M. Mandal and A. Chauhan, Trends in Biotechnl., 21 (2003) 153.
[158] K.F. Reardon, C.S. Gan, A.M. Burja and P.C. Wright, Biochemical Engineering XIV: Frontiers and Adances in Biotechnology, Biological and Biomolecular Enginering, Harrison Hot Springs, BC, Canda, p.11 (2005).
[159] J. Woodward, M. Orr, K. Cordray and E. Greenbaum, Nature, 405 (2000) 1014.
[160] T.M. Bricker, J. Morvant, N. Masri, H.M. Sutton and L.K. Frankel, Biochim. Biophys. Acta, 1409 (1998) 50
[161] Z.Q. Mao and T.N. Veziroglu (eds), Hydrogen Energy Progress XIII (vol 1), Beijing, China, P370-374 (2000).

Bioprocessing for Value-Added Products from Renewable Resources
Shang-Tian Yang (Editor)

Chapter 22. Bioconversion of Whey Lactose into Microbial Exopolysaccharides

Y. Martin Lo, Sanem Argin-Soysal, and Chia-Hua Hsu

Food Bioprocess Engineering Laboratory, Department of Nutrition and Food Science, University of Maryland, 3102 Marie Mount Hall, College Park, Maryland 20742, USA

1. INTRODUCTION

In the last few decades, there has been a growing consensus in the scientific community as well as the dairy and food processing industries to convert dairy coproducts or byproducts into value-added applications. Central to such efforts is the applicability of the end product, which is dependent on its functionalities, whereas the effectiveness of the conversion process is extremely critical from the industry standpoint. To date, products such as whey protein have been successfully recovered from cheese whey and received many commercial applications. However, due in part to its chemical nature, the bioavailability of whey lactose remains an issue, leaving a considerable discrepancy in applying whey lactose as a key carbon source for bioprocessing.

Microbial exopolysaccharides (EPS) such as xanthan gum and alginate continue to receive increasing applications in response to the demands for desirable product texture and enhanced processibility of nutritional and healthful food products. These polymers require simple sugars as building blocks and are produced via natural fermentation processes using specific microorganisms. The inherent specificity of the enzymes involved and the controllability of a bioprocess make bioconversion into EPS an ideal candidate for the effective utilization of whey lactose. It is the objective of this chapter to provide an overview on the bioconversion of whey lactose into microbial EPS, with an emphasis on the challenges facing the fermentation process as well as the criteria to be considered to develop it into an effective and valuable bioconversion process.

2. WHEY LACTOSE

2.1. Source of origin, quantity, and properties

Cheese whey is the greenish-yellow liquid produced following the precipitation and removal of milk casein during cheese making [1, 2]. Commonly considered a waste product that causes serious pollution problems [3], this largest byproduct of the dairy industry represents about 85–95% of the milk volume and retains 55% of milk nutrients, including lactose (4.5–5.0% w/v), soluble proteins (0.6–0.8% w/v), lipids (0.4–0.5% w/v), and mineral salts (8–10% of dried extract). The biochemical oxygen demand (BOD) of whey varies from 30,000 to 50,000 mg O_2 per liter wastewater (ppm) largely depending upon the source of milk and the variety of cheese being made [4]. Depending on how the casein is coagulated, i.e., by enzymatic rennet coagulation or by acid coagulation, cheese whey can be categorized as sweet whey (rennet whey), which is a milk serum containing no calcium, and acid whey, which contains calcium lactate. In addition to differing in composition, sweet whey cannot be converted into acid whey even with post-acidification [2]. Another type of whey is industrial-grade whey, which is obtained when protein coagulation is done with acids other than lactic acid, e.g., hydrochloric acid, sulfuric acid or acetic acid. The composition of fresh whey is shown in Table 1. The total protein in fresh whey includes whey protein, which is defined as the non-casein protein in milk, and the residual proteins from cheese making, such as rennet, the enzyme used for curd formation, and mild lipase, a natural flavoring agent extracted from dairy animals. Citric acid is used to cause the curds (milk solids) to separate from the whey (liquid), and a minute amount can be found in fresh whey.

Table 1
Composition of fresh whey

		Sweet whey	Acid whey
Water	%	93–94	94–95
Dry matter	%	6–6.5	5–6
Lactose	%	4.5–5	3.8–4.3
Lactic acid	%	traces	up to 0.8
Total protein	%	0.8–1.0	0.8–1.0
Whey protein	%	0.6–0.65	0.6–0.65
Citric acid	%	0.1	0.1
Minerals	%	0.5–0.7	0.5–0.7
pH		6.4–6.2	5.0–4.6
°SH*		about 4	20–25

*Titratable acidity (°SH) of a solution is an approximation of the solution's total acidity. It is measured by reacting the acids present with a base such as sodium hydroxide (NaOH) to a chosen end point, close to neutrality, as indicated by an acid sensitive color indicator.

Annual global whey production is estimated at 118 million tons, of which 66% is manufactured in Europe, 25% in North America, and 9% elsewhere [2]. Whey proteins are almost universally recovered via ultrafiltration and sold as concentrates [5]. The permeate containing the lactose accounts for most of the BOD and dissolved salts and hence still constitutes a formidable disposal problem. Therefore, the separation of lactose has been the focus of many studies aimed at reducing the BOD of whey. Many attempts have been made to find various applications for the recovered lactose in the food and pharmaceutical industries due to its multiple functional properties [6]. The isolation of lactose from whey involves several critical steps. Cheese whey is often first concentrated by the dairy plants using various processes such as evaporation, ultrafiltration, nanofiltration, and reverse osmosis to reduce the cost of storing, handling, and transporting large volumes of bulk liquid [3, 4]. Lactose in the concentrated whey is then separated by crystallization. The processing time required for lactose crystallization varies from 12 to 72 h [7, 8]. The quality of the separated lactose is not of pharmaceutical or edible grade, mainly due to the presence of proteins, suspended solids, and minerals [4]. Removing proteins and suspended solids from whey by ultrafiltration and demineralization by electrodialysis are generally recommended [9, 10].

Milk disaccharide lactose (4-O-β-D-galactopyranosyl-D-glucose) can be purified from cheese whey or permeate (50–60% solids) by crystallization [1]. Depending on the raw material, there are two basic recovery methods: crystallization of the lactose in untreated whey and crystallization of the lactose in whey permeate (50–60% solids) in which the protein has been removed by ultrafiltration or other methods. In the manufacturing of non-hygroscopic dry whey products [11], the lactose in condensed whey (a supersaturated solution) is cooled under specific conditions in order to crystallize it prior to drying. The majority of the lactose crystallizes in monohydrate form. If lactose is present in the amorphous state, the resulting powder is hygroscopic. The crystals are harvested and washed to remove the mother liquor and dried. Crude lactose obtained this way contains about 98% lactose. The powder obtained by inducing crystal formation in condensed whey is non-sticking and stable [12]. Edible and USP grades are produced from crude lactose by protein precipitation, decolorization with activated carbon, and subsequent demineralization. Lactose is further refined by recrystallization or by spray drying [13].

2.2. Current applications and limitations

Applications for whey are increasing constantly, in part because of improved separation techniques (ultracentrifugation, ultrafiltration, sterile filtration, reverse osmosis, electrophoresis, chromatography and others) and enzymatic hydrolysis. The industry has developed many uses for lactose as well as for whey itself. Currently, the major uses of whey and whey permeate are in manufacturing dried whey powder and refined lactose. These uses, however, are often aimed at keeping the surplus whey out of sewers rather than at producing a highly desirable product. For example, the manufactured lactose is mainly used in cattle feed [2]. The ultimate goal for the dairy industry should be to turn whey lactose into a profit-generating feedstock for high value-added products.

Lactose purified from cheese whey or permeate is used as a supplement in infant formulas and as an excipient for pharmaceutical products [1]. Its plasticity, light flavor and reduced sweetening power make it apt for use in pill tablets [14]. Although the production of lactose from whey has constantly increased, the amount of purified lactose produced worldwide would require the use of only 5% of the whey available [15, 16]. Hence, alternative uses are being sought. The reduction of lactose to lactitol (4-β-galactopyranosul-D-sorbitol), a non-digestible sweetener with a sweetening power slightly higher than that of lactose, creates an additive in low-calorie diet foods with a caloric value of 2 kcal/g. Its ester, lactitol-palmitate, which has an emulsifying effect, is also used in human nutrition [17]. Lactulose (4-O-β-D-galactopyranosyl-D-fructose), a highly valued disaccharide with worldwide markets in pharmacology, could be synthesized by isomerizing lactose in an alkaline solution. Its sweetness is 48–62% that of sucrose, and it is also used as a Bifidus factor in nutrition [18].

In addition, lactosylurea, a non-protein nitrogen source in ruminant feed containing ammonia below the toxic level [16], and galactose, which has been used to replace sorbitol, which is very expensive [19], can be produced by direct reaction and lactose hydrolysis, respectively [1]. Hydrolyzed lactose solutions possess greater sweetening power than lactose and have applications in both the confectionery and ice-cream industries, replacing saccharose or starch syrup. Sweetness can be increased by converting the glucose in lactose-hydrolyzed whey permeate to fructose with immobilized glucose isomerase [19].

The lactose in whey has also be used in yeast fermentation, in which lactose serves as a nutrient for the participating microorganisms. The metabolic activity of the microorganisms results in the production of CO_2, ethanol and single cell protein as biomass. Enriched whey can be used directly in cattle feeding, or the biomass can be separated and dried as a feed concentrate. A corresponding preparation of the single-cell protein can lead, within certain limitations, to application for human food. Ethanol and CO_2 are used in various other applications, and ethanol is manufactured from whey on a large industrial scale in some countries. After lactose is fermented, the BOD in the residual liquid is reduced to such a level that the liquid can be discharged directly into municipal wastewater systems without any further processing. However, direct fermentation by yeast does not make much technological and economic sense, as apart from *Kluyveromyces fragilis* only a few yeasts can use lactose as a substrate for fermentation [20]. The fermentation process cannot be optimized to achieve a theoretical yield of 0.5 g yeast cells per g of lactose, making the distillation process too expensive [21]. Moreover, the intensive aeration required for yeast growth is difficult to achieve.

The direct fermentation of lactose or of the glucose and galactose obtained from lactose hydrolysis has drawn much attention [22–28]. Intensive studies have also been conducted using recombinant DNA techniques to elucidate the expression of the genes that code for the β-galactosidase and lactose permease system of *Kluyveromyces lactis* in *Saccharomyces cerevisie* [29–31]. It has been shown that, in this way, *S. cerevisiae* could be developed directly on cheese whey, producing high yields of ethanol or other commercially useful fermentation products [32]. However, the recombinant yeasts elaborated up to this point in time are very slow growing and have reduced genetic stability, so yields are low even when these recombinant yeasts are used in specially designed bioreactors [33].

3. MICROBIAL EXOPOLYSACCHARIDES (EPS)

Microbial EPS, different in function from those of higher plants, are the polysaccharides secreted from bacteria to form a layer over the surface of the organism. To distinguish them from any polysaccharides that might be found within the cell, they are hence characterized as exo-polysaccharides [34]. In nature, their functions are thought to be mainly protective, either as a general physical barrier preventing dehydration and the entry of harmful substances, or as a way of binding and neutralizing bacteriophages. In appropriate environments, they may also prevent phagocytosis by other microorganisms or the cells of the immune system. The capsular EPS are often highly immunogenic, and may have evolved their unusual diversity as a way of avoiding antibody responses. Secreted EPS have a role in adhering and penetrating the host, and hence might be involved in pathogenicity. For example, *Pseudomonas aeruginosa*, commonly found in respiratory tract infections, produces alginate that contributes to blockage in the respiratory tract and leads to further infection, and the plant pathogen *Xanthomonas campestris* produces viscous xanthan gum to attach to its host cabbage.

EPS are produced by both Gram-positive and Gram-negative bacteria in varying molecular weights and compositions, which consequently result in different rheological properties. Some species, such as *Agrobacterium* and *Rhizobium*, can synthesize more than one EPS. In general, EPS can be classified as homopolysaccharides (HoPS), which contain only one type of monosaccharide, or as heteropolysaccharides (HePS), which are composed of repeating units containing different monosaccharides and non-sugar molecules [35]. Despite the fact that EPS from a variety of strains are potentially available, only a few, such as xanthan, gellan, curdlan, dextran, and bacterial cellulose, have been commercialized [36]. In the food industry, microbial EPS are classified as hydrocolloids and have enjoyed their traditional roles as thickening, gelling, and suspending agents (Table 2).

Taking into considerations the technical challenges today's leading commercial EPS had to deal with during development and production, the bioconversion of lactose into EPS inevitably inherits many challenges from both biological and engineering standpoints. For instance, the apparent adverse effects caused by diminishing aeration efficiency during xanthan gum fermentation are well acknowledged and have been extensively studied [37, 38]. On the other hand, as seen in the case of utilizing lactose for ethanol production, it is anticipated that not only is the number of microorganisms able to directly metabolize lactose limited, but most of the microorganisms are inhibited by moderate sugar and product concentrations [16]. Since the processes based on microbial cultures on cheese-whey permeate for ethanol production are considered the most profitable alternatives for the transformation of cheese-whey surplus [39], it is of critical importance to understand the metabolic pathways involved in lactose utilization and EPS production. Therefore, to provide a better understanding of the challenges and limitations facing the bioconversion of lactose into EPS, in this section, in-depth discussions on exemplary EPS will be provided, while all aspects concerning the biosynthesis of EPS via lactose fermentation will be closely examined in Section 4.

3.1. Xanthan gum

Xanthan gum, approved by the U.S. Food and Drug Administration (FDA) in 1965 as food grade [40], is known for its distinctive rheological properties, namely high viscosity at low shear, shear-thinning, stability over a broad range of temperatures and pHs, and high resistance to shear degradation [41, 42]. Its high acid stability makes xanthan highly popular in sauces, syrups, toppings, and salad dressings. In drinks, the addition of xanthan with carboxymethylcellulose (CMC) adds 'body' to the liquid and assists with the uniform distribution of fruit pulp etc. The high freeze-thaw stability of xanthan suspensions makes them particularly attractive for the frozen food industry. These properties are also exploited by the chemical and pharmaceutical industries. Moreover, the animal feed industry uses xanthan gum for transporting liquid feeds with added vitamins and other supplements that would otherwise sediment out during transportation or storage.

Table 2
Applications of EPS in food

EPS	Producing Organism(s)	Major Applications	Worldwide Market ($ million)	US Price* ($/lb)
Xanthan	*Xanthomonas campestris*	Thickener, stabilizer, emulsifier, suspending agent	235.0	2.75–4.50
Gellan	*Sphingomonas elodea*	Gelling and suspending agent, stabilizer	15.0	**
Curdlan	*Alcaligenes faecalis* var. *myxogenes*	Gelling agent, texture modifier	N/A	35.00
Alginate	*Azetobacter vinelandii* *A. chroococcum* *Pseudomonas aureginosa*	Thickener and gelling agent	91.0[§]	3.05[§]
Bacterial cellulose	*Acetobacter xylinum* (*Gluconacetobacter xylinus*)	Dispersion and emulsion stability, shape retention	N/A	2.65
Dextran	*Leuconostoc mesenteroids*	Stabilizing agent, texture modifier	N/A	N/A
Dairy-associated EPS	*Streptococcus thermophilus* *Lactobacillus ssp.* *Lactococcus lactis* *Bifidobacterium longum*	Thickener, stabilizer (with the culture)	N/A	***

*Average market price as of August 2005. Pricing may vary depending on volume and contract pricings.
**Gellan gum sells for $25–$30 per lb pure and $5–$6 per lb dilute.
***Regular yogurt cultures sell for $22.71 for 170 grams cup to set 1,000 gal. Heavy body "buttermilk" cultures sell for a premium at $33.68 to $48.96.
[§] The information is based on sodium alginate.
N/A: Not available.

While xanthan is present in a solution alone, however, only transient weak gels are formed. This is because the junction zones are weaker than those in true gel networks, and hence the gels are easily broken down under stress [43]. There has been considerable interest in improving the weak gelation characteristics of xanthan by including galactomannans such as locust bean gum in mixtures. Xanthan gum forms relatively rigid-rod-like structures in solution at ambient temperatures but converts to a random configuration on heating. These rods are able to align themselves–like agarose and the carrageenans–with the unsubstituted regions of galactomannans, such as guar and its derivatives and locust bean gum, to produce fairly rigid mixed gels with applications in food manufacturing. Synergistic interactions between galactomannans with xanthan produces stronger gels with an optimum mixing ratio of ~50:50 by weight. It has been shown that deacetylating the xanthan side chains seems to enhance these synergistic interactions. Xanthan gum can form strong gels when mixed with positively charged polymers because of its anionic nature caused by the presence of pyruvate and glucuronic acid in its structure.

Additional functionalities provided by xanthan gum include: (1) a significant yield value at low concentrations compared to other industrial gums, which explains its ability to stabilize dispersions, such as emulsions or suspensions [44]; (2) completely soluble in cold and hot water; (3) unaffected by enzymes; and, most noteworthy, (4) excellent stability to heat and pH. In particular, the viscosities of xanthan gum solutions remain unchanged from 0°C to 100°C and from a pH of 1 to 13 [45].

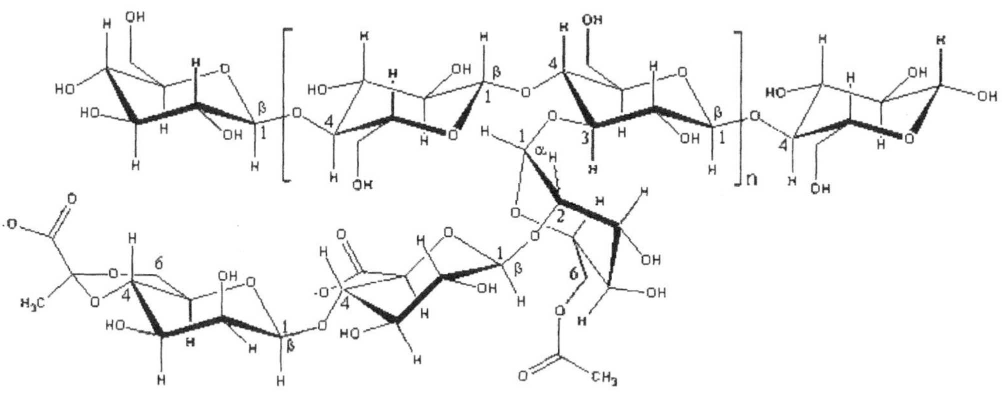

Fig. 1. Chemical structure of xanthan gum.

Produced by the aerobic fermentation of *X. campestris* on glucose, xanthan gum consists of a cellulosic backbone with side chains of two mannose and one glucuronic acid (Fig. 1) [46, 47]. Production of xanthan gum depends on many parameters and variables, including medium composition [48], temperature [49], pH [50], and oxygen transfer [38]. Its average molecular weight of around 6 million Daltons means that it is one of the largest aqueously soluble polysaccharides, making it possible to create extremely viscous solutions. However, xanthan production is greatly hindered by low oxygen transfer rates as the broth becomes viscous [37, 38, 51–54]. Lo and coworkers [38] demonstrated improved xanthan productivity in a novel, centrifugal packed-bed reactor (CPBR), in which innovative aeration strategies were employed [52] in addition to the construction of a fibrous supporting matrix for effective cell immobilization [55]. Primarily attributed to the high cell density achieved on the matrix, the enhanced aeration efficiency at high xanthan concentrations was crucial for reaching higher xanthan productivity.

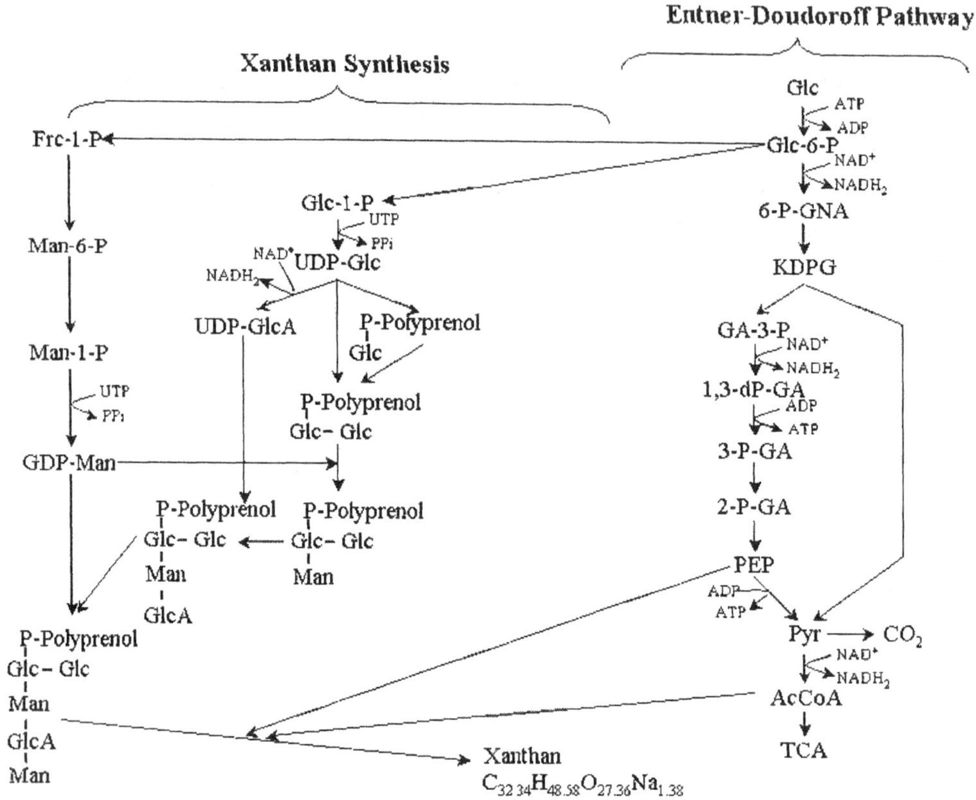

Fig. 2. Metabolic network of xanthan synthesis and glucose metabolism in *X. campestris* NRRL-1459. (*Glc*: glucose, *Fru*: fructose, *Man*: mannose, *GlcA*: glucuronic acid) (adapted from [51])

To date, two discrete systems have been identified in the glucose metabolic pathways of *X. campestris*, one periplasmic and oxidative, the second intracellular and phosphorylative [56]. Proposed as a deviation from the Entner-Doudoroff pathway at glucose-6-phosphate (Glc-6-P) with intermediates derived from phosphoenolpyruvate (PEP) and Acetyl Co A (AcCoA), xanthan synthesis has been proven to demand a significant proportion of the total cellular nicotinamide cofactors [57] and ATP (Fig. 2) [58]. Garcia-Ochoa and colleagues [59] developed a metabolic structured kinetic model based on the work of Pons et al. [60] to describe the evolution of biomass, the consumption of carbon, nitrogen, dissolved oxygen, and the production of xanthan at different temperatures. However, it has been found that carbon flux is dependent upon the distribution between fructose diphosphatase (gluconeogenesis) and fructose diphosphate aldolase (glycolysis), which results in a significant and unavoidable loss of carbon [57]. With at least two enzymes (glucokinase and phosphoglucoisomerase) encoded by genes having yet to be cloned [61], mapping the enzymatic kinetics involved in xanthan biosynthesis remains challenging.

3.2. Gellan gum

Gellan gum was developed as the result of a systematic search for a polysaccharide of the required properties followed by the identification of the organism. It is obtained from cultures of *Sphingomonas elodea* (formerly named *Pseudomonas elodea*) found growing on the *elodea* plant [62, 63]. It has a linear structure with a repeating unit of a tetrasaccharide with one carboxyl group and, in the native state, one acetyl group (Fig. 3). It is therefore sensitive to calcium levels but has rheological properties similar to those of xanthan with a similar charge density. It was clearly intended to compete with xanthan gum, though before permission for food use was obtained it was promoted as an agar substitute, particularly for use in growth media.

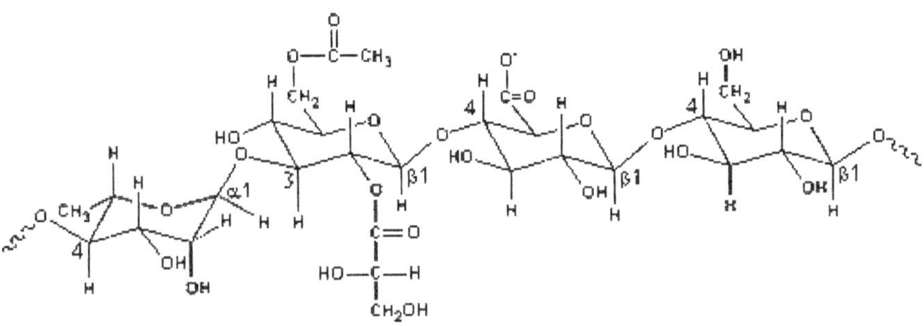

Fig. 3. Chemical structure of gellan gum.

3.3. Curdlan

Curdlan gum is the third fermentation-produced polysaccharide approved for food use in the US. Curdlan is member of the class of molecules known as (1,3)-ß-glucans. These polysaccharides are characterized by repeating glucose subunits joined by a ß linkage between the first and third carbons of the glucose ring. While the primary structure (Fig. 4) is a long chain, curdlan forms more complex tertiary structures due to intramolecular and intermolecular hydrogen bonding. A wide variety of commercial food products already make use of curdlan in Japan, Korea, and Taiwan. Curdlan is produced from the fermentation of an *Alcaligenes faecalis* var. *myxogenes* culture in a medium consisting of glucose, a nitrogen source, and trace amounts of minerals. Raw curdlan formed in the medium is dissolved with alkali, separated from the microorganism, purified, and dried into powder. Curdlan is distinguished for its ability to produce colorless, odorless, and tasteless gels when heated [64]. Other gelling agents often require another step or condition in addition to heating to form gels. For example, carageenan, agar-agar, high methoxyl pectin, gellan gum, and gelatin must be cooled after heating, while sodium alginate and low-methoxyl pectin require the presence of calcium ions. Food ingredients such as lacto albumin, powdered egg whites, and soy protein harden from being heated alone; however, they have characteristic odors and tastes that can limit their usage in foods.

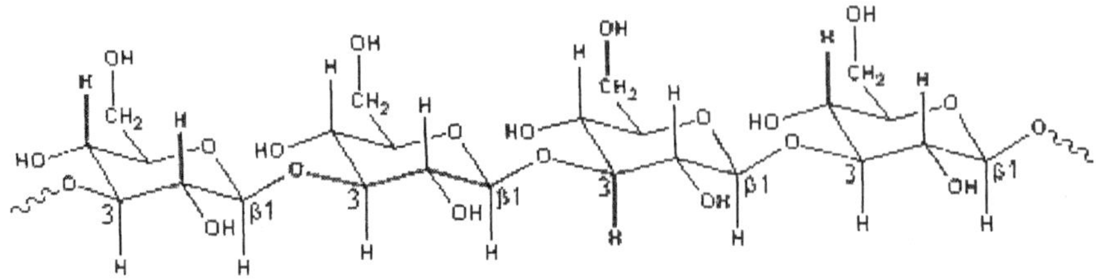

Fig. 4. Chemical structure of curdlan gum.

3.4. Alginate

Although commercial alginates are derived from algal sources, there is a large potential for producing 'tailor-made' alginates from bacterial sources, especially if advantage is taken of the genetic tools for controlling the production of the enzymes that are responsible for the synthesis and epimerization (conversion of D-mannuronic, 'M', to L-guluronic residues, 'G') of the polymeric alginate chain (Fig. 5). There appears to be greater structural diversity (poly-M, poly-G, and poly MG residues) and our understanding of the genes and the enzyme gene produces is much greater for bacterial alginate production compared with the case for seaweed. The main alginate-producing bacteria that have been studied are *Pseudomonas aeruginosa* and *Azotobacter vinelandii* [65]. *P. aeruginosa* has been the subject of particular attention because of its association with respiratory disease and is found in patients suffering from cystic fibrosis. *A. vinelandii* appears to be the most promising in terms of industrial production because of its stable output of alginate. *P. aeruginosa* alginate has no poly-G residues (and hence has a low G-content) whereas *A. vinelandii* can, like seaweed alginate, possess all three block sequences (poly-M, poly-G, and poly-MG residues).

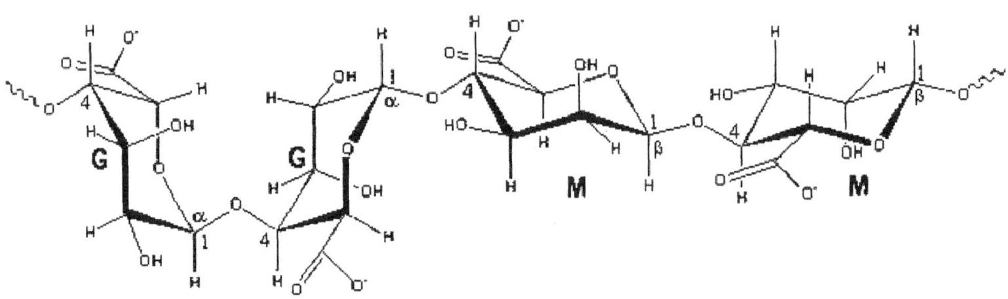

Fig 5. Chemical structure of alginate.

3.5. Bacterial cellulose and dextran

Cellulose, the main component of plant cell walls, can be produced by an acetic acid-producing bacteria *Acetobacter xylinum* (*Gluconacetobacter xylinus*). The production of bacterial cellulose has received great attention because of its wide possible applications [66] to reduce the consumption of cotton and wood, the major resources for all cellulose products. Bacterial cellulose and plant cellulose have the same chemical structure, but different physical and chemical properties. Bacterial cellulose has a diameter approximately 1/100 of that of plant cellulose, and its Young's modulus is almost equivalent to that of aluminum. Moreover, unlike cellulose from plants, which is only 50 or 60% crystalline, bacterial cellulose fibers can be 80% crystalline. The elevated crystallinity enables bacterial cellulose fibers to align in highly ordered structure, making bacterial cellulose more durable in the face of chemical reactions and mechanical pressure. Bacterial cellulose is expected to be a new biodegradable biopolymer.

Bacterial dextrans are produced in substantial quantities by *Leuconostoc mesenteroides* and are familiar to laboratory workers as the basis for cross-linked dextran beads used in gel filtration columns. The product with an average molecular weight of about 60,000 Da is used in medicine as a blood extender, while fractions of defined molecular weights (e.g. the Pharmacia 'T-' series, where 'T500 Dextran' would stand for dextran with an weight-average molecular weight of 500 kDa) serve as polysaccharide standards in molecular weight calibrations [67].

3.6. EPS produced by dairy-associated, acidifying microorganisms

Increasing interest has also been given to dairy-associated acidifying bacteria capable of producing EPS by directly fermenting the lactose within dairy-based products. Some of the bacteria have been found to be candidates in providing probiotic benefits to humans, indicating a great potential for extending their applications to non-dairy products.

3.6.1. EPS produced by lactic acid bacteria

Many different strains of lactic acid bacteria (LAB) produce different types of EPS in varying structures and sizes. EPS from LAB can be classified into two groups: homopolysaccahrides (HoPS) and heteropolysaccharides (HePS). HoPS can further be divided into four main groups: α-D-glucans (dextrans, alternans, mutans), β-D-glucans, fructans (levan produced by *S. salivarius*) and others like polygalactans. HePS are produced by a wide variety of mesophilic and thermophilic LAB [35, 68]. The molecular weight of EPS from LAB varies from 1.0×10^4 to 6.0×10^6, which is quite comparable to that of the other commonly used EPS.

Industrial scale production of EPS from LAB is not very common since the yield of EPS biosynthesis is relatively low. In most cases, the total amount of EPS synthesized does not exceed 1 g/L even under optimal growth conditions, although some higher amounts have been reported for *Lactobacillus returi* strains (4.1 and 4.8 g/L) [69], *S. thermophilus* LY03 (1.5 g/L) [35], and *L. Sakei* 0-1 (1.4 g/L) [70]. The lower energy generated in anaerobic LAB limits EPS formation and results in lower product yields than that from aerobic strains, such as *X. campestris*. Bacterial dextrans produced by *Leuconostoc mesenteroides* are one of the few EPS from LAB that is industrially important since they are widely used in gel filtration columns, in medicine as a blood extender, and to some extent serve as polysaccharide standards in molecular weight calibrations. Although most of the ropy strains produce less than 1 g/L of EPS, the total biocompatible nature of the resulting rheologically active biopolymers favors their possible extensive use in the food industry [71]. Studies have demonstrated that, by adding EPS-producing bacteria such as *Lactobacillus casei* [72, 73] and *L. delbrueckii* ssp. *bulgaricus* RR [74, 75] in dairy products, product viscosity, texture, and mouth feel can be directly improved without the addition of expensive commercial EPS. Moreover, EPS from LAB will enable the production of new products, such as low-milk-solid and low-fat yogurts [76].

The growing interest in EPS produced by bacteria isolated from dairy products [77] is mainly due to the fact that the growth of EPS-producing bacteria can directly affect the rheological properties of fermented dairy products [74, 75, 78]. Moreover, some being produced by probiotic strains, EPS from ropy bacteria are very new but also very promising in the market for healthy products. The value of EPS produced by ropy strains could be greatly increased if more information is gained about their physical and chemical properties and if their rheological determining properties can find applications in nondairy foods [79–81]. Although a higher carbon-to-nitrogen ratio in the medium favors EPS production by bacteria such as *P. acidipropionici*, a higher EPS yield by *S. thermophilus* was obtained when both the carbon and nitrogen concentrations were increased [82], indicating considerable variance between different microorganisms. Therefore, the influence of the substrate's carbon-to-nitrogen ratio on cell growth and EPS production by dairy-related bacteria needs to be specifically determined.

3.6.2. EPS produced by Bifidobacteria

In the quest to incorporate health benefits into dairy products, many yogurt-type products are now enriched with microorganisms such as bifidobacteria, the predominant species in the gut microflora of healthy humans, which has recognized probiotic, nutritional, and therapeutic properties [83–87] and *L. delbrueckii* ssp. *bulgaricus* RR, which is used for EPS production. Bifidobacteria, natural inhabitants of the guts of many animals including humans, are strictly anaerobic (although some strains can tolerate oxygen in the presence of carbon dioxide), gram-positive rods. The growth of bifidobacteria can be enhanced by the addition of promoters. Bovine casein digest and yeast extract were the best promoters when bifidobacteria were grown in synthetic medium. Other growth promoters, including human and bovine milk whey, hog gastric mucin, and bovine serum albumin digest were effective with some species but not with others. Bifidobacteria also grew well when bovine casein digest (20 mg/ml) was used as the nitrogen source. However, the nature of these growth factors has not yet been determined [88].

Bifidobacterium longum is known for its capability to inhibit liver tumors in mice [89] and improve diarrheal conditions by decreasing stool frequency [90]. The production of EPS in the genus of *Bifidobacterium* was not characterized until the last decade [91, 92]. The composition of an acidic EPS (molecular weight > 200 kDa) produced by *B. longum* BB-79 was first reported by Roberts and colleagues [91], where the synthesis of the EPS was affected by the primary carbon source of the culture medium. The acidic nature of EPS, similar to other microbial EPS, has made pH the primary parameter for determining the yield of EPS throughout fermentation [52]. According to Roberts et al. [91] and Andaloussi et al. [92], culture media with an initial pH of 6.0 did show reduced production of EPS throughout the 10-day period, whereas pH in the range 6.0–9.0 had little effect on EPS yields. As indicated by related literature, the time of incubation plays an important role in microbial EPS production [93, 94]. However, the time of incubation appeared to have little effect on the resultant yields of EPS [91].

3.6.3. Effects of the carbon source

Although some researchers are convinced that the nature of the substrate cannot influence the composition of the EPS produced [95, 96], many others have found that changes in the carbon source modified the sugar composition and molecular weight of the EPS and also affected the yield of EPS biosynthesis. Glucose, skim milk, and sucrose are the common carbon sources for EPS production in many ropy strains, while bifidobacteria prefers lactose to other sugars [91]. The majority of EPS-producing microorganisms utilize carbohydrates as their energy source as well as their carbon source for EPS formation [95]. In general, EPS production in defined media was found to be stimulated by limiting nutrients such as nitrogen and by providing excess carbohydrate [93, 97], whereas limiting the carbon source resulted in minimal EPS production [98]. However, the effect of the carbon source on EPS production and sugar composition varies in different microorganisms. For instance, van Geel-Schutten and coworkers [69] reported that the same *Lactobacillus* strains produce varying amounts of EPS with different sugar compositions when they are grown with sucrose, raffinose, and lactose in liquid or on solid media. In another study, the effect of different hexose sugars (glucose, sucrose and lactose) on EPS production by *Agrobacterium*, *Alcaligenes*, *Pseudomonas* and *Xanthomonas* was investigated. Lactose was found to be the best carbon source for *Pseudomonas* sp. for EPS production, whereas sucrose was favored by the other bacterial cultures [99]. The EPS produced by *L. casei* CG11 in the presence of glucose was reported to be different from that formed in the presence of lactose [97]. The sugar composition of EPS produced by *L. bulgaricus* was modified by adding glucose to both milk and milk ultrafiltrate [93]. However, no difference in the amount of EPS produced by *Lactobacillus rhamnosus* strain 9595M was detected when glucose or lactose were used as the carbon source.

In addition, Cerning et al. [97] suggested that sugar concentration had a marked effect on the polymer yield of *L. casei* CG11. Similar observations were found in the fermentation of *L. rhamnosus* C83 [100]: higher EPS yields were obtained when higher initial carbon source concentrations were employed. The optimal production of EPS by *P. acidipropionici* was obtained in a whey-based medium supplemented with 60 g/L of lactose. However, further increasing the lactose concentration to over 60 g/L did not improve EPS production and the utilization of lactose as a carbon source [101]. *B. longum* BB-79 produced the highest amount of EPS by weight (0.466 g/L) when lactose was used as the primary carbon source in liquid media. This represented an approximately 200–300% increase in yield over EPS production in media containing sucrose, fructose, or glucose. The optimum production of EPS in the presence of lactose indicated that EPS can be produced by *B. longum* BB-79 in lactose-containing substrates such as cheese whey. However, to economically maximize the productivity of the EPS, either by increasing the yield or production rate, efforts are needed to determine the optimum conditions for *B. longum* BB-79 growth and EPS production.

3.7. Fermentation conditions and scale-up

Fermentation conditions greatly influence EPS production. For many bacteria, the optimum conditions for EPS production are different from their optimal growth conditions. As discussed earlier in this chapter, the effects of temperature, pH and medium composition on EPS production are well recognized by researchers in the field. It is also known that some microbial EPS can degrade after synthesis. For example, EPS produced by *L. mesenteroides* and *S. thermophilus* showed reduced EPS levels at the later stages of fermentation [102]. For *L. casei*, EPS was found to degrade when pH was controlled at 5.0 [72]. It was suggested that the degradation of EPS may be due to the activation of certain enzymes that are capable of degrading the polymer [98]. However, based on the results obtained in our laboratory, the EPS produced by *B. longum* BB-79 did not decrease during the fermentation, implying that no EPS degrading enzyme was formed in the medium throughout the fermentation, in agreement with the results reported by Roberts et al. [91]. The marked reduction in EPS formation upon prolonged fermentation seems to be dependent on the strain and culture conditions and can be avoided by harvesting the EPS at the appropriate time and under the appropriate pH and temperature during isolation [103].

Most of the previous work on the kinetics of xanthan fermentation focused on the nutritional requirements for xanthan production [104–108]. The specific effects of the carbon and nitrogen source on cell growth and xanthan biosynthesis, respectively, were also studied [109, 110]. It is generally believed that a high nitrogen concentration is required for fast cell growth and high cell density, whereas xanthan biosynthesis is favored by a high concentration ratio of the carbon source to the nitrogen source [111, 112]. Various kinetic models have been constructed based on different nutrient requirements for xanthan biosynthesis and cell growth [113, 114]. However, there is still considerable confusion regarding the underlying mechanisms governing the synergetic effects of carbon and nitrogen substrates.

Diverse observations have been reported on EPS-producing, acidifying bacteria in the literature. For example, in the research by Cerning et al. [98], EPS produced by *L. bulgaricus* contained more galactose than glucose and the EPS composition was significantly affected by its carbon source. A similar effect was observed by Kojic et al. [80]: the sugar composition of the EPS produced by *L. casei* CG11 also depended on the medium composition; however, van den Berg et al. [78] observed that the carbon source did not alter the sugar composition of the EPS produced by *L. sake* 0-1. Therefore, it is reasonable to conclude that the composition of EPS is highly associated with how the microorganism utilizes its carbon source.

Bioreactor design and effective scale-up are of crucial importance in the development of a successful bioprocess. The bioreactor is the heart of a successful bioprocess [115]. In the words of Cooney [116]: "The continued success of biotechnology depends significantly on the development of bioreactors, which represents the focal point for interaction between the life scientists and the process engineer." With the novel design of CPBR, which is superior in mixing, scale-up is the next goal to accomplish based on the understandings of cellular metabolism, optimized cell growth condition, and the rheological properties of EPS. The cost- and time-efficiency of large-scale production is the driving force for scale-up. The statistical results for both individual pieces of equipment and whole-plant construction have risen to the so called "six-tenth factor" [117, 118], which means that, if the capacity of a plant is doubled, the cost will be only $2^{0.6}$ more. Generally, the operation costs decrease with the exponent in the range of 0.3–0.5, which is another reason to scale up. Thus, the scale-up process directly to the of the the affects the production capacity and efficiency of a bioprocess, and is thus relevant to the operating costs.

3.8. Characterization of EPS functionality and structure

Exopolysaccharides produced during fermentation have been found to be responsible for the consistency, texture, color, and flavor of yogurt and other dairy products. If strains produce too many polysaccharides, then the final yogurt product will be slimier. Adding bacteria in the product is important also because consumers are demanding more natural products and the labeling "Live active cultures" suggests health benefits. An example of this is probiotics. Live cultures, such as *Lactobacillus acidophilus,* are being added to dairy products, such as yogurt, for their believed benefits to the intestinal system.

The texture of yogurt is determined by its total solid content, the composition of milk, the nature of the cultures used, and processing. Exopolysaccharides are known to play a key role in the rheological behavior and texture of fermented foods [119]. They improve texture (i.e. increase the viscosity and decrease syneresis), which is important because in some countries, such as France and The Netherlands, the use of stabilizers is not allowed. It is accepted that EPS are the cause of ropiness in dairy products [98]. However, surprisingly, a study based on the analysis of the effect of EPS content on the viscosity of acidified skim milk showed that the apparent viscosity of the system was not related to the concentration of EPS produced [120]. This study also claimed that, before stirring, the permeability of milk gels was affected by the type of yogurt starter. Syneresis, by which a liquid is separated from a gel owing to further coagulation, occurs in gels formed from milk when the whey component in milk separates from the curd [121]. Protons in the protein matrix result in a decrease in NMR relaxation times upon gelation. This could be due to water molecules being confined by the protein network in pores or cavities in a gel-like curd [122]. Relaxation times and the amount of water inside and outside the gel have been shown to be correlated [123].

The stability and properties of the EPS in food products, especially in fermented dairy products, can be highly influenced by interactions with other ingredients, mainly depending on the chemical structure, composition, and purity of the samples. Directly adding protein-rich ingredients to xanthan solution has been reported to affect its rheological behavior [124], and the trend can possibly be anticipated with other EPS. For example, the primary components of the EPS produced by *B. longum* BB-79 were found to be galactose and a sugar, which has been tentatively identified as a lactic acid derivative of a hexose [91]. It has been suggested that the EPS may be synthesized from the same base unit and may have a structure resulting from repeating subunits [92]. Because of the acidic nature of the EPS, the presence of acid in the system is expected to affect the reaction rate of the EPS. Capable of elucidating the structure and dynamics of molecules in solution state in a non-invasive manner, nuclear magnetic resonance (NMR) has been successfully demonstrated as a powerful analytical tool for the structural identification of macromolecules such as proteins [125] and carbohydrates [126]. The invention of the Pulse Field Gradient [127] further enhances its sensitivity, enabling subsequent applicability to oligosaccharides [128] and to various EPS related systems [129, 130]. To enable control of the functional attributes of the EPS, as well as to understand the flux distribution of the substrate (in this case, lactose from whey), it is crucial that its composition, structure, and molecular conformation be fully understood and clearly elucidated.

In various research papers, the quantification and structure of EPS have been evaluated using various methods such as high performance liquid chromatography (HPLC) and NMR spectroscopy. What has consistently been found is that glucose (Glc), galactose (Gal), and sometimes rhamnose (Rha) are part of the EPS structures produced by *Lactobacillus* spp. [61], and EPS of *Streptococcus thermophilus* contain the same and sometimes small amounts of xylose (Xyl), arabinose (Ara), and mannose (Man) [131]. As stated in a previous work, a quantitative analysis of exopolysaccharides using cation exchange HPLC showed a 3:2 ratio of Gal and Glc [132]. It was suggested that the differences in minor sugars present in the polysaccharide structure found along with Gal and Glc may have been due to a difference in growth conditions and/or isolation and purification procedures [98].

Structural analysis of EPS begins with the isolation of the polysaccharides. Initially, there is a precipitation step, and then the sample is dialyzed. The pure polysaccharide can then be quantitatively analyzed using ^1H and ^{13}C NMR spectroscopy or HPLC. Methylation analysis can also be performed to quantitatively analyze polysaccharides [133]. However, NMR and HPLC methods tend to be more accurate and can be automated fairly easily.

4. BIOSYNTHESIS OF EPS VIA LACTOSE FERMENTATION

Microorganisms capable of converting lactose into other products contain an enzyme called lactase that attacks lactose, splitting its molecules into glucose and galactose [134]. Other enzymes then attack the glucose and galactose, converting them into various acids, such as lactic acid. However, only suitable strains could directly ferment lactose into the target product. For example, most yeasts lack the lactose permease system, the membrane lactose carrier that controls the entry of sugar into the cells, as well as β-galactosidase, the intracellular enzyme for lactose hydrolysis, their ability to ferment lactose directly into ethanol is thus greatly hindered [31, 32].

Several dairy associated bacteria use lactose as their major energy source as well as in polysaccharide synthesis. Many *Lactobacillus* and *Pseudomonas* strains, *B. longum*, and dairy propionibacteria have been reported to produce EPS from lactose. Lactose bioconversion gives good yields in some strains; however, with many others EPS production is more favored by other carbon sources such as sucrose and glucose. For instance, *L.*

delbrueckii subsp. *bulgaricus* NCFB 2772 produced larger amounts of EPS when grown on glucose or lactose than when grown on fructose to equal cell densities [135]. *L. delbrueckii* subsp. *bulgaricus* NCFB 2483 showed increasing specific yields of EPS with increasing dilution rates when grown on lactose as well [136]. On the other hand, some other *Lactobacillus* strains, such as *L. reuteri*, produce fewer amounts of EPS when grown on lactose than when sucrose was used as the primary carbon source [69].

Exopolysaccharides produced via fermentation of lactose are potential alternatives to those produced from glucose, since most of these possess already important applications [137]. Dairy propionibacteria as well as LAB are suggested to have the potential to directly utilize whey by lactose fermentation and produce useful polymers for food and non-food uses [138]. While *Rahnella aquatilis* has been shown to be particularly active in metabolizing lactose provided that good aeration conditions are provided [139], other exopolysaccharides are produced in low concentrations via fermentative pathway, and are thus not feasible for industrial applications [137]. Genetically modified strains have been developed in which the *lac* genes are added to the wild-type strain [134]. Unfortunately, the benefits of this approach have yet to be realized industrially.

4.1. EPS biosynthesis via lactose fermentation in dairy associated bacteria

LAB produce a wide variety of EPS where the differentiation can be based on the biosynthesis mechanism and the precursors required [140]. In the case of extracellularly produced HoPS, such as dextran, the polymerization reaction proceeds via extracellular glycosyltransferases, which transfer one monosacharide from a disaccharide to a growing polysaccharide chain. However, many HoPS and HePS are intracellularly produced from sugar nucleotide precursors via glycolysis before being secreted to the environment [34]. The biosynthetic pathway is composed of reactions regarding the sugar transport into the cytoplasm, the synthesis of sugar-1-phosphates, the activation and coupling of sugars, and the export of the EPS [141].

In LAB, many sugars are transported from the periplasm to the cytoplasm via the sugar-specific Phosphoenolpyruvate-phosphotransferase system (PEP-PTS), which is energy efficient as only one molecule of ATP is used. After sugar is taken into the cell, the phosphorylation state of the sugar determines its further role [142]. Lactose may enter the cell either in the phosphorylated state or as a free sugar, depending on the mode of transport (Fig. 6). Once lactose is split into glucose and galactose intracellularly, glucose is metabolized via the glycolytic (homofermentative LAB strains) or phosphoketolase (heterofermentative LAB strains) pathways, whereas galactose is degraded via the tagatose-6-phosphate (galactose-6-phosphate) pathway or Leloir pathway [68, 140]. *L. lactis* strains possess a lactose-specific PEP-PTS sugar transport system that imports extracellular lactose and provides internal lactose-6-phosphate. Lactose-6-phosphate is further hydrolyzed into galactose-6-phosphate and glucose by phospho-β-galactosidase. Glucose is converted into glucose-6-phosphate by the enzyme glucokinase. In galactose negative *L. delbrueckii* subsp. *bulgaricus* and *S. thermophilus* strains, lactose is transported into the cell by a lactose permease, which operates as a lactose-galactose antiporter [143]; the galactose moiety is then released into the medium, leaving only glucose as the energy and carbon source [68, 144].

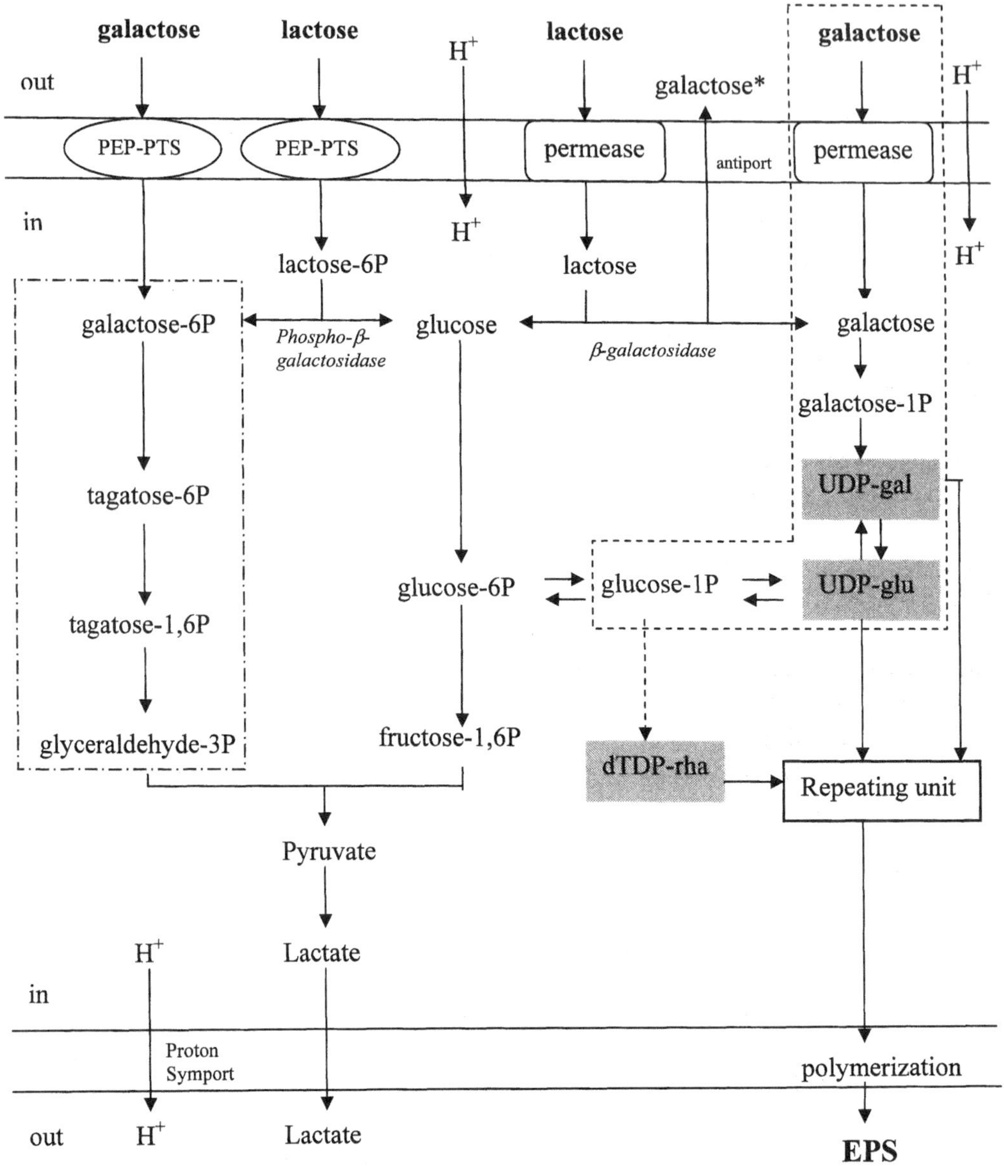

Fig. 6. Schematic diagram of pathways involved in lactose and galactose uptake and dissimilation, as well as exopolysaccharide (EPS) production in lactic acid bacteria. Tagatose-6-phosphate pathway (_ . _ . _) and Leloir pathway (- - - - -) are shown respectively as grouped. (galactose*: in the case of galactose-negative strains, galactose is exported) (adapted and modified from [68] and [140])

HePS are made by the polymerization of repeating unit precursors (UDP-glucose, UDP-galactose and dTDP-rhamnose) formed in the cytoplasm. These are assembled at the membrane by the sequential addition of activated sugars to the growing repeating unit. After a HePS repeating unit is completed, it will be exported through the cell membrane, becoming polymerized into a final HePS [68]. However, the mechanisms of polymerization, chain length determination, and export from the plasma membrane remain unclear.

In bifidobacteria lactose is transported into the cell by proton symport, which couples substrate translocation with the incorporation of a proton, and intracellularly degraded into glucose and galactose by β-galactosidase. Glucose is degraded to produce energy via the unique bifidus pathway that generates 5 ATP as 2 glucose molecules break into 3 acetic acids and 2 lactic acids, while galacatose is converted to glucose-1-phosphate by different enzymes via either the Leloir pathway (during the exponential phase) (Fig. 6) or the pyrophosphorylase pathway (during the stationary phase). Glucose-1-phosphate produced from galactose is converted to glucose-6-phosphate before entering the bifidus pathway [145, 146]. Although the energy mechanism of bifidobacteria has been studied, further research needs to be conducted to identify the mode of lactose utilization during EPS production by *B. longum*.

4.2. Challenges in whey lactose bioconversion

It is generally accepted that processes based on microbial cultures are one of the most profitable alternatives for the transformation of whey lactose [39]. However, a variety of factors, such as the efficacy of waste minimization, the scale of processing and production, and product yields, have to be carefully considered in order to establish a cost-effective, sustainable operation. While the profitability of the traditional uses of EPS as a thickening agent seems to be limited, much potential exists if unique and/or synergistic functionalities of EPS can be identified and characterized. *B. longum* BB-79, an example discussed in this chapter, has added health benefits due to its probiotic nature and appears to be a promising candidate, especially since it has also been shown to digest lactose more effectively than it does other fermentable sugars [91]. Little information, however, exists on the *B. longum* culture conditions that affect the ability of the organism to produce a polymer [75, 91]. Further optimization of the growth environment of *B. longum* is important to achieving maximal EPS production.

While the direct utilization of lactose may not be feasible for all strains of interest, one approach is to hydrolyze lactose into its two monomeric sugars, glucose and galactose, which are readily and efficiently fermented [5]. Two different means have been reported to accomplish the hydrolysis. The glycosidic bond of the disaccharide can be enzymatically hydrolyzed with β-galactosidase. A number of enzyme sources, reactor configurations, and processes have been proposed and tested [147–149]. However, β-galactosidase enzymes are too expensive for producing a product if its commercial value does not justify the cost.

Luckily, a cost-effective method exists; acid-catalyzed hydrolysis [5]. The process has been well characterized for solutions of pure glycosides, and for some dairy effluents [150, 151]. Acid hydrolysis involves heating with simple reagents, but from a mechanistic perspective, the process is quite complex. Monosaccharide products can be further degraded into undesirable chemicals. The number of possible side reactions depends upon, among other things, the permeate composition. As such, an evaluation of acid hydrolysis as a means to generate monosaccharides from lactose in whey permeate must be carried out within the context of the intended use of the hydrolysis products [5]. For example, factors affecting the physiological state of the fermenting organisms must be known. Moreover, adding inorganic acids followed by neutralizing agents will increase the ionic strength and osmotic pressure, which may reduce the activity of the microorganism.

The media used to produce exopolysaccharides requires, besides lactose, enrichment with a nitrogen source. The direct utilization of whey would be a less expensive option if advantage were taken of the nitrogen inventory in the feedstock. Availability of such nitrogen can be further enhanced by hydrolyzing whey proteins as they can stimulate the growth of certain bacteria [152, 153]. The current production of whey protein concentrates also generates considerable quantities of whey permeate that could constitute an alternative feedstock for exopolysaccharide production.

Pintado et al. [137] reported that the low concentration of peptides and free amino acids in plain whey and whey permeate did not permit the active consumption of lactose, whereas a lack of molecular oxygen and the presence of salt played a major role in inhibiting exopolysaccharide production. The qualitative and quantitative profile of the nitrogen fraction has been suggested to affect the metabolism of lactose as well as the production of exopolysaccharide and organic acids. Gorret and coworkers [154] also observed that the anaerobic growth of *Propionibacterium acidipropionici* on milk permeate was only possible if supplemented with yeast extract. The fermentation capacities of the strain were significantly improved by further increasing the supplemented yeast extract.

Lessons learned from the bioconversion of lactose into ethanol have suggested that an efficient option for lactose utilization is to use combined cultures of bacteria and yeast (e.g., lactobacilli and baker's yeast) in a two-stage fermentation [2]. During the first stage, bacteria ferment lactose into lactic acid at pH 4.5–5.0. By adding sulfuric acid or other inorganic acids, the pH can be lowered to the optimal value, which is about 4 for yeast growth. In the second phase, yeast consumes this acid, thus raising the pH value to 6.5. At the same time, yeast cells are autolyzed and provide more nutrients for the other yeast cells. The entire fermentation process takes 10–60 h. However, when the yeast cells uses the mixture of glucose and galactose as a carbon source, it manifests diauxic growth and lower yields in ethanol production, even for strains previously adapted to galactose [16, 155]. Additional disadvantages of these processes are the high price of β-galactosidase and the failure of this enzyme to hydrolyze all the lactose, thus leaving the problems associated with effluent disposal unsolved [17].

5. FUTURE OUTLOOK

The advancement of biotechnology has explored many useful, high-throughput tools for understanding the metabolic pathways associated with EPS production. With the proper design for metabolic flow, the effective utilization of lactose (or glucose and galactose) is within reach. Furthermore, examples in the literature have shown that, by manipulating these pathways it is now possible to specifically control the charge of and/or functional groups on the EPS side chains, which are crucial to the functionality and stability of EPS in food systems. Equally important is the advancement of available techniques for measuring the conformational changes of EPS under different environmental conditions. It is recognized that the conformation of EPS in a solution will greatly affect its rheological properties, hence determining its final applications and values.

6. CONCLUDING REMARKS

Bioconversion is considered the most feasible means to effectively transform milk lactose from a low-value feedstock into value-added products. However, the direct fermentation of lactose by participating microorganisms usually suffers from a limited conversion rate as is found in yeast fermentation. The utilization of lactose as the building blocks for the synthesis of microbial EPS thus appears to be a favorable option. Microbial EPS already have many important applications in food, chemical, and pharmaceutical industries, and they continue to find new uses in emerging markets. A better understanding of the metabolic pathways involved in the bioconversion of lactose to EPS will increase the likelihood for such processes to be profitable.

ACKNOWLEDGEMENT

The authors would like to thank Edgar P. Anders of Anders and Associates, Midland, Michigan for his assistance in acquiring the current U.S. market price of microbial EPS discussed in this chapter.

REFERENCES

[1] M.I.G. Siso, Bioresource Technol., 57 (1996) 1.
[2] E. Spreer, Milk and Dairy Product Technology, Marcel Dekker, New York, 1998.
[3] A.E. Ghaly, M.S.A. Tango, N.S. Mahmoud and A.C. Avery, World J. Microbiol. Biotechnol., 20 (2004) 65.
[4] R. Mukhopadhyay, D. Talukdar, B.P. Chatterjee and A.K. Guha, Process Biochem., 39 (2003) 381.
[5] A. Coté, W.A. Brown, D. Cameron and G.P. van Walsum, J. Dairy Sci., 87 (2004) 1608.
[6] S.T. Yang and E.M. Silva, J. Dairy Sci., 78 (1995) 2541.
[7] T.A. Nickerson, J. Agric. Food Chem., 27 (1979) 672.
[8] P.G. Hobman, J. Dairy Sci., 67 (1984) 2630.
[9] D.W. Houldsworth, J. Soc. Dairy Technol., 33 (1980) 30.
[10] R.R. Zall, J. Dairy Sci., 67 (1984) 2621.
[11] M. Carić, Concentrated and Dried Dairy Products, VCH Publishers, Inc., New York, 1994.
[12] R. Chandan, Dairy-Based Ingredients, Eagan Press, St. Paul, 1997.
[13] T.A. Nickerson, in B.H. Webb (ed.), Byproducts from Milk, AVI Publ. Co., Inc., Westport, page 356, 1970.
[14] V. Yves, Revue Laitiere Française, 372 (1979) 27.
[15] S.G. Coton, J. Soc. Dairy Technol., 33 (1980) 89.
[16] G. Moulin and P. Galzy, Biotechnol. Genet. Eng. Rev., 1 (1984) 347.
[17] T. Sienkiewicz and C.L. Riedel, Whey and Whey Utilization, Th. Mann, Germany, 1990.
[18] K. Dendene, L. Guihaard, S. Nicolas and B. Bariou, J. Chem. Technol. Biotechnol., 61 (1994) 37.

[19] N. Kosaric and Y.J. Asher, Adv. Biochem. Eng., 35 (1985) 25.
[20] A.J. Mawson, Bioresource Technol., 47 (1994) 195.
[21] C.S.F. Tin and A.J. Mawson, Proc. Biochem., 28 (1993) 217.
[22] T.P. Shukla, CRC Crit. Rev. Food Technol., Jan (1975) 325.
[23] B.A. Friend and K.M. Shahani, New Zealand J. Dairy Sci. Technol., 14 (1979) 143.
[24] P.G. Hobman, J. Dairy Sci., 67 (1984) 2630.
[25] V. Gekas and M. Lopez-Leiva, Proc. Biochem., 20 (1985) 2.
[26] C.P. Champagne and J. Goulet, Can. Inst. Food Sci. Technol., 21 (1988) 545.
[27] A.J. Mawson, Biotechnol. Lett., 10 (1988) 503.
[28] A.M. Nolan, N. Barron, T. Brady, T. McAree, D. Smith, L. McHale and A.P. McHale, Biotechnol. Lett., 16 (1994) 849.
[29] K. Sreekrishna and R.C. Dickson, Proc. Nat. Acad. Sci. USA, 82 (1985) 7909.
[30] F. Farahnak, T. Seki, D.Y. Ryu and D. Ogrydziak, Appl. Environ. Microbiol., 51 (1986) 362.
[31] I. Russel, Trends Biotechnol., 4 (1986) 107.
[32] D. Porro, E. Martegani, B.M. Ranza and L. Alberghina, Biotechnol. Lett., 14 (1992) 1085.
[33] Y.S. Jeong, W.R. Vieth and T. Matsuura, Biotechnol. Bioeng., 37 (1991) 587.
[34] I. Sutherland, in R. Whistler (ed.), Microbial Polysaccharides, Academic Press, San Diego, 1993.
[35] L. de Vuyst and B. Degeest, FEMS Microbiol. Rev., 23 (1999) 153.
[36] I. Sutherland, Int. Dairy J., 11 (2001) 663.
[37] F. Garcia-Ochoa, E.G. Castro and V.E. Santos, Enzyme Microb. Technol., 27 (2000) 680.
[38] Y.M. Lo, C.H. Hsu, S.T. Yang and D.B. Min, Bioproc. Biosyst. Eng., 24 (2001) 187.
[39] F.J. Castillo, in H. Verachtert and R. De Mot (eds.), Yeast Biotechnology and Biocatalysis, Marcel Dekker, New York, 1990.
[40] E.L. Meyer, G.G. Fuller, R.C. Clark, and W.M. Kulicke, Macromolecules, 26 (1993) 504.
[41] A.K. Podolsak, C. Tiu, T. Saeki and H. Usui, Polym. Int., 40 (1996) 155.
[42] J.K. Leela and G. Sharma, Bioprocess Eng., 23 (2000) 687.
[43] R. Lapasin and S. Pricl, Rheology of Industrial Polysaccharides: Theory and Applications, Blackie Academic & Professional, London, 1995.
[44] B. Urlacher, O. Noble and A. Imeson, Thickening and Gelling Agents for Food, Blackie Academic & Professional, London, 1997.
[45] J.D. Dziesak, Food Technol., 45 (1991) 116.
[46] P. Jansson, L. Kenne and B. Lindberg, Carbohydr. Res., 45 (1975) 275.
[47] L.D. Melton, L. Mindt and D.A. Rees, Carbohydr. Res., 46 (1976) 245.
[48] Y.M. Lo, S.T. Yang and D.B. Min, Appl. Microbiol. Biotechnol., 47 (1997) 689.
[49] F. Garcia-Ochoa, V.E. Santos and A. Alcon, Enzyme Microb. Technol. 17 (1995) 206.
[50] F. Garcia-Ochoa, V.E. Santos and A. Alcon, Math. Comput. Simul., 42 (1996) 187.
[51] C.H. Hsu and Y.M. Lo, Process Biochem., 38 (2003) 1617.
[52] S.T. Yang, Y.M. Lo and D.B. Min, Biotechnol. Prog., 12 (1996) 630.
[53] Y.M. Lo, S.T. Yang and D.B. Min, J. Membr. Sci., 117 (1996) 237.
[54] C.H. Hsu, Y.F. Chu, S. Argin-Soysal, T.S. Hahm and Y.M. Lo, J. Food Sci., 69 (2004) E441.
[55] S.T. Yang, Y.M. Lo and D. Chattopadhyay, Biotechnol. Prog., 14 (1998) 259.
[56] C. Whitfield, I.W. Sutherland and R.E. Cripps, J. Gen. Microbiol., 128 (1982) 981.
[57] F. Letisse, P. Chevallereau, J.L. Simon and N.D. Lindley, Appl. Microbiol. Biotechnol., 55 (2001) 417.
[58] T.R. Jarman and G.W. Pace, Arch. Microbiol., 137 (1984) 231.
[59] F. Garcia-Ochoa, V.E. Santos and A. Alcon, Enzyme Microb. Technol., 23 (1998) 75.
[60] A. Pons, C.G. Dussap and J.B. Gros, Biotechnol. Bioeng., 33 (1989) 394.
[61] A. Becker, F. Katzen, A. Puhler and L. Ielpi L, Appl. Microbiol. Biotechnol., 50 (1998) 145.
[62] I.W. Sutherland and L. Kennedy, Microbiology, 142 (1996) 867.
[63] K.M. To, Food Ingred. Europe Conf. Proc., 4 (1997) 148.

[64] Y.M. Lo, K.L. Robbins, S. Argin-Soysal and L.N. Sadar, J. Food Sci., 68 (2003) 2057.
[65] W. Sabra, A.P. Zeng and W.D. Deckwer, Appl. Microbiol. Biotechnol., 56 (2001) 315.
[66] S.M.A.S. Keshk and K. Sameshima, Afr. J. Biotechnol., 4 (2005) 478.
[67] M. Tombs and S.E. Harding, An Introducion to Polysaccharide Biotechnology, Taylor and Francis, London, 1998.
[68] L. De Vuyst, F. De Vin, F. Vaningelgem and B. Degeest, Int. Dairy J., 11 (2001) 687.
[69] G.H. van Geel-Schutten, F. Flesch, B. ten Brink, M.R. Smith and L. Dijkhuizen, Appl. Microbiol. Biotechnol., 50 (1998) 697.
[70] D.J.C. van den Berg, G.W. Robijn, A.C. Janssen, M.L.F. Giuseppin, R. Vreeker, J.P. Kamerling, J.F.G. Vliegenthart, A.M. Ledeboer and C.T. Verrips, Appl. Environ. Microbiol., 61 (1995) 2840.
[71] A. Morin, in S. Dumitriu (ed.), Polysaccharides : Structural Diversity and Functional Versatility, Marcel Dekker, New York, 1998.
[72] F. Mozzi, G. Olive, G.S. de Giori and G.F. de Valdez, Biotechnol. Lett., 18 (1996) 435.
[73] F. Mozzi, G. Olive, G.S. de Giori and G.F. de Valdez, Milchwissenschaft, 51 (1996) 670.
[74] S.A. Kimmel and R.F. Roberts, Int. J. Food Microbiol., 40 (1998) 87.
[75] S.A. Kimmel, R.F. Roberts and G.R. Ziegler, Appl. Environ. Microbiol., 64 (1998) 659.
[76] B. Mollet, Cerevisia Belg. J. Brewing Biotechnol., 21 (1996) 63.
[77] B. Degeest, B. Janssens and L. De Vuyst, J. Appl. Microbiol., 91 (2001) 470.
[78] D.J.C. van den Berg, G.W. Robijn, A.C. Janssen, M.L.F. Giuseppin, R. Vreeker, J.P. Kamerling, J.F.G. Vliegenthart, A.M. Ledeboer AM and C.T. Verrips, Appl. Environ. Microbiol., 61 (1995) 2840.
[79] J. Cerning, C. Bouillanne, M.J. Desmazeaud and M. Landon, Biotechnol. Lett., 8 (1986) 625.
[80] M. Kojic, M. Vujcic, A. Banina, P. Cocconcelli, J. Cerning and L. Topisirovic, Appl. Environ. Microbiol., 58 (1992) 4086.
[81] J.A. Teggatz and H.A. Morris, Food Structure, 9 (1990) 133.
[82] L. De Vuyst, F. Vanderveken and S.D.B. van de Ven, J. Appl. Microbiol., 84 (1998) 1059.
[83] T. Mitsuoka, Microbes in the Intestine, Yakult Honsha Co., Tokyo, Japan, 1989.
[84] D.B. Hughes and D.G. Hoover, Food Technol., 45 (1991) 74.
[85] D.G. Hoover, Food Technol., 47 (1993) 120.
[86] J. Ballongue, in S. Salminen and A.V. Wright (eds.), Lactic Acid Bacteria, Marcel Dekker, New York, 1993.
[87] K. Kailasapatha and J. Chin, Immunol. Cell Biol., 78 (2000) 80.
[88] M. Poch and A. Bezkorovainy, J. Dairy Sci., 71 (1988) 3214.
[89] T. Mitzutani and T. Mitsouka, J. Natl. Cancer Inst., 63 (1979) 1365.
[90] R. Takiguchi, M. Ohe and M. Miyamoto, Bifidus-Flores, Fructus Semina, 8 (1985) 7.
[91] C.M. Roberts, W.F. Fett, S.F. Osman, C. Wijey, J.V. O'Connor and D.G. Hoover, J. Appl. Bacteriol., 78 (1995) 463.
[92] S.A. Andaloussi, H. Talbaoui, R. Marczak and R. Bonaly, Appl. Microbiol. Biotechnol., 43 (1995) 995.
[93] J. Cerning, C. Bouillanne and M. Landon, J. Dairy Sci., 75 (1992) 692.
[94] I.W. Sutherland, Adv. Microb. Physiol., 23 (1982) 79.
[95] I.W. Sutherland, in J. Baddiley, N.H. Carey, I.J. Higgins and W.G. Potter (eds.), Biotechnology of Microbial Exopolysaccharides, Cambridge University Press, New York, 1990.
[96] L. Garmar-Nourani, K. Blondeau and M.J. Simonet, J. Appl. Microbiol., 85 (1998) 664.
[97] J. Cerning, C.M.G.C. Renard, J.F. Thibault, C. Bouillanne, M. Landon, M. Desmazeaud and L. Topisirovic, Appl. Environ. Microbiol., 60 (1994) 3914.
[98] J. Cerning, FEMS Microbiol. Rev., 87 (1990) 113.
[99] S.V.N. Vijayendra and R.C. Gupta, Industrial Dairyman, 44 (1992) 595.
[100] L. Garmar-Nourani, K. Blondeau and J.M. Simonet, J. Appl. Bacteriol., 83 (1997) 281.

[101] M. Racin, J. Dumont, C.P. Champagne and A. Morin, J. Appl. Bacteriol., 71 (1991) 233.
[102] K.A. Ludbrook, C.M. Russell and R.I. Greig, J. Food Sci., 62 (1997) 597.
[103] B. Degeest, F. Vaningelgem and L. De Vuyst, Int. Dairy J., 11 (2001) 747.
[104] I.W. Davidson, FEMS Microbiol. Lett., 3 (1978) 347.
[105] J.F. Kennedy, P, Jones and S.A. Barker, Enzyme Microb. Technol., 4 (1982) 39.
[106] P. Souw and A.L. Demain, Appl. Environ. Microbiol., 37 (1979) 567.
[107] I.W. Sutherland, in A.H. Rose (ed.), Advances in Microbial Physiology, Academic, New York, 1982.
[108] M.I. Tait, I.W. Sutherland and A.J. Clarke-Sturman, J. Gen. Microb., 132 (1986) 1483.
[109] C. Nirmala and Purushothaman, Nat. Acad. Sci. Lett., 14 (1991) 71.
[110] M.A. Qadeer and S. Baig, Pakistan J. Sci. Ind. Res., 30 (1987) 886.
[111] F. Garcia-Ochoa, M.A. Garcia-Leon and A. Romero, Chem. Biochem. Eng. Q., 4 (1990) 15.
[112] R.W. Schweickart and A.V. Quinlan, J. Biomech. Eng., 111 (1989) 166.
[113] H.U. Peters, I.S. Suh, A. Schumpe and W.D. Deckwer, Can. J. Chem. Eng., 70 (1992) 742.
[114] O. Vashitz and M. Sheintuch, Biotechnol. Bioeng., 37 (1991) 383.
[115] G. Liden, Bioproc. Biosyst. Eng., 24 (2002) 273.
[116] C.L. Cooney, Science, 219 (1983) 728.
[117] A.J. Hacking, Economic Aspects of Biotechnology, Cambridge University Press, Canbridge, 1986.
[118] B. Reisman, Economic Analysis of Fermentation Processes, CRC Press, Boca Raton, 1988.
[119] W.R. Penny, Chem. Eng., 78 (1971) 89.
[120] M.E. Van Marle and P. Zoon, Netherlands Milk Dairy J., 49 (1995) 47.
[121] M.J. Pearse and A.G. MacKinlay, J. Dairy Sci., 72 (1989) 1401.
[122] D. Auerswald, W. Buchheim and K. Guthy, Milchwissenschaft, 44 (1989) 560.
[123] J. Guillement, P. Marchal, F. Mariette and C. Tellier, J. Agric. Food Chem., 41 (1993) 2259.
[124] C. Noik, F. Monot, D. Ballerini and J. Lecourtier, in P.L. Dubin and P. Tong (eds.), Colloid-Polymer Interactions, American Chemical Society, Columbus, 1993.
[125] A. Bax and S. Grzesiek, Accounts of Chemical Res., 26 (1993) 131.
[126] D. Uhrin, J.R. Brisson and D.R. Bundle, J. Biomolecular NMR, 3 (1993) 367.
[127] C.H. Sotak, D.M. Freeman and R.E. Hurd, J. Magn. Reson., 78 (1988) 355.
[128] A. Otter and D.R. Bundle, J. Magn. Reson., 109 (1995) 194.
[129] M.M. Corsaro, W.D. Grant, S. Grant, C.E. Marciano and M. Parrilli, European J. Biochem., 264 (1999) 554.
[130] W.H.M. van Casteren, C. Dijkema, H.A. Schols, G. Beldman and A.G.J. Voragen, Carbohydr. Res., 324 (2000) 170.
[131] C. Bouillanne, J. Cerning, , M.J. Desmazeaud and M. Landon, Biotechnol. Lett., 10 (1988) 255.
[132] K. Blondeau, C. Bosso, P. Colin-Morel, L. Gamar-Nourani, C. Gey, A. Heyraud, J. Simonet and C. Vanhaverbeke, Carbohydr. Res., 314 (1998) 211.
[133] M.T. Duenas-Chasco, J.L. Espartero, J.L. Gil-Serrano, A. Irastorza-Iribas, M.A. Rodriguez-Carvajal and P. Tejero-Mateo, Carbohydr. Res., 307 (1998) 125.
[134] J.F. Fu, R.Y. Chang and Y.H. Tseng, Appl. Microbiol. Biotechnol., 37 (1992) 225.
[135] G.J. Grobben, J. Sikkema, M.R. Smith and J.A.M. de Bont, J. Appl. Bacteriol., 79 (1995) 103.
[136] A.D. Welman and I.S. Maddox, J. Ind. Microbiol. Biotechnol., 30 (2003) 661.
[137] M.E. Pintado, A.I.E. Pintado and F. X. Malcata, J. Dairy Sci., 82 (1999) 2315.
[138] M. Racine, J. Dumont, C.P. Champagne and A. Morin, J. Appl. Bacteriol., 71 (1991) 233.
[139] J.H. Flatt, T.A. Cooper, J.M. Gonzalez, D.E. Dogger, E.N. Lightfoot and D.C. Cameron, Biotechnol. Prog., 8 (1992) 335.
[140] I.C. Boels, R. van Kranenburg, J. Hugenholtz, M. Kleerebezem and W.M. de Vos, Int. Dairy J., 11 (2001) 723.
[141] A. Laws, Y. Gu and V. Marshall, Biotechnol. Adv., 19 (2001) 597.

[142] P.W. Postma, J.W. Lengeler and G.R. Jacobson, Microbiol. Rev., 57 (1993) 543.
[143] C. Foucaud and B. Poolman, J. Biol. Chem., 267 (1992) 22087.
[144] A.D. Welman, I.S. Maddox and R. Archer, Biotechnol. Lett., 25 (2003) 1515.
[145] F. Krzewinski, C. Brassart, F. Gavini, and S. Bouquelet, Current Microbiol., 32 (1996) 301.
[146] F. Krzewinski, C. Brassart, F. Gavini, and S. Bouquelet, Current Microbiol., 35 (1997) 175.
[147] K. Ovesejevi, V. Graz and F. Batista-Viera, Biotechnol. Tech., 12 (1998) 143.
[148] J. Szczodrak, Acta Biotechnol., 19 (1999) 235.
[149] B. Splechtna, I. Petzelbauer, B. Kuhn, K.D. Kulbe and B. Nidetzky, Appl. Biochem. Biotechnol., 98 (2002) 473.
[150] G.A. Ramsdell and B.H. Webb, J. Dairy Sci., 28 (1945) 677.
[151] T.E. Timell, Can. J. Chem., 42 (1964) 1456.
[152] M. Proulx, P. Ward, S.F. Gauthier and D. Roy, Lait, 74 (1994) 139.
[153] D. St-Gelais, D. Roy, S. Haché and M. Desjardins, J. Dairy Sci., 76 (1993) 3327.
[154] N. Gorret, J.L. Maubois, M. Ghoul and J.M. Engasser, J. Appl. Microbiol., 90 (2001) 779.
[155] R.W. Coughlin and M. Charles, in W.H. Pitcher Jr. (ed.), Immobilized Enzymes for Food Processing, CRC Press, Boca Raton, 1980.

Bioprocessing for Value-Added Products from Renewable Resources
Shang-Tian Yang (Editor)

Chapter 23. Microbial Production of Bioplastics from Renewable Resources

Jian Yu

Hawaii Natural Energy Institute , School of Ocean and Earth Science & Technology, University of Hawaii, 1680 East-West Rd. Honolulu, HI 96822, USA

1. INTRODUCTION

Polyhydroxyalkanoates (PHAs) are a family of biopolyesters synthesized and accumulated by many bacteria as carbon storage materials [1]. They are natural biopolymers synthesized and decomposed via microbial metabolisms, but can be melted and molded like petrochemical thermoplastics. PHAs have attracted academic and industrial attention because of their potential use as biodegradable thermoplastics. Reviews have been made on their molecular biology, physiology, biochemistry, material properties, blending and processing, and biodegradation [1–8]. This chapter is focused on the production of these bioplastics from renewable resources as a potential alternative to petrochemical plastics.

1.1. Petrochemical plastics and their environmental impact

Since the discovery of polyethylene in 1933 and an industrial boom during the 1950's, petrochemical plastics have grown into a major, mature manufacturing industry. The annual production of four major commodity thermoplastic resins (polyethylene, polypropylene, polystyrene and polyvinyl chloride) in 2003 was around 33 million metric tons in the United States, 30 million tons in the EU, and 35 million tons from China, Japan and Korea [9]. These thermoplastics are cheap, durable, lightweight, easy to process, and highly resistant to chemical and biological degradation. These advantages make them widely used in our lives, from improved packaging to new textiles, from kitchenware to vehicles, and from single-use disposable goods to cutting-edge cellular phones, computers and digital storage media. The widespread and increasing use of these durable plastics, however, has raised concerns about their potential adverse impacts on ecosystems. Disposed plastics make up 11% of our trash by weight today, increased from 1% in 1960 [10]. Tons of plastic litter are discarded and finally discharged into the oceans via rivers and municipal drainage [11]. Plastic debris is reduced to microscopic size and accumulates in marine environments [12]. Although its final fate is not yet clear, some harmful effects on marine animals and coral reefs have already been reported [11]. Plastic waste typically makes up 35–55% by volume of the non-biodegradable component of municipal wastes sent to landfill [13]. Reducing landfill volume is a challenge for many municipalities worldwide.

The adverse impact of synthetic plastics on the environment could be significantly reduced if many single-use, nondurable goods were made from biodegradable plastics. Municipal plastic wastes in the US totaled about 25 million tons in 2001, and about 68%, or 17 million tons, were containers, packaging, and nondurable goods, such as diapers, trash bags, utensils, and medical devices [10]. This indicates that more than 50% of durable synthetic resins are being used for nondurable purposes. If the nondurable goods were made from biodegradable plastics, then 50 million tons of plastic wastes that would otherwise accumulate in the environment would be eliminated. Early efforts were made with various blends of petrochemicals and natural polymers, such as polyethylene and starch. This type of blends or composites is not truly biodegradable, but their material properties are, to some extent, deteriorated [14]. The ideal biodegradable thermoplastics, or bioplastics, should have the same or similar thermal and mechanical properties as their synthetic counterparts, but be decomposable into benign products by natural microorganisms when left in the environment. In addition to the environmental benefits, bioplastics can also benefit healthcare and agriculture through such products as controlled-release drugs, pesticides, fertilizers, and biodegradable mulch films [15]. In addition, the biodegradable containers and bags would significantly reduce the cost of the composting business [16].

1.2. Renewable feedstock for bioplastics

Synthetic plastics consume non-renewable fossil fuels that provide both power and raw materials, and approximately 80 million tons of oil and natural gas are consumed every year in the United States alone [17]. In contrast, most biodegradable polymers, including PHAs, are made from renewable feedstocks under relatively mild process conditions. Renewable resources become increasingly important to the manufacturing industry as oil reserves decline and sustainable manufacturing becomes a concern of the industry [18]. One example is polylactic acid (PLA), a biodegradable thermoplastic polymer produced and sold by Cargill under the trade name Natureworks™. About 140,000 tons of the monomer lactic acid are made annually from corn sugar via microbial fermentation. The manufacturer claims that bioprocessing renewable feedstock uses less fossil resources and releases less greenhouse gas than does a petrochemical process [19].

Three types of renewable feedstock are available for PHA production: grains or their components, agriculture and forestry biomass, and industrial wastes. It is beyond the scope of this chapter to evaluate all of the renewable feedstock. Instead, feedstock from corn, biomass, and processing wastes are reviewed as potential raw materials for PHA fermentation. The enabling technologies can be extended to other renewable resources, such as bagasse, straw, and sawdust.

As a raw material in many important food, feed, chemical, and fuel applications, corn is the leading agricultural commodity in the United States. Nearly 10 billion bushels (255 million metric tons) of corn kernel were recently harvested [20]. Almost equal to the weight of kernels, about 208 million metric tons of corn biomass or stover, including stalks, cobs, leaves and husks, were also produced [21]. A portion of the corn biomass (~28 wt%) is left in the fields as fertilizer and for erosion control, but most of it is largely underutilized [22]. The kernel can be directly used for microbial fermentation after dry milling and starch saccharification with amylases or refined into fine products, including glucose, fructose, dextrose, oil, and proteins, via wet milling [21]. Both processes are used in large capacities for ethanol fermentation. With about 2 billion bushels of kernel being wet processed each year, around 5 million tons of corn fiber are discarded as processing waste [21]. Sold as animal feed at \$0.03-0.04 lb^{-1}, this feedstock is underutilized and has a high potential for PHA fermentation because of its availability as fine particles at wet milling sites, thus avoiding transportation and handling costs [23]. Table 1 compares the chemical composition of corn kernel, stover and fiber.

Table 1
Representative composition (weight %) of corn kernel, stover and fiber [21, 23]

Component	Kernel	Stover	Corn fiber
Starch glucan (%)	71.7	-	-
Cellulose glucan (%)	2.7	40.9	37.2
Hemicellulose (%)	6.2[a]	24.3[b]	32.4[c]
Lignin (%)	0.6	11.0	7.8
Protein (%)	9.5	8.9	11.0
Oil, fat (%)	4.3	1.3	2.5
Ash (%)	1.4	7.2	0.6
Others (%)	3.6	6.4	8.5

Note: a. as xylose
b. xylan (21.5%) + arabinan (1.8%) + galactan (1.0%)
c. xylan (17.6%) + arabinan (11.2%) + galactan (3.6)

2. POLYHYDROXYALKANOATES (PHAs)

2.1. Chemical structure and material properties

Since the discovery of poly (3-hydroxybutyrate) (P3HB), the most common polyhydroxyalkanoate, as a carbon storage material in *Bacillus megaterium* in 1926 [24], more than 100 hydroxyalkanoic acids have been identified as the monomers of bacterial polyesters [5].

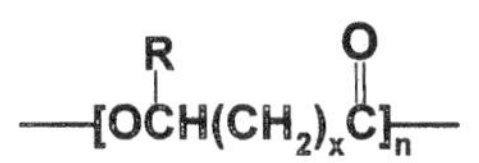

X=1,2,3,4; n= 200 - 12,000
R=H, CH_3, C_2H_5, C_3H_7,...

Fig. 1. The general chemical structure of polyhydroxyalkanoates (PHAs).

Fig. 1 shows the general chemical structure of PHAs. Many PHAs are formed from precursors or structurally related substrates [25]. The hydroxyalkanoate monomers that are supplied via native cell metabolism are usually 3-hydroxyalkanoates (3HAs) and almost all 3HAs are in *R* configuration due to the stereo-specificity of enzymes involved in PHA biosynthesis. The high stereoregularity makes the polyesters optically active, and some PHAs are highly crystalline. The hydroxyalkanoates can be roughly divided into two groups, short-chain-length hydroxyalkanoates (scl-HA) of 3 to 5 carbons and medium-chain-length hydroxyalkanoates (mcl-HA) of 6 to 14 carbons. The size of the side chain (**R**) greatly affects the material properties of PHAs. The small side chains, such as methyl and ethyl groups, of scl-P3HAs result in a stiff material with high crystallinity, high tensile modulus, and low elongation at break, while the large side chains (C3 to C14) of mcl-P3HAs make the material elastic with relatively low crystallinity and melting temperature, but improved elongation at break [26, 27]. Manipulating the side chains and compositions of P3HA copolymers, therefore, can create new polymers with desired material properties. A copolymer of 3-hydroxybutyrate (90 mol%) and 3-hydroxyhexaonate (10 mol%), P(3HB-co-10mol% 3HHx), for example, shows mechanical properties similar to those of low-density polyethylene, very different from P3HB [27].

Table 2 compares the physical properties of petrochemical commodity polymers and a few representative PHA biopolymers. Because of its perfect stereoregularity, P3HB undergoes a detrimental aging process at ambient temperatures, resulting in a brittle material with less practical applications while newly-molded P3HB is ductile [28, 29]. Large cracks appear during secondary crystallization because the reorganization of the initial lamellar crystals tightly constrains the amorphous chains between the crystals [29]. The ductility of P3HB polymers can be improved by introducing large side chains, such as ethyl and propyl groups, into the polyester backbone to disturb or reduce the crystal lattice [27]. Table 2 shows that the propyl group in P(3HB-co-10 mol% 3HHx), P3HB3HHx, is more efficient than the ethyl group in P(3HB-co-10 mol% 3HV), P3HB3HV, in making PHA ductile and increasing the material's elongation at break. The crystal lattice of P3HB3HHx is actually reduced to about half of the P3HB's lattice, while 3-hydroxyvalerate (3HV) barely reduces the P3HB3HV lattice [27]. The 3HV monomers are actually incorporated into P3HB's crystal lattice, a phenomenon called isodimorphism [28].

Molecular size is another important parameter that can be controlled for improved material properties. The weight-average molecular weight (Mw) of P3HB produced by wild-type bacteria such as *Ralstonia eutropha* ranges from 100 kDa to 1,000 kDa with a polydispersity (Mw/Mn) of around 2 [8]. The molecular size is affected to some extent by the carbon source, nutrients, and environmental conditions, such as pH and the nitrogen/carbon ratio [35]. High nutrient concentration increases the molecular size of P3HB, with smaller amounts of P3HB being synthesized and accumulated [36]. A recombinant *E. coli* harboring *R. eutropha* PHA genes (phaCAB) produces an ultra-high molecular weight P3HB (MW 1,100-11,000 kDa) from glucose [37]. The P3HB has a high crystallinity (80%) and reduced solubility in chloroform. After being hot-drawn and annealed, its tensile strength and elongation at break are increased to 175 MPa and 104%, respectively [31].

Table 2

Mechanical properties of synthetic plastics and typical PHAs [30–34]

Property	PP	LDPE	PS	PVC	P3HB	UP3HB	P3HB 3HV	P3HB 3HHx	P3HB 4HB
Melting temp. (°C)	168	123	-[a]	-[a]	177	182	140	127	150
Glass temperature (°C)	-20	-36	90	110	4	4	-1	-1	-7
Crystallinity (%)	60	30	-[a]	-[a]	70	80	60	34	45
Young's modulus (GPa)	1.3	0.4	3.2	3.2	3.5	0.97	0.8	0.5	-[a]
Tensile strength (MPa)	36	20	36	46	43	175	20	21	26
Elongation at break (%)	350	530	2	60	5	104	50	400	444
Notched Izod impact strength (J/m)	50	NB[b]	24	580	60	-[a]	110	-[a]	-[a]

Note: a. Not available; b. No Break.

Abbreviations: PP; polypropylene; LDPE; low-density polyethylene; PS; polystyrene; PVC; polyvinyl chloride; P3HB; poly(3-hydroxybutyrate); UP3HB; ultra-high-molecular-weight poly(3-hydroxybutyrate) (hot drawn and annealed); P3HB3HV; poly(3-hydroxybutyrate-co-10% mol 3-hydroxyvalerate); P3HB3HHx; poly(3-hydroxybutyrate-co-10% mol 3-hydroxyhexanoate); P3HB4HB; poly(3-hydroxybutyrate-co-16% mol 4-hydroxybutyrate).

Incorporating longer monomers, such as 4-hydroxybutyrate (4HB) and 4-hydroxyvalerate (4HV), into the PHA backbone can also change the material properties of PHA polymers [38, 39]. As the 4HB content in poly(3-hydroxybutyrate-co-4-hydroxybutyrate) (P3HB4HB) increases, the crystallinity of the copolymer declines and the material ductility increases [39].. In microbial PHA biosynthesis, precursors of 4HB (e.g., 1,4-butanediol) and 4HV (e.g., 4-ketovaleric acid or levulinic acid) are supplied either as the sole carbon source or as co-substrates with glucose [38].

2.2. PHA physiology

The broad spectrum of PHA-producing bacteria includes different taxonomical groups, such as phototrophic bacteria, archaebacteria, Gram-positive and negative bacteria, and aerobic and anaerobic bacteria [1, 2]. Some microbial species exclusively synthesize short-chain-length PHAs (scl-PHAs) and some make only medium-chain-length PHAs (mcl-PHAs). Few wild types are able to synthesize copolymers of scl- and mcl-hydroxyalkanoates [5]. The microbes synthesize and store the polyesters from a generous supply of carbon source when a complete range of nutrients is not available for cell growth. The biosynthesis of PHA is promoted by deficiency in one or multiple nutrients, including sulfate, magnesium, nitrogen, phosphate, and oxygen [40, 41]. PHA is an ideal carbon storage material due to its low solubility and high molecular weight, causing negligible osmotic stress to the bacterial cells [1]. Once the polyesters are synthesized, they serve as both carbon and energy sources during starvation. The PHA content in most bacteria, however, is low, ranging from 1 to 30 wt% of cell mass [2].

PHAs are stored inside of cells as discrete granules of 0.2–0.5 μm in diameter, as shown in Fig. 2. Suspended in cytoplasm, the inclusions contain about 5 to 10 wt% of water and are largely amorphous [42]. Each granule is surrounded by a phospholipid monolayer membrane in which proteins, including PHA synthase and degradase, are located [43]. Other proteins (phasins) are presumed to be involved in the stabilization of the amorphous hydrophobic PHA inside the hydrophilic cell cytoplasm [44, 45]. One phasin protein from *R. ruber* has two short hydrophobic stretches close to the C-terminus of the protein. These regions have been suggested to be responsible for the binding of the proteins to the PHA inclusions [46]. The proteins and other cellular components around the PHA granules affect the recovery and purification of biopolyesters from the cell mass.

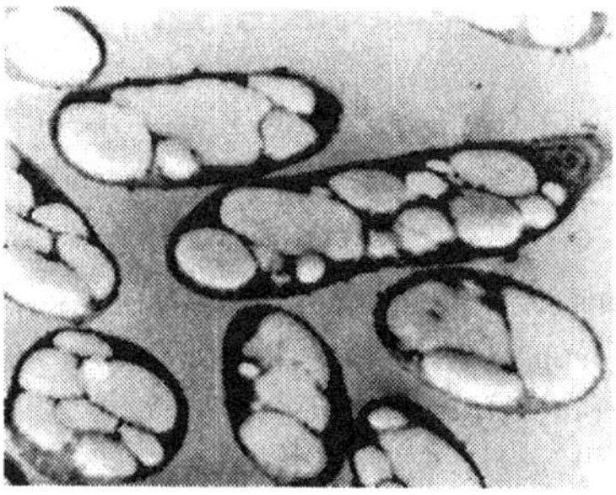

Fig. 2. PHA granules accumulated in *R. eutropha* cells under a transmission electron microscope (University of Hawaii)

2.3. PHA biosynthesis pathways

Starting from two common renewable carbon sources, sugars and oils, scl-PHAs, mcl-PHAs, and copolymer scl-mcl-PHAs are synthesized, depending on microbial strains as shown in Figure 3. *R eutropha* is a representative scl-PHA producer using sugars and short-chain organic acids as the carbon source [47]. Three enzymes are involved in the biosynthesis of P3HB in *R. eutropha* from acetyl-CoA [1]: β-ketothiolase (PhaA) condenses two acetyl-CoA moieties into acetoacetyl-CoA, NADPH-dependent reductase (PhaB) reduces acetoacetyl-CoA into (R)-3-hydroxybutyryl-CoA, and PHA synthase (PhaC) incorporates 3HB into the growing P3HB backbone. In the presence of propionic acid, the cells incorporate 3-hydroxyvalerate into the PHA backbone, forming a copolymer, poly(3-hydroxybutyrate-co-3-hydroxyvalerate), P3HB3HV. A second β-ketothiolase in *R. eutropha* with broad substrate specificity is involved in the condensation of acetyl-CoA and propionyl-CoA into β-ketovaleryl-CoA, the precursor of 3-hydroxyvalerate monomer [48].

Pseudomonas, especially those belonging to rRNA homology group I, are representative of mcl-PHA producers. *P. oleovorans* grows on n-alkane, n-alkene, and long-chain fatty acids and accumulates copolymers of mcl-PHA. Monomer units of carbon-even substrates ranging from C6 to C14 or of carbon-odd substrates ranging from C7 to C13 are supplied via β-oxidation of acyl-CoA [49, 50]. Fatty acids shorter than C6 are not incorporated into the growing PHA backbone by PHA synthase, but are converted into acetyl-CoA, which is utilized via the TCA cycle for energy and as carbon source for cell growth [51]. Conversion of fatty acid β-oxidation intermediates into PHA precursors is presumably completed by enzymes enoyl-CoA hydratase (PhaJ) and 3-ketoacyl-CoA reductase (FabG) [52, 53]. Many *Pseudomonas* spp. can also utilize carbohydrates and other structurally unrelated substrates, such as acetate and lactate, to synthesize mcl-PHAs. The precursors of PHA are provided by de novo fatty acid synthesis and transferred from ACP form to CoA form by enzyme 3-hydroxyacyl-CoA-ACP transferase (PhaG) [54]. *Pseudomonas putida* cultivated on glucose accumulates mcl-PHA containing a major unit of 3-hydroxydecanoate (C10) and minor units of C6, C8, C12, and C14 along with some unsaturated C12 and C14 containing single double bonds [55]. A broad spectrum of PHAs can be synthesized by *Pseudomonas* on related or unrelated substrates, and these specialty biopolyesters may have niche applications.

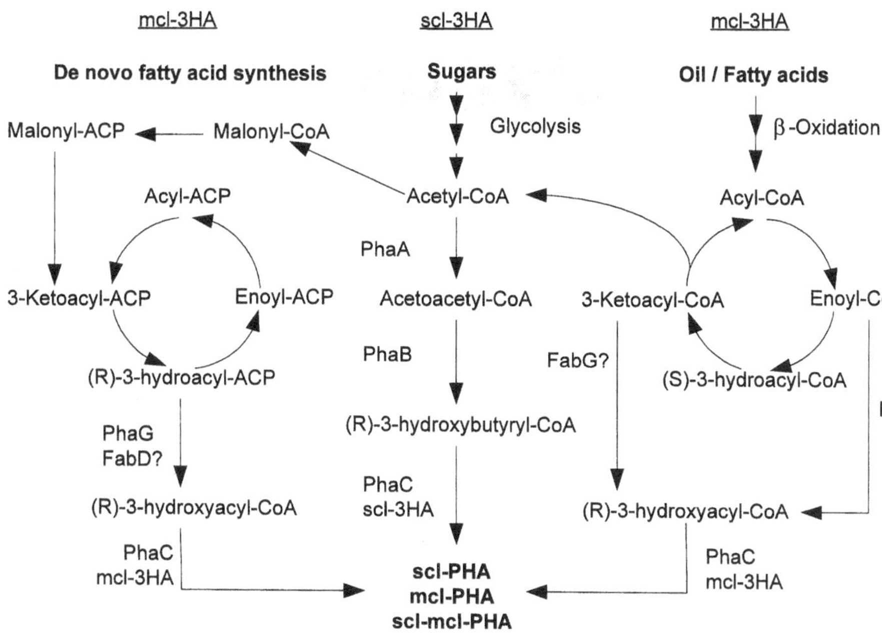

Fig. 3. Biosynthesis pathways of short chain length (scl-)PHA, medium chain length (mcl-)PHA and short-medium-chain length (scl-mcl-)PHA from sugars and oils. PhaA, β-ketothiolase; PhaB, NADPH-dependent acetoacetyl-CoA reductase; PhaC, PHA synthase; PhaG, 3-hydroxyacyl-ACP-CoA transferase; PhaJ, (R)-enoyl-CoA hydratase; FabD, malonyl-CoA-ACP transacylase; FabG, 3-ketoacyl-CoA reductase [3, 7, 8].

3. MICROBIAL PHA FERMENTATION

The amount of PHA produced per volume per time is a technical and economical indicator of microbial fermentation and can be improved by increasing the amount of cell mass per volume per time (cell productivity) and the amount of PHA per cell (specific productivity). The latter also affects polymer recovery after fermentation. The overall volumetric productivity depends on three factors: substrates (including nutrients), the microbial strain, and fermentation technology. To a great extent the microbial strain determines the fermentation performance and PHA properties, but the cost of substrates and bioreactor operation also play an important role in process economy. The carbon sources of most PHA fermentations are pure substrates or their mixtures, including glucose, sucrose, and short or long chain fatty acids. Fed-batch fermentation is a popular method of achieving high cell density and productivity. Efforts have also been made in using cheap raw materials, such as potato processing waste [56] and food scraps [57]. Organic waste streams are usually mixtures of solids and liquids with a complex chemical composition. Therefore, new bioreactor and fermentation technology is needed in order to use this type of raw material. In commercial PHA fermentation, the microbial strain should be able to grow quickly on simple and cheap substrates and nutrients, have high yields of cell mass and PHA but a low oxygen respiration rate, produce the right PHA with the desired molecular size and composition, and accumulate a large amount of polyester in its cells (>50 wt% of dry cell mass). Although many bacteria can synthesize and accumulate PHAs [1, 2], few meet the requirements for large capacity commercial PHA fermentation.

3.1. Pure substrates

Most PHA fermentations in academic research and industrial development were conducted in chemically defined or semi-defined media prepared with pure substances. The results provide invaluable information on metabolism, material properties, and fermentation efficiency in PHA production capacities ranging from several grams to thousands of kilograms.

3.1.1. scl-PHAs synthesis by R. eutropha

Ralstonia eutropha is a representative producer of scl-PHA, usually poly(3-hydroxybutyrate) (P3HB) and poly(3-hydroxybutyrate-co-3-hydroxyvalerate) (P3HB3HV) when one or more essential nutrient is limited. The cells can accumulate PHA polymers in quantities of up to 80 wt% of the dry cell mass. This nonsporulating, Gram-negative aerobe grows on simple carbon sources and mineral salts, including fructose, organic alcohols and acids. A glucose-utilizing mutant was the workhorse of the first industrial production of PHA by ICI in the 1980s [30]. In the presence of glucose and propionic acid (or valeric acid), a precursor of 3-hydroxyvalerate (3HV), *R. eutropha* synthesizes P3HB3HV copolymers with a 3HV content of 5 to 25 mol%, sold under the trade name Biopol^{TM}. Because of the relatively inefficient incorporation of propionic acid, a more expensive substrate than glucose, into the growing PHA backbone, a high acid concentration was maintained in the culture medium, resulting in a high substrate cost and a potential inhibitory effect on cell metabolism [58].

Two-stage fed-batch fermentation was developed for the industrial production of PHA. In the first stage, *R. eutropha* was grown on glucose until reaching a pre-determined cell mass concentration when one essential nutrient (e.g. phosphorus) in the growth media became limited. The second stage of PHA accumulation was started by continuously feeding carbon substrates (glucose and propionic acid) with little or no limiting nutrient(s) in the media [59]. Lack of essential nutrient(s) in this stage should have no adverse effect on PHA biosynthesis, but would limit cell growth. This technology has not been changed very much since it was adopted by ICI in the 1980's. The optimal results of this technique were demonstrated in a laboratory bioreactor fed with solutions of propionic acid and glucose [60]. *R. eutropha* cells grew to a dry mass concentration of 113 g L^{-1} containing 57 wt% P3HB3HV with 14 mol% 3HV. The PHA productivity was around 1.64 g L^{-1} h^{-1}. In P3HB fermentation on glucose only, the strain had a high productivity of 4.6 g P3HB L^{-1} h^{-1} and a high P3HB content of 80 wt%. The relatively low P3HB3HV content and moderate productivity were attributed to the high propionic acid concentration, which was necessary in order to give a desired P3HB3HV composition (10–20 mol% 3HV). An optimal feeding strategy of glucose and propionate improved the P3HB3HV content to 78 wt% of dry cell mass with 16 mol% 3HV [61].

Batch and fed-batch fermentations are widely used in industrial bioprocesses. The latter is more efficient than the former in achieving high concentrations of cells and products because the medium composition can be controlled and substrate inhibition at high initial concentrations is avoided. It uses, however, only a portion of the bioreactor capacity during the feeding period, and its turnaround between two batches takes tens of hours, resulting in high operation costs. A continuous culture system provides an alternative, which consists of two bioreactors in series, the first for cell growth under a balanced nutrient supply and the second for PHA synthesis under limited nitrogen. At a dilution rate of 0.08 h^{-1}, the system produced 1.23 g P3HB L^{-1} h^{-1}, with a yield of 0.36 g P3HB (g glucose)$^{-1}$. The PHA content was 72 wt% of the dry cell mass [62].

3.1.2. scl-PHA synthesis by other strains

Alcaligenes latus is another good strain for scl-PHA production, particularly for the homopolymer P3HB. It accumulates P3HB up to 50–60 wt% of dry cell mass from sucrose during cell growth, resulting in a relatively short fermentation time on an alternative carbohydrate [63]. In batch fermentation, the growth yield ($Y_{x/s}$) was around 0.4 g cell (g sugar)$^{-1}$ and the product yield ($Y_{p/s}$) around 0.24 g P3HB (g sugar)$^{-1}$ [64]. When cell growth was controlled under nitrogen-limited conditions, laboratory fermentation, fed with a high sucrose solution (900 g L^{-1}), reached a high productivity of 4.9 g P3HB L^{-1} h^{-1} and a yield of 0.42 g P3HB (g sucrose)$^{-1}$ [65].

The PHA synthesis gene (phaCAB) of *R. eutropha* has been cloned and expressed in *E. coli* and *Klebsiella* strains in order to utilize other carbohydrates, such as sucrose [66, 67]. A productivity of 1 g PHA L^{-1} h^{-1} was achieved with a recombinant *Klebsiella* on sucrose [67]. A high productivity of 2.8 g P3HB L^{-1} h^{-1} in a fed-batch culture of a recombinant *E. coli* was also achieved under vigorous agitation and oxygen-enriched aeration [68]. An *E. coli* harboring the *A. latus* P3HB gene can produce P3HB on sucrose or P3HB3HV on sucrose and propionic acid, with PHA productivities similar to that of wild types [69, 70]. The precursor of 3-hydroxyvalerate of copolymer P3HB3HV in the recombinant strain is formed from structurally related substrates, not through cell metabolism.

The 3HV precursor of P3HB3HV can also be supplied *in vivo* through cell metabolism on structurally unrelated substrates. A *Pseudomonas* sp. was cultivated on glucose to produce P3HB3HV in the absence of propionic or valeric acids. In a fed-batch culture, the cell mass concentration reached 38 g L^{-1} in 45 h, containing 53 wt% P3HB3HV with 7.5 mol% 3HV. The productivity was around 0.84 g PHA L^{-1} h^{-1} [71].

3.1.3. scl-PHAs containing 4-hydroxyalkanoates

P. putida GPp104 harboring the PHA biosynthesis gene from *T. pfennigii* accumulated a copolyester (3HB: 3HV: 4HV = 0.8: 69.2: 30), which accounted for 42 wt% of its dry cell mass, when cultivated in a mineral medium with levulinic acid as the sole carbon source [72]. In the same medium, recombinant *R. eutropha* accumulated a copolyester (3HB: 3HV: 4HV = 46.4: 51.5: 2.1), which comprised up to 66% of its dry cell mass. In two-stage fed-batch cultivation in a 25-L pilot fermenter, cells were first grown on glucose at 30°C and then shifted to biosynthesis of 4HV-containing polyesters under nitrogen starvation at 35°C on levulinic acid and gluconic acid. In about 100 h (20 h in stage 1), the cell density reached 25g L^{-1} containing 50 wt% PHA and 15 mol% 4HV, or a PHA productivity of 0.125 g L^{-1} h^{-1}. When the culture was scaled up to 600 L, similar fermentation performance was achieved, including 19.7 g L^{-1} of cell mass containing 50 wt% PHA and 34.6 mol% 4HV. The productivity was 0.073 g PHA L^{-1} h^{-1} [72].

3.1.4. scl-mcl-PHAs

Copolymers of scl-mcl-hydroxyalkanoates, such as P3HB3HHx, exhibit good ductility and toughness, similar to those of low density polyethylene (Table 2). *Bacillus megaterium* accumulates PHA copolymers consisting of 95 mol% of 3HB, 3 mol% of 3HHp (3-hydroxyheptanoate) and 2 mol% of an 8-carbon HA [73]. Few wild-type strains, however, can produce the scl-mcl-copolymers, either because their monomer precursors cannot be supplied through native metabolic pathways, or because the PHA polymerase (phaC) has strict substrate specificity. This chain-length limitation is lifted when the PHA synthase gene is expressed in a heterologous host that can provide a wide range of HA monomers. The PHA synthase (phaC) of *R. eutropha* can actually incorporate small amounts of 3-hydroxyhexonate (3HHx) and 3-hydroxyoctonate (3HO) monomers into the PHA backbone [74, 75]. A recombinant strain of PHA-negative *R. eutropha* harboring a PHA synthase gene from *Aeromonas caviae* was able to produce random copolymers of 3HB and 3HHx from vegetable oils and accumulated a high content of PHA (about 80% w/w) [76]. Recently, it was reported that a recombinant *R. eutropha* harboring a PHA gene of *Pseudomonas sp.* could synthesize P3HB3HHx on sugars [77]. Further improvement on PHA yield, content, and the mole composition of copolyesters in the recombinant strains will lead to industrial P3HB3HHx fermentation on renewable resources.

A large-scale fermentation (200 m^3) of *Aeromonas hydrophila* producing P3HB3HHx with 10 mol% 3-hydroxyhexonate on glucose and lauric acid was recently reported [78]. The strain was cultivated on glucose with essential nutrients to a pre-determined cell mass concentration and then directed to PHA synthesis and accumulation on lauric acid under nitrogen or phosphorous limitation. With a PHA productivity of 0.54 g L^{-1} h^{-1}, 1 kg of P3HB3HHx was produced from 2 kg of glucose plus 2 kg of lauric acid. The long chain fatty acid contains more energy than carbohydrates, resulting in a high PHA yield, but the water insoluble substrate also caused operational problems, including long fermentation time and broth foaming.

3.1.5. mcl-PHAs

Pseudomonas oleovorans is a representative mcl-PHA producer that relies on beta-oxidation to convert oils or long chain fatty acids into (R)-3-hydroxyacyl-CoAs. The precursor esters are condensed into PHA chains by two polymerases, phaC1 and phaC2. A genetically engineered *P. oleovorans* with an additional copy of phaC1 in its chromosome was cultivated in a chemostat at a dilution rate of 0.2 h^{-1}. The strain, grown on octanoic acid (C/N = 15), reached a cell concentration of 1.6 g L^{-1}, with 50 wt% of the cell mass being PHA, and a productivity of 0.15 g PHA L^{-1} h^{-1} [79]. This productivity was much lower as compared to continuous cultures of *P. putida* on oleic acid (0.69 g PHA L^{-1} h^{-1}) [80] and of a wild type *P. oleovorans* on octane (0.56 g PHA L^{-1} h^{-1}) [81]. In a pilot culture (25 L) of *P. putida* on technical oleic acid (80–90% w/v), the PHA productivity reached 0.57 g L^{-1} h^{-1} at a PHA yield of 0.56 g (g acid)$^{-1}$ [26].

3.2. Impure substrates

A large amount of organic matter is discarded from agriculture and food processing. Using these organic wastes as cheap feedstock for PHA production not only lowers the substrate cost, but also saves energy and cost in waste disposal. In contrast to pure substrates, such as glucose and organic acids, waste streams usually have complicated compositions. Some components are readily available for fermentation, while some are moderate or poor substrates, or even toxic to microbial cells. Fig. 4 shows the growth of *R. eutropha* on different carbon sources in the concentration range of 0.4 to 4 g L^{-1}. Compared with the control of no carbon substrate in the medium, the carbon substrates are classified as good (glucose, propanol, lactate, and butyrate), moderate (acetate, butanol, and citrate) and poor (ethanol and octanoate). The cells cannot utilize xylose and maltose because they lack the appropriate enzymes, and cell lysis occurs with acetone. In addition, organic waste streams often contain solids of different sizes. The presence of non-PHA solids in the final PHA-containing cell mass is a challenging problem for cost-effective PHA recovery and purification.

3.2.1. Industrial processing wastes

Organic wastes discharged from industrial processes, such as molasses and cheese whey, are readily available for PHA fermentation. Some type of pretreatment may be necessary before the waste stream can be utilized by microbial PHA-producers. Whey is the residual watery portion of milk from cheese manufacturing. The opaque liquid contains about 6% solids and has a high biological oxygen demand (32 g L^{-1}). In a fed-batch culture using whey as the feeding solution, recombinant *E. coli* harboring *R. eutropha*'s PHA gene utilized the organic matter and reached 69 g P3HB L^{-1} in 49 h [82], or a productivity of 1.4 g P3HB L^{-1} h^{-1}. *Azotobacter vinelandii* and its mutants can also accumulate a large amount of scl-PHA (75 wt% of dry cell mass) on beet molasses supplemented with 0.2% fish peptone [83]. Fatty acid mixtures, produced from oily waste of plant and animal origin by steam distillation and saponification, were used as the substrates for mcl-PHA production in *P. putida* [26]. In a 2-L bench-top bioreactor, the PHA productivity reached 0.33 to 0.56 g polymer L^{-1} h^{-1} with a yield of 0.38–0.46 g PHA (g substrate)$^{-1}$.

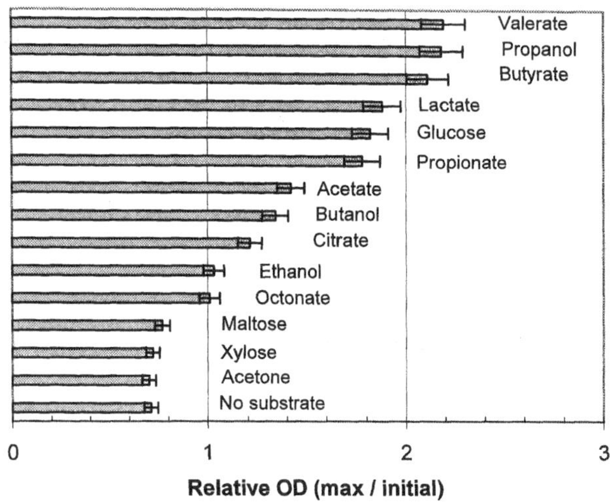

Fig. 4. Growth of *R. eutropha* on different carbon sources under aerobic conditions: 0-4 g substrate L^{-1} (C/N <5) at 30°C for 45 hours. Based on the initial optical density at 620 nm, the relative ODs are calculated from the maximum turbidities measured under the conditions.

3.2.2. Organic wastes

Many bacteria synthesize and accumulate PHAs from the organic matter in wastewater [84, 85] and municipal waste [86, 87]. Like lignocellulosic biomass, a large portion of the organic matter in waste streams is solid or a poor substrate for typical PHA-producers, such as *R. eutropha* (Fig. 4). Pretreatment plus enzymatic or biological conversion is often required to make the carbon source available to the PHA-producing cells [56, 57]. Furthermore, the PHA-producing cells should not be mixed with the non-PHA solids in the raw wastes; otherwise, PHA recovery would be costly. One approach was to use a microbial population under anaerobic conditions to digest and convert the organic matter into soluble carbon substrates, such as butyric and propionic acids. The fermentative intermediates were recovered for PHA fermentation [86, 88]. Although cell mass containing 50–60 wt% PHA was obtained, comparable to that from pure substrate PHA fermentation, the recovery of fermentative acids would be expensive, with high equipment and operation costs. This problem is, to a great extent, resolved by using an integrated biosystem [57]. The system consists of two chambers: one for anaerobic digestion and conversion of organic matter; and another for aerobic PHA synthesis using the fermentative intermediates as the substrates. The two chambers are separated by a barrier through which the soluble fermentative intermediates (plus other nutrients) are transferred from the anaerobic digestion to aerobic PHA synthesis under a driving force created and maintained by microbial activities as shown in Fig. 5. The utilization of fermentative intermediates for PHA biosynthesis at the right side is the sink of carbon flux from the left side.

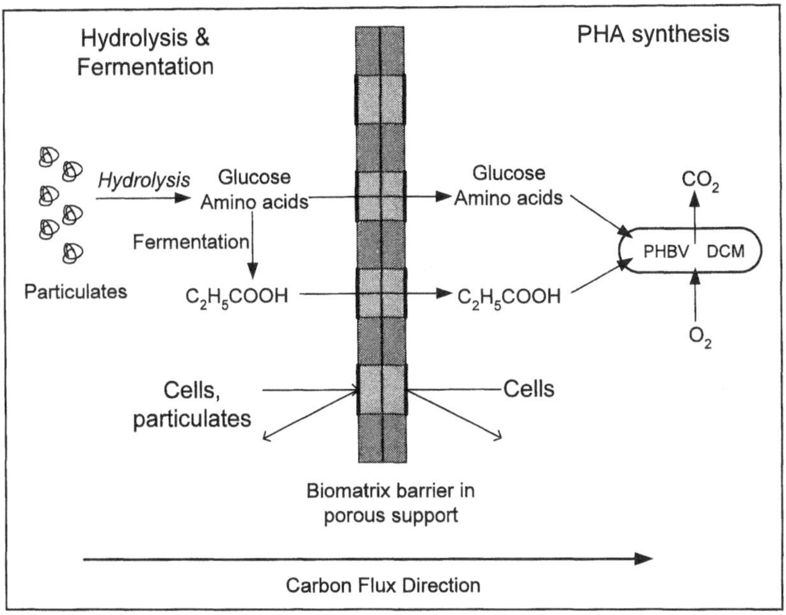

Fig. 5. A schematic integrated biosystem consisting of microbial hydrolysis and conversion of organic matter (left) and aerobic PHA biosynthesis on the soluble intermediates (right). The carbon flux from left to right is driven by concentration gradients across the barrier (40–50 g L^{-1} at the left side and <1 g L^{-1} at the right side). The driving force is created and maintained by two microbial actions.

This biosystem has been demonstrated in a 5-L laboratory facility. A slurry of food scraps was digested from an initial solid content of 17 wt% to 8 wt%, yielding a mixture of lactic (33 g L^{-1}), butyric (28 g L^{-1}), propionic (2 g L^{-1}), and acetic (5 g L^{-1}) acids. At the same time, the acids and other nutrients released from anaerobic digestion of food scraps were transferred through a membrane barrier into an aerobic mineral solution in which they were utilized for cell growth and PHA synthesis. The concentrations of acids were kept at very low level (5–150 mg L^{-1}) in the medium because of fast microbial utilization and PHA synthesis. The cell mass concentration and PHA content reached 22.7 g L^{-1} and 72.6 wt% in 72 h, or a PHA productivity of 0.24 g L^{-1} h^{-1} [57]. The PHA was a copolymer P3HB3HV with 3 mol% 3HV. A 50-L mini-pilot demonstration facility is underway at the University of Hawaii.

4. INTEGRATED BIOPROCESSING OF RENEWABLE FEEDSTOCK TO PHA

Corn biomass (stover and fibers) belongs to a large family of plant materials known as lignocellulose and has a complex structure of cellulose, hemicellulose and lignin (Table 1). Cellulose has a primarily crystalline structure, but can be broken down into glucose by cellulase. Hemicellulose, a polymer of xylose, arabinose, galactose, and other sugars, is not crystalline and thus is more readily hydrolyzed into its component sugars than is cellulose under thermal chemical hydrolysis [89]. Pentose is the primary sugar of hemicellulose hydrolyzate; some glucose (about 10 wt% of the total glucan) and lignin are also released and a portion of sugars underwent further reactions and formed byproducts, including acetic acid, furfural, hydroxymethyl furfural, and levulinic acid [90, 91]. The hydrolytic byproducts at moderate concentrations are inhibitory or toxic to many microbial strains and are removed before use in microbial fermentation, such as ethanol fermentation [89]. They also can be utilized by PHA-producing bacteria for the production of PHA copolymers, such as P3HB4HB from glucose and levulinic acid [72]. Fig. 6 shows a schematic process that integrates the biomass pretreatment and PHA production via simultaneous cellulose hydrolysis and PHA fermentation.

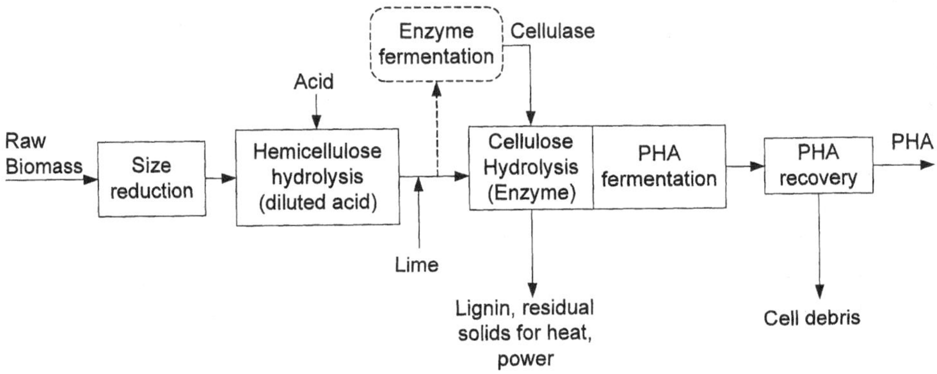

Fig. 6. A schematic process flow sheet of PHA fermentation on lignocellulosic biomass.

4.1. Biomass pretreatment

The conversion of biomass to PHA bioplastics is a capital-intensive process, and PHA should be produced year-around to achieve high returns on capital investment. Corn fiber is a good feedstock candidate because of its year-round availability at the processing facility. Corn stover, however, can be collected only during a limited harvest season. In order to achieve a high yield of glucose from enzymatic hydrolysis, different pretreatment methods are used for different raw materials, including size reduction and thermal chemical hydrolysis of hemicellulose into soluble sugars to expose cellulose to cellulase.

Steam explosion and dilute acid pretreatment are two technologies that can achieve high hemicellulose removal and good cellulose digestibility [92, 93]. Dilute acid hydrolysis recovers more than 80% of hemicellulose sugars; steam treatment recovers less than 65%. In a dilute sulfuric acid solution (0.1 wt%), biomass slurry was treated in a batch or flow-through reactor at 180°C for up to 5 h; more than 90% of xylan and 60% of lignin were removed. The treated biomass had an enzymatic digestibility of greater than 90% [94]. In contrast to the conventional acid concentrations of 0.7 to 3.0 wt%, this much diluted acid treatment relieves the problems associated with acid hydrolysis, including expensive anti-corrosive equipment, base consumption and gypsum formation during neutralization, and byproducts that are potentially toxic to microorganisms in the following fermentation [95]. The toxic compounds generated during dilute acid hydrolysis are mitigated by the addition of $Ca(OH)_2$ (overliming). Typical sugar syrup, containing 76 g L^{-1} of xylose, 14 g L^{-1} of arabinose and mannose, and 13 g L^{-1} of glucose, gypsum, and other residues, is fermentable by yeasts and bacteria [96]. Removal of the toxic byproducts by using ion exchange and activated carbon adsorption is also effective, but adds extra cost [97]. Other technologies including irradiation [98], alkaline delignification [99, 100], ammonia freeze explosion [101], and steam-SO_2 explosion [102] have also been investigated in an effort to reduce the cost of sugar from both hemicellulose and cellulose, but with limited success.

4.2. Simultaneous saccharification and PHA fermentation

Enzymatic hydrolysis following dilute acid pretreatment is the enabling technology for the biomass program of the US Department of Energy [103]. Enzymes provide an almost theoretical yield of glucose with less toxic byproducts than other technologies, such as concentrated acid hydrolysis. Substantial progress has been made in lowering the enzyme cost by a factor of up to ten [103]. The target price of cellulase is \$10 lb^{-1}; this would place it in the same range as that of amylase for starch hydrolysis, making it possible to use commercial cellulase enzymes to prepare cheap sugars from treated cellulose [104]. Otherwise, a fermentation unit is needed to provide cellulase on site by using about 4–9% of the biomass feedstock as shown in Fig. 6. The fungus *Trichoderma reesei* is a popular aerobic producer of cellulase on lignocellulosic sugars [105]. The wild-type strains need a substrate inducer, such as cellulose, to excrete a large amount of cellulase when the easily digestible sugars (e.g. glucose) are unavailable. Many mutant strains of *T. reesei* and other recombinant microbial species have been derived to enhance the yield and rate of cellulase release on different substrates.

Cellulase actually consists of three active components that work synergistically to break cellulose down into glucose [93]: endoglucanase breaks cellulose chains and generates free ends; exoglucanase (or cellobiohydrolase) attacks these ends in order to release cellobiose molecules; and β-glucosidase splits one cellobiose molecule into two molecules of glucose. The release of glucose during cellulose hydrolysis is affected by the condition of the pretreated cellulose, the enzyme and the process conditions [106–108]. Cellulase activity, for example, is inhibited by hydrolytic products, particularly cellobiose and glucose, which results in reduced yield and cellulose hydrolysis rate, eventually completely stopping hydrolysis [109]. Because of this inhibitory effect, the glucose concentration in hydrolytic solutions is limited to only a few percent, a potential problem for high cell-density cultures used in PHA fermentation. Improvement has been made on pretreatment to reduce crystallinity and the levels of lignin, hemicellulose in the pretreated cellulose, and robust enzymes with high activity and low sugar inhibition. The simultaneous enzymatic hydrolysis and microbial fermentation in the same vessel, or simultaneous saccharification and fermentation (SSF), in which the sugars are immediately converted into metabolic products, such as ethanol, to reduce the inhibitory effect of sugars on cellulase is a breakthrough in bioprocessing [110]. As a result, the overall glucose utilization by SSF technology can reach around 90% of the theoretical value, giving a high final concentration of fermentation product for cost-effective recovery. The SSF configuration also reduces capital and operational costs by combining the time-consuming hydrolysis and fermentation in a single vessel and shortening the processing time [110].

4.3. Technical challenges to PHA fermentation

There are three major technical challenges when a PHA-producing bacterium is cultivated in a vessel also used for enzymatic hydrolysis of cellulose. First, the hydrolytic medium contains a large amount of pentose (xylose and arabinose) that must be fully converted to PHA for cost effectiveness, but PHA-producers such as *R. eutropha* cannot utilize pentose (Fig. 4). This problem can be solved by using recombinant *E. coli* or *R. eutropha* that can utilize both pentose and hexose [66, 67, 111]. Second, fungal cellulase usually has a high hydrolytic activity at low pH levels (pH 3–4). The low pH inhibits cell growth and PHA synthesis of most PHA-producing strains. Finally, a solid mixture of PHA-containing cells and residual biomass, such as fibers and lignin, would make polymer recovery very difficult, as discussed in the next section.

One approach is to use microbial species as a converter that can effectively grow in hydrolytic conditions and convert all the sugars into substrates for the PHA-producer [112]. The simple fermentative intermediates can then be fed to the PHA-producer for polymer synthesis. Cost saving can be further achieved if the two vessels are integrated for simultaneous microbial conversion, as shown in Fig. 5. This novel bioreactor configuration has been demonstrated on a laboratory scale to simultaneously convert organic wastes and perform PHA biosynthesis [57].

5. PHA RECOVERY

At the end of PHA fermentation under controlled conditions, the microbial cells accumulate a large amount of PHA polymers: 50–80 wt % of their dry cell mass. Fig. 3 shows the intracellular PHA inclusions (0.2–0.5 μm) in cytoplasm. The dried "plastic" cells can be directly molded into articles that, although not very strong, may have some applications in agriculture and aquaculture [113]. Recovery technologies have been developed to purify the biopolyesters to different extents for a variety of applications. The solid composition of the P3HB inclusions in *B. megaterium* consists of 97.7% polyester, 1.87% protein and 0.46% lipid [114]. A PHA product with a purity of less than 98% may contain non-PHA biomass such as peptidoglycan and protein [115]. The peptidoglycan fragments exhibit various biological activities in mammalian hosts (inflammatory, antitumor, and pyrogenicity), depending upon the size and composition of the fragments [116]. Compared with other bioproducts of small and large molecules, PHA recovery poses a unique challenge because of the native solid state of polymer particles. Two strategies are usually adopted to separate the PHA inclusions from the non-PHA biomass: PHA solution or non-PHA biomass dissolution. In the former, only PHA is dissolved in an appropriate organic solvent, and in the latter, the non-PHA cell components are digested and dissolved by chemical agents.

5.1. PHA solution

Solvent extraction of PHA is widely used in laboratory research to prepare small amounts of biopolyesters. It has also been used at the pilot- [72] and large-scales [78] for recovery of mcl-PHA and scl-mcl-PHA polymers. Unfortunately, only a few solvents can dissolve scl-PHAs, and the most popular ones are halogenated hydrocarbons, such as chloroform and dichloromethane [117]. Although the wet PHA-containing cells can be directly extracted with water immiscible solvents [118], pretreatment of the cell mass, such as water removal at elevated temperature and the extraction of lipids/pigments by PHA-insoluble solvents such as acetone and methanol, is usually performed [119]. Solvent pretreatment not only makes the cells permeable [72], but also removes possible contaminants of the final products. The pretreated cell mass is subjected to extraction in hot chloroform or other appropriate solvents [120], and the dissolved PHA is further separated from the non-soluble cell mass by filtration or centrifugation. A viscous PHA solution is formed even at a relatively low PHA concentration (5% w/v), which causes problems in removing the non-PHA biomass and possible loss of a substantial amount of organic solvents. Precipitating the PHA by adding a PHA-insoluble solvent, such as methanol, facilitates the final separation of PHA from its solution [121]. Solvent extraction gives very pure and almost intact PHA macromolecules. Its major drawbacks include [115]: a large amount of organic solvent needed to make a dilute solution (<5 wt % PHA), lengthy extraction and filtration because of the solution's high viscosity, extra facility and operation costs for solvent recovery, and the possible loss of a large amount of toxic and volatile organic solvents into the environment.

5.2. Non-PHA cellular mass dissolution

For cells containing more than 50 wt % PHA, a highly purified PHA (>98 wt%) can also be prepared by dissolving the non-PHA biomass while the PHA granules are left in solid state. The non-PHA cell mass is generally composed of peptidoglycan, proteins, nucleic acids, lipids and lipopolysaccharides. Cellular organelles and fragments are easily separated from the PHA granules by centrifugation because of the high solid density of PHA (1.2–1.3 $g.cm^{-3}$) [1]. Depending on the dissolving agents and process conditions, the non-PHA biomass can be dissolved in non-selective or selective solutions.

5.2.1. Non-selective dissolution

Dissolution of non-PHA cellular biomass in alkaline hypochlorite solution is a simple and effective process [122], but unfortunately, PHA macromolecules are also seriously degraded, resulting in a low PHA yield and finished polymers of low molecular weight (22 kDa) [123]. Many factors affect the purity, recovery yield, and molecular size of the final PHA polymers, including the concentrations of non-PHA biomass and hypochlorite, temperature, pH, time, and pretreatment of the cell mass. A careful control of these conditions may result in good purity (>90%) and recovery (>90%) while also keeping molecular size above 400 kDa [124]. A highly pure PHA (98%) was also obtained in a strong alkaline solution (pH 13.6) at the expense of PHA recovery (78%) and molecular size (250 kDa).

The non-PHA biomass of recombinant *E. coli* harboring the PHA gene of *R. eutropha* is digested in simple alkaline solution. After treating in a NaOH or KOH solution (0.1 N) at 30°C for 1 h, cells with a P3HB content of 77 wt % can be purified to 91–92 wt% at a yield of 90–93% [125]. The relatively low concentration of base limits the use of this technique to a cell mass with a high initial PHA content. Otherwise, the purity is decreased [26]. A highly pure PHA can be achieved in a strong alkaline solution, but PHA loss and the reduction in molecular size become serious, as in the case of alkaline hypochlorite solutions. Fig. 7 shows the decomposition of P3HB into monomeric hydrolysis products, 3-hydroxybutyric acid and crotonic acid when lyophilized P3HB-containing cells are treated in sodium hydroxide solutions (0.1–1 N) at 70°C.

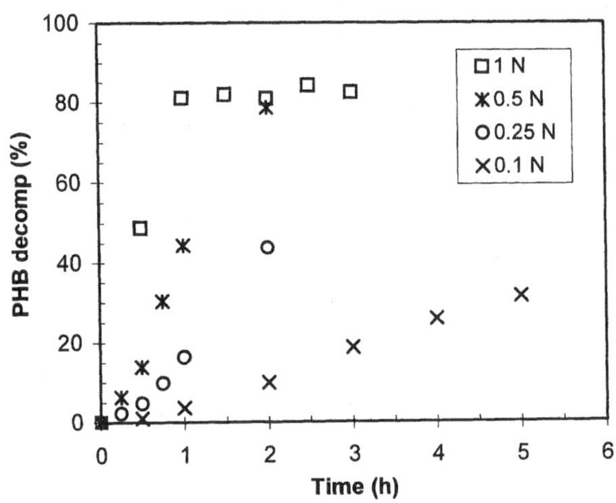

Fig. 7. Decomposition of P3HB granules in lyophilized *R. eutropha* cells (80 wt% P3HB) in an alkaline solution at 70°C. Two monomeric hydrolytic products, 3-hydroxybutyric and crotonic acids, were released and measured during P3HB hydrolysis.

5.2.2. Selective dissolution

In a selective solution, only non-PHA cellular mass is dissolved with little damage to PHA molecules. Proteolytic enzymes have high activities on hydrolysis and dissolution of proteins but little activity on PHA biopolymers [115]. Processing of PHA-containing cell slurry (60 wt% PHA) starts with heat treatment and is followed by enzymatic hydrolysis, surfactant treatment, and finally hydrogen peroxide decolorization. The selective dissolution of non-PHA biomass was chosen by ICI for the recovery of Biopol™ from *R. eutropha* cells [30] because of its small adverse impact on the environment and its production of high molecular-weight PHAs. Multiple steps and the high cost of agents, including enzymes and surfactants, e.g., sodium dodecyl sulfate (SDS), are the major drawbacks of this technique.

Anionic surfactants, such as SDS, can also help to dissolve the non-PHA cellular mass to a great extent with little degradation of the polyesters. *R. eutropha* cells containing 57 wt% of P3HB can be purified to 77–98 wt%, depending on pH (8–10) and the dosage of surfactant (0.25–5 wt%) [124, 125]. In a SDS and EDTA solution, the purity of P3HB3HV is increased from 64 wt% at the end of fermentation to 98%, with a recovery yield of 91% [126]. The technique, however, used a very high dosage of surfactant (0.24 g surfactant per g of cell mass), which would cause problems in wastewater treatment.

6. PHA ECONOMIC ANALYSIS

The annual capacity of early commercial PHA production by ICI was 200 to 300 tons and the price of P3HB3HV (Biopol™) in 1989 was around \$35 kg^{-1} (\$16 lb^{-1}) [30]. It was expected that the price would be below \$11 kg^{-1} (\$5 lb^{-1}) by the mid-1990s' when the capacity was increased to several thousand tons a year. The annual capacity of Biopol™ (Zeneca Bio Products, Billingham, UK) reached 1,000 tons in 1995 and the market price was around \$16 kg^{-1} (\$7.3 lb^{-1}). Compared to current prices of less than \$1.5 kg^{-1} (\$0.7 lb^{-1}) for petrochemical commodity plastics, the high price of PHA has been a major obstacle to its acceptance in the market. However, consumers may be willing to pay a premium rate for the environmental friendliness of PHA bioplastics. It is beyond the scope of this chapter to predict the acceptable price of PHA biopolymers. Although the selling price of synthetic biodegradable polyesters, such as poly-ε-caprolactone (PCL) at \$3–4 kg^{-1}, may serve as a good reference for PHA specialty polymers, the impact of PHAs as environmentally friendly plastics will not be seen until bioplastics are widely accepted at a competitive price. Economic evaluations have been performed to estimate the production costs of PHA in large capacities.

6.1. Process economy

Based on an annual production of 4,300 tons of P3HB by recombinant *E. coli* on glucose supplemented with complex nitrogen sources, the break-even selling price per kilogram of P3HB was \$6.08 in 1995 [13]. Assuming a ten year plant life with a 10% cash flow discount rate and a conservative process design based on available technologies, the analysis determined that the cost was highly sensitive to the P3HB expression level and the recovery strategy and moderately sensitive to the media cost and the cell growth yield. The cost was not affected very much by the maximum cell density achieved, but could be reduced to between \$5.63 kg^{-1} and \$3.59 kg^{-1} by using dairy whey as a partial replacement for glucose as the carbon source. The cost might be further reduced to \$2.67 kg^{-1} by using concentrated dairy whey and a significantly altered processing strategy as claimed by the analysis [13].

P3HB production at 2,850 tons a year was evaluated by cultivating *R. eutropha*, *A. latus*, *M. organophilum*, and recombinant *E. coli* on glucose (\$0.5 kg^{-1}), sucrose (\$0.3 kg^{-1}) and methanol (\$0.2 kg^{-1}), respectively [127]. The production cost ranged from \$5.58 to \$8.16 in 1996, not including the sales cost and investment return. *R. eutropha* gave the lowest production cost even though a relatively long culture time (50 h) on the relatively expensive substrate, glucose was needed. The cost was reduced by 15% to \$4.75 kg^{-1} when the capacity was increased to 100,000 tons a year. The raw material cost was correspondingly increased to around 48% of the total production cost. The fixed investment related costs accounted for 30% of the annual operation cost. In a similar process simulation for 5,000 tons of P3HB a year by *R. eutropha* on glucose, the break-even production cost in 2001 was around \$4.24 kg^{-1} [128].

Based on the above economic evaluations, the production cost of P3HB by *R. eutropha* on glucose using current technologies most likely will be \$4–5 kg^{-1}. P3HB3HV is expected to cost more than P3HB because propionic acid and other co-substrates (\$1–1.5 kg^{-1}) cost more than glucose (\$0.4–0.5 kg^{-1}), and fermentation performance is decreased in the presence of toxic short-chain fatty acids [34, 129]. Few economic evaluations have been performed on production of PHA copolymers. For the production of 5,000 tons a year of poly(3-hydroxybutyrate-co-5mol% 3-hydroxyhexaonate) using recombinant *R. eutropha* cultivated on soybean oil supplemented with mineral salts, the production cost was projected to be \$3.5–4.8 kg^{-1} in 2001 based on laboratory results: 40–50 h cultivation to reach 100–150 g L^{-1} of cell concentration with 80–85 wt% of PHA content and a high polymer yield of 0.7–0.8 g PHA (g oil)$^{-1}$ [128].

6.2. Fermentation cost

In order to obtain high volumetric productivity during PHA fermentation, high cell-density cultures of 80 to 150 g L^{-1} are usually pursued. One major cost factor of PHA fermentation is the supply of sufficient oxygen to the PHA-producing aerobes (such as *Ralstonia*, *Azotobacter*, *E. coli* and *Pseudomonas*) to satisfy their high respiration rate. The effect of oxygen starvation on the aerobes in two-stage fed-batch cultivation is not clear; oxygen starvation is believed to be just another growth-limiting factor during PHA accumulation, much like nitrogen or phosphorous limitation. Different strains may have different sensitivities and responses to oxygen starvation. The reductive step of PHA synthesis in *Azotobacter* appears to serve as an electron sink for the reduction of power accumulated as a consequence of oxygen limitation [130]. At low dissolved oxygen concentrations (1–3% air saturation), P3HB3HV synthesis from glucose and propionate by *R. eutropha* was slow, about 50–70% of the rate under oxygen-sufficient conditions, but the PHA yield per carbon source was increased [131]. With sufficient substrates and nutrients, *R. eutropha* grows exponentially at ~0.3 h^{-1} and consumes 0.3 to 0.4 g O_2 (g cell)$^{-1}$ h^{-1} [35, 132]. Under oxygen starvation (0% air saturation), several distinct metabolites and the intermediates of the tricarboxylic acid cycle were released by *R. eutropha* in response to the accumulation of NADH [132]. Sufficient oxygen supply in the first growth stage, especially at the end of the first stage, when the cell concentration reaches the highest level, seems crucial to the physiology and metabolism of cells in the following PHA accumulation stage. Oxygen starvation at this point irreversibly damages the cells and results in deteriorated substrate utilization and PHA yield [133].

Very high cell density and PHA productivity have been achieved in laboratory bioreactors. However, these good fermentation results are usually obtained with small volumes, vigorous agitation, high volumetric energy inputs, oxygen enrichment, and feeding solutions with very high sugar concentrations (500–900 g L^{-1}). Not only is the sugar concentration in most cheap feedstock low to moderate (~100 g L^{-1}), but also the high capital and operational costs are not applicable to economical fermentation in large capacities. PHA productivity is to a great extent limited by the moderate oxygen transfer rate (OTR) in large bioreactors. The maximum OTR is determined by oxygen partial pressure and k_La, the volumetric oxygen transfer coefficient. The latter depends on volumetric energy input and aeration. A standard laboratory fermenter (15.2 cm in diameter) stirred by a single impeller at 750 rpm and with a superficial air velocity of 18.3 m^3 $(m^2\ h)^{-1}$, has a k_La value of 1,000 h^{-1}, or a maximum OTR of 6–7 g O_2 L^{-1} h^{-1}. In contrast, the k_La value of a large fermenter (58 m^3), operated at a superficial air velocity of 54 m^3 $(m^2\ h)^{-1}$ and with a high energy input of 39 kW m^{-3}, is around 370 h^{-1}, an OTR of 2.4 g O_2 L^{-1} h^{-1} [134]. Since most large fermenters have an energy input of 1–2 kW m^{-3} [127, 128], the aerobic cell respiration rate that can be maintained in the laboratory fermenters may not be satisfied in the large fermenters. Oxygen-enriched air and/or high-pressure operation can increase the partial pressure of oxygen and hence the OTR, but may not be practical in commercial PHA production.

Maintaining a high OTR in large fermenters is costly in terms of both equipment and operation. A moderate productivity (0.5–1.0 kg PHA L^{-1} h^{-1}) may be practical if it can be achieved at much lower capital and operational costs than high density cultures. In this regard, PHA production by transgenic plants gives an example of low investment and operation costs, but also have a low PHA productivity [135, 136]. Microbial PHA fermentation can produce single cells with a high PHA content (60–70 wt%) for easy recovery and purification. Low cost fermenters such as airlift and bubble column bioreactors can provide alternatives to mechanically agitated fermenters because of their simple structure, low equipment and maintenance costs, low energy consumption, around 0.5 kWh (kg oxygen)$^{-1}$, moderate oxygen transfer rate from air (1–2 g O_2 L^{-1} h^{-1}), and moderate PHA productivity (0.5–1 g PHA L^{-1} h^{-1}) [133, 134, 137].

6.3. Substrate cost

The raw materials, including the carbon source, are a major cost factor in microbial PHA production. With a typical P3HB yield of 0.3 g (g glucose)$^{-1}$ and a glucose market price of \$0.4–0.5 kg^{-1} [127, 128], the cost of the carbon source alone is above \$1.3 per kg of P3HB, leaving little room for PHA to compete with petrochemical commodity plastics (<\$1.5 kg^{-1}). It is possible to lower the production cost by using cheap carbon sources, such as molasses, cheese whey, and corn fiber hydrolysate [13, 127]. When lignocellulosic biomass becomes a cheap carbon source for microbial PHA production, the raw biomass delivered from the fields within an 80-km radius can support a plant processing 2000 metric tons of biomass a day at a cost of \$30–35 per ton [138]. The conversion cost of sugars from the biomass using current enzyme technology is around \$0.17 per kg of sugars produced and may be further reduced to \$0.10 kg^{-1} by 2010 [139]. The low price of biomass-derived sugars is the prerequisite for the National Biomass Program's target price of bioethanol in 2010, \$1.07 gal^{-1} [103, 138]. Compared to bioethanol's lower selling price of \$0.35 kg^{-1}, PHAs should have a better profit margin and be competitive with petrochemical plastics.

Unlike pure substrates, cheap carbon feedstock contains low to moderate concentrations of fermentable sugars and a significant amount of organic matters that may be toxic to or unable to be utilized by PHA-producing cells (Fig. 4). Cost-effective pretreatments must be developed in order to effectively utilize the carbon source. Cheap feedstock also contains solids that may not be easily removed before fermentation or digested afterwards with the non-PHA cellular mass. New bioreactor technology, therefore, is needed to use this type of complex feedstock without adding the extra cost of separating and purifying the sugars.

6.4. PHA recovery cost

Solvent extraction method for PHA recovery generally does not require a high purity of PHA-containing cell mass, an advantage over other recovery methods involving the dissolution or digestion of non-PHA biomass. Indeed, PHA can be recovered directly from activated sludge by using solvent extraction [73]. However, the cost of PHA recovery by solvent extraction is quite high, more than half of the total cost as found in a large-scale PHA production [78]. Also, solvent loss during the recovery process has a potentially adverse impact on the environment. PHA recovery by surfactant washing and hypochlorite digestion can save up to 50% of the cost of solvent extraction [127], but chemical degradation of PHA is an unsolved problem and the technique has not been tested in large-scale processing where the process conditions cannot be controlled as accurately as in small-scale applications.

PHA biopolyesters with high molecular weight and high purity may be cost-effectively recovered when hydrolytic enzymes become available at low price. For example, in order to treat cell slurry with a density of 50 g L^{-1} and containing 60 wt% PHA, the dosage of "Alcalase", an industrial enzyme from Novo Industries, is 1.5 AU per 100 g of non-PHA cell mass, or 16.8 g enzyme per kg PHA [115]. At an enzyme price of $48 per kg protein, it costs about $0.81 per kg PHA [104]. Obviously, a technology breakthrough is needed for cost-effective PHA recovery.

7. CONCLUSIONS

New discoveries and progress in the biotechnology and material science of polyhydroxyalkanoates have created new thermoplastic biopolymers that exhibit similar thermal and mechanical properties to those of petrochemical thermoplastics but are truly biodegradable and biocompatible. Microbial fermentation is a feasible technology for the commercial production of bioplastics. Its great impact on many aspects of life will be seen when biodegradable thermoplastics are produced in a cost-effective way from abundant renewable resources, in particular lignocellulosic biomass, the future low-cost carbon source for bio-based industrial products. Compared to bioethanol, which is increasingly important in replacing petroleum-based fuels, PHA has a higher market price and thus should also be competitive with their petrochemical counterparts. Through an integrated bioprocessing, the bioplastics can be co-produced with bioethanol, adding high value to the biorefinery of biomass. This will happen with further improvement in microbial strain, fermentation technology, and PHA recovery.

ACKNOWLEDGEMENT

The author appreciates the support from Hawaii Natural Energy Institute and the School of Ocean and Earth Science & Technology at the University of Hawaii.

REFERENCES

[1] A.J. Anderson and E.A. Dawes, Microbiol. Rev., 54 (1990) 450.
[2] H. Brandl, R.A. Gross, R.W. Lenz and R.C. Fuller, Adv. Biochem. Eng./Biotech., 41 (1990) 77.
[3] D. Byrom (ed), Biomaterials, MacMillan Publishers, Basingstoke, 1991.
[4] H. Verhoogt, B.A. Ramsay and B.D. Favis, Polymer, 35 (1994) 5155.
[5] A. Steinbuchel and H. E. Valentin, FEMS Microbiol. Lett., 128 (1995) 219.
[6] S. Y. Lee, Biotech. Bioeng., 49 (1996) 1.
[7] L.L. Madison and G.W. Huisman, Microbiol. Mol. Biol. Rev., 63 (1999) 21.
[8] K. Sudesh, H. Abe and Y. Doi. Prog. Polym. Sci., 25 (2000) 1503.
[9] Chemical & Engineering News, 82 (2004) 56.
[10] U.S. Environmental Protection Agency, Municipal Solid Waste in the United States 2001, http://www.epa.gov.
[11] J.G.B. Derraik, Mar. Pollut. Bull., 44 (2002) 842.
[12] R.C. Thompson, Y. Olsen, R.P. Mitchell, A. Davis, S.J. Rowland, A.W.G. John, D. McGonigle and A.E. Russell, Science, 304 (2004) 838.
[13] R.J. van Wegen, Y. Ling and A.P.J. Middelberg, Trans IchemE., 76 (1998) 417.
[14] P. P. Klemchuk, Polym. Degrad. Stabil., 27 (1990) 183.
[15] Y. Doi, Microbial Polyester, VCH Publishers, New York, 1990.
[16] R. Tillinger and L.De Baere, 10th Annu. BioEnviron. Polym. Soc., Alburquerque, NM, 2002.
[17] T. Gerngross and S. Slater, Sci. Am., August (2000) 36.
[18] M. Paster, J.L. Pellegrino and T.M. Carole, Industrial Bioproudcts: Today and Tomorrow, http://www.bioproducts-bioenergy.gov/pdfs/BioProductsOpportunitiesReportFinal.pdf. 2003.
[19] Cargill & Dow, http://www.cargilldow.com, 2004.
[20] U.S. Department of Agriculture. Crop Produciton: 2001 Summary. Rep. Cr Pr 3-1(01)c. Washington DC, 2001.
[21] P.J. White and L.A. Johnson, Corn Chemistry and Technology, AACC, St. Paul, MN. 2003.
[22] National Research Council, Biobased Industrial Products: Priorities for Research and Commercilization. National Academy Press, Washington DC, 1999.
[23] M. Gulati, K. Kohlmann, M.R. Ladisch, R. Hespell and R.J. Bothast, Bioresource Technol., 58 (1996) 253.
[24] M. Lemoigne, Bull. Soc. Chim. Biol. (Paris), 8 (1926) 770.
[25] B.H.A. Rehm and A. Steinbuchel, Int. J. Biol. Macromol., 25 (1999) 3.
[26] M.B. Kellerhals, B. Kessler, B. Witholt, A. Tchouboukov and H. Brandl, Macromolecules, 33 (2000) 4690.
[27] Y. Doi, S. Kitamura and H. Abe, Macromolecules, 28 (1995) 4822.
[28] P.J. Barham, P. Barker and S.J. Organ, FEMS Microbiol. Rev., 103 (1992) 289.
[29] G.J.M. de Koning and P.J. Lemstra, Polymer, 34 (1993) 4089.
[30] W.D. Luzier, Proc. Natl. Acad. Sci., 89 (1992) 839.
[31] Y. Aoyagi, Y. Doi and T. Iwata, Polym. Degrad. Stabil., 79 (2003) 209.
[32] J.J. Fischer, Y. Aoyagi, M. Enoki, Y. Doi and T. Iwata, Polym. Degrad. Stabil. 83 (2004) 453.
[33] E. A. Grulke, Polymer Processing Engineering, Prentice Hall, New Jersey, 1994.
[34] T. Tsuge, J. Biosci. Bioeng. 94 (2002) 579.
[35] J. Wang and J. Yu, J. Ind. Microbiol. Biotechnol., 26 (2001) 121.

[36] J. Wang and J. Yu, Process Biochem., 36 (2000) 201.
[37] S. Kusaka, T. Iwata and Y. Doi, J.M.S. Pure Appl. Chem., A35 (1998) 319.
[38] Y. Saito, S. Nakamura, M. Hiramitsu and Y. Doi, Polymer, 39 (1996) 169.
[39] Z. Zhu, P. Dakwa, P. Tapadia, R.W. Whitehouse and S.-Q. Wang, Macromolecules 36 (2003) 4891.
[40] R. Repaske and A.C. Repaske, Appl. Environ. Microbiol., 32 (1976) 585.
[41] A.C. Ward, B.I. Rowley, E.A. Dawes, J. Gen. Microbiol., 102 (1977) 61.
[42] G.J.M. de Koning and P.J. Lemstra, Polymer, 33 (1992) 3292.
[43] C. Lauzier, R.H. Marchssault, P. Smith and H. Chanzy, Polymer, 33 (1992) 823.
[44] E.S. Stuart, A. Tehrani, H.E. Valentin, D. Dennis, R.W. Lenz and R.C. Fuller, J. Biotechnol., 64 (1998) 137.
[45] H.E. Valentin, E.S. Stuart, R.C. Fuller, R.W. Lenz and D. Dennis, J. Biotechnol., 64 (1998) 145.
[46] U. Pieper-Furst, M.H. Madkour and F. Mayer, J. Bacteriol., 177 (1995) 2513.
[47] J. Yu and Y. Si, Biotechnol. Progress, 20 (2004) 1015.
[48] S. Slater, K.L. Houmiel, M.Tran, T.A. Mitsky, N.B. Taylor, S.R. Padgette and K.J. Gruys, J.Bacteriol., 180 (1998) 1979.
[49] H. Brandl, R.A. Gross, R.W. Lenz and R.C. Fuller, Appl. Environ. Microbiol., 54 (1988) 1977.
[50] G.W. Huisman, O. de Leeuw, G. Eggink and B. Witholt, Appl. Environ. Microbiol., 55 (1989) 1949.
[51] G. Du and J. Yu, Biotechnol. Lett., Process Biochem., 38 (2002) 35.
[52] T. Fukui, S. S. Yokomizu, S. Kobayashi and Y. Doi, FEMS Microbiol. Lett., 170 (1999) 69.
[53] K. Taugchi, Y. Aoyagi, H. Matsusaki, T. Fukui and Y. Doi, FEMS Microbiol. Lett., 170 (1999) 69.
[54] B.H.A. Rehm, N. Kruger and A. Steinbuchel, J. Biol. Chem., 273 (1998) 24044.
[55] H. Preusting, A. Nijenhuis and B. Witholt, Macromolecules, 23 (1990) 4220.
[56] D. Rusendi and J.D. Sheppard, Bioresource Technol. 54 (1995) 191.
[57] G. Du and J. Yu, Environ. Sci. Technol., 36 (2002) 5511.
[58] D. Byrom, US Patent No. 5,364,778 (1994).
[59] P. A. Holmes, US Patent 4,477,654 (1984).
[60] B.S. Kim, S.C. Lee, S.Y. Lee, H.N.Chang, Y.K. Chang and S.I. Woo, Enzym. Microb. Technol., 16 (1994) 556.
[61] G. Du, J. Chen, J. Yu and S. Lun, Biochem. Eng. J., 8 (2001) 103.
[62] G. Du, J. Chen, J. Yu and S. Lun, J. Biotechnol., 88 (2001) 59.
[63] O. Hrabak, FEMS Microbiol. Rev., 103 (1992) 251.
[64] E. Grothe, M. Moo-Young and Y. Chisti, Enzym. Microbiol. Technol., 25 (1999) 132.
[65] F. Wang and S. Y. Lee, Appl. Environ. Microbiol., 63 (1997) 3703.
[66] S. Slater, T. Gallaher and D. Dennis, Appl. Environ. Microbiol., 58 (1992) 1089.
[67] H. Zhang, V. Obias, K. Gonyer and D. Dennis, Appl. Environ. Microbiol., 60 (1994) 1198.
[68] F. Wang and S. Y. Lee, Appl. Environ. Microbiol., 63 (1997) 4765.
[69] J.-I. Choi, S.Y. Lee and K. Hang, Appl. Environ. Microbiol., 64 (1998) 4897.
[70] J.-I. Choi and S. Y. Lee, Appl. Environ. Microbiol., 65 (1999) 4363.
[71] H.-J. Son, K.-K. Kim, H.-S. Kim, Y.-K. Kim and S.-J. Lee, J. Ind. Microbiol. Biotechnol., 24 (2000) 36.
[72] V. Gorenflo, G. Schmack, R. Vogel and A. Steinbuchel, Biomacromolecules, 2 (2001) 45.
[73] R.H. Findlay and D.C. White, Appl. Environ. Microbiol., 45 (1983) 71.
[74] D. Dennis, M. McCoy, A. Stang, H.E. Valentin and Z. Wu, J. Biotechnol., 64 (1998) 177.
[75] R.V. Antonio, A. Steinbuchel and BHA Rehm, FEMS Microbiol. Lett., 182 (2000) 111.
[76] T. Fukui and Y. Doi, Appl. Microbiol. Biotechnol., 49 (1998) 333.

[77] K. Matsumoto, S. Nakae, K. Taguchi, H. Matsusaki, M. Seki and Y. Doi, Biomacromolecules, 2 (2001) 934.
[78] G.Q. Chen, G. Zhang, S.J. Park and S.Y. Lee, Appl. Microbiol. Biotechnol., 57 (2001) 50.
[79] M.A. Prieto, M.B. Kellerhals, G.B. Bozato, D. Radnovic, B. Witholt and B. Kessler, Appl. Environ. Microbiol., 65 (1999) 3265.
[80] G.N.M. Huijberts and G. Eggink, Appl. Microbiol. Biotechnol., 46 (1996) 233.
[81] H. Preusting, W. Hazenberg and B. Witholt, Enzym. Microbiol. Technol., 15 (1993) 311.
[82] H.H. Wong and S.Y. Lee, Appl. Microbiol. Biotechnol., 50 (1998) 30.
[83] W.J. Page, FEMS Microbiol. Rev., 103 (1992) 149.
[84] H. Chua, P.H.F. Yu, L.Y. Ho, Appl. Biochem. Biotechnol., 53/65 (1997) 627.
[85] D. Rusendi and J.D. Sheppard, Bioresource Technol., 54 (1995) 191.
[86] S. Lee and J. Yu, Resour. Conserv. Recy., 19 (1997) 151.
[87] H. Takabatake, H. Satoh, T. Mino and T. Matsuo, Water Sci. Technol., 42 (2000) 351.
[88] M.A. Hassan, Y. Shirai, H. Umeki, H. Yamazumi, S. Jin, S. Yamanoto, M. Karim and K. Nakanishi. Biosci. Biotechnol. Biochem., 61 (1997) 1465.
[89] L.R. Lynd, Annu. Rev. Energy Environ. 21 (1996) 403.
[90] E. Palmqvist and B. Hahn-Hagerdal, Bioresource Technol., 74 (2000) 25.
[91] J.Y. Cha and M.A. Hanna, Ind. Crops Products, 16 (2002) 109.
[92] J. R. Baker and R.P. Overend (Eds), Enzymatic conversion of biomass for fuels production. Symp. Ser. 566. ACS, Washington DC. 1994.
[93] C.E. Wyman, ed. Handbook on Bioethanol, Production and Utilization, Taylor & Francis, Washington, DC. 1996.
[94] B. Yang and C.E. Wyman, Biotech. Bioeng., 86 (2004) 88.
[95] Y.Y. Lee, P. Iyer and R.W. Torget, Adv. Biochm. Eng. Biotechnol., 65 (1999) 93.
[96] A. Martinez, M.E. Rodriguez, S.W. York, J.F. Preston and L.O. Ingram, Biotech. Bioeng., 69 (2000) 526.
[97] S.I. Mussatto and I.C. Roberto, Bioresour. Technol., 93 (2004) 1.
[98] A.W. Khan, J. Labrie, J. Mckeown, Rad. Phys. Chem., 29 (1987) 117.
[99] M.A. Millett, A.J. Baker and L.D. Satter, Biotechnol. Bioeng. Symp., 6 (1976) 125.
[100] W.E. Kaar and M.T. Holtzapple, Biomass Bioenergy, 18 (2000) 189.
[101] M.T. Holtzapple, J. Jun, G. Ashok, S.L. Patibandia and B.E. Dale, Appl. Biochem. Biotechnol., 28/29 (1991) 59.
[102] D.J. Schell, R. Torget, A. Power, P.J. Walter, K. Grohmann and N.D. Hinman, Appl. Biochm. Biotechnol., 28/29 (1991) 87.
[103] US Department of Energy, Biomass Program: Multi-Year Technical Plan, http://devafdc.nrel.gov/biogeneral/Program_Review/MYTP.pdf, 2004.
[104] J. Ferrel, http//:www.bioproducts-bioenergy.gov/pdfs/HistoryofOBPandCellulosicEthanol.pdf, 2004.
[105] H. Esterbauer, W. Steiner, I. Labudova, A. Hermann and M. Hayn, Bioresource Technol., 36 (1991) 51.
[106] H.E. Grethlein, Bio/Technolgy, 3 (1985) 155.
[107] V.S. Chang and M.T. Hlotzapple, Appl. Biochem. Biotechnol., 84/86 (2000) 5.
[108] J. Karlsson, J. Medve, F. Tjerneld, Appl. Biochem. Biotechnol., 82 (1999) 243.
[109] M. Takagi, Biotechnol. Bioeng., 26 (1984) 1506.
[110] J.D. Wright, C.E. Wyman and K. Grohmann, Appl. Biochem. Biotechnol., 18 (1987) 75.
[111] L.O. Ingram, T. Conway and F. Alterthum, U.S. Patent No. 5,000,000 (1991).
[112] J. Yu, J. Biotechnol. 86 (2001) 165.
[113] M.N. Baptist and F.X. Werber, US Patent No. 3,107,172 (1963).
[114] R. Griebel, Z. Smith and J.M. Merrick, Biochem. 7 (1968) 3676.
[115] P.A. Holmes and G.B. Lim, US Patent No. 4,910,145 (1992).

[116] P.H. Seidl and K.H. Schleifer, Biological Properties of Peptidoglycans, Gruyter, Berlin, 1986.
[117] M. Terada and R.H. Marchessault, Int. J. Biological Macromolecules, 25 (1999) 207.
[118] W. Baluhut, W. Gierlinger and F. Strempfl, US Patent No. 5,213,976 (1993).
[119] P.J. Senior, L.F. Wright and Barry Alderson, US Patent No. 4,324,907 (1982).
[120] R.M. Lafferty and E. Heinzle, US Patent No. 4,101,533 (1978).
[121] J. Walker and J.R. Whitton, US Patent No. 4,358,583 (1982).
[122] E. Berger, B.A. Ramsay, J.A. Ramsay, C. Chavarie and G. Graunegg, Biotechnol. Tech., 3 (1989) 227.
[123] D.C. Lundgren, R. Alper, C. Schnaitman, R.H. Marchessault, J. Bacteriol., 89 (1965) 245.
[124] B.A. Ramsay, J. Ramsay, E.Berger, C. Chavarie, G. Braunegg, US Patent No. 5,110980 (1992)
[125] J.-I. Choi and S.Y. Lee, Biotechnol. Bioeng., 62 (1999) 546.
[126] Y. Chen, Q. Xu, H. Yang and G. Gu, Procesess Biochem., 36 (2001) 773.
[127] J.-I. Choi and S.Y. Lee, Bioprocess Eng., 17 (1997) 335.
[128] M. Akiyama, T. Tsuge and Y. Doi. Polym. Degrad. Stabil., 80 (2003) 183.
[129] J. Yu and J. Wang, Biotechnol. Bioeng., 73 (2001) 458.
[130] P.J. Senior and E.A. Dawes, Biochem. J., 134 (1973) 134.
[131] G. Lefebvre, M. Rocher and G. Braungegg, Appl. Environ. Microbiol., 63 (1997) 827.
[132] D. Vollbrecht, Eur. J. Appl. Microbiol. Biotechnol., 15 (1982) 117.
[133] L.Z. Tavares, E.S. da Silva and J.G. da C. Pradella, Biochem. Eng. J., 18 (2004) 21.
[134] B. Atkinson and F. Mavituna, Biochemical Engineering and Biotechnology Handbook 2nd Macmillan Publishers Ltd. UK, 1991. p733.
[135] C. Nawrath, Y. Poirier and C. Somerville, Proc. Natl. Acad. Sci. USA, 91 (1994) 12760.
[136] Y. Poirier, C. Nawrath and C. Somerville, Biotechnol., 13 (1995) 142.
[137] M. Moresi, Biotechnol. Bioeng., 23 (1981) 2537.
[138] A. Aden, M. Ruth, K. Ibsen, J. Jechura, K. Neeves, J. Sheehan, B. Wallace, L. Montague, A. Stayton, J. Lukas, http://www.eere.energy.gov/biomass/pdfs/32438.pdf. 2002.
[139] M.F. Ruth and R.J. Wooley, http://devafdc.nrel.gov/pdfs/4913.pdf, 2000.

Bioprocessing for Value-Added Products from Renewable Resources
Shang-Tian Yang (Editor)

Chapter 24. Industrial Applications for Plant Oils and Lipids

Bernie Y. Tao

Department of Agricultural and Biological Engineering, Purdue University, 745 Agricultural Mall Drive, West Lafayette, Indiana, 47907, USA

1. INTRODUCTION

The history and chemistry of the industrial use of natural products and their derivatives are rich in tradition and technology. The chemistry and physics of natural materials encompass some of the most challenging, complex issues facing modern science. Rich in chemical reactivity, stereochemical diversity, and physical structure, natural materials form an extraordinary set of potentially industrially useful products. Many modern products, including plastics, fuels, chemical intermediates and fibers find their origins in natural products from plants and animals. Growing economic, environmental, and political concerns regarding petroleum resources and usage has created an increased interest in utilizing renewable, domestic, raw materials to supplement or replace petroleum-based industrial products. Recent forecasts indicate that efforts to reduce petroleum imports and environmental pressure resulting from petroleum use will spark double-digit annual growth for plant lipid-based chemicals and products over the next several years [1]. Soy-derived chemicals are predicted to see an annual growth of 14.9%, boosting demand above $1 billion by 2007. Methyl esters are anticipated to lead this growth with a 35% annual increase in biodiesel fuel applications.

The term 'oil' is used to describe both petroleum and plant lipids. This is not particularly surprising, since their physical properties are very similar: both are very slippery, non-water soluble, float on water, and easily burned. The chemical structures of vegetable oil and petroleum are also very similar: both are essentially hydrocarbons, although vegetable oil contains some additional oxygen atoms. While petroleum distillates offer a wider range of hydrocarbon structures, plant lipids offer mono-dispersed molecular weight distributions, with mono- or polyunsaturation and partial oxygenation. In fact, the chemical similarities between these materials allows them to be used in many similar products, such as fuels, plastics, lubricants, soaps, detergents, paints, and inks. Amides, esters, and acetates of plant lipids are currently being used as plasticizers, blocking/slip agents and mold-release agents for synthetic polymers. Lipids linked to amines, quaternary ammonium ligands, alcohols, phosphates and sulfur ligands are used as fabric softeners, surfactants, emulsifiers, corrosion inhibitors, anti-static agents, hair conditioners, ink carriers, biodegradable solvents, cosmetic bases, and perfumes. Complexes with aluminum, magnesium or other metal compounds have produced greases and marine lubricating materials. Oxidized lipids are used in the production of urethanes, polyester, and other plastics. In the case of energy fuels, methyl esters are currently used as lubricant additives in diesel internal combustion engines and have been proposed for use in jet aviation fuels.

Most of the technology for the industrial applications of plant lipids was developed following the Great Depression and funded by Henry Ford and the Farm Chemurgic Council. Their goal was to make industrial products from farm crops and thus to help farmers. Prior to World War II, most paint, polymers, and coating applications, such as linoleum and alkyd paints, were based on plant oils.

The industrial revolution of the 18th and 19th century harnessed the power of petroleum and coal to fuel the enormous economic growth of the 20th century. The vast abundance of these resources encouraged their utilization first as fuels and later in novel polymers and chemicals that have significantly changed life in the last century. Because of the limited seasonal availability of plant oils and the rapidly growing transportation and energy industry in the early 20th century, readily available petroleum grew to dominate the industrial energy landscape. With advances in chemistry and engineering, new applications were developed for petroleum in materials: petrochemicals, polymers, synthetic fibers, and coatings.

In the 21st century, our dependency on petroleum has grown so large that it often drives our nation's economic, industrial, and military policies and planning. This dependency has had significant impact on our national economic welfare because most US petroleum is obtained from foreign sources. The enormous utilization of petroleum over the last 50 years has led to the depletion of these non-renewable resources. Anticipation of the limited future availability of petroleum, along with record high prices for gasoline, natural gas, and other petroleum-derived products have spurred interest in plant-based renewable lipids to replace or supplement petrochemical resources.

The energy goals proposed by the US Department of Energy Roadmap for Agriculture Biomass Feedstock Supply in the United States (US Dept. of Energy, November 2003) declared that, by 2030, biomass based fuels will provide 5% of the nation's power, 20% of its transportation fuels, and 25% of its chemicals. This is equivalent to approximately 30% of the current US petroleum consumption in these areas. Domestically produced biofuels would significantly decrease demand for imported oil and strengthen critical rural economies as well as support long-term domestic energy security. Additionally, increasing concerns about global climate change support the use of biofuels, since their use would not produce a net increase in atmospheric carbon dioxide, thus reducing the buildup of atmospheric greenhouse gases. Although petroleum reserves may last for another half-century, development of a bio-based economy that is driven by completely renewable feedstock is inevitable.

For more detailed historical information, the reader may wish to refer to several historical reviews on plant-derived industrial products [2–4]. Reviews of the utilization of soybean-based products have been previously presented [5–7]. Note that since approximately 80% of plant oils are currently used in food or animal feed applications, much of the scientific knowledge of plant lipids and their uses were developed for these applications. An excellent summary of this information can be found in Bailey's Fats and Oils [8].

2. CHEMICAL COMPOSITION OF PLANT LIPIDS

In general, plant lipids include derivatives of fatty acids, such as triacylglycerides (TAGs) and glycerophospholipids (GPLs), a variety of aromatic and hydrocarbon-like compounds, such as sterols, carotenoids and terpenes, and waxes. Of these, TAGs and GPLs are the dominant industrially useful compounds found in most domestic crop plant seeds, such as soybean, corn, and canola.

2.1. Sterols

The main sterol in animal lipids is cholesterol, whereas plant sterols or phytosterols include campersterol, stigmasterol, and sitosterol. Sterols are usually present at 0.1% levels in most plant lipids and are useful in nutritional applications, but have not found any significant industrial uses other than as food additives.

Carotenoids are generally reddish-yellow polyene hydrocarbon compounds, such as carotenes and lycopene or xanthophylls, such as lutein. These are generally very sensitive to and react easily with oxygen due to their high degree of conjugated double bonds. They are widely held as nutritional antioxidants and used in a variety of foods for coloration. Terpenoids are generally the precursors of sterols and do not have any widespread industrial or food applications.

2.2. Glycerophospholipids

Glycerophospholipids (GPLs) are fatty acid diglycerides with a phosphatidyl ester attached to the terminal carbon (see Fig. 1). The terminal ester groups (X) are mainly ethanolamine, choline, serine, or inositol. GPLs are highly amphiphilic and normally are components of cellular or vesicle membranes. The commercial term lecithin is widely used for GPL mixtures, which contain mostly phosphotidyl choline. Lecithins are used in foods as emulsifiers and surface active agents for altering viscosity and crystallization properties. It also has industrial uses as an emulsifying agent in textiles, leather, cosmetics, paints, plastics, concrete release agent, and insecticides. The largest commercial source of lecithin is soybean oil, where it is present at levels of about 1–3%. The global market for lecithin is estimated to be in the range of 130,000 metric tons per year.

2.3. Triacylglycerides

Triacylglycerides (TAGs) are the primary industrial plant lipid used. TAGs consist of three fatty acid molecules (R, R', and R" in Fig. 2) connected to a glycerol molecule via ester linkages. The functional properties of the molecule are determined by its fatty acid composition, geometric configuration, and positional distribution. The fatty acid hydrocarbon chains range in length from 6 carbons to 24 carbons, with C16 and C18 being the predominant lengths found in TAGs from most domestic crops. Although linear, they may also include single or multiple carbon-carbon double bonds.

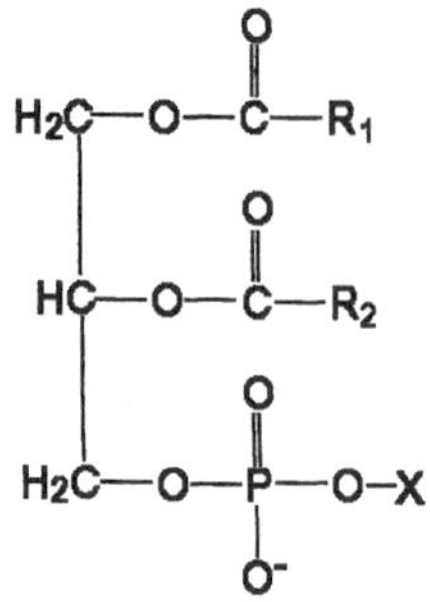

Fig. 1. General structure of glycerophospholipids, R_1 and R_2 indicate fatty acid groups and X indicates an ester-linked group.

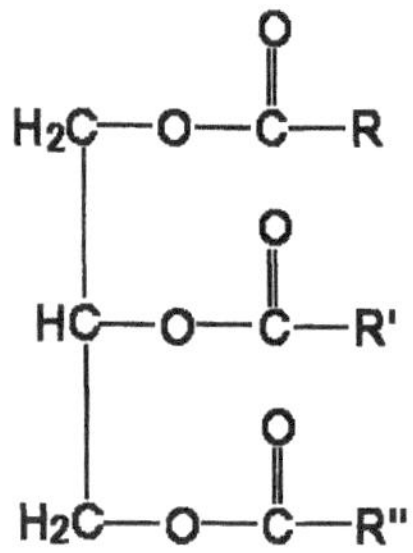

Fig. 2. General structure of triacylglyceride, R, R', and R" indicate fatty acid groups.

The major chemical-structure attributes that have the greatest effects on a TAG's physical properties are the chain length of the fatty acids, their position on the glycerol molecule, and the number and position of carbon double bonds in the fatty acid. Combinations of these structures give TAGs an enormous variety of physical properties, including thermal and crystallographic properties, chemical reactivity, and surface activity. These variations have found extensive applications in a number of industrial applications as well as in biological systems.

Examples of the structures of common C18 fatty acids are given in Fig. 3 below, as C18:0 (stearic acid, octadecanoic acid), C18:1 (oleic acid, 9-octadecenoic acid), and C18:2 (linoleic acid, 9,12-octadecenoic acid), respectively. Note that stearic acid contains only single carbon bonds, and thus its carbon atoms can rotate freely, hence this molecule has no configuration constraints. The inclusion of double bonds within the hydrocarbon chain in oleic and linoleic acids introduces a kink in the linear structure that constrains its configuration and can disrupt simple molecular crystal patterns. In addition to the degree of unsaturation, the stereochemical configuration of the double bonds (*cis* or *trans*) also has a significant effect on the physical and chemical properties of the oil. All naturally occurring fatty acids of plant origin are in the *cis* form. *Trans* fatty acids are generally formed when oils and fats are hydrogenated or heated at a high temperature. The *trans* configuration generally melts at a higher temperature than the *cis* configuration and has been linked to nutritional health issues.

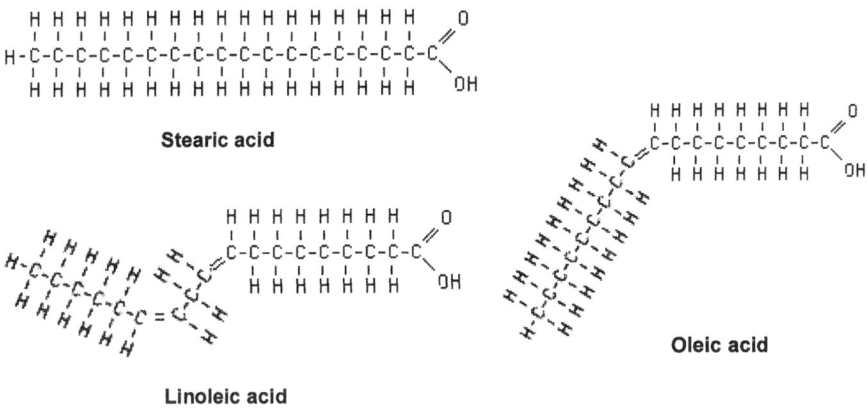

Fig. 3. Structures of common C18 fatty acids: stearic, oleic, and linoleic acids.

In U.S. domestic crops, fatty acid chain length generally ranges from 16 to 20 carbons, with the major oils (soybean, corn, canola) being predominantly C18. Tropical plants, such as coconut and palm, have lower molecular weight fatty acids in the C12–C16 range. Table 1 provides some fatty acid compositions of common vegetable oils.

The positional distribution of fatty acids on the glycerol molecule is another important factor in triacylglyceride physical properties. In general, chain length and the extent of fatty acid saturation are the dominant factors in determining physical properties. However, because of the free rotation of the single bonds among the glycerol and fatty acid carbons and constrained double bond configurations, TAG molecules can form in many different conformations, giving rise to different crystallographic patterns, a property known as polymorphism [9]. These variations in crystalline structure can dramatically affect the physical and thermal properties. For example, the polymorphic behavior of cocoa butter TAG, which is essentially a single chemical isomer, has at least seven different crystalline structures with different melting temperatures. This combines with variations in fatty acid type and distribution on the glycerol molecule to give TAGs an extensive range of crystalline structures. It may be of potential industrial interest to note that similar polymorphic crystallinity in synthetic hydrocarbons are responsible for electronic applications such as liquid crystal displays (LCD) and ferromagnetic liquids, although plant lipids are not yet used in these applications.

Table 1
Typical Fatty Acid Compositions of Common Plant Oils (Source: Excerpted from Bailey's Industrial Oil & Fat Products, Vol. 1, Y. H. Hui ed, pp 24-25, 1996)

Fat source	8 caprylic	10 capric	12 lauric	14 myristic	16 palmitic	18:0 stearic	18:1 oleic	18:2 linoleic	18:3 linolenic	20 arachidic	20:1 gadoleic	22 behenic
Canola					3.9	1.9	64.1	18.7	9.2	0.6	1.0	0.2
Cocoa butter				0.1	25.8	34.5	35.3	2.9		1.1		
Coconut	8.0	6.4	48.5	17.6	8.4	2.56	6.5	1.5		0.1		
Corn					12.2	2.2	27.5	57.0	0.9	0.1		
Linseed					4.8	4.7	19.9	15.9	52.7			
Olive					13.7	2.5	71.1	10.0	0.6	0.9		
Palm			0.3	1.1	45.1	0.1	4.7	38.8	9.4	0.3	0.2	
Peanut				0.1	11.6	3.1	46.5	31.4		1.5	1.4	3.0
Soybean				0.1	11.0	4.0	23.4	53.2	7.8	0.3		0.1

3. CHEMICAL MODIFICATIONS OF FATS AND FATTY ACIDS

Given the extent of organic and physical hydrocarbon chemistry, it is not surprising that industrial oleochemistry employs a wide range of thermal and catalytic reaction technologies as well as separation methods. Currently, the most widely used methods for altering the thermal and crystallization properties of fats are catalytic hydrogenation, hydrolysis, and transesterification, along with a variety of other chemical modifications.

3.1. Hydrogenation

The hydrogenation of unsaturated triglycerides and fatty acids is an important process in which these materials are chemically altered to change their physical and chemical properties. It is normally accomplished by the high-pressure catalytic reduction of unsaturated double bonds with gaseous hydrogen. The predominant catalyst used commercially is nickel, although recently other metallic and organic catalysts have been used to reduce *trans* isomerization reactions for food products. Partial hydrogenation results in a reduction in the number of double bonds and in the isomerization of the normal *cis* configuration of carbon double bonds to form the *trans* conformation. Extensive hydrogenation results in both the elimination of carbon double bonds and the reduction of the carboxylate terminus, producing fatty alcohols.

3.2. Hydrolysis/Transesterification

Hydrolysis splits oils into fatty acids and glycerol by the addition of water, usually by using alkaline or enzymatic catalysts. Partial hydrolysis is also used to make mono- and diacylglycerides that are frequently used as food emulsifiers and can be further chemically modified by linkage to other synthetic moieties to make a variety of surfactants. Esterification employs a variety of acceptor alcohols, most often methanol, to produce methyl esters. This is usually performed using alkaline catalysts at an elevated temperature and pressures in a two-phase liquid reaction. In the case of triglycerides, interesterification is widely used to alter the physical behavior of fats and oils. This involves the rearranging or adding fatty acid groups on the glycerol backbone structure. Alternative transesterification processes have also been researched. One process under investigation utilizes a proprietary immobilized mesoporous iminophosphorane superbase catalyst to convert plant oils and methanol into methyl esters at a high yield at ambient temperatures [10]. This process eliminates the use of soluble catalysts and water washing, thus avoiding environmental problems due to catalyst removal and uses less energy by operating at room temperature. Another alternative process being commercially explored is the BIOX process from Canada. This process, developed at the University of Toronto, uses traditional base-catalyzed transesterification, but uses an inert solvent to co-dissolve methanol and vegetable oil into a single phase. This single phase reaction process operates at room temperature and reaches 95% completion within ten minutes; standard two phase processes require several hours (www.bioxcorp.com).

3.3. Other Chemical Modifications

There are many other chemical reactions done to lipids in combination with these basic reactions to generate specific products. For example, hydrolysis is used in the production of fatty acids which are then further processed by amination, oxidation, cyclization, polymerization, and addition reactions involving sulfur, aromatics, and alkyl branches to produce a wide variety of chemicals. Methyl esters are often hydrogenated to produce fatty alcohols, which are the primary starting point in the production of monoalkylphosphates, alkanoamides, and a variety of surfactants used in detergents and cleaning applications.

The addition of oxygen via reactions with fatty acid carbon double bonds is widely used to introduce new reactivity sites in unsaturated lipids. Oxygenation normally occurs through the exposure of unsaturated triglycerides and fatty acids to air at elevated temperatures. This initially forms oxides from the carbon double bonds which subsequently form epoxides or hydroxyl groups on the hydrocarbon chains. These polyol structures are used as initial reactants in the formation of a variety of polyurethanes, varnishes, and polymers. Recently, ozonation has also been used in the formation of polyols from unsaturated triglycerides and fatty acid derivatives [11]. The oxidation of unsaturated fatty acids can also be accomplished using potassium permanganate, potassium dichromate, chromic acid, sodium hypochlorite, or a variety of transition metals and metal oxides [12]. These reactions may also be used on the native triacylglyceride to produce precursors for cross-linked polymers and coatings.

Recently, alkene metathesis reactions have been explored for altering unsaturated fatty acids and TAGs to form new compounds. This involves catalytically cleaving and reforming carbon-carbon double bonds between molecules. Conjugated linoleate oils, as well as higher molecular weight oligomeric compounds, can be made using ruthenium or rhodium catalysts. This reaction has been used to make a variety of films and polymers using transition metal catalyst complexes [13]. A more detailed description of many of these reactions and industrial applications can be found in Bailey's Industrial Oil & Fat Products [14].

Technologies for separating and purifying TAGs, fatty acids, and esters include crystallization (sometimes referred to as winterization in food oil applications), vacuum distillation, molecular chelation, and chromatography. These are widely used in the food industry to alter the thermal properties of oils or to purify valuable components, such as essential oils for cosmetics and pharmaceuticals. One novel separation technology for methyl esters has been recently developed to provide nearly quantitative separation of saturated from unsaturated esters using urea inclusion processing [15]. Since most vegetable oils are used in edible applications, the process of isolating and purifying oils is well documented and can be found in Bailey's Industrial Oil & Fat Products [16].

4. INDUSTRIAL APPLICATIONS AND TECHNOLOGY

4.1. Coatings and Polymers

Natural oils have been used as waterproofing for millennia, with the earliest records in China, for caulking, boats and furniture. Conjugated oils, such as linseed, tung and some fish oils, are called 'drying oils' because of their ability to cross-link or polymerize on surfaces to form water-proof films. These materials are combined with pigments and used as paints and varnishes, with about 1.2 billion pounds being consumed annually until the 1950's, when synthetic epoxy and urethane resins began to replace them. Today, it is estimated that about 450 million lbs of fatty acids are still used in these applications, mainly as the alkyds and esters of synthetic components.

Modified oils with multiple non-conjugated double bonds, such as soybean and sunflower oil, are called 'semi-drying' oils and are often modified by partial oxidation to form reactive epoxides, which rapidly form cross-links. Highly unsaturated fatty acids and triglycerides are oxygenated to add reactive hydroxyl groups in epoxides, which are then reacted with isocyanates or other cross-linking agents to make polymeric coatings, plastics, and foams. Metallic catalysts based on cobalt and manganese are often added to the oil to accelerate the reaction rate. Most recently, tighter regulatory constraints on volatile organic solvents have renewed interest in drying oils and the use of modified oils in latex paints. Graft copolymers have also been produced from long chain glucans and plant fatty acids. These are used as biodegradable plastics [17–19].

4.2. Printing Inks

Soy oil printing inks are used by more than 90% daily newspapers in the US, a quarter of the 50,000-plus US commercial printers, and is widely used globally. These inks contain unmodified oils and a variety of specialized pigments and resins, depending on the printing application. The amount of soybean oil in the ink also depends on the application, for example, newspaper inks contain 50 – 75% soybean oil, while sheet-fed printing, heat-set and cold-set printing inks contain 20 – 30% soybean oil. One of the major advantages of soy inks is that volatile organic compounds (VOCs) are not released into the air when it dries. Instead of drying by losing solvents, the oil polymerizes, usually catalyzed by heat, pressure, and metallic catalysts. A recent innovation in soy ink is the development of a de-inkable soy resin-based toner for printers, copiers, and fax machines. This product was jointly developed by Battelle (Columbus, OH) and the Ohio Soybean Council and was given a 2003 Annual 100 Innovation Award by R&D magazine (www.rdmag.com). Further information on soy inks can be found at the National Soy Ink Information Center (www.soyink.com).

4.3. Lubricants

The use of natural plant oils as lubricants has roots in historical antiquity. While these materials have excellent lubrication properties, and are biodegradable, they also suffer from limited chemical stability. The presence of multiple double bonds and ester linkages make triglyceride oils reactive in the presence of oxygen and water at elevated temperatures, which has spurred the development of synthetic lubricants.

Modified and unmodified vegetable lipids are used as direct loss lubricants to replace petroleum-based lubricants in applications where the lubricant is lost directly into the environment. In some applications, unmodified oil formulations are used; for examples, underground pump drip oils and railway greases. In other applications, modified methyl esters or TAG are used; for examples, chain saws, underground pumps, fire truck pumps, water utility pumps, and wire rope. Modifications include the use of low unsaturated lipids or chemically modified lipids to remove oxidation potential and thus increase stability. The main benefits of vegetable lipids in these uses are their biodegradability, low toxicity, and decreased flammability. Sources of commercial products are ELM, Inc. (www.elmusa.com), Portec, (www.portec.com), Archer Petroleum, Omaha, NE, and International Lubricants, Inc. Seattle, WA.

For longer-term lubricating applications, natural plant lipids provide excellent lubrication and surface coating, but their viscosity and oxidative reactivity, primarily due to the presence of multiple double bonds and the ester linkages, limit their usage in modern high temperature applications. To overcome some of these difficulties, researchers are exploring chemical modifications of the whole oil and lipid components, using polymerization, ligand addition, partial oxidation, and the addition of anti-oxidant and anti-viscosity modifier formulations.

One promising area for the utilization of modified vegetable lipids is in hydraulic fluids. The commercial availability of a canola-based biodegradable hydraulic fluid in Europe has motivated a joint research project between Caterpillar and Cargill to develop modified soybean oil as a potential lubricant and hydraulic fluid for use in engines. By chemically modifying the double bond structures of soybean oil to reduce its oxidation potential, researchers are creating a new component in lubricating oil base stocks [20]. Their work has targeted uses as machining oils, crankcase oils, and transmission oils. Agro-Management (Colorado Springs, Colorado) has also developed a soybean oil-based lubricant, which also includes some canola oil, for use in automotive engine crankcase.

4.4. Cosmetics/Pharmaceuticals

A variety of oils and modified lipids are used in the cosmetics and pharmaceutical industries. In pharmaceuticals, oils are generally used as dermatological delivery agents, formulated into creams or emulsions to provide more uniform, efficacious application and transport of active agents. The specific oils and lipids allowed in topical formulary are regulated by the FDA (United States Pharmacopoeia 23, National Formulary 18). Among these are safflower oil, soybean oil, almond oil, cocoa butter, corn oil, cottonseed oil, lecithin, olive oil, peanut oil, and sesame oil. Emulsified products may also include surface active agents, such as mono- and di-glycerides.

In cosmetics, natural oils and lipids are used for several purposes. Emolliency, the imparting of softness and flexibility, is caused by the interaction between skin and cosmetic lipids. This can result from moisture retention or addition, if emulsifiers are used. Lubricity, adhesion, gloss, and pigmentation are also important benefits of using lipids in cosmetics. In some cases, there may be biochemical benefits; for example, essential fatty acids, such as linoleic acid, are not synthesized by mammals and are converted to gamma linoleic acid, which is an anti-inflammatory agent. One often cited benefit of the use of natural lipids is the belief that they are safer than synthetic ingredients, perhaps due to their similarity to skin oils. Not surprisingly, there are an abundance of plant and animal lipids used in cosmetics. In addition to common plant lipids, oils from many exotic sources: apricot kernel, hazelnut, rice bran, jojoba, borage seed, mango kernel, black currant seed, kukuinut, primerose, and many others are used. In addition to natural triacylglycerides, glycerophospholipids, and sterols, hydrogenated, esterified, and oxidized lipids are also used. Often, other components from plant sources, such as complex sterol esters or phytosterols, are also included in formulations.

The Food Chemicals Codex and CTFA Cosmetic Ingredient Directory lists most of these compounds.

4.5. Leather Processing

Fats and oils are widely used as lubricants and softening agents in the processing of leather and textiles. Natural oils are combined with sulfonated fats and applied as emulsions to wet leather, altering the surface of its fibers, making them more pliable. Animal fats are usually used, but natural and sulfonated plant oils are also used. While sulfonation increases the surface activity of oils for application to wet leather, unmodified oils can be used after the leather has been dried. Fats provide lubrication and anti-static properties for consumer textile fabrics to protect fibers as they undergo high speed spinning, weaving, and dyeing. A more detailed background of the applications of fats in leather processing can be found in Bailey's Industrial Oil & Fat Products [14].

4.6. Surfactants

The cleansing properties of the metallic salts of fatty acids has been recognized for millennia [21]. The reaction of fatty acids with alkaline metals (sodium and potassium) has been employed for centuries in the production of traditional soaps. Based on records from ancient Babylon about 2800 B.C., soap is probably the first industrial product made from plant oils. Soap making became a well-established industry in Europe during the Middle Ages. It was made from vegetable or animal oils boiled with wood ash, perfumes, and coloring agents. Countries with access to olive oil and other plant oils, such as Italy, Spain and France, became centers of soap manufacturing.

As the chemical structure of oils was discovered in the early 18th century, a variety of chemical reactions were used to create a host of new products. Surfactants made by sulfonating oils and fatty acids demonstrated significantly increased cleaning capability. With the recognition of the similar chemical structures between fatty acids and petrochemical hydrocarbons in the early 1900s, a new host of applications became possible by converting fats to new, useful, industrial products.

The first synthetic detergents were developed in response to shortages of fats during World War I. Detergents had the advantage of working in hard water, which contains high levels of minerals. After World War II, detergent sales surpassed soap sales and today detergents have essentially replaced soap-based products for laundering, dishwashing and household cleaning, although soap still retains a substantial market in personal hygiene products.

The earliest modifications of soaps were the addition of sulfate groups onto primary or secondary carbons in fatty acids. Since the original sulfated fatty acid derivative, the so-called "Turkey red oil" used to dye cloth and fabrics in 1750 [22], a host of alkylated, arylated sulfates and sulfonates have been developed as surfactants. Raw materials shortages during World War I led to the development of alkyl sulfonates as some of the first synthetic surfactants. The postwar growth of the petrochemical industry rapidly led to advances in the development of the current long-chain alkyl-aromatic sulfonates, such as alkyl benzyl sulfonates and other linear olefin detergent derivatives.

In general, sulfate salts require the hydrogenation of fatty acids to corresponding alcohols prior to reaction with sulfuric acid. The addition of sulfate groups to the unsaturated sites produces secondary sulfates. Sulfated glycerides have also been produced, but suffer from a lack of hydrolytic stability. The use of ether linkages vs. ester linkages in ethoxylated, sulfated alkyl aromatics dramatically improves stability.

Alkyl sulfonates via either ester or amide linkages have also been demonstrated to be excellent surfactants. The combination of alkyl aromatic sulfonates with fatty acids and alcohols has also proven to provide excellent wetting.

Cationic surfactants have primarily focused on the use of amines and ammonium salts in combination with aliphatic groups. In addition to possessing excellent stability and high surface activity, these materials are also recognized for their antimicrobial properties and capability as fabric softeners.

A host of chemical variations on the cationic theme have been developed, including the use of polyamines, imidazolines, and amino esters in combination with various alkyl and aromatic ligands. Triethanol amine derivatives have found extensive use in cosmetics, pharmaceuticals and shampoos.

Amphoteric surfactants, containing both cationic and anionic constituents, have also been synthesized [2, 22]. The development of nonionic surfactants derived from soybean lipids and carbohydrates which have excellent emulsifying properties has also been reported [23–25].

While many of these surfactants were originally derived from soybean and other vegetable oils, contemporary production of alkyl detergents is based on paraffinic alcohols and hydrocarbon ligands synthesized from petrochemical materials via Ziegler-Natta catalysis. However, vegetable oil-derived chemical intermediates for detergents, clothes softeners and soaps still have a significant market share, primarily based on their price and availability. Further information on plant-based surfactants is available from a recent review [26].

4.7. Solvents

The methyl esters of vegetable oils are excellent solvents for inks, polymers, and oils, with low volatility and good solubility. They are widely used to replace mineral spirits in the textile screen ink industry and graphics arts industries. Their main benefits are low volatility and flammability, low toxicity, and environmental compatibility. They have also been used in paint removal products in both consumer and industrial applications (AG Environmental Products, Lenexa, KS; Franmar Chemical, Inc., Normal, IL). These products combined methyl esters with more common solvents and cleaners, such as ethanol and citrus oils. These products take advantage of micro-emulsion technology, using non-ionic surfactants combined with methyl or ethyl esters derived from soybean lipids.

Methyl esters have also found applications in the removal and recovery of spilled crude oil and other petroleum products from both coastal and inland spill sites. The product has been listed by the EPA on the National Contingency Plan for oil spill cleanups and is the only shoreline clean-up material licensed by the state of California. In addition to being non-toxic, non-volatile and biodegradable, these vegetable oil solvents serve as carbon sources to stimulate the growth of natural bacteria which break down petroleum contaminants (Cytoculture International Inc., Point Richmond, CA).

Methyl esters have also found use in replacing volatile, regulated solvents to clean greases and other contaminants from metal. These cleaning products may also leave a thin film of non-toxic oils, which help to protect against corrosion (Bioneer Technologies Inc., Flint, MI; Archer Petroleum, Omaha, NE; Soy Environmental Products, Overland Park, KS).

A variety of formulated products based on soybean oil and esterified soybean oil are being used to replace diesel fuel and other volatile petroleum distillates used to lubricate concrete forms, to coat concrete and highways during curing, and to seal or stain concrete (MBI Technologies Inc., Woodburn, IN; AG Environmental Products, Lexena, KS).

4.8. Hydraulic Fluids

Lipid-based hydraulic fluids are commercially available as replacements for petroleum-based fluids. These are generally natural or modified methyl esters of low unsaturation from soy or rapeseed oils containing high oleic acid and erucic acid levels. These fluids are readily biodegradable and nonhazardous, which makes them particularly useful in applications where leakage or spillage will cause environmental damage. European environmental regulations already mandate the use of biodegradable hydraulic fluids in environmentally sensitive areas, such as waterways, farms, and forests, and these have been used for more than a decade, composing about 12% of the European hydraulic fluid market. The U.S. does not have similar environmental regulations.

4.9. Pesticide/Herbicide Adjuvants

Herbicide adjuvant products can use methyl esters and plant lipids as a replacement for petroleum-based components. These components enhance herbicide activity during application and reduce the total pesticide load on the environment, while being biodegradable, non-volatile, low in toxicity and safer for workers. A compendium of methyl ester products and vendors can be found at http://www.stratsoy.uiuc.edu/commproducts.html.

4.10. Glycerin (Glycerol)

The production of methyl esters from TAG also produces glycerin, which finds uses in pharmaceuticals, foods, cosmetics, soaps, plasticizers, and explosives. Recently, glycerin has been used to develop aviation deicers, replacing petroleum or natural gas-derived propylene and ethylene glycols [27, 28].

4.11. Fuels

4.11.1. Biodiesel Fuel/Lubricity Additives

The most recent large-scale industrial use of soybean oil and other vegetable oils is as alternative fuels [29]. The esterification of fatty acids derived from soybean and other vegetable or animal oils with methanol or ethanol, known as biodiesel fuels, can be used either alone or in a mixture with conventional petroleum-derived fuels. The methyl esters of vegetable fatty acids are widely used in diesel engine fuels worldwide, due to lower emissions. In addition to improving the lubricating effect of diesel fuel at low additive levels (1%), methyl esters are also effective diesel fuels, and reduce sulfur oxides, particulate emissions, nitrogen oxide emissions while being environmentally beneficial. European countries have embraced biodiesel fuels more strongly than the U.S., providing significant government subsidies and tax reductions for the use of biodiesel fuels.

Recent life cycle analysis comparisons of petroleum vs. biodiesel methyl esters [30] have calculated that 1.2 units of fossil resources are needed to produce 1 unit of petroleum diesel, while biodiesel yields 3.2 units of fuel product energy for every unit of fossil energy consumed in its life cycle. Emissions data has demonstrated that biodiesel reduces net CO_2 emissions by 78.45%, and also emissions of total particulate matter (32%), CO (35%), and SOx (8%).

4.11.2. Heating Oils

In ancient times, oils from plants and animals were widely used for lighting, such as in oil lamps. For heating purposes, wood and coal were used, but suffered from problems of inconvenience due to fuel loading, storage, and use and the removal of wastes. With the advent of inexpensive petroleum, liquid fuels quickly replaced solid fuels because of their ease in handling and storage. While natural gas has widely replaced liquid home heating fuels due to lower environmental emissions and ease of distribution, petroleum-based liquid home heating oil is still used to heat over 8 million homes in the USA, predominantly in the northeastern US, using approximately 6.6 billion gallons of fuel oil annually. With recent rises in petroleum prices to over $65 per barrel and anticipated future price increases as petroleum resources become less available, recent fuel oil prices have topped $3/gallon. Additional concerns over environmental issues involving sulfur and nitrogen oxide emissions from oil-based home heating systems have sparked a search for alternative fuels to supply this market. In these applications it has been found that degummed vegetable oils [31] and methyl esters [32] can directly replace kerosene like heating fuels. Economically, degummed soybean oil is forecasted to decrease in price over the next decade. The Food and Agricultural Policy Research Institute (FAPRI) expects soybean oil prices to fall steadily over the next decade from an average of $27.97/cwt (1 cwt = 100 lbs) in 2003–04 to $19.45/cwt in 2013–14. If this is correct, the price of soybean oil as a biofuel for heating purposes would be approximately $1.42 per gallon. Soybean methyl ester prices are currently about $2.50 per gallon, making them a very expensive heating fuel alternative. However, recent legislation provides a subsidy of approximately $1/gallon for biodiesel fuel used in mobile transport. If this were applicable to home heating fuels, the cost would be more comparable with vegetable oil prices. Prediction of petroleum fuel prices over the next decade would be highly speculative, but it is probably safe to assume that they will be significantly higher by comparison. In addition to economic advantages, using vegetable oils for home heating has significant environmental benefits, including decreased sulfur oxide emissions and the utilization of this renewable domestic fuel [31, 32].

4.11.3. Aviation Fuels

Petroleum's major use is in mobile transportation fuels. Approximately 75% of this is used for automotive ground transportation fuels, while only 12% is used by aviation. There are a number of promising fuel alternatives for ground transportation, such as ethanol, electricity, natural gas, and hydrogen fuel cells. However, fuel requirements for aviation are much more sensitive than those for ground transportation; for example, while renewable liquid fuels such as ethanol are widely available and economical, ethanol's energy density (2/3 that of petroleum), water solubility, and high volatility make it unattractive in aviation. Vegetable-based lipid fuels are chemically and physically very similar to petroleum fuels, making them an ideal replacement for aviation applications. They are liquids with energy densities similar to jet fuels', and can be used in the current fuel infrastructure with few or no changes. Since they come primarily from renewable, domestic resources, their use will have significant positive implications for homeland security and environmental impact.

One major technical limitation for the use of biodiesel fuels is cold temperature performance. Biodiesel fuels produced from standard domestic sources (soybeans, corn, canola, animal fats) contain significant amounts of long-chain, saturated fatty acids, which result in crystallization at temperatures around the freezing temperature of water. This limits their use in cold environments, such as during winter or at high altitudes. Recently, fractionated soybean methyl esters have been demonstrated as fuel additives for aviation turbine jet fuels [15, 33]. These blends demonstrated cloud points down to -42°C, meeting current ASTM aviation fuel standards. Work by Bist [15] has demonstrated that using urea fractionation, high yields of unsaturated methyl esters can be obtained and blended with Jet A fuel up to a level of 30% (vol.) and used in stationary turbine jet engines.

4.12. Other Uses

Concrete and asphalt release agents employing soybean lipids have been commercialized. These biodegradable compounded materials serve to help release and seal/cure concrete and asphalt from surface molds, replacing inert petrochemical-derived materials (Strategic Market Development, Kankakee, IL; Midwest Biologicals, Inc., Woodburn, IN). A soybean oil-based product called Soapstock is commercially available as a dust suppressant for dust control on gravel roads. Unlike petroleum-based materials previously used for this purpose, the soybean product does not cause environmental water quality problems [34]. Crayons made from soybean oil are currently being commercially produced. Using all natural and environmentally safe materials, Dixon Ticonderoga's Prang Fun Pro TM crayons can replace petroleum-derived crayons as familiar drawing implements for children [35]. Consumer candles produced using hydrogenated soybean oil are also commercially available [36].

5. ISSUES FOR INDUSTRIAL APPLICATIONS OF PLANT LIPIDS

With recent and ongoing research into so many applications, the future holds great promise for developing new technologies for industrial utilization of plant lipids. Many lipid-derived products are meant to replace existing petroleum-based products, such as plastics and fuels. While these may be technologically feasible, other important issues, such as raw materials availability, and of course, economics, will be critical to long term success. Since the majority of plant lipids are currently used in food products, it is likely that significant new industrial uses will require increases in crop production. If this were to occur, changes in production infrastructure, such as segregated storage (industrial vs. food processing systems), year-round availability and storage, may be needed. Additionally, the economics and availability of petroleum will play a significant role in establishing markets for bio-based products.

Regulatory issues involving economics will also be important factors. While some bio-based products have legislative economic support, e.g. biodiesel fuel, other products offer unique benefits over existing alternative materials and also carry significant price differentials. Similarly, opportunities exist in specific niche markets, such as total loss lubricants, biodegradable surfactants, cosmetics, and pharmaceuticals. Based on market volume and product price/profit, these types of products should be a good bet for long term success, since free market forces rather than regulatory support will drive the need for the products.

Some low technology markets that have significant profit margins and product uniqueness already exist, such as children's crayons and candles. Petrochemical paraffins, which carry the stigma of being composed of potentially hazardous carcinogens and heavy metals, can be effectively replaced by vegetable oils that have no aromatics or metals. In such markets involving children and potential safety issues, vegetable lipids can command a reasonable premium price based on their uniqueness and perceived consumer needs.

REFERENCES

[1] Freedonia Focus Report on Soy Products & Markets, The Freedonia Group, Inc., Cleveland, OH (2003).
[2] M. Lager, The Useful Soybean, McGraw-Hill, New York (1945).
[3] T. P. Hilditch, Synthetic Detergents. Chapt. 7, in The Industrial Chemistry of Fats and Waxes', 3rd Ed., Bailliere, Tindall and Cox, London, (1949) 438.
[4] K. S. Markley (ed.), Soybeans and Soybean Products, Interscience Publ., New York (1951).
[5] B. Y. Tao, Chemistry and Industry (1994) 906.
[6] L. Johnson and C. P. Baumel. eds., Industrial Uses of Soy Oil for Tomorrow, Special Report 96, Iowa State University (1995).
[7] W. Shurtleff and A. Aoyaoi, Bibliography of Industrialization of Soybeans, Soyfoods Center, Lafayette. CA (1989).
[8] Y. H. Hui (ed.), Bailey's Industrial Oil & Fat Products, Edible Oil & Fat Products: Products and Application Technology, 5th Ed. Vol. 3, Wiley Interscience, 1996a.
[9] N. Widlak, R. Hartel, and S. Narine, Crystallization and Solidification Properties of Lipids, AOCS Press (2001).
[10] J. G., Verkade, A Powerful Proton Abstracting Agent U.S. Patent 5,052,533 1991
[11] R. Narayan, D. Graiver, and K. Farminer, Ozone-Mediated Transformations of Vegetable Oils., Am. Oil Chem. Soc. Annual Meeting, Cincinnati, OH (2004).
[12] P. Wang and B. Y. Tao, J. Am. Oil Chem. Soc., 75 (1998) 9.
[13] R. C. Larock, Lipid Technol. *15*, (2003) 58.
[14] Y. H. Hui (ed.), Bailey's Industrial Oil & Fat Products, Industrial and Consumer nonedible Products from Oils and Fats, 5th Ed., Vol. 5, , John Wiley & Sons, 1996b.

[15] S. Bist, Development of vegetable lipids derived fatty acid methyl esters as aviation turbine fuel extender, MS thesis, School of Agricultural and Biological Engineering, Purdue University, W. Lafayette, IN, 2004.
[16] Y. H. Hui (ed.), Bailey's Industrial Oil & Fat Products, Edible Oil & Fat Products: Processing Technology, 5th Ed., Vol. 4, John Wiley & Sons, 1996c.
[17] H. Kwatra, J. Caruthers, and B. Y. Tao, Ind. Eng. Chem. Res., 31(2), (1993) 2647.
[18] P. Wang and B. Y. Tao, Appl. Polymer Sci., 52, (1994) 755.
[19] P. Wang, and B. Y. Tao, J. Environ. Polymer Degrad., 3, (1995) 115.
[20] United Soybean Board, Feedstocks, Vol. 1(2) 1997.
[21] F. W. Gibbs, Ann. of Science, 4 (1939) 169.
[22] H. G. Kirschenbauer, Fats and Oils', 2nd ed., Reinhold Pub. Co., New York, (1960) 133.
[23] Y. Chen, Synthesis and Characterization of Maltooligosaccharide Fatty Acid Esters, MS thesis, Dept. of Agricultural and Biological Eng., Purdue University, West Lafayette, IN, 1997.
[24] S. Eastburn and B. Y. Tao, Biotechnol. Adv., 12, (1994) 325.
[25] D. K. Allen and B. Y. Tao, J. Surfactants Detergents, 5 (2002) 245.
[26] SODEOPEC Soaps, Detergents, Oleochemicals, and Personal Care Products, 2004, L. Spitz, ed., AOCS Press
[27] J. A. Smith, Initial testing of biobased aviation deicer fluids, MS thesis, School of Aeronautics and Astronautics Engineering, Purdue University, W. Lafayette, IN 2004.
[28] W. T. Smith, Glycerine-based aviation deicers and anti-icers, MS thesis, School of Agricultural and Biological Engineering, Purdue University, West Lafayette, IN 2004.
[29] B. Y. Tao, Biotechnology (March, 2001) 45.
[30] NREL, Overview of Biodiesel and Petroleum Diesel Life Cycles, NREL/TP-580-24772 (1998).
[31] N. W. VanLanningham, Soybean Oil Containing Triglycerides as a Renewable Component in Residential Heating Applications, MS thesis, Purdue University, West Lafayette, IN 2003.
[32] C. R. Krishna and R.J. McDonald, The Green Fuel Option for the Oil Heat Industry Biofuel Research, The 2003 National Oil Heat Research Alliance Technology Symposium, Boston, MA, Brookhaven National Laboratory.(2003).
[33] R. O. Dunn, American Society of Agricultural Engineers, Vol.44(6) (2001) 1751.
[34] B. F. Haumann, INFORM, 4(12) (1993) 1331.
[35] T. E. Sinwald.and M. J. Howick, Soybean oil marking compositions and methods of making the same, US Patent 5,753,015 (1998).
[36] B. Y. Tao, Vegetable lipid-based composition and candle. US Patent 6,497,735 (2002).

Bioprocessing for Value-Added Products from Renewable Resources
Shang-Tian Yang (Editor)

Chapter 25. Value-Added Products from Animal Manure

Zhiyou Wen*, Wei Liao, Chuanbin Liu, and Shulin Chen

Biomass Processing and Bioproduct Laboratory, Department of Biological Systems Engineering, Washington State University, Pullman, WA 99164, USA

1. INTRODUCTION

The environmentally-friendly disposal and utilization of animal manure is a significant challenge to the livestock industry. During the past decade, the U.S. livestock industry has undergone a substantial structural change featuring a rapid reduction in the number of animal operations and a corresponding increase in herd size on the remaining farms. These concentrated animal operations have created greater environmental concerns because of the amount of animal waste produced at these facilities [1].

Currently, approximately 55 million tons of animal manure is collected each year for subsequent disposal. About 90% of the collected manure is disposed of through land application, with a small amount of the manure being composted prior to disposal. This land application, especially on excessively nutrient loaded land, can cause environmental problems, such as nutrient enrichment of streams and groundwater [2] and air pollution and the emission of NH_3, CO_2, CH_4, N_2O, and other greenhouse gases [3–6]. Composting is an aerobic digestion of organic wastes and can reduce odors and the levels of soluble organics; however, this process takes a relatively long time (4–6 weeks) to attain stabilized materials and the necessary equipment and intensive labor make the process very costly [7]. In addition, composting does not create higher-value chemicals or energy products.

Pyrolysis/combustion and anaerobic digestion are two additional treatment methods for producing energy from animal manure [7]. With an energy content of about 13.4 MJ kg^{-1} [8], collected manure represents an annual renewable energy resource of approximately 7×10^{11} MJ. To date, combustion/ pyrolysis has shown economic potential for recovering energy from animal manure. For example, a laboratory-scale thermochemical process converted swine manure into a raw oil product, producing 2.95-times more energy than the energy needed to bring the feedstock to the desired operating temperature [9]. Also, a power plant built in Suffolk, UK (Fiberpower) utilized poultry litter as fuel for a combustion system. The plant generated a gross output of 14 MW of electricity with a net output of 12.5MW [7]. It is believed that increased commercialization of combustion/pyrolysis technology will depend on the cost of moving the feedstock to the appropriate processing center, the plant capital investment, and the operation cost [7]. Compared with the high temperature and high pressure requirements, and hence, high equipment costs of combustion/pyrolysis systems, anaerobic

*Current Affiliation: Department of Biological Systems Engineering, Virginia Tech, Blacksburg, VA 24061

digestion is capable of being operated at much milder conditions. This technology, however, is limited by low digestion rates and difficulties with the biodegradation of lignocellulosic materials [10-12]. However, the environmental benefits of pyrolysis and anaerobic digestion, such as odor control and greenhouse reduction, may help balance the cost of these manure management processes.

The increasing pollution control requirement for animal operations is challenging the scientific community and the industry to develop alternative animal manure management strategies. The concept of using animal manure as a bioresource for producing value-added products offers such an alternative. Animal manure contains a variety of components including fiber, protein/amino acids, and minerals. Such components can be used as feedstocks for producing value-added products through biological or chemical conversion processes. Such processes can potentially change manure from a disposal problem to a valuable resource for chemicals. The aim of this chapter is to review several examples of utilizing animal manure as feedstock for producing value-added products. First, three types of animal manure, cattle, poultry, and swine manure, were characterized by their fiber, protein/amino acid, and elements content; second, an investigation on conversion of cattle manure fiber into fermentable reducing sugars by acid and enzymatic hydrolysis was discussed; and finally, a fungal-culture process for producing cellulase enzymes from dairy manure was presented. These results, however, are limited to laboratory studies.

2. CHARACTERIZATION OF ANIMAL MANURE

A detailed chemical characterization of animal manure has led to a better understanding of the manure utilization process and has offered insights into improving the process efficiency. However, previous manure characterization reports have been limited to basic elemental composition and the characterization of general classes of chemicals [13]: for example, traditional parameters such as total solids (TS), volatile solids (VS), nitrogen content (N), phosphorus (P), and potassium (K), which provide information for land application purposes, but are insufficient for the study of animal manure as a feedstock for value-added products.

Detailed information on the elemental, fiber composition (cellulose, hemicellulose and lignin), and the protein/amino acid content of various manure sources are critical for producing value-added products. Of particular importance is the quantification of the fiber content because the cellulose and hemicellulose components of the fiber can be converted into reducing sugars for later use in a sugar biorefinery platform. The cellulose, hemicellulose, and lignin content can be determined by the analysis of neutral detergent fiber (NDF), acid detergent fiber (ADF), and acid detergent lignin (ADL) [14]. NDF is used to estimate the total lignocellulosic materials (including cellulose, hemicellulose, and lignin), while ADF is used to estimate the content of lignin and cellulose, and lignin levels can be directly estimated from the ADL value. In addition to fiber, protein is another important component of manure, which potentially can be used to produce amino acids or as a nitrogen source for microbial growth. The protein composition (amino acid content) is determined using an amino acid analyzer; the crude protein content is estimated from the total nitrogen in animal manure. Using the assumption that the proportion of nitrogen in protein is around 16%, the content of crude protein was estimated by multiplying the organic nitrogen content by 6.25. The organic nitrogen, in turn, was calculated by subtracting ammonium-nitrogen from total nitrogen.

Animal manure, including cattle, poultry and swine manure, was recently characterized [15] by, among other factors, its fiber and protein/amino acid composition. The dairy manure was collected at the Dairy Center at Washington State University (WSU) while the beef, feedlot cattle, poultry, and swine manures were collected by the WSU Puyallup Research & Extension Center from the waste piles at several farms within Washington State.

Table 1
Characterization of different types of cattle manure

Parameters	Dairy	Beef	Feedlot
Solid content (% of fresh manure)			
Total solids (dry matter)	13.39	12.56	26.61
Total volatile solids	11.21	9.97	22.78
Elements (% of dry matter)			
Carbon	45.37	43.81	43.56
Nitrogen	3.03	1.94	2.72
Phosphorus	0.48	0.42	0.81
Potassium	2.86	1.44	0.92
Calcium	1.2	1.06	0.69
Magnesium	0.55	0.3	0.34
Sodium	0.47	0.25	0.12
Copper	0.003	0.0002	0.0018
Zinc	0.032	0.0042	0.0087
Iron	0.03	0.059	0.055
Sulfur	0.31	0.25	0.21
Aluminum	0.014	0.017	0.021
Cobalt	0.0009	0.0002	0.0002
Chromium	0.0002	0.0002	0.0001
Manganese	0.051	0.06	0.012
Molybdenum	0.0003	0.0002	0.0008
Nickel	0.001	0.001	-
Vanadium	0.0005	0.0006	-
Fiber content (% of dry matter)			
NDF	52.6	51.5	41.7
AFD	40.4	34.1	20.3
ADL	13.0	12.2	6.1
Cellulose	27.4	21.9	14.2
Hemicellulose	12.2	17.4	21.4
Lignin	13.0	12.2	6.1

NDF: neutral detergent fiber; ADF: acid detergent fiber; ADL: acid detergent lignin.

The total solid content, macro/micro elements, and fiber composition of different types of cattle manure are presented in Table 1. Carbon was the most abundant element, followed by nitrogen, potassium and calcium. The manure also contained phosphorus, magnesium, sodium, sulfur, and trace elements, which together comprised less than 1% of the total dry weight. Fiber (NDF) accounted for a large portion of all three types of cattle manure; demonstrating the potential for its cellulose and hemicellulose components to be degraded to sugars and then fermented into value-added products. The detailed amino acid composition of cattle manures is presented in Table 2. Cattle manure contains a variety of amino acids, among which glutamic acid, aspartic acid, glycine, and alanine were relatively abundant. Amino acids accounted for about 8–10 % of the total dry matter.

Table 2
Amino acid compositions of different cattle manures

Amino acid (% DM)	Dairy	Beef	Feedlot
Taurine	0.06	0.08	0.06
Hydroxyproline	0.08	0.19	0.02
Aspartic acid	0.73	0.82	1.00
Threonine	0.36	0.37	0.57
Serine	0.30	0.35	0.43
Glutamic acid	1.46	1.09	1.93
Proline	0.49	0.44	0.75
Lanthionine	BDL	0.01	0.02
Glycine	0.82	0.66	0.56
Alanine	0.82	0.61	0.72
Cysteine	0.14	0.12	0.23
Valine	0.49	0.44	0.64
Methionine	0.12	0.11	0.23
Isoleucine	0.38	0.33	0.5
Leucine	0.60	0.54	0.82
Tyrosine	0.15	0.21	0.36
Phenylalanine	0.32	0.32	0.58
Hydroxylysine	BDL	0.01	BDL
Histidine	0.09	0.14	0.24
Ornithine	0.03	0.02	0.03
Lysine	0.24	0.37	0.63
Arginine	0.24	0.35	0.48
Tryptophan	BDL	0.05	0.09
Total	7.92	7.61	10.89

BDL – Below Determination Limit

Samples of poultry and swine manures were also analyzed. The poultry manures were collected from starter chick, pullet grower, 17–40 weeks old, and post-molt diet chicken houses. The swine manures were collected from the nursery, finish, phase two, and phase three pigs. The fiber, amino acid, and element contents in these samples are listed in Tables 3 and 4, respectively.

Table 3
Characterization of poultry manures

Parameter (% dry matter)	Different poultry manures			
	Chick starter	Pullet grower	17–40 weeks	Post-molt diet
NDF	31.7	36.4	34.5	31.2
ADF	13.4	14.9	14.3	14.8
Lignin	4.9	7.2	3.2	4.1
Taurine	0.21	0.21	0.06	0.04
Hydroxyproline	0.23	0.24	0.19	0.17
Aspartic acid	1.22	1.31	0.82	1.09
Threonine	0.59	0.74	0.37	0.55
Serine	0.54	0.88	0.35	0.5
Glutamic acid	1.64	1.95	1.09	1.48
Proline	0.72	1.22	0.44	0.54
Lanthionine	0.01	0.01	0.01	0.01
Glycine	2.11	2.53	0.66	1.27
Alanine	1.14	1.23	0.61	0.76
Cysteine	0.31	0.59	0.12	0.18
Valine	0.77	1.02	0.44	0.60
Methionine	0.2	0.22	0.11	0.18
Isoleucine	0.56	0.73	0.33	0.44
Leucine	0.94	1.25	0.54	0.76
Tyrosine	0.33	0.48	0.21	0.31
Phenylalanine	0.53	0.71	0.32	0.45
Hydroxylysine	0.02	0.02	0.01	0.01
Histidine	0.21	0.26	0.14	0.21
Ornithine	0.11	0.13	0.02	0.02
Lysine	0.6	0.73	0.37	0.47
Arginine	0.43	0.76	0.35	0.5
Tryptophan	0.1	0.09	0.05	0.07
Total amino acids	13.52	17.31	7.61	10.61
Nitrogen	6.37	7.74	5.05	4.46
Calcium	4.0	4.0	9.6	6.9
Magnesium	0.66	0.68	0.91	0.96
Sodium	0.77	0.64	0.72	0.6
Potassium	2.7	2.5	3.9	3.8
Phosphorus	2.3	2.6	3.4	3.2
Copper	0.0046	0.0044	0.0059	0.0048
Zinc	0.048	0.04	0.05	0.041
Iron	0.0079	0.0056	0.019	0.025
Sulfur	0.56	0.65	0.76	0.70
Aluminum	0.0055	0.0084	0.019	0.022
Cobalt	0.000041	0.000043	0.000067	0.000054
Chromium	0.000081	0.0001	0.00017	0.00015
Manganese	0.037	0.04	0.054	0.046
Molybdenum	0.00015	0.00015	0.0003	0.00036
Nickel	0.00067	0.00062	0.0012	0.0011

Table 4
Characterization of swine manures

Parameter (% dry matter)	Different swine manure		
	Nursery	Grower	Finisher
NDF	39.2	39.1	37.4
ADF	17.3	18.7	15.8
Lignin	4.1	5.4	2.9
Taurine	0.08	0.06	0.07
Hydroxyproline	0.03	0.09	0.01
Aspartic acid	1.7	2.01	1.61
Threonine	0.82	1	0.77
Serine	0.57	0.77	0.54
Glutamic acid	2.3	2.54	2.15
Proline	0.83	0.95	0.76
Lanthionine	0.03	0	0.04
Glycine	0.95	1.13	0.84
Alanine	1.28	1.37	1.12
Cysteine	0.28	0.39	0.3
Valine	1.11	1.13	1.08
Methionine	0.4	0.49	0.37
Isoleucine	0.91	0.92	0.93
Leucine	1.4	1.65	1.34
Tyrosine	0.63	0.69	0.56
Phenylalanine	0.9	0.97	0.84
Histidine	0.33	0.43	0.29
Ornithine	0.05	0.03	0.08
Lysine	1.1	1.33	1.13
Arginine	0.73	0.88	0.7
Tryptophan	0.13	0.15	0.1
Total amino acids	16.56	18.99	15.63
Nitrogen	4.02	3.63	3.52
Calcium	1.6	4.2	2.9
Magnesium	0.61	0.86	0.62
Sodium	0.28	0.28	0.15
Potassium	1.6	1.6	1.2
Phosphorus	1.5	2.45	1.6
Copper	0.058	0.1786	0.12
Zinc	0.059	0.2167	0.066
Iron	0.025	0.2566	0.051
Sulfur	0.41	0.55	0.37
Aluminum	0.058	0.062	0.1
Cobalt	0.0001	0.00015	0.00015
Chromium	0.0033	0.00037	0.0021
Manganese	0.028	0.074	0.034
Nickel	0.00085	0.0015	0.00063
Vanadium	0.00033	0.00074	0.00032

Since fiber and crude protein are the two major animal manure components, a comparison of these two components in different types of manure has been made and is presented below. As shown in Table 5, the fiber and crude protein contents varied among cattle, swine, and poultry manures. The poultry manures had the highest protein content, followed by the swine and cattle manures. However, the cattle manures had the highest fiber content. Such differences are considered to be caused by the different diets fed to cattle, swine, and poultry, and also the different digestion abilities of the three animals.

Table 5
Comparison of fiber and protein contents (as of % dry matter, DM) in cattle, swine and poultry manures

	Crude protein	Fiber	Hemicellulose	Cellulose	Lignin
Cattle manure					
Dairy	18.1	52.6	12.2	27.4	13.0
Beef	12.1	51.5	17.4	21.9	12.2
Feedlot	17.0	41.7	21.4	14.2	6.1
Swine manure					
Nursery	25.1	39.2	21.9	13.2	4.1
Grower	22.7	39.1	20.4	13.3	5.4
Finisher	22.0	37.4	21.6	12.9	2.9
Poultry manure					
Chick starter	39.8	31.7	18.3	8.5	4.9
Pullet grower	48.4	36.4	21.5	7.7	7.2
17-40 weeks	31.6	34.5	20.2	12.0	2.3
Post-molt	28.0	31.2	16.4	10.7	4.1

In summary, the above results suggest that crude protein and fiber are the two major building blocks for the bioprocessing of animal manure into value-added products. In the following sections, a focus will be placed on manure fiber utilization through a sugar biorefinery platform. Dairy manure will be used as the raw material because of its high fiber content and relatively low protein composition.

3. PRODUCTION OF REDUCING SUGARS FROM CATTLE MANURE LIGNOCELLULOSICS

Cattle manure, especially dairy manure, composes the largest category of animal wastes produced in the US livestock industry. Dairy manure fiber can be converted into reducing sugars, which can then be converted into value-added products, such as xylitol and sorbitol, using a hydrogenation reaction [16] and lactic acid using bacterial or fungal fermentation [17, 18]. Acid hydrolysis and enzymatic hydrolysis of dairy manure fiber have been investigated at the Biomass Processing and Bioproduct Laboratory at WSU, and the results are presented here.

Although the composition of manure fiber is similar to that of most cellulosic materials, such as wood or straw, the structural complexity of this material makes its hydrolysis conditions very different from that of wood or straw. Because the fiber has already undergone some digestion while passing through the animal's digestive tract, some of the more easily digested fiber components have been removed, leaving a more "recalcitrant" component that is not similarly seen in wood or straw. In addition, dairy manure has a nitrogen content of approximately 2.6%, which is considerably higher than the 1% in wheat straw and other fibrous materials [19]. This nitrogen is in the form of indigestible forage proteins, proteins from the metabolism of rumen bacteria, and inorganic nitrogen, such as urine and ammonia [20]. During hydrolysis, ammonia and amino acids from hydrolyzed proteins can react with sugars in the hydrolyzed solution under high-temperature, acidic conditions [16, 21], and ultimately can influence the final sugar yield. To reduce the negative influence of nitrogen on manure fiber hydrolysis, a washing and solid/liquid separation method can be employed.

3.1. Solid/liquid separation

Solid/liquid separation is an effective method for removing nitrogen and protein enriched materials from manure fiber since most nitrogen and proteins are soluble. Manure fiber can be separated from the nitrogen by simply washing with water and subsequently separating the liquid and solid fractions, thus reducing the negative influence of nitrogen on the hydrolysis of fiber.

Table 6
Distribution of solid particles in total dairy manure solid (TS), carbon and nitrogen content in different portions of solid particles[a] [15]

Portion	Solid particles (% of TS)	Carbon (% of portion particles)	Nitrogen (% of portion particles)
>1.68 mm	56.36	43.0	2.40
1.68 mm - 1.19 mm	4.49	39.64	2.94
1.19 mm - 0.84 mm	4.73	40.85	3.03
0.84 mm - 0.42 mm	5.70	41.31	2.97
0.42 mm - 0.125 mm	4.88	39.22	2.96
<0.125 mm (Filtrate)	23.84	37.67	7.12

[a]Experiments were performed by sequentially sieving manure slurry (with 55.45 g of dry matter equivalent) with a set of American Standard sieves (openings: 1.68 mm, 1.19 mm, 0.84 mm, 0.42 mm, and 0.125 mm).

Screening is the easiest way to separate the liquid and solid manure fractions. To determine the proper pore size of the screen to be used in the filtration system, the size distribution of manure solids was tested by passing it through a set of American Standard Sieves. As shown in Table 6, more than 56% of dairy manure solids was larger than 1.68 mm, with the filtrate passing through the 0.125 mm opening containing 23.84% of the total solids. The carbon and nitrogen content of each portion of solid particles was also determined. Particles larger than 1.68 mm contained more carbon than the smaller ones; in contrast, the particles in the filtrate (<0.125 mm) contained more nitrogen than the larger ones. It was also found that particles larger than 1.68 mm accounted for 58% of the total carbon and 37% of the total nitrogen of raw manure, with the remaining total carbon and nitrogen present in the particles smaller than

1.68 mm. The above results indicate that screen with 1.68 mm openings can efficiently separate carbon (mainly fiber) and nitrogen (mainly proteins) into different fractions.

The fiber content of particles larger than 1.68 mm was compared to the overall fiber content of the non-sieved raw manure. As shown in Table 7, the relative proportions of cellulose, hemicellulose, and lignin increased after solid/liquid separation, while the nitrogen and corresponding crude protein content decreased. Further acid hydrolysis of the two types of solid particles shows that the yield of total reducing sugar from the solid after sieving treatment was higher than that from raw manure. Meanwhile, the color of hydrolysate, a direct indication of Maillard byproducts content, was significantly different. The formation of Maillard compounds from raw manure was five times more than that from manure solids after sieving treatment [22]. Based on these results, manure solid separated by a 1.68 mm screen was used in the following acid hydrolysis experiments.

Table 7
Fiber composition of raw dairy manure solid and solid particles separated by American Standard Sieves (1.68 mm openings)

Parameters	Raw manure solid	Solid after separation
Dry matter (%)	15.50	13.26
NDF (% of DM)	48.27	67.11
ADF (% of DM)	35.80	52.23
ADL (% of DM)	13.91	16.56
Cellulose (% of DM)	21.89	35.67
Hemicellulose (% of DM)	12.47	14.88
Lignin (% of DM)	13.91	16.56
N (% of DM)	3.64	2.40
Crude protein (% of DM)	22.75	15

3.2. Acid hydrolysis of manure fiber

Acid hydrolysis, particularly two-stage sulfuric acid hydrolysis, is widely used to treat lignocellulosic materials to obtain sugars [23–27]. In the process, acid first breaks the matrix structure of the fiber into cellulose, hemicellulose, and lignin components [28], and then further reduces these polysaccharides into monosaccharides [29]. This type of application commonly utilizes either concentrated acid hydrolysis at a low temperature or dilute acid hydrolysis at a high temperature [30]. In general, concentrated acid hydrolysis is much more effective than dilute acid hydrolysis [31] and involves two steps: a decrystallization step that breaks down the crystal structure of fiber using sulfuric acid at a concentration greater than 60% and a hydrolysis step using acid with a concentration around 20%–30% to liberate sugars from the decrystallized fiber [32]. It has been reported that a glucose yield of 72–82% can be achieved from mixed wood chips using such a concentrated acid hydrolysis process [33]. However, concentrated acid hydrolysis has a major drawback in its use of highly concentrated acid, which can cause serious environmental concerns [30]. Therefore, a new combination of concentrated and relatively dilute acid processes should be developed for the purpose of combining the shorter reaction time and lower temperatures of concentrated processes with the more complete degradation of lignocellulosics and diminished chemical toxicity in dilute processes.

Fig. 1 summarizes the different procedures used in the acid hydrolysis of manure fiber, i.e., one-stage hydrolysis with decrystallization, one-stage hydrolysis, two-stage hydrolysis, two-stage hydrolysis with alkaline extraction, and two-stage hydrolysis with decrystallization. For all of the procedures, 10% (dry basis) of solid particles after solid/liquid separation (1.68 mm-openings screen) were used.

The sugar yields of the five acid-hydrolysis procedures are presented in Fig. 2. Two-stage hydrolysis resulted in a much higher sugar yield from hemicellulose (arabinose, galactose and xylose) than one-stage hydrolysis. An approximate 100% hemicellulose derived-sugars yield means that hemicellulose was completely hydrolyzed. In terms of cellulose conversion, two-stage hydrolysis with decrystallization converted almost 90% cellulose into sugars, while the cellulose-sugar yield of all the other procedures was less than 35%. This suggests that the crystal structure of manure cellulose is the most difficult part to attack and the critical factor influencing glucose yield.

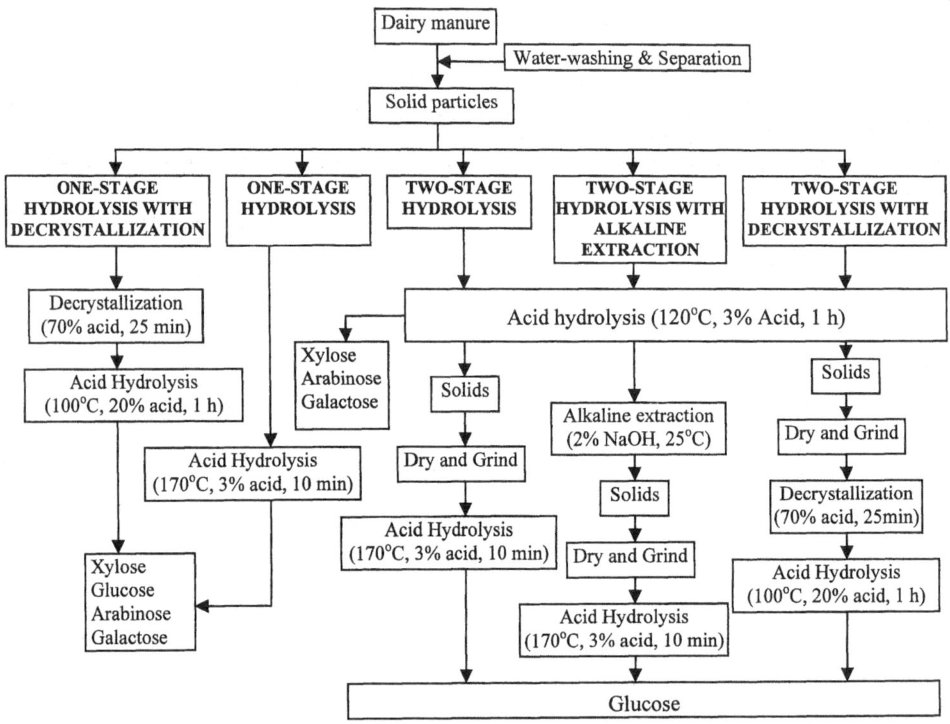

Fig. 1. Different procedures of acid hydrolysis of dairy manure fiber. [15]

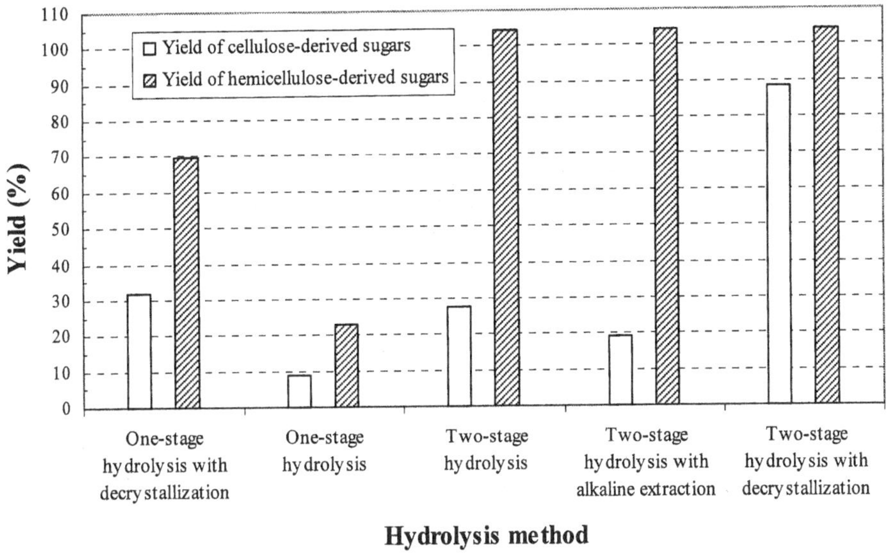

Fig. 2. Comparison of sugar yields of different acid hydrolysis procedures (see Fig.1). [15]

3.3. Enzymatic hydrolysis of manure fiber

Although acid hydrolysis, especially two-stage hydrolysis with decrystallization, gives a very high yield of both hemicellulose and cellulose-derived sugars, this procedure has a number of disadvantages, such as severe environmental and corrosive problems, high acid consumption and recovery costs, and sugar degradation and a consequently low process yield. These factors present major barriers to the economic success of acid hydrolysis [34].

Enzymatic hydrolysis is the preferred method for converting manure fiber into sugars. The reaction is highly specific, has no sugar degradation, occurs under mild conditions (pH ~5 and temperature less than 50°C), and is not corrosive to the environment. Generally, an efficient cellulase system consists of endocellulase (E.C. 3.2.1.4), which cleaves the internal glycosidic bonds, exoglucanase (E.C. 3.2.1.91), which cuts the cellulose chain from either the reducing or non-reducing end, and β-glucosidase (E.C. 3.2.1.21), which hydrolyzes cellubiose to produce glucose.

An enzymatic hydrolysis of manure cellulose has been previously performed [35]. The optimal conditions were determined to be 650 FPU L^{-1} of cellulase, 250 IU L^{-1} of glucosidase, 50 g L^{-1} of substrate, with a supplementation of 0.2% Tween 80 at pH 4.8 and 46°C. Under such conditions, 11.32 g of glucose was production per 100 g manure, corresponding to a ~40% sugar yield. Compared to other lignocellulosic materials, however, such a yield is relatively low. For example, the cellulose conversion of steam-pretreated softwood reached a 69% sugar yield [36], and hydrolysis of rice straw pretreated by acid-catalyzed steam explosion, dilute sulfuric acid, and ammonia fiber explosion resulted in 50–60% carbohydrate conversion [37]. Another interesting yield comparison is to alfalfa, since dairy manure lignocellulosics originate from alfalfa feed. It has been reported that enzymatic saccharification of alfalfa produces 38.8 g of glucose per 100 g of substrate, corresponding to ~60% yield [38]. The reason for the low cellulose conversion of alfalfa-derived manure lignocellulosics may be due to the recalcitrant nature and crystallinity of manure lignocellulosic materials, which is unusually high in crystallinity due to its earlier rumen conversion. Indeed, crystallinity has been identified as the most important structural feature affecting hydrolysis efficiency [39]. In order to obtain a deeper insight into the crystallinity and structural changes undergone by manure fiber, a scanning electronic microscopy technology was used to monitor the micro-structure of manure fiber at different hydrolysis stages.

3.4. Structural change of manure fiber during hydrolysis

The difference in fiber structure between raw manure (as collected) and manure solid particles after water-washing and solid/liquid separation is shown in Fig. 3. The texture of raw manure fiber was rougher and had more substances attached to it than that of washed fiber. The chemical comparison of the two types of manure showed that most of those substances were probably proteins (Table 7). Such "attached" proteins could impede the contact between the acid or enzyme with the substrate and as a result could in part explain the lower sugar yield in raw manure fiber hydrolysis.

Fig. 4A presents the structure of manure fiber after hydrolysis by 3% sulfuric acid at 120°C for an hour before decrystallization. Although hemicellulose was removed by this first step of two-stage hydrolysis [22], the striations on the fiber surface were thin, and the main crystal structure was not destroyed. Further decrystallization (70% sulfuric acid, 25 min, room temperature) of this material showed that the fiber turned into amorphous powders (Fig. 4B). Because the crystal structure was destroyed in decrystallization, a high sugar yield was obtained in this process (Fig. 2). The results suggest that dilute acid treatment can hydrolyze most of the hemicellulose, while the backbone structure of manure, composed mostly of cellulose, could only be degraded after decrystallization by concentrated acid.

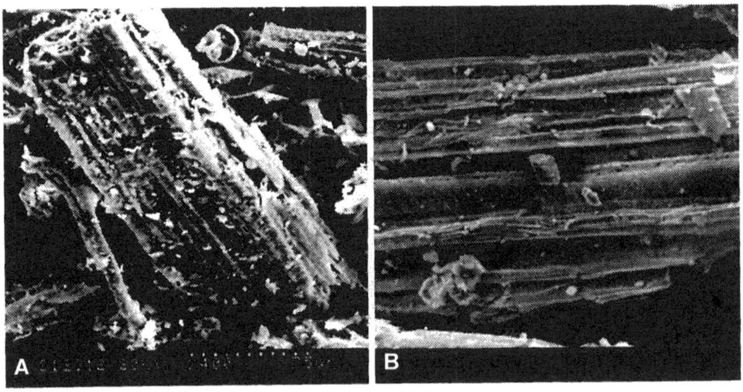

Fig. 3. Scanning electron micrographs of dairy manure fiber (400x). A. Original manure fiber; B. Solid particles after water-washing and solid/liquid separation (with 1.68 mm opening screen). [22]

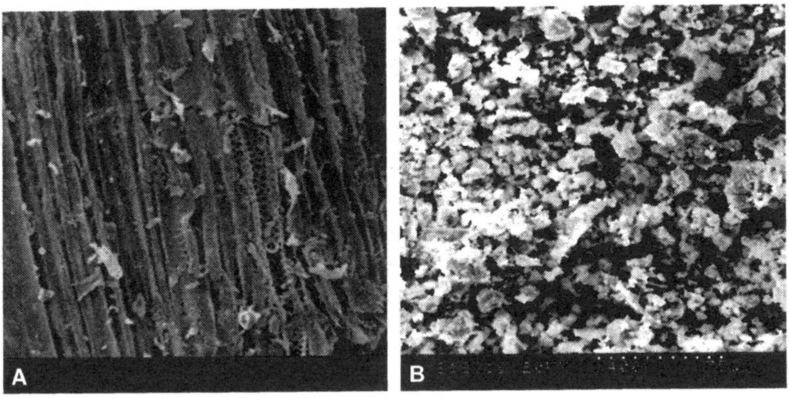

Fig. 4. Scanning electron micrographs of manure fiber (400×). A. right after acid hydrolysis; B. after further treatment with decrystallization. [22]

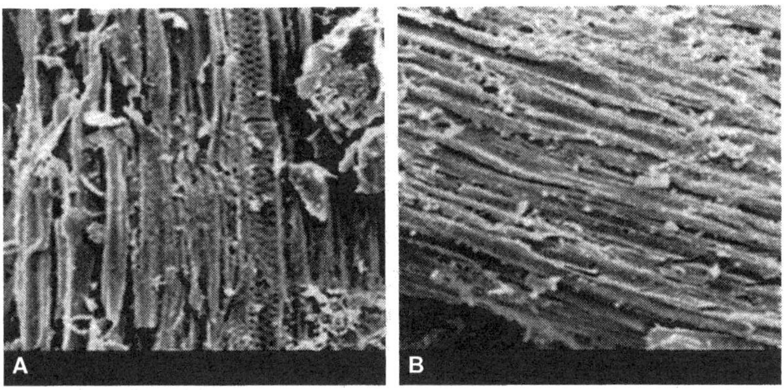

Fig. 5. Scanning electron micrographs of manure fiber before and after enzymatic hydrolysis (400×). A. Solid particles before enzymatic hydrolysis; B. Solid particles after enzymatic hydrolysis.

The structural changes of materials during enzymatic hydrolysis (50°C, pH 4.8 with 650 FPU L^{-1} cellulase and 250 IU L^{-1} β-glucosidase) of manure fiber are presented in Fig. 5. The only difference between the two materials was that the striations on the surface of the manure fibers after hydrolysis were thinner. It is suggested that the main structure was partially degraded, but not as completely destroyed as those in the decrystallization procedure.

4. CELLULOLYTIC ENZYME PRODUCTION FROM DAIRY MANURE

Enzymatic hydrolysis is a preferred method for converting manure fibers into reducing sugars because the reaction is specific and environmentally friendly. The process, however, is currently considered uneconomical because the cost of commercial cellulase enzymes remains very high [40]. Cellulolytic enzymes are produced by a number of bacteria and fungi that can use cellulose as a primary carbon source. Pure, crystalline cellulose, such as Solka Floc, Avicel, and cotton are good cellulase inducers, but are expensive. A number of cellulosic residues can be used as potentially less expensive alternatives. These include wood [41, 42], wastepaper [43], bagasse [44, 45], wheat straw [46, 47], corn cob [48], wheat bran [49], and fruit pomace [50, 51].

The research group at the Biomass Processing and Bioproduct Lab at WSU has developed a process for producing cellulase enzymes from dairy manure by fungal cultures. Dairy manure contains a variety of components that are ideal nutrients for cellulolytic fungi/bacteria cultures (Tables 1 & 2). Indeed, some microorganisms have been used to recover nutrients (such as nitrogen and phosphorus) from manure in order to produce nutritional biomass [52, 53] or value-added products [54].

The rationale for cellulase production from dairy manure and its coupling with manure fiber hydrolysis is shown in Fig. 6. Dairy manure provides not only a carbon source (cellulose), but also a variety of other nutrients, such as nitrogen, phosphorus, and trace elements, for fungal growth. The cellulolytic fungi *Trichoderma* and *Aspergillus* were used as cellulase producers. The culture was performed at 25–30°C with pH 5.5–7.0. The broth containing crude cellulase was used to hydrolyze the manure fiber into reducing sugars that can be further fermented into value-added products. The advantages of this process are the cheap substrate and environmental benefits. In addition, because crude cellulase broth was directly used for hydrolyzing the manure fiber, the high cost of cellulase purification could be avoided.

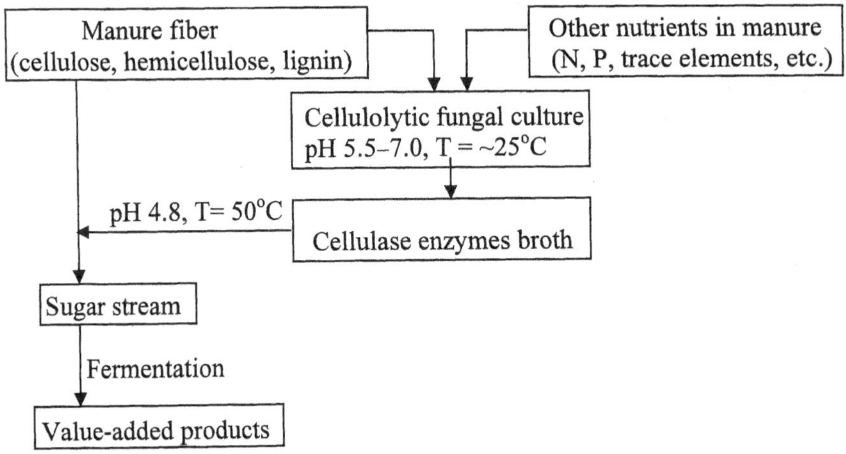

Fig. 6. Flow chart of cellulase production and subsequent manure lignocellulosics utilization.

4.1. Cellulase production by the fungus *T. reesei*

Trichoderma reesei was investigated for producing cellulase from dairy manure. This fungal species has been extensively studied for its cellulase production capability [55–60]. Among various mutants of *T. reesei*, RUT C-30 is of industrial interest because of its high cellulase production level [57] as well as its ability to grow on waste cellulosic materials [42, 43, 60].

T. reesei RUT C-30 was grown in medium containing different concentrations of dairy manure (as collected) with full Mandel salts [55] and 2 mL L^{-1} Tween 80. Filter paper activity increased with manure concentration from 3.35 to 10 g L^{-1} (dry basis) and was maintained at a high level from 10 to 20 g L^{-1} of manure (Fig. 7). The results were consistent with previous reports that used different fungal species and cellulosic materials [42, 46, 62].

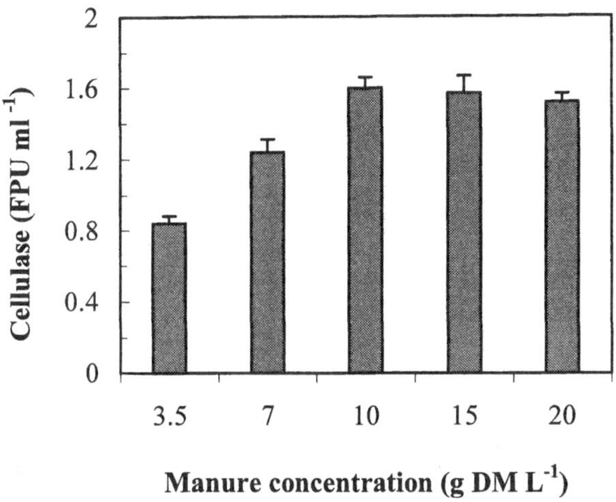

Fig. 7. Effects of manure concentration on cellulase production by the fungus *T. reesei.* Data are means of three replicates and error bars show standard deviation. [65]

As manure is a heterogenic material containing various nutrients, the effects of manure concentration on cellulase production depend not only on the amount of cellulose, but also on other nutrients or ions. The nutrients distribution in manure (at 10 g L^{-1}) and in Mandel salt solution is compared in Table 8. It was found that the amounts of calcium, magnesium, iron, manganese, and zinc contained in the manure were much higher than those in the salt solution and the nitrogen level from manure was lower than (~70% of) that of the salt solution. However, the levels of potassium and phosphorus in manure were much lower (<23%) than those in the salt solution. The results in Table 8 indicate that some salts used in Mandel solutions may not be necessary, as the corresponding nutrients contained in the manure may be sufficient for fungal growth. The possibility of eliminating nutrients from the Mandel salt solution was further tested experimentally. As shown in Table 9, the elimination of nitrogen, calcium, magnesium and trace elements from the salt solution had no negative influence on cellulase production, which remained almost the same as (runs 4 & 5) or became even better (runs 2 & 3) than that of the control. For the KH_2PO_4-eliminated medium (run 1), however, *T. reesei* produced much less cellulase than the control did. This is probably due to the insufficiency of the potassium and phosphorus levels in manure for fungal growth (Table 8). Another reason may be that most manure phosphorus is in the form of organic phosphate and polyphosphates, making its utilization more difficult [13, 62, 63].

Table 8
The distribution of nutrients in dairy manure and salt solution[a] [65]

Nutrient	From Manure	From salt solution	Total concentration in medium	Nutrient ratio of manure to salt solution
Calcium	0.241 g L^{-1}	0.11 g L^{-1}	0.350 g L^{-1}	2.19 : 1
Magnesium	0.097 g L^{-1}	0.029 g L^{-1}	0.126 g L^{-1}	3.34 : 1
Iron	13.4 mg L^{-1}	1.0 mg L^{-1}	14.4 mg L^{-1}	13.4 : 1
Manganese	1.5 mg L^{-1}	0.521 mg L^{-1}	2.021 mg L^{-1}	2.88 : 1
Zinc	1.3 mg L^{-1}	0.317 mg L^{-1}	1.617 mg L^{-1}	4.10 : 1
Cobalt	0.02 mg L^{-1}	0.908 mg L^{-1}	0.928 mg L^{-1}	0.02 : 1
Nitrogen	0.303 g L^{-1}	0.436 g L^{-1}	0.739 g L^{-1}	0.69 : 1
Potassium	0.124 g L^{-1}	0.573 g L^{-1}	0.697 g L^{-1}	0.22 : 1
Phosphorus	0.081 g L^{-1}	0.456 g L^{-1}	0.537 g L^{-1}	0.18 : 1

[a]The calculation was based on 10 g L^{-1} (DM) of manure and the composition of Mandel salt solution [55].

Table 9
Experimental design for eliminating various nutrients from Mandel salt solution and corresponding cellulase activity[a] [65]

Run	Nutrients					Cellulase (FPU mL^{-1})
	KH_2PO_4	$CaCl_2$	$MgSO_4$	$(NH_4)_2SO_4$ & Urea	Trace-elements	
1	-	+	+	+	+	0.797 ± 0.085
2	+	-	+	+	+	1.74 ± 0.037
3	+	+	-	+	+	1.71 ± 0.057
4	+	+	+	-	+	1.61 ± 0.123
5	+	+	+	+	-	1.58 ± 0.109
6	+	+	+	+	+	1.59 ± 0.066

[a] "– " indicates that the corresponding nutrient was eliminated from the salt solution; "+" indicates that the corresponding nutrient was included in Mandel salt solution.

Based on the results in Table 9, a further experiment was conducted by growing the fungal cells in media containing manure (10 g L^{-1}) augmented with KH_2PO_4 (2 g L^{-1}), $CoCl_2$ (2 mg L^{-1}), and Tween 80 (2 mL L^{-1}). Nitrogen, calcium, magnesium, and trace elements (except for cobalt) were simultaneously eliminated from the salt solution. The medium containing manure supplemented with full Mandel nutrients was used as the control. In order to give a detailed cellulase profile of the fungi under this condition, the time course of filter paper activity, CMCase activity, and β-glucosidase activity were monitored, respectively. As shown in Fig. 8, the three enzymes had similar patterns and increased in parallel with each other as time passed. The filter paper activity and β-glucosidase reached their highest activity levels at day 6, while the highest CMCase production occurred near day 8. It was also found that the activities of all three enzyme activities were comparable to the control, suggesting that the medium with reduced nutrients could sufficiently support high cellulase production by *T. reesei* (Fig. 8).

The above work shows that dairy manure was a suitable substrate for cellulase production by *T. reesei.* The optimal culture conditions were determined as follows: 10 g/L manure (dry basis), 2 g $L^{-1}KH_2PO_4$, 2 mg L^{-1} $CoCl_2$, and 2 mL L^{-1} Tween 80; initial medium pH of 5.7 and temperature of 25.5°C. The filter paper activity under these conditions achieved 1.72 FPU mL^{-1}, which was much higher than results obtained using other lignocellulosics residues (Table 10).

The highest filter paper activity and β-glucosidase activity produced by *T. reesei* were 1.72 FPU mL^{-1} and 0.08 IU mL^{-1}, respectively, corresponding to a ratio of β-glucosidase to total cellulase of 0.047. An ideal ratio of β-glucosidase activity to filter paper activity for enzymatic hydrolysis is between 0.12 and 1.5, depending on the source of the enzyme and the type of substrate [41]. For hydrolysis of manure cellulose, however, the optimal ratio of β-glucosidase activity to filter paper activity is around 0.38 [35]. This suggested that the β-glucosidase contained in *T. reesei*-derived cellulase was very low and thus, insufficient to hydrolyze cellubiose to glucose.

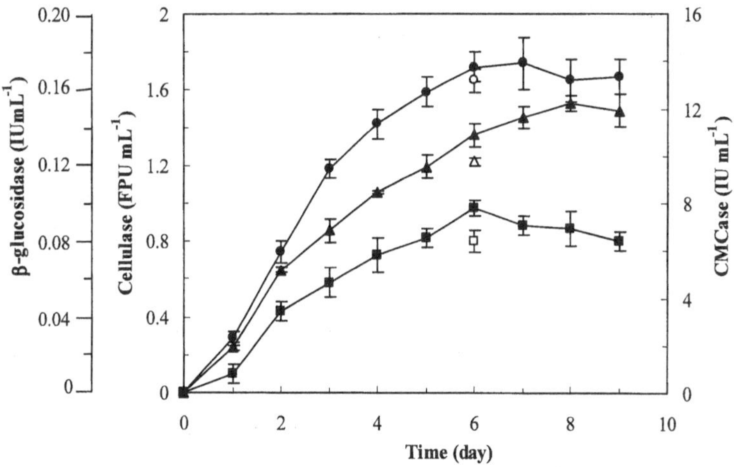

Fig. 8. Time course of cellulase activity, CMCase activity, and β-glucosidase activity in medium containing manure supplemented with KH_2PO_4, $CoCl_2$, and Tween 80. Symbols: •, Total cellulase; ▲, CMCase; ■, β-glucosidase. The open symbols are the control medium containing manure added with Mandel solution [55]. Data are means of three replicates and error bars show standard deviation. [65]

A deficiency of β-glucosidase is common to most strains of *Trichoderma.* Several approaches have been attempted in order to overcome this deficiency. For example, a temperature and pH cycling strategy was applied to culture *T. reesei* RUT-C30 to increase β-glucosidase production [66]. In another study, the mutant *Trichoderma* E12 was grown on microcrystalline cellulose with peanut cake as a nitrogen source for a high C/N ratio. Results for that study showed a well balanced ratio of β-glucosidase activity to filter paper activity

[67]. *Trichoderma* can also be co-cultured with *Aspergillus*, which is a good producer of β-glucosidase [68–72]. Such a mixed culture technique has been studied in our lab [73].

Table 10
Comparison of cellulase production by different fungal species and substrates [65]

Fungal species	Substrate	Cellulase (FPU mL^{-1})	Reference
	Lignocellulosic substrate		
Trichoderma reesei RUT C30	Steam-treated willow	0.66	[42]
Trichoderma reesei RUT C30	Steam-treated willow	1.55	[64]
Trichoderma reesei RUT C30	Wastepaper	0.30	[43]
Trichoderma reesei RUT C30	Dairy manure	1.72	[65]
Chaetomium globosum Kunze	Oil palm fruit fiber	0.95	[61]
Neurospora crassa 4335 (cell-1)	Wheat straw	1.33	[46]
Scytalidium thermophilum 3-A	Apple pomace	0.39	[45]
Scytalidium thermophilum 3-A	Lentil bran	0.23	[45]
Scytalidium thermophilum 3-A	Bagasse	0.21	[45]
	Pure cellulose or reducing sugars		
Trichoderma reesei QM 9414	Acid-swollen cellulose	0.54	[58]
Trichoderma reesei RUT NG14	Acid-swollen cellulose	15	[57]
Trichoderma reesei RUT C30	Acid-swollen cellulose	15	[57]
Trichoderma reesei RUT C30	Solka floc (cellulose)	4.65	[59]
Trichoderma reesei RUT C30	Solka floc (cellulose)	2.10	[60]
Trichoderma reesei RUT C30	Lactose	1.30	[60]
Trichoderma viride QM 6a	Solka floc (cellulose)	3.3	[55]

4.2. Cellulase production by the mixed culture of *Trichoderma reesei* and *Aspergillus phoenicis*

Before performing the mixed culture, β-glucosidase production by pure cultures of *Aspergillus phoenicis* was investigated to determine the optimal culture conditions. Table 11 summarizes the optimal culture conditions for pure cultures of *T. reesei* and *A. phoenicis*. For the two pure cultures, medium composition, temperature, and pH were very similar. In addition, it was found that the response of β-glucosidase activity (by *A. phoenicis*) and filter paper activity (by *T. reesei*) as a function of T and pH did not fall steeply when the values of T and pH changed slightly from their best values [65]. This is a desirable property because it means that the total cellulase production by *T. reesei* and β-glucosidase production by *A. phoenicis* will remain robust even with slight fluctuations in T and pH. Such a property presents the possibility that when T and pH are set at sub-optimal values which are very close to the optimal values, the resulting total cellulase and β-glucosidase will not decrease significantly from their maximum level. In other words, total cellulase from *T. reesei* and β-glucosidase from *A. phoenicis* can be simultaneously maintained at high levels by appropriately controlling pH and T. Based on the above analysis, 27°C and pH 5.5 were selected as operation parameters and enzyme production by the mixed culture was experimentally determined.

Table 11
Summary of optimal conditions for *T. reesei* and *A. phoenicis* [73]

Parameters	Fungal cultures	
	T. reesei	*A. phoenicis*
Manure concentration	10 g L^{-1} (dry basis)	10 – 30 g L^{-1} (dry basis)
Medium composition	2.0 g L^{-1} KH_2PO_4 + 2mL L^{-1} Tween 80 + 2 mg L^{-1} $CoCl_2$	
Temperature	25.5°C	28.2°C
Initial medium pH	5.76	5.14
Enzyme activity	1.71 FPU mL^{-1}	0.69 IU mL^{-1}

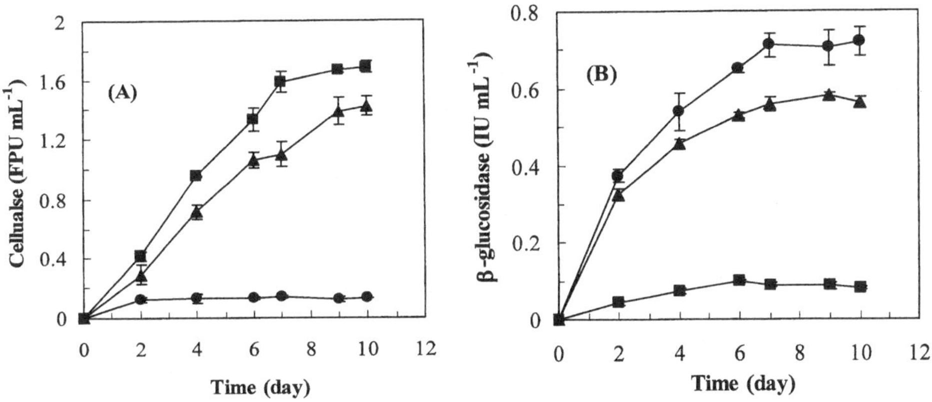

Fig. 9. Time course of cellulase (A) and β-glucosidase (B) production by the pure culture of *T. reesei* (■) and *A. phoenicis* (•) and the mixed culture of the two fungi (▲). Data are means of three replicates and error bars show standard deviation. [73]

The rates of enzyme production by the mixed culture of *T. reesei* and *A. phoenicis* are presented in Fig. 9. The trends for filter paper activity and β-glucosidase activity had similar patterns and increased in parallel as incubation time increased, although *T. reesei* and *A. phoenicis* demonstrated different abilities for producing total cellulase and β-glucosidase. *T. reesei* produced a high level of total cellulase (Fig. 9A) with a very low level of β-glucosidase (Fig. 9B). On the other hand, the total cellulase produced by *A. phoenicis* was very low (Fig. 9A) while its β-glucosidase activity was much higher (Fig. 9B). The mixed culture resulted in a relatively high filter paper activity and β-glucosidase activity, although its filter paper activity was 15% lower than that of the pure *T. reesei* culture (Fig. 9A) and β-glucosidase activity was 18% lower than that of the pure *A. phoenicis* culture (Fig. 9A).

This mixed culture has been studied on various substrates. The enzyme levels produced by the mixed culture are dependent upon the fungal species and substrates used. For example, when *T. reesei* LM-UC4 and *A. phoenicis* QM 329 were grown on bagasse, the filter paper activity and β-glucosidase activity from the mixed culture were much higher than those of the corresponding pure cultures [69]. However, β-glucosidase produced by *A. niger* was higher than the mixed culture of *T. reesei* LM-UC4 and *A. niger* when soymeal was added to the sugarcane bagasse [72]. Similar results were also observed in the mixed culture of *T. reesei* and *A. terrus* on bagasse [74]. When *T. reesei* RUT C30 and *A. phoenicis* were grown on starch substrate, both filter paper activity and β-glucosidase activity were lower than those of each pure culture [68].

The above reports suggest that the cellulase and β-glucosidase production from mixed fungal cultures are species specific and dependent upon the substrates being used. Here, the reduced cellulase and β-glucosidase activities of the mixed culture may be due to the lack of synergism of the enzymes produced from the two fungal species. Although the cellulase level was a little lower than that of the pure *T. reesei* culture, the cellulolytic potential of the mixed culture could be markedly enhanced due to the increased β-glucosidase.

4.3. Hydrolysis of manure cellulose by enzymes derived from dairy manure

To test the effectiveness of the enzymes produced from the mixed fungal culture, hydrolysis of manure cellulose was performed by using enzyme broth from the mixed culture and the pure *T. reesei* culture. Commercial cellulase (Celluclast 1.5 L) was used as control. For each enzyme source used, the produced glucose followed a similar pattern, that is, glucose increased sharply for the first 12 h, and reached the highest levels between 96–132 h (Fig. 10). The glucose produced from mixed culture enzymes was significantly (p <0.01) higher than those obtained from the other two enzyme sources. The high glucose level was due to the high β-glucosidase contained in the mixed culture broth, with the ratio of β-glucosidase activity to filter paper activity being 0.41. The result suggests that the mixed fungal culture is an efficient method of producing cellulolytic enzymes from animal manure and further hydrolyzing manure fiber into sugars.

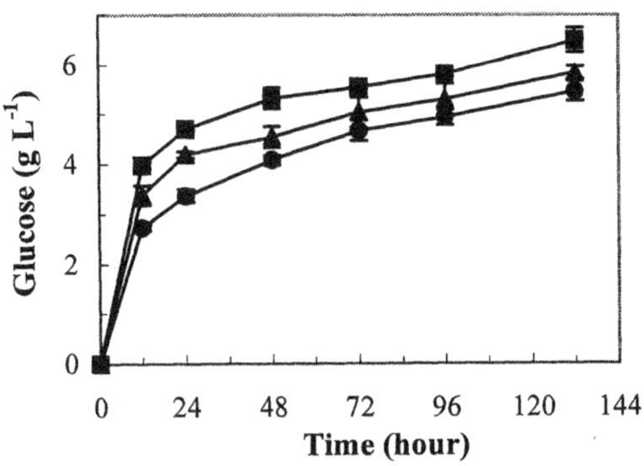

Fig. 10. Glucose concentrations in the hydrolysate during hydrolysis of manure cellulose by using different enzyme sources. ■, Enzymes from mixed culture; ▲, Enzyme from *T. reesei*; •, commercial enzymes. Data are means of three replicates and the error bars show the standard deviation. [73]

5. SUMMARY

The utilization of animal manure for value-added products provides a potential alternative to traditional animal manure management practices, yielding saleable bioproducts while alleviating environmental concerns. Lignocellulosics are a major component of animal manure, especially cattle manure, and is capable of being hydrolyzed into reducing sugars, which can be further converted into various value-added products by biological or chemical processes. However, utilizing cattle manure lignocellulosics to produce fermentable reducing sugars is still not economical. Acid-hydrolysis, especially with concentrated acid-decrystallization, resulted in a very high sugar yield, but causes even more environmental concerns than manure disposal. Enzymatic hydrolysis is an environmentally-friendly method, but is inhibited by the high cost of cellulase enzymes. Producing cellulase from dairy manure is a promising way to reduce the high cellulase cost and provides a new way for utilizing not only manure fiber, but also nitrogen and phosphorus and other nutrients. The hydrolysis efficiency of manure-derived crude cellulase was comparable to commercial enzymes. However, as investigations of cellulase enzymes production from animal manure are still in their infancy, an in-depth understanding of the factors that affect cellulase production is needed. In the future, genetically modified microorganisms may be the most efficient hosts for production of cellulases and reducing sugars from animal manures.

REFERENCES

[1] U.S. Environmental Protection Agency, Managing Manure with Biogas Recovery Systems: Improved Performance at Competitive Costs, EPA-430-F-02-004, 2002.
[2] A.L. Sutton, D.W. Nelson, D.T. Kelly and D.L. Hill, J. Environ. Quality, 15 (1986) 370.
[3] E. Sánchez, R. Borja, P. Weiland, L. Travieso and A. Martín, Bioprocess Eng., 22 (2000) 247.
[4] J.W. Paul, E.G. Beauchamp, X. Zhang, Can. J. Soil Sci., 75 (1993) 539.
[5] A.F. Bouwman and K.W. Van Der Hoek, Atmos. Environ., 31 (1997) 4095.
[6] A.N. Pell, J. Dairy Sci. 80 (1997) 2673.
[7] B.P. Kelleher, J.J. Leahy, A.M. Henihan, T.F. O'Dwyer, D. Sutton and M.J. Leahy, Bioresour. Technol., 83 (2002) 27.
[8] Klass, D.K., Biomass for Renewable Energy, Fuels and Chemicals, Academic Press, San Diego, CA, 1998.
[9] B.J. He, Y. Zhang, T.L. Funk, G.L. Riskowski and Y. Yin, Trans. ASAE, 43 (2000) 1827.
[10] J. Andersson and L. Björnsson, Bioresour. Technol., 85 (2002) 51.
[11] J. Mata-Alvarez, S. Macé and P. Llabrés, Bioresour. Technol., 74 (2000) 3.
[12] J.B. van Lier, A.B. Tilche, K. Ahring, H. Macarie, R. Moletta, M. Dohanyos, L.W. Hulshoff Pol, P. Lens and W. Verstraste, Water Sci. Technol., 43 (2001) 1.
[13] J.P. Fontenot, L.W. Smith and A.L. Sutton, J. Anim. Sci., 57(Suppl. 2) (1983). 221.
[14] H.K. Goering and P. J. van Soest, Agricultural Handbook No. 379, Agricultural Research Service-United State Department of Agriculture, Washington D.C., 1970.
[15] S. Chen, W. Liao, C. Liu, Z. Wen, R.L. Kincaid, J.H. Harrison, D.C. Elliott, M.D. Brown, A.E. Solana and D.J. Stevens, Project Report submitted to the U.S. Department of Energy (Contract DE-AC06-76RL01830), 2003.

[16] O.R. Fennema, Food Chemistry, 3rd edition, Marcel Dekker, New York, 1996.
[17] J.C. Parajo, J.L. Alonso and V. Santos, Process Biochem., 31 (1996) 271.
[18] A.L. Woiciechowski, C.R. Soccol, L.P. Ramos and A. Pandey, Process Biochem., 34 (1999) 949.
[19] S.A. Reinertsen, L.F. Elliot, V.L. Cochran and G.S. Campbell, Soil Biol. Biochem., 16 (1984) 459.
[20] Committee on Animal Nutrition Board on Agriculture National Research Council, National Academy Press, Washington D.C., 1983.
[21] P. Collins and R. Ferrier, Monosaccharides, John Wiley & Sons, New York, 1995.
[22] W. Liao, Y. Liu, C. Liu and S. Chen, Bioresour. Technol., 94 (2004) 33.
[23] C.H. Choi and A.P. Mathews, Bioresour. Technol., 58 (1996) 101.
[24] Kim, J.S., Lee, Y.Y., Torget, R.W., 2001. Cellulose hydrolysis under extremely low sulfuric acid and high-temperature conditions. Appl. Biochem. Biotechnol. 91-93, 331-340.
[25] Q. Nguyen, Internal Report, National Renewable Energy laboratory, Golden, C.O., 1998.
[26] K.C. Patrick Lee, M. Bulls, J. Holmes and J.W. Barrier, Appl. Biochem. Biotechnol., 66 (1997) 1.
[27] J.I. Zerbe and A.J. Baker, Energy from Biomass and Wastes, Institute of Gas Technology, Chicago, 1987.
[28] Forest Products Laboratory, Wood Engineering Handbook (2nd edition), Englewood Cliffs, NJ, 1990.
[29] D. Fengel and G. Wegener, Wood Chemistry, Ultrastructure, Reactions. Walter De Gruyter, Berlin, 1984.
[30] Y. Sun and J. Cheng, Bioresour. Technol., 83 (2002) 1.
[31] K. Grahmann, R. Torget and M. Himmel, Biotechnol. Bioeng. Symp., 15 (1985) 59. .
[32] F. Bayat-makooi, T.M. Singh and I.S. Goldstein, Biotechnol. Bioeng. Symp., 15 (1985) 27.
[33] J. Iranmahboob, F. Nadim and S. Monemi, Biomass Bioenergy, 22 (2002) 401.
[34] J.D. Wright, Chem. Eng. Prog., 8 (1988) 62.
[35] Z. Wen, W. Liao, S. Chen, Bioresour. Technol., 91 (2004) 31.
[36] C. Tengborg, M. Galbe and G. Zacchi, Biotechnol. Prog., 17 (2001) 110.
[37] E.Y. Vlasenko, H. Ding, J.M. Labavitch and S.P. Shoemaker, Bioresour. Technol., 59 (1997) 109.
[38] H.K. Sreenath, R.G. Koegel, A.B. Moldes, T.W. Jeffries and R.J. Straub, Process Biochem., 35 (1999) 33.
[39] Q. Gan, S.J. Allen and G. Taylor, Process Biochem., 38 (2003) 1003.
[40] M. von Sivers and G. Zacchi, Bioresour. Technol., 51 (1995) 43.
[41] S.J.B. Duff and W.D. Murray, W.D., Bioresour. Technol., 55 (1996) 1.
[42] K. Reczey, Z. Szengyel, R. Eklund, G. Zacchi, Bioresour. Technol., 57 (1996) 25.
[43] L.K. Ju and O.A. Afolabi, Biotechnol. Prog., 15 (1999) 91.
[44] M.I. Rajoka and K.A. Malik, Bioresour. Technol., 59 (1997) 21.
[45] Z.B. Ogel, K. Yarangumeli, H. Dundar and I. Ifrij, Enzyme Microb. Technol., 28 (2001) 689.
[46] M.D. Romero, J. Aguado, L. Gonzalez and M. Ladero, Enzyme Microb. Technol., 25 (1999) 244.
[47] E. Kalogeris, P. Christakopoulos, P. Katapodis, A. Alexiou, S. Vlachou, D. Kekos and B. J. Macris, Process Biochem., 38 (2003) 1099.
[48] L. Xia and P. Cen, Process Biochem., 34 (1999) 909.
[49] J.P. Smits, A. Rinzema, J. Tramper, H.M. van Sonsbeek and W. Knol, Appl. Microbol. Biotechnol., 46 (1996) 489.

[50] M.S. Umikalsom, A.B. Ariff, Z.H. Shamsuddin, C.C. Tong, M.A. Hassan, M.I.A. Karim Appl. Microbiol. Biotechnol. 47 (1997) 590.
[51] M.S.Y. Haddadin, I.M. Abu-Reesh, F.A.S. Haddadin and R. K. Robinson, Bioresour. Technol., 78 (2001) 225.
[52] W.W. Mulbry and A.C. Wilkie, J. Appl. Phycol. 13 (2001) 301.
[53] A.C. Wilkie and W.W. Mulbry, Bioresour. Technol., 84 (2002) 81.
[54] M. Potter, F.B. Oppermann-Sanio and A. Steinbuchel, Appl. Environ. Microbiol., 67 (2001) 617.
[55] M. Mandels and J. Weber, Adv. Chem. Ser., 95 (1969) 391.
[56] M. Mandels, J. Weber and R. Parizek, Appl. Microbiol., 21 (1971) 152.
[57] B.S. Montenecourt and D.E. Eveleigh, Adv. Chem. Ser., 181 (1979) 289.
[58] N.J. Gadgil, H.F. Daginawala, T. Chakrabarti and P. Khanna, Enzyme Microb. Technol., 17 (1995) 942.
[59] S. Velkovska, M.R. Marten and D.F. Ollis, J. Biotechnol., 54 (1997) 83.
[60] F.C. Domingues, J.A. Queiroz, J.M.C. Cabral and L.P. Fonseca, Enzyme Microb. Technol., 26 (2000) 401.
[61] M.S. Umikalsom, A.B. Ariff, H.S. Zulkifli, C.C. Tong, M.A. Hassan and M.I.A. Karim Bioresour. Technol., 62 (1997) 1.
[62] M. Henze, P. Harremoes, J.L.C. Jansen and E. Arvin, Wastewater Ttreatment: Biological and Chemical Process, 2nd edition, Springer, Berlin, 1996.
[63] H.B. Moller, S.G. Sommmer and B. K. Ahring, Bioresour. Technol., 85 (2002) 189.
[64] Z. Szengyel and G. Zacchi, Appl. Biochem. Biotechnol., 89 (2000) 31.
[65] Z. Wen, W. Liao and S. Chen, Bioresour. Technol., 96 (2005) 491
[66] S.K. Tangnu, H.W. Blanch and C.R. Wilke, Biotechnol. Bioeng., 23 (1981) 1837.
[67] N. Cochet, Enzyme Microb. Technol., 13 (1991) 104.
[68] S.J.B. Duff, D.C. Cooper and O.M. Fuller, Enzyme Microb. Technol. 9 (1987) 47.
[69] R. Duenas, R.P. Tengerdy and M. Gutierrez-Correa, World J. Microbiol. Biotechnol., 11 (1995) 333.
[70] M. Gutierrez-Correa and R. P. Tengerdy, Biotechnol. Lett., 19 (1997) 665.
[71] B. Flachner, A. Brumbauer and K. Reczey, Enzyme Microb. Technol., 24 (1999) 362.
[72] M. Gutierrez-Correa, L. Portal, P. Moreno and R.P. Tengerdy, Bioresour. Technol., 68 (1999) 173.
[73] Z. Wen, W. Liao and S. Chen, Process. Biochem., (2005) in press.
[74] M.I. Massadeh, W.M.W. Yusoff, O. Omar and J. Kader, Biotechnol. Lett. 23 (2001) 1771.

Index

D

N

O

P

R

FSC
www.fsc.org
MIX
Papier | Fördert gute Waldnutzung
FSC® C083411

Kolibri 360 GmbH
Ferdinand-Jühlke-Straße 7
99095 Erfurt, Deutschland
produktsicherheit@kolibri360.de